DATE DUE

APR 1 1 2005	

BRODART, CO.   Cat. No. 23-221-003

ON THE

SENSATIONS OF TONE

AS A PHYSIOLOGICAL BASIS FOR THE

THEORY OF MUSIC

HERMANN L. F. HELMHOLTZ

The Second English Edition, Translated, thoroughly Revised and Corrected,
rendered conformal to the Fourth (and last) German Edition of 1877,
with numerous additional Notes and a New additional Appendix
bringing down information to 1885, and especially
adapted to the use of Music Students by
ALEXANDER J. ELLIS

With a New Introduction (1954) by
HENRY MARGENAU

DOVER PUBLICATIONS, INC., NEW YORK

This Dover edition, first published in 1954, is an
unabridged and unaltered republication of the second (1885) edition of the Ellis translation of *Die
Lehre von den Tonempfindungen,* as originally published by Longmans & Co. A new Introduction has
been specially written for this Dover edition by
Henry Margenau.

International Standard Book Number: 0-486-60753-4

Manufactured in the United States of America
Dover Publications, Inc.
180 Varick Street
New York, N.Y. 10014

INTRODUCTION

BY HENRY MARGENAU

This book, reprinted more than 90 years after its first publication, is a magnum opus of one of the last great universalists of science. Figures like Helmholtz belong to a dying age in which a full synthetic view of nature was still possible, in which one man could not only unify the practice and teaching of medicine, physiology, anatomy and physics, but also relate these sciences significantly and lastingly to the fine arts. Added to this distinction of the book is yet another, a remarkable circumstance that is unique beyond the historical greatness of the work: its continued usefulness. The *Sensations of Tone* is still required reading for everyone who wishes to prepare himself for work in physiological acoustics, and the musician finds in it unexhausted treasure if he wishes to understand his art.

English readers are particularly fortunate in having available a translation that is excellent almost beyond belief. The clarity of Ellis' prose and his careful choice of technical terms are in no small measure the occasion for the current value of Helmholtz' contributions. This introduction, therefore, wishes to pay more than ordinary tribute to the distinguished translator and annotator whose technical competence in the field of acoustics and whose love of music enhanced the work in a major way. The reader owes Ellis a debt of gratitude for sparing him such literal translations as "clang tint," proposed by Tyndall for the German "Klangfarbe," and rendering it as "quality." This term has now been generally adopted. The other interesting divergence in terminology which arose from Helmholtz' use of the word "Oberton" has not been settled. Tyndall chose the misleading literal version of "overtone" while Ellis advocates "upper partial." There can hardly be a question concerning the greater fitness of the latter phrase.

Since the book speaks for itself, this introduction can serve the reader best by acquainting him with the life of its author and with his astonishing accomplishments in other fields. To exemplify the account I shall append a chronological list of Helmholtz' publications.

Hermann Ludwig Ferdinand Helmholtz was born on August 31, 1821 in Potsdam, the first child of Ferdinand Helmholtz, a teacher in the Gymnasium of that city. He was a sickly boy and, according to his mother, considered unattractive by all. "However," she says, "I was not worried about it; I admired my child, for whenever he opened his eyes he smiled at me and I saw nothing but spirit and intelligence." Illness interfered with the youngster's early training, but when, at the age of 9, he entered the Potsdam Gymnasium he proceeded rapidly, jumped grades and passed into his maturity at the age of 17. He himself, in reflecting upon this period 50 years later, recalls the difficulties he encountered in memorizing unrelated facts, grammar and vocabulary in particular annoyed him. History was his difficult subject; curiously, too, he says he had trouble in distinguishing left from right.

A pronounced interest in natural science developed early and led to the desire on the part of the boy to study physics. His father, however, had five children to support and saw no possibility of financing so expensive an academic course. He applied, therefore, to the medical institute (Friedrich-Wilhelms Institut) in Berlin for a scholarship which would enable Hermann to combine some work in physics with a regular course in medicine. The scholarship was awarded against the pledge that the candidate, after completing his studies, would spend several years as a military physician.

Thus Helmholtz became an "Élève" at the "Royal Medical and Surgical Friedrich-Wilhelms Institute" in Berlin where he spent the years from 1838 to 1842. The freshman schedule at this school comprised forty-eight hours per week. Here he came under the influence of a great teacher, Johannes Müller, whose advice and example largely determined Helmholtz' career. The desire to combine physiology and physics manifestly stems from this early association with a great scientist. During that period, also, a lifelong friendship with Brücke and Du Bois-Reymond was formed. The work at the Institute led to a doctor's degree, acquired at the age of 21 through a dissertation entitled "De Fabrica Systematis Nervosi Evertebratorum," describing a fundamental anatomical discovery demonstrating that nerve fibers originate in ganglion cells. In 1843 the young military doctor entered service as surgeon in the regiment of the royal guards at Potsdam.

It was during this epoch that his thoughts turned increasingly toward the fundamental problem of the day, the principle of conservation of energy. The equivalence of heat and mechanical energy was known; perpetual motion was generally regarded as impossible. Following Liebig, Helmholtz investigated heat phenomena in muscle action. But his ambition rose beyond the wish to establish the conservation law merely in the realm of empirical fact; he set himself the task of deriving it from more fundamental principles and thus to exhibit it in its full generality.

This he accomplished in his famous lecture before the Physical Society of Berlin, on July 23, 1847. The substance of the lecture had been accumulated painstakingly during long hours of hospital duty; it represents one of the greatest scientific documents of all time. Beginning with two basic assumptions, (1) that all matter consists of mass particles and (2) that these interact by central forces, i.e. forces acting along the lines joining the particles, he was able to give a cogent proof of the conservation law. In his reasoning he introduced the idea of "Spann-kraft", the form of energy now generally called potential. The prestige acquired through publication of this paper permitted him to relinquish the duties of a military physician, and in 1848 Helmholtz took his first academic position; he became teacher of anatomy at the Academy for Fine Arts in Berlin. This minor post did not hold him long, for in 1849 he was called to Königsberg as associate (ausserordentlicher) Professor of Physiology.

Having obtained a secure position, he married Olga von Velten, the daughter of a physician, and then turned to his important investigations concerning the speed of nerve impulses. According to the view then current these were so rapid as to escape detection, and astonishment was great when the young professor measured the speed along frog nerves and found it to be about 30m/sec. His discovery of the law of rise of an electric current in an inductive circuit, now named after him, fell into the beginning of the Königsberg period. Indeed it is amusing to find him writing his father in the summer of 1850 that he has become the father of a healthy girl, but that so far as discoveries are concerned he has only a "theorem regarding the rise of electrical currents" to report. Moreover in the same year he invented the ophthalmoscope, that well-known device for illuminating the retina and studying its structure, which brought him fame in the medical world.

Helmholtz' interest in acoustics was aroused for the first time in 1852, apparently by mathematical errors in the publications of Challis, which he corrected. His concern at this time is again with principles, and the subject of acoustics is viewed as a branch of hydrodynamics, the science for which he was later to create the mathematical foundation.

Near the end of his time in Königsberg, he met the man to whom he remained joined in admiration and friendship for the remainder of his life: William Thomson, later Lord Kelvin. The meeting took place in Kreuznach where Thomson had gone seeking a cure for his ailing wife. Helmholtz, 34 at this time, writes this about the occasion. "I expected to find in him, who is one of the foremost mathematical physicists of Europe, a man somewhat older than myself, and I was not a little astonished when a light-blond youth of girlish appearance came toward me. . . . With respect to analytical acumen, clarity of thought and versatility he surpasses every great scientist I have met; indeed I myself feel at times a little stupid in his presence."

In 1855 Helmholtz accepted a call to become full professor of anatomy and physiology in Bonn. His teaching there was a trifle unusual because he mixed fundamental considerations, and indeed a little mathematics, with his anatomy. The minister of education, who received complaints, was ill pleased and felt that "the teaching of anatomy in Bonn was not well cared for." Helmholtz, though never rebuked officially, heard about the complaints and in a letter to Du Bois described their cause: he had once had the temerity of using a cosine in a lecture on physiological optics. Needless to say, such minor infelicities disappeared or were condoned as the magnitude of his scientific achievements was recognized, and the three years at Bonn became a fruitful period.

It was here that his important work on combination tones was done. The occurrence of difference tones arising from two simultaneous simple tones had already been discovered. Helmholtz discovered essentially two experimental facts, the existence of summation tones and the marked dependence of their presence on the strength of the simple tones. His theory, developed in 1856, is well known; it shows how these combination tones originate from the non-linear response of the eardrum or other detector. Two years later there followed the publication of the fundamental work on the physical cause of musical harmony to which a large part of this book is devoted. The author was then thirty-seven years of age.

During this preoccupation with acoustical problems, Helmholtz transferred to a major position as professor of physiology at Heidelberg where he remained from 1858 to 1871. His presence and that of Bunsen and Kirchhoff initiated one of the glorious scientific eras in that beautiful university. Research progressed rapidly; in one year, 1859, he published two remarkable papers, the first dealing with the "timbre" or quality of vowels, the second with air vibrations in open pipes. The paper on vowels develops the relation between quality of sounds and the structure of partial tones; the other presents a mathematical theory of the motion of air near the end of an organ pipe. The solution of this problem had been attempted by well known mathematicians without success. At a later time, Helmholtz described his own method of procedure in a psychologically interesting way. "The pride which I might have experienced over my results in such cases," he said, "was greatly diminished by my realization that success in solving these problems was attained only by way of increasing generalization of favorable instances, by a series of happy conjectures after numerous failures. I was like a mountaineer who, not knowing his path, must climb slowly and laboriously, is forced to turn back frequently because his way is blocked but discovers, sometimes by deliberation and often by accident, new passages which lead him onward for a distance. Finally, when he reaches his goal, he finds to his embarrassment a royal road which would have permitted him easy access by vehicle if he had been clever enough to find the proper start. In my publications, of course, I did not tell the reader of my erratic course but described for him only the wagon road by which he may now reach the summit without labor." It is heartening indeed to have such a confession from a great scientist.

The same productive year, 1859, also dealt two serious personal blows. In June death claimed his father; in December his beloved wife. His own health was impaired to an extent which made scientific work impossible for several months. When he recovered, his attention turned to problems of physiological optics and he conducted investigations which resulted in the publication of his handbook on this subject. For one year he lost himself in his work, producing among other things a careful study of the tempered scale and of certain little-known oriental scales. Then life, and the cares of a father of two small children reasserted themselves, and Helmholtz married in 1861 a young lady of great charm, Anna von Mohl, who remained at his side until his death.

This marriage initiated the period of greatest scientific activity and productivity in the scientist's life. His acoustical researches culminated in the present book, which appeared in 1862; hydrodynamical and electrodynamical problems engaged his attention; indeed the scope of his interest widened to include the theory of knowledge and the axioms of geometry. The impressive range of his genius is recognized in the utterances and writings of his colleagues, Bunsen and Kirchhoff, and in the public esteem he enjoyed. He became "Prorektor" of the University of Heidelberg, invitations to lecture showered down upon him, and tempting offers came from foreign parts. When, in 1870, the professorship of physics in Berlin fell vacant upon the death of Magnus, Helmholtz was free to state his conditions for its acceptance. He required: (1) a salary of 4000 thaler; (2) the promise that a new Institute of Physics would be built; (3) assurance of appointment to the directorship of this institute with authority to permit or refuse its use to others; (4) living quarters in the institute. These conditions were met, and in the spring of 1871 he moved to Berlin.

The period which now begins is one of sustained and general scientific activity, of tremendous successes. It presents a distinguished German professor at the very height of his active career but is in some ways the least interesting from the chronicler's point of view. Helmholtz' main scientific preoccupation was with the problems of the nascent theory of electrodynamics, where his contributions are intertwined with those of Ampère, Maxwell and Hertz. Acoustic researches are mixed with these; indeed in 1878 he published an article entitled "The Telephone and the Quality of Sound." His great work in thermodynamics, too, was done chiefly during this period. One of his papers dealt with the steering of gas-filled balloons, another with thunderstorms. His stature was recognized even by court officials, for it is known that he was once consulted on the arrangement of lightning rods on the Castle of Goslar.

Travels were frequent and extensive and even in those days often paid for by commercial or scientific agencies. He declined two invitations to America because he felt ill at ease at the prospect of writing out popular lectures in English. His letters, to be sure, often contain English phrases, and it is somewhat surprising to find a disposition against visiting the United States in this otherwise adventurous man. Trips to Switzerland, France, Italy, and Spain were undertaken, and diaries indicate the open eye and the critical judgment of the traveler.

In 1877 Helmholtz became Rektor of the University of Berlin, a position which he occupied for one year. His inaugural speech was entitled "On academic freedom in German Universities," and it represents a document of major interest to this day. Ten years later the Physikalisch-Technische Reichsanhalt was founded, and in 1888 Helmholtz became its first president.

His new duties were largely administrative and for a time burdensome. But after one year, when the institute was established and its policies were decided, the eager spirit of its director turned again to research. A congratulatory message from the mathematician Kronecker upon the occasion of Helmholtz' sixty-seventh birthday may have had an influence on the choice of his problem at this time. Kronecker, after playfully commenting on the number 67 as the last of Fermat's three puzzling and critical integers among the first hundred, and predicting smooth sailing for the famous investigator upon passing this numerical hurdle, encourages him to come to the aid of mathematics. "The wealth of your practical experience with sane and interesting problems will give to mathematics a new direction and a new impetus. . . . One-sided and introspective mathematical speculation leads into sterile fields. Therefore, come and join us, esteemed friend, and impress upon pure mathematics the imperishable footprints of your bold and original advances in order that here, too, the paths of the future bear their markings." While Kronecker's sentiment may have outstripped his metaphor, Helmholtz seems nevertheless to have been impressed, and he launched upon a problem, though not of pure mathematics, but exhibiting a goodly measure of mathematical elegance: the motion of air masses.

First he established the existence of discontinuous boundaries between atmospheric layers of different densities. Then he proved the possibility of a transverse wave motion within the boundaries and computed the wave length, arriving at a value of 550m for a certain set of conditions. Later this value was nicely confirmed by observations from balloons. The waves here postulated became of interest in meteorology as they provide a mechanism for the mixing of air masses and the attendant atmospheric circulations. Similar considerations were later applied to water waves. While such researches absorbed the major interest of the aging scientist, he continued an active interest in electrodynamics and took part in controversies regarding the nature of the ether.

At the age of seventy-two, he finally decided to visit America where he was to be sent by the German government as delegate to the Electrical Congress in Chicago (1893). He regarded this journey as a great adventure and viewed it with slight misgivings. Only when he obtained permission to have his wife accompany him, did he willingly accept the assignment. The experiences of the pair are preserved in the letters written by Mrs. Helmholtz to her daughter, many of which throw interesting and amusing side lights upon the cultural life in the United States and also upon the misconceptions of cultured Europeans regarding it. Excerpts from these letters are contained in the three-volume biography (Helmholtz, F. Vieweg and Son, 1903) by Leo Königsberger, to which this brief account is greatly indebted. The return trip was marred by an accident, a fall down the ship's stairs resulting in a considerable loss of blood. He survived this accident, however, though he acquired the scar over his left brow that is in evidence in the beautiful portraits painted by Lenbach during the last year of the scientist's life.

On the 12th of July 1894, Helmholtz suffered what appears to have been a cerebral hemorrhage; he lived in the twilight of delirium and semi-consciousness until he breathed his last, in the midst of a large and devoted family, on September 8, 1894.

HELMHOLTZ' BIBLIOGRAPHY

"On the consumption of matter in muscle action," Müllers Archiv, 1845

"Report on researches from the year 1845 concerning the theory of physiological heat phenomena," Fortschritte der Physik, 1847

"On the Conservation of energy," G. Reimer, Berlin, 1847

"On the development of heat in muscle action," Müllers Archiv, 1847

"On the velocity of propagation of nerve impulses," Comptes Rendus XXX., XXXIII, 1850

"Measurements on the time of twitching of animal muscles and the velocity of propagation of nerve impulses," Müllers Archiv, 1850

"On the methods of measuring smallest time intervals and their application in Physiology," Lecture in the Phys. ökon. Gesellsch. zu Konigsberg, December, 1850

"Description of an ophthalmoscope for the examination of the retina in the living eye," Verlag von A. Forster, Berlin, 1850

"On a new simplest form of the ophthalmoscope," Vierordts Archiv, 1851

"On the course and the duration of electric currents induced by current fluctuations," Poggendorffs Annalen, 1851

"Measurements on the speed of propagation of the impulses in nerves," Zweite Reihe, Müllers Archiv, 1852

"Results of recent investigations on animal electricity," Kieler Allgemeine Monatsschrift, 1852

"On the theory of complex colors," Poggendorffs Annalen, 1852

"On the Nature of Human sense perception," Habilitationsvortrag, 28 June, 1852

"A theorem concerning the distribution of electrical currents in conductors," Berl. Monatsber. 22 July, 1852

"Some laws concerning the distribution of electrical currents in conductors with application to experiments on animal electricity," Poggendorffs Annalen, 1853

"Goethe's Scientific Work," Lecture in the Deutsche Gesellsch. in Konigsberg, 18 January, 1853

"On hitherto unknown changes in the human eye resulting from altered accommodation," Berl. Monatsber. 3 February, 1853

"On the Velocity of Some Processes in Muscles and nerves," Berl. Monatsber. 15 June, 1854

"On the composition of spectral colors," Poggendorffs Annalen, 1855

"On the sensitivity of the human retina to the most refractable rays of sunlight," Poggendorffs Annalen, 1855

"On the accommodation of the eye," Gräfes Archiv für Ophthalmologie, 1855

"Human Vision," Lecture in Konigsberg, 27 February, 1855

"The motions of the chest," Niederrh. Gesellschaft zu Bonn, 12 March, 1856

"The twitching of frog muscles," Niederrh. Sitzungsber. 14 May, 1856

"The explanation of gloss," Niederrh. Sitzungsber. 6 March, 1856

"On Combination tones, or *Tartini* tones," Niederrh. Sitzungsber. May, 1856

"Handbook of Physiological Optics," 1856

"On combination tones," Berl. Monatsber. 22 May, 1856

"The action of arm muscles," Niederrh. Gesellsch. für Heilkunde, 10 December, 1857

"The telestereoscope," Poggendorffs Annalen, 1857

"On integrals of the hydrodynamic equations which correspond to vortex motions," Journal für reine und angew. Mathematik, 1858

"On the subjective after-images in the eye," Niederrh. Sitzungsber. 3 July, 1858

"On after-images," Naturforschervers. in Karlsruhe, September, 1858

"The Physical Cause of harmony and Disharmony," Naturforschervers. in Karlsruhe, September, 1858

"On the quality of vowels," Poggendorffs Annalen, 1859

"On air vibrations in pipes with open ends," Journal f. reine u. angew. Mathematik, 1858

"On Color Blindness," Naturh.-med. Verein in Heidelberg, 11 Nov., 1859

"On Musical temperament," Naturh.-med. Verein in Heidelberg, 23 November, 1860

"On the Arabic-Persian Scale," Naturh.-med. Verein in Heidelberg, 2 July, 1862

"On the motion of the strings of a violin," Proc. of the Glasgow Philosophical Society, 19 December, 1860

"On the Application of the law of the conservation of force to organic nature," Proc. Roy. Inst. 12 April, 1861

"A general transformation method for problems concerning electrical distributions," Lecture in Naturh.-med. Verein in Heidelberg on 8 December, 1861

"The relation between the natural sciences and the totality of the sciences," Prorectoratsrede, 22 November, 1862

"On the form of Horopters," Poggendorffs Annalen, 1864

"The Sensations of tone as physiological foundation for the theory of music," Verlag von Fr. Vieweg u. Sohn, Braunschweig, 1863

"The normal movements of the human eye," Gräfes Archiv für Ophthalmologie, 1863

"On the normal motions of the human eye in relation to binocular vision," Croonian Lecture on 14 April, 1864

"Stereoscopic vision," Lecture in Naturh.-med. Verein in Heidelberg on 6 January, 1856

"On the regulation of Ice," Philosoph. Magazine, 1865

"Recent advances in the theory of vision," Preussische Jahrbücher XXI, 1868

"Mechanics of the small bones of the ear and of the tympanum," Pflügers Archiv fur Physiologie, 1869

"On the discontinuous movement of fluids," Berliner Akademie, 23 April, 1868

"Theory of stationary currents in viscous fluids," Lecture in naturh.-med. Verein, Heidelberg 5 March, 1869

"The axioms of geometry," The Academy Vol. I. 1868

"The physiological action of short electrical pulses in the interior of conducting masses," Naturh.-med. Verein zu Heidelberg, 12 February, 1869

"Electrical oscillations," Naturh.-med Verein zu Heidelberg, 30 April, 1869

"On hay fever," Virchows Archiv für path. Anatomie, 1869

"On acoustical vibrations of the labyrinth of the ear," Lecture in naturh.-med. Verein of Heidelberg, 25 June, 1869

"The theory of electrodynamics," Journ. fur reine u. angew. Mathematik, Bd. 72, 1860

"On the origin of the planetary system," Lecture in Heidelberg on February, 1871

"The velocity of propagation of electrodynamic effects," Berliner Akademie, 25 May, 1871

"The theory of electrodynamics," Journ. fur reine u. angew. Mathematik, Bd. 75, 1873; Journ. fur reine u. angew. Mathematik, Bd 78, 1874

"Galvanic polarization of platinum," Lecture at the Naturforscherversammlung in Leipzig in August, 1872

"Galvanic polarization in gas-free liquids," Berliner Akademie on 16 March, 1876

"A theorem concerning geometrically similar motions of liquids with application to the problem of steering balloons," Berliner Akademie on June 26, 1873

"The theoretical limit for the resolving power of microscopes," Poggendorffs Annalen, 1874

"Theory of anomalous dispersion," Pogendorffs Annalen, Bd. 154, 1875

"Cyclones and thunderstorms," Deutsche Rundschau, 1876

"On academic freedom in German universities," Rectoratsrede, 15 October, 1877

"On galvanic currents caused by differences in concentration; consequences of the mechanical theory of heat," Wiedemanns Annalen, Bd. 3, 1877

"Telephone and quality of sound," Wiedemanns Annalen, Bd. 5, 1878

"Studies on electrical boundary layers," Wiedemanns Annalen, Bd. 7, 1879

"On convective motions near polarized platinum," Wiedemanns Annalen, Bd. 11, 1880

"On the forces acting on the interior of magnetically or dielectrically polarized bodies," Wiedemanns Annalen, Bd. 13, 1881

"An electrodynamic balance," Wiedemanns Annalen, Bd. 14, 1881

"Wissenschaftliche Abhandlungen," Bd. I, 1882; Bd. II, 1883

"The thermodynamics of electrical processes," Berliner Akademie, 2 February, 1882

"Determination of magnetic moments by means of the balance," Berliner Akademie, April 5, 1883

"Static principles of monocyclic systems," Crelle's Journal, Bd. 97, 1884

"On the physical significance of the Principle of least action," Crelle's Journal, Bd. 100, 1886

Report on Sir William Thomson's Mathematical and Physical Papers. Vol. I and II. Nature, Vol. 32, 1885

"On the formation of clouds and thunderstorms," Physikalische Gesellschaft, 22 October, 1886

"Experiment designed to demonstrate the cohesion of liquids," Physikalische Gesellschaft, April 4, 1888

"Further experiments concerning the electrolysis of water," Wiedemanns Annalen, Bd. 34, 1887

"Counting and measuring as viewed from the standpoint of the theory of knowledge," Philosophische Aufsätze, 1887

"On atmospheric motions," Verhandlungen der physikalischen Gesellschaft zu Berlin, 25 October, 1889

"The energies of waves and winds," Wiedemanns Annalen, Bd. 41, 1890

"Attempt to extend the application of Fechner's law in the system of colors," Zeitschrift für Psychologie und Physiologie der Sinnesorgane, Bd. 2, 1891

"The principle of least action in electrodynamics," Berliner Akademie, 12 May, 1892

"Attempt to apply the psychophysical law to the differences in color of tri-chromatic eyes," Zeitschrift für Psychologie und Physiologie der Sinnesorgane, Bd. 3, 1891

"Goethe's anticipations of scientific ideas," Deutsche Rundschau, Bd. 72, 1892

"Electromagnetic theory of color dispersion," Wiedemanns Annalen, Bd. 48, 1892

"Consequences of Maxwell's theory concerning the motions of the pure ether," Berliner Akademie, 6 July, 1893

"On the origin of the correct interpretation of our sense impressions," Zeitschrift für Psychologie und Physiologie der Sinnesorgane, Bd. 7, 1894

TRANSLATOR'S NOTICE

TO THE

SECOND ENGLISH EDITION.

———◦◦◦———

In preparing a new edition of this translation of Professor Helmholtz's great work on the Sensations of Tone, which was originally made from the *third German* edition of 1870, and was finished in June 1875, my first care was to make it exactly conform to the *fourth German* edition of 1877 (the last which has appeared). The numerous alterations made in the fourth edition are specified in the Author's preface. In order that no merely verbal changes might escape me, every sentence of my translation was carefully re-read with the German. This has enabled me to correct several misprints and mistranslations which had escaped my previous very careful revision, and I have taken the opportunity of improving the language in many places. Scarcely a page has escaped such changes.

Professor Helmholtz's book having taken its place as a work which all candidates for musical degrees are expected to study, my next care was by supplementary notes or brief insertions, always carefully distinguished from the Author's by being inclosed in [ ], to explain any difficulties which the student might feel, and to shew him how to acquire an insight into the Author's theories, which were quite strange to musicians when they appeared in the *first German* edition of 1863, but in the twenty-two years which have since elapsed have been received as essentially valid by those competent to pass judgment.

For this purpose I have contrived the Harmonical, explained on pp. 466–469, by which, as shewn in numerous footnotes, almost every point of theory can be illustrated; and I have arranged for its being readily procurable at a moderate charge. It need scarcely be said that my interest in this instrument is purely scientific.

My own Appendix has been entirely re-written, much has been rejected and the rest condensed, but, as may be seen in the Contents, I have added a considerable amount of information about points hitherto little known, such as the Determination and History of Musical Pitch, Non-Harmonic scales, Tuning, &c., and in especial I have given an account of the work recently done on Beats and Combinational Tones, and on Vowel Analysis and Synthesis, mostly since the fourth German edition appeared.

Finally, I wish gratefully to acknowledge the assistance, sometimes very great, which I have received from Messrs. D. J. Blaikley, R. H. M. Bosanquet, Colin Brown, A. Cavaillé-Coll, A. J. Hipkins, W. Huggins, F.R.S., Shuji Isawa, H. Ward Poole, R. S. Rockstro, Hermann Smith, Steinway, Augustus Stroh, and James Paul White, as will be seen by referring to their names in the Index.

ALEXANDER J. ELLIS.

25 Argyll Road, Kensington:
July 1885.

AUTHOR'S PREFACE

TO THE

FIRST GERMAN EDITION.

———•◇•———

IN laying before the Public the result of eight years' labour, I must first pay a debt of gratitude. The following investigations could not have been accomplished without the construction of new instruments, which did not enter into the inventory of a Physiological Institute, and which far exceeded in cost the usual resources of a German philosopher. The means for obtaining them have come to me from unusual sources. The apparatus for the artificial construction of vowels, described on pp. 121 to 126, I owe to the munificence of his Majesty King Maximilian of Bavaria, to whom German science is indebted, on so many of its fields, for ever-ready sympathy and assistance. For the construction of my Harmonium in perfectly natural intonation, described on p. 316, I was able to use the Soemmering prize which had been awarded me by the Senckenberg Physical Society (*die Senckenbergische naturforschende Gesellschaft*) at Frankfurt-on-the-Main. While publicly repeating the expression of my gratitude for this assistance in my investigations, I hope that the investigations themselves as set forth in this book will prove far better than mere words how earnestly I have endeavoured to make a worthy use of the means thus placed at my command.

H. HELMHOLTZ.

HEIDELBERG: *October* 1862.

———————————

AUTHOR'S PREFACE

TO THE

THIRD GERMAN EDITION.

———•◇•———

THE present Third Edition has been much more altered in some parts than the second. Thus in the sixth chapter I have been able to make use of the new physiological and anatomical researches on the ear. This has led to a modification of my view of the action of Corti's arches. Again, it appears that the peculiar articulation between the auditory ossicles called ' hammer ' and ' anvil ' might easily cause within the ear itself the formation of harmonic upper partial tones for simple tones which are sounded loudly. By this means that peculiar series of upper partial tones, on the existence of which the present theory of music is essentially founded, receives a new subjective value, entirely independent of external alterations in the quality of tone. To illustrate the anatomical descriptions, I have been able to add a series of new woodcuts, principally from Henle's Manual of Anatomy, with the author's permission, for which I here take the opportunity of publicly thanking him.

I have made many changes in re-editing the section on the History of Music, and hope that I have improved its connection. I must, however, request the reader to regard this section as a mere compilation from secondary sources; I have neither time nor preliminary knowledge sufficient for original studies in this extremely difficult field. The older history of music to the commencement of Discant, is scarcely more than a confused heap of secondary subjects, while we can only make hypotheses concerning the principal matters in question. Of course, however, every theory of music must endeavour to bring some order into this chaos, and it cannot be denied that it contains many important facts.

For the representation of pitch in just or natural intonation, I have abandoned the method originally proposed by Hauptmann, which was not sufficiently clear in involved cases, and have adopted the system of Herr A. von Oettingen [p. 276], as had already been done in M. G. Guéroult's French translation of this book.

[A comparison of the Third with the Second editions, shewing the changes and additions individually, is here omitted.]

If I may be allowed in conclusion to add a few words on the reception experienced by the Theory of Music here propounded, I should say that published objections almost exclusively relate to my Theory of Consonance, as if this were the pith of the matter. Those who prefer mechanical explanations express their regret at my having left any room in this field for the action of artistic invention and esthetic inclination, and they have endeavoured to complete my system by new numerical speculations. Other critics with more metaphysical proclivities have rejected my Theory of Consonance, and with it, as they imagine, my whole Theory of Music, as too coarsely mechanical.

I hope my critics will excuse me if I conclude from the opposite nature of their objections, that I have struck out nearly the right path. As to my Theory of Consonance, I must claim it to be a mere systematisation of *observed facts* (with the exception of the functions of the *cochlea* of the ear, which is moreover an hypothesis that may be entirely dispensed with). But I consider it a mistake to make the Theory of Consonance the essential foundation of the Theory of Music, and I had thought that this opinion was clearly enough expressed in my book. The essential basis of Music is *Melody*. Harmony has become to Western Europeans during the last three centuries an essential, and, to our present taste, indispensable means of strengthening melodic relations, but finely developed music existed for thousands of years and still exists in ultra-European nations, without any harmony at all. And to my metaphysico-esthetical opponents I must reply, that I cannot think I have undervalued the artistic emotions of the human mind in the Theory of Melodic Construction, by endeavouring to establish the physiological facts on which esthetic feeling is based. But to those who think I have not gone far enough in my physical explanations, I answer, that in the first place a natural philosopher is never bound to construct systems about everything he knows and does not know; and secondly, that I should consider a theory which claimed to have shewn that all the laws of modern Thorough Bass were natural necessities, to stand condemned as having proved too much.

Musicians have found most fault with the manner in which I have characterised the Minor Mode. I must refer in reply to those very accessible documents, the musical compositions of A.D. 1500 to A.D. 1750, during which the modern Minor was developed. These will shew how slow and fluctuating was its development, and that the last traces of its incomplete state are still visible in the works of Sebastian Bach and Handel.

HEIDELBERG : *May* 1870.

AUTHOR'S PREFACE

TO THE

FOURTH GERMAN EDITION.

——◦◦◦——

In the essential conceptions of musical relations I have found nothing to alter in this new edition. In this respect I can but maintain what I have stated in the chapters containing them and in my preface to the third [German] edition. In details, however, much has been remodelled, and in some parts enlarged. As a guide for readers of former editions, I take the liberty to enumerate the following places containing additions and alterations.*

P. 16d, note *.—On the French system of counting vibrations.

P. 18a.—Appunn and Preyer, limits of the highest audible tones.

Pp. 59b to 65b.—On the circumstances under which we distinguish compound sensations.

P. 76a, b, c.—Comparison of the upper partial tones of the strings on a new and an old grand pianoforte.

P. 83, note †.—Herr Clement Neumann's observations on the vibrational form of violin strings.

Pp. 89a to 93b.—The action of blowing organ-pipes.

P. 110b.—Distinction of Ou from U.

Pp. 111b to 116a.—The various modifications in the sounds of vowels.

P. 145a.—The ampullæ and semicircular canals no longer considered as parts of the organ of hearing.

P. 147b.—Waldeyer's and Preyer's measurements adopted.

Pp. 150b to 151d.—On the parts of the ear which perceive noise.

P. 159b.—Koenig's observations on combinational tones with tuning-forks.

P. 176d, note.—Preyer's observations on deepest tones.

P. 179c.—Preyer's observation on the sameness of the quality of tones at the highest pitches.

Pp. 203c to 204a.—Beats between upper partials of the same compound tone condition the preference of musical tones with harmonic upper partials.

Pp. 328c to 329b.—Division of the Octave into 53 degrees. Bosanquet's harmonium.

Pp. 338c to 339b.—Modulations through chords composed of two major Thirds.

P. 365, note †.—Oettingen and Riemann's theory of the minor mode.

P. 372.—Improved electro-magnetic driver of the siren.

P. 373a.—Theoretical formulæ for the pitch of resonators.

P. 374c.—Use of a soap-bubble for seeing vibrations.

Pp. 389d to 396b.—Later use of striking reeds. Theory of the blowing of pipes.

Pp. 403c to 405b.—Theoretical treatment of sympathetic resonance for noises.

P. 417d.—A. Mayer's experiments on the audibility of vibrations.

P. 428c, d.—Against the defenders of tempered intonation.

P. 429.—Plan of Bosanquet's Harmonium.

H. HELMHOLTZ.

Berlin: April 1877.

* [The pages of this edition are substituted for the German throughout these prefaces, and omissions or alterations as respects the first edition of this translation are mostly pointed out in footnotes as they arise.—*Translator.*]

CONTENTS.

PART I. (pp. 7-151.)

ON THE COMPOSITION OF VIBRATIONS.

Upper Partial Tones, and Qualities of Tone.

CHAPTER I. ON THE SENSATION OF SOUND IN GENERAL, pp. 8-25.

CHAPTER II. ON THE COMPOSITION OF VIBRATIONS, pp. 25-36.

PART III. (pp. 234-371.)

THE RELATIONSHIP OF MUSICAL TONES.

Scales and Tonality.

CONTENTS.

LIST OF FIGURES.

LIST OF PASSAGES IN MUSICAL NOTES.

LIST OF TABLES.

Corrigenda.

P. 101d, note, line 12 from bottom, *for* 1. Upper thick *read* 1. Lower thick.

P. 139b *and elsewhere*, cochlean *for* coch'ear *is intentional.*

P. 282d, note, line 10 from bottom, *after* 70·6 cents, *omit* the remainder of the paragraph, and *read* For the possible origin of Villoteau's error see *infrà* p. 520b to 520d'.

P. 329d', note ‡, line 17 from bottom, *for* No. 6 *read* No. 7.

P. 356c, lines 15 and 16 from bottom of text, for $a^1\flat - c + e^1\flat$ read $a^1\flat + c - e^1\flat$.

P. 356d, line 4 from bottom of text, for $c - e_1 - g$ read $c + e_1 - g$.

P. 477, music, line 2, bar 2, *dele* the reference number 8, and the corresponding note below.

P. 478d', last words of lines 7 and 5 from bottom. *for* lightly *and* bad *read* tightly *and* best. *The passage will therefore read* These 24 levers are a quarter of an inch wide, and can play a pianoforte with hammers half the common width, with single strings, but larger and tightly strained, so as to yield the maximum tone, tension nearly to breaking point giving the best tone.'

P. 501, col. 1, *for* 300 cents *read* 330 cents.

P. 519c, No. 130, *for* reosen *read* riosen, *and for additional information on Japanese Scales generally, see* p. 556.

INTRODUCTION.

In the present work an attempt will be made to connect the boundaries of two sciences, which, although drawn towards each other by many natural affinities, have hitherto remained practically distinct—I mean the boundaries of *physical and physiological acoustics* on the one side, and of *musical science and esthetics* on the other. The class of readers addressed will, consequently, have had very different cultivation, and will be affected by very different interests. It will therefore not be superfluous for the author at the outset distinctly to state his intention in undertaking the work, and the aim he has sought to attain. The horizons of physics, philosophy, and art have of late been too widely separated, and, as a consequence, the language, the methods, and the aims of any one of these studies present a certain amount of difficulty for the student of any other ¶ of them; and possibly this is the principal cause why the problem here undertaken has not been long ago more thoroughly considered and advanced towards its solution.

It is true that acoustics constantly employs conceptions and names borrowed from the theory of harmony, and speaks of the ' scale,' ' intervals,' ' consonances,' and so forth; and similarly, manuals of Thorough Bass generally begin with a physical chapter which speaks of 'the numbers of vibrations,' and fixes their 'ratios' for the different intervals; but, up to the present time, this apparent *connection* of acoustics and music has been wholly external, and may be regarded rather as an expression given to the feeling that such a connection must exist, than as its actual formulation. Physical knowledge may indeed have been useful for musical instrument makers, but for the development and foundation of the theory of harmony ¶ it has hitherto been totally barren. And yet the essential facts within the field here to be explained and turned to account, have been known from the earliest times. Even Pythagoras (fl. circa B.C. 540–510) knew that when strings of different lengths but of the same make, and subjected to the same tension, were used to give the perfect consonances of the Octave, Fifth, or Fourth, their lengths must be in the ratios of 1 to 2, 2 to 3, or 3 to 4 respectively, and if, as is probable, his knowledge was partly derived from the Egyptian priests, it is impossible to conjecture in what remote antiquity this law was first known. Later physics has extended the law of Pythagoras by passing from the lengths of strings to the number of vibrations, and thus making it applicable to the tones of all musical instruments, and the numerical relations 4 to 5 and 5 to 6 have been added to the above for the less perfect consonances of the major and minor Thirds, but I am not aware that any real step was ever made towards answering the question : *What have musical consonances to do with the ratios of the first six numbers ?* Musicians, as well as philosophers and physicists, have generally contented themselves with saying in effect that human minds were in some unknown manner so constituted as to discover the numerical relations of musical vibrations, and to have a peculiar pleasure in contemplating simple ratios which are readily comprehensible.

Meanwhile musical esthetics has made unmistakable advances in those points which depend for their solution rather on psychological feeling than on the action of the senses, by introducing the conception of movement in
¶ the examination of musical works of art.   E. Hanslick, in his book ' on the Beautiful in Music ' (*Ueber das musikalisch Schöne*), triumphantly attacked the false standpoint of exaggerated sentimentality, from which it was fashionable to theorise on music, and referred the critic to the simple elements of melodic movement.   The esthetic relations for the structure of musical compositions, and the characteristic differences of individual forms of composition, are explained more fully in Vischer's ' Esthetics ' (*Aesthetik*). In the inorganic world the kind of motion we see, reveals the kind of moving force in action, and in the last resort the only method of recognising and measuring the elementary powers of nature consists in determining the motions they generate, and this is also the case for the motions of bodies or of voices which take place under the influence of human feelings.   Hence
¶ the properties of musical movements which possess a graceful, dallying, or a heavy, forced, a dull, or a powerful, a quiet, or excited character, and so on, evidently chiefly depend on psychological action.   In the same way questions relating to the equilibrium of the separate parts of a musical composition, to their development from one another and their connection as one clearly intelligible whole, bear a close analogy to similar questions in architecture.   But all such investigations, however fertile they may have been, cannot have been otherwise than imperfect and uncertain, so long as they were without their proper origin and foundation, that is, so long as there was no scientific foundation for their elementary rules relating to the construction of scales, chords, keys and modes, in short, to all that is usually contained in works on ' Thorough Bass.'   In this elementary region
¶ we have to deal not merely with unfettered artistic inventions, but with the natural power of immediate sensation.   Music stands in a much closer connection with pure sensation than any of the other arts.   The latter rather deal with what the senses apprehend, that is with the images of outward objects, collected by psychical processes from immediate sensation. *Poetry* aims most distinctly of all at merely exciting the formation of images, by addressing itself especially to imagination and memory, and it is only by subordinate auxiliaries of a more musical kind, such as rhythm, and imitations of sounds, that it appeals to the immediate sensation of hearing.   Hence its effects depend mainly on psychical action.   The *plastic arts*, although they make use of the sensation of sight, address the eye almost in the same way as *poetry* addresses the ear.   Their main purpose s to excite in us the image of an external object of determinate form and colour.   The spectator is essentially intended to interest himself in this

image, and enjoy its beauty; not to dwell upon the means by which it was created. It must at least be allowed that the pleasure of a connoisseur or virtuoso in the constructive art shewn in a statue or a picture, is not an essential element of artistic enjoyment.

It is only in painting that we find colour as an element which is directly appreciated by sensation, without any intervening act of the intellect. On the contrary, in *music*, the sensations of tone are the material of the art. So far as these sensations are excited in music, we do not create out of them any images of external objects or actions. Again, when in hearing a concert we recognise one tone as due to a violin and another to a clarinet, our artistic enjoyment does not depend upon our conception of a violin or clarinet, but solely on our hearing of the tones they produce, whereas the ¶ artistic enjoyment resulting from viewing a marble statue does not depend on the white light which it reflects into the eye, but upon the mental image of the beautiful human form which it calls up. In this sense it is clear that music has a more immediate connection with pure sensation than any other of the fine arts, and, consequently, that the theory of the sensations of hearing is destined to play a much more important part in musical esthetics, than, for example, the theory of *chiaroscuro* or of perspective in painting. Those theories are certainly useful to the artist, as means for attaining the most perfect representation of nature, but they have no part in the artistic effect of his work. In music, on the other hand, no such perfect representation of nature is aimed at; tones and the sensations of tone exist for themselves alone, and produce their effects independently of anything behind ¶ them.

This theory of the sensations of hearing belongs to natural science, and comes in the first place under *physiological acoustics*. Hitherto it is the *physical* part of the *theory of sound* that has been almost exclusively treated at length, that is, the investigations refer exclusively to the motions produced by solid, liquid, or gaseous bodies when they occasion the sounds which the ear appreciates. This *physical acoustics* is essentially nothing but a section of the theory of the motions of elastic bodies. It is physically indifferent whether observations are made on stretched strings, by means of spirals of brass wire, (which vibrate so slowly that the eye can easily follow their motions, and, consequently, do not excite any sensation of sound,) or by means of a violin string, (where the eye can scarcely perceive the vibrations ¶ which the ear readily appreciates). The laws of vibratory motion are precisely the same in both cases; its rapidity or slowness does not affect the laws themselves in the slightest degree, although it compels the observer to apply different methods of observation, the eye for one and the ear for the other. In physical acoustics, therefore, the phenomena of hearing are taken into consideration solely because the ear is the most convenient and handy means of observing the more rapid elastic vibrations, and the physicist is compelled to study the peculiarities of the natural instrument which he is employing, in order to control the correctness of its indications. In this way, although physical acoustics as hitherto pursued, has, undoubtedly, collected many observations and much knowledge concerning the action of the ear, which, therefore, belong to *physiological acoustics*, these results were not the principal object of its investigations; they were merely secondary

and isolated facts.   The only justification for devoting a separate chapter
to acoustics in the theory of the motions of elastic bodies, to which it
essentially belongs, is, that the application of the ear as an instrument
of research influenced the nature of the experiments and the methods of
observation.

But in addition to a *physical* there is a *physiological theory of acoustics*,
the aim of which is to investigate the processes that take place within the
ear itself.   The section of this science which treats of the conduction of the
motions to which sound is due, from the entrance of the external ear to the
expansions of the nerves in the labyrinth of the inner ear, has received
much attention, especially in Germany, since ground was broken by
¶ Johannes Mueller.   At the same time it must be confessed that not many
results have as yet been established with certainty.   But these attempts
attacked only a portion of the problem, and left the rest untouched.
Investigations into the processes of each of our organs of sense, have in
general three different parts.   First we have to discover how the agent
reaches the nerves to be excited, as light for the eye and sound for the ear.
This may be called the *physical* part of the corresponding physiological
investigation.   Secondly we have to investigate the various modes in which
the nerves themselves are excited, giving rise to their various *sensations*,
and finally the laws according to which these sensations result in mental
images of determinate external objects, that is, in *perceptions*.   Hence we
have secondly a specially *physiological* investigation for sensations, and
¶ thirdly, a specially *psychological* investigation for perceptions.   Now whilst
the physical side of the theory of hearing has been already frequently
attacked, the results obtained for its *physiological* and *psychological* sections
are few, imperfect, and accidental.   Yet it is precisely the physiological part
in especial—the theory of the sensations of hearing—to which the theory
of music has to look for the foundation of its structure.

In the present work, then, I have endeavoured in the first place to collect
and arrange such materials for a theory of the *sensations of hearing* as already
existed, or as I was able to add from my own personal investigations.   Of
course such a first attempt must necessarily be somewhat imperfect, and be
limited to the elements and the most interesting divisions of the subject
discussed.   It is in this light that I wish these studies to be regarded.
¶ Although in the propositions thus collected there is little of entirely new
discoveries, and although even such apparently new facts and observations
as they contain are, for the most part, more properly speaking the immediate consequences of my having more completely carried out known
theories and methods of investigation to their legitimate consequences, and
of my having more thoroughly exhausted their results than had heretofore
been attempted, yet I cannot but think that the facts frequently receive new
importance and new illumination, by being regarded from a fresh point of
view and in a fresh connection.

The First Part of the following investigation is essentially physical and
physiological.   It contains a general investigation of the phenomenon of
harmonic *upper partial tones*.   The nature of this phenomenon is established,
and its relation to *quality of tone* is proved.   A series of qualities of tone are
analysed in respect to their harmonic upper partial tones, and it results

that these upper partial tones are not, as was hitherto thought, isolated phenomena of small importance, but that, with very few exceptions, they determine the qualities of tone of almost all instruments, and are of the greatest importance for those qualities of tone which are best adapted for musical purposes. The question of how the ear is able to perceive these harmonic upper partial tones then leads to an hypothesis respecting the mode in which the auditory nerves are excited, which is well fitted to reduce all the facts and laws in this department to a relatively simple mechanical conception.

The Second Part treats of the disturbances produced by the simultaneous production of two tones, namely the *combinational tones* and *beats*. The physiologico-physical investigation shews that two tones can be simul- ¶ taneously heard by the ear without mutual disturbance, when and only when they stand to each other in the perfectly determinate and well-known relations of intervals which form musical consonance. We are thus imme- diately introduced into the field of music proper, and are led to discover the physiological reason for that enigmatical numerical relation announced by Pythagoras. The magnitude of the consonant intervals is independent of the quality of tone, but the harmoniousness of the consonances, and the distinctness of their separation from dissonances, depend on the quality of tone. The conclusions of physiological theory here agree precisely with the musical rules for the formation of chords ; they even go more into par- ticulars than it was possible for the latter to do, and have, as I believe, the authority of the best composers in their favour. ¶

In these first two Parts of the book, no attention is paid to esthetic considerations. Natural phenomena obeying a blind necessity, are alone treated. The Third Part treats of the construction of *musical scales* and *notes*. Here we come at once upon esthetic ground, and the differences of national and individual tastes begin to appear. Modern music has especially developed the principle of *tonality*, which connects all the tones in a piece of music by their relationship to one chief tone, called the tonic. On admitting this principle, the results of the preceding investigations furnish a method of constructing our modern musical scales and modes, from which all arbitrary assumption is excluded.

I was unwilling to separate the physiological investigation from its musical consequences, because the correctness of these consequences must ¶ be to the physiologist a verification of the correctness of the physical and physiological views advanced, and the reader, who takes up my book for its musical conclusions alone, cannot form a perfectly clear view of the meaning and bearing of these consequences, unless he has endeavoured to get at least some conception of their foundations in natural science. But in order to facilitate the use of the book by readers who have no special knowledge of physics and mathematics, I have transferred to appendices, at the end of the book, all special instructions for performing the more complicated experiments, and also all mathematical investigations. These appendices are therefore especially intended for the physicist, and contain the proofs of my assertions.* In this way I hope to have consulted the interests of both classes of readers.

* [The additional Appendix XX. by the Translator is intended especially for the use of musical students.—*Translator.*]

It is of course impossible for any one to understand the investigations thoroughly, who does not take the trouble of becoming acquainted by personal observation with at least the fundamental phenomena mentioned. Fortunately with the assistance of common musical instruments it is easy for any one to become acquainted with harmonic upper partial tones, combinational tones, beats, and the like.* Personal observation is better than the exactest description, especially when, as here, the subject of investigation is an analysis of sensations themselves, which are always extremely difficult to describe to those who have not experienced them.

In my somewhat unusual attempt to pass from natural philosophy into the theory of the arts, I hope that I have kept the regions of physiology ¶ and esthetics sufficiently distinct. But I can scarcely disguise from myself, that although my researches are confined to the lowest grade of musical grammar, they may probably appear too mechanical and unworthy of the dignity of art, to those theoreticians who are accustomed to summon the enthusiastic feelings called forth by the highest works of art to the scientific investigation of its basis. To these I would simply remark in conclusion, that the following investigation really deals only with the analysis of actually existing sensations—that the physical methods of observation employed are almost solely meant to facilitate and assure the work of this analysis and check its completeness—and that this analysis of the sensations would suffice to furnish all the results required for musical theory, even independently of my physiological hypothesis concerning the mechanism of ¶ hearing, already mentioned (p. 5a), but that I was unwilling to omit that hypothesis because it is so well suited to furnish an extremely simple connection between all the very various and very complicated phenomena which present themselves in the course of this investigation.†

* [But the use of the *Harmonical*, described in App. XX. sect. F. No. 1, and invented for the purpose of illustrating the theories of this work, is recommended as greatly superior for students and teachers to any other instrument. —*Translator*.]

† Readers unaccustomed to mathematical and physical considerations will find an abridged account of the essential contents of this book in Sedley Taylor, *Sound and Music*, London, Macmillan, 1873. Such readers will also find a clear exposition of the physical relations of sound in J. Tyndall, *On Sound*, a course of eight lectures, London, 1867, (the last or fourth edition 1883) Longmans, Green, & Co. A German translation of this work, entitled *Der Schall*, edited by H. Helmholtz and G. Wiedemann, was published at Brunswick in 1874.

*** [The marks ¶ in the outer margin of each page, separate the page into 4 sections, referred to as *a*, *b*, *c*, *d*, placed after the number of the page. If any section is in double columns, the letter of the second column is accented, as p. 13*d'*.]

PART I

ON THE COMPOSITION OF VIBRATIONS.

UPPER PARTIAL TONES, AND QUALITIES OF TONE.

CHAPTER I.

ON THE SENSATION OF SOUND IN GENERAL.

SENSATIONS result from the action of an external stimulus on the sensitive apparatus of our nerves. Sensations differ in kind, partly with the organ of sense excited, and partly with the nature of the stimulus employed. Each organ of sense produces peculiar sensations, which cannot be excited by means of any other; the eye gives sensations of light, the ear sensations of sound, the skin sensations of touch. Even when the same sunbeams which excite in the eye sensations of light, impinge on the skin and excite its nerves, they are felt only as heat, not as light. ¶ In the same way the vibration of elastic bodies heard by the ear, can also be felt by the skin, but in that case produce only a whirring fluttering sensation, not sound. The sensation of sound is therefore a species of reaction against external stimulus, peculiar to the ear, and excitable in no other organ of the body, and is completely distinct from the sensation of any other sense.

As our problem is to study the laws of the sensation of hearing, our first business will be to examine how many kinds of sensation the ear can generate, and what differences in the external means of excitement or sound, correspond to these differences of sensation.

The first and principal difference between various sounds experienced by our ear, is that between *noises* and *musical tones*. The soughing, howling, and whistling of the wind, the splashing of water, the rolling and rumbling of carriages, are examples of the first kind, and the tones of all musical instruments of the second. Noises and musical tones may certainly intermingle in very various degrees, and ¶ pass insensibly into one another, but their extremes are widely separated.

The nature of the difference between musical tones and noises, can generally be determined by attentive aural observation without artificial assistance. We perceive that generally, a noise is accompanied by a rapid alternation of different kinds of sensations of sound. Think, for example, of the rattling of a carriage over granite paving stones, the splashing or seething of a waterfall or of the waves of the sea, the rustling of leaves in a wood. In all these cases we have rapid, irregular, but distinctly perceptible alternations of various kinds of sounds, which crop up fitfully. When the wind howls the alternation is slow, the sound slowly and gradually rises and then falls again. It is also more or less possible to separate restlessly alternating sounds in case of the greater number of other noises. We shall hereafter become acquainted with an instrument, called a resonator, which will materially assist the ear in making this separation. On the other hand, a musical tone strikes the ear as a perfectly undisturbed, uniform sound which remains unaltered as long as it exists, and it presents no alternation of various kinds of constituents. To this then corresponds a simple, regular kind of sensation, whereas in a noise many various sensations of musical tone are irregularly mixed up and as it were tumbled about in confusion. We can easily compound noises out of musical tones, as, for example, by simultaneously striking all the keys contained in one or two octaves of a pianoforte. This shews us that musical tones are the simpler and more regular elements of the sensations of hearing, and that we have consequently first to study the laws and peculiarities of this class of sensations.

Then comes the further question : On what difference in the external means of excitement does the difference between noise and musical tone depend ? The normal and usual means of excitement for the human ear is atmospheric vibration. The irregularly alternating sensation of the ear in the case of noises leads us to ¶ conclude that for these the vibration of the air must also change irregularly. For musical tones on the other hand we anticipate a regular motion of the air, continuing uniformly, and in its turn excited by an equally regular motion of the sonorous body, whose impulses were conducted to the ear by the air.

Those regular motions which produce musical tones have been exactly investigated by physicists. They are *oscillations, vibrations,* or swings, that is, up and down, or to and fro motions of sonorous bodies, and it is necessary that these oscillations should be regularly *periodic.* By a *periodic motion* we mean one which constantly returns to the same condition after exactly equal intervals of time. The ength of the equal intervals of time between one state of the motion and its next exact repetition, we call the *length of the oscillation* vibration or swing, or the *period* of the motion. In what manner the moving body actually moves during one period, is perfectly indifferent. As illustrations of periodical motion, take the motion of a clock pendulum, of a stone attached to a string and whirled round in ¶ a circle with uniform velocity, of a hammer made to rise and fall uniformly by its connection with a water wheel. All these motions, however different be their form, are periodic in the sense here. explained. The length of their periods, which in the cases adduced is generally from one to several seconds, is relatively long in comparison with the much shorter periods of the vibrations producing musical tones, the lowest or deepest of which makes at least 30 in a second, while in other cases their number may increase to several thousand in a second.

Our definition of periodic motion then enables us to answer the question proposed as follows :—*The sensation of a musical tone is due to a rapid periodic motion of the sonorous body ; the sensation of a noise to non-periodic motions.*

The musical vibrations of solid bodies are often visible. Although they may be too rapid for the eye to follow them singly, we easily recognise that a sounding string, or tuning-fork, or the tongue of a reed-pipe, is rapidly vibrating between two fixed limits, and the regular, apparently immovable image that we see, notwith- ¶ standing the real motion of the body, leads us to conclude that the backward and forward motions are quite regular. In other cases we can feel the swinging motions of sonorous solids. Thus, the player feels the trembling of the reed in the mouth-piece of a clarinet, oboe, or bassoon, or of his own lips in the mouthpieces of trumpets and trombones.

The motions proceeding from the sounding bodies are usually conducted to our ear by means of the atmosphere. The particles of air must also execute periodically recurrent vibrations, in order to excite the sensation of a musical tone in our ear. This is actually the case, although in daily experience sound at first seems to be some agent, which is constantly advancing through the air, and propagating itself further and further. We must, however, here distinguish between the motion of the individual particles of air—which takes place periodically backwards and forwards within very narrow limits—and the propagation of the sonorous tremor. The latter is constantly advancing by the constant attraction of fresh particles into its sphere of tremor.

This is a peculiarity of all so-called *undulatory motions*. Suppose
be thrown into a piece of calm water.   Round the spot struck there fo
ring of wave, which, advancing equally in all directions, expands to a
increasing circle.   Corresponding to this ring of wave, sound also proce
air from the excited point and advances in all directions as far as the lin
mass of air extend.   The process in the air is essentially identical with tl
surface of the water. The principal difference consists in the spherical propagation
of sound in all directions through the atmosphere which fills all surrounding space,
whereas the waves of the water can only advance in rings or circles on its surface.
The crests of the waves of water correspond in the waves of sound to spherical
shells where the air is condensed, and the troughs to shells where it is rarefied.
On the free surface of the water, the mass when compressed can slip upwards and
so form ridges, but in the interior of the sea of air, the mass must be condensed,
as there is no unoccupied spot for its escape.   ¶

The waves of water, therefore, continually advance without returning.   But
we must not suppose that the particles of water of which the waves are composed
advance in a similar manner to the waves themselves.   The motion of the particles
of water on the surface can easily be rendered visible by floating a chip of wood
upon it.   This will exactly share the motion of the adjacent particles.   Now, such
a chip is not carried on by the rings of wave.   It only bobs up and down and
finally rests on its original spot.   The adjacent particles of water move in the same
manner.   When the ring of wave reaches them they are set bobbing ; when it has
passed over them they are still in their old place, and remain there at rest, while
the ring of wave continues to advance towards fresh spots on the surface of the
water, and sets new particles of water in motion.   Hence the waves which pass
over the surface of the water are constantly built up of fresh particles of water.
What really advances as a wave is only the tremor, the altered form of the surface,
while the individual particles of water themselves merely move up and down ¶
transiently, and never depart far from their original position.

The same relation is seen still more clearly in the waves of a rope or chain.
Take a flexible string of several feet in length, or a thin metal chain, hold it at one
end and let the other hang down, stretched by its own weight alone.   Now, move
the hand by which you hold it quickly to one side and back again.   The excursion
which we have caused in the upper end of the string by moving the hand, will run
down it as a kind of wave, so that constantly lower parts of the string will make a
sidewards excursion while the upper return again into the straight position of rest.
But it is evident that while the wave runs down, each individual particle of the
string can have only moved horizontally backwards and forwards, and can have
taken no share at all in the advance of the wave.

The experiment succeeds still better with a long elastic line, such as a thick
piece of india-rubber tubing, or a brass-wire spiral spring, from eight to twelve feet
in length, fastened at one end, and slightly stretched by being held with the hand ¶
at the other.   The hand is then easily able to excite waves which will run very
regularly to the other end of the line, be there reflected and return.   In this case
it is also evident that it can be no part of the line itself which runs backwards and
forwards, but that the advancing wave is composed of continually fresh particles
of the line.   By these examples the reader will be able to form a mental image of
the kind of motion to which sound belongs, where the material particles of the
body merely make periodical oscillations, while the tremor itself is constantly
propagated forwards.

Now let us return to the surface of the water.   We have supposed that one of
its points has been struck by a stone and set in motion.   This motion has spread
out in the form of a ring of wave over the surface of the water, and having reached
the chip of wood has set it bobbing up and down.   Hence by means of the wave,
the motion which the stone first excited in one point of the surface of the water
has been communicated to the chip which was at another point of the same surface.

..he process which goes on in the atmospheric ocean about us, is of a precisely similar nature. For the stone substitute a sounding body, which shakes the air ; for the chip of wood substitute the human ear, on which impinge the waves of air excited by the shock, setting its movable parts in vibration. The waves of air proceeding from a sounding body, transport the tremor to the human ear exactly in the same way as the water transports the tremor produced by the stone to the floating chip.

In this way also it is easy to see how a body which itself makes periodical oscillations, will necessarily set the particles of air in periodical motion. A falling stone gives the surface of the water a single shock. Now replace the stone by a regular series of drops falling from a vessel with a small orifice. Every separate drop will excite a ring of wave, each ring of wave will advance over the surface of the water precisely like its predecessor, and will be in the same way followed by ¶ its successors. In this manner a regular series of concentric rings will be formed and propagated over the surface of the water. The number of drops which fall into the water in a second will be the number of waves which reach our floating chip in a second, and the number of times that this chip will therefore bob up and down in a second, thus executing a periodical motion, the period of which is equal to the interval of time between the falling of consecutive drops. In the same way for the atmosphere, a periodically oscillating sonorous body produces a similar periodical motion, first in the mass of air, and then in the drumskin of our ear, and the period of these vibrations must be the same as that of the vibration in the sonorous body.

Having thus spoken of the principal division of sound into Noise and Musical Tones, and then described the general motion of the air for these tones, we pass on to the peculiarities which distinguish such tones one from the other. We are acquainted with three points of difference in musical tones, confining our attention ¶ in the first place to such tones as are isolatedly produced by our usual musical instruments, and excluding the simultaneous sounding of the tones of different instruments. Musical tones are distinguished :—

1. By their *force*,
2. By their *pitch*,
3. By their *quality*.

It is unnecessary to explain what we mean by the force and pitch of a tone. By the quality of a tone we mean that peculiarity which distinguishes the musical tone of a violin from that of a flute or that of a clarinet, or that of the human voice, when all these instruments produce the same note at the same pitch.

We have now to explain what peculiarities of the motion of sound correspond to these three principal differences between musical tones.

First, We easily recognise that the *force* of a musical tone increases and diminishes with the extent or so-called *amplitude* of the oscillations of the particles of ¶ the sounding body. When we strike a string, its vibrations are at first sufficiently large for us to see them, and its corresponding tone is loudest. The visible vibrations become smaller and smaller, and at the same time the loudness diminishes. The same observation can be made on strings excited by a violin bow, and on the reeds of reed-pipes, and on many other sonorous bodies. The same conclusion results from the diminution of the loudness of a tone when we increase our distance from the sounding body in the open air, although the pitch and quality remain unaltered ; for it is only the amplitude of the oscillations of the particles of air which diminishes as their distance from the sounding body increases. Hence loudness must depend on this amplitude, and none other of the properties of sound do so.*

---

* Mechanically the force of the oscillations for tones of different pitch is measured by their *vis viva*, that is, by the square of the greatest velocity attained by the oscillating particles. But the ear has different degrees of sensibility for tones of different pitch, so that no measure can be found for the intensity of the sensation of sound, that is, for the loudness of sound, which will hold all pitches. [See the addition to a footnote on p. 75*d*, referring especially to this passage.—*Translator.*]

The second essential difference between different musical tones consists in their *pitch*. Daily experience shews us that musical tones of the same pitch can be produced upon most diverse instruments by means of most diverse mechanical contrivances, and with most diverse degrees of loudness. All the motions of the air thus excited must be periodic, because they would not otherwise excite in us the sensation of a musical tone. But the sort of motion within each single period may be any whatever, and yet if the length of the periodic time of two musical tones is the same, they have the same pitch. Hence : *Pitch depends solely on the length of time in which each single vibration is executed, or,* which comes to the same thing, *on the number of vibrations completed in a given time.* We are accustomed to take a second as the unit of time, and shall consequently mean by the *pitch number* [or *frequency*] of a tone, the number of vibrations which the particles of a sounding body perform in one second of time.* It is self-evident that we find the periodic time or *vibrational period*, that is length of time which ¶ is occupied in performing a single vibration backwards and forwards, by dividing one second of time by the pitch number.

Musical tones are said to be higher, the greater their pitch numbers, that is, the shorter their vibrational periods.

The exact determination of the pitch number for such elastic bodies as produce audible tones, presents considerable difficulty, and physicists had to contrive many comparatively complicated processes in order to solve this problem for each particular case. Mathematical theory and numerous experiments had to render mutual assistance.† It is consequently very convenient for the demonstration of the fundamental facts in this department of knowledge, to be able to apply a peculiar instrument for producing musical tones—the so-called *siren*—which is constructed in such a manner as to determine the pitch number of the tone produced, by a direct observation. The principal parts of the simplest form of the siren are shewn in fig. 1, after Seebeck. ¶

A is a thin disc of cardboard or tinplate, which can be set in rapid rotation about its axle b by means of a string f f, which passes over a larger wheel. On the margin of the disc there is punched a set of holes at equal intervals : of these

FIG. 1.

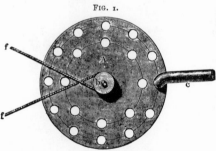

there are twelve in the figure ; one or more similar series of holes at equal distances are introduced on concentric circles, (there is one such of eight holes in the figure), c is a pipe which is directed over one of the holes. Now, on setting the disc in rotation and blowing through the pipe c, the air will pass freeely whenever one of the holes comes under the end of the pipe, but will be checked whenever an unpierced portion ¶ of the disc comes under it. Each hole of the disc, then, that passes the end of the pipe lets a single puff of air escape. Supposing the disc to make a single revolution and the pipe to be directed to the

* [The *pitch number* was called the 'vibrational number' in the first edition of this translation. The pitch *number* of a note is commonly called the *pitch* of the note. By a convenient abbreviation we often write *a′* 440, meaning the note *a′* having the pitch number 440 ; or say that the pitch of *a′* is 440 vib. that is, 440 double vibrations in a second. The second term *frequency*, which I have introduced into the text, as it is much used by acousticians, properly represents *the number of times that any periodically recurring event happens in one second of time*, and, applied to double vibrations, it means the same as pitch number.

The pitch of a musical instrument is the pitch of the note by which it is tuned. But as pitch is properly a *sensation*, it is necessary here to distinguish from this sensation the pitch *number* or *frequency* of vibration by which it is *measured*. The larger the pitch number, the *higher* or *sharper* the pitch is said to be. The lower the pitch number the *deeper* or *flatter* the pitch. These are all metaphorical expressions which must be taken strictly in this sense.—*Translator*.]

† [An account of the more exact modern methods is given in App. XX. sect. B.—*Translator*.]

outer circle of holes, we have twelve puffs corresponding to the twelve holes; but if the pipe is directed to the inner circle we have only eight puffs. If the disc is made to revolve ten times in one second, the outer circle would produce 120 puffs in one second, which would give rise to a weak and deep musical tone, and the inner circle eighty puffs. Generally, if we know the number of revolutions which the disc makes in a second, and the number of holes in the series to which the tube is directed, the product of these two numbers evidently gives the number of puffs in a second. This number is consequently far easier to determine exactly than in any other musical instrument, and sirens are accordingly extremely well adapted for studying all changes in musical tones resulting from the alterations and ratios of the pitch numbers.

The form of siren here described gives only a weak tone. I have placed it first because its action can be most readily understood, and, by changing the disc, it

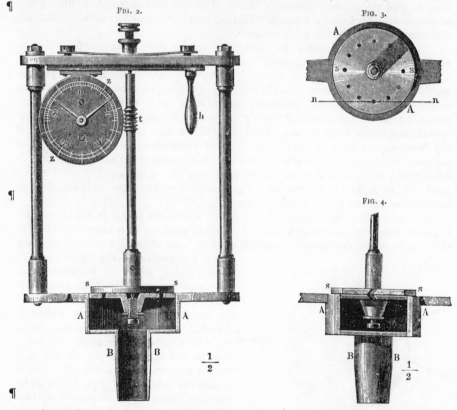

FIG. 2.   FIG. 3.

FIG. 4.

$\frac{1}{2}$ $\frac{1}{2}$

can be easily applied to experiments of very different descriptions. A stronger tone is produced in the siren of Cagniard de la Tour, shewn in figures 2, 3, and 4, above. Here s s is the rotating disc, of which the upper surface is shewn in fig. 3, and the side is seen in figs. 2 and 4. It is placed over a windchest A A, which is connected with a bellows by the pipe B B. The cover of the windchest A A, which lies immediately under the rotating disc s s, is pierced with precisely the same number of holes as the disc, and the direction of the holes pierced in the cover of the chest is oblique to that of the holes in the disc, as shewn in fig. 4, which is a vertical section of the instrument through the line n n in fig. 3. This position of the holes enables the wind escaping from A A to set the disc s s in rotation, and by increasing the pressure of the bellows, as much as 50 or 60 rotations in a second can be produced. Since all the holes of one circle are blown through at the same time in this siren, a much more powerful tone is produced than in Seebeck's, fig. 1 (p. 11c). To record the revolutions, a counter z z is introduced, connected with a toothed wheel which works in the screw t, and advances one tooth for each revolution of the disc s s. By the handle h this counter may be moved slightly to one side, so that the wheelwork and screw may be connected or disconnected at pleasure. If they are connected at the beginning of one second, and disconnected at the beginning of another, the hand of the counter shows how many revolutions of the disc have been made in the corresponding number of seconds.*

Dove † introduced into this siren several rows of holes through which the wind might be directed, or from which it might be cut off, at pleasure. A polyphonic siren of this description with other peculiar arrangements will be figured and described in Chapter VIII., fig. 56.

It is clear that when the pierced disc of one of these sirens is made to revolve with a uniform velocity, and the air escapes through the holes in puffs, the motion of the air thus produced must be *periodic* in the sense already explained. The ¶ holes stand at equal intervals of space, and hence on rotation follow each other at equal intervals of time. Through every hole there is poured, as it were, a drop of air into the external atmospheric ocean, exciting waves in it, which succeed each other at uniform intervals of time, just as was the case when regularly falling drops impinged upon a surface of water (p. 10a). Within each separate period, each individual puff will have considerable variations of form in sirens of different construction, depending on the different diameters of the holes, their distance from each other, and the shape of the extremity of the pipe which conveys the air ; but in every case, as long as the velocity of rotation and the position of the pipe remain unaltered, a regularly periodic motion of the air must result, and consequently the sensation of a musical tone must be excited in the ear, and this is actually the case.

It results immediately from experiments with the siren that two series of the same number of holes revolving with the same velocity, give musical tones of the ¶ same pitch, quite independently of the size and form of the holes, or of the pipe. We even obtain a musical tone of the same pitch if we allow a metal point to strike in the holes as they revolve instead of blowing. Hence it follows firstly that the pitch of a tone depends only on the *number* of puffs or swings, and not on their form, force, or method of production. Further it is very easily seen with this instrument that on increasing the velocity of rotation and consequently the number of puffs produced in a second, the pitch becomes sharper or higher. The same result ensues if, maintaining a uniform velocity of rotation, we first blow into a series with a smaller and then into a series with a greater number of holes. The latter gives the sharper or higher pitch.

With the same instrument we also very easily find the remarkable relation which the pitch numbers of two musical tones must possess in order to form a consonant interval. Take a series of 8 and another of 16 holes on a disc, and blow into both sets while the disc is kept at uniform velocity of rotation. Two ¶ tones will be heard which stand to one another in the exact relation of an Octave. Increase the velocity of rotation ; both tones will become sharper, but both will continue at the new pitch to form the interval of an Octave.‡ Hence we conclude that *a musical tone which is an Octave higher than another, makes exactly twice as many vibrations in a given time as the latter.*

---

* See Appendix I.

† [Pronounce *Dóh-veh*, in two syllables.—*Translator.*]

‡ [When two notes have different pitch numbers, there is said to be an *interval* between them. This gives rise to a sensation, very differently appreciated by different individuals, but. in all cases the *interval is measured by the ratio of the pitch numbers*, and, for some purposes, more conveniently by other numbers called *cents*, derived from these ratios, as explained in App. XX. sect. C. The names of all the intervals usually distinguished are also given in App. XX. sect. D., with the corresponding ratios and cents. These names were in the first place derived from the ordinal number of the note in the scales, or successions of continually sharper notes. The Octave is the eighth note in the major scale. An octave is a set of notes lying within an Octave. Observe that *in this translation* all names of intervals commence with a capital letter, to prevent ambiguity, as almost all such words are also used in other senses.—*Translator.*]

The disc shewn in fig. 1, p. 11c, has two circles of 8 and 12 holes respectively. Each, blown successively, gives two tones which form with each other a perfect Fifth, independently of the velocity of rotation of the disc. Hence, *two musical tones stand in the relation of a so-called Fifth when the higher tone makes three vibrations in the same time as the lower makes two.*

If we obtain a musical tone by blowing into a circle of 8 holes, we require a circle of 16 holes for its Octave, and 12 for its Fifth. Hence the ratio of the pitch numbers of the Fifth and the Octave is 12 : 16 or 3 : 4. But the interval between the Fifth and the Octave is the Fourth, so that we see that *when two musical tones form a Fourth, the higher makes four vibrations while the lower makes three.*

The polyphonic siren of Dove has usually four circles of 8, 10, 12 and 16 holes respectively. The series of 16 holes gives the Octave of the series of 8 holes, and ¶ the Fourth of the series of 12 holes. The series of 12 holes gives the Fifth of the series of 8 holes, and the minor Third of the series of 10 holes. While the series of 10 holes gives the major Third of the series of 8 holes. The four series consequently give the constituent musical tones of a major chord.

By these and similar experiments we find the following relations of the pitch numbers :—

$$1 : 2 \quad \text{Octave}$$
$$2 : 3 \quad \text{Fifth}$$
$$3 : 4 \quad \text{Fourth}$$
$$4 : 5 \quad \text{major Third}$$
$$5 : 6 \quad \text{minor Third}$$

When the fundamental tone of a given interval is taken an Octave higher, the interval is said to be *inverted*. Thus a Fourth is an inverted Fifth, a minor Sixth ¶ an inverted major Third, and a major Sixth an inverted minor Third. The corresponding ratios of the pitch numbers are consequently obtained by doubling the smaller number in the original interval.

From 2 : 3 the Fifth, we thus have   3 : 4 the Fourth
„   4 : 5 the major Third . . .   5 : 8 the minor Sixth
„   5 : 6 the minor Third,   6 : 10 = 3 : 5 the major Sixth.

These are all the consonant intervals which lie within the compass of an Octave. With the exception of the minor Sixth, which is really the most imperfect of the above consonances, the ratios of their vibrational numbers are all expressed by means of the whole numbers, 1, 2, 3, 4, 5, 6.

Comparatively simple and easy experiments with the siren, therefore, corroborate that remarkable law mentioned in the Introduction (p. 1d), according to which the pitch numbers of consonant musical tones bear to each other ratios expressible ¶ by small whole numbers. In the course of our investigation we shall employ the same instrument to verify more completely the strictness and exactness of this law.

Long before anything was known of pitch numbers, or the means of counting them, Pythagoras had discovered that if a string be divided into two parts by a bridge, in such a way as to give two consonant musical tones when struck, the lengths of these parts must be in the ratio of these whole numbers. If the bridge is so placed that $\frac{2}{3}$ of the string lie to the right, and $\frac{1}{3}$ on the left, so that the two lengths are in the ratio of 2 : 1, they produce the interval of an Octave, the greater length giving the deeper tone. Placing the bridge so that $\frac{3}{5}$ of the string lie on the right and $\frac{2}{5}$ on the left, the ratio of the two lengths is 3 : 2, and the interval is a Fifth.

These measurements had been executed with great precision by the Greek musicians, and had given rise to a system of tones, contrived with considerable art. For these measurements they used a peculiar instrument, the *monochord*, consisting of a sounding board and box on which a single string was stretched with a scale below, so as to set the bridge correctly.*

It was not till much later that, through the investigations of Galileo (1638), Newton, Euler (1729), and Daniel Bernouilli (1771), the law governing the motions of strings became known, and it was thus found that the simple ratios of the lengths of the strings existed also for the pitch numbers of the tones they produced, and that they consequently belonged to the musical intervals of the tones of all instruments, and were not confined to the lengths of strings through which the law had been first discovered.

This relation of whole numbers to musical consonances was from all time looked upon as a wonderful mystery of deep significance. The Pythagoreans themselves made use of it in their speculations on the harmony of the spheres. From that time it remained partly the goal and partly the starting point of the strangest and most venturesome, fantastic or philosophic combinations, till in ¶ modern times the majority of investigators adopted the notion accepted by Euler himself, that the human mind had a peculiar pleasure in simple ratios, because it could better understand them and comprehend their bearings. But it remained uninvestigated how the mind of a listener not versed in physics, who perhaps was not even aware that musical tones depended on periodical vibrations, contrived to recognise and compare these ratios of the pitch numbers. To shew what processes taking place in the ear, render sensible the difference between consonance and dissonance, will be one of the principal problems in the second part of this work.

CALCULATION OF THE PITCH NUMBERS FOR ALL THE TONES OF THE MUSICAL SCALE.

By means of the ratios of the pitch numbers already assigned for the consonant intervals, it is easy, by pursuing these intervals throughout, to calculate the ratios ¶ for the whole extent of the musical scale.

The major triad or chord of three tones, consists of a major Third and a Fifth. Hence its ratios are :

$$C : E : G$$
$$1 : \tfrac{5}{4} : \tfrac{3}{2}$$
$$\text{or}\quad 4 : 5 : 6$$

If we associate with this triad that of its dominant $G : B : D$, and that of its sub-dominant $F : A : C$, each of which has one tone in common with the triad of the tonic $C : E : G$, we obtain the complete series of tones for the major scale of C, with the following ratios of the pitch numbers :

$$C : D : E : F : G : A : B : c$$
$$1 : \tfrac{9}{8} : \tfrac{5}{4} : \tfrac{4}{3} : \tfrac{3}{2} : \tfrac{5}{3} : \tfrac{15}{8} : 2.$$
$$[\text{or}\ 24 : 27 : 30 : 32 : 36 : 40 : 45 : 48]$$

¶

In order to extend the calculation to other octaves, we shall adopt the following notation of musical tones, marking the higher octaves by accents, as is usual in Germany,† as follows :

1. *The unaccented or small octave* (the 4-foot octave on the organ‡):—

c   d   e   f   g   a   b

---

* [As the monochord is very liable to error, these results were happy generalisations from necessarily imperfect experiments.—*Translator*.]

† [English works use strokes above and below the letters, which are typographically inconvenient. Hence the German notation is retained.—*Translator*.]

‡ [The note C in the small octave was once omitted by an organ pipe 4 feet in length:

2. *The once-accented octave* (2-foot) :—

3. *The twice-accented octave* (1-foot) :—

And so on for higher octaves. Below the small octave lies the great octave, written with unaccented capital letters; its C requires an organ pipe of eight feet ¶ in length, and hence it is called the 8-foot octave.

4. *Great or 8-foot octave* :—

Below this follows the 16-*foot* or *contra-octave*; the lowest on the pianoforte and most organs, the tones of which may be represented by $C_{,} D_{,} E_{,} F_{,} G_{,} A_{,} B_{,}$, with an inverted accent. On great organs there is a still deeper, 32-foot octave, the tones of which may be written $C_{,,} D_{,,} E_{,,} F_{,,} G_{,,} A_{,,} B_{,,}$, with two inverted accents, but they scarcely retain the character of musical tones. (See Chap. IX.)

¶ Since the pitch numbers of any octave are always twice as great as those for the next deeper, we find the pitch numbers of the higher tones by *multiplying* those of the small or unaccented octave as many times by 2 as its symbol has upper accents. And on the contrary the pitch numbers for the deeper octaves are found by *dividing* those of the great octave, as often as its symbol has lower accents.

$$\text{Thus } c'' = 2 \times 2 \times c = 2 \times 2 \times 2 \, C$$
$$C_{,,} = \tfrac{1}{2} \times \tfrac{1}{2} \times C = \tfrac{1}{2} \times \tfrac{1}{2} \times \tfrac{1}{2} c.$$

For the pitch of the musical scale German physicists have generally adopted that proposed by Scheibler, and adopted subsequently by the German Association of Natural Philosophers (*die deutsche Naturforscherversammlung*) in 1834. This makes the once-accented a′ execute 440 vibrations in a second.* Hence results the

---

thus Bédos (*L'Art du Facteur d'Orgues*, 1766) ¶ made it 4 old French feet, which gave a note a full Semitone flatter than a pipe of 4 English feet. But in modern organs not even so much as 4 English feet are used. Organ builders, however, in all countries retain the names of the octaves as here given, which must be considered merely to determine the place on the staff, as noted in the text, independently of the precise pitch.—*Translator.*]

* The Paris Academy has lately fixed the pitch number of the same note at 435. This is called 870 by the Academy, because French physicists have adopted the inconvenient habit of counting the forward motion of a swinging body as one vibration, and the backward as another, so that the whole vibration is counted as two. This method of counting has been taken from the seconds pendulum, which ticks once in going forward and once again on returning. For symmetrical

backward and forward motions it would be indifferent by which method we counted, but for non-symmetrical musical vibrations which are of constant occurrence, the French method of counting is very inconvenient. The number 440 gives fewer fractions for the first [just] major scale of C, than a′ = 435. The difference of pitch is less than a comma. [The practical settlement of pitch has no relation to such arithmetical considerations as are here suggested, but depends on the compass of the human voice and the music written for it at different times. An Abstract of my *History of Musical Pitch* is given in Appendix XX. sect. H. Scheibler's proposal, named in the text, was chosen, as he tells us (*Der Tonmesser*, 1834, p. 53), as being the mean between the limits of pitch within which Viennese pianofortes at that time rose and fell by heat and cold, which he reckons at ¾ vibration either way. That this proposal had no reference to the

following table for the scale of C major, which will serve to determine the pitch of all tones that are defined by their pitch numbers in the following work.

Notes	Contra Octave $C_{,}$ to $B_{,}$ 16 foot	Great Octave C to B 8 foot	Unaccented Octave c to b 4 foot	Once-accented Octave c' to b' 2 foot	Twice-accented Octave c'' to b'' 1 foot	Thrice-accented Octave c''' to b''' $\frac{1}{2}$ foot	Four-times accented Octave c'''' to b'''' $\frac{1}{4}$ foot
C	33	66	132	264	528	1056	2112
D	37·125	74·25	148·5	297	594	1188	2376
E	41·25	82·5	165	330	660	1320	2640
F	44	88	176	352	704	1408	2816
G	49·5	99	198	396	792	1584	3168
A	55	110	220	440	880	1760	3520
B	61·875	123·75	247·5	495	990	1980	3960 *

The lowest tone on orchestral instruments is the $E_{,}$ of the double bass, making $41\frac{1}{4}$ vibrations in a second.† Modern pianofortes and organs usually go down to $C_{,}$ ¶

expression of the just major scale in whole numbers, is shewn by the fact that he proposed it for an equally tempered scale, for which he calculated the pitch numbers to four places of decimals, and for which, of course, none but the octaves of a' are expressible by whole numbers.—*Translator*.]

* [As it is important that students should be able to hear the exact intervals and pitches spoken of throughout this book, and as it is quite impossible to do so on any ordinary instrument, I have contrived a specially-tuned harmonium, called an Harmonical, fully described in App. XX. sect. F. No. 1, which Messrs. Moore & Moore, 104 Bishopsgate Street, will, in the interests of science, supply to order, for the moderate sum of 165s. The following are the pitch numbers of the first four octaves, the tuning of the fifth octave will be

explained in App. XX. sect. F. The names of the notes are in the notation of the latter part of Chap. XIV. below. Read the sign $D_{,}$ as 'D one,' $E^{,}b$ as 'one E flat,' and $'Bb$ as 'seven B flat.' In playing observe that $D_{,}$ is on the ordinary Db or $C\sharp$ digital, and that $'Bb$ is on the ordinary Gb or $F\sharp$ digital, and that the only keys in which chords can be played are C major and C minor, with the minor chord $D_{,}FA_{,}$ and the natural chord of the Ninth $CE_{,}G'BbD$. The mode of measuring intervals by *ratios* and *cents* is fully explained hereafter, and the results are added for convenience of reference. The pitches of c'' 528, a' 440, $a''b$ 422·4 and $'b'b$ 462, were taken from forks very carefully tuned by myself to these numbers of vibrations, by means of my unique series of forks described in App. XX., at the ¶ end of sect. B.

SCALE OF THE HARMONICAL.

| Notes | Pitch Numbers. | | | | Ratios | | Cents | |
	8 foot	4 foot	2 foot	1 foot	Note to Note	C to Note	Note to Note	C to Note
C	66	132	264	528	—	1 : 1	—	0
					9 : 10		182	
$D_{,}$	$73\frac{1}{3}$	$146\frac{2}{3}$	$293\frac{1}{3}$	$586\frac{2}{3}$	—	9 : 10	—	182
					80 : 81		22	
D	$74\frac{1}{4}$	$148\frac{1}{2}$	297	594	—	8 : 9	—	204
					15 : 16		112	
$E^{,}b$	$79\frac{1}{5}$	$158\frac{2}{5}$	$316\frac{4}{5}$	$633\frac{3}{5}$	—	5 : 6	—	316
					24 : 25		70	
$E_{,}$	$82\frac{1}{2}$	165	330	660	—	4 : 5	—	386
					15 : 16		112	
F	88	176	352	704	—	3 : 4	—	498
					8 : 9		204	
G	99	198	396	792	—	2 : 3	—	702
					15 : 16		112	
$A^{,}b$	$105\frac{3}{5}$	$211\frac{1}{5}$	$422\frac{2}{5}$	$844\frac{4}{5}$	—	5 : 8	—	814
					24 : 25		70	
$A_{,}$	110	220	440	880	—	3 : 5	—	884
					20 : 21		85	
$'Bb$	$115\frac{1}{2}$	231	462	924	—	4 : 7	—	969
					35 : 36		49	
$B^{,}b$	$118\frac{4}{5}$	$237\frac{3}{5}$	$475\frac{1}{5}$	$950\frac{2}{5}$	—	5 : 9	—	1018
					24 : 25		70	
$B_{,}$	$123\frac{3}{4}$	$247\frac{1}{2}$	495	990	—	8 : 15	—	1088
					15 : 16		112	
C	132	264	528	1056	—	1 : 2	—	1200

¶

Translator.]

† [The following account of the actual tones used is adapted from my *History of Musical*

Pitch. $C_{,,}$ commencement of the 32-foot octave, the lowest tone of very large organs, two

with 33 vibrations, and the latest grand pianos even down to $A_{//}$ with $27\frac{1}{2}$ vibrations. On larger organs, as already mentioned, there is also a deeper Octave reaching to $C_{//}$ with $16\frac{1}{2}$ vibrations. But the musical character of all these tones below $E_{/}$ is imperfect, because we are here near to the limit of the power of the ear to combine vibrations into musical tones. These lower tones cannot therefore be used musically except in connection with their higher octaves to which they impart a character of greater depth without rendering the conception of the pitch indeterminate.

Upwards, pianofortes generally reach a'''' with 3520, or even c^v with 4224 vibrations. The highest tone in the orchestra is probably the five-times accented d^v of the piccolo flute with 4752 vibrations. Appunn and W. Preyer by means of small tuning-forks excited by a violin bow have even reached the eight times accented e^{viii} with 40,960 vibrations in a second. These high tones were very painfully unpleasant, and the pitch of those which exceed the boundaries of the musical scale was ¶ very imperfectly discriminated by musical observers.* More on this in Chap. IX.

The musical tones which can be used with advantage, and have clearly distinguishable pitch, have therefore between 40 and 4000 vibrations in a second, extending over 7 octaves. Those which are audible at all have from 20 to 40,000 vibrations, extending over about 11 octaves. This shews what a great variety of different pitch numbers can be perceived and distinguished by the ear. In this respect the ear is far superior to the eye, which likewise distinguishes light of different periods of vibration by the sensation of different colours, for the compass of the vibrations of light distinguishable by the eye but slightly exceeds an Octave.†

Force and *pitch* were the two first differences which we found between musical tones; the third was *quality of tone*, which we have now to investigate. When

Octaves below the lowest tone of the Violoncello. $A_{///}$ the lowest tone of the largest pianos. $C_{/}$ commencement of the 16-foot octave, the lowest note assigned to the Double ¶ Bass in Beethoven's Pastoral Symphony. $E_{/}$, the lowest tone of the German four-stringed Double Bass, the lowest tone mentioned in the text. $F_{/}$, the lowest tone of the English four-stringed Double Bass. $G_{/}$, the lowest tone of the Italian three-stringed Double Bass. $A_{/}$, the lowest tone of the English three-stringed Double Bass. C, commencement of the 8-foot octave, the lowest tone of the Violoncello, written on the second leger line below the bass staff. G, the tone of the third open string of the Violoncello. c, commencement of the 4-foot octave 'tenor C,' the lowest tone of the Vióla, written on the second space of the bass staff. d, the tone of the second open string of the Violoncello. f, the tone signified by the bass or F-clef. g, the lowest tone of the Violin. a, the tone of the highest open string ¶ of the Violoncello. c', commencement of the 2-foot octave, 'middle C,' written on the leger line between the bass and treble staves, the tone signified by the tenor or C-clef. d', the tone of the third open string of the Violin. g', the tone signified by the treble or G-clef. a', the tone of the second open string of the Violin, the 'tuning note' for orchestras. c'', commencement of the 1-foot octave, the usual 'tuning note' for pianos. e'', the tone of the first or highest open string of the Violin. c''', commencement of the $\frac{1}{2}$-foot octave. g''', the usual highest tone of the Flute. c^{iv}, commencement of the $\frac{1}{4}$-foot octave. e^{iv}, the highest tone on the Violin, being the double Octave harmonic of the tone of the highest open string. a^{iv}, the usual highest tone of large pianos. d^v, the highest tone of the piccolo flute. e^{viii}, the highest tone reached by Appunn's forks, see next note.—*Translator.*]

* [Copies of these forks, described in Prof. Preyer's essay 'On the Limits of the Perception of Tone,' (*über die Grenzen der Tonwahrnehmung*, 1876, p. 20), are in the South Kensington Museum, Scientific Collection. I have several times tried them. I did not myself find the tones painful or cutting, probably because there was no beating of inharmonic upper partials. It is best to sound them with two violin bows, one giving the octave of the other. The tones can be easily heard at a distance of more than 100 feet in the gallery of the Museum.—*Translator.*]

† [Assuming the undulatory theory, which attributes the sensation of light to the vibrations of a supposed luminous 'ether,' resembling air but more delicate and mobile, then the phenomena of 'interference' enables us to calculate the lengths of waves of light in empty space, &c., hence the numbers of vibrations in a second, and consequently the ratios of these numbers, which will then clearly resemble the ratios of the pitch numbers that measure musical intervals. Assuming, then, that the yellow of the spectrum answers to the tenor c in music, and Fraunhofer's 'line A' corresponds to the G below it, Prof. Helmholtz, in his *Physiological Optics*, (*Handbuch der physiologischen Optik*, 1867, p. 237), gives the following analogies between the notes of the piano and the colours of the spectrum :—

$F\sharp$, end of the Red.	$f\sharp$,	Violet.
G, Red.	g,	Ultra-violet.
$G\sharp$, Red.	$g\sharp$,	,,
A, Red.	a,	,,
$A\sharp$, Orange-red.	$a\sharp$,	,,
B, Orange.	b,	end of the solar spectrum.
c, Yellow.		The scale therefore extends to about a Fourth beyond the octave. — *Translator.*]
$c\sharp$, Green.		
d, Greenish-blue.		
$d\sharp$, Cyanogen-blue.		
e, Indigo-blue.		
f, Violet.		

we hear notes of the same force and same pitch sounded successively on a pianoforte, a violin, clarinet, oboe, or trumpet, or by the human voice, the character of
the musical tone of each of these instruments, notwithstanding the identity of force
and pitch, is so different that by means of it we recognise with the greatest ease
which of these instruments was used. Varieties of quality of tone appear to be
infinitely numerous. Not only do we know a long series of musical instruments
which could each produce a note of the same pitch ; not only do different individual
instruments of the same species, and the voices of different individual singers shew
certain more delicate shades of quality of tone, which our ear is able to distinguish ;
but notes of the same pitch can sometimes be sounded on the same instrument with
several qualitative varieties. In this respect the ' bowed ' instruments (i.e. those
of the violin kind) are distinguished above all other. But the human voice is still
richer, and human speech employs these very qualitative varieties of tone, in order
to distinguish different letters. The different vowels, namely, belong to the class ¶
of sustained tones which can be used in music, while the character of consonants
mainly depends upon brief and transient noises.

On inquiring to what external physical difference in the waves of sound the
different qualities of tone correspond, we must remember that the amplitude of
the vibration determines the force or loudness, and the period of vibration the
pitch. Quality of tone can therefore depend upon neither of these. The only
possible hypothesis, therefore, is that the quality of tone should depend upon the
manner in which the motion is performed within the period of each single vibration. For the generation of a musical tone we have only required that the motion
should be periodic, that is, that in any one single period of vibration exactly the
same state should occur, in the same order of occurrence as it presents itself in any
other single period. As to the kind of motion that should take place within any
single period, no hypothesis was made. In this respect then an endless variety of
motions might be possible for the production of sound.   ¶

Observe instances, taking first such periodic motions as are performed so slowly
that we can follow them with the eye. Take a pendulum, which we can at any
time construct by attaching a weight to a thread and setting it in motion. The
pendulum swings from right to left with a uniform motion, uninterrupted by jerks.
Near to either end of its path it moves slowly, and in the middle fast. Among
sonorous bodies, which move in the same way, only very much faster, we may
mention tuning-forks. When a tuning-fork is struck or is excited by a violin bow,
and its motion is allowed to die away slowly, its two prongs oscillate backwards
and forwards in the same way and after the same law as a pendulum, only they
make many hundred swings for each single swing of the pendulum.

As another example of a periodic motion, take a hammer moved by a water-
wheel. It is slowly raised by the millwork, then released, and falls down suddenly,
is then again slowly raised, and so on. Here again we have a periodical backwards
and forwards motion ; but it is manifest that this kind of motion is totally different ¶
from that of the pendulum. Among motions which produce musical sounds, that of
a violin string, excited by a bow, would most nearly correspond with the hammer's,
as will be seen from the detailed description in Chap. V. The string clings for a
time to the bow, and is carried along by it, then suddenly releases itself, like the
hammer in the mill, and, like the latter, retreats somewhat with much greater
velocity than it advanced, and is again caught by the bow and carried forward.

Again, imagine a ball thrown up vertically, and caught on its descent with a
blow which sends it up again to the same height, and suppose this operation to be
performed at equal intervals of time. Such a ball would occupy the same time in
rising as in falling, but at the lowest point its motion would be suddenly interrupted,
whereas at the top it would pass through gradually diminishing speed of ascent
into a gradually increasing speed of descent. This then would be a third kind of
alternating periodic motion, and would take place in a manner essentially different
from the other two.

To render the law of such motions more comprehensible to the eye than is possible by lengthy verbal descriptions, mathematicians and physicists are in the habit of applying a graphical method, which must be frequently employed in this work, and should therefore be well understood.

To render this method intelligible suppose a drawing point b, fig. 5, to be fastened to the prong A of a tuning-fork in such a manner as to mark a surface of paper B B. Let the tuning-fork be moved with a uniform velocity in the direction of the upper arrow, or else the paper be drawn under it in the opposite direction, as shewn by the lower arrow. When the fork is not sounding, the point will describe the dotted straight line d c. But if the prongs have been first set in vibration, the point will describe the undulating line d c, for as the prong vibrates, the attached point b will constantly move backwards and forwards, and hence be

¶

FIG. 5.

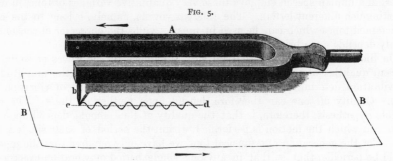

sometimes on the right and sometimes on the left of the dotted straight line d c, as is shewn by the wavy line in the figure. This wavy line once drawn, remains as a permanent image of the kind of motion performed by the end of the fork during
¶ its musical vibrations. As the point b is moved in the direction of the straight line d c with a constant velocity, equal sections of the straight line d c will correspond to equal sections of the time during which the motion lasts, and the distance of the wavy line on either side of the straight line will shew how far the point b has moved from its mean position to one side or the other during those sections of time.

In actually performing such an experiment as this, it is best to wrap the paper over a cylinder which is made to rotate uniformly by clockwork. The paper is wetted, and then passed over a turpentine flame which coats it with lampblack, on which a fine and somewhat smooth steel point will easily trace delicate lines.

FIG. 6.

¶

Fig. 6 is the copy of a drawing actually made in this way on the rotating cylinder of Messrs. Scott and Koenig's *Phonautograph*.

Fig. 7 shews a portion of this curve on a larger scale. It is easy to see the meaning of such a curve. The drawing point has passed with a uniform velocity in the direction e h. Suppose that it has described the section e g in $\frac{1}{10}$ of a second. Divide e g into 12 equal parts, as in the figure, then the point has been $\frac{1}{120}$ of a second in describing the length of any such section horizontally, and the curve shews us on what side and at what distance from the position of rest the vibrating point will be at the end of $\frac{1}{120}$, $\frac{2}{120}$, and so on, of a second, or, generally, at any given short interval of time since it left the point e. We see, in the figure, that after $\frac{1}{120}$ of a second it had reached the height 1, and that it rose gradually till the end of $\frac{3}{120}$ of a second ; then, however, it began to descend gradually till, at the end of $\frac{6}{120} = \frac{1}{20}$ second, it had reached its mean position f, and then it continued descending on the opposite side till the end of $\frac{3}{120}$ of a second and so on.　We can also easily determine where the vibrating point was to be found at the end of any fraction of this hundred-and-twentieth of a second.　A drawing of this kind consequently shews immediately at what point of its path a vibrating particle is to be found at any given instant, and hence gives a complete image of its motion.　If the reader wishes to reproduce the motion of the vibrating point, he has only to cut a narrow vertical slit in a piece of paper, and place it over fig. 6 or fig. 7, so as to shew a very small portion of the curve through the vertical slit, and draw the book slowly but uniformly under the slit, from right to left; the white or black point in the slit will then appear to move backwards and forwards in precisely the same manner as the original drawing point attached to the fork, only of course much more slowly.

We are not yet able to make all vibrating bodies describe their vibrations

¶

FIG. 7.

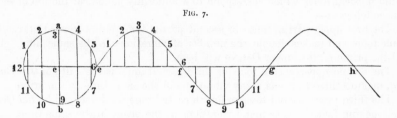

directly on paper, although much progress has recently been made in the methods required for this purpose.　But we are able ourselves to draw such curves for all sounding bodies, when the law of their motion is known, that is, when we know how far the vibrating point will be from its mean position at any given moment of time.　We then set off on a horizontal line, such as e f, fig. 7, lengths corresponding to the interval of time, and let fall perpendiculars to it on ¶ either side, making their lengths equal or proportional to the distance of the vibrating point from its mean position, and then by joining the extremities of these perpendiculars we obtain a curve such as the vibrating body would have drawn if it had been possible to make it do so.

Thus fig. 8 represents the motion of the hammer raised by a water-wheel, or of a point in a string excited by a violin bow.　For the first 9 intervals it rises slowly and uniformly, and during the 10th it falls suddenly down.

FIG. 8.　　　　　　　　　　　　　　FIG. 9.

Fig. 9 represents the motion of the ball which is struck up again as soon as it ¶ comes down.　Ascent and descent are performed with equal rapidity, whereas in fig. 8 the ascent takes much longer time.　But at the lowest point the blow suddenly changes the kind of motion.

Physicists, then, having in their mind such curvilinear forms, representing the law of the motion of sounding bodies, speak briefly of the *form of vibration* of a sounding body, and assert that *the quality of tone depends on the form of vibration.* This assertion, which has hitherto been based simply on the fact of our knowing that the quality of the tone could not possibly depend on the periodic time of a vibration, or on its amplitude (p. 10c), will be strictly examined hereafter.　It will be shewn to be in so far correct that every different quality of tone requires a different form of vibration, but on the other hand it will also appear that different forms of vibration may correspond to the same quality of tone.

On exactly and carefully examining the effect produced on the ear by different forms of vibration, as for example that in fig. 8, corresponding nearly to a violin string, we meet with a strange and unexpected phenomenon, long known indeed to individual musicians and physicists, but commonly regarded as a mere curiosity, its generality and its great significance for all matters relating to musical tones not having been recognised.   The ear when its attention has been properly directed to the effect of the vibrations which strike it, does not hear merely that one musical tone whose pitch is determined by the period of the vibrations in the manner already explained, but in addition to this it becomes aware of a whole series of higher musical tones, which we will call the *harmonic upper partial tones*, and sometimes simply the *upper partials* of the whole musical tone or note, in contra-distinction to the *fundamental* or *prime partial tone* or simply the *prime*, as it may be called, which is the lowest and generally the loudest of all the partial tones, and by the pitch of which we judge of the pitch of the whole *compound musical tone* itself.   The series of these upper partial tones is precisely the same for all com-

¶ pound musical tones which correspond to a uniformly periodical motion of the air. It is as follows :—

The first upper partial tone [or second partial tone] is the upper Octave of the prime tone, and makes double the number of vibrations in the same time.   If we call the prime *C*, this upper Octave will be *c*.

The second upper partial tone [or third partial tone] is the Fifth of this Octave, or *g*, making three times as many vibrations in the same time as the prime.

The third upper partial tone [or fourth partial tone] is the second higher Octave, or *c′*, making four times as many vibrations as the prime in the same time.

The fourth upper partial tone [or fifth partial tone] is the major Third of this second higher Octave, or *e′*, with five times as many vibrations as the prime in the same time.

The fifth upper partial tone [or sixth partial tone] is the Fifth of the second higher Octave, or *g′*, making six times as many vibrations as the prime in the

¶ same time.

And thus they go on, becoming continually fainter, to tones making 7, 8, 9, &c., times as many vibrations in the same time, as the prime tone.   Or in musical notation

Ordinal number of partials															
C	*c*	*g*	*c′*	*e′*	*g′*	⁷*b′*♭	*c″*	*d″*	*e″*	¹¹*f″*	*g″*	¹³*a″*	⁷*b″*♭	*b″*	*c‴*
1	2	3	4	5	6	7	8	9	10	11	12	13	14	15	16
Pitch number 66	132	198	264	330	396	462	528	594	660	726	792	858	924	990	1054 *

where the figures [in the first line] beneath shew how many times the corresponding pitch number is greater than that of the prime tone [and, taking the lowest note to have 66 vibrations, those in the second line give the pitch numbers of all the

¶ other notes].

The whole sensation excited in the ear by a periodic vibration of the air we

* [This diagram has been slightly altered to introduce all the first 16 harmonic partials of *C* 66, (which, excepting 11 and 13, are given on the Harmonical as harmonic notes,) and to shew the notation, symbolising, both in letters and on the staff, the 7th, 11th, and 13th harmonic partials, which are not used in general music.   It is easy to shew on the Harmonical that its lowest note, *C* of this series, contains all these partials, after the theory of the beats of a disturbed unison has been explained in Chap. VIII.   Keep down the note *C*, and touch in succession the notes *c, g, c′, e′, g′*, &c., but in touching the latter press the finger-key such a little way down that the tone of the note is only just audible. This slightly flattens each note, and slow beats can be produced in every case (except, of course, 11 and 13, which are not on the instrument) up to 16.   It should also be observed that the pitch of the beat is very nearly that of the upper (*not* the lower) note in each case.   The whole of these 16 harmonics of *C* 66 (except the 11th and 13th) can be played at once on the Harmonical by means of the harmonical bar, first without and then with the 7th and 14th.   The whole series will be found to sound like a single fine note, and the 7th and 14th to materially increase its richness.   The relations of the partials in this case may be studied from the tables in the footnotes to Chap. X.—*Translator.*]

have called a *musical tone*. We now find that this is *compound*, containing a series of different tones, which we distinguish as the *constitutents* or *partial tones* of the *compound*. The first of these constituents is the *prime partial tone* of the compound, and the rest its *harmonic upper partial tones*. The *number* which shews the *order* of any partial tone in the series shews how many times its vibrational number exceeds that of the prime tone.* Thus, the second partial tone makes twice as many, the third three times as many vibrations in the same time as the prime tone, and so on.

G. S. Ohm was the first to declare that there is only one form of vibration which will give rise to no harmonic upper partial tones, and which will therefore consist solely of the prime tone. This is the form of vibration which we have described above as peculiar to the pendulum and tuning-forks, and drawn in figs. 6 and 7 (p. 10). We will call these *pendular vibrations*, or, since they cannot be analysed into a compound of different tones, *simple vibrations*. In what sense not ¶ merely other musical tones, but all other forms of vibration, may be considered as *compound*, will be shewn hereafter (Chap. IV.). The terms *simple* or *pendular vibration*,† will therefore be used as synonymous. We have hitherto used the expression tone and musical tone indifferently. It is absolutely necessary to distinguish in acoustics first, a *musical tone*, that is, the impression made by *any* periodical vibration of the air ; secondly, a *simple tone*, that is, the impression produced by a *simple* or pendular vibration of the air; and thirdly, a *compound tone*, that is, the impression produced by the *simultaneous* action of *several* simple tones with certain definite ratios of pitch as already explained. A *musical tone* may be either *simple* or *compound*. For the sake of brevity, *tone* will be used in

* [The ordinal number of a partial tone in general, must be distinguished from the ordinal number of an *upper* partial tone in particular. For the same tone the former number is always greater by unity than the latter, because the partials in general include the prime, which is reckoned as the first, and the *upper* partials exclude the prime, which being the *lowest* partial is of course not an *upper* partial at all. Thus the partials generally numbered 2 3 4 5 6 7 8 9 are the same as the upper partials numbered 1 2 3 4 5 6 7 8 respectively. As even the Author has occasionally failed to carry out this distinction in the original German text, and other writers have constantly neglected it, too much weight cannot be here laid upon it. The presence or absence of the word *upper* before the word *partial* must always be care-

fully observed. It is safer never to speak of an upper partial by its ordinal number, but to call the *fifth upper* partial the *sixth* partial, omitting the word *upper* and increasing the ¶ ordinal number by one place. And so in other cases.—*Translator*.]

† The law of these vibrations may be popularly explained by means of the construction in fig. 10. Suppose a point to describe the circle of which c is the centre with a uniform velocity, and that an observer stands at a considerable distance in the prolongation of the line e h, so that he does not see the surface of the circle but only its edge, in which case the point will appear merely to move up and down along its diameter a b. This up and down motion would take place exactly according to the law of pendular vibration. To represent this motion graphi-

FIG. 10.

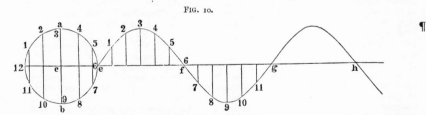

¶

cally by means of a curve, divide the length e g, supposed to correspond to the time of a single period, into as many (here 12) equal parts as the circumference of the circle, and draw the perpendiculars 1, 2, 3, &c., on the dividing points of the line e g in order, equal in length to and in the same direction with, those drawn in the circle from the corresponding points 1, 2, 3, &c. In this way we obtain the curve drawn in fig. 10, which agrees in

form with that drawn by the tuning-fork, fig. 6, p. 20b, but is of a larger size. Mathematically expressed, the distance of the vibrating point from its mean position at any time is equal to the sine of an arc proportional to the corresponding time, and hence the form of simple vibrations are also called the *sine-vibrations* [and the above curve is also known as the *curve of sines*].

the general sense of a *musical tone*, leaving the context or a prefixed qualification to determine whether it is simple or compound. A *compound tone* will often be briefly called a *note*, and a *simple tone* will also be frequently called a *partial*, when used in connection with a compound tone; otherwise, the full expression *simple tone* will be employed. A note has, properly speaking, no single pitch, as it is made up of various partials each of which has its own pitch. By the *pitch of a note or compound tone* then we shall therefore mean the *pitch of its lowest partial* or prime tone. By a *chord or combination of tones* we mean several musical tones (whether simple or compound) produced by different instruments or different parts of the same instrument so as to be heard at the same time. The facts here adduced shew us then that every musical tone in which harmonic upper partial tones can be distinguished, although produced by a single instrument, may really be considered as in itself a chord or combination of various simple tones.*

¶ * [The above paragraph relating to the English terms used in this translation, necessarily differs in many respects from the original, in which a justification is given of the use made by the Author of certain *German* expressions. It has been my object to employ terms which should be thoroughly English, and should not in any way recall the German words. The word *tone* in English is extremely ambiguous. Prof. Tyndall (*Lectures on Sound*, 2nd ed. 1869, p. 117) has ventured to define a *tone* as a *simple tone*, in agreement with Prof. Helmholtz, who in the present passage limits the German word *Ton* in the same way. But I felt that an English reader could not be safely trusted to keep this very peculiar and important class of musical tones, which he has very rarely or never heard separately, invariably distinct from those musical tones ¶ with which he is familiar, unless the word *tone* were uniformly qualified by the epithet *simple*. The only exception I could make was in the case of a *partial tone*, which is received at once as a new conception. Even Prof. Helmholtz himself has not succeeded in using his word *Ton* consistently for a *simple tone* only, and this was an additional warning to me. English musicians have been also in the habit of using *tone* to signify a certain musical interval, and *semitone* for half of that interval, on the equally tempered scale. In this case I write *Tone* and *Semitone* with capital initials, a practice which, as already explained (note, p. 13*d'*,) I have found convenient for the names of all intervals, as *Thirds, Fifths,* &c. Prof. Helmholtz uses the word *Klang* for a *musical tone,* which gene- ¶ rally, but not always, means a *compound tone.* Prof. Tyndall (*ibid.*) therefore proposes to use the English word *clang* in the same sense. But *clang* has already a meaning in English, thus defined by Webster: 'a sharp shrill sound, made by striking together metallic substances, or sonorous bodies, as the *clang* of arms, or any like sound, as the *clang* of trumpets. This word implies a degree of harshness in the sound, or more harshness than *clink*.' Interpreted scientifically, then, *clang* according to this definition, is either *noise* or one of those *musical tones with inharmonic upper partials*, which will be subsequently explained. It is therefore totally unadapted to represent a musical tone in general, for which the simple word *tone* seems eminently suited, being of course originally the tone produced by a *stretched* string. The common word *note*, properly the mark by

which a musical tone is written, will also, in accordance with the general practice of musicians, be used for a *musical tone*, which is generally compound, without necessarily implying that it is one of the few recognised tones in our musical scale. Of course, if *clang* could not be used, Prof. Tyndall's suggestion to translate Prof. Helmholtz's *Klangfarbe* by *clangtint* (*ibid.*) fell to the ground. I can find no valid reason for supplanting the time-honoured expression *quality of tone.* Prof. Tyndall (*ibid.*) quotes Dr. Young to the effect that ' this quality of sound is sometimes called its register, colour, or timbre.' *Register* has a distinct meaning in vocal music which must not be disturbed. *Timbre,* properly a kettledrum, then a helmet, then the coat of arms surmounted with a helmet, then the official stamp bearing that coat of arms (now used in France for a postage label), and then the mark which declared a thing to be what it pretends to be, Burns's ' guinea's stamp,' is a foreign word, often odiously mispronounced, and not worth preserving. *Colour* I have never met with as applied to music, except at most as a passing metaphorical expression. But the difference of tones in *quality* is familiar to our language. Then as to the Partial Tones, Prof. Helmholtz uses *Theiltöne* and *Partialtöne*, which are aptly Englished by *partial simple tones.* The words *simple* and *tone,* however, may be omitted when *partials* is employed, as partials are necessarily both tones and simple. The *constituent tones* of a chord may be either *simple* or *compound.* The *Grundton* or fundamental tone of a compound tone then becomes its *prime tone,* or briefly its *prime.* The *Grundton* or *root* of a *chord* will be further explained hereafter. *Upper* partial (simple) tones, that is, the partials exclusive of the prime, even when *harmonic,* (that is, for the most part, belonging to the first six partial tones,) must be distinguished from the sounds usually called *harmonics* when produced on a violin or harp for instance, for such *harmonics* are not necessarily *simple* tones, but are more generally compounds of *some* of the complete series of partial tones belonging to the musical tone of the whole string, selected by damping the remainder. The fading *harmonics* heard in listening to the sound of a pianoforte string, struck and undamped, as the sound dies away, are also compound and not simple partial tones, but as they have the successive partials for their successive primes, they have the

Now, since quality of tone, as we have seen, depends on the form of vibration, which also determines the occurrence of upper partial tones, we have to inquire how far differences in quality of tone depend on different force or loudness of upper partials. This inquiry will be found to give a means of clearing up our conceptions of what has hitherto been a perfect enigma,—the nature of quality of tone. And we must then, of course, attempt to explain how the ear manages to analyse every musical tone into a series of partial tones, and what is the meaning of this analysis. These investigations will engage our attention in the following chapters.

CHAPTER II.

ON THE COMPOSITION OF VIBRATIONS.

AT the end of the last chapter we came upon the remarkable fact that the human ear is capable, under certain conditions, of separating the musical tone produced by a single musical instrument, into a series of simple tones, namely, the prime partial tone, and the various upper partial tones, each of which produces its own separate sensation. That the ear is capable of distinguishing from each other tones proceeding from different sources, that is, which do not arise from one and the same sonorous body, we know from daily experience. There is no difficulty during a concert in following the melodic progression of each individual instrument or voice, if we direct our attention to it exclusively; and, after some practice, most persons can succeed in following the simultaneous progression of several united parts. This is true, indeed, not merely for musical tones, but also for noises, and for mixtures of music and noise. When several persons are speaking at once, we can generally listen at pleasure to the words of any single one of them, and even understand those words, provided that they are not too much overpowered by the mere loudness of the others. Hence it follows, first, that many different trains of waves of sound can be propagated at the same time through the same mass of air, without mutual disturbance; and, secondly, that the human ear is capable of again analysing into its constituent elements that composite motion of the air which is produced by the simultaneous action of several musical instruments. We will first investigate the nature of the motion of the air when it is produced by several simultaneous musical tones, and how such a compound motion is distinguished from that due to a single musical tone. We shall see that the ear has no decisive test by which it can in all cases distinguish between the effect of a

---

pitch of those partials. But these fading harmonics are not regular compound tones of the kind described on p. 22a, because the lower partials are absent one after another. Both sets of harmonics serve to indicate the existence and place of the partials. But they are no more those upper partial tones themselves, than the original compound tone of the string is its own prime. Great confusion of thought having, to my own knowledge, arisen from confounding such harmonics with upper partial tones, I have generally avoided using the ambiguous substantive harmonic. Properly speaking the harmonics of any compound tone are other compound tones of which the primes are partials of the original compound tone of which they are said to be harmonics. Prof. Helmholtz's term Obertöne is merely a contraction for Oberpartialtöne, but the casual resemblance of the sounds of ober and over, has led Prof. Tyndall to the erroneous translation overtones. The German ober is an adjective meaning upper, but the English preposition over is equivalent to the German preposition über. Compare Oberzahn, an 'upper tooth,' i.e. a tooth in the upper jaw, with Ueberzahn, an 'overtooth,' i.e. one grown over another, a projecting tooth. The continual recurrence of such words as clang, clangtint, overtone, would combine to give a strange un-English appearance to a translation from the German. On the contrary I have endeavoured to put it into as straightforward English as possible. But for those acquainted with the original and with Prof. Tyndall's work, this explanation seemed necessary. Finally I would caution the reader against using overtones for partial tones in general, as almost every one who adopts Prof. Tyndall's word is in the habit of doing. Indeed I have in the course of this translation observed, that even Prof. Helmholtz himself has been occasionally misled to employ Obertöne in the same loose manner. See my remarks in note, p. 23c.—Translator.]

motion of the air caused by several different musical tones arising from different sources, and that caused by the musical tone of a single sounding body. Hence the ear has to analyse the composition of single musical tones, under proper conditions, by means of the same faculty which enabled it to analyse the composition of simultaneous musical tones. We shall thus obtain a clear conception of what is meant by analysing a single musical tone into a series of partial simple tones, and we shall perceive that this phenomenon depends upon one of the most essential and fundamental properties of the human ear.

We begin by examining the motion of the air which corresponds to several simple tones acting at the same time on the same mass of air. To illustrate this kind of motion it will be again convenient to refer to the waves formed on a calm surface of water. We have seen (p. 9*a*) that if a point of the surface is agitated by a stone thrown upon it, the agitation is propagated in rings of waves over the surface

¶ o more and more distant points. Now, throw two stones at the same time on to different points of the surface, thus producing two centres of agitation. Each will give rise to a separate ring of waves, and the two rings gradually expanding, will finally meet. Where the waves thus come together, the water will be set in motion by both kinds of agitation at the same time, but this in no wise prevents both series of waves from advancing further over the surface, just as if each were alone present and the other had no existence at all. As they proceed, those parts of both rings which had just coincided, again appear separate and unaltered in form. These little waves, caused by throwing in stones, may be accompanied by other kinds of waves, such as those due to the wind or a passing steamboat. Our circles of waves will spread out over the water thus agitated, with the same quiet regularity as they did upon the calm surface. Neither will the greater waves be essentially disturbed by the less, nor the less by the greater, provided the waves never break; if that happened, their regular course would certainly be impeded.

¶ Indeed it is seldom possible to survey a large surface of water from a high point of sight, without perceiving a great multitude of different systems of waves, mutually overtopping and crossing each other. This is best seen on the surface of the sea, viewed from a lofty cliff, when there is a lull after a stiff breeze. We first see the great waves, advancing in far-stretching ranks from the blue distance, here and there more clearly marked out by their white foaming crests, and following one another at regular intervals towards the shore. From the shore they rebound, in different directions according to its sinuosities, and cut obliquely across the advancing waves. A passing steamboat forms its own wedge-shaped wake of waves, or a bird, darting on a fish, excites a small circular system. The eye of the spectator is easily able to pursue each one of these different trains of waves, great and small, wide and narrow, straight and curved, and observe how each passes over the surface, as undisturbedly as if the water over which it flits were not agitated at the same time by other motions and other forces. I must own that

¶ whenever I attentively observe this spectacle it awakens in me a peculiar kind of intellectual pleasure, because it bares to the bodily eye, what the mind's eye grasps only by the help of a long series of complicated conclusions for the waves of the invisible atmospheric ocean.

We have to imagine a perfectly similar spectacle proceeding in the interior of a ball-room, for instance. Here we have a number of musical instruments in action, speaking men and women, rustling garments, gliding feet, clinking glasses, and so on. All these causes give rise to systems of waves, which dart through the mass of air in the room, are reflected from its walls, return, strike the opposite wall, are again reflected, and so on till they die out. We have to imagine that from the mouths of men and from the deeper musical instruments there proceed waves of from 8 to 12 feet in length [*c* to *F*], from the lips of the women waves of 2 to 4 feet in length [*c″* to *c′*], from the rustling of the dresses a fine small crumple of wave, and so on; in short, a tumbled entanglement of the most different kinds of motion, complicated beyond conception.

And yet, as the ear is able to distinguish all the separate constituent parts of this confused whole, we are forced to conclude that all these different systems of wave coexist in the mass of air, and leave one another mutually undisturbed. But how is it possible for them to coexist, since every individual train of waves has at any particular point in the mass of air its own particular degree of condensation and rarefaction, which determines the velocity of the particles of air to this side or that? It is evident that at each point in the mass of air, at each instant of time, there can be only one single degree of condensation, and that the particles of air can be moving with only one single determinate kind of motion, having only one single determinate amount of velocity, and passing in only one single determinate direction.

What happens under such circumstances is seen directly by the eye in the waves of water. If where the water shews large waves we throw a stone in, the waves thus caused will, so to speak, cut into the larger moving surface, and this ¶ surface will be partly raised, and partly depressed, by the new waves, in such a way that the fresh crests of the rings will rise just as much above, and the troughs sink just as much below the curved surfaces of the previous larger waves, as they would have risen above or sunk below the horizontal surface of calm water. Hence where a crest of the smaller system of rings of waves comes upon a crest of the greater system of waves, the surface of the water is raised by the sum of the two heights, and where a trough of the former coincides with a trough of the latter, the surface is depressed by the sum of the two depths. This may be expressed more briefly if we consider the heights of the crests above the level of the surface at rest, as positive magnitudes, and the depths of the troughs as negative magnitudes, and then form the so-called algebraical sum of these positive and negative magnitudes, in which case, as is well known, two positive magnitudes (heights of crests) must be added, and similarly for two negative magnitudes (depths of troughs); but when both negative and positive concur, one is to be subtracted ¶ from the other. Performing the addition then in this algebraical sense, we can express our description of the surface of the water on which two systems of waves concur, in the following simple manner : *The distance of the surface of the water at any point from its position of rest is at any moment equal to the [algebraical] sum of the distances at which it would have stood had each wave acted separately at the same place and at the same time.*

The eye most clearly and easily distinguishes the action in such a case as has been just adduced, where a smaller circular system of waves is produced on a large rectilinear system, because the two systems are then strongly distinguished from each other both by the height and shape of the waves. But with a little attention the eye recognises the same fact even when the two systems of waves have but slightly different forms, as when, for example, long rectilinear waves advancing towards the shore concur with those reflected from it in a slightly different direction. In this case we observe those well-known comb-backed waves where ¶ the crest of one system of waves is heightened at some points by the crests of the other system, and at others depressed by its troughs. The multiplicity of forms is here extremely great, and any attempt to describe them would lead us too far. The attentive observer will readily comprehend the result by examining any disturbed surface of water, without further description. It will suffice for our purpose if the first example has given the reader a clear conception of what is meant by *adding waves together*.*

Hence although the surface of the water at any instant of time can assume only one single form, while each of two different systems of waves simultaneously attempts to impress its own shape upon it, we are able to suppose in the above

* The velocities and displacements of the particles of water are also to be added according to the law of the so-called parallelogram of forces. Strictly speaking, such a simple addition of waves as is spoken of in the text, is not perfectly correct, unless the heights of the waves are infinitely small in comparison with their lengths.

sense that the two systems coexist and are superimposed, by considering the actual elevations and depressions of the surface to be suitably separated into two parts, each of which belongs to one of the systems alone.

In the same sense, then, there is also a superimposition of different systems of sound in the air. By each train of waves of sound, the density of the air and the velocity and position of the particles of air, are temporarily altered. There are places in the wave of sound comparable with the crests of the waves of water, in which the quantity of the air is increased, and the air, not having free space to escape, is condensed; and other places in the mass of air, comparable to the troughs of the waves of water, having a diminished quantity of air, and hence diminished density. It is true that two different degrees of density, produced by two different systems of waves, cannot coexist in the same place at the same time; nevertheless the condensations and rarefactions of the air can be (algebraically) ¶ added, exactly as the elevations and depressions of the surface of the water in the former case. Where two condensations are added we obtain increased condensation, where two rarefactions are added we have increased rarefaction; while a concurrence of condensation and rarefaction mutually, in whole or in part, destroy or neutralise each other.

The displacements of the particles of air are compounded in a similar manner. If the displacements of two different systems of waves are not in the same direction, they are compounded diagonally; for example, if one system would drive a particle of air upwards, and another to the right, its real path will be obliquely upwards towards the right. For our present purpose there is no occasion to enter more particularly into such compositions of motion in different directions. We are only interested in the effect of the mass of air upon the ear, and for this we are only concerned with the motion of the air in the passages of the ear. Now the passages of our ear are so narrow in comparison with the length of the waves of ¶ sound, that we need only consider such motions of the air as are parallel to the axis of the passages, and hence have only to distinguish displacements of the particles of air outwards and inwards, that is towards the outer air and towards the interior of the ear. For the magnitude of these displacements as well as for their velocities with which the particles of air move outwards and inwards, the same (algebraical) addition holds good as for the crests and troughs of waves of water.

Hence, *when several sonorous bodies in the surrounding atmosphere, simultaneously excite different systems of waves of sound, the changes of density of the air, and the displacements and velocities of the particles of the air within the passages of the ear, are each equal to the (algebraical) sum of the corresponding changes of density, displacements, and velocities, which each system of waves would have separately produced, if it had acted independently*; * and in this sense we can say that all the separate vibrations which separate waves of sound would ¶ have produced, coexist undisturbed at the same time within the passages of our ear.

After having thus in answer to the first question explained in what sense it is possible for several different systems of waves to coexist on the same surface of water or within the same mass of air, we proceed to determine the means possessed by our organs of sense, for analysing this composite whole into its original constituents.

I have already observed that an eye which surveys an extensive and disturbed surface of water, easily distinguishes the separate systems of waves from each other and follows their motions. The eye has a great advantage over the ear in being able to survey a large extent of surface at the same moment. Hence the eye readily sees whether the individual waves of water are rectilinear or curved, and whether they have the same centre of curvature, and in what direction they

---

* The same is true for the whole mass of external air, if only the addition of the displacements in different directions is made according to the law of the parallelogram of forces.

are advancing. All these observations assist it in determining whether two systems of waves are connected or not, and hence in discovering their corresponding parts. Moreover, on the surface of the water, waves of unequal length advance with unequal velocities, so that if they coincide at one moment to such a degree as to be difficult to distinguish, at the next instant one train pushes on and the other lags behind, so that they become again separately visible. In this way, then, the observer is greatly assisted in referring each system to its point of departure, and in keeping it distinctly visible during its further course. For the eye, then, two systems of waves having different points of departure can never coalesce; for example, such as arise from two stones thrown into the water at different points. If in any one place the rings of wave coincide so closely as not to be easily separable, they always remain separate during the greater part of their extent. Hence the eye could not be easily brought to confuse a compound with a simple undulatory motion. Yet this is precisely what the ear does under similar circum- ¶ stances when it separates the musical tone which has proceeded from a single source of sound, into a series of simple partial tones.

But the ear is much more unfavourably situated in relation to a system of waves of sound, than the eye for a system of waves of water. The ear is affected only by the motion of that mass of air which happens to be in the immediate neighbourhood of its tympanum within the aural passage. Since a transverse section of the aural passage is comparatively small in comparison with the length of waves of sound (which for serviceable musical tones varies from 6 inches to 32 feet),* it corresponds to a single point of the mass of air in motion. It is so small that distinctly different degrees of density or velocity could scarcely occur upon it, because the positions of greatest and least density, of greatest positive and negative velocity, are always separated by half the length of a wave. The ear is therefore in nearly the same condition as the eye would be if it looked at one point of the surface of the water through a long narrow tube, which would permit of ¶ seeing its rising and falling, and were then required to undertake an analysis of the compound waves. It is easily seen that the eye would, in most cases, completely fail in the solution of such a problem. The ear is not in a condition to discover how the air is moving at distant spots, whether the waves which strike it are spherical or plane, whether they interlock in one or more circles, or in what direction they are advancing. The circumstances on which the eye chiefly depends for forming a judgment, are all absent for the ear.

If, then, notwithstanding all these difficulties, the ear is capable of distinguishing musical tones arising from different sources—and it really shews a marvellous readiness in so doing—it must employ means and possess properties altogether different from those employed or possessed by the eye. But whatever these means may be—and we shall endeavour to determine them hereafter—it is clear that the analysis of a composite mass of musical tones must in the first place be closely connected with some determinate properties of the motion of the ¶ air, capable of impressing themselves even on such a very minute mass of air as that contained in the aural passage. If the motions of the particles of air in this passage are the same on two different occasions, the ear will receive the same sensation, whatever be the origin of those motions, whether they spring from one or several sources.

We have already explained that the mass of air which sets the tympanic membrane of the ear in motion, so far as the magnitudes here considered are concerned, must be looked upon as a single point in the surrounding atmosphere. Are there, then, any peculiarities in the motion of a single particle of air which would differ for a single musical tone, and for a combination of musical tones? We have seen that for each single musical tone there is a corresponding periodical

---

* [These are of course rather more than twice the length of the corresponding open flue organ pipes. See Chap. V. sect. 5, and compare p. 26d.—*Translator*.]

motion of the air, and that its pitch is determined by the length of the periodic time, but that the kind of motion during any one single period is perfectly arbitrary, and may indeed be infinitely various. If then the motion of the air lying in the aural passage is not periodic, or if at least its periodic time is not as short as that of an audible musical tone, this fact will distinguish it from any motion which belongs to a musical tone; it must belong either to noises or to several simultaneous musical tones. Of this kind are really the greater number of cases where the different musical tones have been only accidentally combined, and are therefore not designedly framed into musical chords; nay, even where orchestral music is performed, the method of tempered tuning which at present prevails, prevents an accurate fulfilment of the conditions under which alone the resulting motion of the air can be exactly periodic. Hence in the greater number of cases a want of periodicity in the motion might furnish a mark for distinguishing the presence ¶ of a composite mass of musical tones.

But a composite mass of musical tones may also give rise to a *purely periodic motion of the air*, namely, *when all the musical tones which intermingle, have pitch numbers which are all multiples of one and the same old number*, or, which

FIG. 11.

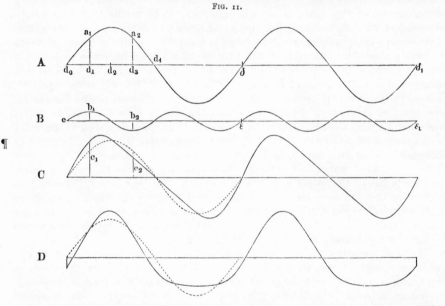

comes to the same thing, when *all these musical tones, so far as their pitch is concerned, may be regarded as the upper partial tones of the same prime tone*. It ¶ was mentioned in Chapter I. (p. 22a, *b*) that the pitch numbers of the upper partial tones are multiples of the pitch number of the prime tone. The meaning of this rule will be clear from a particular example. The curve A, fig. 11, represents a pendular motion in the manner explained in Chapter I. (p. 21b), as produced in the air of the aural passage by a tuning-fork in action. The horizontal lengths in the curves of fig. 11, consequently represent the passing time, and the vertical heights the corresponding displacements of the particles of air in the aural passage. Now suppose that with the first simple tone to which the curve A corresponds, there is sounded a second simple tone, represented by the curve B, an Octave higher than the first. This condition requires that two vibrations of the curve B should be made in the same time as one vibration of the curve A. In A, the sections of the curve $d_0\delta$ and $\delta\delta_1$ are perfectly equal and similar. The curve B is also divided into equal and similar sections e ϵ and ϵ $ϵ_1$ by the points e, ϵ, $ϵ_1$. We could certainly halve each of the sections e ϵ and ϵ $ϵ_1$, and thus obtain equal and similar sections, each of which would then correspond to a single period of B. But by taking sections consisting of two periods of B, we divide B into larger sections, each of which is of the same horizontal length, and hence corresponds to the same duration of time, as the sections of A.

If, then, both simple tones are heard at once, and the times of the points e and d_0, ϵ and δ, ϵ_1 and δ_1 coincide, the heights of the portions of the section of curve e ϵ have to be [algebraically] added to heights of the section of curve $d_0\delta$, and similarly for the sections $\epsilon\,\epsilon_1$ and $\delta\,\delta_1$. The result of this addition is shewn in the curve C. The dotted line is a duplicate of the section $d_0\delta$ in the curve A. Its object is to make the composition of the two sections immediately evident to the eye. It is easily seen that the curve C in every place rises as much above or sinks as much below the curve A, as the curve B respectively rises above or sinks beneath the horizontal line. The heights of the curve C are consequently, in accordance with the rule for compounding vibrations, equal to the [algebraical] sum of the corresponding heights of A and B. Thus the perpendicular c_1 in C is the ¶ sum of the perpendiculars a_1 and b_1 in A and B; the lower part of this perpendicular c_1, from the straight line up to the dotted curve, is equal to the perpendicular a_1, and the upper part, from the dotted to the continuous curve, is equal to the perpendicular b_1. On the other hand, the height of the perpendicular c_2 is equal to the height a_2 diminished by the depth of the fall b_2. And in the same way all other points in the curve C are found.*

It is evident that the motion represented by the curve C is also periodic, and that its periods have the same duration as those of A. Thus the addition of the section $d_0\delta$ of A and e ϵ of B, must give the same result as the addition of the perfectly equal and similar sections $\delta\,\delta_1$ and $\epsilon\,\epsilon_1$, and, if we supposed both curves to be continued, the same would be the case for all the sections into which they would be divided. It is also evident that equal sections of both curves could not continually coincide in this way after completing the addition, unless the curves thus added could be also separated into exactly equal and similar sections of the same ¶ length, as is the case in fig. 11, where two periods of B last as long or have the same horizontal length as one of A. Now the horizontal lengths of our figure represent time, and if we pass from the curves to the real motions, it results that the motion of air caused by the composition of the two simple tones, A and B, is also periodic, just because one of these simple tones makes exactly twice as many vibrations as the other in the same time.

It is easily seen by this example that the peculiar form of the two curves A and B has nothing to do with the fact that their sum C is also a periodic curve. Whatever be the form of A and B, provided that each can be separated into equal and similar sections which have the same horizontal lengths as the equal and similar sections of the other—no matter whether these sections correspond to one or two, or three periods of the individual curves—then any one section of the curve A compounded with any one section of the curve B, will always give a section of the curve C, which will have the same length, and will be precisely equal and ¶ similar to any other section of the curve C obtained by compounding any other section of A with any other section of B.

When such a section embraces several periods of the corresponding curve (as in fig. 11, the sections e ϵ and $\epsilon\,\epsilon_1$ each consist of two periods of the simple tone B,) then the pitch of this second tone B, is that of an upper partial tone of a prime (as the simple tone A in fig. 11), whose period has the length of that principal section, in accordance with the rule above cited.

In order to give a slight conception of the multiplicity of forms producible by comparatively simple compositions, I may remark that the compound curve would

---

* [Readers not used to geometrical constructions are strongly recommended to trace the two curves A and B, and to construct the curve C from them, by drawing a number of perpendiculars to a straight line, and then setting off upon them the lengths of the corresponding perpendiculars in A and B in proper directions, and joining the extremities of the lengths thus found by a curved line. In this way only can a clear conception of the composition of vibrations be rendered sufficiently familiar for subsequent use.— *Translator*.]

receive another form if the curves B, fig. 11, were displaced a little with respect to the curve A before the addition were commenced.   Let B be displaced by being slid to the right until the point e falls under d_1 in A, and the composition will then give the curve D with narrow crests and broad troughs, both sides of the crest being, however, equally steep; whereas in the curve C one side is steeper than the other.   If we displace the curve B still more by sliding it to the right till e falls under d_2, the compound curve would resemble the reflection of C in a mirror: that is, it would have the same form as C reversed as to right and left; the steeper inclination which in C lies to the left would now lie to the right.   Again, if we displace B till e falls under d_3 we obtain a curve similiar to D, fig. 11, but reversed as to up and down, as may be seen by holding the book upside-down, the crests being broad and the troughs narrow.

¶

FIG. 12.

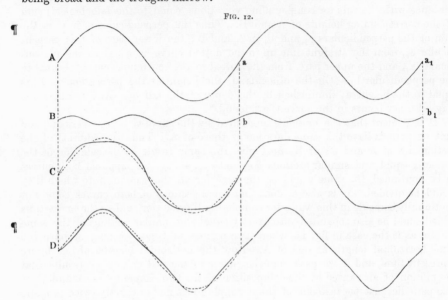

All these curves with their various transitional forms are periodic curves. Other composite periodic curves are shewn at C, D, fig. 12 above, where they are compounded of the two curves A and B, having their periods in the ratio of 1 to 3. The dotted curves are as before copies of the first complete vibration or period of the curve A, in order that the reader may see at a glance that the compound curve is always as much higher or lower than A, as B is higher or lower than the horizontal line.   In C, the curves A and B are added as they stand, but for D the curve B has been first slid half a wave's length to the right, and then the addition ¶ has been effected.   Both forms differ from each other and from all preceding ones. C has broad crests and broad troughs, D narrow crests and narrow troughs.

In these and similar cases we have seen that the compound motion is perfectly and regularly periodic, that is, it is exactly of the same kind as if it proceeded from a single musical tone.   The curves compounded in these examples correspond to the motions of single simple tones.   Thus, the motions shown in fig. 11 (on p. 30b, c) might have been produced by two tuning-forks, of which one sounded an Octave higher than the other.   But we shall hereafter see that a flute by itself when gently blown is sufficient to create a motion of the air corresponding to that shown in C or D of fig. 11.   The motions of fig. 12 might be produced by two tuning-forks of which one sounded the twelfth of the other.   Also a single closed organ pipe of the narrower kind (the stop called *Quintaten**) would give nearly the same motion as that of C or D in fig. 12.

* [The names of the stops on German organs do not always agree with those on English organs.   I find it best, therefore, not to translate them, but to give their explana-

Here, then, the motion of the air in the aural passage has no property by which the composite* musical tone can be distinguished from the single musical tone. If the ear is not assisted by other accidental circumstances, as by one tuning-fork beginning to sound before the other, so that we hear them struck, or, in the other case, the rustling of the wind against the mouthpiece of the flute or lip of the organ pipe, it has no means of deciding whether the musical tone is simple or composite.

Now, in what relation does the ear stand to such a motion of the air? Does it analyse it, or does it not? Experience shews us that when two tuning-forks, an Octave or a Twelfth apart in pitch, are sounded together, the ear is quite able to distinguish their simple tones, although the distinction is a little more difficult with these than with other intervals. But if the ear is able to analyse a composite musical tone produced by two tuning-forks, it cannot but be in a condition to carry out a similar analysis, when the same motion of the air is produced by a ¶ single flute or organ pipe. And this is really the case. The single musical tone of such instruments, proceeding from a single source, is, as we have already mentioned, analysed into partial simple tones, consisting in each case of a prime tone, and one upper partial tone, the latter being different in the two cases.

The analysis of a single musical tone into a series of partial tones depends, then, upon the same property of the ear as that which enables it to distinguish different musical tones from each other, and it must necessarily effect both analyses by a rule which is independent of the fact that the waves of sound are produced by one or by several musical instruments.

The rule by which the ear proceeds in its analysis was first laid down as generally true by G. S. Ohm. Part of this rule has been already enunciated in the last chapter (p. 23*a*), where it was stated that only that particular motion of the air which we have denominated a *simple vibration*, for which the vibrating particles swing backwards and forwards according to the law of pendular motion, ¶ is capable of exciting in the ear the sensation of a single simple tone. *Every motion of the air, then, which corresponds to a composite mass of musical tones, is, according to Ohm's law, capable of being analysed into a sum of simple pendular vibrations, and to each such single simple vibration corresponds a simple tone, sensible to the ear, and having a pitch determined by the periodic time of the corresponding motion of the air.*

The proofs of the correctness of this law, the reasons why, of all vibrational forms, only that one which we have called a simple vibration plays such an important part, must be left for Chapters IV. and VI. Our present business is only to gain a clear conception of what the rule means.

The simple vibrational form is inalterable and always the same. It is only its amplitude and its periodic time which are subject to change. But we have seen in figs. 11 and 12 (p. 30*b* and p. 32*b*) what varied forms the composition of only two simple vibrations can produce. The number of these forms might be greatly in- ¶ creased, even without introducing fresh simple vibrations of different periodic times, by merely changing the proportions which the heights of the two simple

---

tions from E. J. Hopkins's *The Organ, its History and Construction*, 1870, pp. 444-448. In this case Mr. Hopkins, following other authorities, prints the word 'quintaton,' and defines it, in 16 feet tone, as 'double stopped diapason, of rather small scale, producing the Twelfth of the fundamental sound, as well as the ground-tone itself, that is, sounding the 16 and 5⅓ ft. tones,' which means sounding the notes beginning with $C_{,}$, simultaneously with the notes beginning with G, which is called the 5⅓ foot tone, because according to the organ-makers' theory (not practice) the length of the G pipe is ⅓ of the length of the C pipe, and ⅓ of 16 is 5⅓. [See p. 15*d'*, note ‡.] And similarly,

in other cases, 'a pipe for sounding the Twelfth in addition to the fundamental tone.' It seems to be properly the English stop '*Twelfth*, *Octave Quint*, *Duodecima*,' No. 611, p. 141 of Hopkins. —*Translator.*]

* [The reader must distinguish between *single* and *simple* musical tones. A *single* tone may be a *compound* tone inasmuch as it may be compounded of several simple musical tones, but it is *single* because it is produced by *one* sounding body. A *composite* musical tone is necessarily *compound*, but it is called *composite* because it is made up of tones (simple or compound) produced by *several* sounding bodies.— *Translator.*]

vibrational curves A and B bear to each other, or displacing the curve B by other distances to the right or left, than those already selected in the figures. By these simplest possible examples of such compositions, the reader will be able to form some idea of the enormous variety of forms which would result from using more than two simple forms of vibration, each form representing an upper partial tone of the same prime, and hence, on addition, always producing fresh periodic curves. We should be able to make the heights of each single simple vibrational curve greater or smaller at pleasure, and displace each one separately by any amount in respect to the prime,—or, in physical language, we should be able to alter their amplitudes and the difference of their phases; and each such alteration of amplitude and difference of phase in each one of the simple vibrations would produce a fresh change in the resulting composite vibrational form. [See App. XX. sect. M. No. 2.]

¶ The multiplicity of vibrational forms which can be thus produced by the composition of simple pendular vibrations is not merely extraordinarily great: it is so great that it cannot be greater. The French mathematician Fourier has proved the correctness of a mathematical law, which in reference to our present subject may be thus enunciated: *Any given regular periodic form of vibration can always be produced by the addition of simple vibrations, having pitch numbers which are once, twice, thrice, four times, &c., as great as the pitch numbers of the given motion.*

The *amplitudes* of the elementary simple vibrations to which the height of our wave-curves corresponds, and the *difference of phase*, that is, the relative amount of horizontal displacement of the wave-curves, can always be found in every given case, as Fourier has shewn, by peculiar methods of calculation, (which, however, do not admit of any popular explanation,) so that *any given regularly periodic motion can always be exhibited in one single way, and in no other way whatever, as the sum of a certain number of pendular vibrations.*

¶ Since, according to the results already obtained, any regularly periodic motion corresponds to some musical tone, and any simple pendular vibration to a simple musical tone, these propositions of Fourier may be thus expressed in acoustical terms:

Any vibrational motion of the air in the entrance to the ear, corresponding to a musical tone, may be always, and for each case only in one single way, exhibited as the sum of a number of simple vibrational motions, corresponding to the partials of this musical tone.

Since, according to these propositions, any form of vibration, no matter what shape it may take, can be expressed as the sum of simple vibrations, its analysis into such a sum is quite independent of the power of the eye to perceive, by looking at its representative curve, whether it contains simple vibrations or not, and if it does, what they are. I am obliged to lay stress upon this point, because I have by no means unfrequently found even physicists start on the false hypothesis, that the ¶ vibrational form must exhibit little waves corresponding to the several audible upper partial tones. A mere inspection of the figs. 11 and 12 (p. 30b and p. 32b) will suffice to shew that although the composition can be easily traced in the parts where the curve of the prime tone is dotted in, this is quite impossible in those parts of the curves C and D in each figure, where no such assistance has been provided. Or, if we suppose that an observer who had rendered himself thoroughly familiar with the curves of simple vibrations imagined that he could trace the composition in these easy cases, he would certainly utterly fail on attempting to discover by his eye alone the composition of such curves as are shewn in figs. 8 and 9 (p. 21c). In these will be found straight lines and acute angles. Perhaps it will be asked how it is possible by compounding such smooth and uniformly rounded curves as those of our simple vibrational forms A and B in figs. 11 and 12, to generate at one time straight lines, and at another acute angles. The answer is, that an infinite number of simple vibrations are required to generate curves with such discontinuities as are there shewn. But when a great many such curves are combined, and are so chosen that in certain places they all bend in the same direction, and in others in opposite directions, the curvatures mutually strengthen each other in the first case, finally producing an infinitely great curvature, that is, an acute angle, and in the second case they mutually weaken each other, so that ultimately a straight line results. Hence we can generally lay it down as a rule that the force or loudness of the upper partial tones is the greater, the sharper the discontinuities of the atmospheric motion. When the motion alters uniformly and gradually, answering to a vibrational curve proceeding in smoothly curved forms, only the deeper partial tones, which lie nearest to the prime tone, have any perceptible intensity. But where the motion alters by jumps, and hence the vibrational curves shew angles or sudden changes of curvature, the upper partial tones will also have sensible force, although in all these cases the amplitudes decrease as the pitch of the upper partial tones becomes higher.*

We shall become acquainted with examples of the analysis of given vibrational forms into separate partial tones in Chapter V. ¶

The theorem of Fourier here adduced shews first that it is mathematically possible to consider a musical tone as a sum of simple tones, in the meaning we have attached to the words, and mathematicians have indeed always found it convenient to base their acoustic investigations on this mode of analysing vibrations. But it by no means follows that we are obliged to consider the matter in this way. We have rather to inquire, do these partial constituents of a musical tone, such as the mathematical theory distinguishes and the ear perceives, really exist in the mass of air external to the ear? Is this means of analysing forms of vibration which Fourier's theorem prescribes and renders possible, not merely a mathematical fiction, permissible for facilitating calculation, but not necessarily having any corresponding actual meaning in things themselves? What makes us hit upon pendular vibrations, and none other, as the simplest element of all motions producing sound? We can conceive a whole to be split into parts in very different and arbitrary ways. Thus we may find it convenient for a certain calculation to ¶ consider the number 12 as the sum 8 + 4, because the 8 may have to be cancelled, but it does not follow that 12 must always and necessarily be considered as merely the sum of 8 and 4. In another case it might be more convenient to consider 12 as the sum of 7 and 5. Just as little does the mathematical possibility, proved by Fourier, of compounding all periodic vibrations out of simple vibrations, justify us in concluding that this is the only permissible form of analysis, if we cannot in addition establish that this analysis has also an essential meaning in nature. That this is indeed the case, that this analysis has a meaning in nature independently of theory, is rendered probable by the fact that the ear really effects the same analysis, and also by the circumstance already named, that this kind of analysis has been found so much more advantageous in mathematical investigations than any other. Those modes of regarding phenomena that correspond to the most intimate constitution of the matter under investigation are, of course, also always those which lead to the most suitable and evident theoretical treatment. But it ¶ would not be advisable to begin the investigation with the functions of the ear, because these are very intricate, and in themselves require much explanation. In the next chapter, therefore, we shall inquire whether the analysis of compound into simple vibrations has an actually sensible meaning in the external world, independently of the action of the ear, and we shall really be in a condition to shew that certain mechanical effects depend upon whether a certain partial tone

* Supposing n to be the number of the order of a partial tone, and n to be very large, then the amplitude of the upper partial tones decreases: 1) as $\frac{1}{n}$, when the amplitude of the vibrations themselves makes a sudden jump; 2) as $\frac{1}{n.n}$, when their differential quotient makes a sudden jump, and hence the curve has an acute angle; 3) as $\frac{1}{n.n.n}$, when the curvature alters suddenly; 4) when none of the differential quotients are discontinuous, they must decrease at least as fast as e^{-n}.

is or is not contained in a composite mass of musical tones. The existence of partial tones will thus acquire a meaning in nature, and our knowledge of their mechanical effects will in turn shed a new light on their relations to the human ear.

CHAPTER III.

ANALYSIS OF MUSICAL TONES BY SYMPATHETIC RESONANCE.

WE proceed to shew that the simple partial tones contained in a composite mass of musical tones, produce peculiar mechanical effects in nature, altogether independent of the human ear and its sensations, and also altogether independent of ¶ merely theoretical considerations. These effects consequently give a peculiar objective significance to this peculiar method of analysing vibrational forms.

Such an effect occurs in the phenomenon of *sympathetic resonance*. This phenomenon is always found in those bodies which when once set in motion by any impulse, continue to perform a long series of vibrations before they come to rest. When these bodies are struck gently, but periodically, although each blow may be separately quite insufficient to produce a sensible motion in the vibratory body, yet, provided the periodic time of the gentle blows is precisely the same as the periodic time of the body's own vibrations, very large and powerful oscillations may result. But if the periodic time of the regular blows is different from the periodic time of the oscillations, the resulting motion will be weak or quite insensible.

Periodic impulses of this kind generally proceed from another body which is already vibrating regularly, and in this case the swings of the latter in the course ¶ of a little time, call into action the swings of the former. Under these circumstances we have the process called *sympathetic oscillation* or *sympathetic resonance*. The essence of the mechanical effect is independent of the rate of motion, which may be fast enough to excite the sensation of sound, or slow enough not to produce anything of the kind. Musicians are well acquainted with *sympathetic resonance*. When, for example, the strings of two violins are in exact unison, and one string is bowed, the other will begin to vibrate. But the nature of the process is best seen in instances where the vibrations are slow enough for the eye to follow the whole of their successive phases.

Thus, for example, it is known that the largest church-bells may be set in motion by a man, or even a boy, who pulls the ropes attached to them at proper and regular intervals, even when their weight of metal is so great that the strongest man could scarcely move them sensibly, if he did not apply his strength in determinate periodical intervals. When such a bell is once set in motion, it continues, like a ¶ struck pendulum, to oscillate for some time, until it gradually returns to rest, even if it is left quite by itself, and no force is employed to arrest its motion. The motion diminishes gradually, as we know, because the friction on the axis and the resistance of the air at every swing destroy a portion of the existing moving force.

As the bell swings backwards and forwards, the lever and rope fixed to its axis rise and fall. If when the lever falls a boy clings to the lower end of the bell-rope, his weight will act so as to increase the rapidity of the existing motion. This increase of velocity may be very small, and yet it will produce a corresponding increase in the extent of the bell's swings, which again will continue for a while, until destroyed by the friction and resistance of the air. But if the boy clung to the bell-rope at a wrong time, while it was ascending, for instance, the weight of his body would act in opposition to the motion of the bell, and the extent of swing would decrease. Now, if the boy continued to cling to the rope at each swing so long as it was falling, and then let it ascend freely, at every swing the motion of the bell would be only increased in speed, and its swings would gradually become greater and greater, until by their increase the motion imparted on every oscillation of the bell to the walls of the belfry, and the external air would become so great as exactly to be covered by the power exerted by the boy at each swing.

The success of this process depends, therefore, essentially on the boy's applying his force only at those moments when it will increase the motion of the bell. That is, he must employ his strength periodically, and the periodic time must be equal to that of the bell's swing, or he will not be successful. He would just as easily bring the swinging bell to rest, if he clung to the rope only during its ascent, and thus let his weight be raised by the bell.

A similar experiment which can be tried at any instant is the following. Construct a pendulum by hanging a heavy body (such as a ring) to the lower end of a thread, holding the upper end in the hand. On setting the ring into gentle pendular vibration, it will be found that this motion can be gradually and considerably increased by watching the moment when the pendulum has reached its greatest ¶ departure from the vertical, and then giving the hand a very small motion in the opposite direction. Thus, when the pendulum is furthest to the right, move the hand very slightly to the left ; and when the pendulum is furthest to the left, move the hand to the right. The pendulum may be also set in motion from a state of rest by giving the hand similar very slight motions having the same periodic time as the pendulum's own swings. The displacements of the hand may be so small under these circumstances, that they can scarcely be perceived with the closest attention, a circumstance to which is due the superstitious application of this little apparatus as a divining rod. If namely the observer, without thinking of his hand, follows the swings of the pendulum with his eye, the hand readily follows the eye, and involuntarily moves a little backwards or forwards, precisely in the same time as the pendulum, after this has accidentally begun to move. These involuntary motions of the hand are usually overlooked, at least when the observer is not accustomed to exact observations on such unobtrusive influences. By this ¶ means any existing vibration of the pendulum is increased and kept up, and any accidental motion of the ring is readily converted into pendular vibrations, which seem to arise spontaneously without any co-operation of the observer, and are hence attributed to the influence of hidden metals, running streams, and so on.

If on the other hand the motion of the hand is intentionally made in the contrary direction, the pendulum soon comes to rest.

The explanation of the process is very simple. When the upper end of the thread is fastened to an immovable support, the pendulum, once struck, continues to swing for a long time, and the extent of its swings diminishes very slowly. We can suppose the extent of the swings to be measured by the angle which the thread makes with the vertical on its greatest deflection from it. If the attached body at the point of greatest deflection lies to the right, and we move the hand to the left, we manifestly increase the angle between the string and the vertical, and con- ¶ sequently also augment the extent of the swing. By moving the upper end of the string in the opposite direction we should decrease the extent of the swing.

In this case there is no necessity for moving the hand in the same periodic time as the pendulum swings. We might move the hand backwards and forwards only at every third or fifth or other swing of the pendulum, and we should still produce large swings. Thus, when the pendulum is to the right, move the hand to the left, and keep it still, till the pendulum has swung to the left, then again to the right, and then once more to the left, and then return the hand to its first position, afterwards wait till the pendulum has swung to the right, then to the left, and again to the right, and then recommence the first motion of the hand. In this way three complete vibrations, or double excursions of the pendulum, will correspond to one left and right motion of the hand. In the same way one left and right motion of the hand may be made to correspond with seven or more swings of the pendulum. The meaning of this process is always that the motion of the hand must in each case be made at such a time and in such a direction as to be opposed to the deflection of the pendulum and consequently to increase it.

By a slight alteration of the process we can easily make two, four, six, &c., swings of the pendulum correspond to one left and right motion of the hand ; for a sudden motion of the hand at the instant of the pendulum's passage through the vertical has no influence on the size of the swings. Hence when the pendulum lies to the right move the hand to the left, and so increase its velocity, let it swing to the left, watch for the moment of its passing the vertical line, and at that instant return the hand to its original position, allow it to reach the right, and then again the left and once more the right extremity of its arc, and then recommence the first motion of the hand.

We are able then to communicate violent motion to the pendulum by very small periodical vibrations of the hand, having their periodic time exactly as great, ¶ or else two, three, four, &c., times as great as that of the pendular oscillation. We have here considered that the motion of the hand is backwards. This is not necessary. It may take place continuously in any other way we please. When it moves continuously there will be generally portions of time during which it will increase the pendulum's motion, and others perhaps in which it will diminish the same. In order to create strong vibrations in the pendulum, then, it will be necessary that the increments of motion should be permanently predominant, and should not be neutralised by the sum of the decrements.

Now if a determinate periodic motion were assigned to the hand, and we wished to discover whether it would produce considerable vibrations in the pendulum, we could not always predict the result without calculation. Theoretical mechanics would, however, prescribe the following process to be pursued : *Analyse the periodic motion of the hand into a sum of simple pendular vibrations of the hand*—exactly in the same way as was laid down in the last chapter for the periodic motions of ¶ the particles of air,—*then, if the periodic time of one of these vibrations is equal to the periodic time of the pendulum's own oscillations, the pendulum will be set into violent motion, but not otherwise.* We might compound small pendular motions of the hand out of vibrations of other periodic times, as much as we liked, but we should fail to produce any lasting strong swings of the pendulum. Hence the analysis of the motion of the hand into pendular swings has a real meaning in nature, producing determinate mechanical effects, and for the present purpose no other analysis of the motion of the hand into any other partial motions can be substituted for it.

In the above examples the pendulum could be set into sympathetic vibration, when the hand moved periodically at the same rate as the pendulum ; in this case the longest partial vibration of the hand, corresponding to the prime tone of a resonant vibration, was, so to speak, in unison with the pendulum. When three swings of the pendulum went to one backwards and forwards motion of the hand, ¶ it was the third partial swing of the hand, answering as it were to the Twelfth of its prime tone, which set the pendulum in motion. And so on.

The same process that we have thus become acquainted with for swings of long periodic time, holds precisely for swings of so short a period as sonorous vibrations. Any elastic body which is so fastened as to admit of continuing its vibrations for some length of time when once set in motion, can also be made to vibrate sympathetically, when it receives periodic agitations of comparatively small amounts, having a periodic time corresponding to that of its own tone.

Gently touch one of the keys of a pianoforte without striking the string, so as to raise the damper only, and then sing a note of the corresponding pitch forcibly directing the voice against the strings of the instrument. On ceasing to sing, the note will be echoed back from the piano. It is easy to discover that this echo is caused by the string which is in unison with the note, for directly the hand is removed from the key, and the damper is allowed to fall, the echo ceases. The sympathetic vibration of the string is still better shown by putting little paper riders upon it, which are jerked off as soon as the string vibrates. The more exactly the singer hits the pitch of the string, the more strongly it vibrates. A very little deviation from the exact pitch fails in exciting sympathetic vibration.

In this experiment the sounding board of the instrument is first struck by the vibrations of the air excited by the human voice. The sounding board is well known to consist of a broad flexible wooden plate, which, owing to its extensive surface, is better adapted to convey the agitation of the strings to the air, and of the air to the strings, than the small surface over which string and air are themselves directly in contact. The sounding board first communicates the agitations which it receives from the air excited by the singer, to the points where the string is fastened. The magnitude of any single such agitation is of course infinitesimally small. A very large number of such effects must necessarily be aggregated, before any sensible motion of the string can be caused. And such a continuous addition of effects really takes place, if, as in the preceding experiments with ¶ the bell and the pendulum, the periodic time of the small agitations which are communicated to the extremities of the string by the air, through the intervention of the sounding board, exactly corresponds to the periodic time of the string's own vibrations. When this is the case, a long series of such vibrations will really set the string into motion which is very violent in comparison with the exciting cause.

In place of the human voice we might of course use any other musical instrument. Provided only that it can produce the tone of the pianoforte string accurately and sustain it powerfully, it will bring the latter into sympathetic vibration. In place of a pianoforte, again, we can employ any other stringed instrument having a sounding board, as a violin, guitar, harp, &c., and also stretched membranes, bells, elastic tongues or plates, &c., provided only that the latter are so fastened as to admit of their giving a tone of sensible duration when once made to sound.

When the pitch of the original sounding body is not exactly that of the sym- ¶ pathising body, or that which is meant to vibrate in sympathy with it, the latter will nevertheless often make sensible sympathetic vibrations, which will diminish in amplitude as the difference of pitch increases. But in this respect different sounding bodies shew great differences, according to the length of time for which they continue to sound after having been set in action before communicating their whole motion to the air.

Bodies of small mass, which readily communicate their motion to the air, and quickly cease to sound, as, for example, stretched membranes, or violin strings, are readily set in sympathetic vibration, because the motion of the air is conversely readily transferred to them, and they are also sensibly moved by sufficiently strong agitations of the air, even when the latter have not precisely the same periodic time as the natural tone of the sympathising bodies. The limits of pitch capable of exciting sympathetic vibration are consequently a little wider in this case. By the comparatively greater influence of the motion of the air upon light elastic ¶ bodies of this kind which offer but little resistance, their natural periodic time can be slightly altered, and adapted to that of the exciting tone. Massive elastic bodies, on the other hand, which are not readily movable, and are slow in communicating their sonorous vibrations to the air, such as bells and plates, and continue to sound for a long time, are also more difficult to move by the air. A much longer addition of effects is required for this purpose, and consequently it is also necessary to hit the pitch of their own tone with much greater nicety, in order to make them vibrate sympathetically. Still it is well known that bell-shaped glasses can be put into violent motion by singing their proper tone into them ; indeed it is related that singers with very powerful and pure voices, have sometimes been able to crack them by the agitation thus caused. The principal difficulty in this experiment is in hitting the pitch with sufficient precision, and retaining the tone at that exact pitch for a sufficient length of time.

Tuning-forks are the most difficult bodies to set in sympathetic vibration. To

effect this they may be fastened on sounding boxes which have been exactly tuned to their tone, as shewn in fig. 13. If we have two such forks of exactly the same pitch, and excite one by a violin bow, the other will begin to vibrate in sympathy, even if placed at the further end of the same room, and it will continue to sound, after the first has been damped. The astonishing nature of such a case of sympathetic vibration will appear, if we merely compare the heavy and powerful mass of steel set in motion, with the light yielding mass of air which produces the effect by such
¶ small motive powers that they could not stir the lightest spring which was not in tune with the fork. With such forks the time required to set them in full swing by sympathetic action, is also of sensible duration, and the

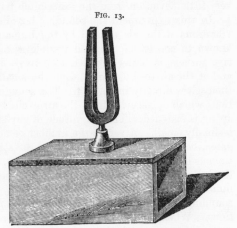

FIG. 13.

slightest disagreement in pitch is sufficient to produce a sensible diminution in the sympathetic effect. By sticking a piece of wax to one prong of the second fork, sufficient to make it vibrate once in a second less than the first—a difference of pitch scarcely sensible to the finest ear—the sympathetic vibration will be wholly destroyed.

After having thus described the phenomenon of sympathetic vibration in general, we proceed to investigate the influence exerted in sympathetic resonance by the different forms of wave of a musical tone.

¶ First, it must be observed that most elastic bodies which have been set into sustained vibration by a gentle force acting periodically, are (with a few exceptions

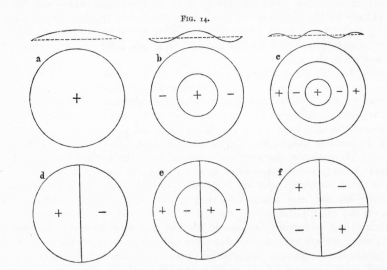

FIG. 14.

¶

to be considered hereafter) always made to swing in pendular vibrations. But they are in general capable of executing several kinds of such vibration, with different periodic times and with a different distribution over the various parts of the vibrating body. Hence to the different lengths of the periodic times correspond different simple tones producible on such an elastic body. These are its so-called *proper tones*. It is, however, only exceptionally, as in strings and the narrower kinds of organ pipes, that these proper tones correspond in pitch with the harmonic upper partial tones of a musical tone already mentioned. They are for the most part inharmonic in relation to the prime tone.

In many cases the vibrations and their mode of distribution over the vibrating bodies can be rendered visible by strewing a little fine sand over the latter. Take, for example, a membrane (as a bladder or piece of thin india-rubber) stretched over a circular ring. In fig. 14 are shewn the various forms which a membrane can assume when it vibrates. The diameters and circles on the surface of the membrane, mark those points which remain at rest during the vibration, and are known as *nodal lines*. By these the surface is divided into a number of compartments which bend alternately up and down, in such a way that while those marked ($+$) rise, those marked ($-$) fall. Over the figures a, b, c, are shewn the forms of a section of the membrane during vibration. Only those forms of motion are drawn which correspond with the deepest and most easily producible tones of the membrane. The number of circles and diameters can be increased at pleasure by ¶ taking a sufficiently thin membrane, and stretching it with sufficient regularity, and in this case the tones would continually sharpen in pitch. By strewing sand on the membrane the figures are easily rendered visible, for as soon as it begins to vibrate the particles of sand collect on the nodal lines.

In the same way it is possible to render visible the nodal lines and forms of vibration of oval and square membranes, and of differently-shaped plane elastic plates, bars, and so on. These form a series of very interesting phenomena discovered by Chladni, but to pursue them would lead us too far from our proper subject. It will suffice to give a few details respecting the simplest case, that of a circular membrane.

In the time required by the membrane to execute 100 vibrations of the form a, fig. 14 (p. 40c), the number of vibrations executed by the other forms is as follows :—

¶

Form of Vibration	Pitch Number	Cents *	Notes nearly
a without nodal lines	100	0	c
b with one circle	229·6	1439	$d' +$
c with two circles	359·9	2217	$b'b +$
d with one diameter	159	805	ab
e with one diameter and one circle . . .	292	1858	$g' -$
f with two diameters	214	1317	$c'\sharp +$

The prime tone has been here arbitrarily assumed as c, in order to note the intervals of the higher tones. Those simple tones produced by the membrane which are slightly higher than those of the note written, are marked ($+$); those lower, by ($-$). In this case there is no commensurable ratio between the prime tone and the other tones, that is, none expressible in whole numbers.

Strew a very thin membrane of this kind with sand, and sound its prime tone strongly in its neighbourhood; the sand will be driven by the vibrations towards ¶ the edge, where it collects. On producing another of the tones of the membrane, the sand collects in the corresponding nodal lines, and we are thus easily able to determine to which of its tones the membrane has responded. A singer who knows how to hit the tones of the membrane correctly, can thus easily make the

* [*Cents* are hundredths of an equal Semitone, and are exceedingly valuable as measures of many, especially unusual, musical intervals. They are fully explained, and the method of calculating them from the Interval Ratios is given in App. XX. sect. C. Here it need only be said that the number of hundreds of cents is the number of *equal*, that is, pianoforte Semitones in the interval, and these may be counted on the keys of any piano, while the units and tens shew the number of hundredths of a Semitone in excess. Wherever *cents* are spoken of in the text, (as in this table), they must be considered as additions by the translator. In the present case, they give the intervals exactly, and not roughly as in the column of notes. Thus, 1439 cents is sharper than 14 Semitones above c, that is, sharper than d' by 39 hundredths of a Semitone, or about ⅖ of a Semitone, and 1858 is flatter than 19 Semitones above c, that is, flatter than g' by 42 hundredths of a Semitone, or nearly ½ a Semitone. —*Translator.*]

sand arrange itself at pleasure in one order or the other, by singing the corresponding tones powerfully at a distance. But in general the simpler figures of the deeper
tones are more easily generated than the complicated figures of the upper tones.
It is easiest of all to set the membrane in general motion by sounding its prime
tone, and hence such membranes have been much used in acoustics to prove the
existence of some determinate tone in some determinate spot of the surrounding
air. It is most suitable for this purpose to connect the membrane with an inclosed
mass of air. A, fig. 15, is a glass bottle,
having an open mouth a, and in place
of its bottom b, a stretched membrane,
consisting of wet pig's bladder, allowed to dry after it has been stretched
and fastened. At c is attached a
¶ single fibre of a silk cocoon, bearing a
drop of sealing-wax, and hanging down
like a pendulum against the membrane.

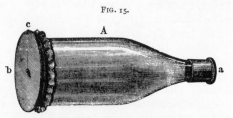

FIG. 15.

As soon as the membrane vibrates, the little pendulum is violently agitated. Such
a pendulum is very convenient as long as we have no reason to apprehend any confusion of the prime tone of the membrane with any other of its proper tones. There
is no scattering of sand, and the apparatus is therefore always in order. But to decide
with certainty what tones are really agitating the membrane, we must after all
place the bottle with its mouth downwards and strew sand on the membrane.
However, when the bottle is of the right size, and the membrane uniformly
stretched and fastened, it is only the prime tone of the membrane (slightly altered
by that of the sympathetically vibrating mass of air in the bottle) which is easily
excited. This prime tone can be made deeper by increasing the size of the membrane, or the volume of the bottle, or by diminishing the tension of the membrane
¶ or size of the orifice of the bottle.

A stretched membrane of this kind, whether it is or is not attached to the bottom of a bottle, will not only be set in vibration by musical tones of the same pitch
as its own proper tone, but also by such musical tones as contain the proper tone
of the membrane among its upper partial tones. Generally, given a number of
interlacing waves, to discover whether the membrane will vibrate sympathetically,
we must suppose the motion of the air at the given place to be mathematically
analysed into a sum of pendular vibrations. If there is one such vibration among
them, of which the periodic time is the same as that of any one of the proper tones
of the membrane, the corresponding vibrational form of the membrane will be superinduced. But if there are none such, or none sufficiently powerful, the membrane
will remain at rest.

In this case, then, we also find that the analysis of the motion of the air into
pendular vibrations, and the existence of certain vibrations of this kind, are deci-
¶ sive for the sympathetic vibration of the membrane, and for this purpose no other
similar analysis of the motion of the air can be substituted for its analysis into
pendular vibrations. The pendular vibrations into which the composite motion of
the air can be analysed, here shew themselves capable of producing mechanical
effects in external nature, independently of the ear, and independently of mathematical theory. Hence the statement is confirmed, that the theoretical view which
first led mathematicians to this method of analysing compound vibrations, is
founded in the nature of the thing itself.

As an example take the following description of a single experiment :—

A bottle of the shape shewn in fig. 15 above was covered with a thin vulcanised india-rubber membrane, of which the vibrating surface was 49 millimetres
(1·93 inches)* in diameter, the bottle being 140 millimetres (5·51 inches) high, and

---

* [As 10 inches are exactly 254 millimetres
and 100 metres, that is, 100,000 millimetres are
3937 inches, it is easy to form little tables for
the calculation of one set of measures from
the other. Roughly we may assume 25 mm.
to be 1 inch. But whenever dimensions are

having an opening at the brass mouth of 13 millimetres (·51 inches) in diameter. When blown it gave $f'\sharp$, and the sand heaped itself in a circle near the edge of the membrane. The same circle resulted from my giving the same tone $f'\sharp$ on an harmonium, or its deeper Octave $f\sharp$, or the deeper Twelfth B. Both $F\sharp$ and D gave the same circle, but more weakly. Now the $f'\sharp$ of the membrane is the prime tone of the harmonium tone $f'\sharp$, the second partial tone of $f\sharp$, the third of B, the fourth of $F\sharp$ and fifth of D.* All these notes on being sounded set the membrane in the motion due to its deepest tone. A second smaller circle, 19 millimetres (·75 inches) in diameter was produced on the membrane by b' and the same more faintly by b, and there was a trace of it for the deeper Twelfth e, that is, for simple tones of which vibrational numbers were $\frac{1}{2}$ and $\frac{1}{3}$ that of b'.†

Stretched membranes of this kind are very convenient for these and similar experiments on the partials of compound tones. They have the great advantage ¶

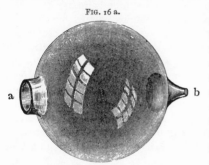

FIG. 16 a.

of being independent of the ear, but they are not very sensitive for the fainter simple tones. Their sensitiveness is far inferior to that of the *res'onātors* which I have introduced. These are hollow spheres of glass or metal, or tubes, with two openings as shewn in figs. 16 a and 16 b. One opening (a) has sharp edges, the other (b) is funnel-shaped, and adapted for insertion into the ear. This smaller end I usually coat with melted sealing wax, and when the wax has cooled down enough not to hurt the finger on being touched, but is still soft, I press the opening into the entrance of my ear. The sealing wax thus moulds itself to the shape of the inner surface of this opening, and when I subsequently use the resonator, it fits easily and is air-tight. ¶ Such an instrument is very like the resonance bottle already described, fig. 15

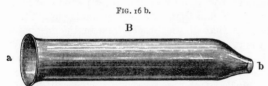

FIG. 16 b.

B

(p. 42a), for which the observer's own tympanic membrane has been made to replace the former artificial membrane.

The mass of air in a resonator, together with that in the aural passage, and with the tympanic membrane or drumskin itself, forms an elastic system which is capable of vibrating in a peculiar manner, and, in especial, the prime tone of the sphere, which is much deeper than any other of its proper tones, can be set into very powerful sympathetic vibration, and then the ear, which is in immediate connection with the air inside the sphere, perceives this augmented tone by direct action. If we stop one ear (which is best done by a plug of sealing wax moulded into the ¶ form of the entrance of the ear),‡ and apply a resonator to the other, most of the tones produced in the surrounding air will be considerably damped; but if the proper tone of the resonator is sounded, it brays into the ear most powerfully.

given in the text in mm. (that is, millimetres), they will be reduced to inches and decimals of an inch.—*Translator.*]

* [As the instrument was tempered, we should have, approximately, for $f\sharp$ the partials $f\sharp$, $f'\sharp$, &c.; for B the partials B, b, $f'\sharp$, &c.; for $F\sharp$ the partials $F\sharp$, $f\sharp$, $c\sharp$, $f'\sharp$, &c.; and for D the partials D, d, a, d', $f'\sharp$, &c. To prevent confusion I have reduced the *upper* partials of the text to ordinary partials, as suggested in p. 23b', note.—*Translator.*]

† [Here the partials of b are b, b', &c., and of e are e, e', b', &c., so that both b and e contain b'.—*Translator.*]

‡ [For ordinary purposes this is quite enough, indeed it is generally unnecessary to stop the other ear at all. But for such experiments as Mr. Bosanquet had to make on beats (see App. XX. section L. art. 4, *b*) he was obliged to use a jar as the resonator, conduct the sound from it through first a glass and then an elastic tube to a semicircular metal tube which reached from ear to ear, to each end of which a tube coated with india-rubber, could be screwed into the ear. By this means, when proper care was taken, all sound but that coming from the resonance jar was perfectly excluded.—*Translator.*]

Hence any one, even if he has no ear for music or is quite unpractised in detecting musical sounds, is put in a condition to pick the required simple tone, even if comparatively faint, from out of a great number of others. The proper tone of the resonator may even be sometimes heard cropping up in the whistling of the wind, the rattling of carriage wheels, the splashing of water. For these purposes such resonators are incomparably more sensitive than tuned membranes. When the simple tone to be observed is faint in comparison with those which accompany it, it is of advantage to alternately apply and withdraw the resonator. We thus easily feel whether the proper tone of the resonator begins to sound when the instrument is applied, whereas a uniform continuous tone is not so readily perceived.

A properly tuned series of such resonators is therefore an important instrument for experiments in which individual faint tones have to be distinctly heard, although accompanied by others which are strong, as in observations on the combinational ¶ and upper partial tones, and a series of other phenomena to be hereafter described relating to chords. By their means such researches can be carried out even by ears quite untrained in musical observation, whereas it had been previously impossible to conduct them except by trained musical ears, and much strained attention properly assisted. These tones were consequently accessible to the observation of only a very few individuals ; and indeed a large number of physicists and even musicians had never succeeded in distinguishing them. And again even the trained ear is now able, with the assistance of resonators, to carry the analysis of a mass of musical tones much further than before. Without their help, indeed, I should scarcely have succeeded in making the observations hereafter described, with so much precision and certainty, as I have been enabled to attain at present.*

It must be carefully noted that the ear does not hear the required tone with augmented force, unless that tone attains a considerable intensity within the mass ¶ of air inclosed in the resonator. Now the mathematical theory of the motion of the air shews that, so long as the amplitude of the vibrations is sufficiently small, the inclosed air will execute pendular oscillations of the same periodic time as those in the external air, and none other, and that only those pendular oscillations whose periodic time corresponds with that of the proper tone of the resonator, have any considerable strength ; the intensity of the rest diminishing as the difference of their pitch from that of the proper tone increases. All this is independent of the connection of the ear and resonator, except in so far as its tympanic membrane forms one of the inclosing walls of the mass of air. Theoretically this apparatus does not differ from the bottle with an elastic membrane, in fig. 15 (p. 42a), but its sensitiveness is amazingly increased by using the drumskin of the ear for the closing membrane of the bottle, and thus bringing it in direct connection with the auditory nerves themselves. Hence we cannot obtain a powerful tone in the resonator except when an analysis of the motion of the external air into ¶ pendular vibrations, would shew that one of them has the same periodic time as the proper tone of the resonator. Here again no other analysis but that into pendular vibrations would give a correct result.

It is easy for an observer to convince himself of the above-named properties of resonators. Apply one to the ear, and let a piece of harmonised music, in which the proper tone of the resonator frequently occurs, be executed by any instruments. As often as this tone is struck, the ear to which the instrument is held, will hear it violently contrast with all the other tones of the chord.

This proper tone will also often be heard, but more weakly, when deeper musical tones occur, and on investigation we find that in such cases tones have been struck which include the proper tone of the resonator among their upper partial tones. Such deeper musical tones are called the *harmonic under tones* of the resonator. They are musical tones whose periodic time is exactly 2, 3, 4, 5, and so on, times as great as that of the resonator. Thus if the proper tone of

---

* See Appendix II. for the measures and different forms of these Resonators.

the resonator is c'', it will be heard when a musical instrument sounds c', f, c, $A\flat$, F, D, C, and so on.* In this case the resonator is made to sound in sympathy with one of the harmonic upper partial tones of the compound musical tone which is vibrating in the external air. It must, however, be noted that by no means all the harmonic upper partial tones occur in the compound tones of every instrument, and that they have very different degrees of intensity in different instruments. In the musical tones of violins, pianofortes, and harmoniums, the first five or six are generally very distinctly present. A more detailed account of the upper partial tones of strings will be given in the next chapter. On the harmonium the un-evenly numbered partial tones (1, 3, 5, &c.) are generally stronger than the evenly numbered ones (2, 4, 6, &c.). In the same way, the upper partial tones are clearly heard by means of the resonators in the singing tones of the human voice, but differ in strength for the different vowels, as will be shewn hereafter.   ¶

Among the bodies capable of strong sympathetic vibration must be reckoned stretched strings which are connected with a sounding board, as on the pianoforte.

The principal mark of distinction between strings and the other bodies which vibrate sympathetically, is that different vibrating forms of strings give simple tones corresponding to the *harmonic* upper partial tones of the prime tone, whereas the secondary simple tones of membranes, bells, rods, &c., are *in*harmonic with the prime tone, and the masses of air in resonators have generally only very high upper partial tones, also chiefly *in*harmonic with the prime tone, and not capable of being much reinforced by the resonator.

The vibrations of strings may be studied either on elastic chords loosely stretched, and not sonorous, but swinging so slowly that their motion may be followed with the hand and eye, or else on sonorous strings, as those of the piano-forte, guitar, monochord, or violin. Strings of the first kind are best made of thin ¶ spirals of brass wire, six to ten feet in length. They should be gently stretched, and both ends should be fastened. A string of this construction is capable of making very large excursions with great regularity, which are easily seen by a large audience. The swings are excited by moving the string regularly backwards and forwards by the finger near to one of its extremities.

A string may be first made to vibrate as in fig. 17, a (p. 46b), so that its appear-ance when displaced from its position of rest is always that of a simple half wave. The string in this case gives a single simple tone, the deepest it can produce, and no other harmonic secondary tones are audible.

But the string may also during its motion assume the forms fig. 17, b, c, d. In this case the form of the string is that of two, three, or four half waves of a simple wave-curve. In the vibrational form b the string produces only the upper Octave of its prime tone, in the form c the Twelfth, and in the form d the second Octave. The dotted lines shew the position of the string at the end of half its ¶ periodic time. In b the point β remains at rest, in c two points γ_1 and γ_2 remain at rest, in d three points δ_1, δ_2, δ_3. These points are called *nodes*. In a swinging spiral wire the nodes are readily seen, and for a resonant string they are shewn by little paper riders, which are jerked off from the vibrating parts and remain sitting on the nodes. When, then, the string is divided by a node into two swinging sections, it produces a simple tone having a pitch number double that of the prime

---

* [The c'' occurs as the 2nd, 3rd, 4th, the 7th being rather flat. The partials are
5th, 6th, 7th, 8th partials of these notes, in fact :—

c' c''
f f' c''
c c' f' c''
$A\flat$ $a\flat$ $e\flat$ $a'\flat$ c''
F f c' f' a' c''
D d a d' $f'\sharp$ a' c''
C c f c' e' f' $b'\flat$ c''.—*Translator.*]

tone. For three sections the pitch number is tripled, for four sections quadrupled, and so on.

To bring a spiral wire into these different forms of vibration, we move it periodically with the finger near one extremity, adopting the period of its slowest swings for a, twice that rate for b, three times for c, and four times for d. Or else we just gently touch one of the nodes nearest the extremity with the finger, and pluck the string half-way between this node and the nearest end. Hence when γ_1 in c, or δ_1 in d, is kept at rest by the finger, we pluck the string at ϵ. The other nodes then appear when the vibration commences.

FIG. 17.

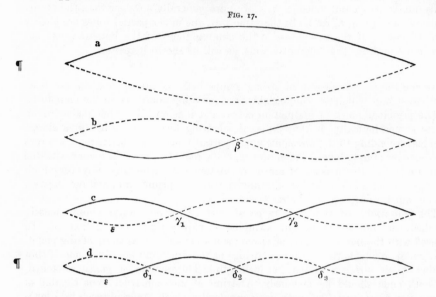

For a sonorous string the vibrational forms of fig. 17 above are most purely produced by applying to its sounding board the handle of a tuning-fork which has been struck and gives the simple tone corresponding to the form required. If only a determinate number of nodes are desired, and it is indifferent whether the individual points of the string do or do not execute simple vibrations, it is sufficient to touch the string very gently at one of the nodes and either pluck the string or rub it with a violin bow. By touching the string with the finger all those simple vibrations are damped which have no node at that point, and only those remain which allow the string to be at rest in that place.

The number of nodes in long thin strings may be considerable. They cease to be formed when the sections which lie between the nodes are too short and stiff to be capable of sonorous vibration. Very fine strings consequently give a greater number of higher tones than thicker ones. On the violin and the lower pianoforte strings it is not very difficult to produce tones with 10 sections ; but with extremely fine wires tones with 16 or 20 sections can be made to sound. [Also compare p. 78d.]

The forms of vibration here spoken of are those in which each point of the string performs pendular oscillations. Hence these motions excite in the ear the sensation of only a single simple tone. In all other vibrational forms of the strings, the oscillations are not simply pendular, but take place according to a different and more complicated law. This is always the case when the string is plucked in the usual way with the finger (as for guitar, harp, zither) or is struck with a hammer (as on the pianoforte), or is rubbed with a violin bow. The resulting motions may then be regarded as compounded of many simple vibrations, which, when taken separately, correspond to those in fig. 17. The multiplicity of such composite forms of motion is infinitely great, the string may indeed be considered as capable of assuming any given form (provided we confine ourselves in all cases to very small deviations from the position of rest), because, according to what was said in Chapter II., any given form of wave can be compounded out of a number of simple waves such as those indicated in fig. 17, a, b, c, d. A plucked, struck, or bowed string therefore allows a great number of harmonic upper partial tones to be heard at the same time as the prime tone, and generally the number increases with the thinness of the string. The peculiar tinkling sound of very fine metallic strings, is clearly due to these very high secondary tones. It is easy to distinguish the upper simple tones up to the sixteenth by means of resonators. Beyond the sixteenth they are too close to each other to be distinctly separable by this means.

Hence when a string is sympathetically excited by a musical tone in its neighbourhood, answering to the pitch of the prime tone of the string, a whole series of different simple vibrational forms will generally be at the same time generated in the string. For when the prime of the musical tone corresponds to the prime of the string all the harmonic upper partials of the first correspond to those of the ¶ second, and are hence capable of exciting the corresponding vibrational forms in the string. Generally the string will be brought into as many forms of sympathetic vibration by the motion of the air, as the analysis of that motion shews that it possesses simple vibrational forms, having a periodic time equal to that of some vibrational form, that the string is capable of assuming. But as a general rule when there is one such simple vibrational form in the air, there are several such, and it will often be difficult to determine by which one, out of the many possible simple tones which would produce the effect, the string has been excited. Consequently the usual unweighted strings are not so convenient for the determination of the pitch of any simple tones which exist in a composite mass of air, as the membranes or the inclosed air of resonators.

To make experiments with the pianoforte on the sympathetic vibrations of strings, select a flat instrument, raise its lid so as to expose the strings, then press down the key of the string (for c' suppose) which you wish to put into sympathetic ¶ vibration, but so slowly that the hammer does not strike, and place a little chip of wood across this c' string. You will find the chip put in motion, or even thrown off, when certain other strings are struck. The motion of the chip is greatest when one of the *under tones* of c' (p. 44d) is struck, as c, F, C, $A\flat$, $F_{,}$, $D_{,}$, or $C_{,}$. Some, but much less, motion also occurs when one of the upper partial tones of c' is struck, as c'', g'', or c''', but in this last case the chip will not move if it has been placed over one of the corresponding nodes of the string. Thus if it is laid across the middle of the string it will be still for c'' and c''', but will move for g''. Placed at one third the length of the string from its extremity, it will not stir for g'', but will move for c'' or c'''. Finally the string c' will also be put in motion when an under tone of one of its upper partial tones is struck ; for example, the note f, of which the third partial tone c'' is identical with the second partial tone of c'. In this case also the chip remains at rest when put on to the middle of the string c', which is its node for c''. In the same way the string c' will move, with the formation of ¶ two nodes, for g', g, or $e\flat$, all which notes have g'' as an upper partial tone, which is also the third partial of c'.*

Observe that on the pianoforte, where one end of the strings is commonly concealed, the position of the nodes is easily found by pressing the string gently on both sides and striking the key. If the finger is at a node the corresponding upper partial tone will be heard purely and distinctly, otherwise the tone of the string is dull and bad.

As long as only one upper partial tone of the string c' is excited, the corresponding nodes can be discovered, and hence the particular form of its vibration determined. But this is no longer possible by the above mechanical method when

---

* [These experiments can of course not be conducted on the usual upright cottage piano. But the experimenter can at least *hear* the tone of c', if c, F, C, &c., are struck and immediately damped, or if c'', g'', c''' are struck and damped. And this sounding of c', although unstruck, is itself a very interesting phenomenon. But of course, as it depends on the ear, it does not establish the results of the text.—*Translator.*]

two upper partial tones are excited, such as c'' and g'', as would be the case if both these notes were struck at once on the pianoforte, because the whole string of c' would then be in motion.

Although the relations for strings appear more complicated to the eye, their sympathetic vibration is subject to the same law as that which holds for resonators, membranes, and other elastic bodies. The sympathetic vibration is always determined by the analysis of whatever sonorous motions exist, into simple pendular vibrations. If the periodic time of one of these simple vibrations corresponds to the periodic time of one of the proper tones of the elastic body, that body, whether it be a string, a membrane, or a mass of air, will be put into strong sympathetic vibration.

These facts give a real objective value to the analysis of sonorous motion into simple pendular vibration, and no such value would attach to any other analysis. ¶ Every individual single system of waves formed by pendular vibrations exists as an independent mechanical unit, expands, and sets in motion other elastic bodies having the corresponding proper tone, perfectly undisturbed by any other simple tones of other pitches which may be expanding at the same time, and which may proceed either from the same or any other source of sound. Each single simple tone, then, can, as we have seen, be separated from the composite mass of tones, by mechanical means, namely by bodies which will vibrate sympathetically with it. Hence every individual partial tone exists in the compound musical tone produced by a single musical instrument, just as truly, and in the same sense, as the different colours of the rainbow exist in the white light proceeding from the sun or any other luminous body. Light is also only a vibrational motion of a peculiar elastic medium, the luminous ether, just as sound is a vibrational motion of the air. In a beam of white light there is a species of motion which *may* be represented as the sum of many oscillatory motions of various periodic times, each of ¶ which corresponds to one particular colour of the solar spectrum. But of course each particle of ether at any particular moment has only *one* determinate velocity, and only *one* determinate departure from its mean position, just like each particle of air in a space traversed by many systems of sonorous waves. The really existing motion of any particle of ether is of course only one and individual ; and our theoretical treatment of it as compound, is in a certain sense arbitrary. But the undulatory motion of light can also be analysed into the waves corresponding to the separate colours, by external mechanical means, such as by refraction in a prism, or by transmission through fine gratings, and each individual simple wave of light corresponding to a simple colour, exists mechanically by itself, independently of any other colour.

We must therefore not hold it to be an illusion of the ear, or to be mere imagination, when in the musical tone of a single note emanating from a musical instrument, we distinguish many partial tones, as I have found musicians inclined ¶ to think, even when they have heard those partial tones quite distinctly with their own ears. If we admitted this, we should have also to look upon the colours of the spectrum which are separated from white light, as a mere illusion of the eye. The real outward existence of partial tones in nature can be established at any moment by a sympathetically vibrating membrane which casts up the sand strewn upon it.

Finally I would observe that, as respects the conditions of sympathetic vibration, I have been obliged to refer frequently to the mechanical theory of the motion of air. Since in the theory of sound we have to deal with well-known mechanical forces, as the pressure of the air, and with motions of material particles, and not with any hypothetical explanation, theoretical mechanics have an unassailable authority in this department of science. Of course those readers who are unacquainted with mathematics, must accept the results on faith. An experimental way of examining the problems in question will be described in the next chapter, in which the laws of the analysis of musical tones by the ear have to be established. The experimental proof there given for the ear, can also be carried out in precisely the same way for membranes and masses of air which vibrate sympathetically, and the identity of the laws in both cases will result from those investigations.*

CHAPTER IV.

ON THE ANALYSIS OF MUSICAL TONES BY THE EAR.

It was frequently mentioned in the preceding chapter that musical tones could be resolved by the ear alone, unassisted by any peculiar apparatus, into a series of partial tones corresponding to the simple pendular vibrations in a mass of air, that ¶ is, into the same constituents as those into which the motion of the air is resolved by the sympathetic vibration of elastic bodies. We proceed to shew the correctness of this assertion.

Any one who endeavours for the first time to distinguish the upper partial tones of a musical tone, generally finds considerable difficulty in merely hearing them.

The analysis of our sensations when it cannot be attached to corresponding differences in external objects, meets with peculiar difficulties, the nature and significance of which will have to be considered hereafter. The attention of the observer has generally to be drawn to the phenomenon he has to observe, by peculiar aids properly selected, until he knows precisely what to look for ; after he has once succeeded, he will be able to throw aside such crutches. Similar difficulties meet us in the observation of the upper partials of a musical tone. I shall first give a description of such processes as will most easily put an untrained ¶ observer into a position to recognise upper partial tones, and I will remark in passing that a musically trained ear will not necessarily hear upper partial tones with greater ease and certainty than an untrained ear. Success depends rather upon a peculiar power of mental abstraction or a peculiar mastery over attention, than upon musical training. But a musically trained observer has an essential advantage over one not so trained in his power of figuring to himself how the simple tones sought for, ought to sound, whereas the untrained observer has continually to hear these tones sounded by other means in order to keep their effect fresh in his mind.

First we must note, that the unevenly numbered partials, as the Fifths, Thirds, Sevenths, &c., of the prime tones, are usually easier to hear than the even ones, which are Octaves either of the prime tone or of some of the upper partials which lie near it, just as in a chord we more readily distinguish whether it contains Fifths and Thirds than whether it has Octaves. The second, fourth, and eighth ¶ partials are higher Octaves of the prime, the sixth partial an Octave above the third partial, that is, the Twelfth of the prime ; and some practice is required for distinguishing these. Among the uneven partials which are more easily distinguished, the first place must be assigned, from its usual loudness, to the third partial, the Twelfth of the prime, or the Fifth of its first higher Octave. Then follows the fifth partial as the major Third of the prime, and, generally very faint, the seventh partial as the minor Seventh† of the second higher Octave of the prime, as will be seen by their following expression in musical notation, for the compound tone c.

---

* Optical means for rendering visible weak sympathetic motions of sonorous masses of air, are described in App. II. These means are valuable for demonstrating the facts to hearers unaccustomed to the observing and distinguishing musical tones.

† [Or more correctly *sub*-minor Seventh ; as the real minor Seventh, formed by taking two Fifths down and then two Octaves up, is sharper by 27 cents, or in the ratio of 63 : 64. —*Translator.*]

	1	2		3	4	5	6	7	8
	c	c'		g'	c''	e''	g''	$^{7}b''\flat$	c'''
[Cents.	0	1200		1902	2400	2786	3102	3369	3600] *

In commencing to observe upper partial tones, it is advisable just before producing the musical tone itself which you wish to analyse, to sound the note you wish to distinguish in it, very gently, and if possible in the same quality of tone as the compound itself. The pianoforte and harmonium are well adapted for these experiments, because they both have upper partial tones of considerable power.

¶   First gently strike on a piano the note g', as marked above, and after letting the digital† rise so as to damp the string, strike the note c, of which g' is the third partial, with great force, and keep your attention directed to the pitch of the g' which you had just heard, and you will hear it again in the compound tone of c. Similarly, first stroke the fifth partial e'' gently, and then c strongly. These upper partial tones are often more distinct as the sound dies away, because they appear to lose force more slowly than the prime. The seventh and ninth partials $b''\flat$ and d''' are mostly weak, or quite absent on modern pianos. If the same experiments are tried with an harmonium in one of its louder stops, the seventh partial will generally be well heard, and sometimes even the ninth.

To the objection which is sometimes made that the observer only imagines he hears the partial tone in the compound, because he had just heard it by itself, I need only remark at present that if e'' is first heard as a partial tone of c on a good piano, tuned in equal temperament, and then e'' is struck on the instrument ¶ itself, it is quite easy to perceive that the latter is a little sharper. This follows from the method of tuning. But if there is a difference in pitch between the two tones, one is certainly not a continuation of the mental effect produced by the other. Other facts which completely refute the above conception, will be subsequently adduced.

A still more suitable process than that just described for the piano, can be adopted on any stringed instrument, as the piano, monochord, or violin. It consists in first producing the tone we wish to hear, as an harmonic, [p. 25d, note] by touching the corresponding node of the string when it is struck or rubbed. The resemblance of the tone first heard to the corresponding partial of the compound is then much greater, and the ear discovers it more readily. It is usual to place a divided scale by the string of a monochord, to facilitate the discovery of the nodes. Those for the third partial, as shewn in Chap. III. (p. 45d), divide the string into three equal parts, those for the fifth into five, and so on. On the piano and violin ¶ the position of these points is easily found experimentally, by touching the string gently with the finger in the neighbourhood of the node, which has been approximatively determined by the eye, then striking or bowing the string, and moving the finger about till the required harmonic comes out strongly and purely. By then sounding the string, at one time with the finger on the node, and at another without, we obtain the required upper partial at one time as an harmonic, and at another in the compound tone of the whole string, and thus learn to recognise the existence of the first as part of the second, with comparative ease. Using thin strings which have loud upper partials, I have thus been able to recognise the

* [The cents, (see p. 41d, note) reckoned from the lowest note, are assigned on the supposition that the harmonies are perfect, as on the Harmonical, not tempered as on the pianoforte. See also diagram, p. 22c.— *Translator.*]

† [The keys played by the fingers on a piano or organ, are best called *digitals* or finger-keys, on the analogy of *pedals* and foot-keys on the organ. The word *key* having another musical sense, namely, the scale in which a piece of music is written, will without prefix be confined to this meaning.—*Translator.*]

partials separately, up to the sixteenth.   Those which lie still higher are too near to each other in pitch for the ear to separate them readily.

In such experiments I recommend the following process.   Touch the node of the string on the pianoforte or monochord with a camel's-hair pencil, strike the note, and immediately remove the pencil from the string.   If the pencil has been pressed tightly on the string, we either continue to hear the required partial as an harmonic, or else in addition hear the prime tone gently sounding with it.   On repeating the excitement of the string, and continuing to press more and more lightly with the camel's-hair pencil, and at last removing the pencil entirely, the prime tone of the string will be heard more and more distinctly with the harmonic till we have finally the full natural musical tone of the string.   By this means we obtain a series of gradual transitional stages between the isolated partial and the compound tone, in which the first is readily retained by the ear.   By applying this last process I have generally succeeded in making perfectly untrained ears ¶ recognise the existence of upper partial tones.

It is at first more difficult to hear the upper partials on most wind instruments and in the human voice, than on stringed instruments, harmoniums, and the more penetrating stops of an organ, because it is then not so easy first to produce the upper partial softly in the same quality of tone.   But still a little practice suffices to lead the ear to the required partial tone, by previously touching it on the piano. The partial tones of the human voice are comparatively most difficult to distinguish for reasons which will be given subsequently.   Nevertheless they were distinguished even by Rameau * without the assistance of any apparatus.   The process is as follows :—

Get a powerful bass voice to sing $e\flat$ to the vowel O, in *sore* [more like *aw* in *saw* than *o* in *so*], gently touch $b'\flat$ on the piano, which is the Twelfth, or third partial tone of the note $e\flat$, and let its sound die away while you are listening to it attentively.   The note $b'\flat$ on the piano will appear really not to die away, ¶ but to keep on sounding, even when its string is damped by removing the finger from the digital, because the ear unconsciously passes from the tone of the piano to the partial tone of the same pitch produced by the singer, and takes the latter for a continuation of the former.   But when the finger is removed from the key, and the damper has fallen, it is of course impossible that the tone of the string should have continued sounding.   To make the experiment for g'' the fifth partial, or major Third of the second Octave above $e\flat$, the voice should sing to the vowel A in *father*.

The resonators described in the last chapter furnish an excellent means for this purpose, and can be used for the tones of any musical instrument.   On applying to the ear the resonator corresponding to any given upper partial of the compound c, such as g', this g' is rendered much more powerful when c is sounded. Now hearing and distinguishing g' in this case by no means proves that the ear alone and without this apparatus would hear g' as part of the compound c.   But ¶ the increase of the loudness of g' caused by the resonator may be used to direct the attention of the ear to the tone it is required to distinguish.   On gradually removing the resonator from the ear, the force of g' will decrease.   But the attention once directed to it by this means, remains more readily fixed upon it, and the observer continues to hear this tone in the natural and unchanged compound tone of the given note, even with his unassisted ear.   The sole office of the resonators in this case is to direct the attention of the ear to the required tone.

By frequently instituting similar experiments for perceiving the upper partial tones, the observer comes to discover them more and more easily, till he is finally able to dispense with any aids.   But a certain amount of undisturbed concentration is always necessary for analysing musical tones by the ear alone, and hence the use of resonators is quite indispensable for an accurate comparison of different

* *Nouveau Système de Musique théorique.*   Paris : 1726.   Préface.

qualities of tones, especially in respect to the weaker upper partials. At least, I must confess, that my own attempts to discover the upper partial tones in the human voice, and to determine their differences for different vowels, were most unsatisfactory until I applied the resonators.

We now proceed to prove that the human ear really does analyse musical tones according to the law of simple vibrations. Since it is not possible to institute an exact comparison of the strength of our sensations for different simple tones, we must confine ourselves to proving that when an analysis of a composite tone into simple vibrations, effected by theoretic calculation or by sympathetic resonance, shews that certain upper partial tones are absent, the ear also does not perceive them.

¶ The tones of strings are again best adapted for conducting this proof, because they admit of many alterations in their quality of tone, according to the manner and the spot in which they are excited, and also because the theoretic or experimental analysis is most easily and completely performed for this case. Thomas Young * first shewed that when a string is plucked or struck, or, as we may add, bowed at any point in its length which is the node of any of its so-called harmonics, those simple vibrational forms of the string which have a node in that point are not contained in the compound vibrational form. Hence, if we attack the string at its middle point, all the simple vibrations due to the evenly numbered partials, each of which has a note at that point, will be absent. This gives the sound of the string a peculiarly hollow or nasal twang. If we excite the string at $\frac{1}{3}$ of its length, the vibrations corresponding to the third, sixth, and ninth partials will be absent; if at $\frac{1}{4}$, then those corresponding to the fourth, eighth, and twelfth partials will fail; and so on.†

This result of mathematical theory is confirmed, in the first place, by analysing the compound tone of the string by sympathetic resonance, either by the ¶ resonators or by other strings. The experiments may be easily made on the pianoforte. Press down the digitals for the notes c and c', without allowing the hammer to strike, so as merely to free them from their dampers, and then pluck the string c with the nail till it sounds. On damping the c string the higher c' will echo the sound, except in the particular case when the c string has been plucked exactly at its middle point, which is the point where it would have to be touched in order to give its first harmonic when struck by the hammer.

If we touch the c string at $\frac{1}{3}$ or $\frac{2}{3}$ its length, and strike it with the hammer, we obtain the harmonic g'; and if the damper of the g' is raised, this string echoes the sound. But if we pluck the c string with the nail, at either $\frac{1}{3}$ or $\frac{2}{3}$ its length, g' is not echoed, as it will be if the c string is plucked at any other spot.

In the same way observations with the resonators shew that when the c string is plucked at its middle the Octave c' is missing, and when at $\frac{1}{3}$ or $\frac{2}{3}$ its length the ¶ Twelfth g' is absent. The analysis of the sound of a string by the sympathetic resonance of strings or resonators, consequently fully confirms Thomas Young's law.

But for the vibration of strings we have a more direct means of analysis than that furnished by sympathetic resonance. If we, namely, touch a vibrating string gently for a moment with the finger or a camel's-hair pencil, we damp all those simple vibrations which have no node at the point touched. Those vibrations, however, which have a node there are not damped, and hence will continue to sound without the others. Consequently, if a string has been made to speak in any way whatever, and we wish to know whether there exists among its simple vibrations one corresponding to the Twelfth of the prime tone, we need only touch one of the nodes of this vibrational form at $\frac{1}{3}$ or $\frac{2}{3}$ the length of the string, in order to reduce to silence all simple tones which have no such node, and leave the Twelfth sounding, if it were there. If neither it, nor any of the sixth, ninth,

---

* London. *Philosophical Transactions*, 1800, vol. i. p. 137.
† See Appendix III.

twelfth, &c., of the partial tones were present, giving corresponding harmonics, the string will be reduced to absolute silence by this contact of the finger.

Press down one of the digitals of a piano, in order to free a string from its damper. Pluck the string at its middle point, and immediately touch it there. The string will be completely silenced, shewing that plucking it in its middle excited none of the evenly numbered partials of its compound tone. Pluck it at $\frac{1}{3}$ or $\frac{2}{3}$ its length, and immediately touch it in the same place; the string will be silent, proving the absence of the third partial tone. Pluck the string anywhere else than in the points named, and the second partial will be heard when the middle is touched, the third when the string is touched at $\frac{1}{3}$ or $\frac{2}{3}$ of its length.

The agreement of this kind of proof with the results from sympathetic resonance, is well adapted for the experimental establishment of the proposition based in the last chapter solely upon the results of mathematical theory, namely, that sympathetic vibration occurs or not, according as the corresponding simple ¶ vibrations are or are not contained in the compound motion. In the last described method of analysing the tone of a string, we are quite independent of the theory of sympathetic vibration, and the simple vibrations of strings are exactly characterised and recognisable by their nodes. If the compound tones admitted of being analysed by sympathetic resonance according to any other vibrational forms except those of simple vibration, this agreement could not exist.

If, after having thus experimentally proved the correctness of Thomas Young's law, we try to analyse the tones of strings by the unassisted ear, we shall continue to find complete agreement.* If we pluck or strike a string in one of its nodes, all those upper partial tones of the compound tone of the string to which the node belongs, disappear for the ear also, but they are heard if the string is plucked at any other place. Thus, if the string c be plucked at $\frac{1}{3}$ its length, the partial tone g' cannot be heard, but if the string be plucked at only a little distance from this point the partial tone g' is distinctly audible. Hence the ear analyses the sound ¶ of a string into precisely the same constituents as are found by sympathetic resonance, that is, into simple tones, according to Ohm's definition of this conception. These experiments are also well adapted to shew that it is no mere play of imagination when we hear upper partial tones, as some people believe on hearing them for the first time, for those tones are not heard when they do not exist.

The following modification of this process is also very well adapted to make the upper partial tones of strings audible. First, strike alternately in rhythmical sequence, the third and fourth partial tone of the string alone, by damping it in the corresponding nodes, and request the listener to observe the simple melody thus produced. Then strike the undamped string alternately and in the same rhythmical sequence, in these nodes, and thus reproduce the same melody in the upper partials, which the listener will then easily recognise. Of course, in order to hear the third partial, we must strike the string in the node of the fourth, and conversely.

The compound tone of a plucked string is also a remarkably striking example ¶ of the power of the ear to analyse into a long series of partial tones, a motion which the eye and the imagination are able to conceive in a much simpler manner. A string, which is pulled aside by a sharp point, or the finger nail, assumes the form fig. 18, A (p. 54a), before it is released. It then passes through the series of forms, fig. 18, B, C, D, E, F, till it reaches G, which is the inversion of A, and then returns, through the same, to A again. Hence it alternates between the forms A and G. All these forms, as is clear, are composed of three straight lines, and on expressing the velocity of the individual points of the strings by vibrational curves, these would have the same form. Now the string scarcely imparts any perceptible portion of its own motion directly to the air. Scarcely any audible tone results when both ends of a string are fastened to immovable supports, as metal bridges, which are again fastened to the walls of a room. The sound of

---

* See Brandt in Poggendorff's *Annalen der Physik*, vol. cxii. p. 324, where this fact is proved.

the string reaches the air through that one of its extremities which rests upon a bridge standing on an elastic sounding board. Hence the sound of the string essentially depends on the motion of this extremity, through the pressure which it exerts on the sounding board. The magnitude of this pressure, as it alters periodically with the time, is shewn in fig. 19, where the height of the line h h corresponds to the amount of pressure exerted on the bridge by that extremity of the string when the string is at rest. Along h h suppose lengths to be set off corresponding to consecutive intervals of time, the vertical ¶ heights of the broken line above or below h h represent the corresponding augmentations or diminutions of pressure at those times. The pressure of the string on the sounding board consequently alternates, as the figure shews, between a higher and a lower value. For some time the greater pressure remains unaltered; then the lower suddenly ensues, and likewise remains for a time unaltered. The letters a to g in fig. 19 correspond to the times at which the string

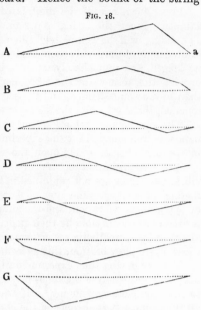

FIG. 18.

assumes the forms A to G in fig. 18. It is this alteration between a greater and a smaller pressure which produces the sound in the air. We cannot but feel astonished that a motion produced by means so simple and so easy to comprehend, ¶ should be analysed by the ear into such a complicated sum of simple tones. For the eye and the understanding the action of the string on the sounding board can be figured with extreme simplicity. What has the simple broken line of fig. 19 to do with wave-curves, which, in the course of one of their periods, shew

FIG. 19.

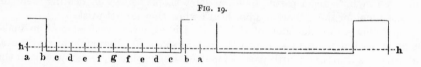

3, 4, 5, up to 16, and more, crests and troughs? This is one of the most striking examples of the different ways in which eye and ear comprehend a periodic motion.

There is no sonorous body whose motions under varied conditions can be so ¶ completely calculated theoretically and contrasted with observation as a string. The following are examples in which theory can be compared with analysis by ear :—

I have discovered a means of exciting simple pendular vibrations in the air. A tuning-fork when struck gives no harmonic upper partial tones, or, at most, traces of them when it is brought into such excessively strong vibration that it no longer exactly follows the law of the pendulum.* On the other hand, tuning-forks have some very high inharmonic secondary tones, which produce that peculiar sharp

---

* [On all ordinary tuning-forks between a and d″ in pitch, I have been able to hear the second partial or Octave of the prime. In some low forks this Octave is so powerful that on pressing the handle of the fork against the table, the prime quite disappears and the Octave only is heard, and this has often proved a source of embarrassment in tuning the forks, or in counting beats to determine pitch numbers. But the prime can always be heard when the fork is held to the ear or over a properly tuned resonance jar, as described in this paragraph. I tune such jars by pouring water in or out until the resonance is strongest, and then I register the height of the water and pitch of the fork for future use on a slip of paper gummed to the side of the jar. I have found that it is not at all necessary to

tinkling of the fork at the moment of being struck, and generally become rapidly inaudible. If the tuning-fork is held in the fingers, it imparts very little of its tone to the air, and cannot be heard unless it is held close to the ear. Instead of holding it in the fingers, we may screw it into a thick board, on the under side of which some pieces of india-rubber tubing have been fastened. When this is laid upon a table, the india-rubber tubes on which it is supported convey no sound to the table, and the tone of the tuning-fork is so weak that it may be considered inaudible. Now if the prongs of the fork be brought near a resonance chamber * of a bottle-form of such a size and shape that, when we blow over its mouth, the air it contains gives a tone of the same pitch as the fork's, the air within this chamber vibrates sympathetically, and the tone of the fork is thus conducted with great strength to the outer air. Now the higher secondary tones of such resonance chambers are also inharmonic to the prime tone, and in general the secondary tones of the chambers correspond neither with the harmonic nor the inharmonic ¶ secondary tones of the forks ; this can be determined in each particular case by producing the secondary tones of the bottle by stronger blowing, and discovering those of the forks with the help of strings set into sympathetic vibration, as will be presently described. If, then, only one of the tones of the fork, namely the prime tone, corresponds with one of the tones of the chamber, this alone will be reinforced by sympathetic vibration, and this alone will be communicated to the external air, and thus conducted to the observer's ear. The examination of the motion of the air by resonators shews that in this case, provided the tuning-fork be not set into too violent motion, no tone but the prime is present, and in such case the unassisted ear hears only a single simple tone, namely the common prime of the tuning-fork and of the chamber, without any accompanying upper partial tones.

The tone of a tuning-fork can also be purified from secondary tones by placing its handle upon a string and moving it so near to the bridge that one of the proper tones of the section of string lying between the fork and the bridge is the same as ¶ that of the tuning-fork. The string then begins to vibrate strongly, and conducts the tone of the tuning-fork with great power to the sounding board and surrounding air, whereas the tone is scarcely, if at all, heard as long as the above-named section is not in unison with the tone of the fork. In this way it is easy to find the lengths of string which correspond to the prime and upper partial tones of the fork, and accurately determine the pitch of the latter. If this experiment is conducted with ordinary strings which are uniform throughout their length, we shield the ear from the inharmonic secondary tones of the fork, but not from the harmonic upper partials, which are sometimes faintly present when the fork is made to vibrate strongly. Hence to conduct this experiment in such a way as to create purely pendular vibrations of the air, it is best to weight one point of the string, if only so much as by letting a drop of melting sealing-wax fall upon it. This causes the upper proper tones of the string itself to be inharmonic to the prime tone, and hence there is a distinct interval between the points where the fork must be placed ¶ to bring out the prime tone and its audible Octave, if it exists.

In most other cases the mathematical analysis of the motions of sound is not nearly far enough advanced to determine with certainty what upper partials will be present and what intensity they will possess. In circular plates and stretched membranes which are struck, it is theoretically possible to do so, but their inhar-

---

put the fork into excessively strong vibration in order to make the Octave sensible. Thus, taking a fork of 232 and another of 468 vibrations, after striking them both, and letting the deeper fork spend most of its energy until I could not see the vibrations with the eye at all, the beats were heard distinctly, when I pressed both on to a table, and continued to be heard even after the forks themselves were separately inaudible. See also Prof. Helmholtz's experiments on a fork of 64 vibrations at the close

of Chap. VII., and Prof. Preyer's in App. XX. sect. L. art. 4, c. The conditions according to Koenig that tuning-forks should have no upper partials are given in App. XX. sect. L. art. 2, a.—*Translator.*]

* Either a bottle of a proper size, which can readily be more accurately tuned by pouring oil or water into it, or a tube of pasteboard quite closed at one end, and having a small round opening at the other. See the proper sizes of such resonance chambers in App. IV.

monic secondary tones are so numerous and so nearly of the same pitch that most observers would probably fail to separate them satisfactorily. On elastic rods, however, the secondary tones are very distant from each other, and are inharmonic, so that they can be readily distinguished from each other by the ear. The following are the proper tones of a rod which is free at both ends; the vibrational number of the prime tone taken to be c, is reckoned as 1 :—

	Pitch Number	Cents *	Notation
Prime tone	1·0000	0	c
Second proper tone	2·7576	1200 + 556	f' +0·2
Third proper tone	5·4041	2400 + 521	f'' +0·1
Fourth proper tone	13·3444	3600 + 886	a''' −0·1

¶ The notation is adapted to the equal temperament, and the appended fractions are parts of the interval of a complete tone.

Where we are unable to execute the theoretical analysis of the motion, we can, at any rate, by means of resonators and other sympathetically vibrating bodies, analyse any individual musical tone that is produced, and then compare this analysis, which is determined by the laws of sympathetic vibration, with that effected by the unassisted ear. The latter is naturally much less sensitive than one armed with a resonator; so that it is frequently impossible for the unarmed ear to recognise amongst a number of other stronger simple tones those which the resonator itself can only faintly indicate. On the other hand, so far as my experience goes, there is complete agreement to this extent: the ear recognises without resonators the simple tones which the resonators greatly reinforce, and perceives no upper partial tone which the resonator does not indicate. To verify this conclusion, I performed numerous experiments, both with the human voice and the harmonium, and they all confirmed it.†

¶ By the above experiments the proposition enunciated and defended by G. S. Ohm must be regarded as proved, viz. that *the human ear perceives pendular vibrations alone as simple tones, and resolves all other periodic motions of the air into a series of pendular vibrations, hearing the series of simple tones which correspond with these simple vibrations.*

Calling, then, as already defined (in pp. 23, 24 and note), the sensation excited in the ear by any periodical motion of the air a *musical tone*, and the sensation excited by a simple pendular vibration a *simple tone*, the rule asserts that *the sensation of a musical tone is compounded out of the sensations of several simple tones.* In particular, we shall henceforth call the sound produced by a single sonorous body its (simple or compound) *tone*, and the sound produced by several musical instruments acting at the same time a *composite tone*, consisting generally of several (simple or compound) tones. If, then, a single note is sounded on a

---

¶ * [For cents see note p. 41d. As a Tone is 200 ct., 0·1 Tone = 20 ct., these would give for the Author's notation f' + 40 ct., f'' + 20 ct., a''' − 10 ct., whereas the column of cents shews that they are more accurately f' + 56 ct., f' + 21 ct., a''' − 14 ct. For convenience, the cents for Octaves are separated, thus 1200 + 556 in place of 1756, but this separation is quite unnecessary. The cents again shew the intervals of the inharmonic partial tones without any assumption as to the value of the prime. By a misprint in all the German editions, followed in the first English edition, the second proper tone was made f' − 0·2 in place of f' + 0·2.—*Translator.*]

† [In my 'Notes of Observations on Musical Beats,' *Proceedings of the Royal Society*, May 1880, vol. xxx. p. 531, largely cited in App. XX. sect. B. No. 7, I showed that I was able to determine the pitch numbers of deep reed tones, by the beats (Chap. VIII.) that their upper partials made with the primes of a set of Scheibler's tuning-forks. The correctness of the process was proved by the fact that the results obtained from different partials of the same reed tone, which were made to beat with different forks, gave the same pitch numbers for the primes, within one or two hundredths of a vibration in a second. I not only employed such low partials as 3, 4, 5 for one tone, and 4, 5, 6 for others, but I determined the pitch number 31·47, by partials 7, 8, 9, 10, 11, 12, 13, and the pitch number 15·94 by partials 25 and 27. The objective reality of these extremely high upper partials, and their independence of resonators or resonance jars, was therefore conclusively shewn. On the Harmonical the beats of the 16th partial of C 66, with c''', when slightly flattened by pressing the note lightly down, are very clear.—*Translator.*]

musical instrument, as a violin, trumpet, organ, or by a singing voice, it must be called in exact language a *tone* of the instrument in question. This is also the ordinary language, but it did not then imply that the tone might be *compound*. When the tone is, as usual, a compound tone, it will be distinguished by this term, or the abridgment, a *compound* ; while *tone* is a general term which includes both simple and compound tones.* The prime tone is generally louder than any of the upper partial tones, and hence it alone generally determines the *pitch* of the compound. The tone produced by any sonorous body reduces to a *single* simple tone in very few cases indeed, as the tone of tuning-forks imparted to the air by resonance chambers in the manner already described. The tones of wide-stopped organ pipes when gently blown are almost free from upper partials, and are accompanied only by a rush of wind.

It is well known that this union of several simple tones into one compound tone, which is naturally effected in the tones produced by most musical instruments, ¶ is artificially imitated on the organ by peculiar mechanical contrivances. The tones of organ pipes are comparatively poor in upper partials. When it is desirable to use a stop of incisive penetrating quality of tone and great power, the wide pipes (*principal register* and *weitgedackt* †) are not sufficient; their tone is too soft, too defective in upper partials ; and the narrow pipes (*geigen-register* and *quintaten* ‡) are also unsuitable, because, although more incisive, their tone is weak. For such occasions, then, as in accompanying congregational singing, recourse is had to the *compound stops*.§ In these stops every key is connected with a larger or smaller series of pipes, which it opens simultaneously, and which give the prime tone and a certain number of the lower upper partials of the compound tone of the note in question. It is very usual to connect the upper Octave with the prime tone, and after that the Twelfth. The more complex compounds (*cornet* §) give the first six partial tones, that is, in addition to the two Octaves of the prime tone and its Twelfth, the higher major Third, and the Octave of the Twelfth. This is as much ¶ of the series of upper partials as belongs to the tones of a major chord. But to prevent these compound stops from being insupportably noisy, it is necessary to reinforce the deeper tones of each note by other rows of pipes, for in all natural tones which are suited for musical purposes the higher partials decrease in force as they rise in pitch. This has to be regarded in their imitation by compound stops. These compound stops were a monster in the path of the old musical theory, which was acquainted only with the prime tones of compounds ; but the practice of organ-builders and organists necessitated their retention, and when they are suitably arranged and properly applied, they form a very effective musical apparatus.

* [Here, again, as on pp. 23, 24, I have, in the translation, been necessarily obliged to deviate slightly from the original. *Klang*, as here defined, embraces *Ton* as a particular case. I use *tone* for the general term, and *compound tone* and *simple tone* for the two particular cases. Thus, as presently mentioned in the text, the *tone* produced by a tuning-fork held over a proper resonance chamber we know, on analysis, to be *simple*, but before analysis it is to us only a (musical) *tone* like any other, and hence in this case the Author's *Klang* becomes the Author's *Ton*. I believe that the language used in my translation is best adapted for the constant accurate distinction between compound and simple tones by English readers, as I leave nothing which runs counter to old habits, and by the use of the words simple and compound, constantly recall attention to this newly discovered and extremely important relation.—*Translator*.]

† [*Principal*—double open diapason. *Grossgedackt*—double stopped diapason. Hopkins, *Organ*, p. 444-5.—*Translator*.]

‡ [' *Geigen Principal*—violin or crisp-

toned diapason, eight feet.' Hopkins, *Organ*, p. 445. 'A manual stop of eight feet, producing a pungent tone very like that of the Gamba, except that the pipes, being of larger scale, speak quicker and produce a fuller tone. Examples of the stop exist at Doncaster, ¶ the Temple Church, and in the Exchange Organ at Northampton.' *Ibid.* p. 138. For *quintaten*, see supra, p. 33d, note.—*Translator*.]

§ [As described in Hopkins, *Organ*, p. 142, these are the *sesquialtera* ' of five, four, three, or two ranks of open metal pipes, tuned in Thirds, Fifths, and Octaves to the Diapason.' The *mixture*, consisting of five to two ranks of open metal pipes smaller than the last, is in England the second, in Germany the first, compound stop (p. 143). The *Furniture* of five to two sets of small open pipes, is variable. 1) The *Cornet, mounted* has five ranks of very large and loudly voiced pipes, 2) the *echo*, is similar, but light and delicate, and is inclosed in a box. In German organs the *cornet* is also a pedal reed stop of four and two feet (*ibid.*).—*Translator*.]

The nature of the case at the same time fully justifies their use. The musician is bound to regard the tones of all musical instruments as compounded in the same way as the compound stops of organs, and the important part this method of composition plays in the construction of musical scales and chords will be made evident in subsequent chapters.

We have thus been led to an appreciation of upper partial tones, which differs considerably from that previously entertained by musicians, and even physicists, and must therefore be prepared to meet the opposition which will be raised. The upper partial tones were indeed known, but almost only in such compound tones as those of strings, where there was a favourable opportunity for observing them ; but they appear in previous physical and musical works as an isolated accidental phenomenon of small intensity, a kind of curiosity, which was certainly occasionally adduced, in order to give some support to the opinion that nature had pre-¶ figured the construction of our major chord, but which on the whole remained almost entirely disregarded. In opposition to this we have to assert, and we shall prove the assertion in the next chapter, that upper partial tones are, with a few exceptions already named, a general constituent of all musical tones, and that a certain stock of upper partials is an essential condition for a good musical quality of tone. Finally, these upper partials have been erroneously considered as weak, because they are difficult to observe, while, in point of fact, for some of the best musical qualities of tone, the loudness of the first upper partials is not far inferior to that of the prime tone itself.

There is no difficulty in verifying this last fact by experiments on the tones of strings. Strike the string of a piano or monochord, and immediately touch one of its nodes for an instant with the finger ; the constituent partial tones having this node will remain with unaltered loudness, and the rest will disappear. We might also touch the node in the same way at the instant of striking, and thus obtain the ¶ corresponding constituent partial tones from the first, in place of the complete compound tone of the note. In both ways we can readily convince ourselves that the first upper partials, as the Octave and Twelfth, are by no means weak and difficult to hear, but have a very appreciable strength. In some cases we are able to assign numerical values for the intensity of the upper partial tones, as will be shewn in the next chapter. For tones not produced on strings this *à posteriori* proof is not so easy to conduct, because we are not able to make the upper partials speak separately. But even then by means of the resonator we can appreciate the intensity of these upper partials by producing the corresponding note on the same or some other instrument until its loudness, when heard through the resonator, agrees with that of the former.

The difficulty we experience in hearing upper partial tones is no reason for considering them to be weak ; for this difficulty does not depend on their intensity, but upon entirely different circumstances, which could not be properly estimated ¶ until the advances recently made in the physiology of the senses. On this difficulty of observing the upper partial tones have been founded the objections which A. Seebeck * has advanced against Ohm's law of the decomposition of a musical tone ; and perhaps many of my readers who are unacquainted with the physiology of the other senses, particularly with that of the eye, might be inclined to adopt Seebeck's opinions. I am therefore obliged to enter into some details concerning this difference of opinion, and the peculiarities of the perceptions of our senses, on which the solution of the difficulty depends.

Seebeck, although extremely accomplished in acoustical experiments and observations, was not always able to recognise upper partial tones, where Ohm's law required them to exist. But we are also bound to add that he did not apply the methods already indicated for directing the attention of his ear to the upper partials in question. In other cases when he did hear the theoretical upper

---

* In Poggendorff's *Annalen der Physik*, vol. lx. p. 449, vol. lxiii. pp. 353 and 368.—*Ohm*, *ibid.* vol. lix. p. 513, and vol. lxii. p. 1.

partials, they were weaker than the theory required. He concluded that the definition of a simple tone as given by Ohm was too limited, and that not only pendular vibrations, but other vibrational forms, provided they were not too widely
separated from the pendular, were capable of exciting in the ear the sensation of
a single simple tone, which, however, had a variable quality. He consequently
asserted that when a musical tone was compounded of several simple tones, part
of the intensity of the upper constituent tones went to increase the intensity of
the prime tone, with which it fused, and that at most a small remainder excited in
the ear the sensation of an upper partial tone. He did not formulate any determinate law, assigning the vibrational forms which would give the impression of
a simple and those which would give the impression of a compound tone. The
experiments of Seebeck, on which he founded his assertions, need not be here
described in detail. Their object was only to produce musical tones for which
either the intensity of the simple vibrations corresponding to the upper partials ¶
could be theoretically calculated, or in which these upper partials could be
rendered separately audible. For the latter purpose the siren was used. We have
just described how the same object can be attained by means of strings. Seebeck
shews in each case that the simple vibrations corresponding to the upper partials
have considerable strength, but that the upper partials are either not heard at all,
or heard with difficulty in the compound tone itself. This fact has been already
mentioned in the present chapter. It may be perfectly true for an observer who
has not applied the proper means for observing upper partials, while another, or
even the first observer himself when properly assisted, can hear them perfectly well.*

Now there are many circumstances which assist us first in separating the
musical tones arising from different sources, and secondly, in keeping together the
partial tones of each separate source. Thus when one musical tone is heard for
some time before being joined by the second, and then the second continues after
the first has ceased, the separation in sound is facilitated by the succession of time. ¶
We have already heard the first musical tone by itself, and hence know immediately what we have to deduct from the compound effect for the effect of this first
tone. Even when several parts proceed in the same rhythm in polyphonic music,
the mode in which the tones of different instruments and voices commence, the
nature of their increase in force, the certainty with which they are held, and the
manner in which they die off, are generally slightly different for each. Thus the
tones of a pianoforte commence suddenly with a blow, and are consequently
strongest at the first moment, and then rapidly decrease in power. The tones of
brass instruments, on the other hand, commence sluggishly, and require a small
but sensible time to develop their full strength. The tones of bowed instruments
are distinguished by their extreme mobility, but when either the player or the
instrument is not unusually perfect they are interrupted by little, very short,
pauses, producing in the ear the sensation of scraping, as will be described more
in detail when we come to analyse the musical tone of a violin. When, then, such ¶
instruments are sounded together there are generally points of time when one or
the other is predominant, and it is consequently easily distinguished by the ear.
But besides all this, in good part music, especial care is taken to facilitate the
separation of the parts by the ear. In polyphonic music proper, where each part
has its own distinct melody, a principal means of clearly separating the progression of each part has always consisted in making them proceed in different rhythms
and on different divisions of the bars ; or where this could not be done, or was at
any rate only partly possible, as in four-part chorales, it is an old rule, contrived
for this purpose, to let three parts, if possible, move by single degrees of the scale,
and let the fourth leap over several. The small amount of alteration in the pitch
makes it easier for the listener to keep the identity of the several voices distinctly
in mind.

* [Here from ' Upper partial tones,' p. 94, to ' former analysis,' p. 100 of the 1st English
edition are omitted, in accordance with the 4th German edition.—*Translator*.]

All these helps fail in the resolution of musical tones into their constituent partials. When a compound tone commences to sound, all its partial tones commence with the same comparative strength; when it swells, all of them generally swell uniformly; when it ceases, all cease simultaneously. Hence no opportunity is generally given for hearing them separately and independently. In precisely the same manner as the naturally connected partial tones form a single source of sound, the partial tones in a compound stop on the organ fuse into one, as all are struck with the same digital, and all move in the same melodic progression as their prime tone.

Moreover, the tones of most instruments are usually accompanied by characteristic irregular noises, as the scratching and rubbing of the violin bow, the rush of wind in flutes and organ pipes, the grating of reeds, &c. These noises, with which we are already familiar as characterising the instruments, materially
¶ facilitate our power of distinguishing them in a composite mass of sounds. The partial tones in a compound have, of course, no such characteristic marks.

Hence we have no reason to be surprised that the resolution of a compound tone into its partials is not quite so easy for the ear to accomplish, as the resolution of composite masses of the musical sounds of many instruments into their proximate constituents, and that even a trained musical ear requires the application of a considerable amount of attention when it undertakes the former problem.

It is easy to see that the auxiliary circumstances already named do not always suffice for a correct separation of musical tones. In uniformly sustained musical tones, where one might be considered as an upper partial of another, our judgment might readily make default. This is really the case. G. S. Ohm proposed a very instructive experiment to shew this, using the tones of a violin. But it is more suitable for such an experiment to use simple tones, as those of a stopped organ pipe. The best instrument, however, is a glass bottle of the form
¶ shewn in fig. 20, which is easily procured and prepared for the experiment. A little rod c supports a guttapercha tube a in a proper position. The end of the tube, which is directed towards the bottle, is softened in warm water and pressed flat, forming a narrow chink, through which air can be made to rush over the mouth of the bottle. When the tube is fastened by an india-rubber pipe to the nozzle of a bellows, and wind is driven over the bottle, it produces a hollow obscure sound, like the vowel *oo* in *too*, which is freer from upper partial tones than even the tone of a stopped pipe, and is only accompanied by a slight
¶ noise of wind. I find that it is easier to keep the pitch unaltered in this instrument while the pressure of the wind is slightly changed, than in stopped pipes. We deepen the tone by partially shading the orifice of the bottle with a little wooden plate; and we sharpen it by

FIG. 20.

pouring in oil or melted wax. We are thus able to make any required little alterations in pitch. I tuned a large bottle to *b♭* and a smaller one to *b′♭* and united them with the same bellows, so that when used both began to speak at the same instant. When thus united they gave a musical tone of the pitch of the deeper *b♭*, but having the quality of tone of the vowel *oa* in *toad*, instead of *oo* in *too*. When, then, I compressed first one of the india-rubber tubes and then the other, so as to produce the tones alternately, separately, and in connection, I was at last able to hear them separately when sounded together, but I could not continue to hear them separately for long, for the upper tone gradually fused with the lower. This fusion takes place even when the upper tone is somewhat stronger than the lower. The alteration in the quality of tone which takes place during this fusion is characteristic. On producing the upper tone first and then letting the lower sound with it, I found that I at first continued to hear the upper tone with its full force, and the under tone sounding below it in its natural quality of *oo* in *too*. But by degrees, as my recollection of the sound of the isolated upper tone died away, it seemed to become more and more indistinct and weak, while the lower tone appeared to become stronger, and sounded like *oa* in *toad*. This weakening of the upper and strengthening of the lower tone was also observed by Ohm on the violin. As Seebeck remarks, it certainly does not always occur, and probably depends on the liveliness of our recollections of the tones as heard separately, and the greater or less uniformity in the simultaneous production of the tones. But where the experiment succeeds, it gives the best proof of the essential dependence of the result on varying activity of attention. With the tones ¶ produced by bottles, in addition to the reinforcement of the lower tone, the altera- tion in its quality is very evident and is characteristic of the nature of the process. This alteration is less striking for the penetrating tones of the violin.*

This experiment has been appealed to both by Ohm and by Seebeck as a corroboration of their different opinions. When Ohm stated that it was an ' illusion of the ear ' to apprehend the upper partial tones wholly or partly as a reinforcement of the prime tone (or rather of the compound tone whose pitch is determined by that of its prime), he certainly used a somewhat incorrect expression, although he meant what was correct, and Seebeck was justified in replying that the ear was the sole judge of auditory sensations, and that the mode in which it apprehended tones ought not to be called an ' illusion.' However, our experiments just described shew that the judgment of the ear differs according to the liveliness of its recollection of the separate auditory impressions here fused into one whole, and according to the intensity of its attention. Hence we can certainly appeal from ¶ the sensations of an ear directed without assistance to external objects, whose interests Seebeck represents, to the ear which is attentively observing itself and is suitably assisted in its observation. Such an ear really proceeds according to the law laid down by Ohm.

Another experiment should be adduced. Raise the dampers of a pianoforte so that all the strings can vibrate freely, then sing the vowel *a* in *father, art*, loudly to any note of the piano, directing the voice to the sounding board ; the sym- pathetic resonance of the strings distinctly re-echoes the same *a*. On singing *oe* in *toe*, the same *oe* is re-echoed. On singing *a* in *fare*, this *a* is re-echoed. For *ee* in *see* the echo is not quite so good. The experiment does not succeed so well if the damper is removed only from the note on which the vowels are sung. The vowel character of the echo arises from the re-echoing of those upper partial tones which characterise the vowels. These, however, will echo better and more clearly when their corresponding higher strings are free and can vibrate sym- ¶ pathetically. In this case, then, in the last resort, the musical effect of the resonance is compounded of the tones of several strings, and several separate partial tones combine to produce a musical tone of a peculiar quality. In addition to the vowels of the human voice, the piano will also quite distinctly imitate the quality of tone produced by a clarinet, when strongly blown on to the sounding board.

Finally, we must remark, that although the pitch of a compound tone is, for

---

* [A very convenient form of this experi- ment, useful even for lecture purposes, is to employ two tuning-forks, tuned as an Octave, say c' and c'', and held over separate resonance jars. By removing first one and then the other, or letting both sound together, the above effects can be made evident, and they even remain when the Octave is not tuned perfectly true. The tone is also brighter and unaccompanied by any windrush. By pressing the handle of the deeper fork on the table, we can excite its other upper partials, and thus produce a third quality of tone, which can be readily appre- ciated; thus, simple c', simple c' + simple c'', compound c'.—*Translator*.]

musical purposes, determined by that of its prime, the influence of the upper partial tones is by no means unfelt. They give the compound tone a brighter and higher effect. Simple tones are dull. When they are compared with compound tones of the same pitch, we are inclined to estimate the compound as belonging to a higher Octave than the simple tones. The difference is of the same kind as that heard when first the vowel *oo* in *too* and then *a* in *tar* are sung to the same note. It is often extremely difficult to compare the pitches of compound tones of different qualities. It is very easy to make a mistake of an Octave. This has happened to the most celebrated musicians and acousticians. Thus it is well known that Tartini, who was celebrated as a violinist and theoretical musician, estimated all combinational tones (Chap. XI.) an Octave too high, and, on the other hand, Henrici * assigns a pitch too low by an Octave to the upper partial tones of tuning-forks.†

¶   The problem to be solved, then, in distinguishing the partials of a compound tone is that of analysing a given aggregate of sensations into elements which no longer admit of analysis. We are accustomed in a large number of cases where sensations of different kinds or in different parts of the body, exist simultaneously, to recognise that they are distinct as soon as they are perceived, and to direct our attention at will to any one of them separately. Thus at any moment we can be separately conscious of what we see, of what we hear, of what we feel, and distinguish what we feel in a finger or in the great toe, whether pressure or a gentle touch, or warmth. So also in the field of vision. Indeed, as I shall endeavour to shew in what follows, we readily distinguish our sensations from one another when we have a precise knowledge that they are composite, as, for example, when we have become certain, by frequently repeated and invariable experience, that our present sensation arises from the simultaneous action of many independent stimuli, each of which usually excites an equally well-known individual sensation. This induces
¶ us to think that nothing can be easier, when a number of different sensations are simultaneously excited, than to distinguish them individually from each other, and that this is an innate faculty of our minds.

Thus we find, among others, that it is quite a matter of course to hear separately the different musical tones which come to our senses collectively, and expect that in every case when two of them occur together, we shall be able to do the like.

The matter is very different when we set to work at investigating the more unusual cases of perception, and at more completely understanding the conditions under which the above-mentioned distinction can or cannot be made, as is the case in the physiology of the senses. We then become aware that two different kinds or grades must be distinguished in our becoming conscious of a sensation. The lower grade of this consciousness, is that where the influence of the sensation in question makes itself felt only in the conceptions we form of external things and processes, and assists
¶ in determining them. This can take place without our needing or indeed being able to ascertain to what particular part of our sensations we owe this or that relation of our perceptions. In this case we will say that the impression of the sensation in question is *perceived synthetically*. The second and higher grade is when we immediately distinguish the sensation in question as an existing part of the sum of the sensations excited in us. We will say then that the sensation is *perceived analytically*.‡ The two cases must be carefully distinguished from each other.

---

* Poggd. *Ann.*, vol. xcix. p. 506. The same difficulty is mentioned by Zamminer (*Die Musik und die musikalischen Instrumente*, 1855, p. 111) as well known to musicians.

† [Here the passage from ' The problem to be solved,' p. 62*b*, to ' from its simple tones,' p. 65*b*, is inserted in this edition from the 4th German edition.—*Translator.*]

‡ [Prof. Helmholtz uses Leibnitz's terms *percipirt* and *appercipirt*, alternating the latter with *wahrgenommen*, and then restricting the meaning of this very common German word. It appeared to me that it would be clearer to an English reader not to invent new words or restrict the sense of old words, but to use *perceived* in both cases, and distinguish them (for *percipirt* and *appercipirt* respectively) by the adjuncts *synthetically* and *analytically*, the use of which is clear from the explanations given in the text.-- *Translator.*]

Seebeck and Ohm are agreed that the upper partials of a musical tone are perceived synthetically. This is acknowledged by Seebeck when he admits that their action on the ear changes the force or quality of the sound examined. The dispute turns upon whether in all cases they can be perceived analytically in their individual existence ; that is, whether the ear when unaided by resonators or other physical auxiliaries, which themselves alter the mass of musical sound heard by the observer, can by mere direction and intensity of attention distinguish whether, and if so in what force, the Octave, the Twelfth, &c., of the prime exists in the given musical sound.

In the first place I will adduce a series of examples which shew that the difficulty felt in analysing musical tones exists also for other senses. Let us begin with the comparatively simple perceptions of the sense of taste. The ingredients of our dishes and the spices with which we flavour them, are not so complicated that they could not be readily learned by any one. And yet there are ¶ very few people who have not themselves practically studied cookery, that are able readily and correctly to discover, by the taste alone, the ingredients of the dishes placed before them. How much practice, and perhaps also peculiar talent, belongs to wine tasting for the purpose of discovering adulterations is known in all wine-growing countries. Similarly for smell ; indeed the sensations of taste and smell may unite to form a single whole. Using our tongues constantly, we are scarcely aware that the peculiar character of many articles of food and drink, as vinegar or wine, depends also upon the sensation of smell, their vapours entering the back part of the nose through the gullet. It is not till we meet with persons in whom the sense of smell is deficient that we learn how essential a part it plays in tasting. Such persons are constantly in fault when judging of food, as indeed any one can learn from his own experience, when he suffers from a heavy cold in the head without having a loaded tongue.

When our hand glides unawares along a cold and smooth piece of metal we ¶ are apt to imagine that we have wetted our hand. This shews that the sensation of wetness to the touch is compounded out of that of unresisting gliding and cold, which in one case results from the good heat-conducting properties of metal, and in the other from the cold of evaporation and the great specific heat of water. We can easily recognise both sensations in wetness, when we think over the matter, but it is the above-mentioned illusion which teaches us that the peculiar feeling of wetness is entirely resolvable into these two sensations.

The discovery of the stereoscope has taught us that the power of seeing the depths of a field of view, that is, the different distances at which objects and their parts lie from the eye of the spectator, essentially depends on the simultaneous synthetical perceptions of two somewhat different perspective images of the same objects by the two eyes of the observer. If the difference of the two images is sufficiently great it is not difficult to perceive them analytically as separate. For example, if we look intently at a distant object and hold one of ¶ our fingers slightly in front of our nose we see two images of our finger against the background, one of which vanishes when we close the right eye, the other belonging to the left. But when the differences of distance are relatively small, and hence the differences of the two perspective images on the retina are so also, great practice and certainty in the observation of double images is necessary to keep them asunder, yet the synthetical perception of their differences still exists, and makes itself felt in the apparent relief of the surface viewed. In this case also, as well as for upper partial tones, the ease and exactness of the analytical perception is far behind that of the synthetical perception.

In the conception which we form of the direction in which the objects viewed seem to lie, a considerable part must be played by those sensations, mainly muscular, which enable us to recognise the position of our body, of the head with regard to the body, and of the eye with regard to the head. If one of these is altered, for example, if the sensation of the proper position of the eye is changed by pressing a finger against the eyeball or by injury to one of the muscles of the eye, our per-ception of the position of visible objects is also changed. But it is only by such occasional illusions that we become aware of the fact that muscular sensations form part of the aggregate of sensations by which our conception of the position of a visible object is determined.

The phenomena of mixed colours present considerable analogy to those of com-pound musical tones, only in the case of colour the number of sensations reduces to three, and the analysis of the composite sensations into their simple elements is still more difficult and imperfect than for musical tones. As early as 1686 R. Waller mentions in the *Philosophical Transactions* the reduction of all colours to the mixture of three fundamental colours, as something already well known. This view could in earlier times only be founded on sensations and experiments arising from the = mixture of pigments. In recent times we have discovered better methods, ¶ by mixing light of different colours, and hence have confirmed the correctness of that hypothesis by exact measurements, but at the same time we have learned that this confirmation only succeeds within a certain limit, conditioned by the fact that no kind of coloured light exists which can give us the sensation of a single one of the fundamental colours with exclusive purity. Even the most saturated and purest colours that the external world presents to us in the prismatic spectrum, may by the development of secondary images of the complementary colours in the eye be still freed as it were from a white veil, and hence cannot be considered as abso-lutely pure. For this reason we are unable to shew objectively the absolutely pure fundamental colours from a mixture of which all other colours without exception can be formed. We only know that among the colours of the spectrum scarlet-red, yellow-green, and blue-violet approach to them nearer than any other objective colours.* Hence we are able to compound out of these three colours almost all the colours that usually occur in different natural bodies, but we cannot produce the ¶ yellow and blue of the spectrum in that complete degree of saturation which they reach when purest within the spectrum itself. Our mixtures are always a little whiter than the corresponding simple colours of the spectrum. Hence it follows that we never see the simple elements of our sensations of colour, or at least see them only for a very short time in particular experiments directed to this end, and consequently cannot have any such exact or certain image in our recollection, as would indisputably be necessary for accurately analysing every sensation of colour into its elementary sensations by inspection. Moreover we have relatively rare opportunities of observing the process of the composition of colours, and hence of recognising the constituents in the compound. It certainly appears to me very characteristic of this process, that for a century and a half, from Waller to Goethe, every one relied on the mixtures of pigments, and hence believed green to be a mixture of blue and yellow, whereas when sky-blue and sulphur-yellow beams of light, *not pigments*, are mixed together, the result is white. To this very cir-¶ cumstance is due the violent opposition of Goethe, who was only acquainted with the colours of pigments, to the assertion that white was a mixture of variously coloured beams of light. Hence we can have little doubt that the power of dis-tinguishing the different elementary constituents of the sensation is originally absent in the sense of sight, and that the little which exists in highly educated observers, has been attained by specially conducted experiments, through which of course, when wrongly planned, error may have ensued.

On the other hand every individual has an opportunity of experimenting on the

---

* [In his *Physiological Optics*, p. 227, Prof. Helmholtz calls scarlet-red or vermilion the part of the spectrum before reaching Fraunhofer's line *C*. He does not use *span-grün*, (= *Grün-span* or verdigris, literally 'Spanish green') in his *Optics*, but talks of green-yellow between the lines *E* and *b*, and he says, on p. 844, that Maxwell took as one of the fundamental colours ' a green near the line *E*,' hence I translate *span-grün* by ' yellow-green.' Maxwell's blue or third colour was between the lines *F* and *G*, but twice as far from the latter as the former. This gives the colour which Prof. H. in his *Optics* calls ' cya-nogen blue,' or Prussian blue. The violet proper does not begin till after the line *G*. It is usual to speak of these three colours, vaguely, as Red, Green, and Blue.—*Translator*.]

composition of two or more musical sounds or noises on the most extended scale, and the power of analysing even extremely involved compounds of musical tones, into the separate parts produced by individual instruments, can readily be acquired by any one who directs his attention to the subject. But the ultimate simple elements of the sensation of tone, simple tones themselves, are rarely heard alone. Even those instruments by which they can be produced, as tuning-forks before resonance chambers, when strongly excited, give rise to weak harmonic upper partials, partly within and partly without the ear, as we shall see in Chapters V. and VII. Hence in this case also, the opportunities are very scanty for impressing on our memory an exact and sure image of these simple elementary tones. But if the constituents to be added are only indefinitely and vaguely known, the analysis of the sum into those parts must be correspondingly uncertain. If we do not know with certainty how much of the musical tone under consideration is to be attributed to its prime, we cannot but be uncertain as to what belongs to the ¶ partials. Consequently we must begin by making the individual elements which have to be distinguished, individually audible, so as to obtain an entirely fresh recollection of the corresponding sensation, and the whole business requires undisturbed and concentrated attention. We are even without the ease that can be obtained by frequent repetitions of the experiment, such as we possess in the analysis of musical chords into their individual tones. In that case we hear the individual tones sufficiently often by themselves, whereas we rarely hear simple tones and may almost be said never to hear the building up of a compound from its simple tones.

The results of the preceding discussion may be summed up as follows :—

1.) The upper partial tones corresponding to the simple vibrations of a compound motion of the air, are perceived synthetically, even when they are not always perceived analytically.

2.) But they can be made objects of analytical perception without any other ¶ help than a proper direction of attention.

3.) Even in the case of their not being separately perceived, because they fuse into the whole mass of musical sound, their existence in our sensation is established by an alteration in the quality of tone, the impression of their higher pitch being characteristically marked by increased brightness and acuteness of quality.

In the next chapter we shall give details of the relations of the upper partials to the quality of compound tones.

CHAPTER V.

ON THE DIFFERENCES IN THE QUALITY OF MUSICAL TONES.

TOWARDS the close of Chapter I. (p. 21d), we found that differences in the quality of musical tones must depend on the form of the vibration of the air. The ¶ reasons for this assertion were only negative. We had seen that force depended on amplitude, and pitch on rapidity of vibration : nothing else was left to distinguish quality but vibrational form. We then proceeded to shew that the existence and force of the upper partial tones which accompanied the prime depend also on the vibrational form, and hence we could not but conclude that musical tones of the same quality would always exhibit the same combination of partials, seeing that the peculiar vibrational form which excites in the ear the sensation of a certain quality of tone, must always evoke the sensation of its corresponding upper partials. The question then arises, can, and if so, to what extent can the differences of musical quality be reduced to the combination of different partial tones with different intensities in different musical tones? At the conclusion of last chapter (p. 60d), we saw that even artificially combined simple tones were capable of fusing into a musical tone of a quality distinctly different from that of either of its constituents, and that consequently the existence of a new upper partial really altered the quality of a tone. By this means we gained a clue to the hitherto enigmatical nature of quality of tone, and to the cause of its varieties.

There has been a general inclination to credit quality with all possible peculiarities of musical tones that were not evidently due to force and pitch. This was correct to the extent that quality of tone was merely a negative conception. But very slight consideration will suffice to shew that many of these peculiarities of musical tones depend upon the way in which they begin and end. The methods of attacking and releasing tones are sometimes so characteristic that for the human voice they have been noted by a series of different letters. To these belong the explosive consonants B, D, G, and P, T, K. The effects of these letters are produced by opening the closed, or closing the open passage through the mouth. For B and P the closure is made by the lips, for D and T by the tongue and upper teeth,* for G and K by the back of the tongue and soft palate. The series of the *mediae*
¶ B, D, G is distinguished from that of the *tenues* P, T, K, by the glottis being sufficiently narrowed, when the closure of the former is released, to produce voice, or at least the rustle of whisper, whereas for the latter or *tenues* the glottis is wide open,† and cannot sound. The *mediae* are therefore accompanied by voice, which is capable of commencing at the beginning of a syllable an instant before the opening of the mouth, and of lasting at the end of a syllable a moment after the closure of the mouth, because some air can be still driven into the closed cavity of the mouth and the vibration of the vocal chords in the larynx can be still maintained. On account of the narrowing of the glottis the influx of air is more moderate, and the noise of the wind less sharp for the *mediae* than the *tenues*, which, being spoken with open glottis, allow of a great deal of wind being forced at once from the chest.‡ At the same time the resonance of the cavity of the mouth, which, as we shall more clearly understand further on, exercises a great influence on the vowels, varies its pitch, corresponding to the rapid alterations in the magnitude of its volume ¶ and orifice, and this brings about a corresponding rapid variation in the quality of the speech sound.

As with consonants, the differences in the quality of tone of struck strings, also partly depends on the rapidity with which the tone dies away. When the strings have little mass (such as those of gut), and are fastened to a very mobile sounding board (as for a violin, guitar, or zither), or when the parts on which they rest or which they touch are but slightly elastic (as when the violin strings, for example, are pressed on the finger board by the soft point of the finger), their vibrations rapidly disappear after striking, and the tone is dry, short, and without ring, as in the *pizzicato* of a violin. But if the strings are of metal wire, and hence of greater weight and tension, and if they are attached to strong heavy bridges which cannot be much shaken, they give out their vibrations slowly to the

---

* [This is true for German, and most Continental languages, and for some dialectal ¶ English, especially in Cumberland, Westmoreland, Yorkshire, Lancashire, the Peak of Derbyshire, and Ireland, but even then only in connection with the trilled R. Throughout England generally, the tip of the tongue is quite free from the teeth, except for TH in *thin* and *then*, and for T and D it only touches the hard palate, seldom advancing so far as the root of the gums.—*Translator.*]

† [This again is true for German, but not for English, French, or Italian, and not even for the adjacent Slavonic languages. In these languages the glottis is quite closed for both the mediae and the tenues in ordinary speech, but the voice begins for the mediae *before* releasing the closure of the lips or tongue and palate, and for the tenues *at the moment of release*. Although in giving vowel sounds, &c., I have generally contented myself with translating the same into English symbols and

examples, it seemed better in the present case, where the author was speaking especially of the phenomena of speech to which he was personally accustomed, to leave the text unaltered and draw attention to English peculiarities in footnotes.—*Translator.*]

‡ [Observe again that this description of the rush of wind accompanying P, T, K, although true for German habits of speech, is not true for the usual English habits, which require the windrush between the opening of the mouth and sounding of the vowel to be entirely suppressed. The English result is a gliding vowel sound preceding the true vowel on commencing a syllable, and following the vowel on ending one. The difference between English P and German P is precisely the same (as I have verified by actual observation) as that between the simple Sanscrit tenuis P, and the postaspirated Sanscrit Ph, as now actually pronounced by cultivated Bengalese. See my *Early English Pronunciation*, p. 1136, col. 1.—*Translator.*]

air and the sounding board; their vibrations continue longer, their tone is more durable and fuller, as in the pianoforte, but is comparatively less powerful and penetrating than that of gut strings, which give up their tone more readily when struck with the same force. Hence the *pizzicato* of bowed instruments when well executed is much more piercing than the tone of a pianoforte. Pianofortes with their strong and heavy supports for the strings have, consequently, for the same thickness of string, a less penetrating but a much more lasting tone than those instruments of which the supports for the strings are lighter.

It is very characteristic of brass instruments, as trumpets and trombones, that their tones commence abruptly and sluggishly. The various tones in these instruments are produced by exciting different upper partials through different styles of blowing, which serve to throw the column of air into vibrating portions of different numbers and lengths similar to those on a string. It always requires a certain amount of effort to excite the new condition of vibration in place of the old, but when once established it is maintained with less exertion. On the other hand, the transition from one tone to another is easy for wooden wind instruments, as the flute, oboe, and clarinet, where the length of the column of air is readily changed by application of the fingers to the side holes and keys, and where the style of blowing has not to be materially altered.

These examples will suffice to shew how certain characteristic peculiarities in the tones of several instruments depend on the mode in which they begin and end. When we speak in what follows of *musical quality of tone*, we shall disregard these peculiarities of beginning and ending, and confine our attention to the peculiarities of the musical tone which continues uniformly.

But even when a musical tone continues with uniform or variable intensity, it is mixed up, in the general methods of excitement, with certain noises, which express greater or less irregularities in the motion of the air. In wind instruments where the tones are maintained by a stream of air, we generally hear more or less whizzing and hissing of the air which breaks against the sharp edges of the mouthpiece. In strings, rods, or plates excited by a violin bow, we usually hear a good deal of noise from the rubbing. The hairs of the bow are naturally full of many minute irregularities, the resinous coating is not spread over it with absolute evenness, and there are also little inequalities in the motion of the arm which holds the bow and in the amount of pressure, all of which influence the motion of the string, and make the tone of a bad instrument or an unskilful performer rough, scraping, and variable. We shall not be able to explain the nature of the motions of the air and sensations of the ear which correspond to these noises till we have investigated the conception of *beats*. Those who listen to music make themselves deaf to these noises by purposely withdrawing attention from them, but a slight amount of attention generally makes them very evident for all tones produced by blowing or rubbing. It is well known that most consonants in human speech are characterised by the maintenance of similar noises, as F, V; S, Z; TH in *thin* and in *then*; the Scotch and German guttural CH, and Dutch G. For some the tone is made still more irregular by trilling parts of the mouth, as for R and L. In the case of R the stream of air is periodically entirely interrupted by trilling the uvula* or the tip of the tongue; and we thus obtain an intermitting sound to which these interruptions give a peculiar jarring character. In the case of L the soft side edges of the tongue are moved by the stream of air, and, without completely interrupting the tone, produce inequalities in its strength.

Even the vowels themselves are not free from such noises, although they are kept more in the background by the musical character of the tones of the voice. Donders first drew attention to these noises, which are partly identical with those which are produced when the corresponding vowels are indicated in low voiceless

---

* [In the northern parts of Germany and of France, and in Northumberland, but not otherwise in England, except as an organic defect. There are also many other trills, into which, as into other phonetic details, it is not necessary to enter.—*Translator.*]

speech. They are strongest for *ee* in *see*, the French *u* in *vu* (which is nearly the same as the Norfolk and Devon *oo* in *too*), and for *oo* in *too*. For these vowels they can be made audible even when speaking aloud.* By simply increasing their force the vowel *ee* in *see* becomes the consonant *y* in *yon*, and the vowel *oo* in *too* the consonant *w* in *wan*.† For *a* in *art*, *a* in *at*, *e* in *met*, *there*, and *o* in *more*, the noises appear to me to be produced in the glottis alone when speaking gently, and to be absorbed into the voice when speaking aloud.‡ It is remarkable that in speaking, the vowels *a* in *art*, *a* in *at*, and *e* in *met*, *there*, are produced with less musical tone than in singing. It seems as if a feeling of greater compression in the larynx caused the tuneful tone of the voice to give way to one of a more jarring character which admits of more evident articulation. The greater intensity thus given to the noises, appears in this case to facilitate the characterisation of the peculiar vowel quality. In singing, on the contrary, we try to favour the musical ¶ part of its quality and hence often render the articulation somewhat obscure.§

Such accompanying noises and little inequalities in the motion of the air, furnish much that is characteristic in the tones of musical instruments, and in the vocal tones of speech which correspond to the different positions of the mouth ; but besides these there are numerous peculiarities of quality belonging to the musical tone proper, that is, to the perfectly regular portion of the motion of the air. The importance of these can be better appreciated by listening to musical instruments or human voices, from such a distance that the comparatively weaker noises are no longer audible. Notwithstanding the absence of these noises, it is generally possible to discriminate the different musical instruments, although it must be acknowledged that under such circumstances the tone of a French horn may be occasionally mistaken for that of the singing voice, or a violoncello may be confused with an harmonium. For the human voice, consonants first disappear at a distance, because they are characterised by noises, but M, N, and the vowels ¶ can be distinguished at a greater distance. The formation of M and N in so far resembles that of vowels, that no noise of wind is generated in any part of the cavity of the mouth, which is perfectly closed, and the sound of the voice escapes through the nose. The mouth merely forms a resonance chamber which alters the quality of tone. It is interesting in calm weather to listen to the voices of men who are descending from high hills to the plain. Words can no longer be recognised, or at most only such as are composed of M, N, and vowels, as *Mamma*, *No*, *Noon*. But the vowels contained in the spoken words are easily distinguished. Wanting the thread which connects them into words and sentences, they form a strange series of alternations of quality and singular inflections of tone.

In the present chapter we shall at first disregard all irregular portions of the motion of the air, and the mode in which sounds commence or terminate, directing our attention solely to the musical part of the tone, properly so called, which corresponds to a uniformly sustained and regularly periodic motion of the air, ¶ and we shall endeavour to discover the relations between the quality of the sound

---

* [At the Comédie Française I have heard M. Got pronounce the word *oui* and Mme. Provost-Ponsin pronounce the last syllable of *hachis* entirely without voice tones, and yet make them audible throughout the theatre.— *Translator.*]

† [That this is not the whole of the phenomenon is shewn by the words *ye, woo*. The whole subject is discussed at length in my *Early English Pronunciation*, pp. 1092–1094, and 1149–1151.—*Translator.*]

‡ [By 'speaking gently' (*leise*) seems to be meant either speaking absolutely without voice, that is with an open glottis, or in a whisper, with the glottis nearly closed. For voice the glottis is quite closed, and this is indicated by 'speaking aloud' (*beim lauten Sprechen*). It would lead too far to discuss

the important phonetic observations in the text.—*Translator.*]

§ [These observations must not be considered as exhausting the subject of the difference between the singing and the speaking voice, which requires a peculiar study here merely indicated. See my *Pronunciation for Singers* (Curwen) and *Speech in Song* (Novello). The difference between English and German habits of speaking and singing must also be borne in mind, and allowed for by the reader. The English vowels given in the text are not the perfect equivalents of Prof. Helmholtz's German sounds. The noises which accompany the vowels are not nearly so marked in English as in German, but they differ very much locally, even in England.— *Translator.*]

and its composition out of individual simple tones. The peculiarities of quality of sound belonging to this division, we shall briefly call its *musical quality*.

The object of the present chapter is, therefore, to describe the different composition of musical tones as produced by different instruments, for the purpose of shewing how different modes of combining the upper partial tones correspond to characteristic varieties of musical quality. Certain general rules will result for the arrangement of the upper partials which answer to such species of musical quality as are called, *soft, piercing, braying, hollow* or *poor, full* or *rich, dull, bright, crisp, pungent*, and so on. Independently of our immediate object (the determination of the physiological action of the ear in the discrimination of musical quality, which is reserved for the following chapter), the results of this investigation are important for the resolution of purely musical questions in later chapters, because they shew us how rich in upper partials, good musical qualities of tone are found to be, and also point out the peculiarities of musical quality ¶ favoured on those musical instruments, for which the quality of tone has been to some extent abandoned to the caprice of the maker.

Since physicists have worked comparatively little at this subject I shall be forced to enter somewhat more minutely into the mechanism by which the tones of several instruments are produced, than will be, perhaps, agreeable to many of my readers. For such the principal results collected at the end of this chapter will suffice. On the other hand, I must ask indulgence for leaving many large gaps in this almost unexplored region, and for confining myself principally to instruments sufficiently well known for us to obtain a tolerably satisfactory view of the source of their tones. In this inquiry lie rich materials for interesting acoustical work. But I have felt bound to confine myself to what was necessary for the continuation of the present investigation.

1. *Musical Tones without Upper Partials.*　　　　　　　¶

We begin with such musical tones as are not decomposable, but consist of a single simple tone. These are most readily and purely produced by holding a struck tuning-fork over the mouth of a resonance tube, as has been described in the last chapter (p. 54*d*).* These tones are uncommonly soft and free from all shrillness and roughness. As already remarked, they appear to lie comparatively deep, so that such as correspond to the deep tones of a bass voice produce the impression of a most remarkable and unusual depth. The musical quality of such deep simple tones is also rather dull. The simple tones of the soprano pitch sound bright, but even those corresponding to the highest tones of a soprano voice are very soft, without a trace of that cutting, rasping shrillness which is displayed by most instruments at such pitches, with the exception, perhaps, of the flute, for which the tones are very nearly simple, being accompanied with very few and faint upper partials. Among vowels, the *oo* in *too* comes nearest to a simple tone, but even this vowel is not entirely free from upper partials. On comparing the ¶ musical quality of a simple tone thus produced with that of a compound tone in which the first harmonic upper partial tones are developed, the latter will be found to be more tuneful, metallic, and brilliant. Even the vowel *oo* in *too*, although the dullest and least tuneful of all vowels, is sensibly more brilliant and less dull than a simple tone of the same pitch. The series of the first six partials of a compound tone may be regarded musically as a major chord with a very predominant fundamental tone, and in fact the musical quality of a compound tone possessing these partials, as, for example, a fine singing voice, when heard beside a simple tone, very distinctly produces the agreeable effect of a consonant chord.

Since the form of simple waves of known periodic time is completely given when their amplitude is given, simple tones of the same pitch can only differ in force and not in musical quality. In fact, the difference of quality remains

---

* On possible sources of disturbance, see Appendix IV.

perfectly indistinguishable, whether the simple tone is conducted to the external air in the preceding methods by a tuning-fork and a resonance tube of any given material, glass, metal, or pasteboard, or by a string, provided only that we guard against any chattering in the apparatus.

Simple tones accompanied only by the noise of rushing wind can also be produced, as already mentioned, by blowing over the mouth of bottles with necks (p. 60c). If we disregard the friction of the air, the proper musical quality of such tones is really the same as that produced by tuning-forks.

2. *Musical Tones with Inharmonic Upper Partials.*

Nearest to musical tones without any upper partials are those with secondary tones which are inharmonic to the prime, and such tones, therefore, in strictness, ¶ should not be reckoned as musical tones at all. They are exceptionally used in artistic music, but only when it is contrived that the prime tone should be so much more powerful than the secondary tones, that the existence of the latter may be ignored. Hence they are placed here next to the simple tones, because musically they are available only for the more or less good simple tones which they represent. The first of these are *tuning-forks* themselves, when they are struck and applied to a sounding board, or brought very near the ear. The [inharmonic] upper partials of tuning-forks lie very high. In those which I have examined, the first made from 5·8 to 6·6 as many vibrations in the same time as the prime tone, and hence lay between its third diminished Fifth and major Sixth. The pitch numbers of these high upper partial tones were to one another as the squares of the odd numbers. In the time that the first upper partial would execute $3 \times 3 = 9$ vibrations, the next would execute $5 \times 5 = 25$, and the next $7 \times 7 = 49$, and so on. Their pitch, therefore, increases with extraordinary rapidity, and they are usually all ¶ inharmonic with the prime, though some of them may exceptionally become harmonic. If we call the prime tone of the fork c, the next succeeding tones are nearly $a''\flat$, d'^{v}, $c'^{\mathrm{v}}\sharp$.* These high secondary tones produce a bright inharmonic clink, which is easily heard at a considerable distance when the fork is first struck, whereas when it is brought close to the ear, the prime tone alone is heard. The ear readily separates the prime from the upper tones and has no inclination to fuse them. The high simple tones usually die off rapidly, while the prime tone remains audible for a long time. It should be remarked, however, that the mutual relations of the proper tones of tuning-forks differ somewhat according to the form of the fork, and hence the above indications must be looked upon as merely approximate. In theoretical determinations of the upper partial tones, each prong of the fork may be regarded as a rod fixed at one end.

The same relations hold for straight elastic rods, which, as already mentioned, when struck, give rather high inharmonic upper partial tones. When such a rod ¶ is firmly supported at the two nodal lines of its prime tone, the continuance of that tone is favoured in preference to the other higher tones, and hence the latter disturb the effect very slightly, more especially as they rapidly die away after the rod has been struck. Such rods, however, are not suitable for real artistic music,

---

* [On calculating the number of cents (as in App. XX. sect. C.), we find that the first tone mentioned, which vibrates from 5·8 to 6·6 as fast as the prime, makes an interval with it of from 3043 to 3267 ct., so that if the prime is called c, the note lies between $g''\flat + 43$, and $a'' - 33$, where $g''\flat$ and a'' are the third diminished Fifth and major Sixth of the prime c mentioned in the text. This Prof. Helmholtz calls $a'''\flat$, or 3200 cents. Then the interval between this partial and the next is 9 : 25 or 1769 ct., and hence the interval with the prime is between 4812 and 5036 cents, or lies between $c'^{\mathrm{v}} + 12$ and $d'^{\mathrm{v}} + 36$, and hence it is called d'^{v} in the text. The interval to the next tone is 25 : 49 or 1165 cents. Adding this to the former numbers the interval with the prime must be between 5977 and 6201 cents, or between $b'^{\mathrm{v}} + 77$ and $d'^{\mathrm{v}} - 3$, for which in the text $c'^{\mathrm{v}}\sharp$ is selected. The indeterminacy arises from the difficulty of finding the pitch of the first inharmonic upper partial. The intervals between that and the next upper partials are 9 : 25 or 1769 ct., 9 : 49 or 2934 ct., 9 : 81 or 3699 ct., and so on. The word 'inharmonic' has been inserted in the text, as tuning-forks have also generally harmonic upper partials. See p. 54d, note.—*Translator.*]

although they have lately been introduced for military and dance music on account of their penetrating qualities of tone.   Glass rods or plates, and wooden rods, were formerly used in this way for the *glass harmonicon* and the *straw-fiddle* or *wood harmonicon*.   The rods were inserted between two pairs of intertwisted strings, which grasped them at their two nodal lines.   The wooden rods in the German *straw-fiddle* were simply laid on straw cylinders.   They were struck with hammers of wood or cork.

The only effect of the material of the rods on the quality of tone in these cases, consists in the greater or less length of time that it allows the proper tones at different pitches to continue.   These secondary tones, including the higher ones, usually continue to sound longest in elastic metal of fine uniform consistency, because its greater mass gives it a greater tendency to continue in any state of motion which it has once assumed, and among metals the most perfect elasticity is found in steel, and the better alloys of copper and zinc, or copper and tin.   In ¶ slightly alloyed precious metals, their greater specific gravity lengthens the duration of the tone, notwithstanding their inferior elasticity.   Superior elasticity appears to favour the continuance of the higher proper tones, because imperfect elasticity and friction generally seem to damp rapid more quickly than slow vibrations.   Hence I think that I may describe the general characteristic of what is usually called a *metallic* quality of tone, as the comparatively continuous and uniform maintenance of higher upper partial tones.   The quality of tone for glass is similar; but as it breaks when violently agitated, the tone is always weak and

FIG. 21.

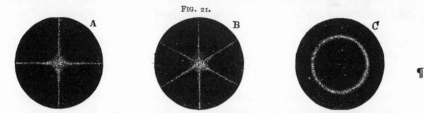

¶

soft, and it is also comparatively high, and dies rapidly away, on account of the smaller mass of the vibrating body.   In wood the mass is small, the internal structure comparatively rough, being full of countless interstices, and the elasticity also comparatively imperfect, so that the proper tones, especially the higher ones, rapidly die away.   And for this reason the *straw-fiddle* or wood harmonicon is perhaps more satisfactory to a musical ear, than harmonicons formed of steel or glass rods or plates, with their piercing inharmonic upper partial tones,—at least so far as simple tones are suitable for music at all, of which I shall have to speak later on.*

For all of these instruments which have to be struck, the hammers are made of wood or cork, and covered with leather.   This renders the highest upper partials much weaker than if only hard metal hammers were employed.   Greater ¶ hardness of the striking mass produces greater discontinuities in the original motion of the plate.   The influence exerted by the manner of striking will be considered more in detail, in reference to strings, where it is also of much importance.

According to Chladni's discoveries, *elastic plates*, cut in circular, oval, square, oblong, triangular, or hexagonal forms, will sound in a great number of different vibrational forms, usually producing simple tones which are mutually inharmonic. Fig. 21 gives the more simple vibrational forms of a circular plate.   Much more complicated forms occur when several circles or additional diameters appear as nodal lines, or where both circles and diameters occur.   Supposing the vibrational form A to give the tone *c*, the others give the following proper tones :—

* [In Java the principal music is produced by harmonicons of metal or wooden rods and kettle-shaped gongs. The wooden harmonicons are frequent also in Asia and Africa.   In Java the rods are laid on the edges of boat-shaped vessels, like old fashion cheese-trays, and kept in position by nails passing loosely through holes.   See App. XX. sect. K. —*Translator.*]

Number of Nodal Circles	Number of Diameters					
	0	1	2	3	4	5
0			c	d'	c''	$g''-g''\sharp$
1	$g\sharp$	$b'\flat$	g''			
2	$g'\sharp +$					

This shews that many proper tones of nearly the same pitch are produced by a plate of this kind. When a plate is struck, those proper tones which have no node at the point struck, will all sound together. To obtain a particular determinate tone it is of advantage to support the plate in points which lie in the nodal lines of that tone ; because those proper tones which have no node in those points will then die off more rapidly. For example, if a circular plate is supported at ¶ 3 points in the nodal circle of fig. 21, C (p. 71c), and is struck exactly in its middle, the simple tone called $g\sharp$ in the table, which belongs to that form, will be heard, and all those other proper tones which have diameters as some of their nodal lines * will be very weak, for example c, d', c'', g'', $b'\flat$ in the table. In the same way the tone $g''\sharp$ with two nodal circles, dies off immediately, because the points of support fall on one of its ventral segments, and the first proper tone which can sound loudly at the same time is that corresponding to three nodal circles, one of its nodal lines being near to that of No. 2. But this is 3 Octaves and more than a whole Tone higher than the proper tone of No. 2, and on account of this great interval does not disturb the latter. Hence a disc thus struck gives a tolerably good musical tone, whereas plates in general produce sounds composed of many inharmonic proper tones of nearly the same pitch, giving an empty tin-kettle sort of quality, which cannot be used in music. But even when the disc is properly supported the tone dies away rapidly, at least in the case of glass plates, because ¶ contact at many points, even when nodal, sensibly impedes the freedom of vibration.

The sound of *bells* is also accompanied by inharmonic secondary tones, which, however, do not lie so close to one another as those of flat plates. The vibrations which usually arise have 4, 6, 8, 10, &c., nodal lines extending from the vertex of the bell to its margin, at equal intervals from each other. The corresponding proper tones for glass bells which have approximatively the same thickness throughout, are nearly as the squares of the numbers 2, 3, 4, 5. so that if we call the lowest tone c, we have for the

Number of nodal lines . . .	4	6	8	10	12
Tones	c	$d' +$	c''	$g''\sharp -$	$d''' +$
Cents	0	1404	2400	3173	3804

The tones, however, vary with the greater or less thickness of the wall of the ¶ bell towards the margin, and it appears to be an essential point in the art of casting bells, to make the deeper proper tones mutually harmonic by giving the bell a certain empirical form. According to the observations of the organist Gleitz,† the bell cast for the cathedral at Erfurt in 1477 has the following proper tones: E, e, $g\sharp$, b, e', $g'\sharp$, b', $c''\sharp$. The [former] bell of St. Paul's, London, gave a and $c'\sharp$. Hemony of Zütphen, a master in the seventeenth century, required a good bell to have three Octaves, two Fifths, one major and one minor Third. The deepest tone is not the strongest. The body of the bell when struck gives a deeper tone than the ' sound bow,' but the latter gives the loudest tone. Probably other vibrational forms of bells are also possible in which nodal circles are formed

---

* Provided that the supported points do not happen to belong to a system of diameters making equal angles with each other.

† 'Historical Notes on the Great Bell and the other Bells in Erfurt Cathedral'

(*Geschichtliches über die grosse Glocke und die übrigen Glocken des Domes zu Erfurt*). Erfurt, 1867.—See also Schafhäutl in the *Kunst- und Gewerbeblatt für das Königreich Bayern*, 1868, liv. 325 to 350; 385 to 427.

parallel to the margin. But these seem to be produced with difficulty and have not yet been examined.

If a bell is not perfectly symmetrical in respect to its axis, if, for example, the wall is a little thicker at one point of its circumference than at another, it will give, on being struck, two different tones of very nearly the same pitch, which will 'beat' together. Four points on the margin will be found, separated from each other by quarter-circles, in which only one of these tones can be heard without accompanying beats, and four others, half-way between the pairs of the others, where the second tone only sounds. If the bell is struck elsewhere both tones are heard, producing beats, and such beats may be perceived in most bells as their tone dies gradually away.

Stretched *membranes* have also inharmonic proper tones of nearly the same pitch. For a circular membrane, of which the deepest tone is *c*, these are, in a vacuum and arranged in order of pitch, as follows :— ¶

Number of Nodal Lines		Tone
Diameters	Circles	
0	0	*c*
1	0	*a*♭
2	0	*c'*♯ + 0·1 *
0	1	*d'* + 0·2
1	1	*g'* − 0·2
0	2	*b'*♭ + 0·1

These tones rapidly die out. If the membranes sound in air,† or are associated with an air chamber, as in the *kettledrum*, the relation of the proper tones may be altered. No detailed investigations have yet been made on the secondary tones of the kettledrum. The kettledrum is used in artistic music, but only to mark ¶ certain accents. It is tuned, indeed, but only to prevent injury to the harmony, not for the purpose of filling up chords.

The common character of the instruments hitherto described is, that, when struck they produce inharmonic upper partial tones. If these are of nearly the same pitch as the prime tone, their quality of sound is in the highest degree unmusical, bad, and tinkettly. If the secondary tones are of very different pitch from the prime, and weak in force, the quality of sound is more musical, as for example in tuning-forks, harmonicons of rods, and bells ; and such tones are applicable for marches and other boisterous music, principally intended to mark time. But for really artistic music, such instruments as these have always been rejected, as they ought to be, for the inharmonic secondary tones, although they rapidly die away, always disturb the harmony most unpleasantly, renewed as they are at every fresh blow. A very striking example of this was furnished by a company of bell-ringers, said to be Scotch, that lately travelled about Germany, and performed all ¶ kinds of musical pieces, some of which had an artistic character. The accuracy and skill of the performance was undeniable, but the musical effect was detestable, on account of the heap of false secondary tones which accompanied the music, although care was taken to damp each bell as soon as the proper duration of its note had expired, by placing it on a table covered with cloth.

Sonorous bodies with inharmonic partials, may be also set in action by violin bows, and then by properly damping them in a nodal line of the desired tone, the secondary tones which lie near it can be prevented from interfering. One simple tone then predominates distinctly, and it might consequently be used for musical purposes. But when the violin bow is applied to any bodies with inharmonic upper partial tones, as tuning-forks, plates, bells, we hear a strong scratching

---

* [These decimals represent tenths of a tone, or 20 cents for the first place. As there can be no sounds in a vacuum, these notes are merely used to conveniently symbolise numbers of vibrations in a second.—*Translator.*]

† See *J. Bourget*, L'Institut, xxxviii., 1870, pp. 189, 190.

sound, which on investigation with resonators, is found to consist mainly of these same inharmonic secondary tones of such bodies, not sounding continuously but only in short irregular fits and starts. Intermittent tones, as I have already noted, produce the effect of grating or scratching. It is only when the body excited by the violin bow has harmonic upper partials, that it can perfectly accommodate itself to every impulse of the bow, and give a really musical quality of tone. The reason of this is that any required periodic motion such as the bow aims at producing, can be compounded of motions corresponding to harmonic upper partial tones, but not of other, inharmonic vibrations.

3. *Musical Tones of Strings.*

We now proceed to the analysis of musical tones proper, which are characterised ¶ by harmonic upper partials. These may be best classified according to their mode of excitement: 1. By striking. 2. By bowing. 3. By blowing against a sharp edge. 4. By blowing against elastic tongues or vibrators. The two first classes comprehend stringed instruments alone, as longitudinally vibrating rods, the only other instruments producing harmonic upper partial tones, are not used for musical purposes. The third class embraces flutes and the flute or flue pipes of organs; the fourth all other wind instruments, including the human voice.

Strings excited by Striking.—Among musical instruments at present in use, this section embraces the *pianoforte, harp, guitar,* and *zither*; among physical, the *monochord,* arranged for an accurate examination of the laws controlling the vibrations of strings; the *pizzicato* of bowed instruments must also be placed in this category. We have already mentioned that the musical tones produced by strings which are struck or plucked, contain numerous upper partial tones. We have the advantage of possessing a complete theory for the motion of plucked ¶ strings, by which the force of their upper partial tones may be determined. In the last chapter we compared some of the conclusions of this theory with the results of experiment, and found them agree. A similarly complete theory may be formed for the case of a string which has been struck in one of its points by a hard sharp edge. The problem is not so simple when soft elastic hammers are used, such as those of the pianoforte, but even in this case it is possible to assign a theory for the motion of the string which embraces at least the most essential features of the process, and indicates the force of the upper partial tones.*

The force of the upper partial tones in a struck string, depends in general on :—

1. The *nature of the stroke.*
2. The *place struck.*
3. The *density, rigidity,* and *elasticity* of the *string.*

First, as to the *nature of the stroke.* The string may be *plucked,* by drawing ¶ it on one side with the finger or a point (the plectrum, or the ring of the zither-player), and then letting it go. This is a usual mode of exciting a string in a great number of ancient and modern stringed instruments. Among the modern, I need only mention the *harp, guitar,* and *zither.* Or else the string may be struck with a hammer-shaped body, as in the pianoforte.† I have already remarked that the strength and number of the upper partial tones increases with the number and abruptness of the discontinuities in the motion excited. This fact determines the various modes of exciting a string. When a string is plucked, the finger, before quitting it, removes it from its position of rest throughout its whole length. A discontinuity in the string arises only by its forming a more or less acute angle at the place where it wraps itself about the finger or point. The angle is more acute for a sharp point than for the finger. Hence the sharp point produces a shriller tone with a greater number of high tinkling upper partials, than the finger. But

---

* See Appendix V.

† [I have here omitted a few words in which, by an oversight, the *spinet* was said to be struck by a hammer-shaped body. See pp. 77c and 78d'.—*Translator.*]

in each case the intensity of the prime tone exceeds that of any upper partial. If the string is struck with a sharp-edged metallic hammer which rebounds instantly, only the one single point struck is directly set in motion. Immediately after the blow the remainder of the string is at rest. It does not move until a wave of deflection rises, and runs backwards and forwards over the string. This limitation of the original motion to a single point produces the most abrupt discontinuities, and a corresponding long series of upper partial tones, having intensities,* in most cases equalling or even surpassing that of the prime. When the hammer is soft and elastic, the motion has time to spread before the hammer rebounds. When thus struck the point of the string in contact with such a hammer is not set in motion with a jerk, but increases gradually and continuously in velocity during the contact. The discontinuity of the motion is consequently much less, diminishing as the softness of the hammer increases, and the force of the higher upper partial tones is correspondingly decreased.   ¶

We can easily convince ourselves of the correctness of these statements by opening the lid of any pianoforte, and, keeping one of the digitals down with a weight, so as to free the string from the damper, plucking the string at pleasure with a finger or a point, and striking it with a metallic edge or the pianoforte hammer itself. The qualities of tone thus obtained will be entirely different. When the string is struck or plucked with hard metal, the tone is piercing and tinkling, and a little attention enables us to hear a multitude of very high partial tones. These disappear, and the tone of the string becomes less bright, but softer, and more harmonious, when we pluck the string with the soft finger or strike it with the soft hammer of the instrument. We also readily recognise the different loudness of the prime tone. When we strike with metal, the prime tone is scarcely heard and the quality of tone is correspondingly *poor*. The peculiar quality of tone commonly termed *poverty*, as opposed to *richness*, arises from the upper partials being comparatively too strong for the prime tone. The prime tone is ¶ heard best when the string is plucked with a soft finger, which produces a rich and yet harmonious quality of tone. The prime tone is not so strong, at least in the middle and deeper octaves of the instrument, when the strings are struck with the pianoforte hammer, as when they are plucked with the finger.

This is the reason why it has been found advantageous to cover pianoforte hammers with thick layers of felt, rendered elastic by much compression. The outer layers are the softest and most yielding, the lower are firmer. The surface of the hammer comes in contact with the string without any audible impact ; the lower layers give the elasticity which throws the hammer back from the string. If you remove a pianoforte hammer and strike it strongly on a wooden table or against a wall, it rebounds from them like an india-rubber ball. The heavier the hammer and the thicker the layers of felt—as in the hammers for the lower octaves—the longer must it be before it rebounds from the string. The hammers for the upper octaves are lighter and have thinner layers of felt. Clearly the makers of these ¶ instruments have here been led by practice to discover certain relations of the elasticity of the hammer to the best tones of the string. The make of the hammer has an immense influence on the quality of tone. Theory shews that those upper partial tones are especially favoured whose periodic time is nearly equal to twice

---

* When *intensity* is here mentioned, it is always measured objectively, by the *vis viva*, or *mechanical equivalent of work* of the corresponding motion. [Mr. Bosanquet (*Academy*, Dec. 4, 1875, p. 580, col. 1) points out that p. 10*d*, note, and Chap. IX., paragraph 3, shew this measure to be inadmissible, and adds : 'if we admit that in similar organ pipes similar proportions of the wind supplied are employed in the production of tone, the mechanical energy of notes of given intensity varies inversely as the vibration number,' i.e. as the pitch number. Messrs. Preece and Stroh, *Proc. R. S.*, vol. xxviii. p. 366, think that 'loudness does not depend upon amplitude of vibration only, but upon the quantity of air put in vibration ; and, therefore, there exists an absolute physical magnitude in acoustics analogous to that of quantity of electricity or quantity of heat, and which may be called quantity of sound,' and they illustrate this by the effect of differently sized discs in their automatic phonograph there described. See also App. XX. sect. M. No. 2.—*Translator.*]

the time during which the hammer lies on the string, and that, on the other hand, those disappear whose periodic time is 6, 10, 14, &c., times as great.*

It will generally be advantageous, especially for the deeper tones, to eliminate from the series of upper partials, those which lie too close to each other to give a good compound tone, that is, from about the seventh or eighth onwards. Those with higher ordinal numbers are generally relatively weak of themselves. On examining a new grand pianoforte by Messrs. Steinway of New York, which was remarkable for the evenness of its quality of tone, I find that the damping resulting from the duration of the stroke falls, in the deeper notes, on the ninth or tenth partials, whereas in the higher notes, the fourth and fifth partials were scarcely to be got out with the hammer, although they were distinctly audible when the string was plucked by the nail.† On the other hand upon an older and much used grand piano, which originally shewed the principal damping in the neighbourhood of the

¶ seventh to the fifth partial for middle and low notes, the ninth to the thirteenth partials are now strongly developed. This is probably due to a hardening of the hammers, and certainly can only be prejudicial to the quality of tone. Observations on these relations can be easily made in the method recommended on p. 52b, c. Put the point of the finger gently on one of the nodes of the tone of which you wish to discover the strength, and then strike the string by means of the digital. By moving the finger till the required tone comes out most purely and sounds the longest, the exact position of the node can be easily found. The nodes which lie near the striking point of the hammer, are of course chiefly covered by the damper, but the corresponding partials are, for a reason to be given presently, relatively weak. Moreover the fifth partial speaks well when the string is touched at two-fifths of its length from the end, and the seventh at two-sevenths of that length. These positions are of course quite free of the damper. Generally we find all the partials which arise from the method of striking used, when we keep on striking

¶ while the finger is gradually moved over the length of the string. Touching the shorter end of the string between the striking point and the further bridge will thus bring out the higher partials from the ninth to the sixteenth, which are musically undesirable.

The method of calculating the strength of the individual upper partials, when the duration of the stroke of the hammer is given, will be found further on.

Secondly as to the *place struck*. In the last chapter, when verifying Ohm's law for the analysis of musical tones by the ear, we remarked that whether strings are plucked or struck, those upper partials disappear which have a node at the point excited. Conversely; those partials are comparatively strongest which have a maximum displacement at that point. Generally, when the same method of striking is successively applied to different points of a string, the individual upper partials increase or decrease with the intensity of motion, at the point of excitement, for the corresponding simple vibrations of the string. The composition of

¶ the musical tone of a string can be consequently greatly varied by merely changing the point of excitement.

Thus if a string be struck in its middle, the second partial tone disappears,

---

* [The following paragraph on p. 123 of the 1st English edition has been omitted, and the passage from 'It will generally be advantageous,' p. 76a, to 'found further on,' p. 76c, has been inserted, both in accordance with the 4th German edition.—*Translator*.]

† [As Prof. Helmholtz does not mention the striking distance of the hammer, I obtained permission from Messrs. Steinway & Sons, at their London house, to examine the c, c' and c'' strings of one of their grand pianos, and found the striking distance to be $\frac{2}{17}$, $\frac{2}{18}$, and $\frac{2}{17}$ of the length of the string respectively. I did not measure the other strings, but I observed that the striking distances varied several times. I got out the 7th and 9th harmonic of c, but on account of difficulties due to the over-stringing and over-barring of the instrument and other circumstances I did not pursue the investigation. Mr. A. J. Hipkins informs me that on another occasion he got out of the c' string, struck at $\frac{1}{9}$ the length, the 6th, 7th, 8th, and 9th harmonics, as in the experiments mentioned in the next footnote, 'the 6th and 7th beautifully strong, the 8th and 9th weaker but clear and unmistakable.' He struck with the hammer always. Observe the 9th harmonic of a string struck with a pianoforte hammer at its node, or $\frac{1}{9}$ its length.—*Translator*.]

because it has a node at that point. But the third partial tone comes out forcibly, because as its nodes lie at $\frac{1}{3}$ and $\frac{2}{3}$ the length of the string from its extremities, the string is struck half-way between these two nodes. The fourth partial has its nodes at $\frac{1}{4}$, $\frac{2}{4}$ ($=\frac{1}{2}$), and $\frac{3}{4}$ the length of the string from its extremity. It is not heard, because the point of excitement corresponds to its second node. The sixth, eighth, and generally the partials with even numbers disappear in the same way, but the fifth, seventh, ninth, and the other partials with odd numbers are heard. By this disappearance of the evenly numbered partial tones when a string is struck at its middle, the quality of its tone becomes peculiar, and essentially different from that usually heard from strings. It sounds somewhat hollow or nasal. The experiment is easily made on any piano when it is opened and the dampers are raised. The middle of the string is easily found by trying where the finger must be laid to bring out the first upper partial clearly and purely on striking the digital.

If the string is struck at $\frac{1}{3}$ its length, the third, sixth, ninth, &c., partials ¶ vanish. This also gives a certain amount of hollowness, but less than when the string is struck in its middle. When the point of excitement approaches the end of the string, the prominence of the higher upper partials is favoured at the expense of the prime and lower upper partial tones, and the sound of the string becomes poor and tinkling.

In pianofortes, the point struck is about $\frac{1}{7}$ to $\frac{1}{9}$ the length of the string from its extremity, for the middle part of the instrument. We must therefore assume that this place has been chosen because experience has shewn it to give the finest musical tone, which is most suitable for harmonies. The selection is not due to theory. It results from attempts to meet the requirements of artistically trained ears, and from the technical experience of two centuries.* This gives particular

* [As my friend Mr. A. J. Hipkins, of Broadwoods', author of a paper on the 'History of the Pianoforte,' in the *Journal of the Society of Arts* (for March 9, 1883, with additions on Sept. 21, 1883), has paid great attention to the archæology of the pianoforte, and from his position at Messrs. Broadwoods' has the best means at his disposal for making experiments, I requested him to favour me with his views upon the subject of the striking place and harmonics of pianoforte strings, and he has obliged me with the following observations:—
'Harpsichords and spinets, which were set in vibration by quill or leather plectra, had no fixed point for plucking the strings. It was generally from about $\frac{1}{5}$ to $\frac{1}{7}$ of the vibrating length, and although it had been observed by Huyghens and the Antwerp harpsichord-maker Jan Couchet, that a difference of quality of tone could be obtained by varying the plucking place on the same string, which led to the so-called lute stop of the 18th century, no attempt appears to have been made to gain a uniform striking place throughout the scale. Thus in the latest improved spinet, a Hitchcock, of early 18th century, in my possession, the striking place of the c's varies from $\frac{1}{2}$ to $\frac{1}{4}$, and in the latest improved harpsichord, a Kirkman of 1773, also in my possession, the striking distances vary from $\frac{1}{2}$ to $\frac{1}{10}$ and for the lute stop from $\frac{1}{9}$ to $\frac{1}{20}$ of the string, the longest distances in the bass of course, but all without apparent rule or proportion. Nor was any attempt to gain a uniform striking place made in the first pianofortes. Stein of Augsburg (the favourite pianoforte-maker of Mozart, and of Beethoven in his virtuoso time) knew nothing of it, at least in his early instruments. The great length of the bass strings as carried out on the single belly-bridge copied from the harpsichord, made a

reasonable striking place for that part of the scale impossible.
'John Broadwood, about the year 1788, was ¶ the first to try to equalise the scale in tension and striking place. He called in scientific aid, and assisted by Signor Cavallo and the then Dr. Gray of the British Museum, he produced a divided belly bridge, which shortening the too great length of the bass strings, permitted the establishment of a striking place, which, in intention, should be proportionate to the length of the string throughout. He practically adopted a ninth of the vibrating length of the string for his striking place, allowing some latitude in the treble. This division of the belly-bridge became universally adopted, and with it an approximately rational striking place.
'Carl Kützing (*Das Wissenschaftliche der Fortepiano-Baukunst*, 1844, p. 41) was enabled to propound from experience, that $\frac{1}{8}$ of the length of the string was the most suitable ¶ distance in a pianoforte for obtaining the best quality of tone from the strings. The love of noise or effect has, however, inclined makers to shorten distances, particularly in the trebles. Kützing appears to have met with $\frac{1}{4}$th in some instances, and Helmholtz has adopted that very exceptional measure for his table on p. 79c. I cannot say I have ever met with a striking place of this long distance from the wrestplank-bridge. The present head of the firm of Broadwood (Mr. Henry Fowler Broadwood) has arrived at the same conclusions as Kützing with respect to the superiority of the $\frac{1}{8}$th distance, and has introduced it in his pianofortes. At $\frac{1}{8}$th the hammers have to be softer to get a like quality of tone; an equal system of tension being presupposed.
'According to Young's law, which Helmholtz by experiment confirms, the impact of

interest to the investigation of the composition of musical tones for this point of excitement. An essential advantage in the choice of this position seems to be that the seventh and ninth partial tones disappear or at least become very weak. These are the first in the series of partial tones which do not belong to the major chord of the prime tone. Up to the sixth partial we have only Octaves, Fifths, and major Thirds of the prime tone; the seventh is nearly a minor Seventh, the ninth a major Second of the prime. Hence these will not fit into the major chord. Experiments on pianofortes shew that when the string is struck by the hammer and touched at its nodes, it is easy to bring out the six first partial tones (at least on the strings of the middle and lower octaves), but that it is either not possible to bring out the seventh, eighth, and ninth at all, or that we obtain at best very weak and imperfect results. The difficulty here is not occasioned by the incapacity of the string to form such short vibrating sections, for if instead of striking ¶ the digital we pluck the string nearer to its end, and damp the corresponding nodes, the seventh, eighth, ninth, nay even the tenth and eleventh partial may be clearly and brightly produced. It is only in the upper octaves that the strings are too short and stiff to form the high upper partial tones. For these, several instrument-makers place the striking point nearer to the extremity, and thus obtain a brighter and more penetrating tone. The upper partials of these strings, which their stiffness renders it difficult to bring out, are thus favoured as against the prime tone. A similarly brighter tone, but at the same time a thinner and poorer one, can be obtained from the lower strings by placing a bridge nearer the striking point, so that the hammer falls at a point less than $\frac{1}{7}$ of the effective length of the string from its extremity.

While on the one hand the tone can be rendered more tinkling, shrill, and acute, by striking the string with hard bodies, on the other hand it can be rendered duller, that is, the prime tone may be made to outweigh the upper partials, by ¶ striking it with a soft and heavy hammer, as, for example, a little iron hammer covered with a thick sheet of india-rubber. The strings of the lower octaves then produce a much fuller but duller tone. To compare the different qualities of tone thus produced by using hammers of different constructions, care must be taken always to strike the string at the same distance from the end as it is struck by the proper hammer of the instrument, as otherwise the results would be mixed up with the changes of quality depending on altering the striking point. These circumstances are of course well known to the instrument-makers, because they have

the hammer abolishes the node of the striking place, and with it the partial belonging to it throughout the string. I do not find, however, that the hammer striking at the $\frac{1}{8}$th eliminates the 8th partial. It is as audible, when touched as an harmonic, as the 9th and higher partials. It is easy, on a long string of say ¶ from 25 to 45 inches, to obtain the series of upper partials up to the fifteenth. On a string of 45 inches I have obtained as far as the 23rd harmonic, the diameter of the wire being 1·17 mm. or ·07 inches, and the tension being 71 kilogrammes or 156·6 lbs. The partials diminish in intensity with the reduction of the vibrating length; the 2nd is stronger than the 3rd, and the 3rd than the 4th, &c. Up to the 7th a good harmonic note can always be brought out. After the 8th, as Helmholtz says, the higher partials are all comparatively weak and become gradually fainter. To strengthen them we may use a narrower harder hammer. To hear them with an ordinary hammer it is necessary to excite them by a firm blow of the hand upon the finger-key and to continue to hold it down. They sing out quite clearly and last a very sensible time. On removing the stop imme-

diately after production, they last much longer and are much brighter.

'I do not think the treble strings are from shortness and stiffness incapable of forming high proper tones. If it were so the notes would be of a very different quality of tone to that which they are found to have. Owing to the very acute pitch of these tones our ears cannot follow them, but their existence is proved by the fact that instrument-makers often bring their treble striking place very near the wrestplank-bridge in order to secure a brilliant tone effect, or ring, by the preponderance of these harmonics.

'The clavichord differs entirely from hammer and plectrum keyboard instruments in the note being started from the end, the tangent (brass pin) which stops the string being also the means of exciting the sound. But the thin brass wires readily break up into segments of short recurrence, the bass wires, which are most indistinct, being helped in the latest instruments by lighter octave strings, which serve to make the fundamental tones apparent.' See also the last note, p. 76d', and App. XX. sect. N.—*Translator*.]

themselves selected heavier and softer hammers for the lower, and lighter and harder for the upper octaves. But when we see that they have not given more than a certain weight to the hammers and have not increased it sufficiently to reduce the intensity of the upper partial tones still further, we feel convinced that a musically trained ear prefers that an instrument to be used for rich combinations of harmony should possess a quality of tone which contains upper partials with a certain amount of strength. In this respect the composition of the tones of pianoforte strings is of great interest for the whole theory of music. In no other instrument is there so wide a field for alteration of quality of tone; in no other, then, was a musical ear so unfettered in the choice of a tone that would meet its wishes.

As I have already observed, the middle and lower octaves of pianoforte strings generally allow the six first partial tones to be clearly produced by striking the digital, and the three first of them are strong, the fifth and sixth distinct, but much ¶ weaker. The seventh, eighth, and ninth are eliminated by the position of the striking point. Those higher than the ninth are always very weak. For closer comparison I subjoin a table in which the intensities of the partial tones of a string for different methods of striking have been theoretically calculated from the formulæ developed in the Appendix V. The effect of the stroke of a hammer depends on the length of time for which it touches the string. This time is given in the table in fractions of the periodic time of the prime tone. To this is added a calculation for strings plucked by the finger. The striking point is always assumed to be at $\frac{1}{7}$ of the length of the string from its extremity.

Theoretical Intensity of the Partial Tones of Strings.

Number of the Partial Tone	Excited by Plucking	Struck by a hammer which touches the string for				Struck by a perfect hard Hammer
		$\frac{3}{7}$ c''	$\frac{3}{10}$ g'	$\frac{3}{14}$ $C_i - c'$	$\frac{3}{20}$	
		of the periodic time of the prime tone				
1	100	100	100	100	100	100
2	81·2	99·7	189·4	249	285·7	324·7
3	56·1	8·9	107·9	242·9	357·0	504·9
4	31·6	2·3	17·3	118·9	259·8	504·9
5	13	1·2	0	26·1	108·4	324·7
6	2·8	0·01	0·5	1·3	18·8	100·0
7	0	0	0	0	0	0

For easier comparison the intensity of the prime tone has been throughout assumed as 100. I have compared the calculated intensity of the upper partials with their force on the grand pianoforte already mentioned, and found that the first series, under $\frac{3}{7}$, suits for about the neighbourhood of c''. In higher parts of ¶ the instrument the upper partials were much weaker than in this column. On striking the digital for c'', I obtained a powerful second partial and an almost inaudible third. The second column, marked $\frac{3}{10}$, corresponded nearly to the region of g', the second and third partials were very strong, the fourth partial was weak. The third column, inscribed $\frac{3}{14}$, corresponds with the deeper tones from c' downwards; here the four first partials are strong, and the fifth weaker. In the next column, under $\frac{3}{20}$, the third partial tone is stronger than the second; there was no corresponding note on the pianoforte which I examined. With a perfectly hard hammer the third and fourth partials have the same strength, and are stronger than all the others. It results from the calculations in the above table that pianoforte tones in the middle and lower octaves have their fundamental tone weaker than the first, or even than the two first upper partials. This can also be confirmed by a comparison with the effects of plucked strings. For the latter the second partial is weaker than the first; and it will be found that the prime tone is much more distinct in the tones of pianoforte strings when plucked by the finger, than when struck by the hammer.

Although, as is shewn by the mechanism of the upper octaves on pianofortes, it is possible to produce a compound tone in which the prime is predominant, makers have preferred arranging the method of striking the lower strings in such a way as to preserve the five or six first partials distinctly, and to give the second and third greater intensity than the prime.

Thirdly, as regards the *thickness and material of the strings.* Very rigid strings will not form any very high upper partials, because they cannot readily assume inflections in opposite directions within very short sections. This is easily observed by stretching two strings of different thicknesses on a monochord and endeavouring to produce their high upper partial tones. We always succeed much better with the thinner than with the thicker string. To produce very high upper ¶ partial tones, it is preferable to use strings of extremely fine wire, such as gold lace makers employ, and when they are excited in a suitable manner, as for example by plucking or striking with a metal point, these high upper partials may be heard in the compound itself. The numerous high upper partials which lie close to each other in the scale, give that peculiar high inharmonious noise which we are accustomed to call 'tinkling.' From the eighth partial tone upwards these simple tones are less than a whole Tone apart, and from the fifteenth upwards less than a Semitone. They consequently form a series of dissonant tones. On a string of the finest iron wire, such as is used in the manufacture of artificial flowers, 700 centimetres (22·97 feet) long, I was able to isolate the eighteenth partial tone. The peculiarity of the tones of the zither depends on the presence of these tinkling upper partials, but the series does not extend so far as that just mentioned, because the strings are shorter.

Strings of gut are much lighter than metal strings of the same compactness, ¶ and hence produce higher partial tones. The difference of their musical quality depends partly on this circumstance and partly on the inferior elasticity of the gut, which damps their partials, especially their higher partials, much more rapidly. The tone of plucked cat-gut strings (*guitar, harp*) is consequently much less tinkling than that of metal strings.

4. *Musical Tones of Bowed Instruments.*

No complete mechanical theory can yet be given for the motion of strings excited by the violin-bow, because the mode in which the bow affects the motion of the string is unknown. But by applying a peculiar method of observation, proposed in its essential features by the French physicist *Lissajous*, I have found it possible to observe the vibrational form of individual points in a violin string, and from this observed form, which is comparatively very simple, to calculate the ¶ whole motion of the string and the intensity of the upper partial tones.

Look through a hand magnifying glass consisting of a strong convex lens, at any small bright object, as a grain of starch reflecting a flame, and appearing as a fine point of light. Move the lens about while the point of light remains at rest, and the point itself will appear to move. In the apparatus I have employed, which is shewn in fig. 22 opposite, this lens is fastened to the end of one prong of the tuning-fork G, and marked L. It is in fact a combination of two achromatic lenses, like those used for the object-glasses of microscopes. These two lenses may be used alone as a doublet, or be combined with others. When more magnifying power is required, we can introduce behind the metal plate A A, which carries the fork, the tube and eye-piece of a microscope, of which the doublet then forms the object-glass. This instrument may be called a *vibration microscope.* When it is so arranged that a fixed luminous point may be clearly seen through it, and the fork is set in vibration, the doublet L moves periodically up and down in pendular vibrations. The observer, however, appears to see the luminous point itself vibrate, and, since the separate vibrations succeed each other so rapidly that the impression on the eye cannot die away during the time of a whole vibration, the path of the luminous point appears as a fixed straight line, increasing in length with the excursions of the fork.*

The grain of starch which reflects the light to be seen, is then fastened to the resonant body whose vibrations we intend to observe, in such a way that the grain moves backwards and forwards horizontally, while the doublet moves up and down vertically.   When both motions take place at once, the observer sees the real horizontal motion of the luminous point combined with its apparent vertical motion, and the combination results in an apparent curvilinear motion.   The field of vision in the microscope then shews an apparently steady and unchangeable bright

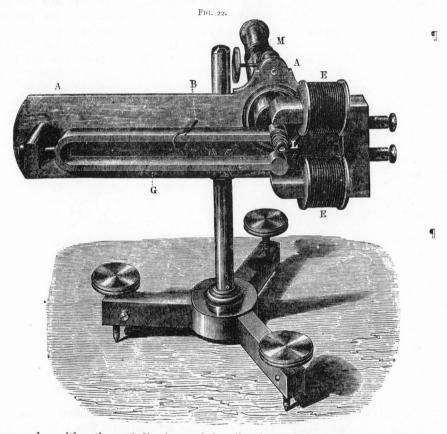

FIG. 22.

curve, when either the periodic times of the vibrations of the grain of starch and of the tuning-fork are exactly equal, or one is exactly two or three or four times as great as the other, because in this case the luminous point passes over exactly the same path every one or every two, three, or four vibrations.   If these ratios of the vibrational numbers are not exactly perfect, the curves alter slowly, and the effect to the eye is as if they were drawn on the surface of a transparent cylinder which slowly revolved on its axis.   This slow displacement of the apparent curves is not disadvantageous, as it allows the observer to see them in different positions.   But if the ratio of the pitch numbers of the observed body and of the fork differs too

---

* The end of the other prong of the fork is thickened to counterbalance the weight of the doublet. The iron loop B which is clamped on to one prong serves to alter the pitch of the fork slightly; we flatten the pitch by moving the loop towards the end of the prong.

E is an electro-magnet by which the fork is kept in constant uniform vibration on passing intermittent electrical currents through its wire coils, as will be described more in detail in Chapter VI.

much from one expressible by small whole numbers, the motion of the curve is too rapid for the eye to follow it, and all becomes confusion.

If the vibration microscope has to be used for observing the motion of a violin string, the luminous point must be attached to that string. This is done by first blackening the required spot on the string with ink, and when it is dry, rubbing it over with wax, and powdering this with starch so that a few grains remain sticking. The violin is then fixed with its strings in a vertical direction opposite the microscope, so that the luminous reflection from one of the grains of starch can be clearly seen. The bow is drawn across the strings in a direction parallel to the prongs of the fork. Every point in the string then moves horizontally, and on setting the fork in motion at the same time, the observer sees the peculiar vibrational curves already mentioned. For the purposes of observation I used the ¶ a' string, which I tuned a little higher, as $b'\flat$, so that it was exactly two Octaves higher than the tuning-fork of the microscope, which sounded $B\flat$.

In fig. 23 are shewn the resulting vibrational curves as seen in the vibration microscope. The straight horizontal lines in the figures, a to a, b to b, c to c

FIG. 23.

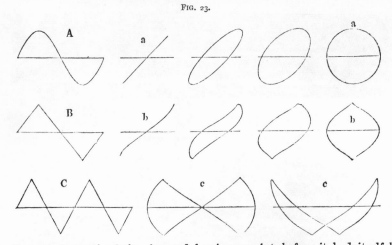

shew the apparent path of the observed luminous point, before it had itself been set in vibration; the curves and zigzags in the same figures, shew the apparent path of the luminous point when it also was made to move. By their side, in A, B, C, the same vibrational forms are exhibited according to the methods used in Chapters I. and II., the lengths of the horizontal line being directly proportional to the corresponding lengths of *time*, whereas in figures a to a, b to b, c to c, the horizontal lengths are proportional to the *excursions* of the vibrating microscope. ¶ A, and a to a, shew the vibrational curves for a tuning-fork, that is for a simple pendular vibration; B and b to b those of the middle of a violin string in unison with the fork of the vibration microscope; C and c, c, those for a string which was tuned an Octave higher. We may imagine the figures a to a, b to b, and c to c, to be formed from the figures A, B, C, by supposing the surface on which these are drawn to be wrapped round a transparent cylinder whose circumference is of the same length as the horizontal line. The curve drawn upon the surface of the cylinder must then be observed from such a point, that the horizontal line which when wrapped round the cylinder forms a circle, appears perspectively as a single straight line. The vibrational curve A will then appear in the forms a to a, B in the forms b to b, C in the forms c to c. When the pitch of the two vibrating bodies is not in an exact harmonic ratio, this imaginary cylinder on which the vibrational curves are drawn, appears to revolve so that the forms a to a, &c., are assumed in succession.

It is now easy to rediscover the forms A, B, C, from the forms a to a, b to b, and c to c, and as the former give a more intelligible image of the motion of the string than the latter, the curves, which are seen as if they were traced on the surface of a cylinder, will be drawn as if their trace had been unrolled from the cylinder into a plane figure like A, B, C. The meaning of our vibrational curves will then precisely correspond to the similar curves in preceding chapters. When four vibrations of the violin string correspond to one vibration of the fork (as in our experiments, where the fork gave $B\flat$ and the string $b'\flat$, p. 82a), so that four waves seem to be traced on the surface of the imaginary cylinder, and when moreover they are made to rotate slowly and are thus viewed in different positions, it is not at all difficult to draw them from immediate inspection as if they had been rolled off on to a plane, for the middle jags have then nearly the same appearance on the cylinder as if they were traced on a plane.

The figures 23 B and 23 C (p. 82b), immediately give the vibrational forms for the middle of a violin string, when the bow bites well, and the prime tone of the ¶ string is fully and powerfully produced. It is easily seen that these vibrational forms are essentially different from that of a simple vibration (fig. 23, A). When the point is taken nearer the ends of the string the vibrational figure is shewn in fig. 24, A, and the two sections $\alpha\beta$, $\beta\gamma$, of any wave, are to one another as the two sections of the string which lie on either side of the observed point. In the figure

FIG. 24.

this ratio is 3 : 1, the point being at $\frac{1}{4}$ the length of the string from its extremity. Close to the end of the string the form is as in fig. 24, B. The short lengths of line in the figure have been made faint because the corresponding motion of the ¶ luminous point is so rapid that they often become invisible, and the thicker lengths are alone seen.*

These figures shew that every point of the string between its two extremities vibrates with a constant velocity. For the middle point, the velocity of ascent is equal to that of descent. If the violin bow is used near the right end of the string descending, the velocity of descent on the right half of the string is less than that of ascent, and the more so the nearer to the end. On the left half of the string the converse takes place. At the place of bowing the velocity of descent appears to be equal to that of the violin bow. During the greater part of each vibration the string here clings to the bow, and is carried on by it ; then it suddenly detaches itself and rebounds, whereupon it is seized by other points in the bow and again carried forward.†

Our present purpose is chiefly to determine the upper partial tones. The vibrational forms of the individual points of the string being known, the intensity ¶ of each of the partial tones can be completely calculated. The necessary mathematical formulæ are developed in Appendix VI. The following is the result of the calculation. When a string excited by a violin bow speaks well, all the upper partial tones which can be formed by a string of its degree of rigidity, are present, and their intensity diminishes as their pitch increases. The amplitude and the intensity of the second partial is one-fourth of that of the prime tone, that of the

* [Dr. Huggins, F.R.S., on experimenting, finds it probable that under the bow, the relative velocity of descent to that of the rebound of the string, or ascent, is influenced by the tension of the hairs of the bow.— *Translator.*]

† These facts suffice to determine the complete motion of bowed strings. See Appendix VI. A much simpler method of observing the vibrational form of a violin string has been given by Herr Clem. Neumann in the *Proceedings (Sitzungsberichte) of the I. and R. Academy* at Vienna, mathematical and physical class, vol. lxi. p. 89. He fastened bits of wire in the form of a comb to the bow itself. On looking through this grating at the string the observer sees a system of rectilinear zigzag lines. The conclusions as to the mode of motion of the string agree with those given above.

third partial a ninth, that of the fourth a sixteenth, and so on. This is the same scale of intensity as for the partial tones of a string plucked in its middle, with this exception, that in the latter case the evenly numbered partials all disappear, whereas they are all present when the bow is used. The upper partials in the compound tone of a violin are heard easily and will be found to be strong in sound if they have been first produced as so-called harmonics on the string, by bowing lightly while gently touching a node of the required partial tone. The strings of a violin will allow the harmonics to be produced as high as the sixth partial tone with ease, and with some difficulty even up to the tenth. The lower tones speak best when the string is bowed at from one-tenth to one-twelfth the length of the vibrating portion of the string from its extremity. For the higher harmonics where the sections are smaller, the strings must be bowed at about one-fourth or one-sixth of their vibrating length from the end.*

The prime in the compound tones of bowed instruments is comparatively more powerful than in those produced on a pianoforte or guitar by striking or plucking the strings near to their extremities; the first upper partials are comparatively weaker; but the higher upper partials from the sixth to about the tenth are much more distinct, and give these tones their cutting character.

The fundamental form of the vibrations of a violin string just described, is, when the string speaks well, tolerably independent of the place of bowing, at least in all essential features. It does not in any respect alter, like the vibrational form of struck or plucked strings, according to the position of the point excited. Yet there are certain observable differences of the vibrational figure which depend upon the bowing point. Little crumples are usually perceived on the lines of the vibrational figure, as in fig. 25, which increase in breadth and height the further the bow is removed from the extremity of the string. When we bow at a node of one of the higher upper partials which is near the bridge, these crumples are simply reduced by the absence of that part of the normal motion of the string which depends on the partial tones having a node at that place. When the observation on the vibrational form is made at one of the other nodes belonging to the deepest tone which is eliminated, none of these crumples are seen. Thus if the string is bowed at $\frac{1}{7}$th, or $\frac{6}{7}$ths, or $\frac{5}{7}$ths, or $\frac{4}{7}$ths, &c., of its length from the bridge, the vibrational figure is simple, as in fig. 24 (p. 83*b*). But if we observe between two nodes, the crumples appear as in fig. 25. Variations in the quality of tone partly depend upon this condition. When the violin bow is brought too near the finger board, the end of which is $\frac{1}{5}$th the length of the string from the bridge, the 5th or 6th partial tone, which is generally distinctly audible, will be absent. The tone is thus rendered duller. The usual place of bowing is at about $\frac{1}{10}$th of the length of the string; for *piano* passages it is somewhat further from the bridge and for *forte* somewhat nearer to it. If the bow is brought near the bridge, and at the same time but lightly pressed, another alteration of quality occurs, which is readily seen on the vibrational figure. A mixture is formed of

FIG. 25.

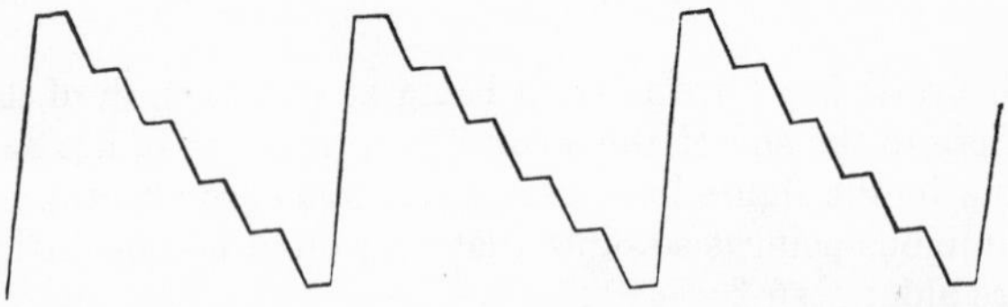

* [The position of the finger for producing the harmonic is often slightly different from that theoretically assigned. Dr. Huggins, F.R.S., kindly tried for me the position of the Octave harmonic on the four strings of his Stradivari, a mark with Chinese white being made under his finger on the finger board. Result, 1st and 4th string exact, 2nd string 3 mm., and 3rd string 5 mm. too near the nut, out of 165 mm. the actual half length of the strings. These differences must therefore be due to some imperfections of the strings themselves. Dr. Huggins finds that there is a space of a quarter of an inch at any point of which the Octave harmonic may be brought out, but the quality of tone is best at the points named above.—*Translator*.]

the prime tone and first harmonic of the string. By light and rapid bowing, namely at about $\frac{1}{20}$th of the length of the string from the bridge, we sometimes obtain the upper Octave of the prime tone by itself, a node being formed in the middle of the string. On bowing more firmly the prime tone immediately sounds. Intermediately the higher Octave may mix with it in any proportion. This is immediately recognised in the vibrational figure. Fig. 26 gives the corresponding series of forms. It is seen how a fresh crest appears on the longer side of the front of a wave, jutting out at first slightly, then more strongly, till at length the crest of the new waves are as high as those of the old, and then the vibrational number has doubled, and the pitch has passed into the Octave above. The quality of the lowest tone of the string is rendered softer and brighter, but less full and powerful when the intermixture commences. It is interesting to observe the vibrational figure while little changes are made in the style of bowing, and note how the resulting slight changes of quality are immediately rendered evident by ¶ very distinct changes in the vibrational figure itself.

The vibrational forms just described may be maintained in a uniformly steady and unchanged condition by carefully uniform bowing. The instrument has then an uninterrupted and pure musical quality of tone. Any scratching of the bow is immediately shewn by sudden jumps, or discontinuous displacements and changes in the vibrational figure. If the scratching continues, the eye has no longer time to perceive a regular figure. The scratching noises of a violin bow must therefore be regarded as irregular interruptions of the normal vibrations of the string, making them to recommence from a new starting point. Sudden jumps in the

FIG. 26.

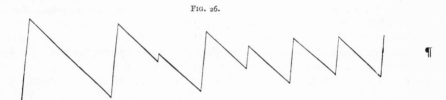

¶

vibrational figure betray every little stumble of the bow which the ear alone would scarcely observe. Inferior bowed instruments seem to be distinguished from good ones by the frequency of such greater or smaller irregularities of vibration. On the string of my monochord, which was only used for the occasion as a bowed instrument, great neatness of bowing was required to preserve a steady vibrational figure lasting long enough for the eye to apprehend it; and the tone was rough in quality, accompanied by much scratching. With a very good modern violin made by Bausch it was easier to maintain the steadiness of the vibrational figure for some time; but I succeeded much better with an old Italian violin of Guadanini, which was the first one on which I could keep the vibrational figure steady enough ¶ to count the crumples. This great uniformity of vibration is evidently the reason of the purer tone of these old instruments, since every little irregularity is immediately felt by the ear as a roughness or scratchiness in the quality of tone.

An appropriate structure of the instrument, and wood of the most perfect elasticity procurable, are probably the important conditions for regular vibrations of the string, and when these are present, the bow can be easily made to work uniformly. This allows of a pure flow of tone, undisfigured by any roughness. On the other hand, when the vibrations are so uniform the string can be more vigorously attacked with the bow. Good instruments consequently allow of a much more powerful motion of the string, and the whole intensity of their tone can be communicated to the air without diminution, whereas the friction caused by any imperfection in the elasticity of the wood destroys part of the motion. Much of the advantages of old violins may, however, also depend upon their age, and especially their long use, both of which cannot but act favourably on the elasticity of the wood.   But the art of bowing is evidently the most important condition of all. How delicately this must be cultivated to obtain certainty in producing a very perfect quality of tone and its different varieties, cannot be more clearly demonstrated than by the observation of vibrational figures.   It is also well known that great players can bring out full tones from even indifferent instruments.

The preceding observations and conclusions refer to the vibrations of the strings of the instrument and the intensity of their upper partial tones, solely in so far as they are contained in the compound vibrational movement of the string.   But partial tones of different pitches are not equally well communicated to the air, and hence do not strike the ear of the listener with precisely the same degrees of intensity as those they possess on the string itself.   They are communicated to the air by means of the sonorous body of the instrument.   As we have had already occasion to remark, vibrating strings do not directly communicate any ¶ sensible portion of their motion to the air.   The vibrating strings of the violin, in the first place, agitate the bridge over which they are stretched.   This stands on two feet over the most mobile part of the ' belly ' between the two ' f holes.' One foot of the bridge rests upon a comparatively firm support, namely, the ' soundpost,' which is a solid rod inserted between the two plates, back and belly, of the instrument.   It is only the other leg which agitates the elastic wooden plates, and through them the included mass of air.*

An inclosed mass of air, like that of the violin, vióla, and violoncello, bounded by elastic plates, has certain proper tones which may be evoked by blowing across the openings, or ' f holes.'   The violin thus treated gives c' according to Savart, who examined instruments made by Stradivari (Stradiuarius).†   Zamminer found the same tone constant on even imperfect instruments.   For the violoncello Savart found on blowing over the holes $F,$ and Zamminer $G.$‡   According to Zamminer the sound-box of the vióla (tenor) is tuned to be a Tone ¶ deeper than that of the violin.§   On placing the ear against the back of a violin and playing a scale on the pianoforte, some tones will be found to penetrate the ear with more force than others, owing to the resonance of the instrument.   On a

---

* [This account is not quite sufficient. Neither leg of the bridge rests exactly on the sound-post, because it is found that this position materially injures the quality of tone. The sound-post is a little in the rear of the leg of the bridge on the e'' string side. The position of the sound-post with regard to the bridge has to be adjusted for each individual instrument. Dr. William Huggins, F.R.S., in his paper ' On the Function of the Sound-post, and on the Proportional Thickness of the Strings of the Violin,' read May 24, 1883, *Proceedings of the Royal Society*, vol. xxxv.
¶ pp. 241–248, has experimentally investigated the whole action of the sound-post, and finds that its main function is to convey vibrations from the belly to the back of the violin, in addition to those conveyed by the sides. The (apparently ornamental) cuttings in the bridge of the violin, sift the two sets of vibrations, set up by the bowed string at right angles to each other and ' allow those only or mainly to pass to the feet which would be efficient in setting the body of the instrument into vibration.' As the peculiar shape of the instrument rendered strewing of sand unavailable, Dr. Huggins investigated the vibrations by means of a ' touch rod,' consisting of ' a small round stick of straight grained deal a few inches long; the forefinger is placed on one end and the other end is put lightly in contact with the vibrating surface. The finger soon becomes very sensitive to small differences of

agitation transmitted by the rod.' In short, the touch rod acts as a sound-post to the finger. The place of least vibration of the belly is exactly over the sound-post and of the back at the point under the sound-post. On removing the sound-post, or covering its ends with a sheet of india-rubber, which did not diminish the support, the tone was poor and thin. But an external wooden clamp connecting belly and back in the places where the sound-post touches them, restored the tone.— *Translator.*]

† [Zamminer, *Die Musik*, 1855, vol. i. p. 37, says c' of 256 vib.—*Translator.*]

‡ [Zamminer, *ibid.* p. 41, and adds that judging from the violin the resonance should be F♯.—*Translator.*]

§ [The passage referred to has not been found.   But Zamminer says, p. 40, ' The length of the box of a violin is 13 Par. inches, and of the vióla 14 inches 5 lines. Exactly in inverse ratio stand the pitch numbers 470 (a misprint for 270 most probably) and 241, which were found by blowing over the wind-holes of the two instruments.' Now the ratio 13 : $14\frac{5}{12}$ gives 182 cents, and the ratio 241 : 270 gives 197 cents, which are very nearly, though not ' exactly ' the same. This, however, makes the resonance of the violin 270 vib. and not 256 vib., and agrees with the next note. I got a good resonance with a fork of 268 vib. from Dr. Huggins's violoncello by Nicholas about A.D. 1792.—*Translator.*]

violin made by Bausch two tones of greatest resonance were thus discovered, one between c' and $c'\sharp$ [between 264 and 280 vib.], and the other between a' and $b'\flat$ [between 440 and 466 vib.]. For a vióla (tenor) I found the two tones about a Tone deeper, which agrees with Zamminer's calculation.*

The consequence of this peculiar relation of resonance is that those tones of the strings which lie near the proper tones of the inclosed body of air, must be proportionably more reinforced. This is clearly perceived on both the violin and violoncello, at least for the lowest proper tone, when the corresponding notes are produced on the strings. They sound particularly full, and the prime tone of these compounds is especially prominent. I think that I heard this also for a' on the violin, which corresponds to its higher proper tone.

Since the lowest tone on the violin is g, the only upper partials of its musical tones which can be somewhat reinforced by the resonance of the higher proper tone of its inclosed body of air, are the higher octaves of its three deepest notes. ¶ But the prime tones of its higher notes will be reinforced more than their upper partials, because these prime tones are more nearly of the same pitch as the proper tones of the body of air. This produces an effect similar to that of the construction of the hammer of a piano, which favours the upper partials of the deep notes, and weakens those of the higher notes. For the violoncello, where the lowest string gives C, the stronger proper tone of the body of air is, as on the violin, a Fourth or a Fifth higher than the pitch of the lowest string. There is consequently a similar relation between the favoured and unfavoured partial tones, but all of

---

* [Through the kindness of Dr. Huggins, F.R.S., the Rev. H. R. Haweis, and the violin-makers, Messrs. Hart, Hill & Withers, I was in 1880 enabled to examine the pitch of the resonance of some fine old violins by Duiffo-prugcar (Swiss Tyrol, Bologna, and Lyons 1510-1538), Amati (Cremona 1596-1684), Ruggieri (Cremona 1668-1720), Stradivari (Cremona 1644-1737), Giuseppe Garneri (known as 'Joseph,' Cremona 1683-1745), Lupot (France 1750-1820). The method adopted was to hold tuning-forks, of which the exact pitch had been determined by Scheibler's forks, in succession over the widest part of the f hole on the g string side of the violin (furthest from the sound-post) and observe what fork excited the maximum resonance. My forks form a series proceeding by 4 vib. in a second, and hence I could only tell the pitch within 2 vib., and it was often extremely difficult to decide on the fork which gave the best resonance. By far the strongest resonance lay between 268 and 272 vib., but one early Stradivari (1696) had a fine resonance at 264 vib. There was also a secondary but weaker maximum resonance at about 252 vib. The 256 vib. was generally decidedly inferior. Hence we may take 270 vib. as the primary maximum, and 252 vib. as the secondary. The first corresponds to the highest English concert pitch $c'' = 540$ vib., now used in London, and agrees with the lower resonance of Bausch's instrument mentioned in the text. The second, which is 120 cents, or rather more than an equal Semitone flatter, gives the pitch which my researches shew was common over all Europe at the time (see App. XX. sect. H.). But although the low pitch was prevalent, a high pitch, a great Semitone (117 ct.) higher, was also in use as a 'chamber pitch.' A violin of Mazzini of Brescia (1560-1640) belonging to the eldest daughter of Mr. Vernon Lushington, Q.C., had the same two maximum resonances, the higher being decidedly the superior. I did not examine for the higher or a' pitches named in the text. Mr. Healey (of the Science and Art Department, South Kensington) thought his violin (supposed to be an Amati) sounded best at the low pitch $c'' = 504$. Subsequently, I examined a fine instrument, bearing inside it the label 'Petrus Guarnerius Cremonensis fecit, Mantuæ sub titulo S. Theresiæ, anno 1701,' in the possession of Mr. A. J. Hipkins, who knew ¶ it to be genuine. I tried this with a series of forks, proceeding by differences of about 4 vibrations from 240 to 560. It was surprising to find that every fork was to a certain extent reinforced, that is, in no case was the tone quenched, and in no case was it reduced in strength. But at 260 vib. there was a good, and at 264 a better resonance; perhaps 262 may therefore be taken as the best. There was no secondary low resonance, but there were two higher resonances, one about 472, (although 468 and 476 were also good,) and another at 520 (although 524 and 528 were also good). As this sheet was passing through the press I had an opportunity of trying the resonance of Dr. Huggins's Stradivari of 1708, figured in Grove's *Dictionary of Music*, iii. 728, as a specimen of the best period of Stradi- ¶ vari's work. The result was essentially the same as the last; every fork was more or less reinforced; there was a subordinate maximum at 252 vib.; a better at from 260 to 268 vib.; very slight maxima at 312, 348, 384, 412, 420, 428 (the last of which was the best, but was only a fair reinforcement), 472 to 480, but 520 was decidedly best, and 540 good. No one fork was reinforced to the extent it would have been on a resonator properly tuned to it, but no one note was deteriorated. Dr. Huggins says that 'the strong feature of this violin is the great equality of all four strings and the persistence of the same fine quality of tone throughout the entire range of the instrument.'—*Translator.*]

them are a Twelfth lower than on the violin. On the other hand, the most favoured partial tones of the vióla (tenor) corresponding nearly with b', do not lie between the first and second strings, but between the second and third; and this seems to be connected with the altered quality of tone on this instrument. Unfortunately this influence cannot be expressed numerically. The maximum of resonance for the proper tones of the body of air is not very marked; were it otherwise there would be much more inequality in the scale as played on these bowed instruments, immediately on passing the pitch of the proper tones of their bodies of ¶ air. We must consequently conjecture that their influence upon the relative intensity of the individual partials in the musical tones of these instruments is not very prominent.

5. *Musical Tones of Flute or Flue Pipes.*

In these instruments the tone is produced by driving a stream of air against an opening, generally furnished with sharp edges, in some hollow space filled with air. To this class belong the bottles described in the last chapter, and shewn in fig. 20 (p. 60c), and especially flutes and the majority of organ pipes. For flutes, the resonant body of air is included in ¶ its own cylindrical bore. It is blown with the mouth, which directs the breath against the somewhat sharpened edges of its mouth hole. The construction of organ pipes will be seen from the two adjacent figures. Fig. 27, A, shews a square wooden pipe, cut open long-wise, and B the external appearance of a round tin pipe. R R in each shews the tube which incloses the sonorous body of air, a b is the *mouth* where it is blown, terminating in a sharp *lip*. In fig. 27, A, we see the air chamber or *throat* K into which the air is first driven from the bellows, and whence it can only escape through the narrow slit c d, which directs it ¶ against the edge of the lip. The wooden pipe A as here drawn is *open*, that is its extremity is uncovered, and it produces a tone with a wave of air *twice* as long as the tube R R. The other pipe, B, is *stopped*, that is, its upper extremity is closed. Its tone has a wave *four times* the length of the tube R R, and hence an Octave deeper than an open pipe of the same length.*

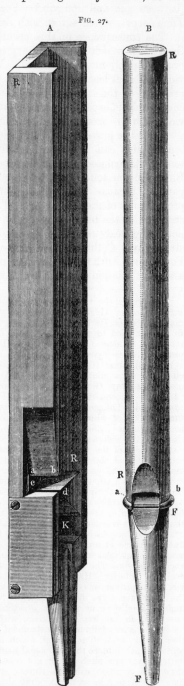

FIG. 27.

Any air chambers can be made to give a musical tone, just like organ pipes, flutes, the bottles previously described, the windchests of violins, &c., provided they have a sufficiently narrow opening,

---

* [These relations are only approximate, as is explained below. The mode of excitement by the lip of the pipe makes them inexact. Also they take no notice of the 'scale' or diameters and depths of the pipes, or of the force of the wind, or of the tempera-

furnished with somewhat projecting sharp edges, by directing a thin flat stream of air across the opening and against its edges.*

The motion of air that takes place in the inside of organ pipes, corresponds to a system of plane waves which are reflected backwards and forwards between the two ends of the pipe. At the stopped end of a cylindrical pipe the reflexion of every wave that strikes it is very perfect, so that the reflected wave has the same intensity as it had before reflexion. In any train of waves moving in a given direction, the velocity of the oscillating molecules in the condensed portion of the wave takes place in the same direction as that of the propagation of the waves, and in the rarefied portion in the opposite direction. But at the stopped end of a pipe its cover does not allow of any forward motion of the molecules of air in the direction of the length of the pipe. Hence the incident and reflected wave at this place combine so as to excite opposite velocities of oscillation of the molecules of air, and consequently by their superposition the velocity of the molecules of air at ¶ the cover is destroyed. Hence it follows that the phases of pressure in both will agree, because opposite motions of oscillation and opposite propagation, result in accordant pressure.

Hence at the stopped end there is no motion, but great alteration of pressure. The reflexion of the wave takes place in such a manner that the phase of conden-sation remains unaltered, but the direction of the motion of oscillation is reversed.

The contrary takes place at the open end of pipes, in which is also included the opening of their mouths. At an open end where the air of the pipe communi-cates freely with the great outer mass of air, no sensible condensation can take place. In the explanation usually given of the motion of air in pipes, it is assumed that both condensation and rarefaction vanish at the open ends of pipes, which is approximatively but not exactly correct. If there were exactly no alteration of density at that place, there would be complete reflexion of every incident wave at the open ends, so that an equally large reflected wave would be generated with ¶ an opposite state of density, but the direction of oscillation of the molecules of air in both waves would agree. The superposition of such an incident and such a

---

ture of the air. The following are adapted from the rules given by M. Cavaillé-Coll, the celebrated French organ-builder, in *Comptes Rendus*, 1860, p. 176, supposing the tempera-ture to be 59° F. or 15° C., and the pressure of the wind to be about 3¼ inches, or 8 centi-metres (meaning that it will support a column of water of that height in the wind gauge). The pitch numbers, for *double* vibrations, are found by dividing 20,080 when the dimensions are given in inches, and 510,000 when in millimetres by the following numbers : (1) for *cylindrical open* pipes, 3 times the length added to 5 times the diameter ; (2) for *cylindri-cal stopped* pipes, 6 times the length added to 10 times the diameter ; (3) for *square open* pipes, 3 times the length added to 6 times the depth (clear internal distance from mouth to back) ; (4) for *square stopped* pipes, 6 times the length added to 12 times the depth.

This rule is always sufficiently accurate for cutting organ pipes to their approximate length, and piercing them to bring out the Octave harmonic, and has long been used for these purposes in M. Cavaillé-Coll's factory. The rule is not so safe for the square wooden as for the cylindrical metal pipes. The pitch of a pipe of known dimensions ought to be first ascertained by other means. Then this pitch number multiplied by the divisors in (3) and (4) should be used in place of the 20,080 or 510,000 of the rule, for all similar pipes.

As to strength of wind, as pressure varies

from 2¾ to 3¼ inches, the pitch number increases by about 1 in 300, but as pressure varies from 3¼ to 4 inches, the pitch number increases by about 1 in 440, the whole increase of pressure from 2¾ to 4 inches increases the pitch number by 1 in 180.

For temperature, I found by numerous observations at very different temperatures that the following practical rule is sufficient for reducing the pitch number observed at one temperature to that due to another. It is not quite accurate, for the air blown from the bellows is often lower than the external tem-perature. Let P be the pitch number observed ¶ at a given temperature, and d the difference of temperature in degrees Fahr. Then the pitch number is $P \times (1 \pm \cdot 00104\, d)$ according as the temperature is higher or lower. The practical operation is as follows : supposing $P = 528$, and $d = 14$ increase of temperature. To 528 add 4 in 100, or 21·12, giving 549·12. Divide by 1000 to 2 places of decimals, giving ·55. Multiply by $d = 14$, giving 7·70. Adding this to 528, we get 535·7 for the pitch number at the new temperature.—*Translator.*]

* [Here the passage from 'These edges,' p. 140, to 'resembling a violin,' p. 141 of the 1st English edition, has been omitted, and the passage from 'The motion of air,' p. 89a, to 'their corners are rounded off,' p. 93b, has been inserted in accordance with the 4th German edition.—*Translator.*]

reflected wave would indeed leave the state of density unaltered at the open ends, but would occasion great velocity in the oscillating molecules of air.

In reality the assumption made explains the essential phenomena of organ pipes. Consider first a pipe with two open ends. On our exciting a wave of condensation at one end, it runs forward to the other end, is there reflected as a wave of rarefaction, runs back to the first end, is here again reflected with another alteration of phase, as a wave of condensation, and then repeats the same path in the same way a second time. This repetition of the same process therefore occurs, after the wave in the tube has passed once forwards and once backwards, that is twice through the whole length of the tube. The time required to do this is equal to double the length of the pipe divided by the velocity of sound. This is the duration of the vibration of the deepest tone which the pipe can give.

¶ Suppose now that at the time when the wave begins its second forward and backward journey, a second impulse in the same direction is given, say by a vibrating tuning-fork. The motion of the air will then receive a reinforcement, which will constantly increase, if the fresh impulses take place in the same rhythm as the forward and backward progression of the waves.

Even if the returning wave does not coincide with the first following similar impulse of the tuning-fork, but only with the second or third or fourth and so on, the motion of the air will be reinforced after every forward and backward passage.

A tube open at both ends will therefore serve as a resonator for tones whose pitch number is equal to the velocity of sound (332 metres) * divided by twice the length of the tube, or some multiple of that number. That is to say, the tones of strongest resonance for such a tube will, as in strings, form the complete series of harmonic upper partials of its prime.

¶ The case is somewhat different for pipes stopped at one end. If at the open end, by means of a vibrating tuning-fork, we excite an impulse of condensation which propagates itself along the tube, it will run on to the stopped end, will be there reflected as a wave of condensation, return, will be again reflected at the open end with altered phase as a wave of rarefaction, and only after it has been again reflected at the stopped end with a similar phase, and then once more at the open end with an altered phase as a condensation, will a repetition of the process ensue, that is to say, not till after it has traversed the length of the pipe four times. Hence the prime tone of a stopped pipe has twice as long a period of vibration as an open pipe of the same length. That is to say, the stopped pipe will be an Octave deeper than the open pipe. If, then, after this double forward and backward passage, the first impulse is renewed, there will arise a reinforcement of resonance.

¶ Partials † of the prime tone can also be reinforced, but only those which are unevenly numbered. For since at the expiration of half the period of vibration, the prime tone of the wave in the tube renews its path with an opposite phase of density, only such tones can be reinforced as have an opposite phase at the expiration of half the period of vibration. But at this time the second partial has just completed a whole vibration, the fourth partial two whole vibrations, and so on.

* [This is 1089·3 feet in a second, which is the mean of several observations in free air ; it is usual, however, in England to take the whole number 1090 feet, at freezing. At 60° F. it is about 1120 feet per second. Mr. D. J. Blaikley (see note p. 97d), in two papers read before the Physical Society, and published in the *Philosophical Magazine* for Dec. 1883, pp. 447-455, and Oct. 1884, pp. 328-334, as the means of many observations on the velocity of sound in dry air at 32° F., *in tubes*, obtained

for diameter	·45	·75	1·25	2·08	3·47 English inches.
pitch various, velocity	1064·26	1072·53	1078·71	1081·78	1083·13 ,, feet.
pitch 260 vib., velocity	1062·12	1072·47	1078·73	1082·51	1084·88 ,, ,,

The velocity in tubes is therefore always less than in free air.—*Translator*.]

† [The original says 'upper partials' (*Obertöne*), but the *upper* partials which are unevenly numbered are the 1st, 3rd, 5th, &c., and these are really the 2nd, 4th, 6th, &c., (that is, the evenly numbered) partial tones, (see foot-note p. 23c,) but it is precisely the latter which are *not* excited in the present case. This is only mentioned as a warning to those who faultily use the faulty expression 'overtones' indifferently for both partials and *upper* partials.—*Translator*.]

These therefore have the same phases, and cancel their effect on the return of the wave with an opposite phase. Hence the tones of strongest resonance in stopped pipes correspond with the series of unevenly numbered partials of its fundamental tone. Supposing its pitch number is n, then $3n$ is the Twelfth of n, that is the Fifth of $2n$ the higher Octave, and $5n$ is the major Third of $4n$ the next higher Octave, and $7n$ the [sub] minor Seventh of the same Octave, and so on.

Now although the phenomena follow these rules in the principal points, certain deviations from them occur because there is not precisely no change of pressure at the open ends of pipes. From these ends the motion of sound communicates itself to the uninclosed air beyond, and the waves which spread out from the open ends of the tubes have relatively very little alteration of pressure, but are not entirely without some. Hence a part of every wave which is incident on the open end of the pipe is *not* reflected, but runs out into the open air, while the remainder or greater portion of the wave is reflected, and returns into the tube. The re- ¶ flexion is the more complete, the smaller are the dimensions of the opening of the tube in comparison with the wave-length of the tone in question.

Theory * also, agreeing with experiment, shews that the phases of the reflected part of the wave are the same as they would be if the reflexion did not take place at the surface of the opening itself but at another and somewhat different plane. Hence what may be called the *reduced length* of the pipe, or that answering to the pitch, is somewhat different from the real length, and the difference between the two depends on the form of the mouth, and not on the pitch of the notes produced unless they are so high and hence their wave-lengths so short, that the dimensions of the opening cannot be neglected in respect to them.

For cylindrical pipes of circular section, with ends cut at right angles to the length, the distance of the plane of reflexion from the end of the pipe is theoretically determined to be at a distance of 0·7854 the radius of the circle.† For a wooden pipe of square section, of which the sides were 36 mm. (1·4 inch) internal ¶ measure, I found the distance of the plane of reflexion 14 mm. (·55 inch).‡

Now since on account of the imperfect reflexion of waves at the open ends of organ pipes (and respectively at their mouths) a part of the motion of the air must escape into the free air at every vibration, any oscillatory motion of its mass of air must be speedily exhausted, if there are no forces to replace the lost motion. In fact, on ceasing to blow an organ pipe scarcely any after sound is observable. Nevertheless the wave is frequently enough reflected forward and backward for its pitch to become perceptible on tapping against the pipe.

The means usually adopted for keeping them continually sounding, is *blowing*. In order to understand the action of this process, we must remember that when

* See my paper in *Crelle's Journal for Mathematics*, vol. lvii.

† Mr. Bosanquet (*Proc. Mus. Assn.* 1877–8, p. 65) is reported as saying: 'Lord Rayleigh and himself had gone fully into the matter, and had come to the conclusion that this correction was much less than Helmholtz supposed. Lord Rayleigh adopted the figure ·6 of this amount, whilst he himself adopted ·55.' See papers by Lord Rayleigh and Mr. Bosanquet in *Philosophical Magazine*. Mr. Blaikley by a new process finds ·576, which lies between the other,two, see his paper in *Phil. Mag.* May 1879, p. 342.

‡ The pipe was of wood, made by Marloye, the additional length being 302 mm. (11·9 in.), corresponding exactly with half the length of wave of the pipe. The position of the nodal plane in the inside of the pipe was determined by inserting a wooden plug of the same diameter as of the internal opening of the pipe at its open end, until the pitch of the pipe, which had now become a closed one, was exactly the same as that of the open pipe before the inser-

tion of the plug. [The sameness of the pitch is determined by seeing that each makes the same number of beats with the same fork.] The nodal surface lay 137 mm. (5·39 inch) from the end of the pipe, while a quarter of ¶ a wave-length was 151 mm. (5·94 inch). At the mouth end of the pipe, on the other hand, 83 mm. (3·27 inch) were wanting to complete the theoretical length of the pipe. [The additional piece being half the length of the wave, the pitch of the pipe before and after the addition of this piece remains the same, by which property the length of the additional piece is found. The length of the pipe from the bottom of the mouth to the open end was 205 mm. = 8·07 inch ; the node, as determined, was 137 mm. = 5·39 inch from the open end, and 68 mm. = 2·68 inch from the bottom of the mouth. These lengths had to be increased by 14 mm. = ·55 in. and 83 mm. = 3·27 in. respectively, to make up each to the quarter length of the wave 151 mm. = 5·95 inch.—*Translator.*]

air is blown out of such a slit as that which lies below the lip of the pipe, it breaks through the air which lies at rest in front of the slit in a thin sheet like a blade or lamina, and hence at first does not draw any sensible part of that air into its own motion. It is not till it reaches a distance of some centimetres [a centimetre is nearly four-tenths of an inch] that the outpouring sheet splits up into eddies or vortices, which effect a mixture of the air at rest and the air in motion. This blade-shaped sheet of air in motion can be rendered visible by sending a stream of air impregnated with smoke or clouds of salammoniac through the mouth of a pipe from which the pipe itself is removed, such as is commonly found among physical apparatus. Any blade-shaped gas flame which comes from a split burner is also an example of a similar process. Burning renders visible the limits between the outpouring sheet of gas and the atmosphere. But the flame does not render the continuation of the stream visible.

¶   Now the blade-shaped sheet of air at the mouth of the organ pipe is wafted to one side or the other by every stream of air which touches its surface, exactly as this gas flame is. The consequence is that when the oscillation of the mass of air in the pipe * causes the air to enter through the ends of the pipe, the blade-shaped stream of air arising from the mouth is also inclined inwards, and drives its whole mass of air into the pipe.† During the opposite phase of vibration, on the other hand, when the air leaves the ends of the pipe the whole mass of this blade of air is driven outwards. Hence it happens that exactly at the times when the air in the pipe is most condensed, more air still is driven in from the bellows, whence the condensation, and consequently also the equivalent of work of the vibration of the air is increased, while at the periods of rarefaction in the pipe the wind of the bellows pours its mass of air into the open space in front of the pipe. We must remember also that the blade-shaped sheet of air requires time in order to traverse the width of the mouth of the pipe, and is during this time exposed to the action ¶ of the vibrating column of air in the pipe, and does not reach the lip (that is the line where the two paths, inwards and outwards, intersect) until the end of this time. Every particle of air that is blown in, consequently reaches a phase of vibration in the interior of the pipe, which is somewhat later than that to which it was exposed in traversing the opening. If the latter motion was inwards, it encounters the following condensation in the interior of the pipe, and so on.

This mode of exciting the tone conditions also the peculiar quality of tone of these organ pipes. We may regard the blade-shaped stream of air as very thin in comparison with the amplitude of the vibrations of air. The latter often amount to 10 or 16 millimetres (·39 to ·63 inches), as may be seen by bringing small flames of gas close to this opening. Consequently the alternation between the periods of time for which the whole blast is poured into the interior of the pipe, and those for which it is entirely emptied outside, is rather sudden, in fact almost instantaneous. Hence it follows ‡ that the oscillations excited by blowing are of ¶ a similar kind; namely, that for a certain part of each vibration the velocity of the particles of air in the mouth and in free space, have a constant value directed outwards, and for a second portion of the same, a constant value directed inwards. With stronger blowing that directed inwards will be more intense and of shorter duration; with weaker blowing, the converse may take place. Moreover, the pressure in the mass of air put in motion in the pipe must also alternate between two constant values with considerable rapidity. The rapidity of this change will, however, be moderated by the circumstance that the blade-shaped sheet of air is not infinitely thin, but requires a short time to pass over the lip of the pipe, and

* [It has, however, not been explained how that 'oscillation' commences. This will be alluded to in the additions to App. VII. sect. B. —*Translator.*]

† [The amount of air which enters as compared with that which passes over the lip out-side the pipe is very small. A candle flame held at the end of the pipe only pulsates; held a few inches from the lip, along the edge of the pipe, it is speedily extinguished.—*Translator.*]

‡ See Appendix VII. [especially sect. B, II.].

that secondly the higher upper partials, whose wave-lengths only slightly exceed the diameter of the pipe, are as a general rule imperfectly developed.

The kind of motion of the air here described is exactly the same as that shewn in fig. 23 (p. 82*b*), B and C, fig. 24 (p. 83*b*), A and B, for the vibrating points of a violin string. Organ-builders have long since remarked the similarity of the quality of tone, for the narrower cylindrical-pipe stops when strongly blown, as shewn by the names : *Geigenprincipal, Vióla di Gamba, Violoncello, Violon-bass.**

That these conclusions from the mechanics of blowing correspond with the facts in nature, is shewn by the experiments of Messrs. Toepler & Boltzmann,† who rendered the form of the oscillation of pressure in the interior of the pipe optically observable by the interference of light passed through a node of the vibrating mass of air. When the force of the wind was small they found almost a simple vibration (the smaller the oscillation of the air-blade at the lip, the more completely the discontinuities disappear). But when the force of the wind was greater they found ¶ a very rapid alternation between two different values of pressure, each of which remained almost unaltered for a fraction of a vibration.

Messrs. Mach and J. Hervert's ‡ experiments with gas flames placed before the end of an open pipe to make the vibrations visible, shew that the form of motion just described really occurs at the ends of the pipes. The forms of vibration which they deduced from the analysis of the forms of the flames correspond with those of a violin string, except that, for the reason given above, their corners are rounded off.

By using resonators I find that on narrow pipes of this kind the partial tones may be clearly heard up to the sixth.

For wide open pipes, on the other hand, the adjacent proper tones of the tube are all somewhat sharper than the corresponding harmonic tones of the prime, and hence these tones will be much less reinforced by the resonance of the tube. Wide pipes, having larger masses of vibrating air and admitting of being much more strongly blown without jumping up into an harmonic, are used for the great body ¶ of sound on the organ, and are hence called *principalstimmen.*§ For the above reasons they produce the prime tone alone strongly and fully, with a much weaker retinue of secondary tones. For wooden 'principal' pipes, I find the prime tone and its Octave or first upper partial very distinct; the Twelfth or second upper partial is. but weak, and the higher upper partials no longer distinctly perceptible. For metal pipes the fourth partial was also still perceptible. The quality of tone in these pipes is fuller and softer than that of the *geigenprincipal.** When flute or flue stops of the organ, and the German flute are blown softly, the upper partials lose strength at a greater rate than the prime tone, and hence the musical quality becomes weak and soft.

Another variety is observed on the pipes which are conically narrowed at their

* [*Geigenprincipal*—violin or crisp-toned diapason, 8 feet,—violin principal, 4 feet. See supra p. 91*d*, note. *Violoncello*—'crisp-toned open stop, of small scale, the Octave to the violone, 8 feet.' *Violon-bass*—this fails in Hopkins, but it is probably his ' *violone*— double bass, a unison open wood stop, of much smaller scale than the Diapason, and formed of pipes that are a little wider at the top than at the bottom, and furnished with ears and beard at the mouth ; the tone of the Violone is crisp and resonant, like that of the orchestral Double Bass ; and its speech being a little slow, it has the Stopped Bass always drawn with it, 16 feet.' *Gamba* or *viol da gamba*— ' bass viol. unison stop, of smaller scale, and thinner but more pungent tone than the violin diapason, 8 feet, . . . one of the most highly esteemed and most frequently disposed stops in Continental organs ; the German gamba is usually composed of cylindrical pipes.' In England till very recently it was made exclu-

sively conical with a bell top. From Hopkins on the *Organ*, pp. 137, 445, &c.—*Translator.*]

† Poggendorff's *Annal.*, vol. cxli. pp. 321- ¶ 352.

‡ Poggendorff's *Annal.*, vol. cxlvii. pp. 590- 604.

§ [Literally ' principal voices or parts ; ' may probably be best translated ' principal work ' or ' diapason-work,' including ' all the open cylindrical stops of Open Diapason measure, or which have their scale deduced from that of the Open Diapason ; such stops are the chief, most important or "*principal*," as they are also most numerous in an organ. The Unison and Double Open Diapasons, Principal, Fifteenth and Octave Fifteenth ; the Fifth, Twelfth, and Larigot ; the Tenth and Tierce ; and the Mixture Stops, when of full or proportional scale, belong to the Diapason-work.' From Hopkins on the *Organ*, p. 131.—*Translator.*]

upper end, in the *salicional, gemshorn,* and *spitzflöte* stops.* Their upper opening has generally half the diameter of the lower section; for the same length the *salicional* pipe has the narrowest, and the *spitzflöte* the widest section. These pipes have, I find, the property of rendering some higher partial tones, from the Fifth to the Seventh, comparatively stronger than the lower. The quality of tone is consequently poor, but peculiarly bright.

The *narrower stopped cylindrical pipes* have proper tones corresponding to the unevenly numbered partials of the prime, that is, the third partial or Twelfth, the fifth partial or higher major Third, and so on. For the *wider* stopped pipes, as for the wide open pipes, the next adjacent proper tones of the mass of air are distinctly higher than the corresponding upper partials of the prime, and consequently these upper partials are very slightly, if at all, reinforced. Hence wide stopped pipes, especially when gently blown, give the prime tone almost alone, and they were ¶ therefore previously adduced as examples of simple tones (p. 60c). Narrow stopped pipes, on the other hand, let the Twelfth be very distinctly heard at the same time with the prime time; and have hence been called *quintaten (quintam tenentes).*† When these pipes are strongly blown they also give the fifth partial, or higher major Third, very distinctly. Another variety of quality is produced by the *rohrflöte.*‡ Here a tube, open at both ends, is inserted in the cover of a stopped pipe, and in the examples I examined, its length was that of an open pipe giving the fifth partial tone of the prime tone of the stopped pipe. The fifth partial tone is thus proportionably stronger than the rather weak third partial on these pipes, and the quality of tone becomes peculiarly bright. Compared with open pipes the quality of tone in stopped pipes, where the evenly numbered partial tones are absent, is somewhat hollow; the wider stopped pipes have a dull quality of tone, especially when deep, and are soft and powerless. But their softness offers a very effective contrast to the more cutting qualities of the narrower open pipes and the ¶ noisy *compound stops,* of which I have already spoken (p. 57b), and which, as is well known, form a compound tone by uniting many pipes corresponding to a prime and its upper partial tones.

Wooden pipes do not produce such a cutting windrush as metal pipes. Wooden sides also do not resist the agitation of the waves of sound so well as metal ones, and hence the vibrations of higher pitch seem to be destroyed by friction. For these reasons wood gives a softer, but duller, less penetrating quality of tone than metal.

It is characteristic of all pipes of this kind that they speak readily, and hence admit of great rapidity in musical divisions and figures, but, as a little increase of force in blowing distinctly alters the pitch, their loudness of tone can scarcely be changed. Hence on the organ *forte* and *piano* have to be produced by stops, which regulate the introduction of pipes with various qualities of tone, sometimes more, sometimes fewer, now the loud and cutting, now the weak and soft. The means of expression on this instrument are therefore somewhat limited, but, on the other ¶ hand, it clearly owes part of its magnificent properties to its power of sustaining tones with unaltered force, undisturbed by subjective excitement.

---

* [*Salicional*—'reedy Double Dulciana, 16 feet and 8 feet, octave salicional, 4 feet.' The Dulciana is described as 'belonging to the Flute-work, ... the pipes much smaller in scale than those of the open diapason ... tone peculiarly soft and gentle' (Hopkins, p. 113). *Gemshorn,* literally 'chamois horn;' in Hopkins, 'Goat-horn, a unison open metal stop, more conical than the Spitz-Flöte, 8 feet.' 'A member of the Flute-work and met with of 8, 4, or 2 feet length in Continental organs. The pipes of this stop are only ⅓ the diameter at the top that they are at the mouth; and the tone is consequently light, but very clear and travelling' (*ibid.* p. 140). *Spitzflöte*—'Spire or taper flute, a unison open metal stop formed of pipes with conical bodies, 8 feet.' 'This stop is found of 8, 4, and 2 feet length in German organs. In England it has hitherto been made chiefly as a 4-feet stop; i.e. of principal pitch. The pipes of the Spitz-flute are slightly conical, being about ¼ narrower at top than at the mouth, and the tone is therefore rather softer than that of the cylindrical stop, but of very pleasing quality' (*ibid.* p. 140).—*Translator.*]

† [See supra p. 33d, note.—*Translator.*]

‡ [*Rohrflöte*—'Double Stopped Diapason of metal pipes with chimneys, 16 feet, Reed-flute, Metal Stopped Diapason, with reeds, tubes or chimneys, 8 feet. Stopped Metal Flute, with reeds, tubes or chimneys, 4 feet' (Hopkins, pp. 444, 445).—*Translator.*]

6. *Musical Tones of Reed Pipes.*

The mode of producing the tones on these instruments resembles that used for the siren : the passage for the air being alternately closed and opened, its stream is separated into a series of individual pulses. This is effected on the siren, as we have already seen, by means of a rotating disc pierced with holes. In reed instruments, elastic plates or tongues are employed which are set in vibration and thus alternately close and open the aperture in which they are fastened. To these belong—

1. *The reed pipes of organs and the vibrators of harmoniums.* Their tongues, shewn in perspective in fig. 28, A, and in section in fig. 28, B, are thin oblong metal plates, z z, fastened to a brass block, a a, in which there is a hole, b b, ¶ behind the tongue and of the same shape. When the tongue is in its position of rest, it closes the hole completely, with the exception of a very fine chink all round its margin. When in motion it oscillates between the positions marked z_1 and z_2 in fig. 28, B. In the position z_1 there is an aperture for the stream of air to enter, in the direction shewn by the arrow, and this is closed when the tongue has reached the other extreme position z_2. The tongue shewn is a *free vibrator* or *anche libre*, such as is now universally employed. These tongues are slightly smaller than the corresponding opening, so that they can bend inwards without touching the edges of the hole.* Formerly, *striking vibrators* ¶ or *reeds* were employed, which on each oscillation struck against their frame. But as these produced a harsh quality of tone and an uncertain pitch they have gone out of use.†

FIG. 28.

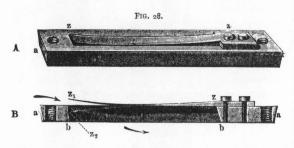

* [The quality of tone produced by the free reed can be greatly modified by comparatively slight changes. If the reed is quite flat, the end not turning up, as it does in fig. 28, above, no tone can be produced. If the size of the slit round the edges be enlarged, by forcing a thin plate of steel between the spring and the flange, and then withdrawing it, the quality of tone is permanently changed. Another change is produced by curving the middle part up and then down in a curve of contrary flexure. Another change results from curving the ends of the reed up as in 'American organs'—a species of harmonium. One of the earliest free reed instruments is the Chinese ' shêng,' which Mr. Hermann Smith thus describes from his own specimen. See also App. XX. sect. K. 'The body of the instrument is in the form and size of a teacup with a tightly fitting cover, pierced with a series of holes, arranged in a circle, to receive a set of small pipe-like canes, 17 in number, and of various lengths, of which 13 are capable of sounding and 4 are mute, but necessary for structure. The lower end of each pipe is fitted with a little free reed of very delicate workmanship, about half an inch long, and stamped in a thin metal plate, having its tip slightly loaded with beeswax, which is also used for keeping the reed in position. One peculiarity to be noticed is that the reed is quite level with the face of the plate, a condition in which modern free reeds would not speak. But this singular provision is made to ensure speaking either by blowing or suction. The corners of the reeds are rounded off, and thus a little space is left between the tip of the reed and the frame for the passage of air, an arrangement quite adverse to the speaking of harmonium reeds. In each pipe the integrity of the column of air is broken by a hole in the side, a short distance above the cup. By this strange contrivance not a single pipe will sound to the wind blown into the cup from a flexible tube, until its side hole has been covered by the finger of the player, and then the pipe gives a note corresponding to its full ¶ speaking length. Whatever be the speaking length of the pipe the hole is placed at a short distance above the cup. Its position has no relation to nodal distance, and it effects its purpose by breaking up the air column and preventing it from furnishing a proper reciprocating relation to the pitch of the reed.' The instrument thus described is the 'sing' of Barrow (*Travels in China*, 1804, where it is well figured as 'a pipe, with unequal reeds or bamboos '), and 'le petit cheng' of Père Amiot (*Mémoires concernant l'histoire . . . des Chinois* . . ., 1780, vol. vi., where a 'cheng' of 24 pipes is figured.—*Translator.*]

† [It will be seen by App. VII. to this edition, end of sect. A., that Prof. Helmholtz has somewhat modified his opinion on this point, in consequence of the information I

The mode in which tongues are fastened in the reed stops of organs is shewn in fig. 29, A and B below.   A bears a resonant cup above; B is a longitudinal section; p p is the air chamber into which the wind is driven; the tongue l is fastened in the groove r, which fits into the block s; d is the tuning wire, which presses against the tongue, and being pushed down shortens it and hence sharpens its pitch, and, conversely, flattens the pitch when pulled up.   Slight variations of pitch are thus easily produced.*

¶ 2. The tongues of *clarinets*, *oboes*, and *bassoons* are constructed in a somewhat similar manner and are cut out of elastic reed plates.   The clarinet has a single wide tongue which is fastened before the corresponding opening of the mouthpiece like the metal tongues previously described, and would strike the frame if its excursions were long enough.   But its

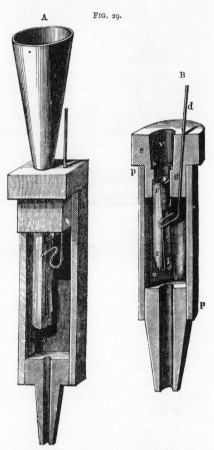

FIG. 29.

obtained from some of the principal English organ-builders, which I here insert from p. 711 of the first edition of this translation:—Mr. Willis tells me that he never uses free reeds, that no power can be got from them, and that he looks upon them as 'artificial toys.' Messrs. J. W. Walker & Sons say that they ¶ have also never used free reeds for the forty or ᴵᴵ more years that they have been in business, and consider that free reeds have been superseded by striking reeds.   Mr. Thomas Hill informs me that free reeds had been tried by his father, by M. Cavaillé-Coll of Paris, and others, in every imaginable way, for the last thirty or forty years, and were abandoned as 'utterly worthless.'   But he mentions that Schulze (of Paulenzelle, Schwartzburg) told him that he never saw a striking reed till he came over to England in 1851, and that Walcker (of Ludwigsburg, Wuertemberg) had little experience of them, as Mr. Hill learnt from him about twenty years ago.   Mr. Hill adds, however, that both these builders speedily abandoned the free reed, after seeing the English practice of voicing striking reeds. This is corroborated by Mr. Hermann Smith's ¶ statement (1875) that Schulze, in 1862, built the great organ at Doncaster with 94 stops, of which only the Trombone and its Octave had free reeds (see Hopkins on the *Organ*, p. 530, for an account of this organ); and that two years ago he built an organ of 64 stops and 4,052 pipes for Sheffield, with not one free reed; also that Walcker built the great organ for Ulm cathedral, with 6,500 pipes and 100 stops, of which 34 had reeds, and out of them only 2 had free reeds; and that more recently he built as large a one for Boston Music Hall, without more free reeds; and again that Cavaillé-Coll quite recently built an organ for Mr. Hopwood of Kensington of 2,252 pipes and 40 stops, of which only one —the *Musette*—had free reeds.   He also says that Lewis, and probably most of the London organ-builders not previously mentioned, have never used the free reed.   The harshness of the

striking reed is obviated in the English method of voicing, according to Mr. H. Smith, by so curving and manipulating the metal tongue, that instead of coming with a discontinuous 'flap' from the fixed extremity down on to the slit of the tube, it 'rolls itself' down, and hence gradually covers the aperture.   The art of curving the tongue so as to produce this effect is very difficult to acquire; it is entirely empirical, and depends upon the keen eye and fine touch of the 'artist,' who notes lines and curves imperceptible to the uninitiated observer, and foresees their influence on the production of quality of tone.   Consequently, when an organ-builder has the misfortune to lose his 'reed-voicer,' he has always great difficulty in replacing him.—*Translator.*]

* [It should be observed that fig. 29, A, shews a *free* reed, and fig. 29, B, a *striking* reed; and that the tuning wire is right in fig. 29, B, because it presses the reed against the edges of its groove and hence shortens it, but it is wrong in fig. 29, A, for the reed being free would strike against the wire and rattle.   For free reeds a clip is used which grasps the reed on both sides and thus limits its vibrating length.

Fig. 28, p. 95*b*, shews the vibrator of an harmonium, not of an organ pipe.   The figures are the same as in all the German editions.— *Translator.*]

excursions are small, and the pressure of the lips brings it just near enough to make the chink sufficiently small without allowing it to strike.   For the oboe and bassoon two reeds or tongues of the same kind are placed opposite each other at the end of the mouthpiece.   They are separated by a narrow chink, and by blowing are pressed near enough to close the chink whenever they swing inwards.

3. *Membranous tongues.*—The peculiarities of these tongues are best studied on those artificially constructed.   Cut the end of a wooden or gutta-percha tube obliquely on both sides, as shewn in fig. 30, leaving two nearly rectangular points standing between the two edges which are cut obliquely. Then gently stretch strips of vulcanised india-rubber over the two oblique edges, so as to leave a small slit between them, and fasten them with a thread.   A reed mouthpiece is thus constructed ¶ which may be connected in any way with tubes or other air chambers.   When the membranes bend inwards the slit is closed; when outwards, it is open.   Membranes which are fastened in this oblique manner speak much better than those which are laid at right angles to the axis of the tube, as Johannes Müller proposed, for in the latter case they require to be bent outwards by the air before they can begin to open and shut alternately.   Membranous tongues of the kind proposed may be blown either in the direction of the arrows or in the opposite direction.   In the first case they open the slit when they move towards the air chamber, that is, towards the further end of the conducting tube.   Tongues of this kind I distinguish as *striking inwards*. When blown they always give deeper tones than they would do if allowed to vibrate freely, that is, without being connected with an air chamber.   The tongues of organ pipes, harmoniums, and wooden wind instruments already mentioned, ¶ are likewise always arranged to strike inwards.   But both membranous and metal tongues may be arranged so as to act against the stream of air, and hence to open when they move towards the outer opening of the instrument.   I then say that they *strike outwards*.   The tones of tongues which strike outwards are always sharper than those of isolated tongues.

FIG. 30.

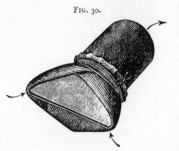

Only two kinds of membranous tongues have to be considered as musical instruments : the human *lips* in *brass instruments*, and the human *larynx* in *singing*.

The lips must be considered as very slightly elastic membranous tongues, loaded with much inelastic tissue containing water, and they would consequently vibrate very slowly, if they could be brought to vibrate by themselves.   In brass instruments they form membranous tongues which strike outwards, and consequently by the above rule produce tones sharper than their proper tones.   But as they offer very slight resistance, they are readily set in motion, by the alternate pressure of the vibrating column of air, when used with brass instruments.* ¶

* [Mr. D. J. Blaikley (manager of Messrs. Boosey & Co.'s Military Musical Instrument Manufactory, who has studied all such instruments theoretically as well as practically, and read many papers upon them, to some of which I shall have to refer) finds that this statement does not represent his own sensations when playing the horn. 'The lips,' he says, ' do not vibrate throughout their whole length, but only through a certain length determined by the diameter of the cup of the mouthpiece. Probably also the vibrating length can be modified by the mere pinch, at least this is the sensation I experience when sounding high notes on a large mouthpiece. The compass (about 4 octaves) possible on a given mouthpiece is much greater than that of any one register of the voice, and the whole range of brass instruments played thus with the lips is about one octave greater than the whole range of the human voice from basso profundo to the highest soprano. That the lips, acting as the vocal chords do, can themselves vibrate rapidly when supported by the rim of a mouthpiece, may be proved, for if such a rim, unconnected with any resonating tube, be held against the lips, various notes of the scale can be produced very faintly, the difficulty being to maintain steadiness of pitch (*Philos. Mag.*, Aug. 1878, p. 2). *The office of the air in the tube in relation to the lips* (leaving out of consideration its work as a resonant body, intensifying and modifying the tone) *is to act as a pendulum governor in facilitating the maintenance* (not the origination) *of a*

In the larynx, the elastic vocal chords act as membranous tongues. They are stretched across the windpipe, from front to back, like the india-rubber strips in fig. 30 (p. 97*a*), and leave a small slit, the glottis, between them. They have the advantage over all artificially constructed tongues of allowing the width of their slit, their tension, and even their form to be altered at pleasure with extraordinary rapidity and certainty, at the same time that the resonant tube formed by the opening of the mouth admits of much variety of form, so that many more qualities of tone can be thus produced than on any instrument of artificial construction. If the vocal chords are examined from above with a laryngoscope, while producing a tone, they will be seen to make very large vibrations for the deeper breast voice, shutting the glottis tightly whenever they strike inwards.

The pitch of the various reeds or tongues just mentioned is altered in very different manners. The metal tongues of the *organ* and *harmonium* are always ¶ intended to produce one single tone apiece. On the motion of these comparatively heavy and stiff tongues, the pressure of the vibrating air has very small influence, and their pitch within the instrument is consequently not much different from that of the isolated tongues. There must be at least one tongue for each note on such instruments.

In *wooden wind instruments*, a single tongue has to serve for the whole series of notes. But the tongues of these instruments are made of light elastic wood, which is easily set in motion by the alternating pressure of the vibrating column of air, and swings sympathetically with it. Such instruments, therefore, in addition to those very high tones, which nearly correspond to the proper tones of their tongues, can, as theory and experience alike shew, also produce deep tones of a very different pitch,* because the waves of air which arise in the tube of the instrument excite an alternation in the pressure of air adjacent to the tongue itself sufficiently powerful to make it vibrate sensibly. Now in a vibrating column of ¶ air the alteration of pressure is greatest where the velocity of the particles of air is smallest ; and since the velocity is always null, that is a minimum, at the end of a closed tube, such as a stopped organ pipe, and the alteration of pressure in that place is consequently a maximum, the tones of these reed pipes must be the same as those which the resonant tube alone would produce, if it were stopped at the place where the tongue is placed, and were blown as a stopped pipe. In musical practice, then, such tones of the instrument as correspond to the proper tones of the tongue are not used at all, because they are very high and screaming, and their pitch cannot be preserved with sufficient steadiness when the tongue is wet. The only tones produced are considerably deeper than the proper tone of the tongue, and have their pitches determined by the length of the column of air, which corresponds to the proper tones of the stopped pipe.

The *clarinet* has a cylindrical tube, the proper tones of which correspond to the third, fifth, seventh, &c., partial tone of the prime. By altering the style of ¶ blowing, it is possible to pass from the prime to the Twelfth or the higher major Third. The acoustic length of the tube may also be altered by opening the side

---

periodic vibration of the lips. Prof. Helmholtz does not say above what produces the alternate pressure, and I can conceive no source for it but a periodic vibration of the lips of a time suited to the *particular* note required.' The depth of the cup is also important :—' The shallower and more " cup-like " the cup,' says Mr. Blaikley, ' the greater the strength of the upper partials. Compare the deep and narrow cup of the French horn with weak upper partials, and the wide and shallow cup of the trumpet with strong upper partials.'—(MS. communications.) Mr. Blaikley kindly sounded for me the same instrument with different mouthpieces or cups, to shew the great difference of quality they produce. In the great bass bombardon on which he produced a tone of 40 vib., the tone was, even at that depth, remarkably rich and fine, owing to the large and deep cup extinguishing the beating upper partials. Mr. Blaikley also drew my attention to the fact that where the tube opens out into the cup, there must be no sharp shoulder, but that the edge must be carefully rounded off, otherwise there is a great loss of power to the blower. In the case of the French horn the cup is very long and almost tapers into the tube.—*Translator*.]

* See Helmholtz, *Verhandlungen des naturhistorischen medicinischen Vereins zu Heidelberg*, July 26, 1861, in the *Heidelberger Jahrbücher*. Poggendorff's *Annalen*, 1861. [Reproduced in part in App. VII. sect. B., I.]

holes of the clarinet, in which case the vibrating column of air is principally that between the mouthpiece and the uppermost open side hole.*

The *oboe* (hautbois) and *bassoon* (fagotto) have conical tubes which are closed up to the vertex of their cone, and have proper tones that are the same as those of open tubes of the same length. Hence the tones of both of these instruments nearly correspond to those of open pipes. By overblowing they give the Octave, Twelfth, second Octave, and so on, of the prime tone. Intermediate tones are produced by opening side holes.

The older *horns* and *trumpets* consist of long conical bent tubes, without keys or side holes.† They can produce such tones only as correspond to the proper tones of the tube, and these again are the natural harmonic upper partials of the prime. But as the prime tone of such a long tube is very deep, the upper partial tones in the middle parts of the scale lie rather close together, especially in the extremely long tubes of the horn,‡ so that they give most of the degrees of the scale. ¶

* [Mr. D. J. Blaikley obligingly furnished me with the substance of the following remarks on clarinets, and repeated his experiments before me in May 1884. The ordinary form of the clarinet is not wholly cylindrical. It is slightly constricted at the mouthpiece and provided with a spreading bell at the other end. The modification of form by key and finger holes also must not be neglected. On a cylindrical pipe played with the lips, the evenly numbered partials are quite inaudible. When a clarinet mouthpiece was added I found traces of the 4th and 6th partials beating with my forks. But on the clarinet with the bell, the 2nd, 4th, and 6th partials were distinct, and I could obtain beats from them with my forks. Mr. Blaikley brought them out (1) by bead and diaphragm resonators tuned to them (fig. 15, p. 42*a*), which I also witnessed, (2) by an irregularly-shaped tubular resonator sunk gradually in water, on which I also heard them, (3) by beats with an harmonium with a constant blast, which I also heard. On the cylindrical tube all the unevenly numbered partials are in tune when played as primes of independent harmonic notes. On the clarinet only the 3rd partial, or 2nd proper tone, can be used as the prime of an independent harmonic tone. The 3rd, 4th, and 5th proper tones of the instrument, are sufficiently near in pitch to the 5th, 7th, and 9th partials of the fundamental tone for these latter to be greatly strengthened by resonance, but the agreement is not close enough to allow of the higher proper tones being used as the primes of independent harmonic compound tones. Hence practically only the 3rd harmonics, or Twelfths, are used on the clarinet. The following table of the relative intensity of the ¶ partials of a *B*♭ clarinet was given by Mr. Blaikley in the *Proc. of the Mus. Assn.* for 1877-8, p. 84:—

PARTIALS—B♭ CLARINETS.

Pitch	1	2	3	4	5	6	7	8, &c.
f′	*f*	Just discernible	*f*	*p*	*mf*	*p*	...	...
b♭	*f*		*f*	*p*	*mf*	...	*mf*	*pp*
a	*f*		*f*	*p*	*mf*	...	*mf*	*pp*
g	*f*		*f*	...	*mf*	*mf*	*p*	*pp*
f	*f*		*f*	...	*mf*	*p*	*mf*	*pp*
e♭	*f*		*mf*	...	*p*	*mf*	*mf*	*pp*
d	*f*		*mf*	*p*	*mf*	*p*	*p*	*pp*

Where *f* means forte, *mf* mezzoforte, *p* piano, *pp* pianissimo.—*Translator*.]

† [Such brass tubes are first worked unbent from cylindrical brass tubes, by putting solid steel cores of the required form inside, and then drawing them through a hole in a piece of lead, which yields enough for the tube to pass through, but presses the brass firmly enough against the core to make the tube assume the proper form. Afterwards the tube is filled with lead, and then bent into the required coils, after which the lead is melted out. The instruments are also not conical in the strict sense of the word, but 'approximate in form to the hyperbolic cone, where the axis of the instrument is an asymptote, and the vertex is at a great or even an infinite distance from the bell end.' From information furnished by Mr. Blaikley.—*Translator*.]

‡ The tube of the *Waldhorn* [foresthorn,

hunting horn of the Germans, answering to our French horn] is, according to Zamminer [p. 312], 13·4 feet long. Its proper prime tone ¶ is *E*,♭. This and the next *E*♭ are not used, but only the other tones, *B*♭, *e*♭, *g*, *b*♭, *d′*♭−, *e′*♭, *f′*, *g′*, *a′*♭+, *b′*♭, &c. [Mr. Blaikley kindly sounded for me the harmonics 8, 9, 10, 11, 12, 13, 14 on an *E*,♭ French horn. The result was almost precisely 320, 360, 400, 440, 480, 520, 560 vib., that is the exact harmonics for the prime tone 40 vib. to which it was tuned, the pitch of English military musical instruments being as nearly as possible *c′* 269, *e′*♭ 319·9, *a′* 452·4. This scale was not completed because the 15th and 16th harmonics 600 and 640 vib. would have been too high for me to measure. Expressed in cents we may compare this scale with just intonation thus:—

Notes	.	.	*e′*♭	*f′*	*g′*	*a′*♭	*b′*♭	*c″*	*d″*♭	*d″*	*e″*♭
Just cents	.	0,	204,	386,	498,	702,	884,	996,	1088,	1200	
Harmonic cents	.	0,	204,	386,	551,	702,	841,	969,	1088,	1200	
Harmonics, No.	.	8,	9,	10,	11,	12,	13,	14,	15,	16.	

The trumpet is restricted to these natural tones. But by introducing the hand into the bell of the French horn and thus partly closing it, and by lengthening the tube of the trombone,* it was possible in some degree to supply the missing tones and improve the faulty ones. In later times trumpets and horns have been frequently supplied with keys † to supply the missing tones, but at some expense of power in the tone and the brilliancy in its quality. The vibrations of the air in these instruments are unusually powerful, and require the resistance of firm, smooth, unbroken tubes to preserve their strength. In the use of brass instruments, the different form and tension of the lips of the player act only to determine which of the proper tones of the tube shall speak; the pitch of the individual tones is almost ‡ entirely independent of the tension of the lips.

¶ On the other hand, in the *larynx* the tension of the vocal chords, which here form the membranous tongues, is itself variable, and determines the pitch of the tone. The air chambers connected with the larynx are not adapted for materially altering the tone of the vocal chords. Their walls are so yielding that they cannot allow the formation of vibrations of the air within them sufficiently powerful to force the vocal chords to oscillate with a period which is different from that required by their own elasticity. The cavity of the mouth is also far too short, and generally too widely open for its mass of air to have material influence on the pitch.

In addition to the tension of the vocal chords (which can be increased not only by separating the points of their insertion in the cartilages of the larynx, but also by voluntarily stretching the muscular fibres within them), their thickness seems also to be variable. Much soft watery inelastic tissue lies underneath the elastic fibrils proper and the muscular fibres of the vocal chords, and in the breast voice this probably acts to weight them and retard their vibrations. The head voice is probably produced by drawing aside the mucous coat below the chords, ¶ thus rendering the edge of the chords sharper, and the weight of the vibrating part less, while the elasticity is unaltered.§

Hence the Fourth $a'b$ was 53 cents (33 : 32) too sharp, and the Sixth c'' was 43 cents (40 : 39) too flat, and they were consequently unusable without modification by the hand. The minor Seventh $d''b$ was too flat by 27 cents (64 : 63), but unless played in (intended) unison against the just form, it produces a better effect. 'In trumpets, strictly so called,' says Mr. Blaikley, 'a great portion of the length is cylindrical and the bell curves out hyperbolically, the two lowest partials are not required as a rule and are not strictly in tune, so the series of partials may be taken as about ·75, 1·90, 3, 4, 5, 6, 7, 8, &c., all the upper notes being brought into tune by modifications in the form of the bell in a good instru- ¶ ment.' The length of the French horn varies with the 'crook' which determines its pitch. The following contains the length in English inches for each crook, as given by Mr. Blaikley : Bb (alto) 108, $A♮$ 114½, Ab 121½, G 128¾, F 144½, $E♮$ 153, Eb 162, $D♮$ 171½, C 192¾, Bb (basso) 216¾, hence the length varies from 9 ft. to 18 ft. ¾ inch. By a curious error in all the German editions, Zamminer is said to make the length of the Eb Waldhorn 27 feet, or the length of the wave of the *lowest* note, in place of his 13·4 feet. Zamminer, however, says that the instrument is named from the *Octave above* the lowest note, and that hence the wave-length of *this Octave* is the length of the horn.—*Translator.*]

* [A large portion of the trombone is composed of a double narrow cylindrical tube on which another slides, so that the length of the

trombone can be altered at will, and chosen to make its harmonics produce a just scale. Some trumpets also are made with a short slide worked by two fingers one way, and returning to its position by a spring. Such instruments are sometimes used by first-rate players, such as Harper, the late celebrated trumpeter, and his son. But, as Mr. Blaikley informed me, an extremely small percentage of the trumpets sold have slides. At present the piston brass instruments have nearly driven all slides, except the trombone, out of the field. —*Translator.*]

† [The keys are nearly obsolete, and have been replaced by pistons which open valves, and thus temporarily increase the length of the tube, so as to make the note blown 1, 2, or 3 Semitones flatter. These can also be used in combination, but are then not so true. This is tantamount to an imperfect slide action. Instruments of this kind are now much used in all military bands, and are made of very different sizes and pitches.— *Translator.*]

‡ [But by no means 'quite.' It is possible to blow out of tune, and to a small extent temper the harmonics.—*Translator.*]

§ [On the subject of the registers of the human voice and its production generally, see Lennox Browne and Emil Behnke, *Voice, Song, and Speech* (Sampson Low, London, 1883, pp. 322). This work contains not merely accurate drawings of the larynx in the different registers, but 4 laryngoscopic photographs from Mr. Behnke's own larynx. A *register*

We now proceed to investigate the *quality of tone* produced on reed pipes, which is our proper subject. The sound in these pipes is excited by intermittent pulses of air, which at each swing break through the opening that is closed by the tongue of the reed. A freely vibrating tongue has far too small a surface to communicate any appreciable quantity of sonorous motion to the surrounding air ; and it is as little able to excite the air inclosed in pipes. The sound seems to be really produced by pulses of air, as in the siren, where the metal plate that opens and closes the orifice does not vibrate at all. By the alternate opening and closing of a passage, a continuous influx of air is changed into a periodic motion, capable of affecting the air. Like any other periodic motion of the air, the one thus produced can also be resolved into a series of simple vibrations. We have already remarked that the number of terms in such a series will increase with the discontinuity of the motion to be thus resolved (p. 34*d*). Now the motion of the air which passes through a siren, or past a vibrating tongue, is discontinuous in a very high ¶ degree, since the individual pulses of air must be generally separated by complete pauses during the closures of the opening. Free tongues without a resonance tube, in which all the individual simple tones of the vibration which they excite in the air are given off freely to the surrounding atmosphere, have consequently always a very sharp, cutting, jarring quality of tone, and we can really hear with either armed or unarmed ears a long series of strong and clear partial tones up to the 16th or 20th, and there are evidently still higher partials present, although it is difficult or impossible to distinguish them from each other, because they do not lie so much as a Semitone apart.* This whirring of dissonant partial tones makes the musical quality of free tongues very disagreeable.† A tone thus produced also shews that it is really due to puffs of air. I have examined the vibrating tongue of a reed pipe, like that in fig. 28 (p. 95*b*), when in action with the vibration microscope of Lissajous, in order to determine the vibrational form of the tongue, and I found that the tongue performed perfectly regular simple vibra- ¶ tions. Hence it would communicate to the air merely a simple tone and not a compound tone, if the sound were directly produced by its own vibrations.

The intensity of the upper partial tones of a free tongue, unconnected with a resonance tube, and their relation to the prime, are greatly dependent on the

---

is defined as ' a series of tones produced by the same mechanism ' (p. 163). The names of the registers adopted are those introduced by the late John Curwen of the Tonic Sol-fa movement. They depend on the appearance of the glottis and vocal chords, and are as follows :

1. Lower thick, 2. Upper thick (both ' chest voice '), 3. Lower thin (' high chest ' voice in men), 4. Upper thin (' falsetto ' in women), 5. Small (' head voice ' in women). The extent of the registers are stated to be (p. 171)

	1. lower thick.	2. upper thick.	3. lower thin.
MEN	E to a,	b to f',	g' to c''
WOMEN	e to c',	d' to f',	g' to c''

	1. lower thick.	2. upper thick.	3. lower thin.
WOMEN ONLY,		d'' to f'',	g'' to f''
		4. upper thin.	5. small.

The mechanism is as follows (pp. 163-171) :—

1. Upper thick. The hindmost points of the pyramids (arytenoid cartilages) close together, an elliptical slit between the vocal ligaments (or chords), which vibrate through their whole length, breadth, and thickness fully, loosely, and visibly. The lid (epiglottis) is low.

2. Upper thick. The elliptical chink disappears and becomes linear. The lid (epiglottis) rises ; the vocal ligaments are stretched.

3. Lower thin. The lid (epiglottis) is more raised, so as to shew the cushion below it, the whole larynx and the insertions of the vocal ligaments in the shield (thyroid) cartilage. The vocal ligaments are quite still, and their vibrations are confined to the thin inner edges. The vocal ligaments are made *thinner* and transparent, as shown by illumination from below. Male voices cease here.

4. Upper thin. An elliptical slit again forms between the vocal ligaments. When this is used by men it gives the *falsetto* arising from the *upper* thin being carried *below* its true place. This slit is gradually reduced in size as the contralto and soprano voices ascend. ¶

5. Small. The back part of the glottis contracts for at least two-thirds of its length, the vocal ligaments being pressed together so tightly that scarcely any trace of a slit remains, and no vibrations are visible. The front part opens as an oval chink, and the edges of this vibrate so markedly that the outline is blurred. The drawings of the two last registers (pp. 168–169) were made from laryngoscopic examination of a lady.

Reference should be made to the book itself for full explanations, and the reader should especially consult Mr. Behnke's admirable little work *The Mechanism of the Human Voice* (Curwen, 3rd ed., 1881, pp. 125).—*Translator.*]

* [See footnote † p. 56*d'*.—*Translator.*]

† [The cheap little mouth harmonicons exhibit this effect very well.—*Translator.*]

nature of the tongue, its position with respect to its frame, the tightness with which it closes, &c. Striking tongues which produce the most discontinuous pulses of air, also produce the most cutting quality of tone.* The shorter the puff of air, and hence the more sudden its action, the greater number of high upper partials should we expect, exactly as we find in the siren, according to Seebeck's investigations. Hard, unyielding material, like that of brass tongues, will produce pulses of air which are much more disconnected than those formed by soft and yielding substances. This is probably the reason why the singing tones of the human voice are softer than all others which are produced by reed pipes. Nevertheless the number of upper partial tones in the human voice, when used in emphatic *forte*, is very great, and they reach distinctly and powerfully up to the four-times accented [or quarter-foot] Octave (p. 26a). To this we shall have to return.

¶ The tones of tongues are essentially changed by the addition of resonance tubes, because they reinforce and hence give prominence to those upper partial tones which correspond to the proper tones of these tubes.† In this case the resonance tubes must be considered as closed at the point where the tongue is inserted.‡

A brass tongue such as is used in organs, and tuned to *b♭*, was applied to one of my larger spherical resonators, also tuned to *b♭*, instead of to its usual resonance tube. After considerably increasing the pressure of wind in the bellows, the tongue spoke somewhat flatter than usual, but with an extraordinarily full, beautiful, soft tone, from which almost all upper partials were absent. Very little wind was used, but it was under high pressure. In this case the prime tone of the compound was in unison with the resonator, which gave a powerful resonance, and consequently the prime tone had also great power. None of the higher partial tones could be reinforced. The theory of the vibrations of air in the sphere further ¶ shews that the greatest pressure must occur in the sphere at the moment that the tongue opens. Hence arose the necessity of strong pressure in the bellows to overcome the increased pressure in the sphere, and yet not much wind really passed.

If instead of a glass sphere, resonant tubes are employed, which admit of a greater number of proper tones, the resulting musical tones are more complex. In the clarinet we have a cylindrical tube which by its resonance reinforces the uneven partial tones.§ The conical tubes of the oboe, bassoon, trumpet, and French horn, on the other hand, reinforce all the harmonic upper partial tones of the compound up to a certain height, determined by the incapacity of the tubes to resound for waves of sound that are not much longer than the width of the opening. By actual trial I found only unevenly numbered partial tones, distinct to the seventh inclusive, in the notes of the clarinet,§ whereas on other instruments, which have conical tubes, I found the evenly numbered partials also. I have not yet had an opportunity of making observations on the further differences of quality in ¶ the tones of individual instruments with conical tubes. This opens rather a wide field for research, since the quality of tone is altered in many ways by the style of blowing, and even on the same instrument the different parts of the scale, when they require the opening of side holes, shew considerable differences in quality. On wooden wind instruments these differences are striking. The opening of side holes is by no means a complete substitute for shortening the tube, and the reflection of the waves of sound at the points of opening is not the same as at the free open end of the tube. The upper partials of compound tones produced by a tube limited by an open side hole, must certainly be in general materially deficient in harmonic purity, and this will also have a marked influence on their resonance.**

* [But see footnote † p. 95*d'*.—*Translator*.]

† [A line has been here cancelled in the translation which had been accidentally left standing in the German, as it refers to a remark on the passage which formerly followed p. 89, l. 2, but was cancelled in the 4th German edition.—*Translator*.]

‡ See Appendix VII.

§ [But see note * p. 99*b*.—*Translator*.]

** [The theory of side holes is excessively complicated and has not been as yet worked

7. *Vowel Qualities of Tone.*

We have hitherto discussed cases of resonance, generated in such air chambers as were capable of reinforcing the prime tone principally, but also a certain number of the harmonic upper partial tones of the compound tone produced. The case, however, may also occur in which the lowest tone of the resonance chamber applied does not correspond with the prime, but only with some one of the upper partials of the compound tone itself, and in these cases we find, in accordance with the principles hitherto developed, that the corresponding upper partial tone is really more reinforced than the prime or other partials by the resonance of the chamber, and consequently predominates extremely over all the other partials in the series. The quality of tone thus produced has consequently a peculiar character, and more or less resembles one of the vowels of the human voice. For the vowels of speech are in reality tones produced by membranous tongues (the vocal ¶ chords), with a resonance chamber (the mouth) capable of altering in length, width, and pitch of resonance, and hence capable also of reinforcing at different times different partials of the compound tone to which it is applied.*

In order to understand the composition of vowel tones, we must in the first place bear in mind that the source of their sound lies in the vocal chords, and that when the voice is heard, these chords act as membranous tongues, and like all tongues produce a series of decidedly discontinuous and sharply separated pulses of air, which, on being represented as a sum of simple vibrations, must consist of a very large number of them, and hence be received by the ear as a very long series of partials belonging to a compound musical tone. With the assistance of resonators it is possible to recognise very high partials, up to the sixteenth, when one of the brighter vowels is sung by a powerful bass voice at a low pitch, and, in the case of a strained forte in the upper notes of any human voice, we can hear, more clearly than on any other musical instrument, those high upper partials ¶ that belong to the middle of the four-times accented Octave (the highest on modern pianofortes, see note, p. 18*d*), and these high tones have a peculiar relation to the ear, to be subsequently considered. The loudness of such upper partials, especially those of highest pitch, differs considerably in different individuals. For cutting bright voices it is greater than for soft and dull ones. The quality of tone in cutting screaming voices may perhaps be referred to a want of sufficient smoothness or straightness in the edges of the vocal chords, to enable them to close in a straight narrow slit without striking one another. This circumstance would give the larynx more the character of striking tongues, and the latter have a much more cutting quality than the free tongues of the normal vocal chords. Hoarseness in voices may arise from the glottis not entirely closing during the vibrations of the vocal chords. At any rate, when alterations of this kind are made in artificial membranous tongues, similar results ensue. For a strong and yet soft quality of voice it is necessary that the vocal chords should, even when ¶ most strongly vibrating, join rectilinearly at the moment of approach with perfect tightness, effectually closing the glottis for the moment, but without overlapping

out scientifically. ‘The general principles,’ writes Mr. Blaikley, ‘are not difficult of comprehension ; the difficulty is to determine quantitatively the values in each particular case.’ The paper by Schafhäutl (writing under the name of Pellisov), ‘Theorie gedeckter cylindrischer und conischer Pfeifen und der Querflöten,’ Schweiger, *Journ.* lxviii. 1833, is disfigured by misprints so that the formulæ are unintelligible, and the theory is also extremely hazardous. But they are the only papers I have found, and are referred to by Theobald Boehm, *Ueber den Flötenbau*, Mainz, 1847. An English version of this, by himself, made for Mr. Rudall in 1847, has recently been edited with additional letters by W. S. Broadwood, and published by Rudall, Carte, & Co., makers of his flutes. See also Victor Mahillon, *Étude sur le doigté de la Flûte Boehm*, 1882, and a paper by M. Aristide Cavaillé-Coll, in *L'Écho Musical* for 11 Jan. 1883 —*Translator*.]

* The theory of vowel tones was first enunciated by Wheatstone in a criticism, unfortunately little known, on Willis's experiments. The latter are described in the *Transactions of the Cambridge Philosophical Society*, vol. iii. p. 231, and Poggendorff's *Annalen der Physik*, vol. xxiv. p. 397. Wheatstone's report upon them is contained in the *London and Westminster Review* for October 1837.

or striking against each other. If they do not close perfectly, the stream of air will not be completely interrupted, and the tone cannot be powerful. If they overlap, the tone must be cutting, as before remarked, as those arising from striking tongues. On examining the vocal chords in action by means of a laryngoscope, it is marvellous to observe the accuracy with which they close even when making vibrations occupying nearly the entire breadth of the chords themselves.*

There is also a certain difference in the way of putting on the voice in speaking and in singing, which gives the speaking voice a much more cutting quality of tone, especially in the open vowels, and occasions a sensation of much greater pressure in the larynx. I suspect that in speaking the vocal chords act as striking tongues.†

¶ When the mucous membrane of the larynx is affected with catarrh, the laryngoscope sometimes shews little flakes of mucus in the glottis. When these are too great they disturb the motion of the vibrating chords and make them irregular, causing the tone to become unequal, jarring, or hoarse. It is, however, remarkable what comparatively large flakes of mucus may lie in the glottis without producing a very striking deterioration in the quality of tone.

It has already been mentioned that it is generally more difficult for the unassisted ear to recognise the upper partials in the human voice, than in the tones of musical instruments. Resonators are more necessary for this examination than for the analysis of any other kind of musical tone. The upper partials of the human voice have nevertheless been heard at times by attentive observers. Rameau had heard them at the beginning of last century. And at a later period Seiler of Leipzig relates that while listening to the chant of the watchman during a sleepless night, he occasionally heard at first, when the watchman was at a distance, the Twelfth of the melody, and afterwards the prime tone. The reason of this difficulty ¶ is most probably that we have all our lives remarked and observed the tones of the human voice more than any other, and always with the sole object of grasping it as a whole and obtaining a clear knowledge and perception of its manifold changes of quality.

We may certainly assume that in the tones of the human larynx, as in all other reed instruments, the upper partial tones would decrease in force as they increase in pitch, if they could be observed without the resonance of the cavity of the mouth. In reality they satisfy this assumption tolerably well, for those vowels which are spoken with a wide funnel-shaped cavity of the mouth, as A [*a* in *art*], or Ä [*a* in *bat* lengthened, which is nearly the same as *a* in *bare*]. But this relation is materially altered by the resonance which takes place in the cavity of the mouth. The more this cavity is narrowed, either by the lips or the tongue, the more distinctly marked is its resonance for tones of determinate pitch, and the more therefore does this resonance reinforce those partials in the compound tone produced by ¶ the vocal chords, which approach the favoured pitch, and the more, on the contrary, will the others be damped. Hence on investigating the compound tones of the human voice by means of resonators, we find pretty uniformly that the first six to eight partials are clearly perceptible, but with very different degrees of force according to the different forms of the cavity of the mouth, sometimes screaming loudly into the ear, at others scarcely audible.

Under these circumstances the investigation of the resonance of the cavity of the mouth is of great importance. The easiest and surest method of finding the tones to which the air in the oral cavity is tuned for the different shapes it assumes

* [Probably these observations were made on the 'upper thick' register, because the chords are then more visible. It is evident that these theories do not apply to the lower thick, upper thin, and small registers, and scarcely to the lower thin, as described above, footnote p. 101c.—*Translator.*]

† [The German habit of beginning open vowels with the 'check' or Arabic hamza, which is very marked, and instantly characterises his nationality, is probably what is here alluded to, as occasioning a sensation of much greater pressure. This does not apply in the least to English speakers.—*Translator.*]

in the production of vowels, is that which is used for glass bottles and other spaces filled with air. That is, tuning-forks of different pitches have to be struck and held before the opening of the air chamber—in the present case the open mouth —and the louder the proper tone of the fork is heard, the nearer does it correspond with one of the proper tones of the included mass of air.* Since the shape of the oral cavity can be altered at pleasure, it can always be made to suit the tone of any given tuning-fork, and we thus easily discover what shape the mouth must assume for its included mass of air to be tuned to a determinate pitch.

Having a series of tuning-forks at command, I was thus able to obtain the following results :—

The pitch of strongest resonance of the oral cavity depends solely upon the vowel for pronouncing which the mouth has been arranged, and alters considerably for even slight alterations in the vowel quality, such, for example, as occur in the different dialects of the same language. On the other hand, the proper tones of ¶ the cavity of the mouth are nearly independent of age and sex. I have in general found the same resonances in men, women, and children. The want of space in the oral cavity of women and children can be easily replaced by a great closure of its opening, which will make the resonance as deep as in the larger oral cavities of men.†

The vowels can be arranged in three series, according to the position of the parts of the mouth, which may be written thus, in accordance with Du Bois-Reymond the elder ‡ :—

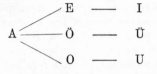

The vowel A [*a* in *father*, or Scotch *a* in *man*] forms the common origin of all three series. With this vowel corresponds a funnel-shaped resonance cavity, ¶

---

* [See note * p. 87*b*, on determining violin resonance. One difficulty in the case of the mouth is that there is a constant tendency to vary the shape of the oral cavity. Another, as shewn at the end of the note cited, is that the same irregular cavity, such as that of the mouth, often more or less reinforces a large number of different tones. As it was important for my phonetic researches, I have made many attempts to determine my own vowel resonances, but have hitherto failed in all my attempts.—*Translator.*]

† [Easily tried by more or less covering the top of a tumbler with the hand, till it resounds to any fork from *c'* to *d''* or higher. —*Translator.*]

‡ *Norddeutsche Zeitschrift*, edited by de la Motte Fouqué, 1812. *Kadmus oder allgemeine Alphabetik*, von F. H. du Bois-Reymond, Berlin, 1862, p. 152. [This is the arrangement usually adopted. But in 1867 Mr. Melville Bell, an orthoepical teacher of many years' standing, who had been led professionally to pay great attention to the shapes of the mouth necessary to produce certain sounds, in his *Visible Speech; the Science of Universal Alphabetics* (London: Simpkin, Marshall & Co., 4to., pp. x. 126, with sixteen lithographic tables), proposed a more elaborate method of classifying vowels by the shape of the mouth. He commenced with 9 positions of the tongue, consisting of 3 in which the middle, or as he terms it, ' front ' of the tongue was raised, highest for *ea* in *seat*, not so high for *a* in *sate*, and lowest for *a* in *sat*; 3 others in which the *back*, instead of the middle, of the tongue

was raised, highest for *oo* in *snood*, lower for *o* in *node*, and lowest for *aw* in *gnawed* (none of which three are determined by the position of the tongue alone), and 3 intermediate positions, where the whole tongue is raised almost evenly at three different elevations. These 9 lingual positions might be accompanied with the ordinary or with increased distension of the pharynx, giving 9 primary and 9 'wide' vowels. And each of the 18 vowels, thus produced, could be 'rounded,' that is, modified by shading the mouth in various degrees with the lips. He thus obtains 36 distinct vowel cavities, among which almost all those used for vowel qualities in different nations may be placed. Subsequent research has shewn how to extend this arrangement materially. See ¶ my *Early English Pronunciation*, part iv., 1874, p. 1279. Also see generally my *Pronunciation for Singers* (Curwen, 1877, pp. 246) and *Speech in Song* (Novello, 1878, pp. 140). German vowels differ materially in quality from the English, and consequently complete agreement between Prof. Helmholtz's observations and those of any Englishman, who repeats his experiments, must not be expected. I have consequently thought it better in this place to leave his German notation untranslated, and merely subjoin in parentheses the nearest English sounds. For the table in the text we may assume A to = *a* in *father*, or else Scotch *a* in *man* (different sounds), E to = *e* in *there*, I to = *i* in *machine*, O to = *o* in *more*, U to = *u* in *sure*; and Ö to = *eu* in French *peu* or else in *peuple* (different sounds), and Ü to = *u* in French *pu*. —*Translator.*]

enlarging with tolerable uniformity from the larynx to the lips. For the vowels of the lower series, O [o in *more*] and U [oo in *poor*], the opening of the mouth is contracted by means of the lips, more for U than for O, while the cavity is enlarged as much as possible by depression of the tongue, so that on the whole it becomes like a bottle without a neck, with rather a narrow mouth, and a single unbroken cavity.* The pitch of such a bottle-shaped chamber is lower the larger its cavity and the narrower its mouth. Usually only one upper partial with strong resonance can be clearly recognised; when other proper tones exist they are comparatively very high, or have only weak resonance. In conformity with these results, obtained with glass bottles, we find that for a very deep hollow U [oo in *poor* nearly], where the oral cavity is widest and the mouth narrowest, the resonance is deepest and answers to the unaccented *f*. On passing from U to O [o in *more* nearly] the resonance gradually rises; and for a full, ringing, pure O the pitch is *b′♭*. The ¶ position of the mouth for O is peculiarly favourable for resonance, the opening of the mouth being neither too large nor too small, and the internal cavity sufficiently spacious. Hence if a *b′♭* tuning-fork be struck and held before the mouth while O is gently uttered, or the O-position merely assumed without really speaking, the tone of the fork will resound so fully and loudly that a large audience can hear it. The usual *a′* tuning-fork of musicians may also be used for this purpose, but then it will be necessary to make a somewhat duller O, if we wish to bring out the full resonance.

On gradually bringing the shape of the mouth from the position proper to O, through those due to Oª [nearly *o* in *cot*, with rather more of the O sound], and A° [nearly *au* in *caught*, with rather more of the A sound] into that for A [Scotch *a* in *man*, with rather more of an O quality in it than English *a* in *father*], the resonance gradually rises an Octave, and reaches *b″♭*. This tone corresponds with the North German A; the somewhat brighter A [*a* in *father*] of the English and Italians rises up to *d‴*, or a major Third higher. It is particularly remarkable what ¶ little differences in pitch correspond to very sensible varieties of vowel quality in the neighbourhood of A; and I should therefore recommend philologists who wish to define the vowels of different languages to fix them by the pitch of loudest resonance.†

For the vowels already mentioned I have not been able to detect any second proper tone, and the analogy of the phenomena presented by artificial resonance chambers of similar shapes would hardly lead us to expect any of sensible loudness.

---

* [This depressed position of the tongue answers better for English *aw* in *saw* than for either *o* in *more* or *oo* in *poor*. For the *o* in *more* the tongue is slightly more raised, especially at the back, while for English *oo* the back of the tongue is almost as high as for *k*, and greatly impedes the oral cavity. If, however, the tongue be kept in the position for *aw* by sound- ¶ ing this vowel, and, while sounding it steadily, the lips be gradually contracted, the sound will be found to pass through certain obscure qualities of tone till it suddenly comes out clearly as a sound a little more like *aw* than *o* in *more* (really the Danish *aa*), and then again passing through other obscure phases, comes out again clearly as a deep sound, not so bright as our *oo* in *poor*, but more resembling the Swedish *o* to which it will reach if the tongue be slightly raised into the A position. It is necessary to bear these facts in mind when following the text, where U is only almost, not quite = *oo* in *poor*, which is the long sound of *u* in *pull*, and is duller than *oo* in *pool* or French *ou* in *poule*.—*Translator*.]

† [Great difficulties lie in the way of carry- ing out this recommendation. The ear of philologists and even of those who are readily able to discriminate vowel sounds, is frequently not acute for differences of pitch. The deter- mination of the pitch even under favour- able circumstances is not easy, especially, as it will be seen, for the higher pitches. Without mechanical appliances even good ears are deceived in the Octave. The differences of pitch noted by Helmholtz, Donders, Merkel, and Koenig, as given on p. 109*d*, probably point to fundamental differences of pronunciation, and shew the desirability of a very extensive series of experiments being carried out with special apparatus, by an operator with an extremely acute musical ear, on speakers of various nationalities and also on various speakers of the same nationality. Great diffi- culty will even then be experienced on account of the variability of the same speaker in his vowel quality for differences of pitch and expression, the want of habit to maintain the position of the mouth unmoved for a sufficient length of time to complete an observation satisfactorily, and, worst of all, the involuntary tendency of the organs to accommodate them- selves to the pitch of the fork presented. Com- pare note * p. 105*c*.—*Translator*.]

Experiments hereafter described shew that the resonance of this single tone is sufficient to characterise the vowels above mentioned.

The second series of vowels consists of A, Ä, E, I. The lips are drawn so far apart that they no longer contract the issuing stream of air, but a fresh constriction is formed between the front (middle) parts of the tongue and the hard palate, the space immediately above the larynx being widened by depressing the root of the tongue, and hence causing the larynx to rise simultaneously. The form of the oral cavity consequently resembles a bottle with a narrow neck. The belly of the bottle is behind, in the pharynx, and its neck is the narrow passage between the upper surface of the tongue and the hard palate. In the above series of letters, Ä, E, I, these changes increase until for I the internal cavity of the bottle is greatest and the neck narrowest. For Ä [the broadest French ê, broader than e in there, and nearly as broad as a in bat lengthened, with which the name of their city is pronounced by the natives of Bath], the whole channel is, however, tolerably wide, ¶ so that it is quite easy to see down to the larynx when the laryngoscope is used. Indeed this vowel gives the very best position of the mouth for the application of this instrument, because the root of the tongue, which impedes the view when A is uttered, is depressed, and the observer can see over and past it.

When a bottle with a long narrow neck is used as a resonance chamber, two simple tones are readily discovered, of which one can be regarded as the proper tone of the belly, and the other as that of the neck of the bottle. Of course the air in the belly cannot vibrate quite independently of that in the neck, and both proper tones in question must consequently be different, and indeed somewhat deeper than they would be if belly and neck were separate and had their resonance examined independently. The neck is approximately a short pipe open at both ends. To be sure, its inner end debouches into the cavity of the bottle instead or the open air, but if the neck is very narrow, and the belly of the bottle very wide, the latter may be looked upon in some respect as an open space with regard to the ¶ vibrations of the air inclosed in the neck. These conditions are best satisfied for I, in which the length of the channel between tongue and palate, measured from the upper teeth to the back edge of the bony palate, is about 6 centimetres [2·36 inches]. An open pipe of this length when blown would give e'''', while the observations made for determining the tone of loudest resonance for I gives nearly d'''', which is as close an agreement as could possibly have been expected in such an irregularly shaped pipe as that formed by the tongue and palate.

In accordance with these experiments the vowels Ä, E, I have each a higher and a deeper resonance tone. The higher tones continue the ascending series of the proper tones of the vowels U, O, A. By means of tuning-forks I found for Ä a tone between g''' and $a'''\flat$, and for E the tone $b'''\flat$. I had no fork suitable for I, but by means of the whistling noise of the air, to be considered presently (p. 108b), its proper tone was determined with tolerable exactness to be d''''.

The deeper proper tones which are due to the back part of the oral cavity are ¶ rather more difficult to discover. Tuning-forks may be used, but the resonance is comparatively weak, because it must be conducted through the long narrow neck of the air chamber. It must further be remembered that this resonance only occurs during the time that the corresponding vowel is gently whispered, and disappears as soon as the whisper ceases, because the form of the chamber on which the resonance depends then immediately changes. The tuning-forks after being struck must be brought as close as possible to the opening of the air chamber which lies behind the upper teeth. By this means I found d'' for Ä and f' for E. For I, direct observation with tuning-forks was not possible; but from the upper partial tones, 1 conclude that its proper tone is as deep as that of U, or near f. Hence, when we pass from A to I, these deeper proper tones of the oral cavity sink, and the higher ones rise in pitch.*

---

* [Mr. Graham Bell, the inventor of the Telephone, son of the Mr. Melville Bell already mentioned (p. 105d, note), was in the habit of bringing out this fact by placing his mouth in

For the third series of vowels from A through Ö [French *eu* in *peu*, or the deeper *eu* in *peuple*], towards Ü [French *u* in *pu*, which is rather deeper than the German sound], we have the same internal positions of the mouth as in the last-named series of vowels.   For Ü the mouth is placed in nearly the same position as for a vowel lying between E and I, and for Ö as for an E which inclines towards Ä.   In addition to the constriction between the tongue and palate as in the second series, we have also a constriction of the lips, which are made into a sort of tube, forming a front prolongation of that made by the tongue and palate.   The air chamber of the mouth, therefore, in this case also resembles a bottle with a neck, but the neck is longer than for the second series of vowels.   For I the neck was 6 centimetres (2·36 inches) long, for Ü, measured from the front edge of the upper teeth to the commencement of the soft palate, it is 8 centimetres (3·15 inches). The pitch of the higher proper tone corresponding to the resonance of the neck
¶ must be, therefore, about a Fourth deeper than for I.   If both ends were free, a pipe of this length would give *b‴*, according to the usual calculation.   In reality it resounded for a fork lying between *g‴* and *a‴♭*, a divergence similar to that found for I, and also probably attributable to the back end of the tube debouching into a wider but not quite open space.   The resonance of the back space has to be observed in the same way as for the I series.   For Ö it is *f′*, the same as for E, and for Ü it is *f*, the same as for I.

The fact that the cavity of the mouth for different vowels is tuned to different pitches was first discovered by Donders,* not with the help of tuning-forks, but by the whistling noise produced in the mouth by whispering.   The cavity of the mouth thus reinforces by its resonance the corresponding tones of the windrush, which are produced partly in the contracted glottis,† and partly in the forward contracted passages of the mouth.   In this way it is not usual to obtain a complete musical tone ; this only happens, without sensible change of the vowel, for Ü and
¶ U, when a real whistle is produced.   This, however, would be a fault in speaking. We have rather only such a degree of reinforcement of the noise of the air as occurs in an organ pipe, which does not speak well, either from a badly-constructed lip or an insufficient pressure of wind.   A noise of this kind, although not brought up to being a complete musical tone, has nevertheless a tolerably determinate pitch, which can be estimated by a practised ear.   But, as in all cases where tones of very different qualities have to be compared, it is easy to make a mistake in the Octave.   However, after some of the important pitches have been determined by

the required positions and then tapping against a finger placed just in front of the upper teeth, for the higher resonance, and placed against the neck, just above the larynx, for the lower. He obligingly performed the experiment several times privately before me, and the successive alterations and differences in their direction
¶ were striking.   The tone was dull and like a wood harmonica.   Considerable dexterity seemed necessary to produce the effect, and I could not succeed in doing so.   He carried out the experiment much further than is suggested in the text, embracing the whole nine positions of the tongue in his father's vowel scheme, and obtaining a double resonance in each case. This fact is stated, and the various vowel theories appreciated in Mr. Graham Bell's paper on 'Vowel Theories' read before the American National Academy of Arts and Sciences, April 15, 1879, and printed in the *American Journal of Otology*, vol. i. July 1879.—*Translator*.]

* *Archiv für die Holländischen Beiträge für Natur- und Heilkunde von* Donders *und* Berlin, vol. i. p. 157.   Older incomplete observations of the same circumstance in Samuel Reyher's *Mathesis Mosaica*, Kiel, 1619.—

Chr. Hellwag, *De Formatione Loquelae Diss., Tubingae*, 1710.—Flörcke, *Neue Berliner Monatsschrift*, Sept. 1803, Feb. 1804.--Olivier *Ortho - epo - graphisches Elementar - Werk*, 1804, part iii. p. 21.

† In whispering, the vocal chords are kept close, but the air passes through a small triangular opening at the back part of the glottis between the arytenoid cartilages.   [According to Czermak (*Sitzungsberichte*, Wiener Akad., Math.-Naturw. Cl. April 29, 1858, p. 576) the vocal chords as seen through the laryngoscope are not quite close for whisper, but are nicked in the middle.   Merkel (*Die Funktionen des menschlichen Schlund- und Kehlkopfes. . . . nach eigenen pharyngo- und laryngoskopischen Untersuchungen*, Leipzig, 1862, p. 77) distinguishes two kinds of whispering : (1) the loud, in which the opening between the chords is from ½ to ¾ of a line wide, producing no resonant vibrations, and that between the arytenoids is somewhat wider ; (2) the gentle, in which the vowel is commenced as in loud speaking, with closed glottis, and, after it has begun, the back part of the glottis is opened, while the back chords remain close and motionless.--*Translator*.]

tuning-forks, and others, as Ü and Ö, by allowing the whisper to pass into a regular whistle, the rest are easily determined by arranging them in a melodic progression with the first.   Thus the series :—

Clear A	Ä	E	I
[a in *father*] d'''	[a in *mat*] g'''	[e in *there*] $b'''b$	[i in *machine*] d''''

forms an ascending minor chord of g in the second Inversion $^6_4$, [with the Fifth in the bass,] and can be readily compared with the same melodic progression on the pianoforte.   I was able to determine the pitch for clear A, Ä, and E by tuning-forks, and hence to fix that for I also.*

* The statements of Donders differ slightly from mine, partly because they have reference to Dutch pronunciation, while mine refer to the North German vowels; and partly because Donders, not having been assisted by tuning-forks, was not always able to determine with certainty to what Octave the noises he heard should be assigned. ¶

Vowel	Pitch according to Donders	Pitch according to Helmholtz
U	f'	f
O	d'	$b'b$
A	$b'b$	$b''b$
Ö	g ?	$c'''\sharp$
Ü	a''	g''' to $a'''b$
E	$c'''\sharp$	$b'''b$
I	f'''	d''''

[The extreme divergence of results obtained by different investigators shews the inherent difficulties of the determination, which (as already indicated) arise partly from different values attributed to the vowels, partly from the difficulty of retaining the form of the mouth steadily for a sufficient time, partly from the wide range of tones which the same cavity of the mouth will more or less reinforce, partly from the difficulty of judging of absolute pitch in general, and especially of the absolute pitch of a scarcely musical whisper, and other causes. In C. L. Merkel's *Physiologie der menschlichen Sprache* (Leipzig, 1866), p. 47, a table is given of the results of Reyher, Hellwag, Flörcke, and Donders (the latter differing materially from that just given by Prof. Helmholtz), and on Merkel's p. 109, he adds his last results.   These are reproduced in the following table with the notes, and their pitch to the ¶ nearest vibration, taking a' 440, and supposing equal temperament. To these I add the results of Donders, as just given, and of Helmholtz, both with pitches similarly assumed. Koenig (*Comptes Rendus*, April 25, 1870) also gives his pitches with exact numbers, reckoned as Octaves of the 7th harmonic of c' 256, and hence called bb, although they are nearer the a of this standard. Reference should also be made to Dr. Koenig's paper on ' Manometric Flames' translated in the *Philosophical Magazine*, 1873, vol. xlv. pp. 1–18, 105–114. Lastly, Dr. Moritz Trautmann (*Anglia*, vol. i. p. 590) very confidently gives results utterly different from all the above, which I subjoin with the pitch as before.   I give the general form of

TABLE OF VOWEL RESONANCES. ¶

Observer.	U	O	A	Ä	E	I	Ü	Ö
1. Reyher . .	c 131	$d\sharp$ 156	a 220 c' 262 }	$d\sharp$ 156	f' 349	c''523		
2. Hellwag .	c 131	$c\sharp$ 139	$f\sharp$ 185	a 220	b 247	c' 262	bb 233	$g\sharp$ 208
3. Flörcke . .	c 131	g 196	c' 262	g' 392	a' 440	c'' 523	g' 392	e' 330
4. Donders according to } Helmholtz .	f'349	d' 294	$b'b$ 466	g''' 1568 $+d''$ 587	$c'''\sharp$ 1109	f'''' 1397	a'' 880	g 196?
5. Donders according to } Merkel .	e 165 } f 175 }	e 165	b 247		c' 262	f''698	a' 440	g 196
6. HELMHOLTZ .	U,f175 Ou.f'349	$b'b$ 466	$b''b$ 932	g''' 1568 $+d''$ 587	b''' 1976 $+f'$ 349	d'''' 2349 $+f$ 175	g''' 1568 $+f$ 175	$c'''\sharp$ 1109 $+f'$ 349
7. Merkel . .	d 147	$f\sharp$ 185	A',a 220 O^a, g 196	d'' 587 A', b 247	E',d'' 587 E', e'' 659	a'' 880	a' 440	$f'\sharp$ 370 or d' 294
8. Koenig, 7th harmonics .	bb 224	$b'b$ 448	b'' 896		$b'''b$ 1792 $b''''b$ 3584	$b'''b$ 3584		
9. Trautmann .	f''698	O',c'''1047 O', a'' 880	f'''' 1397	=E' ?	E',a'''1760 E',c''''2093	f'''' 2794	b''' 1976	Ö,g'''1568 Ö',a''''1760

For U it is also by no means easy to find the pitch of the resonance by a fork, as the smallness of the opening makes the resonance weak. Another phenomenon has guided me in this case. If I sing the scale from *c* upwards, uttering the vowel U for each note, and taking care to keep the quality of the vowel correct, and not allowing it to pass into O,* I feel the agitation of the air in the mouth, and even on the drums of both ears, where it excites a tickling sensation, most powerfully when the voice reaches *f*. As soon as *f* is passed the quality changes, the strong agitation of the air in the mouth and the tickling in the ears cease. For the note *f* the phenomenon in this case is the same as if a spherical resonance chamber were placed before a tongue of nearly the same pitch as its proper tone. In this case also we have a powerful agitation of the air within the sphere and a sudden alteration of quality of tone, on passing from a deeper pitch of the mass of air through that of the tongue to a higher. The resonance of the mouth for U is thus fixed at *f* with more certainty than by means of tuning-forks. But we often meet with a U of higher resonance, more resembling O, which I will represent by the French Ou. Its proper tone may rise as high as *f*.† The resonance of the cavity of the mouth for different vowels may then be expressed in notes as follows :

The mode in which the resonance of the cavity of the mouth acts upon the quality of the voice, is then precisely the same as that which we discovered to exist for artificially constructed reed pipes. All those partial tones are reinforced which coincide with a proper tone of the cavity of the mouth, or have a pitch sufficiently near to that of such a tone, while the other partial tones will be more or less damped. The damping of those partial tones which are not strengthened is the more striking the narrower the opening of the mouth, either between the lips as for U, or between the tongue and palate as for I and Ü.

These differences in the partial tones of the different vowel sounds can be easily and clearly recognised by means of resonators, at least within the once and twice accented Octaves [264 to 1056 vib.]. For example, apply a *b′♭* resonator to the ear, and get a bass voice, that can preserve pitch well and form its vowels with purity, to sing the series of vowels to one of the harmonic under tones of *b′♭*, such as *b♭*, *e♭*, *B♭*, *G♭*, *E♭*. It will be found that for a pure, full-toned O the *b′♭* of the resonator will bray violently into the ear. The same upper partial tone is still very powerful for a clear Ä and a tone intermediate between A and Ö, but is weaker for A, E, Ö, and weakest of all for U and I. It will also be found that the resonance of O is materially weakened if it is taken too dull, approaching U,

the vowel at the head of each column, and when the writer distinguishes different forms I add them immediately before the resonance note. Thus we have Helmholtz's *Ou* between U and O ; Merkel's O' between O and A, his obscure A', E' and clear A', E'; Trautmann's O' = Italian open O, and (as he says) English *a* in *all* (which is, however, slightly different), O' ordinary *o* in Berliner *ohne*, E' Berlin *Schnee*, E' French *père* (the same as Ä ?), Ö' Berlin *schön*, French *peu*, Ö' French *leur*. Of course this is far from exhausting the list of vowels in actual use.—*Translator.*]

* [That is, according to the previous directions, to keep the tongue altogether depressed, in the position for *aw* in *gnaw*, which is not natural for an Englishman, so that for English *oo* in *too* we may expect the result to be materially different.—*Translator.*]

† [Prof. Helmholtz may mean the Swedish *o*, see note * p. 106*d*. The following words immediately preceding the notes, which occur in the 3rd German edition, appear to have been accidentally omitted in the 4th. They are, however, retained as they seem necessary. —*Translator.*]

or too open, becoming A°.　But if the $b''\flat$ resonator be used, which is an Octave higher, it is the vowel A that excites the strongest sympathetic resonance; while O, which was so powerful with the $b'\flat$ resonator, now produces only a slight effect.

For the high upper partials of Ä, E, I, no resonators can be made which are capable of sensibly reinforcing them. We are, then, driven principally to observations made with the unassisted ear. It has cost me much trouble to determine these strengthened partial tones in the vowels, and I was not acquainted with them when my previous accounts were published.* They are best observed in high notes of women's voices, or the falsetto of men's voices. The upper partials of high notes in that part of the scale are not so nearly of the same pitch as those of deeper notes, and hence they are more readily distinguished. On $b'\flat$, for example, women's voices could easily bring out all the vowels, with a full quality of tone, but at higher pitches the choice is more limited. When $b'\flat$ is sung, then, the Twelfth f'' is heard for the broad Ä, the double Octave $b'''\flat$ for E, the high Third d'''' for I, ¶ all clearly, the last even piercingly. [See table on p. 124, note.] †

Further, I should observe, that the table of notes given on the preceding page, relates only to those kinds of vowels which appear to me to have the most characteristic quality of tone, but that in addition to these, all intermediate stages are possible, passing insensibly from one to the other, and are actually used partly in dialects, partly by particular individuals, partly in peculiar pitches while singing, or to give a more decided character while whispering.

It is easy to recognise, and indeed is sufficiently well known, that the vowels with a single resonance from U through O to clear A can be altered in continuous succession. But I wish further to remark, since doubts have been thrown on the deep resonance I have assigned to U, that when I apply to my ear a resonator tuned to f', and, singing upon f or $B\flat$ as the fundamental tone, try to find the vowel resembling U which has the strongest resonance, it does not answer to a dull U, but to a U on the way to O.‡　　　　　　　　　　　　　　　　¶

Then again transitions are possible between the vowels of the A—O—U series and those of the A—Ö—Ü series, as well as between the last named and those of the A—E—I series. I can begin on the position for U, and gradually transform the cavity of the mouth, already narrowed, into the tube-like forms for Ö and Ü, in which case the high resonance becomes more distinct and at the same time higher, the narrower the tube is made. If we make this transition while applying a resonator between $b'\flat$ and $b''\flat$ to the ear, we hear the loudness of the tone increase at a certain stage of the transition, and then diminish again. The higher the resonator, the nearer must the vowel approach to Ö or Ü. With a proper position of the mouth the reinforced tone may be brought up to a whistle. Also in a gentle whisper, where the rustle of the air in the larynx is kept very weak, so that with vowels having a narrow opening of the mouth it can be scarcely heard, a strong fricative noise in the opening of the mouth is often required to make the vowel audible. That is to say, we then make the vowels more like their related ¶ consonants, English W and German J [English Y].

Generally speaking the vowels § with double resonance admit of numerous modifications, because any high pitch of one of the resonances may combine with any low pitch of the other. This is best studied by applying a resonator to the ear and trying to find the corresponding vowel degrees in the three series which reinforce its tone, and then endeavouring to pass from one of these to the other in such a way that the resonator should have a reinforced tone throughout.

---

* *Gelehrte Anzeigen der Bayerischen Akademie der Wissenschaften,* June 18, 1859.

† [The passage 'In these experiments' to 'too deep to be sensible,' pp. 166-7 of the 1st English edition, is here cancelled, and p. 111*b*, 'Further, I should observe,' to p. 116*a*, 'high tones of A, E, I,' inserted in its place from the 4th German edition.—*Translator.*]

‡ [An U sound verging towards O is generally conceived to be *duller*, not *brighter*, by English writers, but here U is taken as the dullest vowel. This remark is made merely to prevent confusion with English readers.— *Translator.*]

§ [Misprinted *Consonanten* in the German. —*Translator.*]

Thus the resonator $b'\flat$ answers to O, to an Äö and to an E which resembles Ä, and these sounds may pass continuously one into the other.

The resonator f' answers to the transition Ou—Ö—E. The resonator d'' to Oa—Äö—Ä. In a similar manner each of the higher tones may be connected with various deeper tones. Thus assuming a position of the mouth which would give e''' for whistling, we can, without changing the pitch of the fricative sound in the mouth, whisper a vowel inclining to Ö or inclining to Ü, by allowing the fricative sound in the larynx to have a higher or deeper resonance in the back part of the mouth.*

¶ In comparing the strength of the upper partials of different vowels by means of resonators, it is further to be remembered, that the reinforcement by means of the resonance of the mouth affects the prime tone of the note produced by the voice, as well as the upper partials. And as it is especially the vibrations of the prime, which by their reaction on the vocal chords retain these in regular vibratory motion, the voice speaks much more powerfully, when the prime itself receives such a reinforcement. This is especially observable in those parts of the scale which the singer reaches with difficulty. It may also be noted with reed pipes having metal tongues. When a resonance pipe is applied to them tuned to the tone of the tongue, or a little higher, extraordinarily powerful and rich tones are produced, by means of strong pressure but little wind, and the tongue oscillates in large excursions either way. The pitch of a metal tongue becomes a little flatter than before. This is not perceived with the human voice because the singer is able to regulate the tension of the vocal chords accordingly. Thus I find distinctly that at $b'\flat$, the extremity of my falsetto voice, I can sing powerfully an O, an Ä, and an A on the way to Ö, which have their resonance at this pitch, whereas U, if it is not made to come very near O, and I, are dull and uncertain, with the expenditure of more air than in the former case. Regard must be had to this circumstance in ¶ experiments on the strength of upper partials, because those of a vowel which speaks powerfully, may become proportionally too powerful, as compared with those of a vowel which speaks weakly. Thus I have found that the high tones of the soprano voice which lie in the reinforcing region of the vowel A at the upper extremity of the doubly-accented [or one-foot] Octave, when sung to the vowel A, exhibit their higher Octave more strongly than is the case for the vowels E and I, which do not speak so well although the latter have their strong resonance at the upper end of the thrice-accented [or six-inch] Octave.

It has been already remarked (p. 39c) that the strength and amplitude of sympathetic vibration is affected by the mass and boundaries of the body which vibrates sympathetically. A body of considerable mass which can perform its vibrations as much as possible without any hindrance from neighbouring bodies, and has not its motion damped by the internal friction of its parts, after it has once been excited, can continue to vibrate for a long time, and consequently, if it ¶ has to be set in the highest degree of sympathetic vibration, the oscillations of the exciting tone must, for a comparatively long time, coincide with those proper vibrations excited in itself. That is to say, the highest degree of sympathetic resonance can be produced only by using tones which lie within very narrow limits of pitch. This is the case with tuning-forks and bells. The mass of air in the cavity of the mouth, on the other hand, has but slight density and mass, its walls, so far as they are composed of soft parts, are not capable of offering much resistance, are imperfectly elastic, and when put in vibration have much internal friction to stop their motion. Moreover the vibrating mass of air in the cavity of the mouth communicates through the orifice of the mouth with the outer air, to which it rapidly gives off large parts of the motion it has received. For this reason a

* This appears to me to meet the objections which were made by Herr G. Engel, in Reichart's and Du Bois-Reymond's *Archiv.*, 1869, pp. 317-319. Herr J. Stockhausen drew my attention to the habit of using such deviations from the usual qualities of vowels in syllables which are briefly uttered.

vibratory motion once excited in the air filling the cavity of the mouth is very rapidly extinguished, as any one may easily observe by filliping his cheek with a finger when the mouth is put into different vowel positions.   We thus very easily distinguish the pitch of the resonance for the various transitional degrees from O towards U in one direction and towards A in the other.   But the tone dies away rapidly.   The various resonances of the cavity of the mouth can also be made audible by rapping the teeth.   Just for this reason a tone, which oscillates approximately in agreement with the few vibrations of such a brief resonance tone, will be reinforced by sympathetic vibration to an extent not much less than another tone which exactly coincides with the first; and the range of tones which can thus be sensibly reinforced by a given position of the mouth, is rather considerable.*   This is confirmed by experiment.   When I apply a $b'\flat$ resonator to the right, and an f'' resonator to the left ear and sing the vowel O on $B\flat$, I find a reinforcement not only of the 4th partial $b'\flat$ which answers to the proper tone of the ¶ cavity of the mouth, but also, very perceptibly, though considerably less, of f'', the 6th partial, also.   If I then change O into an A, until f'' finds its strongest resonance, the reinforcement of $b'\flat$ does not entirely disappear although it becomes much less.

The position of the mouth from O to O_a appears to be that which is most favourable for the length of its proper tone and the production of a resonance limited to a very narrow range of pitch.   At least, as I have before remarked, for this position the reinforcement of a suitable tuning-fork is most powerful, and tapping the cheek or the lips gives the most distinct tone.   If then for O the reinforcement by resonance extends to the interval of a Fifth, the intervals will be still greater for the other vowels.   With this agree experiments.   Apply any resonator to the ear, take a suitable under tone, sing the different vowels to this under tone, and let one vowel melt into another.   The greatest reinforcements by resonance take place on that vowel or those vowels, for which one of the characteristic tones in ¶ the diagram p. 100b coincides with the proper tone of the resonator.   But more or less considerable reinforcement is also observed for such vowels as have their characteristic tones at moderate differences of pitch from the proper tone of the resonator, and the reinforcement will be less the greater these differences of pitch.

By this means it becomes possible in general to distinguish the vowels from each other even when the note to which they are sung is not precisely one of the harmonic under tones of the vowels.   From the second partial tone onwards, the intervals are narrow enough for one or two of the partials to be distinctly reinforced by the resonance of the mouth.   It is only when the proper tone of the cavity of the mouth falls midway between the prime tone of the note sung by the voice and its higher Octave, or is more than a Fifth deeper than that prime tone, that the characteristic resonance will be weak.

Now in speaking, both sexes choose one of the deepest positions of their voice. Men generally choose the upper half of the great (or eight-foot) Octave; and ¶ women the upper half of the small (or four-foot) Octave.†   With the exception of U, which admits of fluctuations in its proper tone of nearly an Octave, all these pitches of the speaking voice have the corresponding proper tones of the cavity of the mouth situated within sufficiently narrow intervals from the upper partials of the speaking tone to create sensible resonance of one or more of these partials, and thus characterise the vowel.‡   To this must be added that the speaking voice, probably through great pressure of the vocal ligaments upon one another, converting

* On this subject see Appendix X., and the corresponding investigation in the text in Part I. Chap. VI. therein referred to.

† [That is both use their 'lower thick' register, as described in the note p. 101d, but are an Octave apart.—Translator.]

‡ [Observe here that the quality of the vowel tone is not made to consist in the identity of certain of its partials with exact pitches but in their coming near enough to those pitches to receive reinforcement, and that the character of a vowel quality of tone, like that of all qualities of tone, depends not on the absolute pitch, but on the relative force of the upper partials.   As Prof. Helmholtz's theory has often been grievously misunderstood, I

them into striking reeds, has a jarring quality of tone, that is, possesses stronger upper partials than the singing voice.

In singing, on the other hand, especially at higher pitches, conditions are less favourable for the characterisation of vowels. Every one knows that it is generally much more difficult to understand words when sung than when spoken, and that the difficulty is less with male than with female voices, each having been equally well cultivated. Were it otherwise, 'books of the words' at operas and concerts would be unnecessary. Above f', the characterisation of U becomes imperfect even if it is closely assimilated to O. But so long as it remains the only vowel of indeterminate sound, and the remainder allow of sensible reinforcement of their upper partials in certain regions, this negative character will distinguish U. On the other hand a soprano voice in the neighbourhood of f'' should not be able to clearly distinguish U, O, and A; and this agrees with my own experience. On singing the three vowels ¶ in immediate succession, the resonance f''' for A will, however, be still somewhat clearer in the cavity of the mouth when tuned for $b''\flat$, than when it is tuned to $b'\flat$ for O. The soprano voice will in this case be able to make the A clearer, by elevating the pitch of the cavity of the mouth towards d''' and thus making it approach to f''. The O, on the other hand, can be separated from U by approaching O_a, and giving the prime more decisive force. Nevertheless these vowels, if not sung in immediate succession, will not be very clearly distinguished by a listener who is unacquainted with the mode of pronouncing the vowels that the soprano singer uses.*

A further means of helping to discriminate vowels, moreover, is found in commencing them powerfully. This depends upon a general relation in bodies excited to sympathetic vibration. Thus, if we excite sympathetic vibration in a suitable body with a tone somewhat different from its proper tone, by commencing it suddenly with great power, we hear at first, in addition to the exciting tone which is rein ¶ forced by resonance, the proper tone of the sympathetically vibrating body.† But the latter soon dies away, while the first remains. In the case of tuning-forks with large resonator, we can even hear beats between the two tones. Apply a $b'\flat$ resonator to the ear, and commence singing the vowel O powerfully on g, of which the upper partials g' and d'' have only a weak lasting resonance in the cavity of the mouth, and you may hear immediately at the commencement of the vowel, a short sharp beat between the $b'\flat$ of the cavity of the mouth and of the resonator. On selecting another vowel, this $b'\flat$ vanishes, which shews that the pitch of the cavity of the mouth helps to generate it. In this case then also the sudden commencement of the tones g' and d'' belonging to the compound tone of the voice, excites the inter mediate proper tone $b'\flat$ of the cavity of the mouth, which rapidly fades. The same thing may be observed for other pitches of the resonator used, when we sing notes, powerfully commenced, which have upper partials that are not reinforced by the resonator, provided that a vowel is chosen with a characteristic pitch which ¶ answers to the pitch of the resonator. Hence it results that when any vowel in any pitch is powerfully commenced, its characteristic tone becomes audible as a short beat. By this means the vowel may be distinctly characterised at the moment of commencement, even when it becomes intermediate on long con tinuance. But for this purpose, as already remarked, an exact and energetic com mencement is necessary. How much such a commencement assists in rendering the words of a singer intelligible is well known. For this reason also the vocal isation of the briefly uttered words of a reciting *parlando*, is more distinct than that of sustained song.‡

draw particular attention to the point in this place. See also the table which I have added in a footnote on p. 124d.—*Translator*.]

  * [In my *Pronunciation for Singers* (Cur wen, 1877), and my *Speech in Song* (Novello, 1878) I have endeavoured to give a popular explanation of the alterations which a singer

may make in the vowels in English, German, French and Italian, at different pitches, so as to remain intelligible.—*Translator*.]

  † See the mathematical statement of this pro cess in App. IX., remarks on equations 4 to 4b.

  ‡ The facts here adduced meet, I think, the objections brought against my vowel theory by

Moreover vowels admit of other kinds of alterations in their qualities of tone, conditioned by alterations of their characteristic tones within certain limits. Thus the resonating capability of the cavity of the mouth may undergo in general alterations in strength and definition, which would render the character of the various vowels and their difference from one another in general more or less conspicuous or obscure. Flaccid soft walls in any passage with sonorous masses of air, are generally prejudicial to the force of the vibrations. Partly too much of the motion is given off to the outside through the soft masses, partly too much is destroyed by friction within them. Wooden organ pipes have a less energetic quality of tone than metal ones, and those of pasteboard a still duller quality, even when the mouthpiece remains unaltered. The walls of the human throat, and the cheeks, are, however, much more yielding than pasteboard. Hence if the tone of the voice with all its partials is to meet with a powerful resonance and come out unweakened, these most flaccid parts of the passage for our voice, must be as much as possible ¶ thrown out of action, or else rendered elastic by tension, and in addition the passage must be made as short and wide as possible. The last is effected by raising the larynx. The soft wall of the cheeks can be almost entirely avoided, by taking care that the rows of teeth are not too far apart. The lips, when their co-operation is not necessary, as it is for Ö and Ü, may be held so far apart that the sharp firm edges of the teeth define the orifice of the mouth. For A the angles of the mouth can be drawn entirely aside. For O they can be firmly stretched by the muscles above and below them (*levator anguli oris* and *triangularis menti*), which then feel like stretched cords to the touch, and can be thus pressed against the teeth, so that this part of the margin of the orifice of the mouth is also made sharp and capable of resisting.

In the attempt to produce a clear energetic tone of the voice we also become aware of the tension of a large number of muscles lying in front of the throat, both those which lie between the under jaw and the tongue-bone and help to form ¶ the floor of the cavity of the mouth (*mylohyoideus, geniohyoideus*, and perhaps also *biventer*), and likewise those which run down near the larynx and air tubes, and draw down the tongue-bone (*sternohyoideus, sternothyroideus* and *thyrohyoideus*). Without the counteraction of the latter, indeed, considerable tension of the former would be·impossible. Besides this a contraction of the skin on both sides of the larynx which takes place at the commencement of the tone of the voice, shews that the *omohyoideus* muscle, which runs obliquely down from the tongue-bone backwards to the shoulder-blade, is also stretched. Without its co-operation the muscles arising from the under jaw and breast-bone would draw the larynx too far forwards. Now the greater part of these muscles do not go to the larynx at all, but only to the tongue-bone, from which the larynx is suspended. Hence they cannot directly assist in the formation of the voice, so far as this depends upon the action of the larynx. The action of these muscles, so far as I have been able to observe it on myself, is also much less when I utter a dull guttural A, than when I endeavour to ¶ change it into a ringing, keen and powerfully penetrating A. Ringing and keen, applied to a quality of tone, imply many and powerful upper partials, and the stronger they are, of course the more marked are the differences of the vowels which their own differences condition. A singer, or a declaimer, will occasionally interpose among his bright and rich tones others of a duller character as a contrast. Sharp characterisation of vowel quality is suitable for energetic, joyful or vigorous frames of mind ; indifferent and obscure quality of tone for sad and troubled, or taciturn states. In the latter case speakers like to change the proper tone of the vowels, by drawing the extremes closer to a middle Äö (say the short German E [the final

Herr E. v. Quanten (Poggendorff's *Annal.*, vol. cliv. pp. 272 and 522), so far as they do not rest upon misconceptions. [In the 1st edition of this translation, during the printing of which v. Quanten's first paper appeared, I added an article, pp. 724–741, with especial reference to it. In consequence of the new matter added by Prof. Helmholtz in his 4th German edition here followed, this article is omitted from the present edition.—*Translator.*]

English obscure A in *idea*]), and hence select somewhat deeper tones in place of the high tones of A, E, I.

A peculiar circumstance must also be mentioned which distinguishes the human voice from all other instruments and has a peculiar relation to the human ear. Above the higher reinforced partial tones of I, in the neighbourhood of e'''' up to g'''' [2640 to 3168 vib.] the notes of a pianoforte have a peculiar cutting effect, and we might be easily led to believe that the hammers were too hard, or that their mechanism somewhat differed from that of the adjacent notes. But the phenomenon is the same on all pianofortes, and if a very small glass tube or sphere is applied to the ear, the cutting effect ceases, and these notes become as soft and weak as the rest, but another and deeper series of notes now becomes stronger and more cutting. Hence it follows that the human ear by its own resonance favours the tones between e'''' and g'''', or, in other words, that it is tuned to one of these pitches.*

¶ These notes produce a feeling of pain in sensitive ears. Hence the upper partial tones which have nearly this pitch, if any such exist, are extremely prominent and affect the ear powerfully. This is generally the case for the human voice when it is strained, and will help to give it a screaming effect. In powerful male voices singing *forte*, these partial tones sound like a clear tinkling of little bells, accompanying the voice, and are most audible in choruses, when the singers shout a little. Every individual male voice at such pitches produces dissonant upper partials. When basses sing their high e', the 7th partial tone † is d'''', the 8th e'''', the 9th $f''''\sharp$, and the 10th $g''''\sharp$. Now, if e'''' and $f''''\sharp$ are loud, and d'''' and $g''''\sharp$, though weaker, are audible, there is of course a sharp dissonance. If many voices are sounding together, producing these upper partials with small differences of pitch, the result is a very peculiar kind of tinkling, which is readily recognised a second time when attention has been once drawn to it. I have not noticed any difference of effect for different vowels in this case, but the tinkling ceases as soon

¶ as the voices are taken *piano*; although the tone produced by a chorus will of course still have considerable power. This kind of tinkling is peculiar to human voices; orchestral instruments do not produce it in the same way either so sensibly or so powerfully. I have never heard it from any other musical instrument so clearly as from human voices.

The same upper partials are heard also in soprano voices when they sing *forte*; in harsh, uncertain voices they are tremulous, and hence shew some resemblance to the tinkling heard in the notes of male voices. But I have heard them brought out with exact purity, and continue to sound on perfectly and quietly, in some steady and harmonious female voices, and also in some excellent tenor voices. In the melodic progression of a voice part, I then hear these high upper partials of the four-times accented Octave, falling and rising at different times within the compass of a minor Third, according as different upper partials of the notes sung enter the region for which our ear is so sensitive. It is certainly remarkable that

¶ it should be precisely the human voice which is so rich in those upper partials for which the human ear is so sensitive. Madame E. Seiler, however, remarks that dogs are also very sensitive for the high e'''' of the violin.

This reinforcement of the upper partial tones belonging to the middle of the four-times accented Octave, has, however, nothing to do with the characterisation of vowels. I have mentioned it here, merely because these high tones are readily remarked in investigations into the vowel qualities of tone, and the observer must not be misled to consider them as peculiar characteristics of individual vowels. They are simply a characteristic of strained voices.

The humming tone heard when singing with closed mouth, lies nearest to U.

---

* I have lately found that my right ear is most sensitive for f'''', and my left for c''''. When I drive air into the passage leading to the tympanum, the resonance descends to $c''''\sharp$ and $g'''\sharp$. The chirp of the cricket corresponds precisely to the higher resonance, and on merely applying a short paper tube to the entrance of my ear, this chirp is rendered extraordinarily weak.

† [The first six partial tones are e', e'', b'', c''', $g'''\sharp$, b''', the seventh is 27 cents flatter than d''''.--*Translator.*]

This hum is used in uttering the consonants M, N and Nᵍ. The size of the exit of the air (the nostrils) is in this case much smaller in comparison with the resonant chamber (the internal nasal cavity) than the opening of the lips for U in comparison with the corresponding resonant chamber in the mouth. Hence, in humming, the peculiarities of the U tone are much enhanced. Thus although upper partials are present, even up to a considerably high pitch, yet they decrease in strength as they rise in pitch much faster than for U. The upper Octave is tolerably strong in humming, but all the higher partial tones are weak. Humming in the N-position differs a little from that in the M-position, by having its upper partials less damped than for M. But it is only at the instant when the cavity of the mouth is opened or closed that a clear difference exists between these conso-nants. We cannot enter into the details of the composition of the sound of the other consonants, because they produce noises which have no constant pitch, and are not musical tones, to which we have here to confine our attention.

The theory of vowel sounds here explained may be confirmed by experiments with artificial reed pipes, to which proper resonant chambers are attached. This was first done by Willis, who attached reed pipes to cylindrical chambers of variable length, and produced different tones by increasing the length of the resonant tube. The shortest tubes gave him I, and then E, A, O, up to U, until the tube exceeded the length of a quarter of a wave. On further increasing the length the vowels returned in converse order. His determination of the pitch of the resonant pipes agrees well with mine for the deeper vowels. The pitch found by Willis for the higher vowels was relatively too high, because in this case the length of the wave was smaller than the diameter of the tubes, and consequently the usual calcula-tion of pitch from the length of the tubes alone was no longer applicable. The vowels E and I were also far from accurately resembling those of the voice, because the second resonance was absent, and hence, as Willis himself states, they could not be well distinguished.*

Vowel	In the Word	Pitch, Willis	Pitch, Helmholtz	Length of Tube in Inches
O	No	c''	c''	4·7
A°	Nought	$e''b$	$e''b$	3·8
	Paw	g''	g''	3·05
A	Part	$d'''b$	$d'''b$	2·2
	Pad	f'''		1·8
E	Pay	d''''	$b'''b$	1·0
	Pet	c'''''	c''''	0·6
I	See	g'''''	d''''	0·38 (?)

The vowels are obtained much more clearly and distinctly with properly tuned resonators, than with cylindrical resonance chambers. On applying to a reed pipe which gave $b\flat$, a glass resonator tuned to $b\flat$, I obtained the vowel U; changing the resonator to one tuned for $b'\flat$, I obtained O; the $b''\flat$ resonator gave a rather close A, and the d''' resonator a clear A. Hence by tuning the applied chambers in the same way we obtain the same vowels quite independently of the form of the chamber and nature of its walls. I also succeeded in producing various grades of

* [Probably the first treatise on phonology in which Willis's experiments were given at length, and the above table cited, with Wheat-stone's article from the *London and Westmin-ster Review*, which was kindly brought under my notice by Sir Charles Wheatstone himself, was my *Alphabet of Nature*, London, 1845. The table includes U exemplified by *but, boot*, with an indefinite length of pipe. The word *pad* is misprinted *paa* in all the German editions of Helmholtz (even the 4th, which appeared after the correction in my translation), and as he therefore could not separate its A from that in *part*, he gives no pitch. It is really the nearest English representative of the German . The sounds in *nought, paw*, which Sir John Her-schel, when citing Willis (Art. ' Sound,' in *Encyc. Metropol.*, par. 375), could not distin-guish, were probably meant for the broad Italian open O, or English *o* in *more*, and the English *aw* in *maw* respectively. The length of the pipe in inches is here added from Willis's paper. I have heard Willis's experiments repeated by Wheatstone.—*Translator.*]

Ä, Ö, E, and I with the same reed pipe, by applying glass spheres into whose external opening glass tubes were inserted from 6 to 10 centimetres (2·36 to 3·94 inches) in length, in order to imitate the double resonance of the oral cavity for these vowels.

Willis has also given another interesting method for producing vowels. If a toothed wheel, with many teeth, revolve rapidly, and a spring be applied to its teeth, the spring will be raised by each tooth as it passes, and a tone will be produced having its pitch number equal to the number of teeth by which it has been struck in a second. Now if one end of the spring is well fastened, and the spring be set in vibration, it will itself produce a tone which will increase in pitch as the spring diminishes in length. If then we turn the wheel with a constant velocity, and allow a watch spring of variable length to strike against its teeth, we shall obtain for a long spring a quality of tone resembling U, and as we shorten the

¶ spring other qualities in succession like O, A, E, I, the tone of the spring here playing the part of the reinforced tone which determines the vowel. But this imitation of the vowels is certainly much less complete than that obtained by reed pipes. The reason of this process also evidently depends upon our producing compound tones in which certain upper partials (which in this case correspond with the proper tones of the spring itself) are more reinforced than others.

Willis himself advanced a theory concerning the nature of vowel tones which differs from that I have laid down in agreement with the whole connection of all other acoustical phenomena. Willis imagines that the pulses of air which produce the vowel qualities, are themselves tones which rapidly die away, corresponding to the proper tone of the spring in his last experiment, or the short echo produced by a pulse or a little explosion of air in the mouth, or in the resonance chamber of a reed pipe. In fact something like the sound of a vowel will be heard if we only tap against the teeth with a little rod, and set the cavity of the mouth in the posi-

¶ tion required for the different vowels. Willis's description of the motion of sound for vowels is certainly not a great way from the truth; but it only assigns the mode in which the motion of the air ensues, and not the corresponding reaction which this produces in the ear. That this kind of motion as well as all others is actually resolved by the ear into a series of partial tones, according to the laws of sympathetic resonance, is shewn by the agreement of the analysis of vowel qualities of tone made by the unarmed ear and by the resonators. This will appear still more clearly in the next chapter, where experiments will be described shewing the direct composition of vowel qualities from their partial tones.

Vowel qualities of tone consequently are essentially distinguished from the tones of most other musical instruments by the fact that the loudness of their partial tones does not depend solely upon their numerical order but preponderantly upon the absolute pitch of those partials. Thus when I sing the vowel A to the note $E\flat$,* the reinforced tone $b''\flat$ is the 12th partial of the compound tone;

¶ and when I sing the same vowel A to the note $b'\flat$, the reinforced tone is still $b''\flat$, but is now the 2nd partial of the compound tone sung.†

From the examples adduced to shew the dependence of quality of tone from the mode in which a musical tone is compounded, we may deduce the following general rules :—

　　1. *Simple Tones*, like those of tuning-forks applied to resonance chambers and wide stopped organ pipes, have a very soft, pleasant sound, free from all roughness, but wanting in power, and dull at low pitches.

　　2. *Musical Tones*, which are accompanied by a moderately loud series of the

---

* [$E\flat$ has for 2nd partial $e\flat$, for 3rd $b\flat$, and hence for 6th $b'\flat$, and for 12th, $b''\flat$.— *Translator.*]

† [See App. XX. sec. M. No. 1, for Jenkin and Ewing's analysis of vowel sounds by means of the Phonograph. *Translator.*]

lower partial tones, up to about the sixth partial, are more harmonious and musical. Compared with simple tones they are rich and splendid, while they are at the same time perfectly sweet and soft if the higher upper partials are absent. To these belong the musical tones produced by the pianoforte, open organ pipes, the softer piano tones of the human voice and of the French horn. The last-named tones form the transition to musical tones with high upper partials; while the tones of flutes, and of pipes on the flue-stops of organs with a low pressure of wind, approach to simple tones.

3. If only the unevenly numbered partials are present (as in narrow stopped organ pipes, pianoforte strings struck in their middle points, and clarinets), the quality of tone is *hollow*, and, when a large number of such upper partials are present, *nasal*. When the prime tone predominates the quality of tone is *rich*; but when the prime tone is not sufficiently superior in strength to the upper partials, the quality of tone is *poor*. Thus the quality of tone in the wider open ¶ organ pipes is richer than that in the narrower; strings struck with pianoforte hammers give tones of a richer quality than when struck by a stick or plucked by the finger; the tones of reed pipes with suitable resonance chambers have a richer quality than those without resonance chambers.

4. When partial tones higher than the sixth or seventh are very distinct, the quality of tone is *cutting* and *rough*. The reason for this will be seen hereafter to lie in the dissonances which they form with one another. The degree of harshness may be very different. When their force is inconsiderable the higher upper partials do not essentially detract from the musical applicability of the compound tones; on the contrary, they are useful in giving character and expression to the music. The most important musical tones of this description are those of bowed instruments and of most reed pipes, oboe (hautbois), bassoon (fagotto), harmonium, and the human voice. The rough, braying tones of brass instruments are extremely penetrating, and hence are better adapted to give the impression of great power ¶ than similar tones of a softer quality. They are consequently little suitable for artistic music when used alone, but produce great effect in an orchestra. Why high dissonant upper partials should make a musical tone more penetrating will appear hereafter.

CHAPTER VI.

ON THE APPREHENSION OF QUALITIES OF TONE.

UP to this point we have not endeavoured to analyse given musical tones further than to determine the differences in the number and loudness of their partial tones. Before we can determine the function of the ear in apprehending qualities of tone, ¶ we must inquire whether a determinate relative strength of the upper partials suffices to give us the impression of a determinate musical quality of tone or whether there are not also other perceptible differences in quality which are independent of such a relation. Since we deal only with *musical tones*, that is with such as are produced by exactly periodic motions of the air, and exclude all irregular motions of the air which appear as noises, we can give this question a more definite form. If we suppose the motion of the air corresponding to the given musical tone to be resolved into a sum of pendular vibrations of air, such individual pendular vibrations will not only differ from each other in force or amplitude for different forms of the compound motion, but also in their relative position, or, according to physical terminology, in their *difference of phase*. For example, if we superimpose the two pendular vibrational curves A and B, fig. 31 (p. 120a), first with the point e of B on the point d_0 of A, and next with the point e of B on the point d^1 of A, we obtain the two entirely distinct vibrational curves

C and D. By further displacement of the initial point e so as to place it on d_2 or d_3 we obtain other forms, which are the inversions of the forms C and D, as has been already shewn (supra, p. 32a). If, then, musical quality of tone depends solely on the relative force of the partial tones, all the various motions C, D, &c., must

FIG. 31.

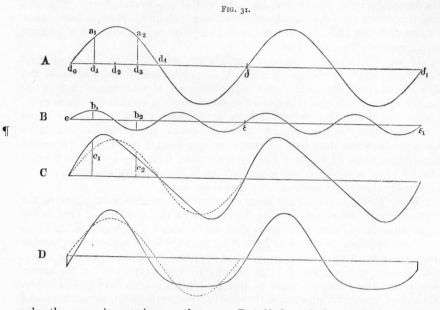

make the same impression on the ear. But if the relative position of the two waves, that is the difference of phase, produces any effect, they must make different impressions on the ear.

Now to determine this point it was necessary to compound various musical tones out of simple tones artificially, and to see whether an alteration of quality ensued when force was constant but phase varied. Simple tones of great purity, which can have both their force and phase exactly regulated, are best obtained from tuning-forks having the lowest proper tone reinforced, as has been already described (p. 54d), by a resonance chamber, and communicated to the air. To set the tuning-forks in very uniform motion, they were placed between the limbs of a little electro-magnet, as shewn in fig. 32, opposite. Each tuning-fork was screwed into a separate board d d, which rested upon pieces of india-rubber tubing e e that were cemented below it, to prevent the vibrations of the fork from being directly communicated to the table and hence becoming audible. The limbs b b of the electro-magnet are surrounded with wire, and its pole f is directed to the fork. There are two clamp screws g on the board d d which are in conductive connection with the coils of the electro-magnet, and serve to introduce other wires which conduct the electric current. To set the forks in strong vibration the strength of these streams must alternate periodically. These are generated by a separate apparatus to be presently described (fig. 33, p. 122b, c).

When forks thus arranged are set in vibration, very little indeed of their tone is heard, because they have so little means of communicating their vibrations to the surrounding air or adjacent solids. To make the tone strongly audible, the resonance chamber i, which has been previously tuned to the pitch of the fork, must be brought near it. This resonance chamber is fastened to another board k, which slides in a proper groove made in the board d d, and thus allows its opening to be brought very near to the fork. In the figure the resonance chamber is shewn at a distance from the fork in order to exhibit the separate parts distinctly ; when in use, it is brought as close as possible to the fork. The mouth of the resonance chamber can be closed by a lid l attached to a lever m. By pulling the string n the lid is withdrawn from the opening and the tone of the fork is communicated to the air with great force. When the thread is let loose, the lid is brought over the mouth of the chamber by the spring p, and the tone of the fork is no longer heard. By partial opening of the mouth of the chamber, the tone of the fork can be made to receive any desired intermediate degree of strength. The whole of the strings which open the various resonance chambers belonging to a series of such forks are attached to a keyboard in such a way that by pressing a key the corresponding chamber is opened.

At first I had eight forks of this kind, giving the tones $B\flat$ and its first seven harmonic upper partials, namely $b\flat$, f', $b'\flat$, d'', f'', $a''\flat$,* and $b''\flat$. The prime tone $B\flat$ corresponds to the pitch in which bass voices naturally speak. Afterwards I had forks made of the pitches d''', f''', $a'''\flat$* and $b'''\flat$, and assumed $b\flat$ for the prime of the compound tone.

To set the forks in motion, intermittent electrical currents had to be conducted ¶ through the coils of the electro-magnet, giving as many electrical shocks as the

FIG. 32.

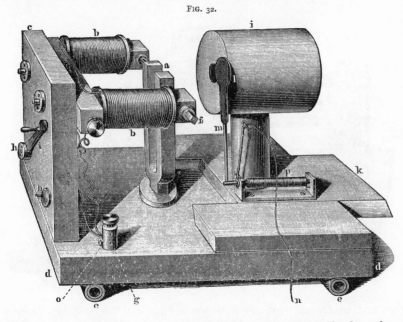

lowest forks made vibrations in a second, namely 120. Every shock makes the iron of the electro-magnet b b momentarily magnetic, and hence enables it to attract the prongs of the fork, which are themselves rendered permanently magnetic. The prongs of the lowest fork $B\flat$ are thus attracted by the poles of the electro- ¶ magnet, for a very short time, once in *every* vibration; the prongs of the second for $b\flat$, which moves twice as fast, once every *second* vibration, and so on. The vibrations of the forks are therefore both excited and kept up as long as the electric currents pass through the apparatus. The vibrations of the lower forks are very powerful, those of the higher proportionally weaker.

The apparatus shewn in fig. 33 (p. 122 b, c) serves to produce intermittent currents of exactly determinate periodicity. A tuning-fork a is fixed horizontally between the limbs b b of an electro-magnet; at its extremities are fastened two platinum wires c c, which dip into two little cups d filled half with mercury and half with alcohol, forming the upper extremities of brass columns. These columns have clamping screws i to receive the wires, and stand on two boards f, g, which turn about an axis, as at f, and which can each be somewhat raised or lowered by a thumb-

* [These being 7th harmonics $^7a''\flat$ and   and $a'''\flat$, in the justly intoned scale of eb.—
$^7a'''\flat$ are 27 cents flatter than the $a''\flat$ Translator.]

screw, as at g, so as to make the points of the platinum wires c c exactly touch the mercury below the alcohol in the cups d. A third clamping screw e is in conductive connection with the handle of the tuning-fork. When the fork vibrates, and an electric current passes through it from i to e, the current will be broken every time that the end of the fork a rises above the surface of the mercury in the cup d, and re-made every time the platinum wire dips again into the mercury. This intermittent current being at the same time conducted through the electromagnet b b, fig. 33, the latter becomes magnetic every time it passes, and thus keeps up the vibrations of the fork a, which is itself magnetic. Generally only one of the cups d is used for conducting the current. Alcohol is poured over the mercury to prevent the latter from being burned by the electrical sparks which arise when the stream is interrupted. This method of interrupting the current was invented by Neef, who used a simple vibrating spring in place of the tuning-fork, as may be generally seen in the induction apparatus so much used for medical purposes. But the vibrations of a spring communicate themselves to all adjacent

FIG. 33.

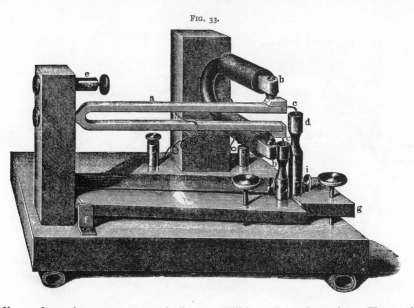

bodies and are for our purposes both too audible and too irregular. Hence the necessity of substituting a tuning-fork for the spring. The handle of a well worked symmetrical tuning-fork is extremely little agitated by the vibrations of the fork and hence does not itself agitate the bodies connected with it, so powerfully as the fixed end of a straight spring. The tuning-fork of the apparatus in fig. 33 must be in exact unison with the prime tone B♭. To effect this I employ a little clamp of thick steel wire h, placed on one of the prongs. By slipping this towards the free end of the prong the tone is deepened, and by slipping it towards the handle of the fork, the tone is raised.*

When the whole apparatus is in action, but the resonance chambers are closed, all the forks are maintained in a state of uniform motion, but no sound is heard, beyond a gentle humming caused by the direct action of the forks on the air. But on opening one or more resonance chambers, the corresponding tones are heard with sufficient loudness, and are louder as the lid is more widely opened. By this means it is possible to form, in rapid succession, different combinations of the prime

* This apparatus was made by Fessel in Cologne. More detailed descriptions of its separate parts, and instructions for the experiments to be made by its means, are given in Appendix VIII. [This apparatus was exhibited by R. Koenig (see Appendix II.) in the International Exhibition of 1872 in London. --*Translator*.]

tone with one or more harmonic upper partials having different degrees of loudness, and thus produce tones of different qualities.

Among the natural musical tones which appear suitable for imitation with forks, the vowels of the human voice hold the first rank, because they are accompanied by comparatively little extraneous noise and shew distinct differences of quality which are easy to seize. Most vowels also are characterised by comparatively low upper partials, which can be reached by our forks; E and I alone somewhat exceed these limits. The motion of the very high forks is too weak for this purpose when influenced only by such electrical currents as I was able to use without disturbance from the noise of the electric sparks.

The first series of experiments was made with the eight forks Bb to $b''b$. With these U, O, Ö, and even A could be imitated; the last not very well because of my not possessing the upper partials c''' and d''', which lie immediately above its characteristic tone $b''b$, and are sensibly reinforced in the natural sound of this ¶ vowel. The prime tone Bb of this series, when sounded alone, gave a very dull U, much duller than could be produced in speech. The sound became more like U when the second and third partial tones bb and f' were allowed to sound feebly at the same time. A very fine O was produced by taking $b'b$ strong, and bb, f', d'' more feebly; the prime tone Bb had then, however, to be somewhat damped. On suddenly changing the pressure on the keys and hence the position of the lids before the resonance chambers, so as to give Bb strong, and all the upper partials weak, the apparatus uttered a good clear U after the O.

A or rather A° [nearly o in *not*] was produced by making the fifth to the eighth partial tones as loud as possible, and keeping the rest under.

The vowels of the second and third series, which have higher characteristic tones, could be only imperfectly imitated by bringing out their reinforced tones of the lower pitch. Though not very clear in themselves they became so by contrast on alternation with U and O. Thus a passably clear Ä was obtained by giving loudness ¶ chiefly to the fourth and fifth tones, and keeping down the lower ones, and a sort of E by reinforcing the third, and letting the rest sound feebly. The difference between O and these two vowels lay principally in keeping the prime tone Bb and its Octave bb much weaker for Ä and E than for O.*

To extend my experiments to the brighter vowels, I afterwards added the forks d''', f''', $a'''b$, $b'''b$, the two upper ones of which, however, gave a very faint tone, and I chose bb as the prime tone in place of Bb. With these I got a very good Ä and A, and at least a much more distinct E than before. But I could not get up to the high characteristic tone of I.

In this higher series of forks, the prime tone bb, when sounded alone, reproduced U. The same prime bb with moderate force, accompanied with a strong Octave $b'b$, and a weaker Twelfth f'', gave O, which has the characteristic tone $b'b$. A was obtained by taking bb, $b'b$, and f'' moderately strong, and the characteristic tones $b''b$ and d''' very strong. To change A into Ä it was necessary to increase ¶ somewhat the force of $b'b$ and f'' which were adjacent to the characteristic tone d'', to damp $b''b$, and bring out d''' and f''' as strongly as possible. For E the two deepest tones of the series, bb and $b'b$, had to be kept moderately loud, as being adjacent to the deeper characteristic tone f', while the highest f''', $a'''b$, $b'''b$ had to be made as prominent as possible. But I have hitherto not succeeded so well with this as with the other vowels, because the high forks were too weak, and because perhaps the upper partials which lie above the characteristic tone $b'''b$ could not be entirely dispensed with.†

* The statements in the *Münchener gelehrte Anzeigen* for June 20, 1859, must be corrected accordingly. At that time I did not know the higher upper partials of E and I, and hence made the O too dull to distinguish it from the imperfect E.

† [The following tabular statement of the above results will serve to shew their relations more clearly. In the first line are placed the notes of the forks and the numbers of the corresponding partials. The letters *pp*, *p*, *mf*, *f*, *ff* below them are the usual musical indications of force, *pianissimo*, *piano*, *mezzo forte*, *forte*, *fortissimo*. Where no such mark is

In precisely the same way as the vowels of the human voice, it is possible to imitate the quality of tone produced by organ pipes of different stops, if they have not secondary tones which are too high, but of course the whizzing noise, formed by breaking the stream of air at the lip, is wanting in these imitations. The tuning-forks are necessarily limited to the imitation of the purely musical part of the tone. The piercing high upper partials, required for imitating reed instruments, were absent, but the nasality of the clarinet was given by using a series of unevenly numbered partials, and the softer tones of the horn by the full chorus of all the forks.

But though it was not possible to imitate every kind of quality of tone by the present apparatus, it sufficed to decide the important question as to the effect of altered difference of phase upon quality of tone. As I particularly observed at the beginning of this chapter, this question is of fundamental importance for the ¶ theory of auditory sensation. The reader who is unused to physical investigations must excuse some apparently difficult and dry details in the explanation of the experiments necessary for its decision.

The simple means of altering the phases of the secondary tones consists in bringing the resonance chambers somewhat out of tune by narrowing their apertures, which weakens the resonance, and at the same time alters the phase. If the resonance chamber is tuned so that the simple tone which excites its strongest resonance coincides with the simple tone of the corresponding fork, then, as the mathematical theory shews,* the greatest velocity of the air at the mouth of the chamber in an outward direction, coincides with the greatest velocity of the ends of the fork in an inward direction. On the other hand, if the chamber is tuned to be slightly deeper than the fork, the greatest velocity of the air slightly precedes, and if it is tuned slightly higher, that greatest velocity slightly lags behind the greatest velocity of the fork. The more the tuning is altered, the ¶ greater will be the difference of phase, till at last it reaches the duration of a quarter of a vibration. The magnitude of the difference of phase agrees during this change precisely with the strength of the resonance, so that to a certain degree we are able to measure the former by the latter. If we represent the strength of the sound in the resonance chamber when in unison with the fork by 10, and divide the periodic time of a vibration, like the circumference of a circle, into 360

added the partial is not mentioned in the text. For the second series of experiments the forks of corresponding pitches are kept under the old ones, but the whole are now numbered as partials of $b\flat$.

First Forks	1 $B\flat$	2 $b\flat$	3 f'	4 $b'\flat$	5 a''	6 f''	7 $a''\flat$	8 $b''\flat$	10	12	14	16
Vowels U	f	pp	pp									
O	mf	p	p	f	p							
A°	p	p	p	p	ff	ff	ff	ff				
Ä	p	pp	p	f	f							
E	p	pp	f	p		p	p	p				

Second Forks		1 $b\flat$		2 $b'\flat$		3 f''		4 $b''\flat$	5 d'''	6 f'''	7 $a'''\flat$	8 $b'''\flat$
Vowels U		f										
O		mf										
A°		mf		mf		p		f	f	ff		
Ä				mf		mf		f	p	ff	ff	
E		mf				f					ff	ff

See Appendix XX. sect. M. No. 2, for Messrs. Preece and Stroh's new method of vowel synthesis.—*Translator.*]

* See the first part of Appendix IX.

degrees, the relation between the strength of the resonance and the difference of phase is shewn by the following table :—

Strength of Resonance	Difference of Phase in angular degrees
10	0°
9	35° 54'
8	50° 12'
7	60° 40'
6	68° 54'
5	75° 31'
4	80° 48'
3	84° 50'
2	87° 42'
1	89° 26'

This table shews that a comparatively slight weakening of resonance by altering the tuning of the chamber occasions considerable differences of phase, but that when the weakening is considerable there are relatively slight changes of phase. We can take advantage of this circumstance when compounding the vowel sounds by means of the tuning-forks to produce every possible alteration of phase. It is only necessary to let the lid shade the mouth of the resonance chamber till the strength of the tone is perceptibly diminished. As soon as we have learned how to estimate roughly the amount of diminution of loudness, the above table gives us the corresponding alteration of phase. We are thus able to alter the vibrations of the tones in question to any amount, up to a quarter of the periodic time of a vibration. Alterations of phase to the amount of half the periodic time are produced by sending the electric current through the electro-magnets of the corresponding fork in an opposite direction, which causes the ends of the fork to be repelled instead of attracted by the electro-magnets on the passage of the current, and thus sets the fork vibrating in the contrary direction. This counter-excitement of the fork, however, by repelling currents, must not be continued too long, as the magnetism of the fork itself would otherwise gradually diminish, whereas attracting currents strengthen it or maintain it at a maximum. It is well known that the magnetism of masses of iron that are violently agitated is easily altered.

After a tone has been compounded, in which some of the partials have been weakened and at the same time altered in phase by the half-shading of the apertures of their corresponding resonance chambers, we can re-compound the same tone by an equal amount of weakening in the same partials, but without shading the aperture, and therefore without change of phase, by simply leaving the mouths of the chambers wide open, and increasing their distances from the exciting forks, until the required amount of enfeeblement of sound is attained.

For example, let us first sound the forks $B\flat$ and $b\flat$, with fully opened resonance chambers, and perfect accord. They will vibrate as shewn by the vibrational forms fig. 31, A and B (p. 120a), with the points e and d_0 coincident, and produce at a distance the compound vibration represented by the vibrational curve C. But by closing the resonance chamber of the fork $B\flat$ we can make the point e on the curve B coincide with the points between d_0 and d_1 on the curve A. To make e coincide with d_1, the loudness of $B\flat$ must be made about three-quarters of what it would be if the mouth of the chamber were unshaded. The point e can be made to coincide with d_4 by reversing the current in the electro-magnets and fully opening the mouth of the resonance chamber ; and then by imperfectly opening the chamber of $B\flat$ the point e can be made to move towards δ. On the other hand, an imperfect opening of the chamber $b\flat$ will make e recede from coincidence with δ (which is the same thing as coincidence with d_0) or with d_1, towards d_4 or d_3 respectively. The proportions of loudness may be made the same in all these cases, without any alteration of phase, by removing the corresponding chambers to the proper distance from its forks without shading its mouth.

In this manner every possible difference of phase in the tones of two chambers can be produced. The same process can of course be applied to any required number of forks. I have thus experimented upon numerous combinations of tone with varied differences of phase, and I have never experienced the slightest difference in the quality of tone. So far as the quality of tone was concerned, I found that it was entirely indifferent whether I weakened the separate partial tones by shading the mouths of their resonance chambers, or by moving the chamber itself to a sufficient distance from the fork. Hence the answer to the proposed question is: *the quality of the musical portion of a compound tone depends solely on the number and relative strength of its partial simple tones, and in no respect on their differences of phase.**

¶ The preceding proof that quality of tone is independent of difference of phase, is the easiest to carry out experimentally, but its force lies solely in the theoretical proposition that phases alter contemporaneously with strength of tone when the mouths of the resonance chambers are shaded, and this proposition is the result of mathematical theory alone. We cannot make vibrations of air directly visible. But by a slight change in the experiment it may be so conducted as to make the alteration of phase immediately visible. It is only necessary to put the tuning-forks themselves out of tune with their resonance chambers, by attaching little lumps of wax to the prongs. The same law holds for the phases of a tuning-fork kept in vibration by an electric current, as for the resonance chambers themselves. The phase gradually alters by a quarter period, while the strength of the tone of the fork is reduced from a maximum to nothing at all, by putting it out of tune. The phase of the motion of the air retains the same relation to the phase of the vibration of the fork, because the pitch, which is

¶ determined by the number of interruptions of the electrical current in a second, is not altered by the alteration of the fork. The change of phase in the fork can be observed directly by means of Lissajou's vibration microscope, already described and shewn in fig. 22 (p. 80*d*). Place the prongs of the fork and the microscope of this instrument horizontally, and the fork to be examined vertically; powder the upper end of one of its prongs with a little starch, direct the microscope to one of the grains of starch, and excite both forks by means of the electrical currents of the interrupting fork (fig. 33, p. 122*b*). The fork of Lissajou's instrument is in unison with the interrupting fork. The grain of starch vibrates horizontally, the object-glass of the microscope vertically, and thus, by the composition of these two motions, curves are generated, just as in the observations on violin strings previously described.

When the observed fork is in unison with the interrupting fork, the curve becomes an oblique straight line (fig. 34, 1), if both forks pass through their

¶

FIG. 34.

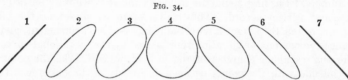

position of rest at the same moment. As the phase alters, the straight line passes through a long oblique ellipse (2, 3), till on the difference of phase becoming a quarter of a period, it develops into a circle (4); and then as the difference of phase increases, it passes through oblique ellipses (5, 6) in another direction, till it reaches another straight line (7), on the difference becoming half a period.

If the second fork is the upper Octave of the interrupting fork, the curves

---

* [The experiments of Koenig with the wave-siren, explained in App. XX. sect. L. art. 6, shew that this law requires a slight modification. Moreover Koenig contends that the 'apparent exception' of p. 127*c*, is an 'actual' one (*ibid.*).—*Translator.*]

1, 2, 3, 4, 5, in fig. 35, shew the series of forms. Here 3 answers to the case when both forks pass through their position of rest at the same time ; 2 and 4 differ from that position by $\frac{1}{12}$, and 1 and 5 by $\frac{1}{4}$ of a wave of the higher fork.

If we now bring the forks into the most perfect possible unison with the interrupting fork, so that both vibrate as strongly as possible, and then alter their

FIG. 35.

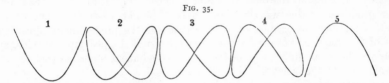

tuning a little by putting on or removing pieces of wax, we also see one figure of the microscopic image gradually passing into another, and can thus easily assure our- ¶ selves of the correctness of the law already cited. Experiments on quality of tone are then conducted by first bringing all the forks as exactly as possible to the pitches of the harmonic upper partial tones of the interrupting fork, next removing the resonance chambers to such distances from the forks as will give the required relations of strength, and finally putting the forks out of tune as much as we please by sticking on lumps of wax. The size of these lumps should be previously so regulated by microscopical observation as to produce the required difference of phase. This, however, at the same time weakens the vibrations of the forks, and hence the strength of the tones must be restored to its former state by bringing the resonance chambers nearer to the forks.

The result in these experiments, where the forks are put out of tune, is the same as in those where the resonance chambers were put out of tune. There is no perceptible alteration of quality of tone. At least there is no alteration so marked as to be recognisable after the expiration of the few seconds necessary ¶ for resetting the apparatus, and hence certainly no such change of quality as would change one vowel into another.

An apparent exception to this rule must here be mentioned. If the forks $B\flat$ and $b\flat$ are not perfectly tuned as Octaves, and are brought into vibration by rubbing or striking, an attentive ear will observe very weak beats which appear like small changes in the strength of the tone and its quality. These beats are certainly connected with the successive entrance of the vibrating forks on varying difference of phase. Their explanation will be given when combinational tones are considered, and it will then be shewn that these slight variations of quality are referable to changes in the strength of one of the simple tones.

Hence we are able to lay down the important law that *differences in musical quality of tone depend solely on the presence and strength of partial tones, and in no respect on the differences in phase under which these partial tones enter into composition.* It must be here observed that we are speaking only of musical ¶ quality as previously defined. When the musical tone is accompanied by un-musical noises, such as jarring, scratching, soughing, whizzing, hissing, these motions are either not to be considered as periodic at all, or else correspond to high upper partials, of nearly the same pitch, which consequently form strident dissonances. We were not able to embrace these in our experiments, and hence we must leave it for the present doubtful whether in such dissonating tones difference of phase is an element of importance. Subsequent theoretic considerations will lead us to suppose that it really is.

If we wish only to imitate vowels by compound tones without being able to distinguish the differences of phase in the individual constituent simple tones, we can effect our purpose tolerably well with organ pipes. But we must have at least two series of them, loud open and soft stopped pipes, because the strength of tone cannot be increased by additional pressure of wind without at the same time changing the pitch. I have had a double row of pipes of this kind made by Herr

Appunn in Hanau, giving the first sixteen partial tones of $B\flat$. All these pipes stand on a common windchest, which also contains the valves by which they can be opened or shut. Two larger valves cut off the passage from the windchest to the bellows. While these valves are closed, the pipe valves are arranged for the required combination of tones, and then one of the main valves of the windchest is opened, allowing all the pipes to sound at once. The character of the vowel is better produced in this way by short jerks of sound, than by a long continued sound. It is best to produce the prime tone and the predominant upper partial tones of the required vowels on both the open and stopped pipes at once, and to open only the weak stopped pipes for the next adjacent tones, so that the strong tone may not be too isolated.

The imitation of the vowels by this means is not very perfect, because, among other reasons, it is impossible to graduate the strength of tone on the different pipes ¶ so delicately as on the tuning-forks, and the higher tones especially are too screaming. But the vowel sounds thus composed are perfectly recognisable.

We proceed now to consider the part played by the ear in the apprehension of quality of tone. The assumption formerly made respecting the function of the ear, was that it was capable of distinguishing both the pitch number of a musical tone (which gives the pitch), and also the *form* of the vibrations (on which the difference of quality depends). This last assertion was based simply on the exclusion of all other possible assumptions. As it could be proved that sameness of pitch always required equal pitch numbers, and as loudness visibly depended upon the amplitude of the vibrations, the quality of tone must necessarily depend on something which was neither the number nor the amplitude of the vibrations. There was nothing left us but form. We can now make this view more definite. The experiments just described shew that waves of very different forms (as fig. 31, C, D, p. 120*a*, and fig. 12, C, D, p. 22*b*), may have the same quality of tone, and ¶ indeed, for every case, except the simple tone, there is an infinite number of forms of wave of this kind, because any alteration of the difference of phase alters the form of wave without changing the quality of tone. The only decisive character of a quality of tone, is that the motion of the air which strikes the ear when resolved into a sum of pendulum vibrations gives the same degree of strength to the same simple vibration.

Hence the ear does not distinguish the different forms of waves in themselves, as the eye distinguishes the different vibrational curves. The ear must be said rather to decompose every wave form into simpler elements according to a definite law. It then receives a sensation from each of these simpler elements as from an harmonious tone. By trained attention the ear is able to become conscious of each of these simpler tones separately. And what the ear distinguishes as different qualities of tone are only different combinations of these simpler sensations.

The comparison between ear and eye is here very instructive. When the ¶ vibrational motion is rendered visible, as in the vibration microscope, the eye is capable of distinguishing every possible different form of vibration one from another, even such as the ear cannot distinguish. But the eye is not capable of directly resolving the vibrations into simple vibrations, as the ear is. Hence the eye, assisted by the above-named instrument, really distinguishes the *form of vibration*, as such, and in so doing distinguishes *every different* form of vibration. The ear, on the other hand, does *not* distinguish *every* different form of vibration, but only such as when resolved into pendular vibrations, give different constituents. But on the other hand, by its capability of distinguishing and feeling these very constituents, it is again superior to the eye, which is quite incapable of so doing.

This analysis of compound into simple pendular vibrations is an astonishing property of the ear. The reader must bear in mind that when we apply the term 'compound' to the vibrations produced by a single musical instrument, the 'composition' has no existence except for our auditory perceptions, or for mathematical theory. In reality, the motion of the particles of the air is not at all compound, it is quite simple, flowing from a single source. When we turn to external nature for an analogue of such an analysis of periodical motions into simple motions, we find none but the phenomena of sympathetic vibration. In reality if we suppose the dampers of a pianoforte to be raised, and allow any musical tone to impinge powerfully on its sounding board, we bring a set of strings into sympathetic vibration, namely *all* those strings, and *only* those, which correspond with the simple tones contained in the given musical tone. Here, then, we have, by a purely mechanical process, a resolution of air waves precisely similar to that performed by the ear. The air wave, quite simple in itself, brings a certain number of strings into sympathetic vibration, and the sympathetic vibration of these strings depends on the same law as the sensation of harmonic upper partial tones in the ear.*

There is necessarily a certain difference between the two kinds of apparatus, because the pianoforte strings readily vibrate with their upper partials in sympathy, and hence separate into several vibrating sections. We will disregard this peculiarity in making our comparison. It would besides be easy to make an instrument in which the strings would not vibrate sensibly or powerfully for any but their prime tones, by simply loading the strings slightly in the middle. This would make their higher proper tones inharmonic to their primes.

Now suppose we were able to connect every string of a piano with a nervous fibre in such a manner that this fibre would be excited and experience a sensation every time the string vibrated. Then every musical tone which impinged on the instrument would excite, as we know to be really the case in the ear, a series of sensations exactly corresponding to the pendular vibrations into which the original motion of the air had to be resolved. By this means, then, the existence of each partial tone would be exactly so perceived, as it really is perceived by the ear. The sensations of simple tones of different pitch would under the supposed conditions fall to the lot of different nervous fibres, and hence be produced quite separately, and independently of each other.

Now, as a matter of fact, later microscopic discoveries respecting the internal construction of the ear, lead to the hypothesis, that arrangements exist in the ear similar to those which we have imagined. The end of every fibre of the auditory nerve is connected with small elastic parts, which we cannot but assume to be set in sympathetic vibration by the waves of sound.

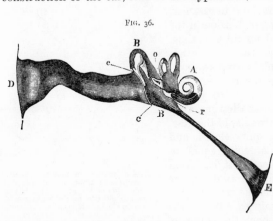

FIG. 36.

The construction of the ear may be briefly described as follows:—The fine ends of the fibres of the auditory nerves are expanded on a delicate membrane in a cavity filled with fluid. Owing to its involved form this cavity is known as the *labyrinth* of the ear. To conduct the vibrations of the air with sufficient force into the fluid of the labyrinth is the office of a second portion of the ear, the *tympănum* or *drum* and the parts within it. Fig. 36 above is a

---

* [Raise the dampers of a piano, and utter the vowel A (*ah*) sharply and loudly, directing it well on to the sound board, pause a second and the vowel will be echoed from the strings. Re-damp, raise the dampers and cry U (*oo*) as before, and that will also be echoed. Re-damp, raise the dampers and cry I (*ee*), and that again will be echoed. The other vowels may be tried in the same way. The echo, though imperfect, is always true enough to surprise a hearer to whom it is new, even if the pitch of the vowel is taken at hazard. It will be improved if the vowels are sung loudly to notes of the piano. The experiment is so easy to make and so fundamental in character, that it should be witnessed by every student.— *Translator*.]

diagrammatic section, of the size of life, shewing the cavities belonging to the auditory apparatus. A is the labyrinth, B B the cavity of the *tympănum* or drum, D the funnel-shaped entrance into the *meătus* or external auditory passage, narrowest in the middle and expanding slightly towards its upper extremity. This *meătus*, in the ear or passage, is a tube formed partly of cartilage or gristle and partly of bone, and it is separated from the tympănum or drum, by a thin circular membrane, the *membrāna tympănī* or *drumskin** c c, which is rather laxly stretched in a bony ring. The *drum* (tym'pănum) B lies between the outer passage (meătus) and the labyrinth. The drum is separated from the labyrinth by bony walls, pierced with two holes, closed by membranes. These are the so-called *windows* (*fenes'trae*) of the labyrinth. The upper one o, called the *oval* window (*fenes'tra ōvālis*), is connected with one of the ossicles or little bones of the ear called the stirrup. The lower or *round* window r (*fenes'tra rotun'da*) has no ¶ connection with these ossicles.

The drum of the ear is consequently completely shut off from the external passage and from the labyrinth. But it has free access to the upper part of the pharynx or throat, through the so-called *Eustachian†* tube E, which in Germany is termed a *trumpet*, because of the trumpet-like expansion of its pharyngeal extremity and the narrowness of its opening into the drum. The end which opens into the drum is formed of bone, but the expanded pharyngeal end is formed of thin flexible cartilage or gristle, split along its upper side. The edges of the split are closed by a sinewy membrane. By closing the nose and mouth, and either condensing the air in the mouth by pressure, or rarefying it by suction, air can be respectively driven into or drawn out of the drum through this tube. At the entrance of air into the drum, or its departure from it, we feel a sudden jerk in the ear, and hear a dull crack. Air passes from the pharynx to the drum, or from the drum to the pharynx only at the moment of making the motion of swallowing. ¶ When the air has entered the drum it remains there, even after nose and mouth are opened again, until we make another motion of swallowing. Then the air leaves the drum, as we perceive by a second cracking in the ear, and the cessation of the feeling of tension in the drumskin which had remained up till that time. These experiments shew that the tube is not usually open, but is opened only during swallowing, and this is explained by the fact that the muscles which raise the *vēlum palātī* or soft palate, and are set in action on swallowing, arise partly from the cartilaginous extremity of the tube. Hence the drum is generally quite closed, and filled with air, which has a pressure equal to ¶ that of the external air, because it has from time to time, that is whenever we swallow, the means of equalising itself with the same by free communication. For a strong pressure of the air, the tube opens even without the action of swallowing, and its power of resistance seems to be very different in different individuals.

In two places, this air in the drum is likewise separated from the fluid of the labyrinth merely by a thin stretched membrane, which closes the two *windows of the*

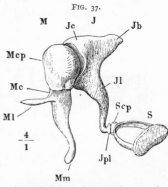

FIG. 37.

Ossicles of the ear in mutual connection, seen from the front, and taken from the right side of the head, which has been turned a little to the right round a vertical axis. M hammer or *malleus*. J anvil or *incus*. S stirrup or *stapes*. Mcp head, Mc neck, Ml long process or *proces'sus grā'cilis*, Mm handle or *manū'brium* of the hammer.—Jc body, Jb short process, Jl long process, Jpl orbicular process or *os orbĭcŭlāre* or *proces'sus lentĭcŭlāris*, of the anvil.—Scp head or *capĭt'ŭlum* of the stirrup.

---

* [In common parlance the *drumskin* of the ear, or *tympanic membrane*, is spoken of as the drum itself. Anatomists as well as drummers distinguish the membranous cover (drumskin) which is struck, from the hollow cavity (drum) which contains the resonant air.

The quantities of the Latin words are marked, as I have heard musicians give them incorrectly.—*Translator.*]

† [Generally pronounced *yoo-stāi'-kĭ-ăn*, but sometimes *yoo-stāi'-shĭ-ăn.—Translator.*]

labyrinth, already mentioned, namely, the *oval* window (o, fig. 36, p. 129c) and the round window (r). Both of these membranes are in contact on their outer side with the air of the drum, and on their inner side with the water of the labyrinth. The membrane of the round window is free, but that of the oval window is connected with the drumskin of the ear by a series of three little bones or auditory ossicles, jointed together. Fig. 37 shews the three ossicles in their natural connection, enlarged four diameters. They are the *hammer* (mal'leus) M, the *anvil* (in'cūs) J, and the stirrup (sta'pes*) S. The hammer is attached to the drumskin, and the stirrup to the membrane of the oval window.

The hammer, shewn separately in fig. 38, has a thick, rounded upper extremity, the *head* cp, and a thinner lower extremity, the *handle* m. Between these two is

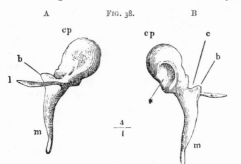

FIG. 38.

A   B

cp   cp   c

b   b

l

m $\frac{4}{1}$ m

Right hammer, A from the front, B from behind. cp head.
c neck, b short, l long process, m handle. * Surface of
the joint.

a contraction c, the *neck*. At the back of the head is the surface of the *joint*, by means of which it fits on to ¶ the anvil. Below the neck, where the handle begins, project two processes, the *long* l, also called *processus Foliānus* and *pr. grăcǐlis*, and the short b, also called *pr. bre'vis*. The long process has the proportionate length shewn in the figure, in children only; in adults it appears to be absorbed down to a little stump. It is directed forwards, and is covered by the bands which fasten the hammer

in front. The *short process* b, on the other hand, is directed towards the drumskin, and presses its upper part a little forwards. From the point of this process b to the point of the handle m the hammer is attached to the upper portion of the ¶

FIG. 39.   FIG. 40.

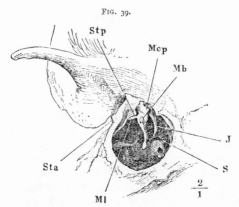

Left temporal bone of a newly-born child, with the auditory
ossicles *in situ*. Sta, spīna tympănica antĕrior. Stp,
spīna tympănica postĕrior. Mcp, head of the hammer.
Mb short, Ml long process of hammer. J anvil. S
stirrup.

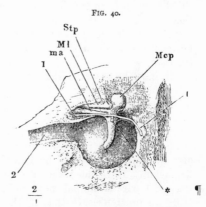

Right drumskin with the hammer, seen from the
inside. The inner layer of the fold of mucous
membrane belonging to the hammer (see
below) is removed. Stp, spīna tympănica
post. Mcp, head of the hammer. Ml, long
process of hammer. ma, ligāmen'tum mallĕi
ant. ı chorda tympănī. 2 Eustachian tube.
* Tendon of the M. tensor tympănī, cut
through close to its insertion.

drumskin, in such a manner that the point of the handle draws the drumskin considerably towards the inner part of the ear.

Fig. 39 above shews the hammer in its natural position as seen from without, after the drumskin has been removed, and fig. 40 shows the hammer lying against the drumskin as seen from within. The hammer is fastened along

---

* [*Stapes* is usually called *stāï'-pēez*. It is
not a classical word, and is usually received as   a contraction for *stătǐpĕs* or foot-rest, also not
classical.— *Translator*.]

the upper margin of the drumskin by a fold of mucous membrane, within which run a series of rather stiff bundles of tendinous fibres. These straps arise in a line which passes from the processus Folianus (fig. 38, 1), above the contraction of the neck, towards the lower end of the surface of the joint for the anvil, and in elderly people is developed into a prominent ridge of bone. The tendinous bands or ligaments are strongest and stiffest at the front and back end of this line of insertion. The front portion of the ligament, lig. mallei anterius (fig. 40, ma), surrounds the processus Folianus, and is attached partly to a bony spine (figs. 39 and 40, Stp) of the osseous ring of the drum, which projects close to the neck of the hammer, and partly to its under edge, and partly falls into a bony fissure which leads towards the articulation of the jaw. The back portion of the same ligament, on the other hand, is attached to a sharp-edged bony ridge projecting inwards from the drumskin, and parallel to it, a little above the opening, through ¶ which a traversing nerve, the chorda tympănī (fig. 40, 1, 1, p. 131c), enters the bone. This second bundle of fibres may be called the lig. mallĕī posterĭus. In fig. 39 (p. 131c) the origin of this ligament is seen as a little projection of the ring to which the drumskin is attached. This projection bounds towards the right the upper edge of the opening for the drumskin, which begins to the left of Stp, exactly at the place where the long process of the anvil makes its appearance in the figure. These two ligaments, front and back, taken together form a moderately tense sinewy chord, round which the hammer can turn as on an axis. Hence even when the two other ossicles have been carefully removed, without loosening these two ligaments, the hammer will remain in its natural position, although not so stiffly as before.

The middle fibres of the broad ligamentous band above mentioned pass outwards towards the upper bony edge of the drumskin. They are comparatively short, and are known as lig. mallei externum. Arising above the line of the axis of the ¶ hammer, they prevent the head from turning too far inwards, and the handle with the drumskin from turning too far outwards, and oppose any down-dragging of the ligament forming the axis. The first effect is increased by a ligament (lig. mallei superius) which passes from the processus Folianus, upwards, into the small slit, between the head of the hammer and the wall of the drum, as shewn in fig. 40 (p. 131c).

It must be observed that in the upper part of the channel of the Eustachian tube, there is a *muscle for tightening the drumskin* (m. tensor tympănī), the tendon of which passes obliquely across the cavity of the drum and is attached to the upper part of the handle of the hammer (at *, fig. 40, p. 131c). This muscle must be regarded as a moderately tense elastic band, and may have its tension temporarily much increased by active contraction. The effect of this muscle is also principally to draw the handle of the hammer inwards, together with the drumskin. But since its point of attachment is so close to the ligamentous axis, ¶ the chief part of its pull acts on this axis, stretching it as it draws it inwards. Here we must observe that in the case of a rectilinear inextensible cord, which is moderately tense, such as the ligamentous axis of the hammer, a slight force which pulls it sideways, suffices to produce a very considerable increase of tension. This is the case with the present arrangement of stretching muscles. It should also be remembered that quiescent muscles not excited by innervation, are always stretched elastically in the living body, and act like elastic bands. This elastic tension can of course be considerably increased by the innervation which brings the muscles into action, but such tension is never entirely absent from the majority of our muscles.

The *anvil*, which is shown separately in fig. 41, resembles a double tooth with two fangs; the surface of its joint with the hammer (at *, fig. 41), replacing the masticating surface. Of the two roots of the tooth which are rather widely separated, the upper, directed backwards, is called the *short process* b; the other, thinner and directed downwards, the *long process of the anvil* l. At the tip of the latter is the knob which articulates with the stirrup. The tip of the short process, on the other hand, by means of a short ligament and an imperfectly

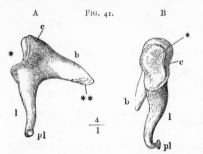

FIG. 41.

Right anvil. A medial surface. B front view. c body. b short, l long process. pl proc. lenticu-lāris or os orbiculāre. ○ Articulation with the head of the hammer. ○○ Surface resting on the wall of the drum.

developed joint at its under surface, is connected with the back wall of the cavity of the drum, at the spot where this passes backwards into the air cavities of the mastoid process behind the ear. The joint between anvil and hammer is a curved depression of a rather irregular form, like a saddle. In its action it may be compared with the joints of the well-known Breguet watchkeys, which have rows of interlocking teeth, offering scarcely any resistance to revolution in one direction, but allowing no revolution what- ¶ ever in the other. Interlocking teeth of this kind are developed upon the under side of the joint between hammer and anvil.

The tooth on the hammer projects towards the drumskin, that of the anvil lies inwards; and, conversely, towards the upper end of the hollow of the joint, the anvil projects outwards, and the hammer inwards. The consequence of this arrangement is that when the hammer is drawn inwards by the handle, it bites the anvil firmly and carries it with it. Conversely, when the drumskin, with the hammer, is driven outwards, the anvil is not obliged to follow it. The interlocking teeth of the surfaces of the joint then separate, and the surfaces glide over each other with very little friction. This arrangement has the very great advantage of preventing any possibility of the stirrup's being torn away from the oval window, when the air in the auditory passage is considerably rarefied. There is also no danger from driving in the hammer, as might happen when the air in the auditory ¶ passage was condensed, because it is powerfully opposed by the tension of the drumskin, which is drawn in like a funnel.

When air is forced into the cavity of the drum in the act of swallowing, the contact of hammer and anvil is loosened. Weak tones in the middle and upper regions of the scale are then not heard much more weakly than usual, but stronger tones are very sensibly damped. This may perhaps be explained by supposing that the adhesion of the articulating surfaces suffices to transfer weak motions from one bone to the other, but that strong impulses cause the surfaces to slide over one another, and hence the tones due to such impulses must be enfeebled.

Deep tones are damped in this case, whether they are strong or weak, perhaps because these always require larger motions to become audible.*

Another important effect on the apprehension of tone, which is due to the above arrangement in the articulation of hammer and anvil, will have to be considered in relation to combinational tones. [See p. 158b.] ¶

Since the attachment of the tip of the short process of the anvil lies sensibly inwards and above the ligamentous axis of the hammer, the head of the hammer separates from the articulating surface between hammer and anvil, when the head is driven outwards, and therefore the handle and drumskin are driven inwards. The consequence is that the ligaments holding the anvil against the hammer, and on the tip of the short process of the anvil, are sensibly stretched, and hence the tip is raised from its osseous support. Consequently in the normal position of the ossicles for hearing, the anvil has no contact with any other bone but the hammer, and both bones are held in position only by stretched ligaments, which are tolerably tight, so that only the revolution of the hammer about its ligamentous axis remains comparatively free.

The third ossicle, the *stirrup*, shewn separately in fig. 42, has really a most striking resemblance to the implement after which it has been named. The foot B

* On this point see Part II. Chapter IX.

is fastened into the membrane of the oval window, and fills it all up, with the exception of a narrow margin. The head cp, has an articulating hole for the tip of the long process of the anvil (processus lenticulāris, or os orbiculāre). The joint is surrounded by a lax membrane. When the drumskin is normally drawn inwards, the anvil presses on the stirrup, so that no tighter ligamentous fastening of the joint is necessary. Every increase in the push on the hammer

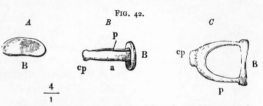

FIG. 42.

Right stirrup : seen, *A* from within, *B* from front, *C* from behind.
B foot. cp, head or capitulum. a Front, p back limb.

arising from the drumskin also occasions an increase in the push of the stirrup ¶ against the oval window ; but in this action the upper and somewhat looser margin of its foot is more displaced than the under, so that the head rises slightly ; this motion again causes a slight elevation of the tip of the long process in the anvil, in the direction conditioned by its position, inwards and underneath the ligamentous axis of the hammer.

The excursions of the foot of the stirrup are always very small, and according to my measurements* never exceed one-tenth of a millimetre ($\cdot$00394 or about $\frac{1}{254}$ of an inch). But the hammer when freed from anvil and stirrup, with its handle moving outwards, and sliding over the articulating surface of the anvil, can make excursions at least nine times as great as it can execute when acting in connection with anvil and stirrup.

The first advantage of the apparatus belonging to the drum of the ear, is that the whole sonorous motion of the comparatively wide surface of the drumskin (vertical diameter 9 to 10 millimetres, [or 0·35 to 0·39 inches,] just over one-third of an ¶ inch ; horizontal diameter, 7·5 to 9 millimetres, [or 0·295 to 0·35 inches,] that is about five-sixths of the former dimensions) is collected and transferred by the ossicles to the relatively much smaller surface of the oval window or of the foot of the stirrup, which is only 1·5 to 3 millimetres [0·06 to 0·12 inches] in diameter. The surface of the drumskin is hence 15 to 20 times larger than that of the oval window.

In this transference of the vibrations of air into the labyrinth it is to be observed that though the particles of air themselves have a comparatively large amplitude of vibration, yet their density is so small that they have no very great moment of inertia, and consequently when their motion is impeded by the drumskin of the ear, they are not capable of presenting much resistance to such an impediment, or of exerting any sensible pressure against it. The fluid in the labyrinth, on the other hand, is much denser and heavier than the air in the auditory passage, and for moving it rapidly backwards and forwards as in sonorous oscillations, a far greater exertion of ¶ pressure is required than was necessary for the air in the auditory passage. On the other hand the amplitude of the vibrations performed by the fluid in the labyrinth are relatively very small, and extremely minute vibrations will in this case suffice to give a vibratory motion to the terminations and appendages of the nerves, which lie on the very limits of microscopic vision.

The mechanical problem which the apparatus within the drum of the ear had to solve, was to transform a motion of great amplitude and little force, such as impinges on the drumskin, into a motion of small amplitude and great force, such as had to be communicated to the fluid in the labyrinth.

A problem of this sort can be solved by various kinds of mechanical apparatus, such as levers, trains of pulleys, cranes, and the like. The mode in which it is solved by the apparatus in the drum of the ear, is quite unusual, and very peculiar.

---

* Helmholtz, 'Mechanism of the Auditory Ossicles,' in Pflueger's *Archiv für Physiologie* vol. i. pp. 34-43. In this paper an attempt is made to prove the correctness of the account of this mechanism given in the text.

A leverage is certainly employed, but only to a moderate extent. The tip of the handle of the hammer, on which the pull of the drumskin first acts, is about once and a half as far from the axis of rotation as that point of the anvil which presses on the stirrup (see fig. 39, p. 131c). The handle of the hammer consequently forms the longer arm of a lever, and the pressure on the stirrup will be once and a half as great as that which drives in the hammer.

The chief means of reinforcement is due to the form of the drumskin. It has been already mentioned that its middle or *navel* (umbĭlĭcus) is drawn inwards by the handle, so as to present a funnel shape. But the meridian lines of this funnel drawn from the navel to the circumference, are not straight lines; they are slightly convex on the outer side. A diminution of pressure in the auditory passage increases this convexity, and an augmentation diminishes it. Now the tension caused in an inextensible thread, having the form of a flat arch, by a force acting perpendicular to its convexity, is very considerable. It is well known that a sensible force ¶ must be exerted to stretch a long thin string into even a tolerably straight horizontal line. The force is indeed very much greater than the weight of the string which pulls the string from the horizontal position.* In the case of the drumskin, it is not gravity which prevents its radial fibres from straightening themselves, but partly the pressure of the air, and partly the elastic pull of the circular fibres of the membrane. The latter tend to contract towards the axis of the funnel-shaped membrane, and hence produce the inflection of the radial fibres towards this axis. By means of the variable pressure of air during the sonorous vibrations of the atmosphere this pull exerted by the circular fibres is alternately strengthened and weakened, and produces an effect on the point where the radial fibres are attached to the tip of the handle of the hammer, similar to that which would happen if we could alternately increase and diminish the weight of a string stretched horizontally, for this would produce a proportionate increase and decrease in the pull exerted by the hand which stretched it. ¶

In a horizontally stretched string such as has been just described, it should be further remarked that an extremely small relaxation of the hand is followed by a considerable fall in the middle of the string. The relaxation of the hand, namely, takes place in the direction of the chord of the arc, and easy geometrical considerations shew that chords of arcs of the same length and different, but always very small curvature, differ very slightly indeed from each other and from the lengths of the arcs themselves.† This is also the case with the drumskin. An extremely little yielding in the handle of the hammer admits of a very considerable change in the curvature of the drumskin. The consequence is that, in sonorous vibrations, the parts of the drumskin which lie between the inner attachment of this membrane to the hammer and its outer attachment to the ring of the drum, are able to follow the oscillations of the air with considerable freedom, while the motion of the air is transmitted to the handle of the hammer with much diminished amplitude but much increased force. After this, as the motion passes from the ¶ handle of the hammer to the stirrup, the leverage already mentioned causes a second and more moderate reduction of the amplitude of vibration with corresponding increase of force.

We now proceed to describe the innermost division of the organ of hearing, called the *labyrinth*. Fig. 43 (p. 134c) represents a cast of its cavity, as seen from different positions. Its middle portion, containing the *oval window* Fv (fenestra vestĭbŭlī) that receives the foot of the stirrup, is called the *vestibule of the labyrinth*.

---

* [The following quatrain, said to have been unconsciously produced by Vince, as a corollary to one of the propositions in his 'Mechanics,' will serve to impress the fact on a non-mathematical reader:—

Hence no force, however great,
Can stretch a cord, however fine,
Into a horizontal line,
So as to make it truly straight.—*Translator*.]

† The amount of difference varies as the square of the depth of the arc. If the length of the arc be l, and the distance of its middle from the chord be s, the chord is shorter than the arc by the length $\frac{8}{3} \cdot \frac{s^2}{l}$.

From the vestibule proceeds forwards and underwards, a spiral canal, the *snail-shell* or cochlĕa, at the entrance to which lies the *round window* Fc (fenestra cochlĕae), which is turned towards the cavity of the drum.   Upwards and backwards, on the other hand, proceed three semicircular canals from the vestibule, the *horizontal, front vertical*, and *back vertical semicircular canals*, each of which debouches with both its mouths in the vestibule, and each of which has at one end a bottle-shaped enlargement, or *ampulla* (ha, vaa, vpa).   The aquaeductus vestĭbŭlī shewn in the figure, Av, appears (from Dr. Fr. E. Weber's investigations) to form a communication between the water of the labyrinth, and the spaces for lymph within the cranium.   The rough places Tsf and * are casts of canals which introduce nerves.

¶ The whole of this cavity of the labyrinth is filled with fluid, and surrounded by the extremely hard close mass of the petrous bone, so that there are only two yielding spots on its walls, the two windows, the oval Fv, and the round Fc.   Into the first, as already described, is fastened the foot of the stirrup, by a narrow membranous margin.   The second is closed by a membrane.   When the stirrup is driven against the oval window, the whole mass of fluid in the labyrinth is necessarily driven against the round window, as the only spot where its walls can give way.   If, as Politzer did, we put a finely drawn glass tube as a manometer into the round window, without in other respects injuring the labyrinth, the water in this tube will be driven upwards as soon as a strong pressure of air acts on the

FIG. 43.

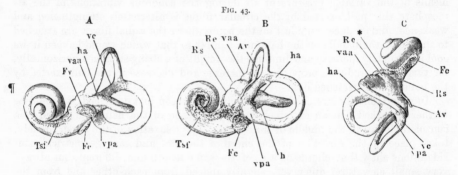

A, left labyrinth from without. B, right labyrinth from within. C, left labyrinth from above. Fc, fenestra cochleae or round window. Fv, fenestra vestĭbŭlī, or oval window. Re, recessus ellipticus. Rs, recessus sphaericus. h, horizontal semicircular canal. ha, ampulla of the same. vaa, ampulla of the front vertical semicircular canal. vpa, ampulla of the back vertical semicircular canal. vc, common limb of the two vertical semicircular canals. Av, cast of the aquaeductus vestĭbŭlī. Tsf, tractus spīrālis forāmĭnōsus. * Cast of the little canals which debouch on the pÿrāmis vestĭbŭlī.

outside of the drumskin and causes the foot of the stirrup to be driven into the oval window.

¶ The terminations of the auditory nerve are spread over fine membranous formations, which lie partly floating and partly expanded in the hollow of the bony labyrinth, and taken together compose the *membranous labyrinth*.   This last has on the whole the same shape as the bony labyrinth.   But its canals and cavities are smaller, and its interior is divided into two separate sections; first the *utrĭcŭlus* with the *semicircular canals*, and second the *saccŭlus* with the *membranous cochlĕa*.   Both, the utriculus and the sacculus lie in the vestibule of the bony labyrinth ; the utriculus opposite to the *recessus elliptĭcus* (Re, fig. 43 above), the sacculus opposite to the *recessus sphaerĭcus* (Rs).   These are floating bags filled with water, and only touch the wall of the labyrinth at the point where the nerves enter them.

The form of the utriculus with its membranous semicircular canals is shewn in fig. 44.   The ampullae project much more in the membranous than in the bony semicircular canals.   According to the recent investigations of Rüdinger, the membranous semicircular canals do not float in the bony ones, but are fastened to the convex side of the latter.   In each ampulla there is a pad-like prominence directed inwards, into which fibriles of the auditory nerve enter; and on the utriculus there is a place which is flatter and thickened. The peculiar manner in which the nerves terminate in this place will be described hereafter. Whether these, and the whole apparatus of the semicircular canals, assist in the sensation of hearing, has latterly been rendered very doubtful. [See p. 151*b*.]

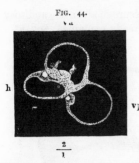

FIG. 44.

Utriculus and membranous semicircular canals (left side) seen from without. va front, vp back vertical, h horizontal semicircular canal.

In the inside of the utriculus is found the auditory *sand*, consisting of little crystals of lime connected by means of a mucous mass with each other and with the thickened places where the nerves are so abundant. In the hollow of the bony vestibule, near the utriculus, and fastened to it, but not communicating with it, lies the sacculus, also provided with a similar thickened ¶ spot full of nerves. A narrow canal connects it with the canal of the membranous cochlea. As to the cavity of the cochlea, we see by fig. 43 opposite, that it is exactly similar to the shell of a garden snail; but the canal of the cochlea is divided into two almost completely separated galleries, by a transverse partition, partly bony and partly membranous. These galleries communicate only at the vertex of the cochlea through a small opening, the *hĕlĭcotrēma*, bounded by the *hāmŭlus* or hook-shaped termination of its central axis or *mŏdĭ'ŏlus*. Of the two galleries into which the cavity of the bony cochlea is divided, one communicates directly with the vestibule and is hence called the *vestibule gallery* (scāla vestĭbŭlī). The other gallery is cut off from the vestibule by the membranous partition, but just at its base, where it begins, is the round window, and the yielding membrane, which closes this, allows the fluid in the gallery to exchange vibrations with the air in the drum. Hence this is called the *drum gallery* (scāla tympănī). ¶

Finally, it must be observed that the membranous partition of the cochlea is not a single membrane, but a membranous canal (ductus cochleāris). Its inner margin is turned towards the central axis or mŏdĭŏlus, and attached to the rudimentary bony partition (lāmĭna spīrālis). A part of the opposite external surface is attached to the inner surface of the bony gallery. Fig. 45 shews the bony parts of a cochlea which has been laid open, and fig. 46 (p. 138*a*), a transverse section of the canal (which is imperfect on the left hand at bottom). In both figures Ls denotes the bony part of the partition, and in fig. 46 v and b are the two unattached parts of the membranous canal. The transverse ¶ section of this canal is, as the figure shews, nearly triangular, so that an angle of the triangle near Lls is attached to the edge of the bony partition. The commencement of the ductus cochlearis at the base of the cochlea, communicates, as already stated, by means of a narrow membranous canal with the sacculus in the vestibule. Of the two unattached strips of its membranous walls, the one facing the vestibule gallery is a soft membrane, incapable of offering much resistance—*Reissner's membrane* (membrāna vestibulāris, v, fig. 46, p. 138*a*); but the other, the *membrāna băsĭlāris* (b), is a firm, tightly stretched, elastic membrane, striped radially, corresponding to its radial fibres. It splits easily in the direction of these fibres, shewing that it is but loosely

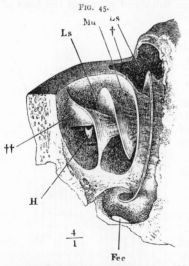

FIG. 45.

Bony cochlea (right side) opened in front. Md, mŏdĭŏ.us. Ls, lāmĭna spīrālis. H, hāmŭus. Fec, fenestra cochleae. † Section of the partition of the cochlea. †† Upper extremity of the same.

connected in a direction transverse to them. The terminations of the nerves of the cochlea and their appendages, are attached to the membrāna bǎsĭlāris, as is shewn by the dotted lines in fig. 46.

When the drumskin is driven inwards by increased pressure of air in the auditory passage, it also forces the auditory ossicles inwards, as already explained, and as a consequence the foot of the stirrup penetrates deeper into the oval window. The fluid of the labyrinth, being surrounded in all other places by firm bony walls, has only one means of escape,—the round window with its yielding membrane. To reach it, the fluid of the labyrinth must either pass

¶ through the hĕlĭcŏtrēma, the narrow opening at the vertex of the cochlea, flowing over from the vestibule gallery into the drum gallery, or, as it would probably not have sufficient time to do this in the case of sonorous vibrations, press the membranous partition of the cochlea against the drum gallery. The converse action must take place when the air in the auditory passage is rarefied.

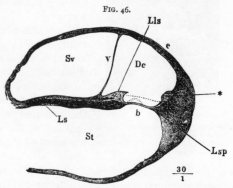

FIG. 46.

Transverse section of a spire of a cochlea which has been softened in hydrochloric acid. Ls, lāmĭna spĭrālis. Lls, limbus lāmĭnae spĭrālis. Sv, scāla vestĭbŭlĭ. St, scala tympănĭ. Dc, ductus cochleāris. Lsp, lĭgāmentum spĭrāle. v, membrāna vestibulāris. b, membrāna bǎsĭlāris. e, outer wall of the ductus cochleāris. * its fillet. The dotted lines shew sections of the membrāna tectōria and the auditory rods.

Hence the sonorous vibrations of the air in the outer auditory passage are finally transferred to the membranes of the labyrinth, more especially those of the cochlea, and to the expansions of the nerves upon them.

¶ The terminal expansions of these nerves, as I have already mentioned, are connected with very small elastic appendages, which appear adapted to excite the nerves by their vibrations.

The nerves of the vestibule terminate in the thickened places of the bags of the membranous labyrinth, already mentioned (p. 137a), where the tissue has a greater and almost cartilaginous consistency. One of these places provided with nerves, projects like a fillet into the inner part of the ampulla of each semicircular canal, and another lies on each of the little bags in the vestibule. The nerve fibres here enter between the soft cylindrical cells of the fine cuticle (ĕpĭthēlĭum) which covers the internal surface of the fillets. Projecting from the internal surface

¶ of this epithelium in the ampullae, Max Schultze discovered a number of very peculiar, stiff, elastic hairs, shewn in fig. 47. They are much longer than the vibratory hairs of the ciliated epithe'lium (their length is $\frac{1}{25}$ of a Paris line, [or ·00355 English inch,] in the ray fish), brittle, and running to a very fine point. It is clear that fine stiff hairs of this kind are extremely well adapted for moving sympathetically with the motion of the fluid, and hence for producing mechanical irritation in the nerve fibres which lie in the soft epithelium between their roots.

According to Max Schultze, the corresponding

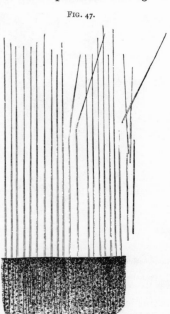

FIG. 47.

thickened fillets in the vestibules, where the nerves terminate, have a similar soft epithelium, and have short hairs which are easily destroyed. Close to these surfaces, which are covered with nerves, lie the calcareous concretions, called *auditory stones* (ŏtŏlĭths), which in fishes form connected convexo-concave solids, shewing on their convex side an impression of the nerve fillet. In human beings, on the other hand, the otoliths are heaps of little crystalline bodies, of a longish angular form, lying close to the membrane of the little bags, and apparently attached to it. These otoliths seem also extremely well suited for producing a mechanical irritation of the nerves whenever the fluid in the labyrinth is suddenly agitated. The fine light membrane, with its interwoven nerves, probably instantly follows the motion of the fluid, whereas the heavier crystals are set more slowly in motion, and hence also yield up their motion more slowly, and thus partly drag and partly squeeze the adjacent nerves. This would satisfy the same conditions of exciting nerves, as Heidenhain's *tetănomōtor*. By this instrument the nerve which acts on a muscle is exposed to the action of a very rapidly oscillating ivory hammer, which at every blow squeezes without bruising the nerve. A powerful and continuous excitement of the nerve is thus produced, which is shewn by a ¶ powerful and continuous contraction of the corresponding muscle. The above parts of the ear seem to be well suited to produce similar mechanical excitement.

The construction of the cochlea is much more complex. The nerve fibres enter through the axis or modiolus of the cochlea into the bony part of the partition, and then come on to the membranous part. Where they reach this, peculiar formations were discovered quite recently (1851) by the Marchese Corti, and have been named after him. On these the nerves terminate.

The expansion of the cochlean nerve is shewn in fig. 48. It enters through the axis (2) and sends out its fibres in a radial direction from the axis through the

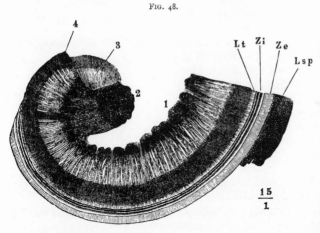

FIG. 48.

bony partition (1, 3, 4), as far as its margins. At this point the nerves pass under the commencement of the membrāna basilāris, penetrate this in a series of openings, and thus reach the ductus cochleāris and those nervous, elastic formations which lie on the inner zone (Zi) of the membrane.

The margin of the bony partition (a to b, fig. 49, p. 140*a*), and the inner zone of the membrana basilaris (a a′) are shewn after Hensen. The under side of the figure * corresponds with the scala tympāni, the upper with the ductus cochleāris. Here h ¶ at the top and k at the bottom, are the two plates of the bony partition, between which the expansion of the nerve b proceeds. The upper side of the bony partition bears a fillet of close ligamentous tissue (Z, fig. 49, also shewn at Lls, fig. 46, p. 138*a*), which, on account of the toothlike impressions on its upper side, is called the toothed layer (*zo'na denticula'ta*), and which carries a peculiar elastic pierced membrane, Corti's membrane, M.C. fig. 49. This membrane is stretched parallel to the membrana basilāris as far as the bony wall on the outer side of the duct, and is there attached a little above the other. Between these two membranes lie the parts in and on which the nerve fibres terminate.

Among these *Corti's arches* (over g in fig. 49) are relatively the most solid formations. The series of these contiguous arches consists of two series of *rods*

* [As the engraving would have been too wide for the page if placed in its proper horizontal position, it has been printed vertically; the *left* side consequently corresponds to the *upper*, and its *right* to the *under* side.—*Translator.*]

or *fibres*, an external and an internal. A single pair of these is shewn in fig. 50, A, below, and a short series of them in fig. 50, B, attached to the membrana basilaris, and at † also connected with the pierced tissue, into which fit the terminal cells of the nerves (fig. 49, c), which will be more fully described presently. These formations are shewn in fig. 51, (p. 141*b*, *c*), as seen from the vestibule gallery ; a is the *denticulated layer*, c the openings for the nerves on the internal margin of the membrana basilaris, its external margin being visible at u u ; d is the inner series of Corti's rods, e the outer ; over these, between e and x is seen the pierced membrane, against which lie the terminal cells of ¶ the nerves.

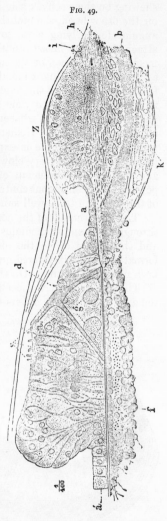

FIG. 49.

The fibres of the first, or outer series, are flat, somewhat S-shaped formations, having a swelling at the spot where they rise from the membrane to which they are attached, and ending in a kind of articulation which serves to connect them with the second or inner series. In fig. 51, p. 141, at d will be seen a great number of these ascending fibres, lying beside each other in regular succession. In the same way they may be seen all along the membrane of the cochlea, close together, so that there must be many thousands of them. Their sides lie close together, and even seem to be connected, leaving however occasional gaps in the line ¶ of connection, and these gaps are probably traversed by nerve fibres. Hence the fibres of the first series as a whole form a stiff layer, which endeavours to erect itself when the natural fastenings no longer resist, but allows the membrane on which they stand to crumple up between the attachments d and e of Corti's arches.

The fibres of the second, or inner series, which form the descending part of the arch e, fig. 50, below, are smooth, flexible, cylindrical threads with thickened ends. The upper extremity forms a kind of joint to connect them with the fibres of the first series, the lower extremity is enlarged in a bell shape and is attached closely to the membrane at the base. In the microscopic preparations they gene-

¶

FIG. 50.

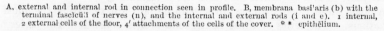

A, external and internal rod in connection seen in profile. B, membrana basi'aris (b) with the terminal fascicul'l of nerves (n), and the internal and external rods (i and e). 1 internal, 2 external cells of the floor, 4' attachments of the cells of the cover. * * epithelium.

rally appear bent in various ways ; but there can be no doubt that in their natural condition they are stretched with some degree of tension, so that they pull down the upper jointed ends of the fibres of the first series. The fibres of the first series arise from the inner margin of the membrane, which can be relatively little agitated, but the fibres of the second series are attached nearly in the middle of the membrane, and this is precisely the place where its vibrations will have the greatest excursions. When the pressure of the fluid in the drum gallery of the labyrinth is increased by driving the foot of the stirrup against the oval window, the membrane at the base of the arches will sink downwards, the fibres of the second series be more tightly stretched, and perhaps the corresponding places of the fibres of the first series be bent a little downwards. It does not, however, seem probable that the fibres of the first series themselves move to any great extent, for their lateral connections are strong enough to make them hang together in masses like a membrane, when they have been released from their attachment in anatomical preparations. On reviewing the whole arrangement, there can be no doubt that Corti's organ is an apparatus adapted for receiving the vibrations of the membrana ¶

FIG. 51.

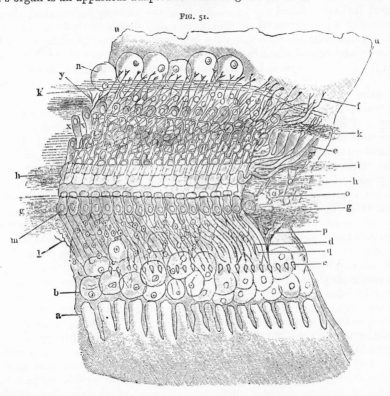

basilaris, and for vibrating of itself, but our present knowledge is not sufficient to determine with accuracy the manner in which these vibrations take place. For this purpose we require to estimate the stability of the several parts and the degree of tension and flexibility, with more precision than can be deduced from such observations as have hitherto been made on isolated parts, as they casually group themselves under the microscope.

Now Corti's fibres are wound round and covered over with a multitude of very delicate, frail formations, fibres and cells of various kinds, partly the finest terminational runners of nerve fibres with appended nerve cells, partly fibres of ligamentous tissue, which appear to serve as a support for fixing and suspending the nerve formations.

The connection of these parts is best shewn in fig. 49 opposite. They are grouped like a pad of soft cells on each side of and within Corti's arches. The most important of them appear to be the cells c and d, which are furnished with hairs, precisely resembling the ciliated cells in the ampullae and utriculus. They appear to be directly connected by fine varicose nerve fibres, and constitute the most constant part of the cochlean formations; for with birds and reptiles, where the structure of the cochlea is much simpler, and even Corti's arches are absent, these little ciliated cells are always to be found, and their hairs are so placed as to strike against Corti's membrane during the vibration of the membrana basilaris. The cells at a and a', fig. 49 (p. 140), which appear in an enlarged condition at b and n in fig. 51 (p. 141), seem to have the character of an epithelium. In fig. 51 there will also be observed bundles and nets of fibres, which may be partly merely supporting fibres of a ligamentous nature, and may partly, to judge by their appearance as strings of beads, possess the character of bundles of the finest fibriles of nerves. But these parts are all so frail and delicate that there is still much doubt as to their connection and office.

¶   The essential result of our description of the ear may consequently be said to consist in having found the terminations of the auditory nerves everywhere connected with a peculiar auxiliary apparatus, partly elastic, partly firm, which may be put in sympathetic vibration under the influence of external vibration, and will then probably agitate and excite the mass of nerves. Now it was shewn in Chap. III., that the process of sympathetic vibration was observed to differ according as the bodies put into sympathetic vibration were such as when once put in motion continued to sound for a long time, or soon lost their motion, p. 39c. Bodies which, like tuning-forks when once struck, go on sounding for a long time, are susceptible of sympathetic vibration in a high degree notwithstanding the difficulty of putting their mass in motion, because they admit of a long accumulation of impulses in themselves minute, produced in them by each separate vibration of the exciting tone. But precisely for this reason there must be the exactest agreement between the pitches of the proper tone of the fork and of the exciting tone, because other-
¶ wise subsequent impulses given by the motion of the air could not constantly recur in the same phase of vibration, and thus be suitable for increasing the subsequent effect of the preceding impulses. On the other hand if we take bodies for which the tone rapidly dies away, such as stretched membranes or thin light strings, we find that they are not only susceptible of sympathetic vibration, when vibrating air is allowed to act on them, but that this sympathetic vibration is not so limited to a particular pitch, as in the other case, and they can therefore be easily set in motion by tones of different kinds. For if an elastic body on being once struck and allowed to sound freely, loses nearly the whole of its motion after ten vibrations, it will not be of much importance that any fresh impulses received after the expiration of this time, should agree exactly with the former, although it would be of great importance in the case of a sonorous body for which the motion generated by the first impulse would remain nearly unchanged up to the time that the second impulse was applied. In the latter case the second impulse could not increase the
¶ amount of motion, unless it came upon a phase of the vibration which had precisely the same direction of motion as itself.

The connection between these two relations can be calculated independently of the nature of the body put into sympathetic vibration,* and as the results are important to enable us to form a judgment on the state of things going on in the ear, a short table is annexed. Suppose that a body which vibrates sympathetically has been set into its state of maximum vibration by means of an exact unison, and that the exciting tone is then altered till the sympathetic vibration is reduced to $\frac{1}{10}$ of its former amount. The amount of the required difference of pitch is given in the first column in terms of an equally tempered Tone, [which is $\frac{1}{6}$ of an Octave]. Now let the same sonorous body be struck, and let its sound be allowed to die away gradually. The number of vibrations which it has made by the time that its intensity is reduced to $\frac{1}{10}$ of its original amount is noted, and given in the second column.

* The mode of calculation is explained in Appendix X.

Difference of Pitch, in terms of an equally tempered Tone, necessary to reduce the intensity of sympathetic vibration to $\frac{1}{10}$ of that produced by perfect unisonance	Number of vibrations after which the intensity of tone in a sonorous body whose sound is allowed to die out, reduces to $\frac{1}{10}$ of its original amount
1. One eighth of a Tone	38·00
2. One quarter of a Tone	19·00
3. One Semitone	9·50
4. Three quarters of a Tone	6·33
5. A whole Tone	4·75
6. A Tone and a quarter	3·80
7. A tempered minor Third or a Tone and a half. . . .	3·17
8. A Tone and three quarters	2·71
9. A tempered major Third or two whole Tones . . .	2·37

Now, although we are not able exactly to discover how long the ear and its individual parts, when set in motion, will continue to sound, yet well-known experiments allow us to form some sort of judgment as to the position which the parts of the ear must occupy in the scale exhibited in this table. Thus, there cannot possibly be any parts of the ear which continue to sound so long as a tuning-fork, for that would be patent to the commonest observation. But even if there were any parts in the ear answering to the first degree of our table, that is requiring 38 vibrations to be reduced to $\frac{1}{10}$ of their force,—we should recognise this in the deeper tones, because 38 vibrations last $\frac{1}{3}$ of a second for A, $\frac{1}{6}$ for a, $\frac{1}{12}$ for a', &c., and such a long endurance of sensible sound would render rapid musical passages impossible in the unaccented and once-accented Octaves. Such a state of things would disturb musical effect as much as the strong resonance of a vaulted room, or as raising the dampers on a piano. When making a shake, we can readily strike 8 or 10 notes in a second, so that each tone separately is struck from 4 to 5 times. If, then, the sound of the first tone had not died off in our ear before the end of the second sound, at least to such an extent as not to be sensible when the latter was sounding, the tones of the shake, instead of being individually distinct, would merge into a continuous mixture of both. Now shakes of this kind, with 10 tones to a second, can be clearly and sharply executed throughout almost the whole scale, although it must be owned that from A downwards, in the great and contra Octaves they sound bad and rough, and their tones begin to mix. Yet it can be easily shewn that this is not due to the mechanism of .the instrument. Thus if we execute a shake on the harmonium, the keys of the lower notes are just as accurately constructed and just as easy to move as those of the higher ones. Each separate tone is completely cut off with perfect certainty at the moment the valve falls on the air passage, and each speaks at the moment the valve is raised, because during so brief an interruption the tongues remain in a state of vibration. Similarly for the violoncello. At the instant when the finger which makes the shake falls on the string, the latter must commence a vibration of a different periodic time, due to its length; and the instant that the finger is removed, the vibration belonging to the deeper tone must return. And yet the shake in the bass is as imperfect on the violoncello as on any other instrument. Runs and shakes can be relatively best executed on a pianoforte because, at the moment of striking, the new tone sounds with great but rapidly decreasing intensity. Hence, in addition to the inharmonic noise produced by the simultaneous continuance of the two tones, we also hear a distinct prominence given to each separate tone. Now, since the difficulty of shaking in the bass is the same for all instruments, and for individual instruments is demonstrably independent of the manner in which the tones are produced, we are forced to conclude that the difficulty lies in the ear itself. We have, then, a plain indication that the vibrating parts of the ear are not damped with sufficient force and rapidity to allow of successfully effecting such a rapid alternation of tones.

Nay more, this fact further proves that *there must be different parts of the ear which are set in vibration by tones of different pitch and which receive the sensation of these tones.* Thus, it might be supposed that as the vibratory mass of the whole ear, the drumskin, auditory ossicles, and fluid in the labyrinth, were vibrating at the same time, the inertia of this mass was the cause why the sonorous vibrations in the ear were not immediately extinguished. But this hypothesis would not suffice to explain the fact observed. For an elastic body set into sympathetic vibration by any tone, vibrates sympathetically in the pitch number of the exciting tone ; but as soon as the exciting tone ceases, it goes on sounding in the pitch number of its own proper tone. This fact, which is derived from theory, may be perfectly verified on tuning-forks by means of the vibration microscope.

If, then, the ear vibrated as a single system, and were capable of continuing its vibration for a sensible time, it would have to do so with its own pitch number, which is totally independent of the pitch number of the former exciting tone. The consequence is that shakes would be equally difficult upon both high and
¶ low tones, and next that the two tones of the shake would not mix with each other, but that each would mix with a third tone, due to the ear itself. We became acquainted with such a tone in the last chapter, the high f'''', p. 116*a*. The result, then, under these circumstances would be quite different from what is observed.

Now if a shake of 10 notes in a second, be made on A, of which the vibrational number is 110, this tone would be struck every $\frac{1}{5}$ of a second. We may justly assume that the shake would not be clear, if the intensity of the expiring tone were not reduced to $\frac{1}{10}$ of its original amount in this $\frac{1}{5}$ of a second. In this case, after at least 22 vibrations, the parts of the ear which vibrate sympathetically with A must descend to at least $\frac{1}{10}$ of their intensity of vibration as their tone expires, so that their power of sympathetic vibration cannot be of the first degree in the table on p. 143*a*, but may belong to the second, third, or some other higher degree. That the degree cannot be any much higher one, is shewn in the first place by the fact that shakes and runs begin to be difficult even on tones which do
¶ not lie much lower. This we shall see by observations on beats subsequently detailed. We may on the whole assume that the parts of the ear which vibrate sympathetically have an amount of damping power corresponding to the third degree of our table, where the intensity of sympathetic vibration with a Semitone difference of pitch is only $\frac{1}{10}$ of what it is for a complete unison. Of course there can be no question of exact determinations, but it is important for us to be able to form at least an approximate conception of the influence of damping on the sympathetic vibration of the ear, as it has great significance in the relations of consonance. Hence when we hereafter speak of individual parts of the ear vibrating sympathetically with a determinate tone, we mean that they are set into strongest motion by that tone, but are also set into vibration less strongly by tones of nearly the same pitch, and that this sympathetic vibration is still sensible for the interval of a Semitone. Fig. 52 may serve to give a general conception of the law by which
¶ the intensity of the sympathetic vibration decreases, as the difference of pitch increases. The horizontal line a b c represents a portion of the musical scale, each of the lengths a b and b c standing for a whole (equally tempered) Tone. Suppose that the body which vibrates sympathetically has been tuned to the tone b and that the vertical line b d represents the maximum

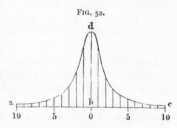

FIG. 52.

of intensity of tone which it can attain when excited by a tone in perfect unison with it. On the base line, intervals of $\frac{1}{10}$ of a whole Tone are set off, and the vertical lines drawn through them shew the corresponding intensity of the tone in the body which vibrates sympathetically, when the exciting tone differs from a unison by the corresponding interval. The following are the numbers from which fig. 52 was constructed :—

Difference of Pitch	Intensity of Sympathetic Vibration	Difference of Pitch	Intensity of Sympathetic Vibration
0·0	100	0·6	7·2
0·1	74	0·7	5·4
0·2	41	0·8	4·2
0·3	24	0·9	3·3
0·4	15	Whole Tone	2·7
Semitone	10		

Now we cannot precisely ascertain what parts of the ear actually vibrate sympathetically with individual tones.* We can only conjecture what they are at present in the case of human beings and mammals. The whole construction of the partition of the cochlea, and of Corti's arches which rest upon it, appears most suited for executing independent vibrations. We do not need to require of them the power of continuing their vibrations for a long time without assistance. ¶

But if these formations are to serve for distinguishing tones of different pitch, and if tones of different pitch are to be equally well perceived in all parts of the scale, the elastic formations in the cochlea, which are connected with different nerve fibres, must be differently tuned, and their proper tones must form a regularly progressive series of degrees through the whole extent of the musical scale.

According to the recent anatomical researches of V. Hensen and C. Hasse, it is probably the breadth of the membrana basilaris in the cochlea, which determines the tuning.† At its commencement opposite the oval window, it is comparatively narrow, and it continually increases in width as it approaches the apex of the cochlea. The following measurements of the membrane in a newly born child, from the line where the nerves pass through on the inner edge, to the attachment to the ligamentum spirale on the outer edge, are given by V. Hensen:

Place of Section	Breadth of Membrane or Length of Transverse Fibres	
	Millimetres	Inches
0·2625 mm. [=0·010335 in.] from root . . .	0·04125	·00162
0·8626 mm. [=0·033961 in.] from root . . .	0·0825	·00325
Middle of the first spire	0·169	·00665
End of first spire	0·3	·01181
Middle of second spire	0·4125	·01624
End of second spire	0·45	·01772
At the hamulus	0·495	·01949

The breadth therefore increases more than twelvefold from the beginning to the end.

Corti's rods also exhibit an increase of size as they approach the vertex of the cochlea, but in a much less degree than the membrana basilaris. The following are Hensen's measurements :— ¶

	at the round window		at the hamulus	
	mm.	inch	mm.	inch
Length of inner rod	0·048	0·00189	0·0855	0·00337
Length of outer rod	0·048	0·00189	0·098	0·00386
Span of the arch . . .	0·019	0·00075	0·085	0·00335

* [Here the passage, 'The particles of auditory sand,' to 'used for musical tones,' on pp. 217-18 of the 1st English edition has been cancelled, and the passage 'We can only conjecture,' to 'without assistance,' on p. 145a added in its place from the 4th German edition. —*Translator*.]

† In the 1st [German] edition of this book (1863), which was written at a time when the more delicate anatomy of the cochlea was just beginning to be developed, I supposed that the different degrees of stiffness and tension in Corti's rods themselves might furnish the reason of their different tuning. By Hensen's measures of the breadth of the membrana basilaris (*Zeitschrift für wissensch. Zoologie*, vol. xiii. p. 492) and Hasse's proof that Corti's rods are absent in birds and amphibia, far more definite foundations for forming a judgment have been furnished, than I then possessed.

Hence it follows, as Henle has also proved, that the greatest increase of breadth falls on the outer zone of the basilar membrane, beyond the line of the attachment of the outer rods. This increases from 0·023 mm. [=·000905 in.] to 0·41 mm. [=·016142 inch] or nearly twentyfold.

In accordance with these measures, the two rows of Corti's rods are almost parallel and upright near to the round window, but they are bent much more strongly towards one another near the vertex of the cochlea.

It has been already mentioned that the membrana basilaris of the cochlea breaks easily in the radial direction, but that its radial fibres have considerable tenacity. This seems to me to furnish a very important mechanical relation, namely that this membrane in its natural connection admits of being tightly stretched in the transverse direction from the modiolus to the outer wall of the cochlea, but can have only little tension in the direction of its length, because it ¶ could not resist a strong pull in this direction.

Now the mathematical theory of the vibration of a membrane with different tensions in different directions shews that it behaves very differently from a membrane which has the same tension in all directions.* On the latter, vibrations produced in one part, spread uniformly in all directions, and hence if the tension were uniform it would be impossible to set one part of the basilar membrane in vibration, without producing nearly as strong vibrations (disregarding individual nodal lines) in all other parts of the membrane.

But if the tension in direction of its length is infinitesimally small in comparison with the tension in direction of the breadth, then the radial fibres of the basilar membrane may be approximatively regarded as forming a system of stretched strings, and the membranous connection as only serving to give a fulcrum to the pressure of the fluid against these strings. In that case the laws of their motion would be the same as if every individual string moved independently ¶ of all the others, and obeyed, by itself, the influence of the periodically alternating pressure of the fluid of the labyrinth contained in the vestibule gallery. Consequently any exciting tone would set that part of the membrane into sympathetic vibration, for which the proper tone of one of its radial fibres that are stretched and loaded with the various appendages already described, corresponds most nearly with the exciting tone; and thence the vibrations will extend with rapidly diminishing strength on to the adjacent parts of the membrane. Fig. 52, on p. 144d, might be taken to represent, on an exaggerated scale of height, a longitudinal section of that part of the basilar membrane in which the proper tone of the radial fibres of the membrane are nearest to the exciting tone.

The strongly vibrating parts of the membrane would, as has been explained in respect to all bodies which vibrate sympathetically, be more or less limited, according to the degree of damping power in the adjacent parts, by friction against the fluid in the labyrinth and in the soft gelatinous parts of the nerve fillet.

¶   Under these circumstances the parts of the membrane in unison with higher tones must be looked for near the round window, and those with the deeper, near the vertex of the cochlea, as Hensen also concluded from his measurements. That such short strings should be capable of corresponding with such deep tones, must be explained by their being loaded in the basilar membrane with all kinds of solid formations; the fluid of both galleries in the cochlea must also be considered as weighting the membrane, because it cannot move without a kind of wave motion in that fluid.

The observations of Hasse shew that Corti's arches do not exist in the cochlea of birds and amphibia, although the other essential parts of the cochlea, as the basilar membrane, the ciliated cells in connection with the terminations of the nerves, and Corti's membrane, which stands opposite the ends of these ciliae, are all present. Hence it becomes very probable that Corti's arches play only a secondary part in the function of the cochlea. Perhaps we might look for the effect

* See Appendix XI.

of Corti's arches in their power, as relatively firm objects, of transmitting the vibrations of the basilar membrane to small limited regions of the upper part of the relatively thick nervous fillet, better than it could be done by the immediate communication of the vibrations of the basilar membrane through the soft mass of this fillet.   Close to the outside of the upper end of the arch, connected with it by the stiffer fibriles of the membrana reticulāris, are the ciliated cells of the nervous fillet (see c in fig. 49, p. 140).   In birds, on the other hand, the ciliated cells form a thin stratum upon the basilar membrane, and this stratum can readily receive limited vibrations from the membrane, without communicating them too far sideways.

According to this view Corti's arches, in the last resort, will be the means of transmitting the vibrations received from the basilar membrane to the terminal appendages of the conducting nerve.   In this sense the reader is requested hereafter to understand references to the vibrations, proper tone, and intonation of ¶ Corti's arches ; the intonation meant is that which they receive through their connection with the corresponding part of the basilar membrane.

According to Waldeyer there are about 4500 outer arch fibres in the human cochlea.   If we deduct 300 for the simple tones which lie beyond musical limits, and cannot have their pitch perfectly apprehended, there remain 4200 for the seven octaves of musical instruments, that is, 600 for every Octave, 50 for every Semitone [that is, 1 for every 2 cents] ; certainly quite enough to explain the power of distinguishing small parts of a Semitone.*   According to Prof. W. Preyer's investigations,† practised musicians can distinguish with certainty a difference of pitch arising from half a vibration in a second, in the doubly accented Octave.   This would give 1000 distinguishable degrees of pitch in the Octave, from 500 to 1000 vibrations in the second.   Towards the limits of the scale the power to distinguish differences diminishes.   The 4200 Corti's arches appear then, in this respect, to be enough to apprehend distinctions of this ¶ amount of delicacy.   But even if it should be found that many more than 4200 degrees of pitch could be distinguished in the Octave, it would not prejudice our assumption.   For if a simple tone is struck having a pitch between those of two adjacent Corti's arches, it would set them both in sympathetic vibration, and that arch would vibrate the more strongly which was nearest in pitch to the proper tonė.   The smallness of the interval between the pitches of two fibres still distinguishable, will therefore finally depend upon the delicacy with which the different forces of the vibrations excited can be compared.   And we have thus also an explanation of the fact that as the pitch of an external tone rises continuously, our sensations also alter continuously and not by jumps, as must be the case if only one of Corti's arches were set in sympathetic motion at once.

To draw further conclusions from our hypothesis, when a simple tone is presented to the ear, those Corti's arches which are nearly or exactly in unison with it will be strongly excited, and the rest only slightly or not at all.   Hence every ¶ simple tone of determinate pitch will be felt only by certain nerve fibres, and

---

* [A few lines of the 1st English edition have here been cancelled, and replaced by others from the 4th German edition.—*Translator.*]

† [*Ueber die Grenzen der Tonwahrnehmung* (On the limits of the perception of tone), June 1876. Rearranged in English by the Translator in the *Proceedings of the Musical Association* for 1876-7, pp. 1-32, under the title of 'On the Sensitiveness of the Ear to Pitch and Change of Pitch in Music.' On p. 11 of this arrangement it is stated that, including Delezenne's results,

at vib.	a difference of	or interval of	
120	·418 vib.	6 cents	} was per-
440	·364 ,,	1·4 ,,	} ceived.

at vib.	a difference of	or interval of	
500	·300 vib.	1·0 cents	} was per-
1000	·500 ,,	·9 ,,	} ceived.

but on the other hand

at vib.	a difference of	or interval of	
60	·200 vib.	6 cents	} was
110	·091 ,,	1·4 ,,	} not
250	·150 ,,	1·0 ,,	} per-
400	·200 ,,	·9 ,,	} ceived,

the intervals perceived, or not perceived, being the same, but the pitches different. And generally throughout the scale a difference of $\frac{1}{5}$ vib. is *not* heard, but

from $G\sharp$	to $g'\sharp$	$\frac{2}{3}$ vib.	
and from	a to c''	$\frac{1}{3}$,,	} is heard.
and from	c to c''	$\frac{1}{2}$,,	

—*Translator.*]

simple tones of different pitch will excite different fibres. When a compound musical tone or chord is presented to the ear, all those elastic bodies will be excited, which have a proper pitch corresponding to the various individual simple tones contained in the whole mass of tones, and hence by properly directing attention, all the individual sensations of the individual simple tones can be perceived. The chord must be resolved into its individual compound tones, and the compound tone into its individual harmonic partial tones.

This also explains how it is that the ear resolves a motion of the air into pendular vibrations and no other. Any particle of air can of course execute only one motion at one time. That we considered such a motion mathematically as a sum of pendular vibrations, was in the first instance merely an arbitrary assumption to facilitate theory, and had no meaning in nature. The first meaning in nature that we found for this resolution came from considering sympathetic ¶ vibration, when we discovered that a motion which was not pendular, could produce sympathetic vibrations in bodies of those different pitches, which corresponded to the harmonic upper partial tones. And now our hypothesis has also reduced the phenomenon of hearing to that of sympathetic vibration, and thus furnished a reason why an originally simple periodic vibration of the air produces a sum of different sensations, and hence also appears as compound to our perceptions.

The sensation of different pitch would consequently be a sensation in different nerve fibres. The sensation of a quality of tone would depend upon the power of a given compound tone to set in vibration not only those of Corti's arches which correspond to its prime tone, but also a series of other arches, and hence to excite sensation in several different groups of nerve fibres.

Physiologically it should be observed that the present assumption reduces sensations which differ qualitatively according to pitch and quality of tone, to a ¶ difference in the nerve fibres which are excited. This is a step similar to that taken in a wider field by Johannes Müller in his theory of the specific energies of sense. He has shewn that the difference in the sensations due to various senses, does not depend upon the actions which excite them, but upon the various nervous arrangements which receive them. We can convince ourselves experimentally that in whatever manner the optic nerve and its expansion, the retina of the eye, may be excited, by light, by twitching, by pressure, or by electricity, the result is never anything but a sensation of light, and that the tactual nerves, on the contrary, never give us sensations of light or of hearing or of taste. The same solar rays which are felt as light by the eye, are felt by the nerves of the hand as heat; the same agitations which are felt by the hand as twitterings, are tone to the ear.

Just as the ear apprehends vibrations of different periodic time as tones of different pitch, so does the eye perceive luminiferous vibrations of different periodic time as different colours, the quickest giving violet and blue, the mean green and ¶ yellow, the slowest red. The laws of the mixture of colours led Thomas Young to the hypothesis that there were three kinds of nerve fibres in the eye, with different powers of sensation, for feeling red, for feeling green, and for feeling violet. In reality this assumption gives a very simple and perfectly consistent explanation of all the optical phenomena depending on colour. And by this means the qualitative differences of the sensations of sight are reduced to differences in the nerves which receive the sensations. For the sensations of each individual fibre of the optic nerve there remains only the quantitative differences of greater or less irritation.

The same result is obtained for hearing by the hypothesis to which our investigation of quality of tone has led us. The qualitative difference of pitch and quality of tone is reduced to a difference in the fibres of the nerve receiving the sensation, and for each individual fibre of the nerve there remains only the quantitative differences in the amount of excitement.

The processes of irritation within the nerves of the muscles, by which their contraction is determined, have hitherto been more accessible to physiological investigation than those which take place in the nerves of sense. In those of the muscle, indeed, we find only quantitative differences of more or less excitement, and no qualitative differences at all. In them we are able to establish, that during excitement the electrically active particles of the nerves undergo determinate changes, and that these changes ensue in exactly the same way whatever be the excitement which causes them. But precisely the same changes also take place in an excited nerve of sense, although their consequence in this case is a sensation, while in the other it was a motion; and hence we see that the mechanism of the process of irritation in the nerves of sense must be in every respect similar to that in the nerves of motion. The two hypotheses just explained really reduce the processes in the nerves of man's two principal senses, notwithstanding their apparently involved qualitative differences of sensations, to the same simple scheme with which we are familiar in the nerves of motion. Nerves have been often and not unsuitably compared to telegraph wires. Such a wire conducts one ¶ kind of electric current and no other; it may be stronger, it may be weaker, it may move in either direction; it has no other qualitative differences. Nevertheless, according to the different kinds of apparatus with which we provide its terminations, we can send telegraphic despatches, ring bells, explode mines, decompose water, move magnets, magnetise iron, develop light, and so on. So with the nerves. The condition of excitement which can be produced in them, and is conducted by them, is, so far as it can be recognised in isolated fibres of a nerve, everywhere the same, but when it is brought to various parts of the brain, or the body, it produces motion, secretions of glands, increase and decrease of the quantity of blood, of redness and of warmth of individual organs, and also sensations of light, of hearing, and so forth. Supposing that every qualitatively different action is produced in an organ of a different kind, to which also separate fibres of nerve must proceed, then the actual process of irritation in individual nerves may always be precisely the same, just as the electrical current in the tele- ¶ graph wires remains one and the same notwithstanding the various kinds of effects which it produces at its extremities. On the other hand, if we assume that the same fibre of a nerve is capable of conducting different kinds of sensation, we should have to assume that it admits of various kinds of processes of irritation, and this we have been hitherto unable to establish.

In this respect then the view here proposed, like Young's hypothesis for the difference of colours, has a still wider signification for the physiology of the nerves in general.

Since the first publication of this book, the theory of auditory sensation here explained, has received an interesting confirmation from the observations and experiments made by V. Hensen * on the auditory apparatus of the Crustaceae. These animals have bags of auditory stones (otoliths), partly closed, partly opening outwards, in which these stones float freely in a watery fluid and are supported by hairs of a peculiar formation, attached to the stones at one end, and, ¶ partly, arranged in a series proceeding in order of magnitude, from larger and thicker to shorter and thinner. In many crustaceans also we find precisely similar hairs on the open surface of the body, and these must be considered as auditory hairs. The proof that these external hairs are also intended for hearing, depends first on the similarity of their construction with that of the hairs in the bags of otoliths; and secondly on Hensen's discovery that the sensation of hearing remained in the *Mysis* (opossum shrimp) when the bags of otoliths had been extirpated, and the external auditory hairs of the antennae were left.

Hensen conducted the sound of a keyed bugle through an apparatus formed on the model of the drumskin and auditory ossicles of the ear into the water of a little box in which a specimen of Mysis was fastened in such a way as to allow the external auditory hairs of the tail to be observed. It was then seen that certain tones of the horn set certain hairs into strong vibration, and other tones

* *Studien über das Gehörorgan der Deca-* and Kölliker's *Zeitschrift für wissenschaftliche* *poden*, Leipzig, 1863. Reprinted from Siebold   *Zoologie*, vol. xiii.

other hairs. Each hair answered to several notes of the horn, and from the notes mentioned we can approximatively recognise the series of under tones of one and the same simple tone. The results could not be quite exact, because the resonance of the conducting apparatus must have had some influence.

Thus one of these hairs answered strongly to $d\sharp$ and $d'\sharp$, more weakly to g, and very weakly to G. This leads us to suppose that it was tuned to some pitch between d'' and $d''\sharp$. In that case it answered to the second partial of d' to $d'\sharp$, the third of g to $g\sharp$, the fourth of d to $d\sharp$, and the sixth of G to $G\sharp$. A second hair answered strongly to $a\sharp$ and the adjacent tones, more weakly to $d\sharp$ and $A\sharp$. Its proper tone therefore seems to have been $a\sharp$.

¶ By these observations (which through the kindness of Herr Hensen I have myself had the opportunity of verifying) the existence of such relations as we have supposed in the case of the human cochlea, have been directly proved for these Crustaceans, and this is the more valuable, because the concealed position and ready destructibility of the corresponding organs of the human ear give little hope of our ever being able to make such a direct experiment on the intonation of its individual parts.*

So far the theory which has been advanced refers in the first place only to the lasting sensation produced by regular and continued periodical oscillations. But as regards *the perception of irregular motions of the air, that is, of noises*, it is clear that an elastic apparatus for executing vibrations could not remain at absolute rest in the presence of any force acting upon it for a time, and even a momentary motion or one recurring at irregular intervals would suffice, if only powerful enough, to set it in motion. The peculiar advantage of resonance over proper tone depends precisely on the fact that disproportionately weak individual impulses, provided that they succeed each other in correct rhythm, are capable of producing comparatively considerable motions. On the other hand, momentary

¶ but strong impulses, as for example those which result from an electric spark, will set every part of the basilar membrane into an almost equally powerful initial motion, after which each part would die off in its own proper vibrational period. By that means there might arise a simultaneous excitement of the whole of the nerves in the cochlea, which although not equally powerful would yet be proportionately gradated, and hence could not have the character of a determinate pitch. Even a weak impression on so many nerve fibres will produce a clearer impression than any single impression in itself. We know at least that small differences of brightness are more readily perceived on large than on small parts of the circle of vision, and little differences of temperature can be better perceived by plunging the whole arm, than by merely dipping a finger, into the warm water.

Hence a perception of momentary impulses by the cochlear nerves is quite possible, just as noises are perceived, without giving an especially sensible prominence to any determinate pitch.

¶ If the pressure of the air which bears on the drumskin lasts a little longer, it will favour the motion in some regions of the basilar membrane in preference to other parts of the scale. Certain pitches will therefore be especially prominent. This we may conceive thus : every instant of pressure is considered as a pressure that will excite in every fibre of the basilar membrane a motion corresponding to itself in direction and strength and then die off ; and all motions in each fibre which are thus excited are added algebraically, whence, according to circumstances, they reinforce or enfeeble each other.† Thus a uniform pressure which lasts during the first half vibration, that is, as long as the first positive excursion, increases the excursion of the vibrating body. But if it lasts longer it weakens the effect first produced. Hence rapidly vibrating bodies would be proportionably less excited by such a pressure, than those for which half a vibration lasts as long as, or longer than, the pressure itself. By this means such an

* [From here to the end of this chapter is an addition from the 4th German edition.— *Translator.*]   † See the mathematical expression for this conception at the end of Appendix XI.

impression would acquire a certain, though an ill-defined, pitch. In general the intensity of the sensation seems, for an equal amount of *vis viva* in the motion, to increase as the pitch ascends. So that the impression of the highest strongly excited fibre preponderates.

A determinate pitch, to a more remarkable extent, may also naturally result, if the pressure itself which acts on the stirrup of the drum alternates several times between positive and negative. And thus all transitional degrees between noises without any determinate pitch, and compound tones with a determinate pitch may be produced. This actually takes place, and herein lies the proof, on which Herr S. Exner * has properly laid weight, that such noises must be perceived by those parts of the ear which act in distinguishing pitch.

In former editions of this work I had expressed a conjecture that the auditory ciliae of the ampullae, which seemed to be but little adapted for resonance, and those of the little bags opposite the otoliths, might be especially active in the ¶ perception of noises.

As regards the ciliae in the ampullae, the investigations of Goltz have made it extremely probable that they, as well as the semicircular canals, serve for a totally different kind of sensation, namely for the perception of the turning of the head. Revolution about an axis perpendicular to the plane of one of the semicircular canals cannot be immediately transferred to the ring of water which lies in the canal, and on account of its inertia lags behind, while the relative shifting of the water along the wall of the canal might be felt by the ciliae of the nerves of the ampullae. On the other hand, if the turning continues, the ring of water itself will be gradually set in revolution by its friction against the wall of the canal, and will continue to move, even when the turning of the head suddenly ceases. This causes the illusive sensation of a revolution in the contrary direction, in the well-known form of giddiness. Injuries to the semicircular canals without injuries to the brain produce the most remarkable disturbances of equilibrium in the lower ¶ animals. Electrical discharges through the ear and cold water squirted into the ear of a person with a perforated drumskin, produce the most violent giddiness. Under these circumstances these parts of the ear can no longer with any probability be considered as belonging to the sense of hearing. Moreover impulses of the stirrup against the water of the labyrinth adjoining the oval window are in reality ill adapted for producing streams through the semicircular canals.

On the other hand the experiments of Koenig with short sounding rods, and those of Preyer with Appunn's tuning-forks, have established the fact that very high tones with from 4000 to 40,000 vibrations in a second can be heard, but that for these the sensation of interval is extremely deficient. Even intervals of a Fifth or an Octave in the highest positions are only doubtfully recognised and are often wrongly appreciated by practised musicians. Even the major Third c^v—e^v [4096 : 5120 vibrations] was at one time heard as a Second, at another as a Fourth or a Fifth ; and at still greater heights even Octaves and Fifths were confused. ¶

If we maintain the hypothesis, that every nervous fibre hears in its own peculiar pitch, we should have to conclude that the vibrating parts of the ear which convey these sensations of the highest tones to the ear, are much less sharply defined in their capabilities of resonance, than those for deeper tones. This means that they lose any motion excited in them comparatively soon, and are also comparatively more easily brought into the state of motion necessary for sensation. This last assumption must be made, because for parts which are so strongly damped, the possibility of adding together many separate impulses is very limited, and the construction of the auditory ciliae in the little bags of the otoliths seems to me more suited for this purpose than that of the shortest fibres of the basilar membrane. If this hypothesis is confirmed we should have to regard the auditory ciliae as the bearers of squeaking, hissing, chirping, crackling sensations of sound, and to consider their reaction as differing only in degree from that of the cochlear fibres.†

* *Pflueger, Archiv. für Physiologie*, vol. xiii.   † [See App. XX. sect. L. art. 5.—*Translator.*

PART II.

ON THE INTERRUPTIONS OF HARMONY.

COMBINATIONAL TONES AND BEATS,*
CONSONANCE AND DISSONANCE.

—••—

CHAPTER VII.

COMBINATIONAL TONES.

IN the first part of this book we had to enunciate and constantly apply the proposition that oscillatory motions of the air and other elastic bodies, produced by several sources of sound acting simultaneously, are always the exact sum of the individual motions producible by each source separately. This law is of extreme importance in the theory of sound, because it reduces the consideration of compound cases to those of simple ones. But it must be observed that this law holds ¶ strictly only in the case where the vibrations in all parts of the mass of air and of the sonorous elastic bodies are of *infinitesimally small dimensions* ; that is to say, only when the alterations of density of the elastic bodies are so small that they may be disregarded in comparison with the whole density of the same body ; and in the same way, only when the displacements of the vibrating particles vanish as compared with the dimensions of the whole elastic body. Now certainly in all practical applications of this law to sonorous bodies, the vibrations are always *very small*, and near enough to being *infinitesimally small* for this law to hold with great exactness even for the real sonorous vibrations of musical tones, and by far the greater part of their phenomena can be deduced from that law in conformity with observation. Still, however, there are certain phenomena which result from the fact that this law does *not* hold with perfect exactness for vibrations of elastic bodies, which, though almost always *very small*, are far from being *infinitesimally small*.† One of these phenomena, with which we are here interested, ¶ is the occurrence of *Combinational Tones*, which were first discovered in 1745 by Sorge,‡ a German organist, and were afterwards generally known, although their pitch was often wrongly assigned, through the Italian violinist Tartini (1754), from whom they are often called *Tartini's tones*.§

These tones are heard whenever two musical tones of different pitches are

* [So much attention has recently been paid to the whole subject of this second part —Combinational Tones and Beats—mostly since the publication of the 4th German edition, that I have thought it advisable to give a brief account of the investigations of Koenig, Bosanquet, and Preyer in App. XX. sect. L., and merely add a few footnotes to refer the reader to them where they especially relate to the statements in the text. But the reader should study the text of this second part, so as to be familiar with Prof. Helmholtz's views, before taking up the Appendix.—*Translator.*]

† Helmholtz, on 'Combinational Tones,' in Poggendorff's *Annalen*, vol. xcix. p. 497. *Monatsberichte* of the Berlin Academy, May 22, 1856. From this last an extract is given in Appendix XII.

‡ *Vorgemach musikalischer Composition* (Antechamber of musical composition).

§ [In England they have hence been often called by Tartini's name, *terzi suoni*, or third sounds, resulting from the combination of two.

sounded together, loudly and continuously. The pitch of a combinational tone is generally different from that of either of the generating tones, or of their harmonic upper partials. In experiments, the combinational are readily distinguished from the upper partial tones, by not being heard when only one generating tone is sounded, and by appearing simultaneously with the second tone. Combinational tones are of two kinds. The first class, discovered by Sorge and Tartini, I have termed *differential tones*, because their pitch number is the *difference* of the pitch numbers of the generating tones. The second class of *summational tones*, having their pitch number equal to the *sum* of the pitch numbers of the generating tones, were discovered by myself.

On investigating the combinational tones of two compound musical tones, we find that both the primary and the upper partial tones may give rise to both differential and summational tones. In such cases the number of combinational tones is very great. But it must be observed that generally the differential are ¶ stronger than the summational tones, and that the stronger generating simple tones also produce the stronger combinational tones. The combinational tones, indeed, increase in a much greater ratio than the generating tones, and diminish also more rapidly. Now since in musical compound tones the prime generally predominates over the partials, the differential tones of the two primes are generally heard more loudly than all the rest, and were consequently first discovered. They are most easily heard when the two generating tones are less than an octave apart, because in that case the differential is deeper than either of the two generating tones. To hear it at first, choose two tones which can be held with great force for some time, and form a justly intoned harmonic interval. First sound the low tone and then the high one. On properly directing attention, a weaker low tone will be heard at the moment that the higher note is struck; this is the required combinational tone.* For particular instruments, as the harmonium, the combinational tones can be made more audible by properly tuned resonators. In this ¶ case the tones are generated in the air contained within the instrument. But in other cases, where they are generated solely within the ear, the resonators are of little or no use.

A commoner English name is *grave harmonics*, which is inapplicable, as they are not necessarily graver than both of the generating tones. Prof. Tyndall calls them *resultant tones*. I prefer retaining the Latin expression, first introduced, as Prof. Preyer informs us (*Akustische Untersuchungen*, p. 11), by G. U. A. Vieth (d. 1836 in Dessau) in Gilbert's *Annalen der Physik* 1805, vol. xxi. p. 265, but only for the tones here distinguished as differential, and afterwards used by Scheibler and Prof. Helmholtz. I shall, however, use 'combinational tones' to express all the additional tones which are heard when two notes are sounded at the same time.—*Translator.*]

* [I have found that combinational tones can be made quite audible to a hundred people at once, by means of two flageolet fifes or whistles, blown as strongly as possible. I choose very close dissonant intervals because the great depth of the low tone is much more striking, being very far below anything that can be touched by the instrument itself. Thus g'''' being sounded loudly on one fife by an assistant, I give $f'''\sharp$, when a deep note is instantly heard which, if the interval were pure, would be g, and is sufficiently near to g to be recognised as extremely deep. As a second experiment, the g'''' being held as before, I give first $f'''\sharp$ and then e'''' in succession. If the intervals were pure the combinational tones would jump from g to c'', and in reality, the jump is very nearly the same and quite appreciable.

The differential tones are well heard on the English concertina, for the same reason as on the harmonium. High notes forming Semitones tell well. It is convenient to choose close dissonant intervals for first examples in order to dissipate the old notion that the 'grave harmonic' is necessarily the 'true fundamental bass' of the 'chord.' It is very easy when playing two high generating notes, as g''' and $g'''\sharp$ or the last and a''', to hear at the same time the rattle of the beats (see next chapter) and the deep combinational tones about $F_{\flat}\sharp$ and $G_{\flat}\sharp$, much resembling a thrashing machine two or three fields off. The beats ¶ and the differentials have the same frequency (note p. 11d). See infrà, App. XX. sect. L. art. 5, *f*. The experiment can also be made with $b''c'''$ and $b''b$ b'' on any harmonium. And if all three notes $b''b$, b', c''' are held down together, the ear can perceive the two sets of beats of the upper notes as sharp high rattles, and the beats of the two combinational tones, about the pitch of C, which have altogether a different character and frequency. On the Harmonical, notes $b''c''$ should beat 66, notes b ''b b'' should beat 39·6, and notes '$b''b$ $b''b$ should beat 26·4 in a second, and these should be the pitches of their combinational notes; the two first should therefore beat 26·4 times in a second, and the two last 13·2 times in a second. But the tone 26·4 is so difficult to hear that the beats are not distinct.—*Translator.*]

The following table gives the first differential tones of the usual harmonic intervals :—

Intervals	Ratio of the vibrational numbers	Difference of the same	The combinational tone is deeper than the deeper generating tone by
Octave	1 : 2	1	a Unison
Fifth	2 : 3	1	an Octave
Fourth	3 : 4	1	a Twelfth
Major Third . . .	4 : 5	1	Two Octaves
Minor Third . . .	5 : 6	1	Two Octaves and a major Third
Major Sixth . . .	3 : 5	2	a Fifth
Minor Sixth . . .	5 : 8	3	a major Sixth

or in ordinary musical notation, the generating tones being written as minims and ¶ the differential tones as crotchets—

Octave.   Fifth.   Fourth.   Major Third.   Minor Third.   Major Sixth.   Minor Sixth.

When the ear has learned to hear the combinational tones of pure intervals and sustained tones, it will be able to hear them from inharmonic intervals and in the rapidly fading notes of a pianoforte. The combinational tones from inhar-¶monic intervals are more difficult to hear, because these intervals beat more or less strongly, as we shall have to explain hereafter. The combinational tones arising from such as fade rapidly, for example those of the pianoforte, are not strong enough to be heard except at the first instant, and die off sooner than the generating tones. Combinational tones are also in general easier to hear from the simple tones of tuning-forks and stopped organ pipes than from compound tones where a number of other secondary tones are also present. These compound tones, as has been already said, also generate a number of differential tones by their harmonic upper partials, and these easily distract attention from the differential tones of the primes. Combinational tones of this kind, arising from the upper partials, are frequently heard from the violin and harmonium.

Example.—Take the major Third *c′e′*, ratio of pitch numbers 4 : 5. First difference 1, that is *C*. The first harmonic upper partial of *c′* is *c″*, relative pitch number 8. Ratio of this and *e′*, 5 : 8, difference 3, that is *g*. The first upper partial of *e′* is *e″*, relative pitch number 10; ¶ ratio for this and *c′*, 4 : 10, difference 6, that is *g′*. Then again *c″ e″* have ratio 8 : 10, difference 2, that is *c*. Hence, taking only the first upper partials we have the series of combinational tones 1, 3, 6, 2 or *C, g, g′, c*. Of these the tone 3, or *g*, is often easily perceived.

These multiple combinational tones cannot in general be distinctly heard, except when the generating compound tones contain audible harmonic upper partials. Yet we cannot assert that the combinational tones are absent, where such partials are absent ; but in that case they are so weak that the ear does not readily recognise them beside the loud generating tones and first differential. In the first place theory leads us to conclude that they do exist in a weak state, and in the next place the beats of impure intervals, to be discussed presently, also establish their existence. In this case we may, as Hallstroem suggests,* consider the multiple combinational tones to arise thus : the first differential tone, or *combinational tone of the first order*, by combination with the generating tones themselves, produce other differential tones, or *combinational tones of the second order* ; these again

* Poggendorff's *Annalen*, vol. xxiv. p. 438.

produce new ones with the generators and differentials of the first order, and so on.

Example.—Take two simple tones *c′* and *e′*, ratio 4 : 5, difference 1, differential tone of the first order *C*. This with the generators gives the ratios 1 : 4 and 1 : 5, differences 3 and 4, differential tones of the second order *g*, and *c′* once more. The new tone 3, gives with the generators the ratios 3 : 4 and 3 : 5, differences 1 and 2, giving the differential tones of the third order *C* and *c*, and the same tone 3 gives with the differential of the first order 1, the ratio 1 : 3, difference 2, and hence as a differential of the fourth order *c* once more, and so on. The differential tones of different orders which coincide when the interval is perfect, as it is supposed to be in this example, no longer exactly coincide when the generating interval is not pure ; and consequently such beats are heard, as would result from the presence of these tones. More on this hereafter.

The differential tones of different orders for different intervals are given in the following notes, where the generators are minims, the combinational tones of the ¶ first order crotchets, of the second quavers, and so on. The same tones also occur with compound generators as combinational tones of their upper partials.*

The series are broken off as soon as the last order of differentials furnishes no fresh tones. In general these examples shew that the complete series of harmonic partial tones 1, 2, 3, 4, 5, &c., up to the generators themselves,† is produced.

The second kind of combinational tones, which I have distinguished as *summational*, is generally much weaker in sound than the first, and is only to be heard

* [These examples are best calculated by giving to the notes in the example the numbers representing the harmonics on p. 22c. Thus

Octave, notes 4 : 8. Diff. 8 − 4 = 4.

Fifth, notes 4 : 6. Diff. 6 − 4 = 2.
  2nd order, 4 − 2 = 2, 6 − 2 = 4.

Fourth, notes 6 : 8. Diff. 8 − 6 = 2.
  2nd order, 8 − 2 = 6, 6 − 2 = 4.
  3rd order, 6 − 4 = 2, 6 − 2 = 4.

Major Third, notes 4 : 5. Diff. 5 − 4 = 1.
  2nd. 4 − 1 = 3, 5 − 1 = 4.
  3rd. 4 − 3 = 1, 5 − 3 = 2.
  4th. 4 − 2 = 2, 4 − 1 = 3.

Minor Third, notes 5 : 6. Diff. 6 − 5 = 1.
  2nd. 5 − 1 = 4, 6 − 1 = 5.
  3rd. 5 − 4 = 1, 6 − 4 = 2.
  4th. 4 − 1 = 3, 6 − 2 = 4.
  5th. 6 − 4 = 2, 6 − 3 = 3.

Major Sixth, notes 6 : 10. Diff. 10 − 6 = 4. ¶
  2nd. 10 − 4 = 6, 6 − 4 = 2.
  3rd. 10 − 2 = 8, 6 − 2 = 4.
  4th. 6 − 4 = 2.

Minor Sixth, notes 5 : 8. Diff. 8 − 5 = 3.
  2nd. 5 − 3 = 2, 8 − 3 = 5.
  3rd. 5 − 2 = 3, 8 − 2 = 6.
  4th. 3 − 2 = 1, 5 − 3 = 2.
  5th. 5 − 1 = 4, 8 − 1 = 7.
  6th. 8 − 7 = 5 − 4 = 1, 4 − 2 = 2, 8 − 4 = 4.

The existence of these differential tones of higher orders cannot be considered as completely established.—*Translator.*]

† [See App. XX. sect. L. art. 7, for the influence of such a series on the consonance of simple tones. It is not to be supposed that all these tones are audible. Mr. Bosanquet derives them direct from the generators, see App. XX. sect. L. art. 5, *a.*—*Translator.*]

with decent ease under peculiarly favourable circumstances on the harmonium and polyphonic siren. Scarcely any but the first summational tone can be perceived, having a vibrational number equal to the sum of those of the generators. Of course summational tones may also arise from the harmonic upper partials. Since their vibrational number is always equal to the sum of the other two, they are always higher in pitch than either of the two generators. The following notes will shew their nature for the simple intervals :—

	Octave.	Fifth.	Fourth.	Major Sixth.	Major Third.	Minor Third.	Minor Sixth.*
	2 + 4 = 6.	2 + 3 = 5.	3 + 4 = 7.	3 + 5 = 8.	4 + 5 = 9.	5 + 6 = 11.	5 + 8 = 13.

In relation to music I will here remark at once that many of these summational tones form extremely inharmonic intervals with the generators. Were they not generally so weak on most instruments, they would give rise to intolerable dissonances. In reality, the major and minor Third, and the minor Sixth, sound very badly indeed on the polyphonic siren, where all combinational tones are remarkably loud, whereas the Octave, Fifth, and major Sixth are very beautiful. Even the Fourth on this siren has only the effect of a tolerably harmonious chord of the minor Seventh.

¶   It was formerly believed that the combinational tones were purely subjective, and were produced in the ear itself.† Differential tones alone were known, and these were connected with the beats which usually result from the simultaneous sounding of two tones of nearly the same pitch, a phenomenon to be considered in the following chapters. It was believed that when these beats occurred with sufficient rapidity, the individual increments of loudness might produce the sensation of a new tone, just as numerous ordinary impulses of the air would, and that the frequency of such a tone would be equal to the frequency of the beats. But this supposition, in the first place, does not explain the origin of summational tones, being confined to the differentials ; secondly, it may be proved that under certain conditions the combinational tones exist objectively, independently of the ear which would have had to gather the beats into a new tone ; and thirdly, this supposition cannot be reconciled with the law confirmed by all other experiments, that the only tones which the ear hears, correspond to pendular vibrations of the ¶ air.‡

And in reality a different cause for the origin of combinational tones can be established, which has already been mentioned in general terms (p. 152c). Whenever the vibrations of the air or of other elastic bodies which are set in motion at the same time by two generating simple tones, are so powerful that they can no longer be considered infinitely small, mathematical theory shews that vibrations of the air must arise which have the same frequency as the combinational tones.§

Particular instruments give very powerful combinational tones. The condition

---

* [The notation of the last 5 bars has been altered to agree with the diagram of harmonics of *C* on p. 22c.—*Translator.*]

† [The result of Mr. Bosanquet's and Prof. Preyer's quite recent experiments is to shew that they are so. See App. XX. sect. L. art. 4, *b, c. Translator.*]

‡ For Prof. Preyer's remarks on these objections, and for other objections, see App. XX. sect. L. art. 5, *b, c. —Translator.*]

§ [The tones supposed to arise from beats, and the differential tones thus generated, are essentially distinct, having sometimes the same but frequently different pitch numbers. See App. XX. sect. L. art. 3, *d.—Translator.*]

for their generation is that the same mass of air should be violently agitated by two simple tones simultaneously. This takes place most powerfully in the polyphonic siren,* in which the same rotating disc contains two or more series of holes which are blown upon simultaneously from the same windchest. The air of the windchest is condensed whenever the holes are closed ; on the holes being opened, a large quantity of air escapes, and the pressure is considerably diminished. Consequently the air in the windchest, and partly even that in the bellows, as can be easily felt, comes into violent vibration. If two rows of holes are blown, vibrations arise in the air of the windchest corresponding to both tones, and each row of openings gives vent not to a stream of air uniformly supplied, but to a stream of air already set in vibration by the other tone. Under these circumstances the combinational tones are extremely powerful, almost as powerful, indeed, as the generators. Their objective existence in the mass of air can be proved by vibrating membranes tuned to be in unison with the combinational tones. Such ¶ membranes are set in sympathetic vibration immediately upon both generating tones being sounded simultaneously, but remain at rest if only one or the other of them is sounded. Indeed, in this case the summational tones are so powerful that they make all chords extremely unpleasant which contain Thirds or minor Sixths. Instead of membranes it is more convenient to use the resonators already recommended for investigating harmonic upper partial tones. Resonators are also unable to reinforce a tone when no pendular vibrations actually exist in the air ; they have no effect on a tone which exists only in auditory sensation, and hence they can be used to discover whether a combinational tone is objectively present. They are much more sensitive than membranes, and are well adapted for the clear recognition of very weak objective tones.

The conditions in the harmonium are similar to those in the siren. Here, too, there is a common windchest, and when two keys are pressed down, we have two openings which are closed and opened rhythmically by the tongues. In this case ¶ also the air in the common receptacle is violently agitated by both tones, and air is blown through each opening which has been already set in vibration by the other tongue. Hence in this instrument also the combinational tones are objectively present, and comparatively very distinct, but they are far from being as powerful as on the siren, probably because the windchest is very much larger in proportion to the openings, and hence the air which escapes during the short opening of an exit by the oscillating tongue cannot be sufficient to diminish the pressure sensibly. In the harmonium also the combinational tones are very clearly reinforced by resonators tuned to be in unison with them, especially the first and second differential and the first summational tone.† Nevertheless I have convinced myself, by particular experiments, that even in this instrument the greater part of the force of the combinational tone is generated in the ear itself. I arranged the portvents in the instrument so that one of the two generators was supplied with air by the bellows moved below by the foot, and the second generator was blown by the ¶ reserve bellows, which was first pumped full and then cut off by drawing out the so-called expression-stop, and I then found that the combinational tones were not much weaker than for the usual arrangement. But the objective portion which the resonators reinforce was much weaker. The noted examples given above (pp. 154–5–6) will easily enable any one to find the digitals which must be pressed down in order to produce a combinational tone in unison with a given resonator.

On the other hand, when the places in which the two tones are struck are entirely separate and have no mechanical connection, as, for example, if they come from two singers, two separate wind instruments, or two violins, the reinforcement

* A detailed description of this instrument will be given in the next chapter.

† [The experiments of Bosanquet, App. XX. sect. L. art. 4, b, render it probable that this apparent reinforcement by a resonator arose from imperfect blocking of both ears when using it. See also p. 43d', note.—*Translator.*]

of the combinational tones by resonators is small and dubious. Here, then, there does not exist in the air any clearly sensible pendular vibration corresponding to the combinational tone, and we must conclude that such tones, which are often powerfully audible, are really produced in the ear itself. But analogously to the former cases we are justified in assuming in this case also that the external vibrating parts of the ear, the drumskin and auditory ossicles, are really set in a sufficiently powerful combined vibration to generate combinational tones, so that the vibrations which correspond to combinational tones may really exist objectively in the parts of the ear without existing objectively in the external air. A slight reinforcement of the combinational tone in this case by the corresponding resonator may, therefore, arise from the drumskin of the ear communicating to the air in the resonator those particular vibrations which correspond to the combinational tone.*

Now it so happens that in the construction of the external parts of the ear for ¶ conducting sound, there are certain conditions which are peculiarly favourable for the generation of combinational tones. First we have the unsymmetrical form of the drumskin itself. Its radial fibres, which are externally convex, undergo a much greater alteration of tension when they make an oscillation of moderate amplitude towards the inside, than when the oscillation takes place towards the outside. For this purpose it is only necessary that the amplitude of the oscillation should not be too small a fraction of the minute depth of the arc made by these radial fibres. Under these circumstances deviations from the simple superposition of vibrations arise for very much smaller amplitudes than is the case when the vibrating body is symmetrically constructed on both sides.†

But a more important circumstance, as it seems to me, when the tones are powerful, is the loose formation of the joint between the hammer and anvil (p. 133*b*). If the handle of the hammer is driven inwards by the drumskin, the anvil and stirrup must follow the motion unconditionally. But that is not the case for the ¶ subsequent outward motion of the handle of the hammer, during which the teeth of the two ossicles need not catch each other. In this case the ossicles may *click*. Now I seem to hear this clicking in my own ear whenever a very strong and deep tone is brought to bear upon it, even when, for example, it is produced by a tuning-fork held between the fingers, in which there is certainly nothing that can make any click at all.

This peculiar feeling of mechanical tingling in the ear had long ago struck me when two clear and powerful soprano voices executed passages in Thirds, in which case the combinational tone comes out very distinctly. If the phases of the two tones are so related that after every fourth oscillation of the deeper and every fifth of the higher tone, there ensues a considerable outward displacement of the drumskin, sufficient to cause a momentary loosening in the joint between the hammer and anvil, a series of blows will be generated between the two bones, which would be absent if the connection were firm and the oscillation regular, and these blows ¶ taken together would exactly generate the first differential tone of the interval of a major Third. Similarly for other intervals.

It must also be remarked that the same peculiarities in the construction of a sonorous body which makes it suitable for allowing combinational tones to be heard when it is excited by two waves of different pitch, must also cause a single simple tone to excite in it vibrations corresponding to its harmonic upper partials; the effect being the same as if this tone then formed summational tones with itself.

This result ensues because a simple periodical force, corresponding to pendular vibrations, cannot excite similar pendular vibrations in the elastic body on which it acts, unless the elastic forces called into action by the displacements of the ex-

* [See latter half of Appendix XVI.— *Translator.*]

† See my paper on combinational tones already cited, and Appendix XII. For unsymmetrical vibrating bodies the disturbances are proportional to the first power of the amplitude, whereas for symmetrical ones they are proportional to only the second power of this magnitude, which is very small in both cases.

cited body from its position of equilibrium, are proportional to these displacements themselves. This is always the case so long as these displacements are infinitesimal. But if the amplitude of the oscillations is great enough to cause a sensible deviation from this proportionality, then the vibrations of the exciting tone are increased by others which correspond to its harmonic upper partial tones. That such harmonic upper partials are occasionally heard when tuning-forks are strongly excited, has been already mentioned (p. 54d). I have lately repeated these experiments with forks of a very low pitch. With such a fork of 64 vib. I could, by means of proper resonators, hear up to the fifth partial. But then the amplitude of the vibrations was almost a centimetre ['3937 inch]. When a sharp-edged body, such as the prong of a tuning-fork, makes vibrations of such a length, vortical motions, differing sensibly from the law of simple vibrations, must arise in the surrounding air. On the other hand, as the sound of the fork fades, these upper partials vanish long before their prime, which is itself only very weakly ¶ audible. This agrees with our hypothesis that these partials arise from disturbances depending on the size of the amplitude.

Herr R. Koenig,* with a series of forks having sliding weights by which the pitch might be gradually altered, and provided also with boxes giving a good resonance and possessing powerful tones, has investigated beats and combinational tones, and found that those combinational tones were most prominent which answered to the difference of one of the tones from the partial tone of the other which was nearest to it in pitch; and in this research partial tones as high as the eighth were effective (at least in the number of beats).† He has unfortunately not stated how far the corresponding upper partials were separately recognised by resonators.‡

Since the human ear easily produces combinational tones, for which the principal causes lying in the construction of that organ have just been assigned, it must also form upper partials for powerful simple tones, as is the case for tuning-forks and the masses of air which they excite in the observations described. Hence ¶ we cannot easily have the sensation of a *powerful* simple tone, without having also the sensation of its harmonic upper partials.§

The importance of combinational tones in the construction of chords will appear hereafter. We have, however, first to investigate a second phenomenon of the simultaneous sounding of two tones, the so-called *beats*.

CHAPTER VIII.

ON THE BEATS OF SIMPLE TONES.

WE now pass to the consideration of other phenomena accompanying the simul ¶ taneous sounding of two simple tones, in which, as before, the motions of the air and of the other co-operating elastic bodies without and within the ear may be conceived as an undisturbed coexistence of two systems of vibrations corresponding to the two tones, but where the auditory sensation no longer corresponds to the sum of the two sensations which the tones would excite singly. Beats, which have now to be considered, are essentially distinguished from combinational tones as follows:—In combinational tones the composition of vibrations in the elastic vibrating bodies which are either within or without the ear, undergoes certain disturbances, although the ear resolves the motion which is finally conducted to it,

---

* Poggendorff's *Annal.*, vol. clvii. pp. 177–236.

† [Even with this parenthetical correction, the above is calculated to give an inadequate impression of the results of Koenig's paper, which is more fully described in Appendix XX.

sect. L.—*Translator.*]

‡ [Koenig states that no upper partials could be heard. See Appendix XX. sect. L. art. 2, a.—*Translator.*]

§ [See App. XX. sect. L. art. 1, ii.—*Translator.*]

into a series of simple tones, according to the usual law. In beats, on the contrary, the objective motions of the elastic bodies follow the simple law ; but the composition of the sensations is disturbed. As long as several simple tones of a sufficiently different pitch enter the ear together, the sensation due to each remains undisturbed in the ear, probably because entirely different bundles of nerve fibres are affected. But tones of the same, or of nearly the same pitch, which therefore affect the same nerve fibres, do not produce a sensation which is the sum of the two they would have separately excited, but new and peculiar phenomena arise which we term *interference*, when caused by two perfectly equal simple tones, and *beats* when due to two nearly equal simple tones.

We will begin with *interference*. Suppose that a point in the air or ear is set in motion by some sonorous force, and that its motion is represented by the curve 1, fig. 53. Let

¶ the second motion be precisely the same at the same time and be represented by the curve 2, so that the crests of 2 fall on the crests of 1, and also the troughs of 2 on the troughs of 1. If both motions proceed at once, the whole motion will be their sum, represented by 3, a curve of the same

FIG. 53.

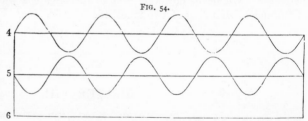

kind but with crests twice as high and troughs twice as deep as those of either of the others. Since the intensity of sound is proportional to the square of the ¶ amplitude, we have consequently a tone not of twice but of four times the loudness of either of the others.

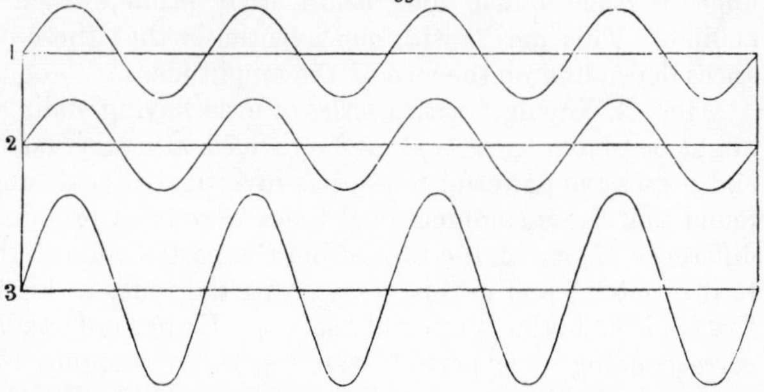

But now suppose the vibrations of the second motion to be displaced by half the periodic time. The curves to be added will stand under one another, as

4 and 5 in fig. 54, and when we come to add to them, the heights of the second curve will be still the same as those of the first, but, being always in the contrary direction, the two will mutually destroy each other, giving as their

FIG. 54.

¶ sum the straight line 6, or no vibration at all. In this case the crests of 4 are added to the troughs of 5, and conversely, so that the crests fill up the troughs, and crests and troughs mutually annihilate each other. The intensity of sound also becomes nothing, and when motions are thus cancelled within the ear, sensation also ceases ; and although each single motion acting alone would excite the corresponding auditory sensation, when both act together there is no auditory sensation at all. One sound in this case completely cancels what appears to be an equal sound. This seems extraordinarily paradoxical to ordinary contemplation because our natural consciousness apprehends sound, not as the motion of particles of the air, but as something really existing and analogous to the sensation of sound. Now as the sensation of a simple tone of the same pitch shews no oppositions of positive and negative, it naturally appears impossible for one positive sensation to cancel another. But the really cancelling things in such a case are the vibrational impulses which the two sources of sound exert on the ear. When it so happens that the vibrational impulses due to one source constantly coincide with opposite ones due to the other, and exactly counterbalance each other, no motion can possibly ensue in the ear, and hence the auditory nerve can experience no sensation.

The following are some instances of sound cancelling sound :—

Put two perfectly similar stopped organ pipes tuned to the same pitch close beside each other on the same portvent. Each one blown separately gives a powerful tone ; but when they are blown together, the motion of the air in the two pipes takes place in such a manner that as the air streams out of one it streams into the other, and hence an observer at a distance hears no tone, but at most the rushing of the air. On bringing the fibre of a feather near to the lips of the pipes, this fibre will vibrate in the same way as if each pipe were blown separately. Also if a tube be conducted from the ear to the mouth of one of the pipes, the tone of that pipe is heard so much more powerfully that it cannot be entirely destroyed by the tone of the other.* ¶

Every tuning-fork also exhibits phenomena of interference, because the prongs move in opposite directions. On striking a tuning-fork and slowly revolving it about its longitudinal axis close to the ear, it will be found that there are four positions in which the tone is heard clearly ; and four intermediate positions in which it is inaudible. The four positions of strong sound are those in which either one of the prongs, or one of the side surfaces of the fork, is turned towards the ear. The positions of no sound lie between the former, almost in planes which make an angle of 45° with the surfaces of the prongs, and pass through the axis of the fork. If in fig. 55, a and b are the ends of the fork seen from above, c, d, e, f will be the four places of strong sound, and the dotted lines

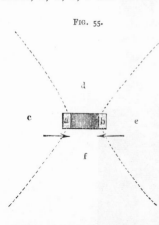

FIG. 55.

the places four of silence. The arrows under a and b shew the mutual motion of the two prongs. Hence while the prong a gives the air about c an impulse in the direction c a, the prong b gives it an ¶ opposite one. Both impulses only partially cancel one another at c, because a acts more powerfully than b. But the dotted lines shew the places where the opposite impulses from a and b are equally strong, and consequently completely cancel each other. If the ear be brought into one of these places of silence and a narrow tube be slipped over one of the prongs a or b, taking care not to touch it, the sound will be immediately augmented, because the influence of the covered prong is almost entirely destroyed, and the uncovered prong therefore acts alone and undisturbed.†

A double siren which I have had constructed is very convenient for the demonstration of these relations.‡ Fig. 56 (p. 162) is a perspective view of this instru- ¶ ment. It is composed of two of Dove's polyphonic sirens, of the kind previously mentioned, p. 13a; a_0 and a_1 are the two windchests, c_0 and c_1 the discs attached to a common axis, on which a screw is introduced at k, to drive a counting apparatus which can be introduced, as described on p. 12b. The upper box a_1 can be turned round its axis, by means of a toothed wheel, in which works a smaller wheel e provided with the driving handle d§. The axis of the box a_1 round which it turns, is a prolongation of the upper pipe g_1, which conducts the wind. On each of the two discs of the siren are four rows of holes, which

* [If a screen of any sort, as the hand, be interposed between the mouths of the pipe, the tone is immediately restored, and then generally remains even if the hand be removed.—*Translator.*]

† [If instead of bringing the tuning-fork to the ear, it be slowly turned before a proper

resonance chamber, the alternation of sound and silence, &c., can be made audible to many persons at once.—*Translator.*]

‡ Constructed by the mechanician Sauerwald in Berlin.

§ [Three turns of the handle cause one turn of the box round its axis.—*Translator.*]

can be either blown separately or together in any combination at pleasure, and at i are the studs for opening and closing the series of holes by a peculiar arrangement.* The lower disc has four rows of 8, 10, 12, 18 holes, the upper of 9, 12, 15, 16. Hence if we call the tone of 8 holes c, the lower disc gives the tones c, e,

FIG. 56.

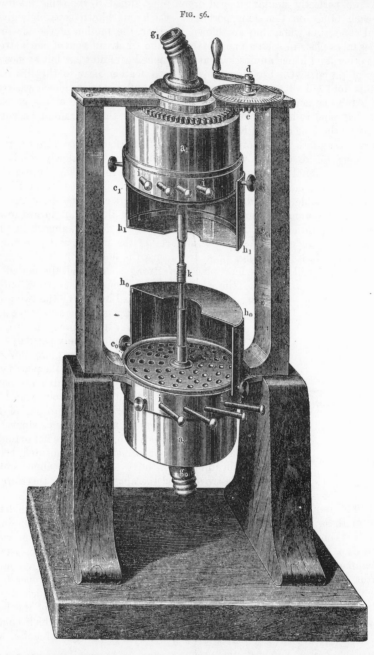

g, d' and the upper d, g, b, c'. We are therefore able to produce the following intervals :—

    1. Unison : gg on the two discs simultaneously.

    2. Octaves : $c\,c'$ and $d\,d'$ on the two.

             * Described in Appendix XIII.

3. Fifths : $c\,g$ and $g\,d'$ either on the lower disc alone or on both discs together.

4. Fourths : $d\,g$ and $g\,c'$ on the upper disc alone or on both together.

5. Major Third : $c\,e$ on the lower alone, and $g\,b$ on the upper alone, or $g\,b$ on both together.

6. Minor Third : eg on the lower, or on both together; $b\,d'$ on both together.

7. Whole Tone [major Tone] : $c\,d$ and $c'\,d'$ on both together [the minor Tone is produced by d and e on both together].

8. Semitone [diatonic Semitone] : $b\,c'$ on the upper.

When both tones are produced from the same disc the objective combinational tones are very powerful, as has been already remarked, p. 157a. But if the tones are produced from different discs, the combinational tones are weak. In the latter case, (and this is the chief point of interest to us at present) the two tones can be made to act together with any desired difference of phase. This is effected by altering the position of the upper box. ¶

We have first to investigate the phenomena as they occur in the unison $g\,g$. The effect of the interference of the two tones in this case is complicated by the fact that the siren produces compound and not simple tones and that the interference of individual partial tones is independent of that of the prime tone and of one another. In order to damp the upper partial tones in the siren by means of a resonance chamber, I caused cylindrical boxes of brass to be made, of which the back halves are shewn at $h_1\,h_1$ and $h_0\,h_0$ fig. 56, opposite. These boxes are each made in two sections, so that they can be removed, and be again attached to the windchest by means of screws. When the tone of the siren approaches the prime tone of these boxes, its quality becomes full, strong and soft, like a fine tone on the French horn ; otherwise the siren has rather a piercing tone. At the same time we use a small quantity of air, but considerable pressure. The circumstances are of the same nature as when a tongue is applied to a resonance chamber of the same pitch. Used in this way the siren is very well adapted for ¶ experiments on interference.

If the boxes are so placed that the puffs of air follow at exactly equal intervals from both discs, similar phases of the prime tone and of all partials coincide, and all are reinforced.

If the handle is turned round half a right angle, the upper box is turned round $\frac{1}{6}$ of a right angle, or $\frac{1}{24}$ of the circumference, that is half the distance between the holes in the series of 12 holes which is in action for g. Hence the difference in the phase of the two primes is half the vibrational period, the puffs of air in one box occur exactly in the middle between those of the other, and the two prime tones mutually destroy each other. But under these circumstances the difference of phase in the upper Octave is precisely the whole of the vibrational period ; that is, they reinforce each other, and similarly all the evenly numbered harmonic upper partials reinforce each other in the same position, and the unevenly numbered ones destroy each other. Hence in the new position the tone is weaker, ¶ because deprived of several of its partials ; but it does not entirely cease ; it rather jumps up an Octave. If we further turn the handle through another half a right angle so that the box is turned through a whole right angle, the puffs of the two discs again agree completely, and the tones reinforce one another. Hence in a complete revolution of the handle there are four positions where the whole tone of the siren appears reinforced, and four intermediate positions where the prime tone and all uneven upper partials vanish, and consequently the Octave occurs in a weaker form accompanied by the evenly numbered upper partials. If we attend to the first upper partial, which is the Octave of the prime, by listening to it through a proper resonator, we find that it vanishes after turning through a quarter of a right angle, and is reinforced after turning through half a right angle, and hence for every complete revolution of the handle it vanishes 8 times, and is reinforced 8 times. The third partial, (or second *upper* partial,) the Twelfth of the prime tone, vanishes in the same time 12 times, the fourth partial 16 times, and so on.

Other compound tones behave like those of the siren. When two tones of the same pitch are sounded together having differences of phase corresponding to half the periodic time, the tone does not vanish, but jumps up an Octave. When, for example, two open organ pipes, or two reed pipes of the same construction and pitch, are placed beside each other on the same portvent, their vibrations generally accommodate themselves in such a manner that the stream of air enters first one and then the other alternately; and while the tone of stopped pipes, which have only unevenly numbered partials, is then almost entirely destroyed, the tone of the open pipes and reed pipes falls into the upper Octave. This is the reason why no reinforcement of tone can be effected on an organ or harmonium by combining tongues or pipes of the same kind, [on the same portvent].

¶ So far we have combined tones of precisely the same pitch; now let us inquire what happens when the tones have slightly different pitch. The double siren just described is also well fitted for explaining this case, for we can slightly alter the pitch of the upper tone by slowly revolving the upper box by means of the handle, the tone becoming flatter when the direction of revolution is the same as that of the disc, and sharper when it is opposite to the same. The vibrational period of a tone of the siren is equal to the time required for a hole in the rotating disc to pass from one hole in the windbox to the next. If, through the rotation of the box, the hole of the box advances to meet the hole of the disc, the two holes come into coincidence sooner than if the box were at rest : and hence the vibrational period is shorter, and the tone sharper. The converse takes place when the revolution is in the opposite direction. These alterations of pitch are easily heard when the box is revolved rather quickly. Now produce the tones of twelve holes on both discs. These will be in absolute unison as long as the upper box is at rest. The two tones constantly reinforce or enfeeble each other according to the position of the upper box. But on setting the upper box in motion, the pitch of

¶ the upper tone is altered, while that of the lower tone, which has an immovable windbox, is unchanged. Hence we have now two tones of slightly different pitch sounding together. And we hear the so-called *beats* of the tones, that is, the intensity of the tone will be alternately greater and less in regular succession.* The arrangement of our siren makes the reason of this readily intelligible. The revolution of the upper box brings it alternately in positions which as we have seen correspond to stronger and weaker tones. When the handle has been turned through a right angle, the windbox passes from a position of loudness through a position of weakness to a position of strength again. Consequently every complete revolution of the handle gives us four beats, whatever be the rate of revolution of the discs, and hence however low or high the tone may be. If we stop the box at the moment of maximum loudness, we continue to hear the loud tone; if at a moment of minimum force, we continue to hear the weak tone.

¶ The mechanism of the instrument also explains the connection between the number of beats and the difference of the pitch. It is easily seen that the number of the puffs is increased by one for every quarter revolution of the handle. But every such quarter revolution corresponds to one beat. Hence *the number of beats in a given time is equal to the difference of the numbers of vibrations executed by the two tones in the same time*. This is the general law which determines the number of beats, for all kinds of tones. This law results immediately from the construction of the siren; in other instruments it can only be verified by very accurate and laborious measurements of the numbers of vibrations.

The process is shewn graphically in fig. 57. Here *c c* represents the series of puffs belonging to one tone, and *d d* those belonging to the other. The distance for *c c* is divided into 18 parts, the same distance is divided into 20 parts for *d d*. At

---

* [The German word *Schwebung*, which might be rendered 'fluctuation,' implies this : The loudest portion only is called the *Stoss*, or 'beat.' But it is not usual to make the distinction in English, where the whole phenomenon alled beats.—*Translator*.]

1, 3, 5, both puffs concur, and the tone is reinforced. At 2 and 4 they are inter-mediate and mutually enfeeble each other. The number of beats for the whole distance is 2, because the difference of the numbers of parts, each of which correspond to a vibration, is also 2.

The intensity of tone varies; swelling from a minimum to a maximum, and lessening from the maximum to the minimum. It is the places of maximum

FIG. 57.

intensity which are properly called *beats*, and these are separated by more or less distinct pauses.

Beats are easily produced on all musical instruments, by striking two notes of ¶ nearly the same pitch. They are heard best from the simple tones of tuning-forks or stopped organ pipes, because here the tone really vanishes in the pauses. A little fluctuation in the pitch of the beating tone may then be remarked.* For the compound tones of other instruments the upper partial tones are heard in the pauses, and hence the tone jumps up an Octave, as in the case of interference already described. If we have two tuning-forks of exactly the same pitch, it is only necessary to stick a little wax on to the end of one, to strike both, and bring them near the same ear or to the surface of a table, or sounding board. To make two stopped pipes beat, it is only necessary to bring a finger slowly near to the lip of one, and thus flatten it. The beats of compound tones are heard by striking any note on a pianoforte out of tune when the two strings belonging to the same note are no longer in unison; or if the piano is in tune it is sufficient to attach a piece of wax, about the size of a pea, to one of the strings. This puts them suffi-ciently out of tune. More attention, however, is required for compound tones ¶ because the enfeeblement of the tone is not so striking. The beat in this case resembles a fluctuation in pitch and quality. This is very striking on the siren according as the brass resonance cylinders (h_0 h_0 and h_1 h_1 of fig. 56, p. 162) are attached or not. These make the prime tone relatively strong. Hence when beats are produced by turning a handle, the decrease and increase of loudness in the tone is very striking. On removing the resonance cylinders, the upper partial tones are relatively powerful, and since the ear is very uncertain when comparing the loudness of tones of different pitch, the alteration of force during the beats is much less striking than that of pitch and quality of tone.

On listening to the upper partials of compound tones which beat, it will be found that these beat also, and that for each beat of the prime tone there are two of the second partial, three of the third, and so on. Hence when the upper partials are strong, it is easy to make a mistake in counting the beats, especially when the beats of the primes are very slow, so that they occur at intervals of a second or two. ¶ It is then necessary to pay great attention to the pitch of the beats counted, and sometimes to apply a resonator.

It is possible to render beats visible by setting a suitable elastic body into sympathetic vibration with them. Beats can then occur only when the two exciting tones lie near enough to the prime tone of the sympathetic body for the latter to be set into sensible sympathetic vibration by both the tones used. This is most easily done with a thin string which is stretched on a sounding board on which have been placed two tuning-forks, both of very nearly the same pitch as the string. On observing the vibrations of the string through a microscope, or attaching a fibril of a goosefeather to the string which will make the same excursions on a magnified scale, the string will be clearly seen to make sympathetic

* See the explanation of this phenomenon which was given me by Mons. G. Guéroult, [the French translator of this work,] in Appen-dix XIV.

vibrations with alternately large and small excursions, according as the tone of the two forks is at its maximum or minimum.

The same effect can be obtained from the sympathetic vibration of a stretched membrane. Fig. 58 is the copy of a drawing made by a vibrating membrane of

FIG. 58.

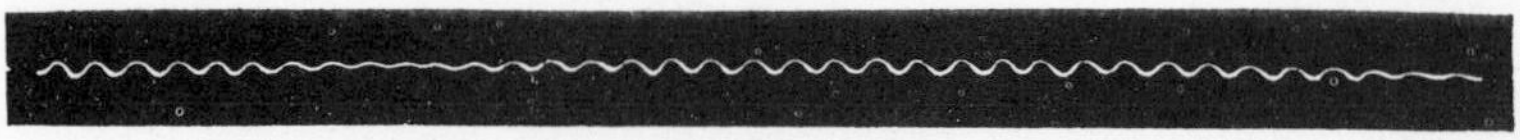

this sort, used in the phonautograph of Messrs. Scott & Koenig, of Paris. The membrane of this instrument, which resembles the drumskin of the ear, carries a small stiff style, which draws the vibrations of the membrane upon a rotating cylinder. In the present case the membrane was set in motion by two organ pipes, that beat. ¶ The undulating line, of which only a part is here given, shews that times of strong vibration have alternated with times of almost entire rest. In this case, then, the beats are also sympathetically executed by the membrane. Similar drawings again have been made by Dr. Politzer, who attached the writing style to the auditory bone (the columella) of a duck, and then produced a beating tone by means of two organ pipes. This experiment shewed that even the auditory bones follow the beats of two tones.*

Generally this must always be the case when the pitches of the two tones struck differ so little from each other and from that of the proper tone of the sympathetic body, that the latter can be put into sensible vibration by both tones at once. Sympathetic bodies which do not damp readily, such as tuning-forks, consequently require two exciting tones which differ extraordinarily little in pitch, in order to shew visible beats, and the beats must therefore be very slow. For bodies readily damped, as membranes, strings, &c., the difference of the exciting ¶ tones may be greater, and consequently the beats may succeed each other more rapidly.

This holds also for the elastic terminal formations of the auditory nerve fibres. Just as we have seen that there may be visible beats of the auditory ossicles, Corti's arches may also be made to beat by two tones sufficiently near in pitch to set the same Corti's arches in sympathetic vibration at the same time. If then, as we have previously supposed, the intensity of auditory sensation in the nerve fibres involved increases and decreases with the intensity of the elastic vibrations, the strength of the sensation must also increase and diminish in the same degree as the vibrations of the corresponding elastic appendages of the nerves. In this case also the motion of Corti's arches must still be considered as compounded of the motions which the two tones would have produced if they had acted separately. According as these motions are directed in the same or in opposite directions they will reinforce or enfeeble each other by (algebraical) addition. It is not till these motions ¶ excite sensation in the nerves that any deviation occurs from the law that each of the two tones and each of the two sensations of tones subsist side by side without disturbance.

We now come to a part of the investigation which is very important for the theory of musical consonance, and has also unfortunately been little regarded by acousticians. The question is: what becomes of the beats when they grow faster and faster? and to what extent may their number increase without the ear being unable to perceive them? Most acousticians were probably inclined to agree with the hypothesis of Thomas Young, that when the beats became very quick they gradually passed over into a combinational tone (the first differential). Young imagined that the pulses of tone which ensue during beats, might have the same

---

* The beats of two tones are also clearly shewn by the vibrating flame described at the end of Appendix II. The flame must be connected with a resonator having a pitch sufficiently near to those of the two generating tones. Even without using the rotating mirror for observing the flames, we can easily recognise the alterations in the shape of the flame which take place isochronously with the audible beats.

effect on the ear as elementary pulses of air (in the siren, for example), and that just as 30 puffs in a second through a siren would produce the sensation of a deep tone, so would 30 beats in a second resulting from any two higher tones produce the same sensation of a deep tone. Certainly this view is well supported by the fact that the vibrational number of the first and strongest combinational tone is actually the number of beats produced by the two tones in a second. It is, however, of much importance to remember that there are other combinational tones (my summational tones), which will not agree with this hypothesis in any respect,* but on the other hand are readily deduced from the theory of combinational tones which I have proposed (in Appendix XII.). It is moreover an objection to Young's theory, that in many cases the combinational tones exist externally to the ear, and are able to set properly tuned membranes or resonators into sympathetic vibration,† because this could not possibly be the case, if the combinational tones were nothing but the series of beats with undisturbed superposition of the two waves. ¶ For the mechanical theory of sympathetic vibration shews that a motion of the air compounded of two simple vibrations of different periodic times, is capable of putting such bodies only into sympathetic vibration as have a proper tone corresponding to one of the two given tones, provided no conditions intervene by which the simple superposition of two wave systems might be disturbed; and the nature of such a disturbance was investigated in the last chapter.‡   Hence we may consider combinational tones as an accessory phenomenon, by which, however, the course of the two primary wave systems and of their beats is not essentially interrupted.

Against the old opinion we may also adduce the testimony of our senses, which teaches us that a much greater number of beats than 30 in a second can be distinctly heard. To obtain this result we must pass gradually from the slower to the more rapid beats, taking care that the tones chosen for beating are not too far apart from each other in the scale, because audible beats are not produced unless ¶ the tones are so near to each other in the scale that they can both make the same elastic appendages of the nerves vibrate sympathetically.§ The number of beats, however, can be increased without increasing the interval between the tones, if both tones are taken in the higher octaves.

The observations are best begun by producing two simple tones of the same pitch, say from the once-accented octave by means of tuning-forks or stopped organ pipes, and slowly altering the pitch of one. This is effected by sticking more and more wax on one of the forks; or more and more covering the mouth of one of the pipes. Stopped organ pipes are also generally provided with a movable plug or lid at the stopped end, in order to tune them; by pulling this out we flatten, by pushing it in we sharpen the tone.**

When a slight difference in pitch has been thus produced, the beats are heard at first as long drawn out fluctuations alternately swelling and vanishing. Slow beats of this kind are by no means disagreeable to the ear. In executing music ¶ containing long sustained chords, they may even produce a solemn effect, or else give a more lively, tremulous or agitating expression. Hence we find in modern

---

* [Prof. Preyer shews, App. XX. sect. L. art. 4, *d*, that summational tones, as suggested by Appunn, may be considered as differential tones of the second order, if such are admitted. —*Translator.*]

† [After the experiments of Prof. Preyer and Mr. Bosanquet, App. XX. sect. L. art. 4, this must be considered as due to some error of observation.—*Translator.*]

‡ [See Bosanquet's theory of 'transformation' in App. XX. sect. L. art. 5, *a.—Translator.*]

§ [Koenig knows no such limitation. See App. XX. sect. L. art. 3.—*Translator.*]

** [A cheap apparatus, useful for demonstration of the following facts, is made with two 'pitch pipes,' each consisting of an extensible stopped pipe, which has the compass of the once-accented octave and is blown as a whistle, the two being connected by a bent tube with a single mouthpiece. By carefully adjusting the lengths of the pipes, I was first able to produce complete destruction of the tone by interference, the sound returning immediately when the mouth of one whistle was stopped by the finger. Then on gradually lengthening one of the pipes the beats began to be heard slowly, and increased in rapidity. The tone being nearly simple the beats are well heard.— *Translator.*]

organs and harmoniums, a stop with two pipes or tongues, adjusted to beat. This imitates the trembling of the human voice and of violins which, appropriately introduced in isolated passages, may certainly be very expressive and effective, but applied continuously, as is unfortunately too common, is a detestable malpractice.

The ear easily follows slow beats of not more than 4 to 6 in a second. The hearer has time to apprehend all their separate phases, and become conscious of each separately, he can even count them without difficulty.* But when the interval between the two tones increases to about a Semitone, the number of beats becomes 20 or 30 in a second, and the ear is consequently unable to follow them sufficiently well for counting. If, however, we begin with hearing slow beats, and then increase their rapidity more and more, we cannot fail to recognise that the sensational impression on the ear preserves precisely the same character, appearing as a series of separate pulses of sound, even when their frequency is so great that we have ¶ no longer time to fix each beat, as it passes, distinctly in our consciousness and count it.†

But while the hearer in this case is quite capable of distinguishing that his ear now hears 30 beats of the same kind as the 4 or 6 in a second which he heard before, the effect of the collective impression of such a rapid beat is quite different. In the first place the mass of tone becomes confused, which I principally refer to the psychological impressions. We actually hear a series of pulses of tone, and are able to recognise it as such, although no longer capable of following each singly or separating one from the other. But besides this psychological consideration, the sensible impression is also unpleasant. Such rapidly beating tones are jarring and rough. The distinctive property of jarring, is the intermittent character of the sound. We think of the letter R as a characteristic example of a jarring tone. It is well known to be produced by interposing the uvula, or else the thin tip of the tongue, in the way of the stream of air passing out of the mouth, ¶ in such a manner as only to allow the air to force its way through in separate pulses, the consequence being that the voice at one time sounds freely, and at another is cut off.‡

Intermittent tones were also produced on the double siren just described by using a little reed pipe instead of the wind-conduit of the upper box, and driving the air through this reed pipe. The tone of this pipe can be heard externally only when the revolution of the disc brings its holes before the holes of the box and open an exit for the air. Hence, if we let the disc revolve while air is driven through the pipe, we obtain an intermittent tone, which sounds exactly like beats arising from two tones sounded at once, although the intermittence is produced by purely mechanical means. Such effects may also be produced in another way on the same siren. Remove the lower windbox and retain only its pierced cover, over which the disc revolves. At the under part apply one extremity of an india-rubber tube against one of the holes in the cover, the other end being conducted ¶ by a proper ear-piece to the observer's ear. The revolving disc alternately opens and closes the hole to which the india-rubber tube has been applied. Hold a tuning-fork in action or some other suitable musical instrument above and near

---

* [See App. XX. sect. B. No. 7, for directions for observing beats.—*Translator.*]

† [The Harmonical is very convenient for this purpose. On the d♭ key is a d_1 one comma lower than d. These dd_1 beat about 9, 18, 36, 73 times in 10 seconds in the different Octaves, the last barely countable. Also e^1♭ and e_1 beat 33, 66, 132, 364 in 10 seconds in the different Octaves. The two first of these sets of beats can be counted, the two last cannot be counted, but will be distinctly perceived as separate pulses. Similarly the beats between all consecutive notes, (except F and G, B and C), can be counted in the lowest Octave, but become rapidly too fast to be followed. As, however, these are not simple tones, the beats are not perfectly clear.—*Translator.*]

‡ [Phonautographic figures of the effect of r, resemble those of fig. 58, p. 166a. Six varieties of these figures are given on p. 19 of Donders's important little tract, on 'The Physiology of Speech Sounds, and especially of those in the Dutch Language' (*De Physiologie der Spraakklanken, in het bijzonder van die der nederlandsche taal.* Utrecht 1870, pp. 24).—*Translator.*]

the rotating disc.   Its tone will be heard intermittently and the number of intermissions can be regulated by altering the velocity of the rotation of the disc.

In both ways then we obtain intermittent tones.   In the first case the tone of the reed pipe as heard in the outer air is interrupted, because it can only escape from time to time.   The intermittent tone in this case can be heard by any number of listeners at once.   In the second case the tone in the outer air is continuous, but reaches the ear of the observer, who hears it through the disc of the siren, intermittently.   It can certainly be heard by one observer only, but then all kinds of musical tones of the most diverse pitch and quality may be employed for the purpose.   The intermission of their tones gives them all exactly the same kind of roughness which is produced by two tones which beat rapidly together.   We thus come to recognise clearly that beats and intermissions are identical, and that either when fast enough produces what is termed a jar or rattle.   ¶

Beats produce intermittent excitement of certain auditory nerve fibres.   The reason why such an intermittent excitement acts so much more unpleasantly than an equally strong or even a stronger continuous excitement, may be gathered from the analogous action of other human nerves.   Any powerful excitement of a nerve deadens its excitability, and consequently renders it less sensitive to fresh irritants.   But after the excitement ceases, and the nerve is left to itself, irritability is speedily re-established in a living body by the influence of arterial blood.   Fatigue and re-freshment apparently supervene in different organs of the body with different velocities ; but they are found wherever muscles and nerves have to operate.   The eye, which has in many respects the greatest analogy to the ear, is one of those organs in which both fatigue and refreshment rapidly ensue.   We need to look at the sun but an instant to find that the portion of the retina, or nervous expansion of the eye, which was affected by the solar light has become less sensitive for other light.   Immediately afterwards on turning our eyes to a uniformly illuminated ¶ surface, as the sky, we see a dark spot of the apparent size of the sun ; or several such spots with lines between them, if we had not kept our eye steady when look-ing at the sun but had moved it right and left.   An instant suffices to produce this effect ; nay, an electric spark, that lasts an immeasurably short time, is fully capable of causing this species of fatigue.

When we continue to look at a bright surface, the impression is strongest at first, but at the same time it blunts the sensibility of the eye, and consequently the impression becomes weaker, the longer we allow the eye to act.   On coming out of darkness into full daylight we feel blinded ; but after a few minutes, when the sensibility of the eye has been blunted by the irritation of the light,—or, as we say, when the eye has grown accustomed to the glare,—this degree of brightness is found very pleasant.   Conversely, in coming from full daylight into a dark vault, we are insensible to the weak light about us, and can scarcely find our way about, yet after a few minutes, when the eye has rested from the effect of the strong light, ¶ we are able to see very well in the semi-dark room.

These phenomena and the like can be conveniently studied in the eye, because individual spots in the eye may be excited and others left at rest, and the sensations of each may be afterwards compared.   Put a piece of black paper on a tolerably well-lighted white surface, look steadily at a point on or near the black paper, and then withdraw the paper suddenly.   The eye sees a secondary image of the black paper on the white surface, consisting of that portion of the white surface where the black paper lay, which now appears brighter than the rest.   The place in the eye where the image of the black paper had been formed, has been rested in com-parison with all those places which had been affected by the white surface, and on removing the black paper this rested part of the eye sees the white surface in its first fresh brightness, while those parts of the retina which had been already fatigued by looking at it, see a decidedly greyer tinge on the whiter surface.

Hence by the continuous uniform action of the irritation of light, this irritation itself blunts the sensibility of the nerve, and thus effectually protects this organ against too long and too violent excitement.

It is quite different when we allow intermittent light to act on the eye, such as flashes of light with intermediate pauses. During these pauses the sensibility is again somewhat re-established, and the new irritation consequently acts much more intensely than if it had lasted with the same uniform strength. Every one knows how unpleasant and annoying is any flickering light, even if it is relatively very weak, coming, for example, from a little flickering taper or rushlight.

The same thing holds for the nerves of touch. Scraping with the nail is far more annoying to the skin than constant pressure on the same place with the same pressure of the nail. The unpleasantness of scratching, rubbing, tickling depends upon the intermittent excitement which they produce in the nerves of touch.

¶   A jarring intermittent tone is for the nerves of hearing what a flickering light is to the nerves of sight, and scratching to the nerves of touch. A much more intense and unpleasant excitement of the organs is thus produced than would be occasioned by a continuous uniform tone. This is even shewn when we hear very weak intermittent tones. If a tuning-fork is struck and held at such a distance from the ear that its sound cannot be heard, it becomes immediately audible if the handle of the fork be revolved by the fingers. The revolution brings it alternately into positions where it can and cannot transmit sound to the ear [p. 161b], and this alternation of strength is immediately perceptible by the ear. For the same reason one of the most delicate means of hearing a very weak, simple tone consists in sounding another of nearly the same strength, which makes from 2 to 4 beats in a second with the first. In this case the strength of the tone varies from nothing to 4 times the strength of the single simple tone, and this increase of strength combines with the alternation to make it audible.

¶   Just as this alternation of strength will serve to strengthen the impression of the very weakest musical tones upon the ear, we must conclude that it must also serve to make the impression of stronger tones much more penetrating and violent, than they would be if their loudness were continuous.

We have hitherto confined our attention to cases where the number of beats did not exceed 20 or 30 in a second. We saw that the beats in the middle part of the scale are still quite audible and form a series of separate pulses of tone. But this does not furnish a limit to their number in a second.

The interval $b'\,c''$ gave us 33 beats in a second, and the effect of sounding the two notes together was very jarring. The interval of a whole tone $b'\flat\,c''$ gives nearly twice as many beats, but these are no longer so cutting as the former. The rule assigns 88 beats in a second to the minor Third $a'\,c''$, but in reality this interval scarcely shews any of the roughness produced by beats from tones at closer intervals. We might then be led to conjecture that the increasing number of beats weakened

¶ their impression and made them inaudible. This conjecture would find an analogy in the impossibility of separating a series of rapidly succeeding impressions of light on the eye, when their number in a second is too large. Think of a glowing stick swung round in a circle. If it executes 10 or 15 revolutions in a second, the eye believes it sees a continuous circle of fire. Similarly for colour-tops, with which most of my readers are probably familiar. If the top be spun at the rate of more than 10 revolutions in a second, the colours upon it mix and form a perfectly unchanging impression of a mixed colour. It is only for very intense light that the alternations of the various fields of colour have to take place more quickly, 20 to 30 times in a second. Hence the phenomena are quite analogous for ear and eye. When the alternation between irritation and rest is too fast, the alternation ceases to be felt, and sensation becomes continuous and lasting.

However, we may convince ourselves that in the case of the ear, an increase of the number of beats in a second is not the only cause of the disappearance of the corresponding sensation. As we passed from the interval of a Semitone $b'\ c''$ to that of a minor Third $a'\ c''$, we not only increased the number of beats, but the width of the interval. Now we can increase the number of beats without increasing the interval by taking it in a higher Octave. Thus taking $b'\ c''$ an Octave higher we have $b''\ c'''$ with 66 beats, and another Octave would give us $b'''\ c''''$ with as many as 132 beats, and these are really audible in the same way as the 33 beats of $b'\ c''$, although they certainly become weaker in the higher positions. Nevertheless the 66 beats of the interval $b''\ c'''$ are much more distinct and penetrating than the same number in the whole Tone $b'\flat c''$, and the 88 of the interval $e'''\ f'''$ are still quite evident, while the 88 of the minor Third $a'\ c''$ are practically inaudible. My assertion that as many as 132 beats in a second are audible will perhaps appear very strange and incredible to acousticians. But the experiment is easy to repeat, and if on an instrument which gives sustained tones, as an organ or harmonium, we strike a series of intervals of a Semitone each, beginning low ¶ down, and proceeding higher and higher, we shall hear in the lower parts very slow beats (B,C gives $4\frac{1}{8}$, $B\ c$ gives $8\frac{1}{4}$, $b\ c'$ gives $16\frac{1}{2}$ beat in a second), and as we ascend the rapidity will increase but the character of the sensation remain unaltered. And thus we can pass gradually from 4 to 132 beats in a second, and convince ourselves that though we become incapable of counting them, their character as a series of pulses of tone, producing an intermittent sensation, remains unaltered. It must be observed, however, that the beats, even in the higher parts of the scale, become much shriller and more distinct, when their number is diminished by taking intervals of quarter tones or less. The most penetrating roughness arises even in the upper parts of the scale from beats of 30 to 40 in a second. Hence high tones in a chord are much more sensitive to an error in tuning amounting to the fraction of a Semitone, than deep ones. While two c' notes which differ from one another by the tenth part of a Semitone, produce about 3 beats in two seconds,* which cannot be observed without considerable attention, ¶ and, at least, gives no feeling of roughness, two c'' notes with the same error give 3 beats in one second, and two c''' notes 6 beats in one second, which becomes very disagreeable. The character of the roughness also alters with the number of beats. Slow beats give a coarse kind of roughness, which may be considered as chattering or jarring ; and quicker ones have a finer but more cutting roughness.

Hence it is not, or at least not solely, the large number of beats which renders them inaudible. The magnitude of the interval is a factor in the result, and consequently we are able with high tones to produce more rapid audible beats than with low tones.

Observation shews us, then, on the one hand, that equally large intervals by no means give equally distinct beats in all parts of the scale. The increasing number of beats in a second renders the beats in the upper part of the scale less distinct. The beats of a Semitone remain distinct to the upper limits of the four-times accented octave [say 4000 vib.], and this is also about the limit for musical ¶ tones fit for the combinations of harmony. The beats of a whole tone, which in deep positions are very distinct and powerful, are scarcely audible at the upper limit of the thrice-accented octave [say at 2000 vib.]. The major and minor Third, on the other hand, which in the middle of the scale [264 to 528 vib.] may be regarded as consonances, and when justly intoned scarcely shew any roughness, are decidedly rough in the lower octaves and produce distinct beats.

On the other hand we have seen that distinctness of beating and the roughness of the combined sounds do not depend solely on the number of beats. For if we could disregard their magnitudes all the following intervals, which by calculation should have 33 beats, would be equally rough :

* [Taking $c' = 264$, a tone one-tenth of a Semitone or 10 cents higher make 265·5 vibrations, and these tones beat $1\frac{1}{2}$ times in a second. The figures in the text have bee altered to these more exact numbers.—*Translator.*]

the Semitone . . $b' c''$ [528–495=33]
the whole Tones . . $e' d'$ [major, 297–264] and $d' e'$ [minor 330–297]
the minor Third . . $e\,g$ [198–165]
the major Third . . $c\,e$ [165–132]
the Fourth . . . $G\,c$ [132–99]
the Fifth . . . $C\,G$ [99–66]

and yet we find that these intervals are more and more free from roughness.*

The roughness arising from sounding two tones together depends, then, in a compound manner on the magnitude of the interval and the number of beats produced in a second. On seeking for the reason of this dependence, we observe that, as before remarked, beats in the ear can exist only when two tones are produced sufficiently near in the scale to set the same elastic appendages of the auditory nerve in sympathetic vibration at the same time. When the two tones produced

¶ are too far apart, the vibrations excited by both of them at once in Corti's organs are too weak to admit of their beats being sensibly felt, supposing of course that no upper partial or combinational tones intervene. According to the assumptions made in the last chapter respecting the degree of damping possessed by Corti's organs (p. 144c), it would result, for example, that for the interval of a whole Tone $c\,d$, such of Corti's fibres as have the proper tone $c\sharp$, would be excited by each of the tones with $\frac{1}{10}$ of its own intensity; and these fibres will therefore fluctuate between the intensities of vibration o and $\frac{4}{10}$. But if we strike the simple tones c and $c\sharp$, it follows from the table there given that Corti's fibres which correspond to the middle between c and $c\sharp$ will alternate between the intensities o and $\frac{12}{10}$. Conversely the same intensity of beats would for a minor Third amount to only 0·194, and for a major Third to only 0·108, and hence would be scarcely perceptible beside the two primary tones of the intensity 1.

Fig. 59, which we used on p. 144d to express the

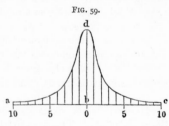

FIG. 59.

¶ intensity of the sympathetic vibration of Corti's fibres for an increasing interval of tone, may here serve to shew the intensity of the beats which two tones excite in the ear when forming different intervals in the scale. But the parts on the base line must now be considered to represent fifths of a *whole* Tone, and not as before of a Semitone. In the present case the distance of the two tones from each other is doubly as great as that between either of them and the intermediate Corti's fibres.

Had the damping of Corti's organs been equally great at all parts of the scale, and had the number of beats no influence on the roughness of the sensation, equal intervals in all parts of the scale would have given equal roughness to the combined tones. But as this is not the case, as the same intervals diminish in roughness

¶ as we ascend in the scale, and increase in roughness as we descend, we must either assume that the damping power of Corti's organs of higher pitch is less than that of those of lower pitch, or else that the discrimination of the more rapid beats meets with certain hindrances in the nature of the sensation itself.

At present I see no way of deciding between these two suppositions; but the former is possibly the more improbable, because, at least with our artificial musical instruments, the higher the pitch of a vibrating body, the more difficulty is experienced in isolating it sufficiently to prevent it from communicating its vibrations to its environment. Very short, high-pitched strings, little metal tongues or plates, &c., yield high tones which die off with great rapidity, whereas it is easy to generate deep tones with correspondingly greater bodies which shall retain their tone for a considerable time. On the other hand the second supposition is favoured by the analogy of another nervous apparatus, the eye. As has been already re-

* [All these intervals can be tried on the Harmonical, but as the tones are compound, the student should listen to the beats of the primes only.—*Translator.*]

marked, a series of impressions of light, following each other rapidly and regularly, excite a uniform and continuous sensation of light in the eye. When the separate luminous irritations follow one another very quickly, the impression produced by each one lasts unweakened in the nerves till the next supervenes, and thus the pauses can no longer be distinguished in sensation. In the eye, the number of separate irritations cannot exceed 24 in a second without being completely fused into a single sensation. In this respect the eye is far surpassed by the ear, which can distinguish as many as 132 intermissions in a second, and probably even that is not the extreme limit. Much higher tones of sufficient strength would probably allow us to hear still more.* It lies in the nature of the thing, that different kinds of apparatus of sensation should shew different degrees of mobility in this respect, since the result does not depend simply on the mobility of the molecules of the nerves, but also depends upon the mobility of the auxiliary apparatus through which the excitement is induced or expressed. Muscles are much less mobile than ¶ the eye ; ten electrical discharges in a second directed through them generally suffice to bring the voluntary muscles into a permanent state of contraction. For the muscles of the involuntary system, of the bowels, the vessels, &c., the pauses between the irritations may be as much as one, or even several seconds long, without any intermission in the continuity of contraction.

The ear is greatly superior in this respect to any other nervous apparatus. It is eminently the organ for small intervals of time, and has been long used as such by astronomers. It is well known that when two pendulums are ticking near one another, the ear can distinguish whether the ticks are or are not coincident, within one hundredth of a second. The eye would certainly fail to determine whether two flashes of light coincided within $\frac{1}{24}$ second ; and probably within a much larger fraction of a second.†

But although the ear shews its superiority over other organs of the body in this respect, we cannot hesitate to assume that, like every other nervous apparatus, ¶ the rapidity of its power of apprehension is limited, and we may even assume that we have approached very near the limit when we can but faintly distinguish 132 beats in a second.

* [In the two high notes g'''' f''''‡ of the flageolet ‧fifes (p. 153d, note), which if justly intoned should give 198 beats in a second, I could hear none, though the tones were very powerful, and the *scream* was very cutting indeed.—In the case of b'' c''', which on the Harmonical are tuned to make 1056 and 990, the rattle of the 66 beats, or thereabouts, is quite distinct, and the differential tone is very powerful at the same time.—*Translator*.]

† [The following is an interesting comparison between eye and ear, and eye and hand. The usual method of observing transits is by counting the pendulum ticks of an astronomical clock, and by observing the distances of the apparent positions of a star before and after passing each bar of the transit instrument at the moments of ticking, to estimate the moment at which it had passed each bar. This is done for five bars and a mean is taken. But a few years ago a chronograph was introduced at Greenwich Observatory, consisting of a uniformly revolving cylinder in which a point pricks a hole every second. Electrical communication being established with a knob on the transit instrument, the observer presses the knob at the moment he sees a star dis-appear behind a bar, and an electrical current causes another point to make a hole between the seconds holes on the chronograph. By applying a scale, the time of transit is thus measured off. A mean, of course, is taken as before. On my asking Mr. Stone (now Astronomer at Oxford, then chief assistant at Greenwich Observatory) as to the relative degree of accuracy of the two methods, he told me that he considered the first gave results to one-tenth, and the second to one-twentieth of a second. It must be remembered that the first method also required a mental estimation which had to be performed in less than a ¶ second, and the result borne in mind, and that this was avoided by the second plan. On the other hand in the latter the sensation had to be conveyed from the eye to the brain, which issued its orders to the hand, and the hand had to obey them. Hence there was an endeavour at performing simultaneously, several acts which could only be successive. Any one will find upon trial that an attempt to merely make a mark at the moment of hearing an expected sound, as, for example, the repeated tick of a common half seconds clock, is liable to great error.—*Translator.*]

CHAPTER IX.

DEEP AND DEEPEST TONES.

BEATS give us an important means of determining the limit of the deepest tones, and of accounting for certain peculiarities of the transition from the sensation of separate pulses of air to a perfectly continuous musical tone, and to this inquiry we now proceed.

The question: what is the smallest number of vibrations in a second which can produce the sensation of a musical tone? has hitherto received very contradictory replies. The estimates of different observers fluctuate between 8 (Savart) ¶ and about 30. The contradiction is explained by the existence of certain difficulties in the experiments.

In the first place it is necessary that the strength of the vibrations of the air for very deep tones should be extremely greater than for high tones, if they are to make as strong an impression on the ear. Several acousticians have occasionally started the hypothesis that, *caeteris paribus*, the strength of tones of different heights is directly proportional to the *vis viva* of the motion of the air, or, which comes to the same thing, to the amount of the mechanical work applied for producing it. But a simple experiment with the siren shews that when equal amounts of mechanical work are applied to produce deep and high tones under conditions otherwise alike, the high tones excite a very much more powerful sensation than the deep ones. Thus, if the siren is blown by a bellows, which makes its disc revolve with increasing rapidity, and if we take care to keep up a perfectly uniform motion of the bellows by raising its handle by the same amount the same ¶ number of times in a minute, so as to keep its bag equally filled, and drive the same amount of air under the same pressure through the siren in the same time, we hear at first, while the revolution is slow, a weak deep tone, which continually ascends, but at the same time gains strength at an extraordinary rate, till when the highest tones producible on my double siren (near to *a''*, with 880 vibrations in a second) are reached, their strength is almost insupportable. In this case by far the greatest part of the uniform mechanical work is applied to the generation of sonorous motion, and only a small part can be lost by the friction of the revolving disc on its axial supports, and the air which it sets into a vortical motion at the same time ; and these losses must even be greater for the more rapid rotation than for the slower, so that for the production of the high tones less mechanical work remains applicable than for the deep ones, and yet the higher tones appear to our sensation extraordinarily more powerful than the deep ones. How far upwards this increase may extend, I have as yet been unable to determine, for the rapidity ¶ of my siren cannot be further increased with the same pressure of air.

The increase of strength with height of tone is of especial consequence in the deepest part of the scale. It follows that in compound tones of great depth, the upper partial tones may be superior to the prime in strength, even though in musical tones of the same description, but of greater height, the strength of the prime greatly predominates. This is readily proved on my double siren, because by means of the beats it is easy to determine whether any partial tone which we hear is the prime, or the second or third partial tone of the compound under examination. For when the series of 12 holes are open in both windboxes, and the handle, which moves the upper windbox, is rotated once, we shall have, as already shown, 4 beats for the primes, 8 for the second partials, and 12 for the third partials. Now we can make the disc revolve more slowly than usual, by allowing a well-oiled steel spring to rub against the edge of one disc with different degrees of pressure, and thus we can easily produce series of puffs which correspond to very deep tones, and then, turning the handle, we can count the beats.

By allowing the rapidity of the revolution of the discs to increase gradually, we find that the first audible tones produced make 12 beats for each revolution of the handle, the number of puffs being from 36 to 40 in the second. For tones with from 40 to 80 puffs, each revolution of the handle gives 8 beats. In this case, then, the upper Octave of the prime is the strongest tone. It is not till we have 80 puffs in a second that we hear the four beats of the primes.

It is proved by these experiments that motions of the air, which do not take the form of pendular vibrations, can excite distinct and powerful sensations of tone, of which the pitch number is 2 or 3 times the number of the pulses of the air, and yet that the prime tone is not heard through them. Hence, when we continually descend in the scale, the strength of our sensation decreases so rapidly that the sound of the prime tone, although its *vis viva* is independently greater than that of the upper partials, as is shewn in higher positions of a musical tone of the same composition, is overcome and concealed by its own upper partials. Even ¶ when the action of the compound tone on the ear is much reinforced, the effect remains the same. In the experiments with the siren the uppermost plate of the bellows is violently agitated for the deep tones, and when I laid my head on it, my whole head was set into such powerful sympathetic vibration that the holes of the rotating disc, which vanish to an eye at rest, became again separately visible, through an optical action similar to that which takes place in stroboscopic discs. The row of holes in action appeared to stand still, the other rows seemed to move partly backwards and partly forwards, and yet the deepest tones were no more distinct than before. At another time I connected my ear by means of a properly introduced tube with an opening leading to the interior of the bellows. The agitation of the drumskin of the ear was so great that it produced an intolerable itching, and yet the deepest tones remained as indistinct as ever.

In order, then, to discover the limit of deepest tones, it is necessary not only to produce very violent agitations in the air but to give these the form of simple ¶ pendular vibrations. Until this last condition is fulfilled we cannot possibly say whether the deep tones we hear belong to the prime tone or to an upper partial tone of the motion of the air.* Among the instruments hitherto employed the wide-stopped organ pipes are the most suitable for this purpose. Their upper partial tones are at least extremely weak, if not quite absent. Here we find that even the lower tones of the 16-foot octave, $C_{\prime}$ to $E_{\prime}$, begin to pass over into a droning noise, so that it becomes difficult for even a practised musical ear to assign their pitch with certainty ; and, indeed, they cannot be tuned by the ear alone, but only indirectly by means of the beats which they make with the tones of the upper octaves. We observe a similar effect on the same deep tones of a piano or harmonium ; they form drones and seem out of tune, although their musical character is on the whole better established than in the pipes, because of their accompanying upper partial tones. In music, as artistically applied in an orchestra, the deepest tone used is, therefore, the $E_{\prime}$ of the [4-stringed German] double bass, with $41\frac{1}{4}$ vibra- ¶ tions in a second, [see p. 18c, note] and I think I may predict with certainty that all efforts of modern art applied to produce good musical tones of a lower pitch must fail, not because proper means of agitating the air cannot be discovered, but because the human ear cannot hear them. The 16-foot $C_{\prime}$ of the organ, with 33 vibrations in a second, certainly gives a tolerably continuous sensation of drone, but does not allow us to give it a definite position in the musical scale. We almost begin to observe the separate pulses of air, notwithstanding the regular form of the motion. In the upper half of the 32-foot octave, the perception of the separate pulses becomes still clearer, and the continuous part of the sensation,

---

* Thus Savart's instrument, where a rotating rod strikes through a narrow slit, is totally unsuitable for making the lowest tone audible. The separate puffs of air are here very short in relation to the whole periodic time of the vibration, and consequently the upper partial tones must be very strongly developed, and the deepest tones, which are heard for 8 to 16 passages of the rod through the hole in a second, can be nothing but upper partials.

which may be compared with a sensation of tone, continually weaker, and in the lower half of the 32-foot octave we can scarcely be said to hear anything but the individual pulses, or if anything else is really heard, it can only be weak upper partial tones, from which the musical tones of stopped pipes are not quite free.

I have tried to produce deep simple tones in another way. Strings which are weighted in their middle with a heavy piece of metal, on being struck give a compound tone consisting of many simple tones which are mutually inharmonic. The prime tone is separated from the nearest upper partials by an interval of several Octaves, and hence there is no danger of confusing it with any of them; besides, the upper tones die away rapidly, but the deeper ones continue for a very long time. A string of this kind * was stretched on a sounding-box having a single opening which could be connected with the auditory passage, so that the air of the sounding-box could escape nowhere else but into the ear. The tones of a string of customary

¶ pitch are under these circumstances insupportably loud. But for $D_{\prime}$, with $37\frac{1}{8}$ vibrations in a second, there was only a very weak sensation of tone, and even this was rather jarring, leading to the conclusion that the ear began even here to feel the separate pulses separately, notwithstanding their regularity. At $B_{\prime\prime}\flat$, with $29\frac{1}{3}$ vibrations in a second, there was scarcely anything audible left. It appears, then, that those nerve fibres which perceive such tones begin as early as at this note to be no longer excited with a uniform degree of strength during the whole time of a vibration, whether it be the phases of greatest velocity or the phases of greatest deviation from their mean position in the vibrating formations in the ear which effect the excitement.†

---

* It was a thin brass pianoforte string. The weight was a copper kreutzer piece, [pronounce *kroitser*; three kreutzers make a penny at Heidelberg, where the experiment was probably tried,] pierced in the middle by a hole
¶ through which the wire passed, and then made to grip the wire immovably by driving a steel point between the hole in the kreutzer and the string.

† Subsequently I obtained two large tuning-forks from Herr Koenig in Paris, with sliding weights on their prongs. By altering the position of the weights, the pitch was changed, and the corresponding number of vibrations was given on a scale which runs along the prongs. One fork gave 24 to 35, the other 35 to 61 vibrations. The sliding weight is a plate, 5 centimetres [nearly 2 inches] in diameter, and forms a mirror. On bringing the ear close to these plates the deep tones are well heard. For 30 vibrations I could still hear a weak drone, for 28 scarcely a trace, although this arrangement made it easily possible to form
¶ oscillations of 9 millimetres [about ⅓ inch] in amplitude, quite close to the ear. Prof. W. Preyer has been thus able to hear down to 24 vib. He has also applied another method (*Physiologische Abhandlungen*, Physiological Treatises, Series 1, part 1, 'On the limits of the perception of tone,' pp. 1–17) by using very deep, loaded tongues, in reed pipes, which were constructed for this purpose by Herr Appunn of Hanau, and gave from 8 to 40 vib. These were set into strong vibration by blowing, and then on interrupting the wind, the dying off of the vibrations was listened to by laying the ear against the box. He states that tones were heard downwards as low as 15 vib. But the proof that the tones heard corresponded with the primes of the pipes depends only on the fact, that the pitch gradually ascended as they passed over into the tones of from 25 to 32 vib., which were more audible, but died off more

rapidly. With extensive vibrations, however, the tongues may have very easily given their point of attachment longitudinal impulses of double the frequency, because when they reached each extremity of their amplitude they might drive back the point of attachment through their flexion, whereas in the middle of the vibration they would draw it forward by the centrifugal force of their weight. Since the power of distinguishing pitch for these deepest tones is extremely imperfect, I do not feel my doubts removed by the judgment of the ear when the estimates are not checked by the counting of beats.

[This check I am fortunately able to supply. A copy of the instrument used by Prof. Preyer is in the South Kensington Museum. It consists of an oblong box, in the lower part of which are the loaded harmonium reeds, not attached to pipes, but vibrating within the box, and governed by valves which can be opened at pleasure. On account of the beats between tongue and tongue taking place in strongly condensed air, they are accelerated, and the nominal pitch, obtained by counting the beats from reed to reed, is not quite the same as the actual pitch (see App. XX. sect. B. No. 6). The series of tones is supposed to proceed from 8 to 32 vib. by differences of 1 vib., from 32 to 64 by differences of 2 vib., and from 64 to 128 by differences of 4 vibs. In November 1879, for another purpose, I determined the pitch of every reed by Scheibler's forks, (see App. XX. sect. B. No. 7) by means of the upper partials of the reeds. For Reeds 8, 9, 10, 11, I used from the 20th to the 30th partial, but I consider only Reed 11 as quite certain. I found it made 10·97 vib. by the 20th, and 10·95 by both the 21st and 24th partials. From Reed 11 upwards I determined every pitch, in many cases by several partials, the result only differing in the second place of decimals. I give the two lowest Octaves, the

Hence although tones of 24 to 28 vib. have been heard, notes do not begin to have a definite pitch till about 40 vibrations are performed in a second. These facts will agree with the hypothesis concerning the elastic appendages to the auditory nerves, on remembering that the deeply intoned fibres of Corti may be set in sympathetic vibration by still deeper tones, although with rapidly decreasing strength, so that sensation of tone, but no discrimination of pitch, is possible. If the most deeply intoned of Corti's fibres lie at greater intervals from each other in the scale, but at the same time their damping power is so great that every tone which corresponds to the pitch of a fibre, also pretty strongly affects the neighbouring fibres, there will be no safe distinction of pitch in their vicinity, but it will proceed continuously without jumps, while the intensity of the sensation must at the same time become small.

Whilst simple tones in the upper half of the 16-foot octave are perfectly con-

only pitches of interest for the present purpose, premising that I consider the three lowest pitches (for which the upper partials lay too close together) and the highest (which had a bad reed) to be very uncertain. ¶

Nominal . . .	8	9	10	11	12	13	14	15	16
Actual . . .	7·91	8·89	9·81	10·95	11·90	12·90	13·93	14·91	15·91
Nominal . . .	17	18	19	20	21	22	23	24	25
Actual . .	16·90	17·91	18·89	19·91	20·91	21·91	22·88	23·97	24·92
Nominal . .	26	27	28	29	30	31	32		
Actual . .	25·92	26·86	27·85	28·84	29·77	30·68	31·47		

There can, therefore be no question as to the real pitch. At Prof. Preyer's request I examined this instrument in Oct. 1877, and he has printed my notes in his *Akustische Untersuchungen*, pp. 6–8. From these I extract the following:—

R means Reed, and R 21··25 means that the two reeds 21 and 25 were sounded together and gave beats.

R 21··25, beat 4 in 1 sec., counted for 20 sec. Hence both of their lowest partials must have been effective.

R 20··24, beat 4 in 1 sec., counted for 10 sec.

R 19··23, beat 4 in 1 sec., counted for 20 sec.

R 17··21, same beats.

R 16··20, same beats quite distinctly.

R 15··19, at first I lost the beats, but afterwards by getting R 15 well into action before R 19 was set on, and keeping on pumping, I got out the 4 in a second quite distinctly. Hence the lowest partial of R 15 was effective.

R 15··17, here also I once heard 4 in a sec., but this must have been from the Octaves.

R 14··16, I was quite unable to distinguish anything in the way of beats, but volleys like a *feu de joie* about a second in length, but impossible to count accurately; they may have been 2 in a sec. and I counted double. At the same time I seemed occasionally to hear a low beat, so low and gentle that I could not count it, and the great exertion of pumping the bellows full enough to keep these two low reeds in action, prevented accurate observation.

R 15 decidedly seemed flatter than R 13, so that I could have only heard the lowest partial of R 15 and the Octave of R 13.

On sounding R 14 and R 15 separately, I seemed to hear from each a very low tone, in quality more like a differential tone than anything else. This could also be heard even with R 13 and R 12, below the thumps, and even in R 11.

At R 8 I heard only the sishing of the escape of wind from the reed, 8 times in a second, as well as I could count, and I also heard beats evidently arising from the higher partials, and also 8 in a second.

At R 9 there was the same kind of sishing and equally rapid beats. But in addition I seemed to hear a faint low tone.

At R 10 there was no mistake as to the existence of such a musical tone.

At R 11 and R 12 it was still more distinct.

At R 13 the tone was very distinct and was quite a good musical tone at R 14, but the sish was still audible. Was this the lowest partial or its Octave ?

R 16 gave quite an organ tone, nothing like a hum or a differential, but the sish and beats remain. I must have heard the lowest partial, ¶ and by continual pumping I was able to keep it in my ear.

R 18·· 20 gave beats of 2 in a sec. very distinctly.

Up to R 25 the sish could be heard at the commencement, but it rapidly disappeared. It feels as if the tone were getting gradually into practice. This effect continued up to R 22, after which the sish was scarcely brought out at all. In fact long before this the sish was made only at the first moment, and was rather a bubble than a sish.

In listening to the very low beats, the beats of the lowest partials as such could not be separated from the general mass of beats, but the 4 in a sec. were quite clear from R 15··19. The lowest pair in which I was distinctly able to hear the bell-like beat of the lowest partials distinct from the general crash was R 30··34. But I fancied I heard it at R 28··32.

Prof. Preyer also, in the same place, details ¶ his experiments with two enormous tuning-forks giving 13·7 and 18·6 vib. The former gave no musical tone at all, though the vibrations were visible for 3 min. and were distinctly separable by touch. The latter had 'an unmistakable dull tone, without droning or jarring.' He concludes: 'Less than 15 vib. in a sec. give no musical tone. At from 16 to 24, say then 20 in the sec. the series of aerial impulses begins to dissolve into a tone, assuming that there are no pauses between them. Above 24 begins the musical character of these bass tones. Herr Appunn,' adds Prof. Preyer, 'informed me that the differential tone of 27·85 vib., generated by the two forks of 111·3 and 83·45 vib., was "surprisingly beautiful" and had a "wondrous effect."'—*Translator.*]

tinuous and musical, yet for aerial vibrations of a different form, for example when compound tones are used, discontinuous pulses of sound are still heard even within this octave. For example, blow the disc of the siren with gradually increasing speed. At first only pulses of air are heard; but after reaching 36 vibrations in a second, weak tones sound with them, which, however, are at first upper partials. As the velocity increases the sensation of the tones becomes continually stronger, but it is a long time before we cease to perceive the discontinuous pulses of air, although these tend more and more to coalesce. It is not till we reach 110 or $117\frac{1}{3}$ vibrations in a second (A or $B\flat$ of the great octave) that the tone is tolerably continuous. It is just the same on the harmonium, where, in the *cor anglais* stop, c with 132 vibrations in a second still jars a little, and in the bassoon stop we observe the same jarring even in c' with 264 vibrations in a second. Generally the same observation can be made on all cutting, snarling, or braying tones, which, as has ¶ been already mentioned, are always provided with a very great number of distinct upper partial tones.

The cause of this phenomenon must be looked for in the beats produced by the high upper partials of such compound tones, which are too nearly of the same pitch. If the 15th and 16th partials of a compound tone are still audible, they form the interval of a Semitone, and naturally produce the cutting beats of this dissonance. That it is really the beats of these tones which cause the roughness of the whole compound tone, can be easily felt by using a proper resonator. If $G_{\prime}$ is struck, having $49\frac{1}{2}$ vibrations in a second, the 15th partial is $f''\sharp$, the 16th g'', and the 17th $g''\sharp$ [nearly], &c. Now when I apply the resonator g'', which reinforces g'' most, and $f''\sharp$, $g''\sharp$ somewhat less, the roughness of the tone becomes extremely more prominent, and exactly resembles the piercing jar produced when $f''\sharp$ and g'' are themselves sounded. This experiment succeeds on the pianoforte, as well as on both stops of the harmonium. It also distinctly succeeds for higher pitches, ¶ as far as the resonators reach. I possess a resonator for g''', and although it only slightly reinforces the tone, the roughness of $G_{\prime}$ with 99 vibrations in a second, was distinctly increased when the resonator was applied.*

Even the 8th and 9th partials of a compound tone, which are a whole Tone apart, cannot but produce beats, although they are not so cutting as those from the higher upper partials. But the reinforcement by resonators does not now succeed so well, because the deeper resonators at least are not capable of simultaneously reinforcing the tones which differ from each other by a whole Tone. For the higher resonators, where the reinforcement is slighter, the interval between the tones capable of being reinforced is greater, and thus by means of the resonators g' and g''' I succeeded in increasing the roughness of the tones G to g (having 99 and 198 vibrations in a second respectively), which is due to the 7th, 8th, and 9th partial tones (f'', g'', a'', and f''', g''', a''' respectively). On comparing the tone of G as heard in the resonators with the tone of the dissonances f'' g'' ¶ and g'' a'' as struck directly, the ear felt their close resemblance, the rapidity of intermittence being nearly the same.

Hence there can no longer be any doubt that motions of the air corresponding to deep musical tones compounded of numerous partials, are capable of exciting at one and the same time a continuous sensation of deep tones and a discontinuous sensation of high tones, and become rough or jarring through the latter.† Herein lies the explanation of the fact already observed in examining qualities of tone, that compound tones with many high upper partials are cutting, jarring, or braying; and also of the fact that they are more penetrating and cannot readily pass unobserved, for an intermittent impression excites our nervous apparatus much more powerfully than a continuous one, and continually forces itself afresh on our

---

* [The student should now perform the experiments on the Harmonical indicated on p. 22d, note.—*Translator*.]

† [This is particularly noticeable on Ap-

punn's Reed pipes of 32 and 64 vib. in the South Kensington Museum. Their musical character is quite destroyed by the loud thumping of the upper partials.—*Translator*.]

perception.* On the other hand simple tones, or compound tones which have only a few of the lower upper partials, lying at wide intervals apart, must produce perfectly continuous sensations in the ear, and make a soft and gentle impression, without much energy, even when they are in reality relatively strong.

We have not yet been able to determine the upper limit of the number of intermittences perceptible in a second for high notes, and have only drawn attention to their becoming more difficult to perceive, and making a slighter impression, as they became more numerous. Hence even when the form of vibration, that is the quality of tone, remains the same, while the pitch is increased, the quality of tone will generally appear to diminish in roughness. The part of the scale adjacent to $f'''\sharp$, for which the ear is peculiarly sensitive, as I have already remarked (p. 116a), must be particularly important, as dissonant upper partials which lie in this neighbourhood cannot but be especially prominent. Now $f''''\sharp$ is the 8th partial of $f'\sharp$ with $366\frac{2}{3}$ vibrations in a second, a tone belonging to the upper tones of a man's and ¶ the lower tones of a woman's voice, and it is the 16th partial of the unaccented $f\sharp$, which lies in the middle of the usual compass of men's voices.† I have already mentioned that when human voices are strained these high notes are often heard sounding with them. When this takes place in the deeper tones of men's voices, it must produce cutting dissonances, and in fact, as I have already observed, when a powerful bass voice is trumpeting out its notes in full strength, the high upper partial tones in the four-times-accented octave are heard, in quivering tinkles (p. 116c). Hence jarring and braying are much more usual and more powerful in bass than in higher voices. For compound tones above $f\sharp$, the dissonances of the higher upper partials in the four-times-accented octave, are not so strong as those of a whole Tone, and as they occur at so great a height they can scarcely be distinct enough to be clearly sensible.

In this way we can explain why high voices have in general a pleasanter tone, and why all singers, male and female, consequently strive to touch high notes. ¶ Moreover in the upper parts of the scale slight errors of intonation produce many more beats than in the lower, so that the musical feeling for pitch, correctness, and beauty of intervals is much surer for high than low notes.

According to the observations of Prof. W. Preyer the difference in the qualities of tone of tuning-forks and reeds entirely disappears when they reach a height of c^{v} 4224, doubtless for the reason he assigns, namely that the upper partials of the reeds fall in the seventh and eighth accented octave, which are scarcely audible.

CHAPTER X.

### BEATS OF THE UPPER PARTIAL TONES.	¶

THE beats hitherto considered, were produced by two simple tones, without any intervention of upper partial or combinational tones. Such beats could only arise when the two given tones made a comparatively small interval with each other. As soon as the interval increased even to a minor Third the beats became indistinct. Now it is well known that beats can also arise from two tones which make a much greater interval with each other, and we shall see hereafter that these beats play a principal part in settling the consonant intervals of our musical scales, and they

* [In Prof. Tyndall's paper 'On the Atmosphere as a Vehicle of Sound,' read before the Royal Society, Feb. 12, 1874, in trying the distance at which intense sounds could be heard at sea, he says (*Philosophical Transactions* for 1874, vol. clxiv. p. 189), ' The influence of " beats " was tried on June 3 [1873] by throwing the horns slightly out of unison; but though the beats rendered the sound characteristic, they did not seem to augment the range.'—*Translator.*]

† [On the compass of voices see App. XX. sect. N. No. 1.—*Translator.*]

must consequently be closely examined. The beats heard when the two generating tones are more than a minor Third apart in the scale, arise from upper partial and combinational tones.* When the compound tones have distinctly audible upper partials, the beats resulting from them are generally clearer and stronger than those due to the combinational tones, and it is much more easy to determine their source. Hence we begin the investigation of the beats occurring in wider intervals with those which arise from the presence of upper partial tones. It must not be forgotten, however, that beats of combinational tones are much more general than these, as they occur with all kinds of musical tones, both simple and compound, whereas of course those due to upper partial tones are only found when such partials are themselves distinct. But since all tones which are useful for musical purposes are, with rare exceptions, richly endowed with powerful upper partial tones, the beats due to these upper partials are relatively of much greater practical importance
¶ than those due to the weak combinational tones.

When two compound tones are sounded at the same time, it is readily seen, from what precedes, that beats may arise whenever any two upper partial tones lie sufficiently near to each other, or when the prime of one tone approaches to an upper partial of the other. The number of beats is of course, as before, the difference of the vibrational numbers of the two partial tones to which the beats are due. When this difference is small, and the beats are therefore slow, they are relatively most distinct to hear and to count and to investigate, precisely as for beats of prime tones. They are also more distinct when the particular partial tones which generate them are loudest. Now, for the tones most used in music, partials with a low ordinal number are loudest, because the intensity of partial tones usually diminishes as their ordinal number increases.

Let us begin, then, with examples like the following, on an organ in its principal or violin stops,† or upon an harmonium :

¶

The minims in these examples denote the prime tones of the notes struck, and the crotchets the corresponding upper partial tones. If the octave C c in the first example is tuned accurately, no beats will be heard. But if the upper note is changed into B as in the second example, or $d\flat$ as in the third, we obtain the same beats as we should from the two tones B c, or c $d\flat$, where the interval is a Semitone. The number of beats ($16\frac{1}{2}$ in a second) is the same in each case, but their intensity is naturally less in the former case, because they are somewhat smothered by the strong deep tone C, and also because c, the second partial of C, has generally less force than its prime. ‡
¶   In examples 4 and 5 beats will be heard on keyed instruments tuned according to the usual system of temperament. If the tempered intonation is exact there will be one beat in a second,§ because the note a'' on the instrument does not exactly

---

* [But as upper partial and combinational tones are both simple, it is always simple tones which beat together, and the laws of Chap. VIII. therefore govern all beats. With a little practice the bell-like sound of the beating partials may be distinguished amid the confused beating of harsh reed tones. It only remains to determine when and how these extra beating tones arise.—*Translator.*]

† [See p. 93, notes * and §. On English organs the open diapason and keraulophon or gamba might be used.—*Translator.*]

‡ [On the Harmonical, instead of varying the Octave in C c by a Semitone up or down,

we can slightly flatten the upper note, by just pressing it down enough to speak, when the beats will arise. Or by using the d and d_1 we can produce mistuned Octaves as D d_1 or D_1 d. And for the Fifth in No. 4 and 5, we can use d' a'' or d' a', or take this mistuned Fifth lower, as d a' or d a, the true Fifth being d_1 a, which may be contrasted with it.—*Translator.*]

§ [Suppose d' has 297, then equally tempered a ought to have 445 vibs. The third partial of d' has therefore $3 \times 297 = 891$ vib., and the Octave of a has $2 \times 445 = 890$ vib., and these two tones beat $891 - 890 =$ once in a second.—*Translator.*]

agree with the note a'', which is the third partial tone of the note d'. On the other hand the note a'' on the instrument exactly coincides with a'', the second partial tone of the note a' in the fifth example, so that on instruments exactly tuned in any temperament the two examples 4 and 5 should give the same number of beats.

Since the first upper partial tone makes exactly twice as many vibrations in a second as its prime, the c on the instrument in Ex. 1, is identical with the first upper partial of the prime tone C, provided c makes twice as many vibrations in a second as C. The two notes C, c, cannot be struck together without producing beats, unless this exact relation is maintained. The least deviation from this exact relation is betrayed by beats. In the fourth example the beats will not cease till we tune a'' on the instrument so as to coincide with the third partial tone of the note d, and this can only happen when the pitch number of a'' is precisely three times that of d'. In the fifth example we have to make the pitch number of a' half as great as that of a'', which is three times that of d'; that is the pitch numbers of d' and a' ¶ must be exactly as $2 : 3$, or beats will ensue. Any deviation from this ratio will be detected at once by beats.

Now we have already shewn that the pitch numbers of two tones which form an Octave are in the ratio $1 : 2$, and those of two which form a Fifth in that of $2 : 3$. These ratios were discovered long ago by merely following the judgment of the ear respecting the most pleasant concord of two tones. The circumstances just stated furnish the reason why these intervals when tuned according to these simple ratios of numbers, and in no other case, will produce an undisturbed concord, whereas very small deviations from this mathematical intonation will betray themselves by that restless fluctuation of tone known as beats. The d' and a' of the last example, if d' tuned as a perfect Fifth below a [that is as d_1 on the Harmonical], make $293\frac{1}{3}$ and 440 vibrations in a second respectively, and their common upper partial a'' makes $3 \times 293\frac{1}{3} = 2 \times 440 = 880$ vibrations in a second. In the tempered intonation d' makes almost exactly $293\frac{2}{3}$ vibrations in a second, and hence its second upper ¶ partial (or third partial) tone makes 881 vib. in the same time, and this extremely small difference is betrayed to the ear by one beat in a second. That imperfect Octaves and Fifths will produce beats, was a fact long known to organ-builders, who made use of it practically to obtain the required just or tempered intonation with greater ease and certainty. Indeed, there is no more sensitive means of proving the correctness of intervals.

Two musical tones, therefore, which stand in the relation of a perfect Octave, a perfect Twelfth, or a perfect Fifth, go on sounding uniformly without disturbance, and are thus distinguished from the next adjacent intervals, imperfect Octaves and Fifths, for which a part of the tone breaks up into distinct pulses, and consequently the two tones do not continue to sound without interruption. For this reason the perfect Octave, Twelfth, and Fifth will be called *consonant intervals* in contradistinction to the next adjacent intervals, which are termed *dissonant*. Although these names were given long ago, long before anything was known about upper partial tones and ¶ their beats, they give a very correct notion of the essential character of the phenomenon which consists in the undisturbed or disturbed coexistence of sounds.

Since the phenomena just described form the essential basis for the construction of normal musical intervals, it is advisable to establish them experimentally in every possible form.

We have stated that the beats heard are the beats of those partial tones of both compounds which nearly coincide. Now it is not always very easy on hearing a Fifth or an Octave which is slightly out of tune, to recognise clearly with the unassisted ear which part of the whole sound is beating. On listening we are apt to feel that the whole sound is alternately reinforced and weakened. Yet an ear accustomed to distinguish upper partial tones, after directing its attention on the common upper partials concerned, will easily hear the strong beats of these particular tones, and recognise the continued and undisturbed sound of the primes. Strike the note d', attend to its upper partial a'', and then strike a tempered Fifth

a'; the beats of a'' will be clearly heard. To an unpractised ear the resonators already described will be of great assistance. Apply the resonator for a'', and the above beats will be heard with great distinctness. If, on the other hand, a resonator, tuned to one of the prime tones d' or a', be employed, the beats are heard much less distinctly, because the continuous part of the tone is then reinforced.

This last remark must not be taken to mean that no other simple tones beat in this combination except a''. On the contrary, there are other higher and weaker upper partials, and also combinational tones which beat, as we shall learn in the next chapter, and these beats coexist with those already described. But the beats of the lowest common upper partials are the most prominent, simply because these beats are the loudest and slowest of all.

Secondly, a direct experimental proof is desirable that the numerical ratios here deduced from the pitch numbers are really those which give no beats. This proof
¶ is most easily given by means of the double siren (fig. 56, p. 162). Set the discs in revolution and open the series of 8 holes on the lower and 16 on the upper, thus obtaining two compound tones which form an Octave. They continue to sound without beats as long as the upper box is stationary. But directly we begin to revolve the upper box, thus slightly sharpening or flattening the tone of the upper disc, beats are heard. As long as the box was stationary, the ratio of the pitch numbers was exactly 1 : 2, because exactly 8 pulses of air escaped on one rotation of the lower, and 16 on one rotation of the upper disc. By diminishing the speed of rotation of the handle this ratio may be altered as slightly as we please, but however slowly we turn it, if it move at all, the beats are heard, which shews that the interval is mistuned.

Similarly with the Fifth. Open the series of 12 holes above, and 18 below, and a perfectly unbroken Fifth will be heard as long as the upper windbox is at rest. The ratio of the vibrational numbers, fixed by the holes of the two series, is exactly
¶ 2 to 3. On rotating the windchest, beats are heard. We have seen that each revolution of the handle increases or diminishes the number of vibrations of the tone due to the 12 holes by 4 (p. 164c). When we have the tone of 12 holes on the lower discs also, we thus obtain 4 beats. But with the Fifth from 12 and 18 holes each revolution of the handle gives 12 beats, because the pitch number of the third partial tone increases on each revolution of the handle by $3 \times 4 = 12$, when that of the prime tone increases by 4, and we are now concerned with the beats of this partial tone.

In these investigations the siren has the great advantage over all other musical instruments, of having its intervals tuned according to their simple numerical relations with mechanical certainty by the method of constructing the instrument, and we are consequently relieved from the extremely laborious and difficult measurements of the pitch numbers which would have to precede the proof of our law on any other musical instrument. Yet the law had been already established by such
¶ measurements, and the ratios were shewn to approximate more and more closely to those of the simple numbers, as the degree of perfection increased, to which the methods of measuring numbers of vibrations and tuning perfectly had been brought.

Just as the coincidences of the two first upper partial tones led us to the natural consonances of the Octave and Fifth, the coincidences of higher upper partials would lead us to a further series of natural consonances. But it must be remarked that in the same proportion that these higher upper partials become weaker, the less perceptible become the beats by which the imperfect are distinguished from the perfect intervals, and the error of tuning is shewn. Hence the delimitation of those intervals which depend upon coincidences of the higher upper partials becomes continually more indistinct and indeterminate as the upper partials involved are higher in order. In the following table the first horizontal line and first vertical column contain the ordinal numbers of the coincident upper partial tones, and at their intersection will be found the name of the corresponding interval between the prime tones, and the ratio of the vibrational numbers of the tones composing it. This numerical ratio always results from the ordinal numbers of the two coincident upper partial tones.

Coincident Partial Tones	1	2	3	4	5
6	2 Octaves and Fifth 1 : 6	Twelfth 1 : 3	Octave 1 : 2	Fifth 2 : 3	Minor Third 5 : 6
5	2 Octaves & Major Third 1 : 5	Major Tenth 2 : 5	Major Sixth 3 : 5	Major Third 4 : 5	
4	Double Octave 1 : 4	Octave 1 : 2	Fourth 3 : 4		
3	Twelfth 1 : 3	Fifth 2 : 3			
2	Octave 1 : 2				

The two lowest lines of this table contain the intervals already considered, the Octave, Twelfth, and Fifth. In the third line from the bottom the 4th partial gives the intervals of the Fourth and double Octave. The 5th partial determines the major Third, either simple or increased by one or two Octaves, and the major Sixth. The 6th partial introduces the minor Third in addition. Here I have stopped, because the 7th partial tone is entirely eliminated, or at least much weakened, on instruments such as the piano, where the quality of tone can be regulated within certain limits.* Even the 6th partial is generally very weak, but an endeavour is made to favour all the partials up to the 5th. We shall return hereafter to the intervals characterised by the 7th partial, and to the minor Sixth, which is determined by the 8th. The following is the order of the consonant ¶ intervals beginning with those distinctly characterised, and then proceeding to those which have their limits somewhat blurred, so to speak, by the weaker beats of the higher upper partial tones :—

1. Octave . . . .   1 : 2
2. Twelfth . . . .   1 : 3
3. Fifth . . . .   2 : 3
4. Fourth . . . .   3 : 4
5. Major Sixth . . . . . .   3 : 5
6. Major Third . . . . . .   4 : 5
7. Minor Third . . . . . .   5 : 6

The following examples in musical notation shew the coincidences of the upper partials. The primes are as before represented by minims, and the upper partials by crotchets. The series of upper partials is continued up to the common tone ¶ only.

Octave.   Twelfth.   Fifth.   Fourth. Maj. Sixth. Maj. Third. Min. Third.
1 : 2   1 : 3   2 : 3   3 : 4   3 : 5   4 : 5   5 : 6

We have hitherto confined our attention to beats arising from intervals which differ but slightly from those of perfect consonances. When the difference is

* [But see Mr. Hipkins' remarks and experiments, supra, p. 77c, note.—*Translator.*]

small the beats are slow, and hence easy both to observe and count. Of course beats continue to occur when the deviation of the two coincident upper partials increases. But as the beats then become more numerous the overwhelming mass of sound of the louder primes conceals their real character more easily than the quicker beats of dissonant primes themselves. These more rapid beats give a rough effect to the whole mass of sound, but the ear does not readily recognise its cause, unless the experiments have been conducted by gradually increasing the imperfection of an harmonic interval, so as to make the beats gradually more and more rapid, thus leading the observer to mark the intermediate steps between the numerable rapid beats on the one hand, and the roughness of a dissonance on the other, and hence to convince himself that the two phenomena differ only in degree.

¶ In the experiments with pairs of simple tones we saw that the distinctness and roughness of their beats depended partly on the magnitude of the interval between the beating tones, and partly upon the rapidity of the beats themselves, so that for high tones this increasing rapidity injured the distinctness of even the beats arising from small intervals, and obliterated them in sensation. At present, as we have to deal with beats of upper partials, which, when their primes lie in the middle region, principally belong to the higher parts of the scale, the rapidity of the beats has a preponderating influence on the distinctness of their definition.

The law determining the number of beats in a second for a given imperfection in a consonant interval, results immediately from the law above assigned for the beats of simple tones. When two simple tones, making a small interval, generate beats, the number of beats in a second is the difference of their vibrational numbers. Let us suppose, by way of example, that a certain prime tone has the pitch number 300. The pitch numbers of the primes which make consonant intervals with it, will be as follows :—

¶

Prime tone = 300	
Upper Octave = 600	Lower Octave = 150
,, Fifth = 450	,, Fifth = 200
,, Fourth = 400	,, Fourth = 225
,, Major Sixth = 500	,, Major Sixth = 180
,, Major Third = 375	,, Major Third = 240
,, Minor Third = 360	,, Minor Third = 250

Now assume that the prime tone has been put out of tune by one vibration in a second, so that its pitch number becomes 301, then calculating the vibrational number of the coincident upper partial tones, and taking their difference, we find the number of beats thus :—

¶

Interval upwards	Beating Partial Tones		Number of Beats
Prime	1 × 300 = 300	1 × 301 = 301	1
Octave	1 × 600 = 600	2 × 301 = 602	2
Fifth	2 × 450 = 900	3 × 301 = 903	3
Fourth	3 × 400 = 1200	4 × 301 = 1204	4
Major Sixth . . .	3 × 500 = 1500	5 × 301 = 1505	5
Major Third . . .	4 × 375 = 1500	5 × 301 = 1505	5
Minor Third . . .	5 × 360 = 1800	6 × 301 = 1806	6

Interval downwards	Beating Partial Tones		Number of Beats
Prime	1 × 300 = 300	1 × 301 = 301	1
Octave	2 × 150 = 300	1 × 301 = 301	1
Fifth	3 × 200 = 600	2 × 301 = 602	2
Fourth	4 × 225 = 900	3 × 301 = 903	3
Major Sixth . . .	5 × 180 = 900	3 × 301 = 903	3
Major Third . . .	5 × 240 = 1200	4 × 301 = 1204	4
Minor Third . . .	6 × 250 = 1500	5 × 301 = 1505	5

Hence the number of beats which arise from putting one of the generating tones out of tune to the amount of one vibration in a second, is always given by the two numbers which define the interval. The smaller number gives the number of beats which arise from increasing the pitch number of the upper tone by 1. The larger number gives the number of beats which arise from increasing the pitch number of the lower tone by 1. Hence if we take the major Sixth $c\,a$, having the ratio 3 : 5, and sharpen a so as to make one additional vibration in a second, we shall have 3 beats in a second; but if we sharpen c so as to make one more vibration in a second, we obtain 5 beats in a second, and so on.

Our calculation and the rule based on it shew that if the amount by which one of the tones is put out of tune remains constant, the number of the beats increases according as the interval is expressed in larger numbers. Hence for Sixths and Thirds the pitch numbers of the tones must be much more nearly in the normal ratio, if we wish to avoid slow beats, than for Octaves and Unisons. On the other ¶ hand a slight imperfection in the tuning of Thirds brings us much sooner to the limit where the beats become too rapid to be distinctly separable. If we change the Unison $c''\,c''$, by flattening one of the tones, into the Semitone $b'\,c''$, on sounding the notes together there results a clear dissonance with 33 beats, the number which, as before observed, seems to give the maximum of harshness. But to obtain 33 beats from fifth $f'\,c''$, it is only necessary to alter c'' by a quarter of a Tone. If it is changed by a Semitone, so that $f'\,c''$ becomes $f'\,b'$, there result 66 beats, and their clearness is already much injured. To obtain 33 beats the c'' must not be changed in the Fifth $c''\,g''$ by more than one-sixth of a Tone, in the Fourth $c''\,f''$ by more than one-eighth, in the major Third $c''\,e''$ and major Sixth $c''\,a''$ by more than one-tenth, and in the minor Third $c''\,e''\flat$ by more than one-twelfth. Conversely, if in each of these intervals the pitch number of c'' be altered by 33, so that c'' becomes b' or $d''\flat$, we obtain the following numbers of beats :—   ¶

The interval of the	becomes	or	and gives beats
Octave . . . . $c''\,c'''$	$b'\,c'''$	$d''\flat\,c'''$	66
Fifth . . . . $c''\,g''$	$b'\,g''$	$d''\flat\,g''$	99
Fourth . . . . $c''\,f''$	$b'\,f''$	$d''\flat\,f''$	132
Major Third . . . $c''\,e''$	$b'\,e''$	$d''\flat\,e''$	165
Minor Third . . . $c''\,e''\flat$	$b'\,e''\flat$	$d''\flat\,e''\flat$	198

Now since 99 beats in a second produce very weak effects even under favourable circumstances for simple tones, and 132 beats in a second seem to lie at the limit of audibility, we must not be surprised if such numbers of beats, produced by the weaker upper partials, and smothered by the more powerful prime tones, no longer produce any sensible effect, and in fact vanish so far as the ear is concerned. Now this relation is of great importance in the practice of music, for in the table it will ¶ be seen that the mistuned Fifth gives the interval $b'\,g''$, which is much used as an imperfect consonance under the name of *minor Sixth*. In the same way we find the major Third $d''\flat f''$ as a mistuned Fourth, and the Fourth $b'e''$ as a mistuned major Third, and so on. That, at least in this part of the scale, the major Third does not produce the beats of a mistuned Fourth, or the Fourth those of a mistuned major Third, is explained by the great number of beats. In point of fact these intervals in this part of the scale give a perfectly uninterrupted sound, without a trace of beats or harshness, when they are tuned perfectly.

This brings us to the investigation of those circumstances which affect the perfection of the consonance for the different intervals. A consonance has been characterised by the coincidence of two of the upper partial tones of the compounds forming the chord. When this is the case the two compound tones cannot generate any slow beats. But it is possible that some other two upper partial tones of these two compounds may be so nearly of the same pitch that they can generate

rapid beats. Cases of this kind occur in the last examples in musical notation (p. 183*d*). Among the upper partials of the major Third *FA* occur *f'* and *e'*, side by side; and among those of the minor Third *FA♭* will be found *a'* and *a'♭*. In each case there is the dissonance of a Semitone, and these must produce the same beats as if they had been given directly as simple prime tones. Now although such beats can produce no very prominent impression, partly on account of their rapidity, partly on account of the weakness of the tones which generate them, and partly because the primes and other partial tones are sounding on at the same time unintermittently, yet they cannot but exert some effect on the harmoniousness of the interval. In the last chapter we found that in certain qualities of tone, where the higher upper partials are strongly developed, sensible dissonances may arise within a single compound tone (p. 178*b*). When two such musical tones are sounded together, there will be not only the dissonances resulting from the higher
¶ upper partial tones in each individual compound, but also those which arise from a partial tone of the one forming a dissonance with a partial tone of the other, and in this way there must be a certain increase in roughness.

An easy method of finding those upper partials in each consonant interval which form dissonances with each other, may be deduced from what has been already stated concerning larger imperfections in tuning consonant intervals (p. 185*c, d*). We thus found that the major Third might be considered as a mistuned Fourth, and the Fourth again as a mistuned Third. On raising the pitch of a compound tone by a Semitone, we raise the pitch of all its upper partial tones by the same amount. Those upper partials which coincide for the interval of a Fourth, separate by a Semitone when by altering the pitch of one generating tone we convert the Fourth into a major Third, and similarly those which coincide for the major Third differ by a Semitone for the Fourth, as will appear in the following example :—

¶

Fourth.   Major Third.   Minor Third.

The 4th and 3rd partial in the Fourth of the first example coincide as *f'*. But if the Fourth *B♭* sinks, as in the second example, to the major Third *A*, its 3rd partial *f'* sinks also to *e'*, and forms a dissonance with the 4th partial *f'* of *F*, which was unaltered. On the other hand the 5th and 4th tone of the two compounds, which in the first example formed the dissonance *a' b'♭*, now coincide as *a'*. In
¶ the same way the consonant unison *a'a'* of the second example appears as the dissonance *a'a'♭* in the third, and the dissonance *c''c''♯* in the second becomes the consonant unison *c''c''* in the third.

Hence *in each consonant interval those upper partials form a dissonance, which coincide in one of the adjacent consonant intervals,** and in this sense we can say, that every consonance is disturbed by the proximity of the consonances next adjoining it in the scale, and that the resulting disturbance is the greater, the

---

* [That is, in intervals which differ from the first by raising or depressing one of its tones by a Semitone (either $\frac{16}{15}$ or $\frac{25}{24}$), as in the table on p. 185*c*, or even a Tone ($\frac{9}{8}$). Thus for the Fifth, $\frac{3}{2} \times \frac{16}{15} = \frac{8}{5}$ a minor Sixth; and $\frac{3}{2} \times \frac{8}{9} = \frac{4}{3}$ a Fourth. For the Fourth, $\frac{4}{3} \times \frac{15}{16} = \frac{5}{4}$ a major Third; and $\frac{4}{3} \times \frac{9}{8} = \frac{3}{2}$ a Fifth. For the major Third $\frac{5}{4} \times \frac{16}{15} = \frac{4}{3}$ a Fourth; and $\frac{5}{4} \times \frac{24}{25} = \frac{6}{5}$ a minor Third. For the minor Third $\frac{6}{5} \times \frac{25}{24} = \frac{5}{4}$ a major Third, and $\frac{6}{5} \times \frac{15}{16} = \frac{9}{8}$ a major Tone. The adjacency of the consonant intervals is best shewn in fig. 60, A (p. 193), where it appears that the order may be taken as; 1) Unison, 2) minor Third, 3) major Third, 4) Fourth, 5) Fifth, 6) minor Sixth, 7) major Sixth, 8) Octave. In the table on p. 187*b*, other intervals, not perfectly consonant, are intercalated among these. --*Translator.*]

lower and louder the upper partials which by their coincidence characterise the disturbing interval, or, in other words, the smaller the number which expresses the ratio of the pitch numbers.

The following table gives a general view of this influence of the different consonances on each other. The partials are given up to the 9th inclusive, and corresponding names assigned to the intervals arising from the coincidence of the higher upper partial tones. The third column contains the ratios of their pitch numbers, which at the same time furnish the number of the order of the coincident partial tones. The fourth column gives the distance of the separate intervals from each other, and the last a measure of the relative strength of the beats resulting from the mistuning of the corresponding interval, reckoned for the quality of tone of the violin.* The degree to which any interval disturbs the adjacent intervals, increases with this last number.†

Intervals	Notation	Ratio of the Pitch Numbers	Relative Distance	Cents in the Intervals	Difference of Cents	Intensity of Influence
Unison . . .	C	1 : 1	—	0	—	100·0
			8 : 9		204	
Second . . .	D	8 : 9	—	204	—	1·4
			63 : 64		27	
Supersecond . .	D+	7 : 8	—	231	—	1·8
			48 : 49		36	
Subminor Third .	E♭ −	6 : 7	—	267	—	2·4
			35 : 36		49	
Minor Third . .	E♭	5 : 6	—	316	—	3·3
			24 : 25		70	
Major Third . .	E	4 : 5	—	386	—	5·0
			35 : 36		49	
Supermajor Third .	E+	7 : 9	—	435	—	1·6
			27 : 28		63	
Fourth . . .	F	3 : 4	—	498	—	8·3
			20 : 21		85	
Subminor Fifth . .	G♭ −	5 : 7	—	583	—	2·8
			14 : 15		119	
Fifth	G	2 : 3	—	702	—	16·7
			15 : 16		112	
Minor Sixth . .	A♭	5 : 8	—	814	—	2·5
			24 : 25		70	
Major Sixth . .	A	3 : 5	—	884	—	6·7
			20 : 21		85	
Subminor Seventh .	B♭ −	4 : 7	—	969	—	3·6
			35 : 36		49	
Minor Seventh . .	B♭	5 : 9	—	1018	—	2·2
			9 : 10		182	
Octave . . .	c	1 : 2	—	1200	—	50·0

The most perfect chord is the *Unison*, for which both compound tones have the same pitch. All its partial tones coincide, and hence no dissonance can occur except such as is contained in each compound separately (p. 178*b*).

It is much the same with the *Octave*. All the partial tones of the higher note of this interval coincide with the evenly numbered partials of the deeper, and reinforce them, so that in this case also there can be no dissonance between two upper partial tones, except such as already exists, in a weaker form, among those of the deeper note. A note accompanied by its Octave consequently becomes brighter in quality, because the higher upper partial tones on which brightness of quality depends, are partly reinforced by the additional Octave. But a similar effect would also be produced by simply increasing the intensity of the lower note without adding the Octave ; the only difference would be, that in the latter case the reinforcement of the different partial tones would be somewhat differently distributed.

The same holds for the *Twelfth* and *double Octave*, and generally for all those

* See Appendix XV.
† [Two columns have been added, shewing the cents in the intervals named, and in the intervals between adjacent notes. See also App. XX. sect. D.—*Translator*.]

cases in which the prime tone of the higher note coincides with one of the partial tones of the lower note, although as the interval between the two notes increases the difference between consonance amd dissonance tends towards obliteration.

The cases hitherto considered, where the prime of one compound tone coincides with one of the partials of the other, may be termed *absolute consonances*. The second compound tone introduces no new element, but merely reinforces a part of the other.

Unison and Octave disturb the next adjacent intervals considerably, in the sense assigned to this expression on p. 186d, so that the minor Second C Db, and the major Seventh C B, which differ from the Unison and Octave by a Semitone respectively, are the harshest dissonances in our scale. Even the major Second C D, and the minor Seventh C Bb, which are a whole Tone apart from the disturbing intervals, must be reckoned as dissonances, although, owing to the greater
¶ interval of the dissonant partial tones, they are much milder than the others. In the higher regions of the scale their roughness is materially lessened by the increased rapidity of the beats. Since the dissonance of the minor Seventh is due to the second partial tone, which in most musical qualities of tone is much weaker than the prime, it is still milder than that of the major Second, and hence lies on the very boundary between dissonance and consonance.

To find additional good consonances we must consequently go to the middle of the Octave, and the first we meet is the *Fifth*. Immediately next to it within the interval of a Semitone there are only the intervals 5 : 7 and 5 : 8 in our table, and these cannot much disturb it, because in all the better kinds of musical tones the 7th and 8th partials are either very weak or entirely absent. The next intervals with stronger upper partials are the Fourth 3 : 4 and the major Sixth 3 : 5. But here the interval is a whole Tone, and if the tones 1 and 2 of the interval of the Octave could produce very little disturbing effect in the minor Seventh, the dis-
¶ turbance by the tones 2 and 3, or by the vicinity of the Fifth to the Fourth and major Sixth must be insignificant, and the reaction of these two intervals with the tones 3 and 4 or 3 and 5 on the Fifth must be entirely neglected. Hence the Fifth remains a perfect consonance, in which there is no sensible disturbance of closely adjacent upper partial tones. It is only in harsh qualities of tone (harmonium, double-bass, violoncello, reed organ pipes) with high upper partial tones, and deep primes, when the number of beats is small, that we remark that the Fifth is somewhat rougher than the Octave.* Hence the Fifth has been acknowledged as a consonance from the earliest times and by all musicians. On the other hand the intervals next adjacent to the Fifth are those which produce the harshest dissonances after those next adjacent to the Octave. Of the dissonant intervals next

* [The above discussion may be rendered easier by the following considerations, which the student should illustrate or hear illustrated on the Harmonical. Take the pitch numbers of the two prime tones which form the Fifth to be 2 and 3, and find those of their upper partials thus, assuming C G to be the two notes.

Nos. of the Partials . .	1	2	3	4	5	6	7	8
Partials of lower note . .	2	4	6	8	10	12	14	16
Lower note	C	c	g	c′	e′	g′	b′b	c″
Fifth or 2 : 3, upper note .		G	g		d′		g′		b′
Partials of upper note . .		3		6		9		12		15
Nos. of the Partials . .		1		2		3		4		5

We see that the principal beating tones are 14 and 15, or b′b b′, the 7th partial of the lower and 5th of the upper; and 15 and 16, or b′ c″, the 5th of the upper and 8th of the lower note, and that these beats are unimportant because the 7th and 8th partials are generally weak; but if they are strong these beats being those of a Semitone and of nearly a Semitone, are very harsh. On the Harmonical it will be found that the 12th C g is faultless, but the 5th C G is decidedly harsh.

The next beating partial tones are 8 and 9, or c′ d′, the 4th partial of the lower and 3rd of the upper note, and these being a whole Tone apart, the beats are not of importance even when strong, and with weak upper partials are insignificant. Similarly for the beats of 9 and 10, or d′ e′, the 3rd partial of the upper and 5th of the lower note. On referring to the text it will be seen that the same intervals are there compared and in the same order as here.—*Translator*.]

the Fifth, those in which the Fifth is flattened, that is which lie between the Fifth and Fourth, and are disturbed firstly by the tones 2 and 3, and secondly by the tones 3 and 4, are more decidedly dissonant than those in which the Fifth is sharpened and which lie between the Fifth and major Sixth, because for the latter the second disturbance arises from the tone 3 and the weaker tone 5.* The intervals between the Fifth and Fourth are consequently always considered dissonant in musical practice. But between the Fifth and major Sixth lies the interval of the *minor Sixth*, which is treated as an imperfect consonance, and owes this preference mainly to its being the inversion of the major Third. On keyed instruments, as the piano, the same keys will strike notes which at one time represent the consonance $C\,A\flat$, and at another the dissonance $C\,G\sharp$.†

Next to the Fifth follow the consonances of the *Fourth* 3 : 4 and the *major Sixth*, the chief disturbance of which arises usually from the Fifth. The Fourth is somewhat further from the Fifth (the interval is 8 : 9) than the major Sixth is ¶ (the interval is 9 : 10), and hence the major Sixth is a less perfect consonance than the Fourth. But close by the Fourth lies the major Third with the 4th and 5th partials coincident, and hence when these partials are strongly developed, the Fourth may lose its advantage over the major Sixth. It is also well known that the old theoretical musicians long disputed as to whether the Fourth should be considered consonant or dissonant. The precedence given to the Fourth over the major Sixth and major Third, is rather due to its being the inversion of the Fifth than to its own inherent harmoniousness. The Fourth, the major Sixth and minor Sixth, are rendered less pleasant by being widened by an Octave (thus becoming the Eleventh, and major and minor Thirteenth), because they then lie near the Twelfth, and consequently the disturbance by the characteristic tones of the Twelfth 1 and 3, is greater, and hence also the adjacent intervals 2 : 5 for the Eleventh, and 2 : 7 for the Thirteenth, are more disturbing than are the 4 : 5 for the Fourth and the 4 : 7 for the Sixth in the lower Octave.‡   ¶

* [Taking the scheme in the last note, and supposing G to be altered first to $G\flat$ and then to $A\flat$, we may write the several schemes thus :

No. of Partials of lower note . .		1	2	3	4	5	6	7	8
Lower Note		C	c	g	c'	e'	g'	$b'\flat$	c''
Fifth or 2 : 3 ⎫		G	g	d'	g'	b'			
Flattened ⎬ forms of the upper note ⎨	$G\flat$	$g\flat$	$d'\flat$	$g'\flat$	$b'\flat$				
Sharpened ⎭		$A\flat$	$a\flat$	$e'\flat$	$a'\flat$	c''			
No. of Partials of upper note . .		1	2	3	4	5			

If the $G\flat$ were made sufficiently flat, we should have its 5th partial $b'\flat$ coinciding with the 7th partial of C, which, however, is never felt as a consonance, and the interval then becomes 5 : 7. This, however, never occurs in musical practice, where the $b'\flat$ from $G\flat$ is always sharper than that from C, but this dissonance is not felt, the $g\flat\,g$ or tones 2 of the upper and 3 of the lower note, and $c'\,d'\flat$ or tones 3 of the upper and 4 of the lower note, producing the chief disturbance. If $A\flat$ is taken sufficiently sharp for its 5th partial c'' to coincide with the eighth of C we have the interval 5 : 8 or minor Sixth. Here again we have the disturbance from $a\flat\,g$ the tones 2 of the upper and 3 of the lower note, but the *second* disturbance is now from $e'\flat\,e'$ or tones

3 of the upper and 5 of the lower note, instead of from $d'\flat\,c'$ or tones 3 of the upper and 4 of the lower note, and as the tone 5 is weaker the disturbance on the whole is weaker. This is the case in musical practice.—*Translator.*]

† [This is the result of equal temperament, in which $A^{1}\flat$, which is 814 cents above C, is confounded with $G_{2}\sharp$, which is only 772 cents ¶ above C, a difference of 42 cents. The interval $c'\,a^{1'}\flat$ can be played on the Harmonical and at that pitch will be found good. The interval $a^{1}\flat\,e'$, which is the same as that of $c'g'_2\sharp$, but a major Third lower, will be found very harsh.—*Translator.*]

‡ [Treating these intervals as in the preceding notes we have :

No. of Partials . . .		1	2	3	4	5	6	7	8
Lower note . . .		C	c	g	c'	e'	g'	$b'\flat$	c''
Fourth or 3 : 4 . .		F	f	c'	f'	a'		c''	
No. of Partials . .		1	2	3	4	5		6	

No. of Partials . . .		1	2	3	4	5	6	7	8
Lower note . . .		C	c	g	c'	e'	g'	$b'\flat$	c''
Eleventh or 3 : 8 . .				f			f'		c''
No. of Partials . .				1			2		3

Next in the order of the consonances come the *major* and *minor Third*. The latter is very imperfectly delimited on instruments which, like the pianoforte, do not strongly develop the 6th partial of the compound tone, because it can then be imperfectly tuned without producing sensible beats.* The minor Third is sensibly exposed to disturbance from the Unison, and the major Third from the Fourth ; and both mutually disturb each other, the minor Third coming off worse than the major.† For the harmoniousness of either interval it is necessary that the disturbing beats should be very rapid. Hence in the upper part of the scale these intervals are pure and good, but in the lower part they are very rough. All antiquity, therefore, refused to accept Thirds as consonances. It was not till the time of Franco of Cologne (at the end of the twelfth century) that they were admitted as *imperfect consonances*. The reason of this may probably be that musical theory was developed among classical nations and in medieval times principally in respect to men's voices, and in the lower part of this scale Thirds are far from good. With this we must connect the fact that the proper intonation of major Thirds was not discovered in early times, and that the *Pythagorean Third*, with its ratio of 64 : 81, was looked upon as the normal form till towards the close of the middle ages.‡

No. of Partials.	1	2	3	4	5	6	7	8
Lower note	C	c	g	c'	e'	g'	$b'b$	c''
Major Sixth or 3 : 5	A		a		e'		a'	c''‡
No. of Partials.	1		2		3		4	5

No. of Partials.	1	2	3	4	5	6	7	8	9	10
Lower note	C	c	g	c'	c'	g'	$b'b$	c''	d''	e''
Major Thirteenth or 3 : 10			a			a'				e''
No. of Partials.			1			2				3

No. of Partials.	1	2	3	4	5	6	7	8
Lower note	C	c	g	c'	e'	g'	$b'b$	c''
Minor Sixth or 5 : 8	Ab		ab	$e'b$		$a'b$		c''
No. of Partials.	1		2	3		4		5

No. of Partials.	1	2	3	4	5	6	7	8	9	10	12	16
Lower note	C	c	g	c'	e'	g'	$b'b$	c''	d''	e''	g''	c'''
Minor Thirteenth or 5 : 16		ab				$a'b$				$e''b$	$a''b$	c'''
No. of Partials.		1				2				3	4	5

These diagrams will make the text immediately intelligible, but as the notes refer to the ordinary notation the fact that f to g in the Fourth is a wider interval than g to a in the major Sixth is not expressed. It is, however, readily seen how much worse is the minor Sixth with g to ab, and that in all these cases the disturbance arises from the 2nd and 3rd partials which coincide for the Fifth. It is also seen how the disturbance is increased in the Eleventh and Thirteenths because one of the disturbing tones then becomes a prime, and hence sounds much louder. See also the table of partials on p. 197c, d. — *Translator.*]

* [As the usual tempered tuning of the piano makes the minor Third greatly too flat, the circumstance mentioned in the text becomes a great advantage on that instrument. On the tempered harmonium even $e'\,g'$, $e''\,g''$ are very harsh, as compared with the same intervals on the Harmonical.—*Translator.*]

† [This will be made clearer by the following diagrams :

No. of Partials.	1	2	3	4	5	6	7	8
Lower note	C	c	g	c'	e'	g'	$b'b$	c''
Major Third or 4 : 5	E	c	b	e'		g'‡	b'	d''
No. of Partials.	1	2	3	4		5	6	

No. of Partials.	1	2	3	4	5	6	7	8
Lower note	C	c	g	c'	e'	g'	$b'b$	c''
Minor Third or 5 : 6	Eb	eb	bb	$e'b$	g'		$b'b$	$d''b$
No. of Partials.	1	2	3	4	5		6	7

The 6th partial of this Eb is not the same as the 7th partial of C, although the notation makes it appear so, but it is sharper in the ratio of 36 : 35, and hence if the partials were not so high would be very disturbing. It is seen that $g'\,g'$♯ are the 6th and 5th partials for the major Third, and $e'b\,e'$ the 4th and 5th for the minor Third; the interval being the same (24 : 25), the disturbance is worse in the latter case, because the partials are lower and hence louder.—*Translator.*]

‡ [The ordinary major Third on the tempered harmonium is very little flatter than this, but still it is much less harsh. The Harmonical does not contain a Pythagorean major Third, 64 : 81, the nearest approach being $'bb : d_1 = 63 : 80$, but it contains a Pythagorean minor Third df, which may be con-

The important influence exercised on the harmoniousness of the consonances, especially the less perfect ones, by the rapidity of the weak beats of the dissonant upper partials, has already been indicated. If we place all the intervals above the same bass note, the number of their beats in a second varies much, and is much greater for the imperfect than for the perfect consonances. But we can give all the intervals hitherto considered such a position in the scale that the number of their beats in a second should be the same. Since we have found that 33 beats in a second produce about the maximum amount of roughness, I have so chosen the position of the intervals in the following examples in musical notation, as to give

trasted with the just minor Third d_1f. The following arrangement of the consonant intervals will show the beating partials in each case, and the exact ratios of their intervals. The number of the partial is subscribed in each case. The beating interval is inoffensive for 5 : 6, but its action becomes sensible for 7 : 8, 8 : 9, and 9 : 10, and for 14 : 15, 15 : 16, 24 : 25 the effect is decidedly bad if the tones are strong enough and the beats slow enough; the strength depends on the lowness of the

ordinal numbers of the beating partials, and the rapidity depends on their position in the scale. This must be taken into consideration, as in fig. 60, p. 193. A prefixed *, †, ‡, || draws attention to the beating partials. The order of the intervals is that of their relative harmoniousness as assigned in my paper 'On the Physical Constitution and Relations of Musical Chords,' in the *Proceedings of the Royal Society*, June 16, 1864, vol. xiii. p. 392, Table VIII., here re-arranged.

Interval	Beating partials
$C\ c$ Octave or 1 : 2, cents 1200	$\{$ 1 2 3 4 5 6 7 8 9 10 $\{$ $\quad 2_1$ 4_2 6_3 8_4 10_5
$C\ G$ Fifth or 2 : 3, cents 702	$\{$ 2_1 4_2 6_3 $*8_4$ $*10_5$ 12_6 $†14_7$ $†16_8$ 18_9 $‡20_{10}$ $\{$ $\quad 3_1$ 6_2 $*9_3$ 12_4 $†15_5$ 18_6 $‡21_7$
$C\ e$ Major Tenth or 2 : 5, cents 1386	$\{$ 2_1 4_2 6_3 8_4 10_5 12_6 $*14_7$ $*16_8$ 18_9 20_{10} $\{$ $\quad 5_1$ 10_2 $*15_3$ 20_4
$C\ g$ Twelfth or 1 : 3, cents 1902	$\{$ 1 2 3 4 5 6 7 8 9 10 $\{$ $\quad 3_1$ 6_2 9_3
$C\ F$ Fourth or 3 : 4, cents 498	$\{$ 3_1 6_2 $*9_3$ 12_4 $†15_5$ 18_6 21_7 24_8 $‡27_9$ $‡30_{10}$ $\{$ 4_1 $*8_2$ 12_3 $†16_4$ 20_5 24_6 $‡28_7$
$C\ A$ Major Sixth or 3 : 5, cents 884	$\{$ 3_1 6_2 $*9_3$ 12_4 15_5 18_6 $†21_7$ $‡24_8$ 27_9 30_{10} $\{$ 5_1 $*10_2$ 15_3 $†20_4$ $‡25_5$ 30_6
$C\ E$ Major Third or 4 : 5, cents 386	$\{$ 4_1 8_2 12_3 $*16_4$ 20_5 $†24_6$ 28_7 32_8 $‡36_9$ 40_{10} $\{$ 5_1 10_2 $*15_3$ 20_4 $†25_5$ 30_6 $‡35_7$ 40_8
$C\ E♭$ Minor Third or 5 : 6, cents 316	$\{$ 5_1 10_2 15_3 20_4 $*25_5$ 30_6 $†35_7$ $‡40_8$ 45_9 $‖50_{10}$ $\{$ 6_1 12_2 18_3 $*24_4$ 30_5 $†36_6$ $‡42_7$ $‖48_8$
$C\ A♭$ Minor Sixth or 5 : 8, cents 814	$\{$ 5_1 10_2 $*15_3$ 20_4 $†25_5$ $‡30_6$ 35_7 40_8 $‖45_9$ $‖50_{10}$ $\{$ 8_1 $*16_2$ $†24_3$ $‡32_4$ 40_5 $‖48_6$
$C\ e♭$ Minor Tenth or 5 : 12, cents 1516	$\{$ 5_1 10_2 15_3 20_4 $*25_5$ 30_6 $†35_7$ 40_8 $‡45_9$ $‡50_{10}$ 55_{11} 60_{12} $\{$ 12_1 $*24_2$ $†36_3$ $‡48_4$ 60_5
$C\ f$ Eleventh or 3 : 8, cents 1698	$\{$ 3_1 6_2 $*9_3$ 12_4 $†15_5$ $†18_6$ 21_7 $‡24_8$ $‡27_9$ $30\ _0$ $\{$ $*8_1$ $†16_2$ $‡24_3$
$C\ a$ Ma. Thrtnth. or 3 : 10, cents 2084	$\{$ 3_1 6_2 $*9_3$ 12_4 15_5 $†18_6$ 21_7 24_3 $‡27_9$ $30\ _0$ $\{$ $*10_1$ $†20_2$ $‡30_3$
$C\ a♭$ Mi. Thrtnth. or 5 : 16, cents 2014	$\{$ 5_1 10_2 $*15_3$ 20_4 25_5 $†30_6$ $†35_7$ 40_3 $‡45_9$ $‡50._0$ $\{$ $*16_1$ $†32_2$ $‡48_3$

See note p. 195 for the intervals depending on 7.

The last four of the above intervals are so rough that they are seldom reckoned as consonances. The order was determined merely by frequently sounding the intervals in just intonation on justly intoned reed instruments, and relates solely to the effect on my own ear. The greater richness of the major Tenth over the Twelfth made me prefer the former. The effect is very much like that of a compound tone, in which the prime is inaudible; even the tones 1 and 3 are supplied partly by combinational tones. Hence when a man's voice accompanies a woman's at a Third below (that is really a tenth) the effect is more agreeable than when another woman sings the real Third below, as long as the Thirds are major; the contrary is the case when the Thirds are minor. In ordinary rules for harmony no distinction is made between Tenths and Thirds, Fourths and Elevenths, &c. The above table shews that the differences are of extreme importance. The dissonant character attributed to the Fourth is apparently due to the Eleventh. As will be seen hereafter, the minor Tenth, the Eleventh, and both Thirteenths ought to be avoided or else treated as dissonances.—*Translator.*]

that number in every case.   The intonation is supposed to be that of the scale of *C major* with just intervals, but $b\flat$ represents the subminor Seventh of c $(4 : 7)$.*

$15_1 : 16_1$ $8_2 : 15_1$ $8_1 : 9_1$ $7_1 : 8_1$ $7_4 : 9_3$ $6_1 : 7_1$ $5_3 : 7_2$

$4_2 : 7_1$ $5_3 : 8_2$ $5_7 : 6_6$ $4_4 : 5_3$ $3_3 : 5_2$ $3_3 : 4_2$ $2_4 : 3_3$

The prime tones of the notes in this example are all partials of $C_{\prime}$, which makes 33 vibrations in a second, and hence their own pitch numbers and those

¶ of their upper partials are multiples of 33 ; consequently the difference of these pitch numbers, which gives the number of beats, must always be 33, 66, or some higher multiple of 33.

In the low positions here assigned the beats arising from the dissonant upper partials are as effective as their intensity will allow, and in this case the Sixths, Thirds, and even the Fourth are considerably rough.   But the major Sixth and major Third shew their superiority over the minor Third and minor Sixth, by descending lower down in the scale, and yet sounding somewhat milder than the others.   It is also a well-known practical rule among musicians to avoid these close intervals in low positions, when soft chords are required, though there was no justification for this rule in any previous theory of chords.

My theory of hearing by means of the sympathetic vibration of elastic appendages to the nerves, would allow of calculating the intensity of the beats of the different intervals, when the intensity of the upper partials in the corre-

¶ sponding quality of tone belonging to the instrument used, is known, and the intervals are so chosen that the number of beats in a second is the same.   But such a calculation would be very different for different qualities of tone, and holds only for such a particular case as may be assumed.

For intervals constructed on the same lower note a new factor comes into play, namely, the number of beats which occur in a second ; and the influence of this factor on the roughness of the sensation cannot be calculated directly by any fixed law.   But to obtain a general graphical representation of the complicated relations which co-operate to produce the effect, I have made such a calculation, knowing that diagrams teach more at a glance than the most complicated descriptions, and have hence constructed figs. 60, A and B (p. 193).   In order to construct them I have been forced to assume a somewhat arbitrary law for the dependence of roughness upon the number of beats.   I chose for this purpose the simplest mathematical formula which would shew that the roughness vanishes when there

¶ are no beats, increases to a maximum for 33 beats, and then diminishes as the number of beats increases.   Next I have selected the quality of tone on the violin in order to calculate the intensity and roughness of the beats due to the upper partials taken two and two together, and from the final results I have constructed figs. 60, A and B, opposite.   The base lines $c'c''$, $c''c'''$ denote those parts of the musical scale which lie between the notes thus named, but the pitch is taken to increase continuously [as when the finger slides down the violin string], and not by separate steps [as when the finger stops off definite lengths of the violin string]. It is further assumed that the notes or compound tones belonging to any individual part of the scale, are sounded together with the note c', which forms the constant lower note of all the intervals.   Fig. 60 A, therefore, shews the roughness of all intervals which are less than an Octave, and fig. 60 B of those which are greater

---

* [The ordinal numbers of the partials which beat 33 times in a second, are here subscribed.   Thus $4_4 : 5_3$ means that the ratio of the primes is $4 : 5$, and that the beating partials are the 4th of 4, and the 3rd of 5, having the ratio 16 : 15.— *Translator.*]

than one Octave, and less than two. Above the base line there are prominences marked with the ordinal numbers of the partials. The height of these prominences at every point of their width is made proportional to the roughness produced by the two partial tones denoted by the numbers, when a note of corresponding pitch is sounded at the same time with the note c'. The roughnesses produced by the different pairs of upper partials are erected one over the other.* It will be seen that the various roughnesses arising from the different intervals encroach on each other's regions, and that only a few narrow valleys remain, corresponding to the position of the best consonances, in which the roughness of the chord is comparatively small. The deepest valleys in the first Octave c' c'' belong to the Octave c', and the Fifth g'; then comes the Fourth f, the major Sixth a', and the major Third e', in the order already found for these intervals. The minor Third $e'\flat$, and the minor Sixth $a'\flat$, have 'cols' rather than valleys, the bottoms of their

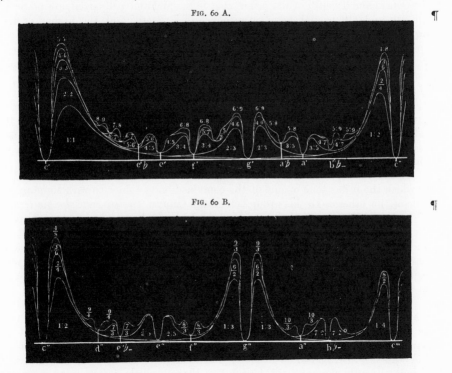

FIG. 60 A.

FIG. 60 B.

depressions lie so high, corresponding to the greater roughness of these intervals. They are almost the same as for the intervals involving 7, as 4 : 7, 5 : 7, 6 : 7.†

In the second Octave as a general rule all those intervals of the first Octave are improved, in which the smaller of the two numbers expressing the ratio was even; thus the Twelfth 1 : 3 or $c'g''$, major Tenth 2 : 5 or $c'e''$, subminor Fourteenth 2 : 7 or $c'b''\flat-$, and subminor Tenth 3 : 7 or $c'e''\flat-$‡ are smoother than the Fifth 2 : 3 or $c'g'$, major Third 4 : 5 or $c'e'$, subminor Seventh 4 : 7 or $c'b'\flat-$, and subminor Third 6 : 7 or $c'e'\flat-$. The other intervals are relatively deteriorated. The Eleventh or $c'f''$ or increased Fourth is distinctly worse than the major Tenth or $c'e''$; the major Thirteenth or $c'a''$, or increased major Sixth, is similarly worse than the subminor Fourteenth $c'b''\flat-$. The minor Third or $c'e'\flat$, when increased to a minor Tenth or $c'e''\flat,$‡ and the minor Sixth or $c'a'\flat$, when increased to a minor

* [The method in which these diagrams were calculated is shewn in the latter part of Appendix XV.—*Translator*.]

† [The interval 4 : 7 is over $b'\flat-$, meaning

$b'\flat$; the interval 5 : 7 is the 'col' between f and g', and the interval 6 : 7 is the next 'col' to the left of $e'\flat$.—*Translator*.]

‡ [By carrying a line down from $e'\flat$ in

Thirteenth or $c'a''\flat$, fare still worse, on account of the increased disturbance of the adjacent intervals. The conclusions here drawn from calculation are easily confirmed by experiments on justly intoned instruments.* That they are also attended to in the practice of musical composition, notwithstanding the theoretical assumption that the nature of a chord is not changed by altering the pitch of any one of its constituents by whole octaves, we shall see further on, when considering chords and their inversions.

It has already been mentioned that peculiarities of individual qualities of tone may have considerable effect in altering the order of the relative harmoniousness of the intervals. The quality of tone in the musical instruments now in use has been of course selected and altered with a view to its employment in harmonic combinations. The preceding investigation of the qualities of tone in our principal musical instruments has shewn that in what are considered good qualities of tone

¶ the Octave and Twelfth of the prime, that is the 2nd and 3rd partial, are powerful, the 4th and 5th partial have only moderate strength, and the higher partials rapidly diminish in force. Assuming such a quality of tone, the results of this chapter may be summed up as follows.

When two musical tones are sounded at the same time, their united sound is generally disturbed by the beats of the upper partials, so that a greater or less part of the whole mass of sound is broken up into pulses of tone, and the joint effect is rough. This relation is called *Dissonance*.

But there are certain determinate ratios between pitch numbers, for which this rule suffers an exception, and either no beats at all are formed, or at least only such as have so little intensity that they produce no unpleasant disturbance of the united sound. These exceptional cases are called *Consonances*.

1. The most perfect *consonances* are those that have been here called *absolute*, in which the prime tone of one of the combined notes coincides with some partial

¶ tone of the other. To this group belong the *Octave, Twelfth,* and *double Octave.*

2. Next follow the *Fifth* and the *Fourth,* which may be called *perfect consonances,* because they may be used in all parts of the scale without any important disturbance of harmoniousness. The Fourth is the less perfect consonance and approaches those of the next group. It owes its superiority in musical practice simply to its being the defect of a Fifth from an Octave, a circumstance to which we shall return in a later chapter.

3. The next group consists of the *major Sixth* and the *major Third,* which may be called *medial consonances.* The old writers on harmony considered them as imperfect consonances. In lower parts of the scale the disturbance of the harmoniousness is very sensible, but in the higher positions it disappears, because the beats are too rapid to be sensible. But each, in good musical qualities of tone, is independently characterised, by the fact that any little defect in its intonation produces sensible beats of the upper partials, and consequently each interval is

¶ sharply separated from all adjacent intervals.

4. The *imperfect consonances,* consisting of the *minor Third* and *minor Sixth,* are not in general independently characterised, because in good musical qualities of tone the partials on which their definition depends are often not found for the minor Third, and are generally absent for the minor Sixth, so that small imperfections in the intonation of these intervals do not necessarily produce beats.†

fig. 60 A, it will be seen that $e''\flat$ belongs to the little depression to the right of the fraction $\frac{7}{3}$ between $e''\flat$—and e''. The slight depression for $a''\flat$ is just under the fraction $\frac{10}{3}$ to the left of a''. The depression for $e'\flat$—is just to the left of that for $e'\flat$.—*Translator.*]

* [The student is strongly recommended to verify all these consonances on the Harmonical, where $b b$—, that is $'b b$, is placed on the $g\flat$ digital. The Harmonical does not contain $e b$—, that is, $^{7}e b$, and hence, in place of

$c'e''\flat$—and $c'e'\flat$—, the student should take the same intervals a Fourth lower, as $g\,^{7}b'\flat$ and $g\,^{7}b\flat$. All the other notes are on the instrument in all the octaves.—*Translator.*]

† [It must be recollected that in the minor Sixth the 2nd and 3rd partials form the Semitone 15 : 16, and the 3rd and 5th form the Semitone 24 : 25 (see note p. 191c), and that the resulting beats, which in good qualities of tone are never absent, will always be more powerful than those which arise from small

They are all less suited for use in lower parts of the scale than the others, and they owe their precedence as consonances over many other intervals which lie on the boundaries of consonance and dissonance, essentially to their being indispensable in the formation of chords, because they are defects of the major Sixth and major Third from the Octave or Fifth. The subminor Seventh 4 : 7 or c'b'♭— is very often more harmonious than the minor Sixth 5 : 8 or c'a'♭, in fact it is always so when the third partial tone of the note is strong as compared with the second, because then the Fifth has a more powerfully disturbing effect on the intervals distant from it by a Semitone, than the Octave on the subminor Seventh, which is rather more than a whole Tone removed from it.* But this subminor Seventh when combined with other consonances in chords produces intervals which are all worse than itself, as 6 : 7, 5 : 7, 7 : 8, &c., and it is consequently not used as a consonance in modern music.†

5. By increasing the interval by an Octave, the Fifth c'g' and major Third ¶ c'e' are improved on becoming the Twelfth c'g'' and major Tenth c'e''. But the

errors of intonation, even in qualities of tone in which an 8th partial is well developed.— *Translator.*]

* [Reverting to the diagrams before given (p. 191c, note), we may compare the effect of these intervals thus :

C Ab — Minor Sixth or 5 : 8, cents 814
C : 5_1 10_2 *15_3 20_4 †25_5 ‡30_6 35_7 40_8 ‖45_9 ‖50_{10}
Ab : 8_1 *16_2 †24_3 ‡32_4 40_5 ‖48_6

C Bb- = C 7Bb — Subminor Seventh or 4 : 7, cents 969
C : 4_1 *8_2 †12_3 16_4 ‡20_5 ‡24_6 28_7 32_8 ‖36_9 40_{10}
Bb : *7_1 †14_2 ‡21_3 28_4 ‖35_5

Hence for the minor Sixth the chief beats arise from the interval 15 : 16, or the 3rd partial of the lower and 2nd of the upper note, that is, from those tones which would coincide for the Fifth, which is what is meant in the text by saying that the interval is disturbed by the Fifth. But in the subminor Seventh the chief disturbance is from 7 : 8, or the prime of the upper and 2nd partial of the lower note, which would coincide for the Octave. The beats from the interval 12 : 14 or 6 : 7 are hardly perceptible, but this is the interval which replaces the 15 : 16 in the minor Sixth, being due to those upper partials which would have coincided for the Fifth. Both CAb and C'Bb can be played on the Harmonical, and the effect in the different Octaves should be compared.—*Translator.*]

† [In fig. 60 A (p. 193b), the bottom of the ¶ valley of 4 : 7 above b'b—, is just a little lower than that of 5 : 7, between f' and g', and that of 6 : 7, which, with that of 7 : 8, lies between c' and e'b. If we take the diagrams for these intervals we have :

C Eb- or G'Bb — Subminor Third or 6 : 7, cents 267
C : 6_1 12_2 18_3 *24_4 †30_5 ‡36_6 42_7 §48_8 ‖54_9 60_{10}
Eb : 7_1 14_2 *21_3 †28_4 ‡35_5 42_6 §49_7 ‖56_8

C Gb- or E'Bb — Subminor Fifth or 5 : 7, cents 583
C : 5_1 10_2 *15_3 †20_4 25_5 30_6 35_7 40_8 45_9 50_{10}
Gb : 7_1 *14_2 †21_3 28_4 35_5 42_6 49_7

C D+ or 7BbC — Supersecond or 7 : 8, cents 231
C : 7_1 14_2 21_3 28_4 *35_5 †42_6 49_7 56_8 63_9 70_{10}
D : 8_1 16_2 24_3 *32_4 †40_5 48_6 56_7 64_8

The second forms in these examples, G'Bb, E'Bb, 'BbG, can be played on the Harmonical. We see, then, that 6 : 7 is disturbed by a continual repetition of this same interval among its lower partials, and also by the intervals 21 : 24 = 7 : 8 from the 3rd and 4th partials, 28 : 30 = 14 : 15 from the 4th and 5th partials, and 35 : 36 from the 5th and 6th partials. On looking at the diagram, fig. 60 A (p. 193c), it will be seen that of these four sources the first is chief, but the others are active. For the subminor Fifth 5 : 7 the great disturbance is from 14 : 15, or the 2nd and 3rd partial, but there is also an active one from 20 : 21, or the 4th and 3rd partial, and these are almost the only ones noted in fig. 60 A. In the Supersecond the continual repetition of the interval 7 : 8 produces the chief effect, but 32 : 35 from the 4th and 5th partials, and 40 : 42 = 20 : 21, from the 5th and 6th partials, also ¶ produce much effect, as shewn in the fig. 60 A. The interval 7 : 9, which is much pleasanter, has not been considered by Prof. Helmholtz, but is available in all Octaves on the Harmonical. Mr. Poole distinguished 5 : 6, 6 : 7, 7 : 9 as the minor, minim, and maxim Third, here called minor, subminor, and super-major Third. There is also the wide (or super) minor Third 14 : 17. I add the analysis of the two last, both of which are on the Harmonical.

7Bb d — Super-major Third or 7 : 9, cents 435
7Bb : 7_1 14_2 21_3 *28_4 †35_5 42_6 49_7 ‡56_8 63_9
d : 9_1 18_2 *27_3 †36_4 45_5 ‡54_6 63_7

7b''b 17d'''b — Super-minor Third or 14 : 17, cents 336
7b''b : 14_1 28_2 42_3 *56_4 †70_5
17d'''b : 17_1 34_2 *51_3 †68_4

In the last there are a quantity of beating partials, but if 17d'''b be kept as here high in the scale, they will not be heard, and the result is really superior to the Pythagorean minor Third 27 : 32, cents 294.—*Translator.*]

Fourth $c'f'$ and major Sixth $c'a'$ become worse as the Eleventh $c'f''$ and major Thirteenth $c'a''$. The minor Third $c'e'\flat$ and minor Sixth $c'a'\flat$, however, become still worse as the minor Tenth $c'e''\flat$ and minor Thirteenth $c'a''\flat$, so that the latter intervals are far less harmonious than the subminor Tenth $3 : 7, c'e''\flat -$ [or $g \, {}^7b'\flat$], and subminor Fourteenth $2 : 7, c'b''\flat -$.

The order of the consonances here proposed is based upon a consideration of the harmoniousness of each individual interval independently of any connection with other intervals, and consequently without any regard to key, scale, and modulation. Almost all writers on musical theory have proposed similar orders for the consonances, agreeing in their general features with each other and with that here deduced from the theory of beats. Thus all put the Unison and Octave first, as the most perfect of all consonances; and next in order comes the Fifth, after which the Fourth is placed by those, who do not include the modulational
¶ properties of the Fourth, but restrict their observation to the independent harmoniousness of the interval. There is great diversity, on the other hand, in the arrangement of the Sixths and Thirds. The Greeks and Romans did not acknowledge these intervals to be consonances at all, perhaps because in the unaccented Octave, within which their music, arranged for men's voices, usually lay, these intervals really sound badly, and perhaps because their ear was too sensitive to endure the trifling increase of roughness generated by compound tones when sounded together in Thirds and Sixths. In the present century, the Archbishop Chrysanthus of Dyrrhachium declares that modern Greeks have no pleasure in polyphonic music, and consequently he disdains to enter upon it in his book on music, and refers those who are curious to know its rules, to the writings of the West.* Arabs are of the same opinion according to the accounts of all travellers.

This rule remained in force even during the first half of the middle ages, when the first attempts were made at harmonies for two voices. It was not till towards
¶ the end of the twelfth century that Franco of Cologne included the Thirds among the consonances. He distinguishes :—

1. *Perfect Consonances :* Unison and Octave.

2. *Medial Consonances :* Fifth and Fourth.

3. *Imperfect Consonances :* Major and minor Thirds.

4. *Imperfect Dissonances :* Major and minor Sixth.

5. *Perfect Dissonances :* Minor Second, augmented Fourth, major and minor Seventh.†

It was not till the thirteenth and fourteenth centuries that musicians began to include the Sixths among the consonances. Philipp de Vitry and Jean de Muris ‡ mention as perfect consonances the Unison, Octave, and Fifth ; as imperfect, the Thirds and Sixths. The Fourth has been cut out. The first author opposes the major Third and major Sixth, as more perfect, to the minor Third and minor Sixth. The same order is found in the *Dodecachordon* of Glareanus, 1557, § who
¶ merely added the intervals increased by an Octave. The reason why the Fourth was not admitted as either a perfect or an imperfect consonant, must be looked for in the rules for the progression of parts. Perfect consonances were not allowed to follow each other between the same parts, still less dissonances ; but imperfect consonances, as the Thirds and Sixths, were permitted to do so. But on the other hand the perfect consonances, Octaves, and Fifths were admitted in chords on which the music paused, as in the closing chord. Here, however, the Fourth of the bass could not occur because it does not occur in the triad of the tonic. Again a succession of Fourths for two voices was not admitted, as the Fourth and Fifth were too closely related for such a purpose. Hence so far as the progression of

---

* Θεωρητικὸν μέγα τῆς Μουσικῆς παρὰ Χρυσάνθου. Τεργέστη, 1832, cited by Coussemaker, *Histoire de l'harmonie*, p. 5.

† Gerbert, *Scriptores ecclesiastici de Musica Sacra*. Saint-Blaise, 1784, vol. iii. p. 11. —Coussemaker, *Histoire de l'harmonie*, Paris,

1852, p. 49.

‡ Coussemaker, *ibid*. p. 66 and p. 68.

§ [This is the date of the abstract by Woneggar of Lithuania, the date of the original work is 1547, ten years earlier.—*Translator*.]

parts was concerned, the Fourth shared the properties of dissonances, and it was at once placed among them ; but it would have been better to have placed it in an intermediate class between perfect and imperfect consonances.   As far as harmoniousness is concerned, there can be no doubt that, for most qualities of tone, the Fourth is much superior to the major Third and major Sixth, and beyond all doubt better than the minor Third and minor Sixth.   But the Eleventh, or Fourth increased by an Octave, sounds far from well when the third partial tone is in any degree strong.*

The dispute as to the consonance or dissonance of the Fourth has been continued to the present day.   As late as 1840, in Dehn's treatise on harmony we find it asserted that the Fourth must be treated and resolved as a dissonance ; but Dehn certainly puts a totally different interpretation on the question in dispute by laying it down that the Fourth of any bass within its key and independently of the intervals with which it is combined, has to be treated as a dissonance.   Otherwise it has been the constant custom in modern music to allow the reduplication of the tonic to occur as the Fourth of the dominant in conjunction with the dominant even in final chords, and it was long so used in these chords, even before Thirds were allowed in them, and in this way it came to be recognised as one of the superior consonances.†

CHAPTER XI.

BEATS DUE TO COMBINATIONAL TONES.

WHEN two or more compound tones are sounded at the same time beats may arise from the combinational tones as well as from the harmonic upper partials.   In Chapter VII. it was shewn that the loudest combinational tone resulting from two

* [See the Eleventh analysed in p. 191c, footnote.—*Translator.*]

† The following general view of the partials of the first 16 harmonics of C 66, (which, with the exception of the 11th and 13th, can be studied on the Harmonical,) will shew generally how they affect each other in any combination. The number of vibrations of each partial of each harmonic is given, whence the beats can be immediately found.

Partials of C	1 C	2 c	3 g	4 c'	5 e'	6 g'	7 ♭b	8 c''	9 d''	10 e''	11 f''	12 g''	13 a''	14 ♭b''	15 b''	16 c'''
1	66															
2	132	132														
3	198	—	198													
4	264	264	—	264												
5	330	—	—	—	330											
6	396	396	396	—	—	396										
7	462	—	—	—	—	—	462									
8	528	528	—	528	—	—	—	528								
9	594	—	594	—	—	—	—	—	594							
10	660	660	—	—	660	—	—	—	—	660						
11	726	—	—	—	—	—	—	—	—	—	726					
12	792	792	792	792	—	792	—	—	—	—	—	792				
13	858	—	—	—	—	—	—	—	—	—	—	—	858			
14	924	924	—	—	—	—	924	—	—	—	—	—	—	924		
15	990	—	990	—	990	—	—	—	—	—	—	—	—	—	990	
16	1056	1056	—	1056	—	—	—	1056	—	—	—	—	—	—	—	1056
17	1122															
18	1188	1188	1188	—	—	1188	—	—	1188							
19	1254															
20	1320	1320	—	1320	1320	—	—	—	—	1320						
21	1386	—	1386	—	—	—	1386									
22	1452	1452	—	—	—	—	—	—	—	—	1452					
23	1518															
24	1584	1584	1584	1584	—	1584	—	—	—	—	—	1584				
25	1650	—	—	—	1650											
26	1716	—	—	—	—	—	—	—	—	—	—	—	1716			
27	1782	—	—	—	—	—	—	—	1782							
28	1848	—	—	—	—	—	—	—	—	—	—	—	—	1848		
30	1980	—	—	—	—	—	—	—	—	1980	—	—	—	—	1980	
32	2112	—	—	—	—	—	—	2112	—	—	—	—	—	—	—	2112

generating tones is that corresponding to the difference of their pitch numbers, or the differential tone of the first order. It is this combinational tone, therefore, which is chiefly effective in producing beats. Even this loudest combinational tone is somewhat weak, unless the generators are very loud; the differential tones of higher orders, and the summational tones, are still weaker. Beats due to such weak tones as those last mentioned cannot be observed unless all other beats which would disturb the observer are absent, as, for instance, in sounding two simple tones, which are entirely free from upper partials. On the other hand the beats of the first differential tones [owing to difference of pitch and quality] can be heard very well at the same time as those due to the harmonic upper partials of compound tones, by an ear accustomed to hear combinational tones.

The *differential tones of the first order* alone, and independently of the combinational tones of higher orders, are capable of causing beats (1) when two ¶ compound tones sound together, (2) when three or more simple or compound tones sound together. On the other hand beats generated by *combinational tones of higher orders* have to be considered when two simple tones sound together.

We commence with the differential tones of compound tones. In the same way that the prime tones in such cases develop combinational tones, any pair of upper partials of the two compounds will also develop combinational tones, but such tones will diminish very rapidly in intensity as the upper partials become weaker. When one or more of these combinational tones nearly coincide with other combinational tones, or the primes or upper partials of the generators, beats ensue. Let us take as an example a slightly incorrectly tuned Fifth, having the pitch numbers 200 and 301, in place of 200 and 300, as in a justly intoned Fifth. We calculate the vibrational numbers of the upper partials by multiplying those of the primes by 1, 2, 3, and so on. We find the vibrational numbers of the differential tones of the first order, by subtracting these numbers from each other, ¶ two and two. The following table contains in the first horizontal line and vertical column the vibrational numbers of the several partials of the two compound tones, and in their intersections the differences of those numbers, which are the pitch numbers of the differential tones due to them.

		Partials of the Fifth			
		301	602	903	
Partia's of the Lower Note	2)0	101	402	703	Combinational Tones
	400	99	202	503	
	600	299	2	303	
	8co	499	198	103	
	1000	699	398	97	

If we arrange these tones by pitch we find the following groups :—

¶

2	97	198	299	398	499	600	699	800	903	1000
	99	200	301	400	503	602	703			
	101	202	303	402						
	103									

The number 2 is too small to correspond to a combinational tone. It only shews the number of beats due to the two upper partials 600 and 602.* In all the other groups, however, tones are found whose vibrational numbers differ by 2, 4, or 6, and hence produce respectively 2, 4, and 6 beats in the same time that the two first-named partials produce 2 beats. The two strongest combinational tones are 101 and 99, and these also are well distinguished from the rest by their low pitch

We observe in this example that the slowest beats due to the combinational tones are the same in number as those due to the upper partials [600 and 602]. This is a general rule and applies to all intervals.†

* [The last three, 800, 903, 1000, are simply non-beating upper partials.—*Translator.*]

† [But the beats of the upper partials are always distinguished by their high pitch.—*Translator.*]

Further it is easy to see that if in our example we replaced 200 and 301, by the numbers 200 and 300 belonging to the perfect Fifth, all the numbers in our table would become multiples of 100, and hence all the different combinational and upper partial tones which now beat would become coincident and not generate any beats. What is here shewn to be the case in this example for the Fifth is also true for all other harmonic intervals.*

The first differential tones of compounds cannot generate beats, except when the upper partials of the same compounds generate them, and the rapidity of the beats is the same in both cases, supposing that the series of upper partials is complete. Hence the addition of combinational tones makes no essential difference in the results obtained in the last chapter on investigating the beats due to the upper partials only. There can be only a slight increase in the strength of the beats.†

But the case is essentially different when two simple tones are sounded together, so that there are no upper partials to consider. If combinational tones were not ¶ taken into account, two simple tones, as those of tuning-forks or stopped organ pipes, could not produce beats unless they were very nearly of the same pitch, and such beats are strong when their interval is a minor or major Second, but weak for a Third and then only recognisable in the lower parts of the scale (p. 171*d*), and they gradually diminish in distinctness as the interval increases, without shewing any special differences for the harmonic intervals themselves. For any larger interval between two simple tones there would be absolutely no beats at all, if there were no upper partial or combinational tones, and hence the consonant intervals discovered in the former chapter would be in no respect distinguished from adjacent intervals; there would in fact be no distinction at all between wide consonant intervals and absolutely dissonant intervals.

Now such wider intervals between simple tones are known to produce beats, although very much weaker than those hitherto considered, so that even for such tones there is a difference between consonances and dissonances, although it is ¶ very much more imperfect than for compound tones. And these facts depend, as Scheibler shewed,‡ on the combinational tones of higher orders.

It is only for the *Octave* that the first differential tone suffices. If the lower note makes 100 vibrations in a second, while the imperfect Octave makes 201, the first differential tone makes $201 - 100 = 101$, and hence nearly coincides with the lower note of 100 vibrations, producing one beat for each 100 vibrations. There is no difficulty in hearing these beats, and hence it is easily possible to distinguish imperfect Octaves from perfect ones, even for simple tones, by the beats produced by the former.§

For the *Fifth*, the first order of differential tones no longer suffices. Take an imperfect Fifth with the ratio 200 : 301; then the differential tone of the first order is 101, which is too far from either primary to generate beats. But it forms an imperfect Octave with the tone 200, and, as just seen, in such a case beats ensue. Here they are produced by the differential tone 99 arising from the tone 101 and ¶

---

* This is proved mathematically in Appendix XVI.

† [The great difference in the pitch of the two sets of beats, which are not necessarily even Octaves of each other, keeps them well apart. The beating partials, in this case 600, 602, and the beating differentials, here 101 and 99, are entirely removed from each other.—*Translator*.]

‡ ['The physical and musical Tonometer, which makes evident to the eye, by means of the pendulum, the absolute vibrations of the tones, and of the principal kinds of combinational tones, as well as the most precise exactness of equally tempered and mathematical chords, invented and executed by Heinrich Scheibler, silk manufacturer in Crefeld.' (*Der*

physikalische und musikalische Tonmesser, &c.)—Essen, G. D. Bädeker, 1834, pp. viii. 80, 5 lithographed Tables (called 3 on title-page), and an engraving of tuning-forks and waves. A most remarkable pamphlet, but unfortunately very obscurely written, as the author says in his preface, 'to write clearly and briefly on a scientific subject is a skill (*Fertigkeit*) I do not possess, and have never attempted.' See also App. XX. sect. B. No. 7. I do not find anywhere that Scheibler attempted to shew that combinational tones existed, especially intermediate ones; he merely assumed them and found the beats.—*Translator*.]

§ [See App. XX. sect. L. art. 3, latter part of *d.*—*Translator*.]

the tone 200, and this tone 99 makes two beats in a second with the tone 101. These beats then serve to distinguish the imperfect from the justly intoned Fifth even in the case of two simple tones. The number of these beats is also exactly as many as if they were beats due to the upper partial tones.* But to observe these beats the two primary tones must be loud, and the ear must not be distracted by any extraneous noise. Under favourable circumstances, however, they are not difficult to hear.†

For an imperfect *Fourth*, having, say, the vibrational numbers 300 and 401, the first differential tone is 101 ; this with the tone 300 produces the differential tone 199 of the second order, and this again with the tone 401 the differential tone 202 of the third order, and this makes 3 beats with the differential tone 199 of the second order, that is, precisely as many beats as would have been generated by the upper partial tones 1200 and 1203, if they had existed. These beats of the Fourth ¶ are very weak even when the primary tones are powerful. Perfect quiet and great attention are necessary for observing them.‡ And after all there may be a doubt whether by strong excitement of the primary tones, weak partials may not have arisen, as we already considered on p. 159*b, c*. §

The beats of an imperfect *major Third* are scarcely recognisable, even under the most favourable conditions. If we take as the vibrational numbers of the primary tones 400 and 501, we have :—

$$501-400=101,\text{ the differential tone of the first order}$$
$$400-101=299, \quad ,, \qquad ,, \quad ,, \quad \text{second} ,,$$
$$501-299=202, \quad ,, \qquad ,, \quad ,, \quad \text{third} ,,$$
$$400-202=198, \quad ,, \qquad ,, \quad \text{fourth} ,,$$

The tones 202 and 198 produce 4 beats. Scheibler succeeded in counting these beats of the imperfect major Third.** I have myself believed that I heard them ¶ under favourable circumstances. But in any case they are so difficult to perceive that they are not of any importance in distinguishing consonance from dissonance.

Hence it follows that two simple tones making various intervals adjacent to the major Third and sounded together will produce a uniform uninterrupted mass of sound, without any break in their harmoniousness, provided that they do not approach a Second too closely on the one hand or a Fourth on the other. My own experiments with stopped organ pipes justify me in asserting that however much this conclusion is opposed to musical dogmas, it is borne out by the fact, provided that really simple tones are used for the purpose.†† It is the same with intervals near to the major Sixth; these also shew no difference as long as they remain sufficently far from the Fifth and Octave. Hence although it is not difficult to tune perfect major and minor Thirds on the harmonium or reed pipes or on the violin, by sounding the two tones together and trying to get rid of the beats, it is perfectly impossible to do so on stopped organ pipes or tuning-forks without the ¶ aid of other intervals. It will appear hereafter that the use of more than two tones will allow these intervals to be perfectly tuned even for simple tones.

Intermediate between the compound tones possessing many powerful upper partials, such as those of reed pipes and violins, and the entirely simple tones of tuning-forks and stopped organ pipes, lie those compound tones in which only the

* [But, as before, the pitch is very different.—*Translator*.]

† [Scheibler, *ibid.* p. 21. I myself succeeded in hearing and counting them.—*Translator*.]

‡ [Scheibler says, p. 24, they are heard as well as for the Fifth. I have not found it so.—*Translator*.]

§ [Supposing the pitch numbers of the mistuned Fourth are 300 and 401, then the beating upper partials would be 1200 and 1203, a very high pitch ; but the beating differentials are 202 and 199, which are so much lower in pitch and so inharmonic to the others that there is no danger of confusing them.—*Translator*.]

** [Scheibler, *ibid.* p. 25, says only ' as beats of this kind are too indistinct,' he uses another method for tuning the major Third. See note *, p. 203*d*. He also calculates the intermediate tones differently. But neither he nor any one seems to have tried to verify their existence, which is doubtful.—*Translator*.]

†† [Or at any rate tones without the 4th partial, which those of stopped organ pipes do not possess.—*Translator*.]

lowest of the upper partials are audible, such as the tones of wide open organ pipes or the human voice when singing some of the obscurer vowels, as *oo* in *too*. For these the partials would not suffice to distinguish all the consonant intervals, but the addition of the first differential tones renders it possible.

A. *Compound Tones consisting of the prime and its Octave.* These cannot delimit Fifths and Fourths by beats of the partials, but are able to do so by those of the first differential tones.

a. *Fifth.* Let the vibrational numbers of the prime tones be 200 and 301, which are accompanied by their Octaves 400 and 602 ; all four tones are then too far apart to beat.　But the differential tones

$$301 - 200 = 101$$
$$400 - 301 = \underline{99}$$
$$\text{Difference} \quad 2$$

give two beats.　The number of these beats again is precisely the same as if they had been produced by the two next upper partials.* Namely

$$2 \times 301 - 3 \times 200 = 2$$

b. *Fourth.* Let the vibrational numbers of the primes be 300 and 401, and of the first upper partials 600 and 802 ; these cannot produce any beats.　But the first differential tones give 3 beats, thus † :—

$$600 - 401 = 199$$
$$802 - 600 = \underline{202}$$
$$\text{Difference} \quad 3$$

For Thirds it would be necessary to take differential tones of the second order into account.

B. *Compound Tones consisting of the prime and Twelfth.* Such tones are produced by the narrow stopped pipes on the organ (*Quintaten*, p. 33*d*, note). These are related in the same way as those which have only the Octave.

a. *Fifth.* Primes 200 and 301, upper partials 600 and 903.　First differential tone

$$903 - 600 = 303$$
$$\text{Fifth} = \underline{301}$$
$$\text{Number of beats} \quad 2$$

b. *Fourth.* Primes 300 and 401, upper partials 900 and 1203.　First differential tone

$$1203 - 900 = 303$$
$$\text{Lower prime} = \underline{300}$$
$$\text{Number of beats} \quad 3$$

Even in this case the beats of the Third cannot be perceived without the help of the weak second differential tones.

C. *Compound Tones having both Octave and Twelfth as audible partials.* Such tones are produced by the wide (wooden) open pipes of the organ (*Principal*, p. 93*d'*, note).　The beats of the upper partials here suffice to delimit the Fifths, but not the Fourths.　The Thirds can now be distinguished by means of the first differential tones.

a. *Major Third.* Primes 400 and 501, with the Octaves 800 and 1002, and Twelfths 1200 and 1503.　First differential tones ‡

$$1002 - 800 = 202$$
$$1200 - 1002 = \underline{198}$$
$$\text{Number of beats} \quad 4$$

---

* [The same in number, but observe that the first set of beats are at pitch 100, and the second at pitch 600.—*Translator.*]

† [At pitch 200, whereas the partials if they existed would beat at pitch 1200.—*Translator.*]

‡ [These are the same two beating tones as calculated on p. 200*b*, but they are quite differently derived.—*Translator.*]

b. *Minor Third.* Primes 500 and 601, Octaves 1202, Twelfths 1500 and 1803. Differential tones *

$$1500 - 1202 = 298$$
$$1803 - 1500 = 303$$

Number of beats 5

c. *Major Sixth.* Primes 300 and 501, Octaves 600 and 1002, Twelfths 900 and 1503. Differential tones

$$600 - 501 = 99$$
$$1002 - 900 = 102$$

Number of beats 3

¶ In fact not only the beats of imperfect Fifths and Fourths, but also those of imperfect major and minor Thirds are easily heard on open organ pipes, and can be immediately used for the purposes of tuning.

Thus, where upper partials, owing to the quality of tone, do not suffice, the combinational tones step in to make every imperfection in the consonant intervals of the Octave, Fifth, Fourth, major Sixth, major and minor Third immediately sensible by means of beats and roughness in the combined sound, and thus to distinguish these intervals from all those adjacent to them.   It is only perfectly simple tones that so far make default in determining the Thirds ; and for them also the beats which disturb the harmoniousness of imperfect Fifths and Fourths, are relatively too weak to affect the ear sensibly, because they depend on differential tones of higher orders.   In reality, as I have already mentioned, two stopped pipes, giving tones which lie between a major and a minor Third apart, will give just as good a consonance as if the interval were exactly either a major or a minor Third. This does not mean that a practised musical ear would not find such an interval ¶ strange and unusual, and hence would perhaps call it false, but the immediate impression on the ear, the simple perception of harmoniousness, considered independently of any musical habits, is in no respect worse than for one of the perfect intervals.†

Matters are very different when more than two simple tones are sounded together.   We have seen that Octaves are precisely limited even for simple tones by the beats of the first differential tone with the lower primary.   Now suppose that an Octave has been tuned perfectly, and that then a third tone is interposed to act as a Fifth.   Then if the Fifth is not perfect, beats will ensue from the first differential tone.

Let the tones forming the perfect Octave have the pitch numbers 200 and 400, and let that of the imperfect Fifth be 301.   The differential tones are

$$400 - 301 = 99$$
$$301 - 200 = 101$$

¶

Number of beats 2

These beats of the Fifth which lies between two Octaves are much more audible than those of the Fifth alone without its Octave.   The latter depend on the weak differential tones of the second order, the former on those of the first orde .   Hence Scheibler some time ago laid down the rule for tuning tuning-forks, first to tune two of them as a perfect Octave, and then to sound them both at once with the Fifth, in order to tune the latter.‡   If Fifth and Octave are both perfect, they also give together the perfect Fourth.

The case is similar, when two simple tones have been tuned to be a perfect

---

* [This was not given for simple tones before, but Scheibler calculates the result in that case, p. 26, and says he could use it still less than for the major Third.—*Translator.*]

† [See Prof. Preyer's theory of consonance for cases where neither partial nor combinational tones are present, App. XX. sect. L. art. 7.—*Translator.*]

‡ [I have been unable to find the passage referred to.—*Translator.*]

Fifth, and we interpose a new tone between them to act as a major Third. Let the perfect Fifth have the pitch numbers 400 and 600. On intercalating the impure major Third with the pitch number 501 in lieu of 500, the differential tones are

$$600 - 501 = 99$$
$$501 - 400 = 101$$

Number of beats	2*

The major Sixth is determined by combining it with the Fourth. Let 300 and 400 be the vibrational numbers of a perfect Fourth, and 501 that of an imperfect major Sixth. The differential tones are

$$501 - 400 = 101$$
$$400 - 300 = 100$$

Number of beats	1	¶

If we tried to intercalate an interval between the tones forming a perfect Fourth, and having the vibrational numbers 300 and 400, it could only be the sub-minor Third with the vibrational number 350. Taking it imperfect and = 351, we have the differential tones

$$400 - 351 = 49$$
$$351 - 300 = 51$$

Number of beats	2

These intervals 8 : 7 and 7 : 6 are, however, too close to be consonances, and hence they can only be used in weak discords (chord of the dominant Seventh).†

The above considerations are also applicable to any single compound tone con- ¶ sisting of several partials. Any two partials of sufficient force will also produce differential tones in the ear. If, then, the partials correspond exactly to the series of harmonic partials, as assigned by the series of smaller whole numbers, all these differentials resulting from partials coincide exactly with the partials themselves, and give no beats. Thus if the prime makes n vibrations in a second, the upper partials make $2n$, $3n$, $4n$, &c., vibrations, and the differences of these numbers are again n, or $2n$, or $3n$, &c. The pitch numbers of the summational tones fall also into this series.

On the other hand, if the pitch numbers of the upper partials are ever so slightly different from those giving these ratios, then the combinational tones will differ from one another and from the upper partials, and the result will be beats. The tone therefore ceases to make that uniform and quiet impression which a compound tone with harmonic upper partials always makes on the ear. How con- siderable this influence is, we may hear from any firmly attached harmonious string after we have fastened a small piece of wax on any part of its length. This, ¶ as theory and experiment alike shew, produces an inharmonic relation of the upper partials. If the piece of wax is very small, then the alteration of tone is also very small. But the slightest mistuning suffices to do considerable harm to the tunefulness of the sound, and renders the tone dull and rough, like a tin kettle.

* [On this was founded Scheibler's method of tuning the perfect major Third (alluded to in p. 200d', note) and also the tempered major Third.

First tune a perfect Fifth, and then an auxiliary Fifth, 2 vib. sharper. Then if the major Third is perfect we have A 220, C♯ 275, E 332 and 275 − 220 = 55, 332 − 275 = 57, and 57 − 55 = 2. Hence the tuning of C♯ must be altered till the differential tones beat 2 in a second.

For the tempered major Third we have, using the perfect Fifth, A 220, C♯ 277·1824, E 330. Then, 277·1824 − 220 = 57·1824, 330 − 277·1824 = 52·8176 and 57·1824 − 52·8176 = 4·3648, and hence the tuning of the inter- mediate fork must be altered till these beats are heard. These are Scheibler's own ex- amples, p. 26, reduced to ordinary double vibrations.—*Translator.*]

† [In actual practice, for the chord of the dominant Seventh the interval is 4 : 7⅑ the in- terval of the just subminor Seventh 4 : 7 not be- ing used, even in just intonation.—*Translator.*]

Herein we find the reason why tones with harmonic upper partials play such a leading part in the sensation of the ear. They are the only sounds which, even when very intense, can produce sensations that continue in undisturbed repose, without beats, corresponding to the purely periodic motion of the air, which is the objective foundation of these tones. I have already stated as a result of the summary which I gave of the composition of musical tones in Chapter V., No. 2, p. 119a, that besides tones with harmonic upper partials, the only others used (and that also generally in a very subordinate manner) are either such as have a section of the series of harmonic upper partials, (like those of well tuned bells), or such as have secondary tones (as those in bars) so very weak and so far distant from their primes, that their differentials have but little force and at any rate do not produce any distinct beats.

¶ Collecting the results of our investigations upon beats, we find that when two or more simple tones are sounded at the same time, they cannot go on sounding without mutual disturbance, unless they form with each other certain perfectly definite intervals. Such an undisturbed flow of simultaneous tones is called a *consonance*. When these intervals do not exist, beats arise, that is, the whole compound tones, or individual partial and combinational tones contained in them or resulting from them, alternately reinforce and enfeeble each other. The tones then do not coexist undisturbed in the ear. They mutually check each other's uniform flow. This process is called *dissonance*.*

Combinational tones are the most general cause of beats. They are the sole cause of beats for simple tones which lie as much as, or more than, a minor Third apart.† For two simple tones they suffice to delimit the Fifth, perhaps the Fourth, but certainly not the Thirds and Sixths. These, however, will be strictly delimited when the major Third is added to the Fifth to form the common major chord, and when the Sixth is united with the Fourth to form the chord of the

¶ Sixth and Fourth, $\frac{6}{4}$.

Thirds, however, are strictly delimited, by means of the beats of imperfect intervals, in a chord of two compound tones, each consisting of a prime and the two next partial tones. The beats of such intervals increase in strength and distinctness, with the increase in number and strength of the upper partial tones in the compounds. By this means the difference between dissonance and consonance, and of perfectly from imperfectly tuned intervals, becomes continually more marked and distinct, increasing the certainty with which the hearer distinguishes the correct intervals, and adding much to the powerful and artistic effect of successions of chords. Finally when the high upper partials are relatively too strong (in piercing and braying qualities of tone) each separate tone will by itself generate intermittent sensations of tone, and any combination of two or more compounds of this description produces a sensible increase of this harshness, while at the same

¶ time the large number of partial and combinational tones renders it difficult for the hearer to follow a complicated arrangement of parts in a musical composition.

These relations are of the utmost importance for the use of different instruments in the different kinds of musical composition. The considerations which determine the selection of the proper instrument for an entire composition or for individual phrases in movements written for an orchestra are very multifarious. First in rank stands mobility and power of tone in the different instruments. On this there is no need to dwell. The bowed instruments and pianoforte surpass all others in mobility, and then follow the flutes and oboes. To these are opposed the trumpets and trombones, which commence sluggishly, but surpass all instruments in power. Another essential consideration is expressiveness, which in general depends on the power of producing with certainty any degree of rapid alterations in loudness at the pleasure of the player. In this respect also bowed instruments,

* [See Prof. Preyer's addendum to this theory in App. XX. sect. L. art. 7. – *Translator*.]   † [But see App. XX. sect. L. art. 3.—*Translator*.]

and the human voice, are pre-eminent. Artificial reed instruments, both of wood and brass, cannot materially diminish their power without stopping the action of the reeds. Flutes and organ pipes cannot greatly alter the force of their tone without at the same time altering their pitch. On the pianoforte the strength with which a tone commences is determined by the player, but not its duration; so that the rhythm can be marked delicately, but real melodic expression is wanting. All these points in the use of the above instruments are easy to observe and have long been known and allowed for. The influence of quality proper was more difficult to define. Our investigations, however, on the composition of musical tones have given us a means of taking into account the principal differences in the effect of the simultaneous action of different instruments and of shewing how the problem is to be solved, although there is still a large field left for a searching investigation in detail.

Let us begin with the simple tones of *wide stopped organ pipes*. In themselves ¶ they are very soft and mild, dull in the low notes, and very tuneful in the upper. They are quite unsuited, however, for combinations of harmony according to modern musical theory. We have already explained that simple tones of this kind discriminate only the very small interval of a Second by strong beats. Imperfect Octaves, and the dissonant intervals in the neighbourhood of the Octave, (the Sevenths and Ninths,) beat with the combinational tones, but these beats are weak in comparison with those due to upper partials. The beats of imperfect Fifths and Fourths are entirely inaudible except under the most favourable conditions. Hence in general the impression made on the ear by any dissonant interval, except the Second, differs very little from that made by consonances, and as a consequence the harmony loses its character and the hearer has no certainty in his perception of the difference of intervals.* If polyphonic compositions containing the harshest and most venturesome dissonances are played upon wide stopped organ pipes, the whole is uniformly soft and harmonious, and for that very reason ¶ also indefinite, wearisome and weak, without character or energy. Every reader that has an opportunity is requested to try this experiment. There is no better proof of the important part which upper partial tones play in music, than the impression produced by music composed of simple tones, such as we have just described.. Hence the wide stopped pipes of the organ are used only to give prominence to the extreme softness and tunefulness of certain phrases in contradistinction to the harsher effect of other stops, or else, in connection with other stops, to strengthen their prime tones. Next to the wide stopped organ pipes as regards quality of tone stand *flutes* and the *flue pipes* on organs (open pipes, blown gently). These have the Octave plainly in addition to the prime, and when blown more strongly even produce the Twelfth. In this case the Octaves and Fifths are more distinctly delimited by upper partial tones; but the definition of Thirds and Sixths has to depend upon combinational tones, and hence is much weaker. The musical character of these pipes is therefore not much unlike that of the wide ¶ stopped pipes already described. This is well expressed by the old joke that nothing is more dreadful to a musical ear than a flute-concerto, except a concerto for two flutes.† But in combination with other instruments which give effect to the connection of the harmony, the flute, from the perfect softness of its tone and its great mobility, is extraordinarily pleasant and attractive, and cannot be replaced by any other instrument. In ancient music the flute played a much more important part than at present, and this seems to accord with the whole ideal of classical art, which aimed at keeping every thing unpleasant from its productions, confining itself to pure beauty, whereas modern art requires more abundant means

---

* [But see Prof. Preyer in App. XX. sect. L. art. 7.—*Translator*.]

† [In the original, ' dass einem musikalischen Ohre nichts schrecklicher sei als ein Flötenconcert, ausgenommen ein Concert von zwei Flöten.' The pun on ' Concert,' first as a *concerto* or peculiar piece of music for one instrument, and secondly as a *concert*, or piece of music for several instruments, cannot be properly rendered in the translation.—*Translator*.]

of expression, and consequently to a certain extent admits into its circle what in itself would be contrary to the gratification of the senses. However this be, the earnest friends of music, even in classical times, contended for the harsher tones of stringed instruments in opposition to the effeminate flute.

The open organ pipes afford a favourable means of meeting the harmonic requirements of polyphonic music, and consequently form the *principal* stops.* They make the lower partials distinctly audible, the wide pipes up to the third, the narrow ones (*geigen principal*†) up to the sixth partial tone. The wider pipes have more power of tone than the narrower; to give them more brightness the 8-foot stops, which contain the ‘principal work,’ are connected with the 4-foot stops, which add the Octave to each note, or the *principal* is connected with the *geigen principal*, so that the first gives power and the second brightness. By this means qualities of tone are produced which contain the first six partial tones in ¶ moderate force, decreasing as the pitch ascends. These give a very distinct feeling for the purity of the consonant intervals, enabling us to distinguish clearly between consonance and dissonance, and preventing the unavoidable but weak dissonances that result from the higher upper partials in the imperfect consonances, from be- coming too marked, but at the same time not allowing the hearer’s appreciation of the progression of the parts to be disturbed by a multitude of loud accessory tones. In this respect the organ has an advantage over all other instruments, as the player is able to mix and alter the qualities of tone at pleasure, and make them suitable to the character of the piece he has to perform.

The narrow *stopped pipes* (*Quintaten*),‡ for which the prime tone is ac- companied by the Twelfth, the *reed-flute* (*Rohrflöte*)§ where the third and fifth partial are both present, the conical open pipes, as the *goat-horn* (*Gemshorn*),** which reinforce certain higher partials †† more than the lower, and so forth, serve only to give distinctive qualities of tone for particular parts, and thus to separate ¶ them from the rest. They are not well adapted for forming the chief mass of the harmony.

Very piercing qualities of tone are produced by the *reed pipes* and *compound stops* ‡‡ on the organ. The latter, as already explained, are artificial imitations of the natural composition of all musical tones, each key bringing a series of pipes into action, which correspond to the first three or first six partial tones of the corresponding note. They can be used only to accompany congregational singing. When employed alone they produce insupportable noise and horrible confusion. But when the singing of the congregation gives overpowering force to the prime tones in the notes of the melody, the proper relation of quality of tone is restored, and the result is a powerful, well-proportioned mass of sound. Without the assistance of these compound stops it would be impossible to control a vast body of sound produced by unpractised voices, such as we hear in [German] churches.

The *human voice* is on the whole not unlike the organ in quality, so far as ¶ harmony is concerned. The brighter vowels, of course, generate isolated high partial tones, but these are so unconnected with the rest that they can have no universal and essential effect on the sound of the chords. For this we must look to the lower partials, which are tolerably uniform for all vowels. But of course in particular consonances the characteristic tone of the vowels may play an important part. If, for example, two human voices sing the major Third *b♭ d′* on the vowel *a* in *father*, the fourth partial of *b♭* (or *b″♭*), and the third partial of *d′* (or *a″*), fall among the tones characteristically reinforced by *A*, and consequently the imperfec- tion of the consonance of a major Third will come out harshly by the dissonance *a″ b″♭*, between these upper partials; whereas if the vowel be changed to *o* in *no*, the dissonance disappears. On the other hand the Fourth *b♭ e′♭* sounds perfectly

---

* [See p. 141d′, note §.—*Translator.*]
† [See p. 141d, note.—*Translator.*]
‡ [See p. 33d, note.—*Translator.*]
§ [See p. 94d′, note.—*Translator.*]

** [See p. 94d, note.—*Translator.*]
†† [Generally the 4th, 6th, and 7th. – *Translator.*]
‡‡ [See p. 57a′, note.—*Translator.*]

well on the vowel *a* in *father*, because the higher note *e'♭* has the same upper partial *b''♭* as the deeper *b♭*. But if *a* in *father* being inclined towards *a* in *fall*, or *a* in *fat*, the upper partials *f''* and *e''♭* or else *d'''* and *e'''♭* might interrupt the consonance. This serves to shew, among other things, that the translation of the words of a song from one language into another is not by any means a matter of indifference for its musical effect.*

Disregarding at present these reinforcements of partials due to the characteristic resonance of each vowel, the musical tones of the human voice are on the whole accompanied by the lower partials in moderate strength, and hence are well adapted for combinations of chords, precisely as the tones of the *principal* stops of the organ. Besides this the human voice has a peculiar advantage over the organ and all other musical instruments in the execution of polyphonic music. The words which are sung connect the notes belonging to each part, and form a clue which readily guides the hearer to discover and pursue the related parts of the whole body ¶ of sound. Hence polyphonic music and the whole modern system of harmony were first developed on the human voice. Indeed, nothing can exceed the musical effect of well harmonised part music perfectly executed in just intonation by practised voices. For the complete harmoniousness of such music it is indispensably necessary that the several musical intervals should be justly intoned, and our present singers† unfortunately seldom learn to take just intervals, because they are accustomed from the first to sing to the accompaniment of instruments which are tuned in equal temperament, and hence with imperfect consonances. It is only such singers as have a delicate musical feeling of their own who find out the correct result, which is no longer taught them.

Richer in upper partials, and consequently brighter in tone than the human voice and the *principal* stops on the organ, are the *bowed instruments*, which consequently fill such an important place in music. Their extraordinary mobility and expressiveness give them the first place in instrumental music, and the moderate ¶ acuteness of their quality of tone assigns them an intermediate position between the softer flutes and the braying brass instruments. There is a slight difference between the different instruments of this class; the tenor and double-bass have a somewhat acuter and thinner quality than the violin and violoncello, that is, they have relatively stronger upper partials. The audible partials reach to the sixth or eighth, according as the bow is brought nearer the finger-board for *piano*, or nearer the bridge for *forte*, and they decrease regularly in force as they ascend in pitch. Hence on bowed instruments the difference between consonance and dissonance is

---

* [Also, it shews how the musical effect of different stanzas in a ballad, though sung to the same written notes, will constantly vary, quite independently of difference of expression. This is often remarkable on the closing cadence of the stanza. As the vowel changes from *a* in *father*, to *a* in *mat*; *e* in *met*, or *i* in *sit*, or again to *o* in *not*, *u* in *but*, and *u* in *put*, the musical result is totally different, though the pitch remains unaltered. To shew the effect of the different vowels throughout a piece of music, I asked a set of about 8 voices to sing, before about 200 others, the first half of *See the conquering hero comes*, first to *lah*, then to *lee*, and then to *loo*. The difference of *effect* was almost ludicrous. Much has to be studied in the relation of the qualities of vowels to the effect of the music. In this respect, too, the *pitch* chosen for the tonic will be found of great importance.—*Translator.*]

† [This refers to Germany, not to the English Tonic Solfaists, nor to the English madrigal singers. On Dec. 27, 1869, at a meeting of the Tonic Solfa College I had an unusual opportunity of contrasting the effect of just

and tempered intonation in the singing of the same choir. It was a choir of about 60 mixed voices, which had gained the prize at the International Exhibition at Paris in 1867, and had been kept well together ever since. After singing some pieces without accompaniment ¶ and hence in the just intonation to which the singers had been trained, and with the most delightful effect of harmony, they sang a piece with a pianoforte accompaniment. Of course the pianoforte itself was inaudible among the mass of sound produced by sixty voices. But it had the effect of perverting their intonation, and the whole charm of the singing was at once destroyed. There was nothing left but the everyday singing of an ordinary choir. The disillusion was complete and the effect most unsatisfactory as a conclusion. If the same piece of music or succession of chords in C major or C minor, w'thout any modulation, be played first on the Harmonical and then be contrasted with an ordinary tempered harmonium, the same kind of difference will be felt, but not so strongly.—*Translator.*]

clearly and distinctly marked, and the feeling for the justness of the intervals very certain ; indeed it is notorious that practised violin and violoncello players have a very delicate ear for distinguishing differences of pitch. On the other hand the piercing character of the tones is so marked, that soft song-like melodies are not well suited for bowed instruments, and are better given to flutes and clarinets in the orchestra. Full chords are also relatively too rough, since those upper partials which form dissonant intervals in every consonance, are sufficiently strong to make the dissonance obtrusive, especially for Thirds and Sixths. Moreover, the imperfect Thirds and Sixths of the tempered musical scale are on bowed instruments very perceptibly different in effect from the justly intoned Thirds and Sixths when the player does not know how to substitute the pure intervals for them, as the ear requires. Hence in compositions for bowed instruments, slow and flowing progressions of chords are introduced by way of exception only, because they are not ¶ sufficiently harmonious ; quick movements and figures, and arpeggio chords are preferred, for which these instruments are extremely well adapted, and in which the acute and piercing character of their combined sounds cannot be so distinctly perceived.

The beats have a peculiar character in the case of bowed instruments. Regular, slow, numerable beats seldom occur. This is owing to the minute irregularities in the action of the bow on the string, already described, to which is due the well-known scraping effect so often heard. Observations on the vibrational figure shew that every little scrape of the bow causes the vibrational curve to jump suddenly backwards or forwards, or in physical terms, causes a sudden alteration in the phase of vibration. Now since it depends solely on the difference of phase whether two tones which are sounded at the same time mutually reinforce or enfeeble each other, every minutest catching or scraping of the bow will also affect the flow of the beats, and when two tones of the same pitch are played, every jump in the phase will ¶ suffice to produce a change in the loudness, just as if irregular beats were occurring at unexpected moments. Hence the best instruments and the best players are necessary to produce slow beats or a uniform flow of sustained consonant chords. Probably this is one of the reasons why quartetts for bowed instruments, when executed by players who can play solo pieces pleasantly enough, sometimes sound so intolerably rough and harsh that the effect bears no proper ratio to the slight roughness which each individual player produces on his own instrument.* When I was making observations on vibrational figures, I found it difficult to avoid the occurrence of one or two jumps in the figure every second. Now in solo-playing the tone of the string is thus interrupted for almost inappreciably minute instants, which the hearer scarcely perceives, but in a quartett when a chord is played for which all the notes have a common upper partial tone, there would be from four to eight sudden and irregular alterations of loudness in this common tone every second, and this could not pass unobserved. Hence for good combined performance, ¶ a much greater evenness of tone is required than for solo-playing.†

The *pianoforte* takes the first place among stringed instruments for which the strings are struck. The previous analysis of its quality of tone shews that its deeper octaves are rich, but its higher octaves relatively poor, in upper partial tones. In the lower octaves, the second or third partial tone is often as loud as the prime, nay, the second partial is often louder than the prime. The consequence is that

* [To myself, one of the principal reasons for the painful effect here alluded to, which is unfortunately so extremely well known, is the fact that the players not having been taught the nature of just intonation, do not accommodate the pitches of the notes properly. When quartett players are used to one another they overcome this difficulty. But when they learn thus, it is a mere accommodation of the different intervals by ear to the playing of (say) the leader. (See App. XX. sect. G. art. 7.) The real relations of the just tones are in fact not gene- rally known. If the music notes could be previously marked by duodenals, in the way suggested in App. XX. sect. E. art. 26, much of this difficulty might be avoided from the first. But the marking would require a study not yet commenced by the greater number of musicians.—*Translator.*]

† [On violins combinational tones are strong. I have been told that violinists watch for the Octave differential tone, in tuning their Fifths.—*Translator.*]

the dissonances near the Octave (the Sevenths and Ninths) are almost as harsh as the Seconds, and that diminished and augmented Twelfths and Fifths are rather rough. The 4th, 5th, and 6th partial tones, on the other hand, on which the Thirds depend, decrease rapidly in force, so that the Thirds are relatively much less dis· tinctly delimited than the Octaves, Fifths, and Fourths. This last circumstance is important, because it makes the sharp Thirds of the equal temperament much more endurable upon the piano than upon other instruments with a more piercing quality of tone, whereas the Octaves, Fifths, and Fourths are delimited with great distinctness and certainty. Notwithstanding the relatively large number of upper partial tones on the pianoforte, the impression produced by dissonances is far from being so penetrating as on instruments of long-sustained tones. On the piano the note is powerful only at the moment when it is struck, and rapidly decreases in strength, so that the beats which characterise the dissonances have not time to become sensible during the strong commencement of the tone; they do not even begin ¶ until the tone is greatly diminished in intensity. Hence in the modern music written for the pianoforte, since the time that Beethoven shewed how the characteristic peculiarities of the instrument were to be utilised in compositions, we find an accumulation and reduplication of dissonant intervals which would be perfectly insupportable on other instruments. The great difference becomes very evident when an attempt is made to play recent compositions for the piano on the harmonium or organ.

That instrument-makers, led solely by practised ears, and not by any theory, should have found it most advantageous to arrange the striking place of the hammer so that the 7th partial tone entirely disappears, and the 6th is weak although actually present,* is manifestly connected with the structure of our system of musical tones. The 5th and 6th partial tones serve to delimit the minor Third, and in this way almost all the intervals treated as consonances in modern music are determined on the piano by coincident upper partials; the Octave, Fifth, ¶ and Fourth by relatively loud tones; the major Sixth and major Third by weak ones; and the minor Third by the weakest of all. If the 7th partial tone were also present, the subminor † Seventh 4 : 7, as $c''7b'\flat$, would injure the harmoniousness of the minor Sixth; the Subminor Fifth 5 : 7, as $e''7b'\flat$, that of the Fifth and Fourth; and the subminor Third 6 : 7, as $g''7b'\flat$, that of the minor Third; without any gain in the more accurate determination of new intervals suitable for musical purposes.

Mention has already been made of a further peculiarity in the selection of quality of tone on the pianoforte, namely that its upper notes have fewer and weaker upper partial tones than the lower. This difference is much more marked on the piano than on any other instrument, and the musical reason is easily assigned. The high notes are usually played in combination with much lower notes, and the relation between the two groups of notes is given by the high upper partials of the deeper tones. When the interval between the bass and treble amounts to two or ¶ three Octaves, the second Octave, higher Third and Fifth of the bass note, are in the close neighbourhood of the treble, and form direct consonances and dissonances with it, without any necessity for using the upper partials of the treble note. Hence the only effect of upper partials on the highest notes of the pianoforte would be to give them shrillness, without any gain in respect to musical definition. In actual practice the construction of the hammers on good instruments causes the notes of the highest Octaves to be only gently accompanied by their second partials. This makes them mild and pleasant, with a flute-like tone. Some instrument-makers, however, prefer to make these notes shrill and piercing, like the piccolo flute, by transferring the striking place to the very end of the highest strings. This contrivance succeeds in increasing the force of the upper partial

---

* [But see Mr. Hipkins' observations on pp. 77, 78, note.—*Translator*.]

† [For these terms see the table on p. 187.

The 7th partial was very distinct on the pianos Mr. Hipkins examined. See also App. XX. Sect. N.—*Translator*.]

tones, but gives a quality of tone to these strings which does not suit the character of the others, and hence certainly detracts from their charm.

In many other instruments, where their construction does not admit of such absolute control over the quality of tone as on the pianoforte, attempts have been made to produce similar varieties of quality in the high notes, by other means. In the bowed instruments this purpose is served by the resonance box, the proper tones of which lie within the deepest Octaves of the scale of the instrument. Since the partial tones of the sounding strings are reinforced in proportion to their proximity to the partial tones of the resonance box, this resonance will assist the prime tones of the higher notes, as contrasted with their upper partials, much more than it will do so for the deep notes. On the contrary, the deepest notes of the violin will have not only their prime tones, but also their Octaves and Fifths favoured by the resonance ; for the deeper proper tone of the resonance box ¶ lies between the prime and 2nd partial, and its higher proper tone between the 2nd and 3rd partials. A similar effect is attained in the *compound stops* of the organ, by making the series of upper partial tones, which are represented by distinct pipes, less extensive for the higher than for the lower notes in the stop. Thus each digital opens six pipes for the lower octaves, answering to the first six partial tones of its note ; but in the two upper octaves, the digital opens only three or even two pipes, which give the Octave and Twelfth, or merely the Octave, in addition to the prime.

There is also a somewhat similar relation in the human voice, although it varies much for the different vowels. On comparing the higher and lower notes which are sung to the same vowel, it will be found that the resonance of the cavity of the mouth generally reinforces relatively high upper partials of the deep notes of the bass, whereas for the soprano, where the note sung comes near to the characteristic pitch of the vowel, or even exceeds it, all the upper partials become much ¶ weaker. Hence in general, at least for the open vowels, the audible upper partials of the bass are much more numerous than those of the soprano.

We have still to consider the artificial reed instruments, that is the *wind instruments of wood and brass*. Among the former the clarinet, among the latter the horn are distinguished for the softness of their tones, whereas the bassoon and hautbois in the first class, and the trombone and trumpet in the second represent the most penetrating qualities of tone used in music.

Notwithstanding that the keyed horns used for so-called concerted music have a far less braying quality of tone than trumpets proper, which have no side holes, yet the number and the force of their upper partial tones are far too great for the harmonious effect of the less perfect consonances, and the chords on these instruments are very noisy and harsh, so that they are only endurable in the open air. In artistic orchestral music, therefore, trumpets and trombones, which on account of their penetrative power cannot be dispensed with, are seldom employed for ¶ harmonies, except for a few and if possible perfect consonances.

The *clarinet* is distinguished from all other orchestral wind instruments by having no evenly numbered partial tones.* To this circumstance must be due many remarkable deviations in the effect of its chords from those of other instruments. When two clarinets are playing together all of the consonant intervals will be delimited by combinational tones alone, except the major Sixth 3 : 5, and the Twelfth 1 : 3. But the differential tones of the first order, which are the strongest among all combinational tones, will always suffice to produce the beats of imperfect consonances. Hence it follows that in general the consonances of two clarinets have but little definition, and must be proportionately agreeable. This is really the case, except for the minor Sixth and minor Seventh, which are too near the major Sixth, and for the Eleventh and minor Thirteenth, which are too near the Twelfth. On the other hand, when a clarinet is played in combination with a violin or oboe, the majority of consonances will have a perceptibly different effect according as

---

* [But see Mr. Blaikley's observations, supra, p. 99b, note.—*Translator.*]

the clarinet takes the upper or the lower note of the chord. Thus the major Third d' $f\sharp$ will sound better when the clarinet takes d' and the oboe $f\sharp$, so that the 5th partial of the clarinet coincides with the 4th of the oboe. The 3rd and 4th and the 5th and 6th partials, which are so disturbing in the major Third,* cannot here be heard, because the 4th and 6th partials do not exist on the clarinet. But if the oboe takes d' and the clarinet $f\sharp$, the coincident 4th partial will be absent, and the disturbing 3rd and 5th present. For the same reason it follows that the Fourth and minor Third will sound better when the clarinet takes the upper tone. I have made experiments of this kind with the clarinet and a bright stop of the harmonium, which possessed the evenly numbered partial tones, and was tuned in just intonation † and not in equal temperament. When $b\flat$ was played on the clarinet, and $e'\flat$, d', $d'\flat$, in succession on the harmonium, the major Third $b\flat$ d' sounded better than the Fourth $b\flat$ $e'\flat$, and much better than the minor Third $b\flat$ $d'\flat$. If, retaining $b\flat$ on the clarinet, I played f, $g\flat$, g in succession ¶ on the harmonium, the major Third $g\flat$ $b\flat$ was rougher, not merely than the Fourth f $b\flat$, but even than the minor Third g $b\flat$.

This example, to which I was led by purely theoretical considerations that were immediately confirmed by experiment, will serve to shew how the use of exceptional qualities of tone will affect the order of agreeableness of the consonances which was settled for those usually heard.

Enough has been said to shew the readiness with which we can now account for numerous peculiarities in the effects of playing different musical instruments in combination. Further details are rendered impossible by the want of sufficient preliminary investigations, especially into the exact differences of various qualities of tone. But in any case it would lead us too far from our main purpose to pursue a subject which has rather a technical than a general interest.

¶

CHAPTER XII.

CHORDS.

WE have hitherto examined the effect of sounding together only two tones which form a determinate interval. It is now easy to discover what will happen when more than two tones are combined. The simultaneous production of more than two separate compound tones is called a *chord*. We will first examine the harmoniousness of chords in the same sense as we examined the harmoniousness of any two tones sounded together. That is, we shall in this section deal exclusively with the isolated effect of the chord in question, quite independently of any musical connection, mode, key, modulation, and so on. The first problem is to determine *under what conditions chords are consonant*, in which case they are termed *concords*. It is quite clear that the first condition of a concord is that each tone of it should ¶ form a consonance with each of the other tones; for if any two tones formed a dissonance, beats would arise destroying the tunefulness of the chord. Concords of three tones are readily found by taking two consonant intervals to any one fundamental tone as c, and then seeing whether the new third interval between the two new tones, which is thus produced, is also consonant. If this is the case each one of the three tones forms a consonant interval with each one of the other two, and the chord is consonant, or is a concord.‡

Let us confine ourselves in the first place to intervals which are less than an Octave. The consonant intervals within these limits, we have found to be : 1) the Fifth c g, $\frac{3}{2}$; 2) the Fourth $c f$, $\frac{4}{3}$; 3) the major Sixth c a, $\frac{5}{3}$; 4) the major Third c e, $\frac{5}{4}$; 5) the minor Third c $e\flat$, $\frac{6}{5}$; 6) the minor Sixth c $a\flat$, $\frac{8}{5}$; to which we may

---

* [See table on p. 191, note.—*Translator.*]

† [Try the Harmonical and clarinet.—*Translator.*]

‡ [If two tones each consonant with a third are dissonant with each other, I call the result a 'con-dissonant triad.' See App. XX. sect. E. art. 5.—*Translator.*]

add 7) the subminor or natural Seventh $c^7b\flat$, $\frac{7}{4}$, which approaches to the minor Sixth in harmoniousness. The following table gives a general view of the chords contained within an Octave. The chord is supposed to consist of the fundamental tone C, some one tone in the first horizontal line, and some one tone of the first vertical column. Where the line and column corresponding to these two selected tones intersect, is the name of the interval which these two latter tones form with each other. This name is printed in *italics* when the interval is *consonant*, and in Roman letters when dissonant, so that the eye sees at a glance what *concords* are thus produced. [Under the name, the equivalent interval in cents has been inserted by the Translator.]

C \newline 0	$G\,\frac{3}{2}$ \newline 702	$F\,\frac{4}{3}$ \newline 498	$A\,\frac{5}{3}$ \newline 884	$E\,\frac{5}{4}$ \newline 386	$E\flat\,\frac{6}{5}$ \newline 316	$A\flat\,\frac{8}{5}$ \newline 814
$G\,\frac{3}{2}$ \newline 702						
$F\,\frac{4}{3}$ \newline 498	major Second \newline $\frac{9}{8}$ \newline 204					
$A\,\frac{5}{3}$ \newline 884	major Second \newline $\frac{10}{9}$ \newline 182	*major Third* \newline $\frac{5}{4}$ \newline 386				
$E\,\frac{5}{4}$ \newline 386	*minor Third* \newline $\frac{6}{5}$ \newline 316	minor Second \newline $\frac{16}{15}$ \newline 112	*Fourth* \newline $\frac{4}{3}$ \newline 498			
$E\flat\,\frac{6}{5}$ \newline 316	*major Third* \newline $\frac{5}{4}$ \newline 386	major Second \newline $\frac{10}{9}$ \newline 182	superfluous Fourth \newline $\frac{25}{18}$ \newline 568	minor Second \newline $\frac{25}{24}$ \newline 70		
$A\flat\,\frac{8}{5}$ \newline 814	minor Second \newline $\frac{16}{15}$ \newline 112	*minor Third* \newline $\frac{6}{5}$ \newline 316	minor Second \newline $\frac{25}{24}$ \newline 70	diminished Fourth \newline $\frac{32}{25}$ \newline 428	*Fourth* \newline $\frac{4}{3}$ \newline 498	
$^7B\flat\,\frac{7}{4}$ \newline 969	subminor Third \newline $\frac{7}{6}$ \newline 267	sub Fourth \newline $\frac{21}{16}$ \newline 471	subminor Second \newline $\frac{21}{20}$ \newline 85	subminor Fifth \newline $\frac{7}{5}$ \newline 583	sub Fifth \newline $\frac{35}{24}$ \newline 653	submajor Second \newline $\frac{35}{32}$ \newline 155

From this it follows that the only consonant *triads* or chords of three notes, that can possibly exist within the compass of an Octave are the following :—

1) $C\ E\ G$ 2) $C\ E\flat\ G$
3) $C\ F\ A$ 4) $C\ F\ A\flat$
5) $C\ E\flat\ A\flat$ 6) $C\ E\ A$.*

The two first of these triads are considered in musical theory as the fundamental triads from which all others are deduced. They may each be regarded as composed of two Thirds, one major and the other minor, superimposed in different orders. The chord $C\ E\ G$, in which the major Third is below, and the minor above, is a *major triad*. It is distinguished from all other major triads by having its tones in the closest position, that is, forming the smallest intervals with each other. It is hence considered as the *fundamental chord* or basis of all other major chords. The triad $C\ E\flat\ G$, which has the minor Third below, and the major above, is the *fundamental chord* of all *minor triads*.

* [The reader ought to hear the whole set of triads that could be formed from the table, at least all exclusive of those formed by the last line. The ordinary tuning of the harmonium, organ, and piano does not permit this. But they can all (inclusive of those formed by the last line) be played on the Harmonical.— *Translator.*]

The next two chords, $C\ F\ A$ and $C\ F\ A\flat$, are termed, from their composition, *chords of the Sixth and Fourth,* written $\frac{6}{4}$ [C to F being a Fourth, and C to A a major, but C to $A\flat$ a minor sixth]. If we take $G_{\prime}$ instead of C for the fundamental or bass tone, these chords of the Fourth and Sixth become $G_{\prime}\ C\ E$ and $G_{\prime}\ C\ E\flat$. Hence we may conceive them as having been formed from the fundamental major and minor triads $C\ E\ G$ and $C\ E\flat\ G$, by transposing the Fifth G an Octave lower, when it becomes $G_{\prime}$.

The two last chords, $C\ E\flat\ A\flat$ and $C\ E\ A$, are termed *chords of the Sixth and Third,* or simply *chords of the Sixth,* written $\frac{6}{3}$ [C to E being a major Third, and C to $E\flat$ a minor Third; and C to A a major Sixth, and C to $A\flat$ a minor Sixth]. If we take E as the bass note of the first, and $E\flat$ as that of the second, they become $E\ G\ c$, $E\flat\ G\ c$, respectively. Hence they may be considered as the transpositions ¶ or *inversions* of a fundamental major and a fundamental minor chord, $C\ E\ G$, $C\ E\flat\ G$, in which the bass note C is transposed an Octave higher and becomes c.

Collecting these inversions, the six consonant triads will assume the following form [the numbers shewing their correspondence with the forms on p. 212d] :—

1)	$C\ E\ G$	2)	$C\ E\flat\ G$
5)	$E\ G\ c$	6)	$E\flat\ G\ c$
3)	$G\ c\ e$	4)	$G\ c\ e\flat$

We must observe that although the natural or subminor seventh $^{7}B\flat$ forms a good consonance with the bass note C, a consonance which is indeed rather superior than inferior to the minor Sixth $C\ A\flat$, yet it never forms part of any triad, because it would make worse consonances with all the other intervals consonant to C than it does with C itself. The best triads which it can produce are $C\ E\ ^{7}B\flat = 4 : 5 : 7$, and $C\ G\ ^{7}B\flat = 4 : 6 : 7$. In the first of these occurs the interval $E\ ^{7}B\flat = 5 : 7$, ¶ (between a Fourth and Fifth,) in the latter the subminor Third $G\ ^{7}B\flat = 6 : 7$.* On the other hand the minor Sixth makes a perfect Fourth with the minor Third, so that this minor Sixth remains the worst interval in the chords of the Sixth and Third, and of the Sixth and Fourth, for which reason these triads can still be considered as consonant. This is the reason why the natural or subminor Seventh is never used as a consonance in harmony, whereas the minor Sixth can be employed, although, considered independently, it is not more harmonious than the subminor Seventh.

The triad $C\ E\ A\flat$, to which we shall return, [Chap. XVII. Dissonant Triads, No. 4] is very instructive for the theory of music. It must be considered as a dissonance, because it contains the diminished Fourth $E\ A\flat$, having the interval ratio $\frac{32}{25}$. Now this diminished Fourth $E\ A\flat$ is so nearly the same as a major Third $E\ G\sharp$, that on our keyed instruments, the organ and pianoforte, the two intervals are not distinguished. We have in fact

$$E\ A\flat = \tfrac{32}{25} = \tfrac{5}{4}\cdot\tfrac{128}{125}$$

or, approximatively $\qquad (E\ A\flat) = (E\ G\sharp)\cdot\tfrac{43}{42}$ †

On the pianoforte it would seem as if this triad, which for practical purposes may be written either $C\ E\ A\flat$ or $C\ E\ G\sharp$, must be consonant, since each one of its tones forms with each of the others an interval which is considered as consonant on the piano, and yet this chord is one of the harshest dissonances, as all musicians are agreed, and as any one can convince himself immediately. On a justly intoned instrument [as the Harmonical] the interval $E\ A\flat$ is immediately recognised as dissonant. This chord is well adapted for shewing that the original meaning of the intervals asserts itself even with the imperfect tuning of the piano, and determines the judgment of the ear.‡

* [Add the consonance $G'B\flat\ D = 6 : 7 : 9$. —*Translator.*]

† [$E\ A\flat$ has 428 cents, and $E\ G\sharp$ has 386

cents, difference 42 cents, the great diĕsis. See App. XX. sect. D.—*Translator.*]

‡ [Inserting the values of the intervals in

The harmonious effect of the various inversions of triads already found depends in the first place upon the greater or less perfection of the consonance of the several intervals they contain.   We have found that the Fourth is less agreeable than the Fifth, and that minor are less agreeable than major Thirds and Sixths.   Now the triad

C E G has a Fifth,   a major Third, and a minor Third
E G C ,,   a Fourth, a minor Third, and a minor Sixth
G C E ,,   a Fourth, a major Third, and a major Sixth

C $E\flat$ G ,,   a Fifth,   a minor Third, and a major Third
$E\flat$ G C ,,   a Fourth, a major Third, and a major Sixth
G C $E\flat$,,   a Fourth, a minor Third, and a minor Sixth

¶ For just intervals the Thirds and Sixths decidedly disturb the general tunefulness more than the Fourths, and hence the major chords of the Sixth and Fourth are more harmonious than those in the fundamental position, and these again than the chords of the Sixth and Third.   On the other hand the *minor* chords of the Sixth and Third are more agreeable than those in the fundamental position, and these again are better than the minor chords of the Sixth and Fourth.   This conclusion will be found perfectly correct for the middle parts of the scale, provided the intervals are all justly intoned.   The chords must be struck separately, and not connected by any modulation.   As soon as modulational connections are allowed, as for example in a concluding cadence, the tonic feeling, which finds repose in the tonic chord, disturbs the power of observation, which is here the point of importance.   In the lower parts of the scale either major or minor Thirds are more disagreeable than Sixths.

Judging merely from the intervals we should expect that the minor triad C $E\flat$ G would sound as well as the major C E G, as each has a Fifth, a major ¶ and a minor Third.   This is, however, far from being the case.   The minor triad is very decidedly less harmonious than the major triad, in consequence of the combinational tones, which must consequently be here taken into consideration. In treating of the relative harmoniousness of the consonant intervals we have seen that combinational tones may produce beats when two intervals are compounded, even when each interval separately produced no beats at all, or at least none distinctly audible (pp. 200b-204b).

Hence we must determine the combinational tones of the major and minor triads.   We shall confine ourselves to the combinational tones of the first order produced by the primes and the first upper partial tones.   In the following examples the primes are marked as minims, the combinational tones resulting from these primes are represented by crotchets, those from the primes and first upper partials by quavers and semiquavers.   A downwards sloping line, when placed before a note, shews that it represents a tone slightly deeper than that ¶ of the note in the scale which it precedes.

1.) Major Triads with their Combinational Tones :*

cents, the two chords, $A^1\flat$ 386 C 386 E_1 and C 386 E_1 386 $G_2\sharp$ are seen to be identical, but when the first is inverted to C 386 E_1 428 $A^1\flat$ it becomes different from the other. Both, however, remain harshly dissonant. On tempered instruments of course they become identical C 400 E 400 $G\sharp$, C 400 E 400 $A\flat$, and are very harsh. The definition of consonant triads does not apply to tempered chords, in none of which are any of the intervals purely consonant.—*Translator.*]

* [As all the differentials must be harmonics of C 66, if we represent this note by 1, the harmonics and hence differentials will all be contained in the series

1	2	3	4	5	6	7	8	9	10	11	12	13	14	15	16
C	c	g	c'	e'	g'	$^7b'b$	c''	d''	e''	$^{11}f''$	g''	$^{13}a''$	$^7b''b$	b''	c'''

First Chord.—The notes will then be 4, 5, 6, represented by minims, and their Octaves 8, 10, 12, which are not given in notes.

1) Crotchets, $5-4=6-5=1$, $6-4=2$.
2) (none)
3) Quavers, $12-10=2$, $8-5=3$.
4) Semiquavers, $12-5=7$, $12-4=8$.

Second Chord.—Notes 5, 6, 8; Octaves 10, 12, 16.

1) Crotchets, $6-5=1$, $8-6=2$, $8-5=3$.
2) Quavers, $10-8=12-10=2$, $12-8=4$.
3) Semiquavers, $12-5=7$, $16-6=10$, (but this is also an audible partial,) $16-5=11$,

2.) Minor Triads with their Combinational Tones :*

¶

In the *major triads* the combinational tones of the first order, and even the deeper combinational tones of the second order (written as crotchets and quavers) are merely doubles of the tones of the triad in deeper Octaves. The higher combinational tones of the second order (written as semiquavers) are extremely weak, because, other conditions being the same, the intensity of combinational tones decreases as the interval between the generating tones increases, with which again the high position of these combinational tones is connected. I have always ¶

which being more than half an equal Semi-tone (51 cents) above equally tempered *f″* is represented on the staff as a flattened *f″♯*.

Third Chord.—Notes 6, 8, 10; Octaves 12, 16, 20.

   1) Crotchets, $10-8=8-6=2$, $10-6=4$.
   2) Quavers, $12-10=2$, $12-8=4$.
   3) Semiquavers, $20-6=14$.

How far these higher notes marked by semiquavers are effective, except possibly when they beat with each other, or with some partials of the original notes, remains to be proved.—*Translator.*]

* [In minor chords the case is different. On referring to the list of harmonics in the last note, it will be seen that the only minor chord is 10, 12, 15 or *e″ g″ b″*, and this is the chord upon the major Third above the third Octave of the fundamental. Hence in the example where the chord taken is *c′ e′b g′* and its inversions, the harmonics must be formed on $A_{\prime\prime}b$ which is the same interval below *c′*. The list of harmonics in these examples is therefore

1	2	3	4	5	6	7	8	9	10	11	12	13	14	15	16
$A_{\prime\prime}b$	$A_{\prime}b$	$E_{\prime}b$	$A_{\prime}b$	c	$e b$	$^7g b$	$a b$	$b b$	c'	$^{11}d'b$	$e'b$	$^{13}f'$	$^7g'b$	g'	$a'b$

18	20	21	22	24	25	26	27	28	30	32	33	39	40
$b'b$	c''	$^7d''b$	$^{11}d''b$	$e''b$	e''	$^{13}f''$	f''	$^7g''b$	g''	$a''b$	$^{11}a''b$	$^{13}c'''$	c'''

The omitted harmonics are not used in this investigation, though differentials of higher orders occur up to the 48th harmonic.

First Chord.—Notes 10, 12, 15; Octaves 20, 24, 30.

   1) Crotchets, $12-10=2$, $15-12=3$, $15-10=5$.
   2) Quavers, $20-15=5$, $20-12=8$, $24-15=9$.
   3) Semiquavers, $24-10=14$, $30-12=18$, $30-10=20$.

Second Chord.—Notes 12, 15, 20; Octaves 24, 30, 40.

   1) Crotchets, $15-12=3$, $20-15=5$, $20-12=8$.
   2) Quavers, $24-20=4$, $24-15=9$, $30-20=10$.
   3) Semiquavers, $30-12=18$, $40-15=25$, $40-12=28$.

Third Chord.—Notes 15, 20, 24; Octaves ¶ 30, 40, 48.

   1) Crotchets, $24-20=4$, $20-15=5$, $24-15=9$.
   2) Quavers, $30-24=6$, $30-20=10$, $40-24=16$.
   3) Semiquavers, $40-15=25$, $48-20=28$, $48-15=33$. This I have here represented as $^{11}a''b$ because it is the Twelfth above $^{11}d'b$, but in the text it is called a flattened *a″* because it is almost the one-sixth of $C''=528$. In fact on the Harmonical, $\frac{5}{3} \times 528 = 880$, and $A_{\prime\prime}b$ would be $\frac{4}{5} \cdot C_{\prime} = \frac{4}{5} \cdot 33 = 26\cdot3$, so that $33 \times 26\cdot3 = 867\cdot9$ vibrations. The interval 880 : 867·9 has 24 cents, and hence *a″* is more than a comma too sharp. The same observation applies as in the last footnote regarding the audible effect of the high notes, when not beating with each other, or with audible partials. *Translator.*]

been able to hear the deeper combinational tones of the second order, written as quavers, when the tones have been played on an harmonium, and the ear was assisted by the proper resonators :* but I have not been able to hear those written with semiquavers. They have been added merely to make the theory complete. Perhaps they might be occasionally heard from very loud musical tones having powerful upper partials. But they may be certainly neglected in all ordinary cases.

For the *minor triads*, on the other hand, the combinational tones of the first order, which are easily audible, begin to disturb the harmonious effect. They are not near enough indeed to beat, but they do not belong to the harmony. For the fundamental triad, and that of the Sixth and Third [the two first chords], these combinational tones, written as crotchets, form the major triad $A\flat$ C $E\flat$, and for the triad of the Sixth and Fourth [the third chord], we find entirely new tones, ¶ $A\flat$, $B\flat$, which have no relations with the original triad.† The combinational tones of the second order, however (written as quavers), are sometimes partly above and generally partly below the prime tones of the triad, but so near to them, that beats must arise ; whereas in the corresponding major triads the tones of this order fit perfectly into the original chord. Thus for the fundamental minor triad in the example, c' $e'\flat$ g', the deeper combinational tones of the second order give the dissonances $a\flat$ $b\flat$ c', and similarly for the triad of the Sixth and Third, $e'\flat$ g' c''. And for the triad of the Sixth and Fourth g' c'' $e''\flat$ we find the dissonances $b\flat$ c' and g' $a'\flat$. This disturbing action of the combinational tones on the harmoniousness of minor triads is certainly too slight to give them the character of dissonances, but they produce a sensible increase of roughness, in comparison with the effect of major chords, for all cases where just intonation is employed, that is, where the mathematical ratios of the intervals are preserved. In the ordinary tempered intonation of our keyed instruments, the roughness due to the ¶ combinational tones is proportionally less marked, because of the much greater roughness due to the imperfection of the consonances. Practically I attribute more importance to the influence of the more powerful deep combinational tones of the first order, which, without increasing the roughness of the chord, introduce tones entirely foreign to it, such as those of the $A\flat$ and $E\flat$ major triads in the case of the C minor triads. The foreign element thus introduced into the minor chord is not sufficiently distinct to destroy the harmony, but it is enough to give a mysterious, obscure effect to the musical character and meaning of these chords, an effect for which the hearer is unable to account, because the weak combinational tones on which it depends are concealed by other and louder tones, and are audible only to a practised ear.‡ Hence minor chords are especially adapted to express mysterious obscurity or harshness.§ F. T. Vischer, in his *Esthetics* (vol. iii. § 772), has carefully examined this character of the minor mode, and shewn how it suits many degrees of joyful and painful excitement, and that all shades of ¶ feeling which it expresses agree in being to some extent 'veiled' and obscure.

Every minor Third and every Sixth when associated with its principal com-

---

* [See note † on p. 157d.—*Translator.*]

† [From the list of harmonics on p. 215c it will be seen that these tones occur as lower harmonics of the tone whence the minor chords are derived.—*Translator.*]

‡ [The Author is of course always speaking of chords in just intonation. When tempered, as on the harmonium, even the major chords are accompanied by unrelated combinational tones, sufficiently close to beat and sufficiently loud for Scheibler to have laid down a rule for counting the beats in order to verify the correctness of the tempered tuning (see p. 203d). But still the different effects of the two chords are very marked.—*Translator.*]

§ [The English names *major* and *minor*

were chosen because the first Third in the fundamental position is *major* in the first case and *minor* in the second. In German the terms are *dur* and *moll*, that is, *hard* and *soft*.] It is well known that the names *dur* and *moll* are not connected with the hard or soft character of the pieces of music written in these modes, but are historically derived from the angular form of ♮ and the rounded form of ♭, which were the *B durum* and *B molle* of the medieval musical notation. [The probable origin of the forms b ♮ h ♯ is given from observations on the p'ates in Gaforius's *Theoricum Opus Harmonicae Disciplinae*, 1480, the earliest printed book on music, in a footnote, infrà p. 312d. *Translator.*]

binational tone, becomes at once a major chord. C is the combinational tone of the minor Third e' g'; c of the major Sixth g e', and g of the minor Sixth e' c''.* Since, then, these dyads naturally produce consonant triads, if any new tone is added which does not suit the triads thus formed, the contradiction is necessarily sensible.

Modern harmonists are unwilling to acknowledge that the minor triad is less consonant than the major. They have probably made all their experiments with tempered instruments, on which, indeed, this distinction may perhaps be allowed to be a little doubtful. But on justly intoned instruments † and with a moderately piercing quality of tone, the difference is very striking and cannot be denied. The old musicians, too, who composed exclusively for the voice, and were consequently not driven to enfeeble consonances by temperament, shew a most decided feeling for that difference. To this feeling I attribute the chief reason for their avoidance of a minor chord at the close. The medieval composers down to Sebastian Bach ¶ used for their closing chords either exclusively major chords, or doubtful chords without the Third; and even Handel and Mozart occasionally conclude a minor piece of music with a major chord. Of course other considerations, besides the degree of consonance, have great weight in determining the final chord, such as the desire to mark the prevailing tonic or key-note with distinctness, for which purpose the major chord is decidedly superior. More upon this in Chapter XV.

After having examined the consonant triads which lie within the compass of an Octave, we proceed to those with wider intervals. We have found in general that consonant intervals remain consonant when one of their tones is transposed an Octave or two higher or lower at pleasure, although such transposition has some effect on its degree of harmoniousness. It follows, then, that in all the consonant chords which we have hitherto found, any one of the tones may be transposed some Octaves higher or lower at pleasure. If the three intervals of the triad were consonant before, they will remain so after transposition. We have already seen ¶ how the chords of the Sixth and Third, and of the Sixth and Fourth, were thus obtained from the fundamental form. It follows further that when larger intervals are admitted, no consonant triads can exist which are not generated by the transposition of the major and minor triads. Of course if such other chords could exist, we should be able by transposition of their tones to bring them within the compass of an Octave, and we should thus obtain a new consonant triad within this compass, whereas our method of discovering consonant triads enabled us to determine every one that could lie within that compass. It is certainly true that slightly dissonant chords which lie within the compass of an Octave are sometimes rendered smoother by transposing one of their tones. Thus the chord $1 : \frac{7}{6} : \frac{7}{4}$, or C, $^7E\flat$, $^7B\flat$,‡ is slightly dissonant in consequence of the interval $1 : \frac{7}{6}$; the interval $1 : \frac{7}{4}$, or subminor Seventh, does not sound worse than the minor Sixth; the interval $\frac{7}{6} : \frac{7}{4}$ is a perfect Fifth. Now transposing the tone $^7E\flat$, an Octave higher to $^7e\flat$, and thus transforming the chord into $1 : \frac{7}{4} : \frac{7}{3}$, we obtain $1 : \frac{7}{3}$ in ¶ place of $1 : \frac{7}{6}$, and this is much smoother, indeed it is better than the minor Tenth of our minor scale $1 : \frac{12}{5}$,§ and a chord thus composed, which I have had carefully tuned on the harmonium, although its unusual intervals produced a strange effect, is not rougher in sound than the worst minor chord, that of the Sixth and Fourth. This chord, C, $^7B\flat$, $^7e\flat$, is also much injured by the unsuitable combinational tones $G_{\prime}$ and F.** Of course it would not be worth while to introduce such strange

---

* [For $e' : g' = 5 : 6$, diff. $6 - 5 = 1$ or C; $g : e' = 3 : 5$, diff. $5 - 3 = 2$ or c; $e' : c'' = 5 : 8$, diff. $8 - 5 = 3$ or g.—*Translator.*]

† See Chapter XVI. for remarks upon just and tempered intonation, and for a justly intoned instrument suitable for such experiments. [The Harmonical can also be used. See App. XX. sect. F. for this and other instruments.]

‡ [See these intervals examined in p. 195, note *.—*Translator.*]

§ [The intervals $6 : 7 = g' : ^7b'b$, $3 : 7 = g : ^7b'b$, and $5 : 12 = e : g'$ can be tried and compared on the Harmonical.—*Translator.*]

** [The ratios are $12 : 21 : 28$, and $21 - 12 = 9$, but $9 : 12 = 3 : 4$, hence if 12 is C, 9 is $G_{\prime}$. Again $28 - 12 = 16$, $12 : 16 = 3 : 4$ and hence 16 is F.—*Translator.*]

tones as $^7B\flat$, $^7e\flat$, into the scale for the sake of a chord which in itself is not superior to the worst of our present consonant chords, and for which the tones could not be transposed without greatly deteriorating its effect.*

The transposition of some tones in a consonant triad, for the purpose of widening their intervals, affects their harmoniousness in the first place by changing the intervals. Major Tenths, as we found in Chapter X. p. 195*b*, sound better than major Thirds, but minor Tenths worse than minor Thirds, the major and minor Thirteenth worse than the minor Sixth (p. 196*a*). The following rule embraces all the cases :—*Those intervals in which the smaller of the two numbers expressing the ratios of the pitch numbers is* EVEN, *are* IMPROVED *by having one of their tones transposed by an Octave,* because the numbers expressing the ratio are thus diminished.

¶

The *Fifth*	. .	2 : 3	becomes the *Twelfth*	. . 2 : 6	= 1 : 3
The *major Third*	.	4 : 5	,, major Tenth	. 4 : 10	= 2 : 5
The *subminor Third*		6 : 7	,, subminor Tenth	6 : 14	= 3 : 7.

Those intervals in which the smaller of the two numbers expressing the ratio of the vibrational numbers is ODD, *are* MADE WORSE *by having one of their tones transposed by an Octave*, as the Fourth 3 : 4 [which becomes the Eleventh 3 : 8], the minor Third 5 : 6 [which becomes the minor Tenth 5 : 12], and the Sixths [major] 3 : 5, and [minor] 5 : 8 [which become the Thirteenths, major 3 : 10 and minor 5 : 16].

Besides this the principal combinational tones are of essential importance. An example of the first combinational tones of the consonant intervals within the compass of an Octave is given below, the primary tones being represented by minims and the combinational tones by crotchets, as before.†

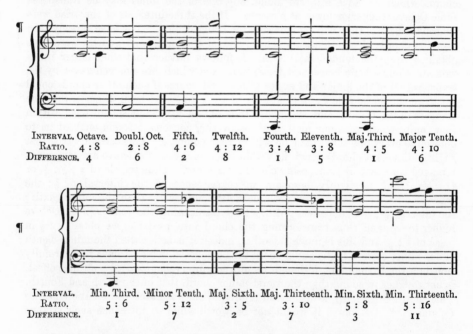

INTERVAL.	Octave.	Doubl. Oct.	Fifth.	Twelfth.	Fourth.	Eleventh.	Maj. Third.	Major Tenth.
RATIO.	4 : 8	2 : 8	4 : 6	4 : 12	3 : 4	3 : 8	4 : 5	4 : 10
DIFFERENCE.	4	6	2	8	1	5	1	6

INTERVAL.	Min. Third.	Minor Tenth.	Maj. Sixth.	Maj. Thirteenth.	Min. Sixth.	Min. Thirteenth.
RATIO.	5 : 6	5 : 12	3 : 5	3 : 10	5 : 8	5 : 16
DIFFERENCE.	1	7	2	7	3	11

The upwards sloping line prefixed to f'' denotes a degree of sharpening of about a quarter of a Tone [53 cents] ; and the downwards sloping line prefixed to $b'\flat$ flattens it [by 27 cents] to the subminor Seventh of *c*. Below the notes are added

---

* [They are, however, insisted on by Poole, see App. XX. sect. F. No. 6.—*Translator.*]

† Some of the bars and numbers have been changed to make all agree with the footnote to

p. 214*d*. All these notes and their combinational notes can by this means be played on the Harmonical.—*Translator.*]

the names of the intervals, the numbers of the ratios, and the differences of those numbers, giving the pitch numbers of the several combinational tones.

We find in the first place that the combinational tones of the *Octave, Fifth, Twelfth, Fourth,* and *major Third* are merely transpositions of one of the primary tones by one or more Octaves, and therefore introduce no foreign tone. Hence these five intervals can be used in all kinds of consonant triads, without disturbing the effect by the combinational tones which they introduce. In this respect the major Third is really superior to the major Sixth and the Tenth in the construction of chords, although its independent harmoniousness is inferior to that of either.

The *double Octave* introduces the Fifth as a combinational tone. Hence if the fundamental tone of a chord is doubled by means of the double Octave, the chord is not injured. But injury would ensue if the Third or Fifth of the chord were doubled in the double Octave.

Then we have a series of intervals which are made into complete major triads ¶ by means of their combinational tones, and hence produce no disturbance in major chords, but are injurious to minor chords. These are the *Eleventh, minor Third, major Tenth, major Sixth,* and *minor Sixth.*

But the *minor Tenth,* and the *major* and *minor Thirteenth* cannot form part of a chord without injuring its consonance by their combinational tones.

We proceed to apply these considerations to the construction of triads.

1. Major Triads.

Major triads can be so arranged that the combinational tones remain parts of the chord. This gives the most perfectly harmonious positions of these chords. To find them, remember that no minor Tenths and no [major or minor] Thirteenths are admissible, so that the minor Thirds and [both major and minor] Sixths must be in their closest position. By taking as the uppermost tone first the Third, then ¶ the Fifth, and lastly the fundamental tone, we find the following positions of these chords, within a compass of two Octaves, in which the combinational tones (here written as crotchets as usual) do not disturb the harmony.

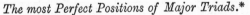

The most Perfect Positions of Major Triads.*

When the Third lies uppermost, the Fifth must not be more than a major Sixth below, as otherwise a [major] Thirteenth would be generated. But the fundamental tone can be transposed. Hence when the Third is uppermost the only two positions which are undisturbed are Nos. 1 and 2. When the Fifth lies uppermost, the Third must be immediately under it, or otherwise a minor Tenth

---

* [Calculation according to list of harmonics p. 214*d*, footnote.

1) Chord 4, 6, 10. Differentials $6-4=2$, $10-6=4$, $10-4=6$, which is also one of the tones.

2) Chord 6, 8, 10. Differentials $8-6=$ $10-8=2$, $10-6=4$.

3) Chord 4, 10, 12. Differentials $12-10$ $=2$, $10-4=6$, $12-4=8$.

4) Chord 8, 10, 12. Differentials $10-8$ $=12-10=2$, $12-8=4$.

5) Chord 3, 5, 8. Differentials $5-3=2$, $8-5=3$, (which is also one of the tones,) $8-3$ $=5$, (which is also one of the tones).

6) Chord 5, 6, 8. Differentials $6-5=1$, $8-6=2$, $8-5=3$.

These chords should be studied on the Harmonical, and the combinational tones listened for, and afterwards the tones played as substantive notes.—*Translator.*]

would be produced ; but the fundamental tone may be transposed. Finally, when the fundamental tone is uppermost, the major Third can lie only in the position of a minor Sixth below it, but the Fifth may be placed at pleasure. Hence it follows that the only possible positions of the major chord which will be entirely free from disturbance by combinational tones, are the six here presented, among which we find the three *close* positions Nos. 2, 4, 6 already mentioned [p. 215a], and three new ones Nos. 1, 3, 5. Of these new positions two (Nos. 1, 3) have the fundamental tone in the bass, just as in the primary form, and are considered as *open* positions of that form, while the third (No. 5) has the Fifth in the bass, just as in the chord of the Sixth and Fourth [of which it is also considered as an *open* position]. The chord of the Sixth [and Third] (No. 6), on the other hand, admits of no opener position [if it is to remain perfectly free from combinational disturbance].

¶ The order of these chords in respect to harmoniousness of the intervals is, perhaps, the same as that presented above. The three intervals of No. 1 (the Fifth, major Tenth, and major Sixth) are the best, and those of No. 6 (the Fourth, minor Third, and minor Sixth) are relatively the most unfavourable of the intervals that occur in these chords.

The remaining positions of the major triads present individual unsuitable combinational tones, and on justly intoned instruments are unmistakably rougher than those previously considered, but this does not make them dissonant, it merely puts them in the same category as minor chords. We obtain all of them which lie within the compass of two Octaves, by making the transpositions forbidden in the last cases. They are as follows, in the same order as before, No. 7 being made from No. 1, and so on :—

*The less Perfect Positions of Major Triads.**

Musicians will immediately perceive that these positions of the major triad are much less in use. The combinational tone $^7b'\flat$, gives the positions 7 to 10 some-

---

* [Calculation in continuation of the last note.

7) Chord 3, 4, 10. Differentials $4-3=1$, $10-4=6$, $10-3=7$.

¶ 8) Chord 3, 8, 10. Differentials $10-8=2$, $8-3=5$, $10-3=7$, which gives the interval 7 : 8 with the tone 8.

9) Chord 4, 5, 12. Differentials $5-4=1$, $12-4=8$, $12-5=7$, the two last differential tones being 7 : 8.

10) Chord 5, 8, 12. Differentials $8-5=3$, $12-8=4$, $12-5=7$, which gives the interval 7 : 8 with the tone 8.

11) Chord 5, 6, 16. Differentials $6-5=1$, $16-6=10$, $16-5=11$, which two last form the dissonant trumpet interval 11 : 10 of 165 cents or about three-quarters of an equal tone.

12) Chord 5, 12, 16. Differentials $16-12=4$, $12-5=7$, $16-5=11$, which forms the same dissonant trumpet interval 11 : 10, but this time with one of the tones, and therefore more harshly.

All these 12 chords should be well studied on the Harmonical, and for the first 10, the differential tones can be played also as substantive notes (remembering that $^7B\flat$ is on the $G\flat$ digital), which will enable the student to acquire a better idea of the roughness. The tones 11 and 13 could not be introduced among the first 4 Octaves on the Harmonical without incurring the important losses of f'' and a''. But if we take the chords an Octave higher we can play $^{11}f'''$ and $^{13}a'''$.

The chords should also be played in lower and higher positions, not only as Octaves of those given, but from the other major chords on the Harmonical as $FA_1C_2\,GB_1D$, $A^1\flat CE^1\flat$, $E^1\flat\,G\,B^1\flat$. Particular attention should be paid to the contrasting of the positions 1 and 7, 2 and 8, 3 and 9, 4 and 10, 5 and 11, 6 and 12. Unless the ear acquires the habit of attending to these differences it will not properly form the requisite conceptions of major chords. For future purposes the results should also be contrasted with those obtained by playing the same chords on a tempered instrument, —if possible of the same pitch, A 440.—*Translator.*]

thing of the character of the chord of the dominant Seventh in the key of F' major, $c\ e\ g\ b\flat$. The two last, 11 and 12, are much the least pleasing ; indeed they are decidedly rougher than the better positions of the minor chord.

2. Minor Triads.

No minor chord can be obtained perfectly free from false combinational tones, because its Third can never be so placed relatively to the fundamental tone, as not to produce a combinational tone unsuitable to the minor chord. If only one such tone is admitted, the Third and Fifth of the minor chord must lie close together and form a major Third, because in any other position they would produce a second unsuitable combinational tone. The fundamental tone and the Fifth must never be so placed as to form an Eleventh, because in that case the resulting combinational tone would make them into a major triad. These conditions can be fulfilled ¶ by only three positions of the minor chord, as follows :—

The most Perfect Positions of Minor Triads.*

The remaining positions which do not sound so well are :—   ¶

The less Perfect Positions of Minor Triads.†

* [Calculation according to the list of harmonics on p. 215c, footnote.

1) Chord 24, 30, 40. Differentials $30-24 = 6$, $40-30 = 10$, $40-24 = 16$.

2) Chord 20, 24, 30. Differentials $24-20 = 4$, $30-24 = 6$, $30-20 = 10$.

3) Chord 10, 24, 30. Differentials $30-24 = 6$, $24-10 = 14$, $30-10 = 20$.

These can also be studied on the Harmonical, and the differentials to Nos. 1 and 2 can be played as substantive tones. Not so No. 3, but the effect may be felt by playing the chord a major Third higher as $eg'b'$, being the 10, 24, 30 harmonics of $C_{\prime\prime}$ and giving the differentials G, $'b\flat b$, e' which can be played as substantive tones, but being so low will make the effect very rough.—*Translator.*]

† [Calculation in continuation of the last note.

4) Chord 12, 15, 40. Differentials $15-12 = 3$, $40-15 = 25$, $40-12 = 28$.

5) Chord 12, 30, 40. Differentials $40-30 = 10$, $40-12 = 18$, $40-12 = 28$.

6) Chord 15, 20, 24. Differentials $24-20 = 4$, $20-15 = 5$, $24-15 = 9$.

7) Chord 12, 20, 30. Differentials $20-12 = 8$, $30-20 = 10$, $30-12 = 18$, where 18 forms the dissonance $20 : 18 = 10 : 9$ with the tone 20.

8) Chord 10, 15, 24. Differentials $15-10 = 5$, $24-15 = 9$, $24-10 = 14$, which forms the dissonant interval $15 : 14$ with one of the tones 15.

9) Chord 10, 12, 30. Differentials $12-10 = 2$, $30-12 = 18$, $30-10 = 20$, the two last form together the dissonance $20 : 18 = 10 : 9$.

10) Chord 15, 20, 48 referred to $A_{\prime\prime}b$. Interpret by taking the Octaves below the numbers in p. 215c, note. Differentials $20-15 = 5 = C$; $48-20 = 28 = {}^7g'b$, $48-15 = 33 = {}^{11}a'b$, see p. 215d', note, towards the end of the observations on the *Third Chord*.

The positions Nos. 4 to 10 each produce two unsuitable combinational tones, one of which necessarily results from the fundamental tone and its [minor] Third; the other results in No. 4 from the Eleventh G C, and in the rest from the transposed major Third $E\flat$ G. The two last positions, Nos. 11 and 12, are the worst of all, because they give rise to three unsuitable combinational tones [two of which beat with original tones].

The influence of the combinational tones may be recognised by comparing the different positions. Thus the position No. 3, with a minor Tenth c' $e''\flat$ and major Third $e''\flat$ g'', sounds unmistakably better than the position No. 7, with major Tenth $e'\flat$ g'' and major Sixth $e'\flat$ c'', although the two latter intervals when struck separately sound better than the two first. The inferior effect of chord No. 7 is consequently solely due to the second unsuitable combinational tone, $b'\flat$.

¶ This influence of bad combinational tones is also apparent from a comparison with the major chords. On comparing the minor chords Nos. 1 to 3, each of which has only one bad combinational tone, with the major chords Nos. 11 and 12, each of which has two such tones, those minor chords will be found really pleasanter and smoother than the major. Hence in these two classes of chords it is not the major and minor Third, nor the musical mode, which decides the degree of harmoniousness, it is wholly and solely the combinational tones.

Four Part Chords or Tetrads.

It is easily seen that all consonant tetrads must be either major or minor triads to which the Octave of one of the tones has been added.* For every consonant tetrad must admit of being changed into a consonant triad by removing one of its tones. Now this can be done in four ways, so that, for example, the tetrad C E G c gives the four following triads:—

¶ C E G, C E c, E G c, C G c.

Any such triad, if it is not merely a dyad, or interval of two tones, with the Octave of one added, must be either a major or a minor triad, because there are no other consonant triads. But the only way of adding a fourth tone to a major or minor triad, on condition that the result should be consonant, is to add the Octave of one of its tones. For every such triad contains two tones, say C and G, which form either a direct or inverted Fifth. Now the only tones which can be combined with C and G so as to form a consonance are E and $E\flat$; there are no others at all. But E and $E\flat$ cannot be both present in the same consonant chord,

11) Chord 15, 40, 48 referred to $A_{,,,}\flat$ as in last chord.

Differentials $48-40=8=A\flat$, $40-15=25$ $=e'$, $48-15=33=$$^{11}a'\flat$ as in last chord, which see.

¶ 12) Chord 15, 24, 40 referred to $A_{,,,}\flat$. Differentials $24-15=9=B\flat$, $40-24=16=a\flat$, $40-15=25=e'$ where the differentials 16, 25 form the dissonant intervals 16 : 15, 25 : 24 with the two tones 15 and 24 respectively. All these chords can be studied on the Harmonical, and their differentials can be played as substantive tones in Nos. 6, 7, and 12. No. 8 can be taken a major Third higher as in chord No. 3 of the last note, that is as e' b' g'' giving the differentials e, d, $^7b\flat$ which can be played. Also No. 9 may be played as e' g' b'' giving differentials c, d'', e''. Nos. 4 and 5 do not admit of such treatment because $e'''\flat$ is not on the instrument. Nos. 10 and 11 cannot be so played because $^{11}a'\flat$ is not on the instrument. In fact it is the 33rd harmonic of $A_{,,,}\flat=13\cdot15$, and this (see footnote p. 215d', remarks on *Third Chord*,) $=33 \times 13\cdot15=433\cdot95$ vib.; whereas $a=440$, and hence is too sharp by the interval

$440 : 433\cdot95$ or 24 cents, rather more than a comma.

The student should try all the minor chords not only in different positions in Octaves, but with all the other minor chords on the Harmonical, namely, $F'A^1\flat C$, $GB^1\flat D$, D_1FA_1 (which contrast with the dissonance DFA_1 for future purposes), A_1CE_1, E_1GB_1, also in different Octaves, till the ear learns to distinguish these 12 different forms.

Finally the 12 forms of the major should be contrasted with the corresponding 12 forms of the minor triad, for the three possible cases FA_1C and $FA^1\flat C$; CE_1G and $CE^1\flat G$; GB_1D and $GB^1\flat D$. To merely read over these pages by eye instead of studying them by ear is useless, and ordinary tempered instruments only impede instead of assisting the investigator.—*Translator.*]

* [That is, if we exclude the harmonic Seventh from consideration, as on p. 195d, those who admit it (as Mr. Poole, App. XX. sect. F. No. 6) consider $CE_1G^7B\flat$ to be a perfectly consonant tetrad.—*Translator.*]

and hence every consonant chord of four or more parts, which contains C and G, must either contain E and some of the Octaves of C, E, G, or else Eb and some of the Octaves of C, Eb, G.

Every consonant chord of three or more parts will therefore be either a major or a minor chord, and may be formed from the fundamental position of the major and minor triad, by transposing or adding the Octaves above or below some or all of its three tones.

To obtain the perfectly harmonious positions of major tetrads, we have again to be careful that no minor Tenths and no [major or minor] Thirteenths occur. Hence the Fifth may not stand more than a minor Third above, or a Sixth below the Third of the chord; and the fundamental tone must not be more than a Sixth above the Third. When these rules are carried out, the avoidance of the minor Thirteenths is effected by not taking the double Octave of the Third and Fifth. These rules may be briefly enunciated as follows : *Those major chords are ¶ most harmonious in which the fundamental tone or the Fifth does not lie more than a Sixth above the Third, or the Fifth does not lie more than a Sixth above or below it.* The fundamental tone, on the other hand, may be as far *below* the Third as we please.

The corresponding positions of the major tetrads are found by combining any two of the more perfect positions of the major triads which have two tones in common, as follows, where the lower figures refer to the positions of the major triads already given.

*The most Perfect Positions of Major Tetrads within the Compass of Two Octaves.**

We see that *chords of the Sixth and Third* must lie quite close, as No. 7 ; † and that *chords of the Sixth and Fourth* ‡ must not have a compass of more than an Eleventh, but may occur in all the three positions (Nos. 5, 6, 11) in which it can be constructed within this compass. Chords which have the fundamental tone in the bass can be handled most freely.

It will not be necessary to enumerate the less perfect positions of major tetrads. ¶ They cannot have more than two unsuitable combinational tones, as in the 12th position of the major triads, p. 220c. The major triads of C can only have the false combinational tones marked ^{7}bb and ^{11}f, [that is, with pitch numbers bearing to that of C the ratios 7 : 1, or 11 : 1].

Minor tetrads, like the corresponding triads, must at least have one false combinational tone. There is only one single position of the minor tetrad which has only one such tone. It is No. 1 in the following example, and is compounded of the positions Nos. 1 and 2 of the minor triads on p. 221b. But there may be as

---

* [These major tetrads can all be played on the Harmonical, and should be tried in every position of Octaves and for all the major chords on the instrument, namely $F A, C$, $C E, G$, $G B, D$, $A^{1}b C E^{1}b$, $E^{1}b G B^{1}b$, till the ear is perfectly familiar with the different forms and the student can tell them at once and designate them by their number in this list on hear-

ing another person play them.—*Translator.*]

† [This chord has the Third both lowest and highest and is marked $\frac{6}{3}$, but is more commonly marked 6.—*Translator.*]

‡ [These chords have the Fifth lowest and are marked $\frac{6}{4}$.—*Translator.*]

many as 4 false combinational tones, as, for example, on combining positions Nos. 10 and 11 of the minor triads, p. 221c.

Here follows a list of the minor tetrads which have not more than two false combinational tones, and which lie within the compass of two Octaves. The false combinational tones only are noted in crotchets, and those which suit the chord are omitted.

Best Positions of Minor Tetrads.*

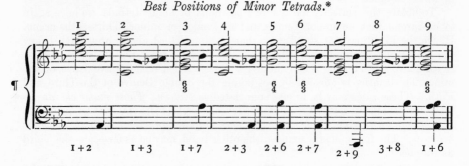

The chord of the Sixth and Fourth [marked $\frac{6}{4}$] occurs only in its closest position, No. 5; but that of the Sixth and Third [marked $\frac{6}{3}$] is found in three positions (Nos. 3, 6, and 9), namely, in all positions where the compass of the chord does not exceed a Tenth; the fundamental chord occurs three times with the Octave of the fundamental note added (Nos. 1, 2, 4), and twice with the Octave of the Fifth added (Nos. 7, 8).

In musical theory, as hitherto expounded, very little has been said of the ¶ influence of the transposition of chords on harmonious effect. It is usual to give as a rule that close intervals must not be used in the bass, and that the intervals should be tolerably evenly distributed between the extreme tones. And even these rules do not appear as consequences of the theoretical views and laws usually given, according to which a consonant interval remains consonant in whatever part of the scale it is taken, and however it may be inverted or combined with others. They rather appear as practical exceptions from general rules. It was left to the musician himself to obtain some insight into the various effects of the various positions of chords by mere use and experience. No rule could be given to guide him.

The subject has been treated here at such length in order to shew that a right view of the cause of consonance and dissonance leads to rules for relations which previous theories of harmony could not contain. The propositions we have enun-

---

¶   * [Calculation of the combinational tones, by the list of harmonics in p. 215c.

1) Chord 20, 24, 30, 40.   Differentials
$24-20=4=A\flat$, $40-24=16=a'\flat$.

2) Chord 10, 24, 30, 40.   Differentials
$24-10=14={}^{7}g'\flat$, $40-24=16=a'\flat$.

3) Chord 12, 15, 20, 30.   Differentials
$20-12=8=a\flat$, $30-12=18=b'\flat$.

4) Chord 10, 20, 24, 30.   Differentials
$24-20=4=A\flat$, $24-10=14={}^{7}g'\flat$.

5) Chord 15, 20, 24, 30.   Differentials
$24-20=4=A\flat$, $24-15=9=b\flat$.

6) Chord 12, 20, 24, 30.   Differentials
$24-20=4=A\flat$, $20-12=8=a\flat$, $30-12=18$ $=b'\flat$.

7) Chord 10, 12, 15, 30.   Differenti ls
$12-10=2=A_{/}\flat$, $30-12=18=b'\flat$.

8) Chord 10, 15, 24, 30.   Differentials
$24-15=9=b\flat$, $24-10=14={}^{7}g'\flat$.

9) Chord 12, 15, 20, 24.   Differentials

$24-20=4=A\flat$, $20-12=8=a\flat$, $24-15=9$ $=b\flat$.

These chords should all be studied on the Harmonical. With the exception of Nos. 2, 4, 7, 8 the differentials can also be played on it as substantive tones. But they can be transposed. Thus No. 2 may be played as e g' b' e'' giving the differentials ${}^{7}b\flat$, c'. No. 4 will become e' e'' g'' b'' giving the differential ${}^{7}b'\flat$, which can be played. No. 7 becomes e' g' b' b'' giving the differentials C and d''. No. 8 becomes e' b' g'' b'' giving the differentials d' and ${}^{7}b'\flat$. These chords should also be studied in all the minor forms on the Harmonical, not only in different Octaves, but on all the minor chords on that instrument, viz. $D_{/} F A_{1}$, $A_{/} C E_{1}$, $E_{/} G B_{1}$, $F A^{1}\flat C$, $C E^{1}\flat G$, $G B^{1}\flat D$, till the ear recognises the form, and the student can name the number of the position to another person's playing.—*Translator*.]

ciated agree, however, with the practice of the best composers, of those, I mean, who studied vocal music principally, before the great development of instrumental music necessitated the general introduction of tempered intonation, as any one may easily convince himself by examining those compositions which aimed at producing an impression of perfect harmoniousness. Mozart is certainly the composer who had the surest instinct for the delicacies of his art. Among his vocal compositions the *Ave verum corpus* is particularly celebrated for its wonderfully pure and smooth harmonies. On examining this little piece as one of the most suitable examples for our purpose, we find in its first clause, which has an extremely soft and sweet effect, none but major chords, and chords of the dominant Seventh. All these major chords belong to those which we have noted as being the more perfectly harmonious. Position 2 occurs most frequently, and then 8, 10, 1, and 9 [of p. 223c]. It is not till we come to the final modulation of this first clause that we meet with two minor chords, and a major chord in an ¶ unfavourable position. It is very striking, by way of comparison, to find that the second clause of the same piece, which is more veiled, longing, and mystical, and laboriously modulates through bolder transitions and harsher dissonances, has many more minor chords, which, as well as the major chords scattered among them, are for the most part brought into unfavourable positions, until the final chord again restores perfect harmony.

Precisely similar observations may be made on those choral pieces of Palestrina, and of his contemporaries and successors, which have a simple harmonic construction without any involved polyphony. In transforming the Roman church music, which was Palestrina's task, the principal weight was laid on harmonious effect in contrast to the harsh and unintelligible polyphony of the older Netherland * system, and Palestrina and his school have really solved the problem in the most perfect manner. Here also we find an almost uninterrupted flow of consonant chords, with some dominant Sevenths, or dissonant passing notes, charily interspersed. Here ¶ also the consonant chords wholly, or almost wholly, consist of those major and minor chords which we have noted as being in the more perfect positions. Only in the final cadence of a few clauses, on the contrary, in the midst of more powerful and more frequent dissonances, we find a predominance of the unfavourable positions of the major and minor chords. Thus that expression which modern music endeavours to attain by various discords and an abundant introduction of dominant Sevenths, was obtained in the school of Palestrina by the much more delicate shading of various inversions and positions of consonant chords. This explains the harmoniousness of these compositions, which are nevertheless full of deep and tender expression, and sound like the songs of angels with hearts affected but undarkened by human grief in their heavenly joy. Of course such pieces of music require fine ears both in singer and hearer, to let the delicate gradation of expression receive its due, now that modern music has accustomed us to modes of expression so much more violent and drastic.   ¶

The great majority of major tetrads in Palestrina's *Stabat mater* are in the positions 1, 10, 8, 5, 3, 2, 4, 9 [of p. 223c], and of minor tetrads in the positions 9, 2, 4, 3, 5, 1 [of p. 224a]. For the major chords one might almost think that some theoretical rule led him to avoid the bad intervals of the minor Tenth and the [major or minor] Thirteenth. But this rule would have been entirely useless for minor chords. Since the existence of combinational tones was not then known, we can only conclude that his fine ear led him to this practice, and that the judgment of that ear exactly agreed with the rules deduced from our theory.

These authorities may serve to lead musicians to allow the correctness of my arrangement of consonant chords in the order of their harmoniousness. But any one can convince himself of their correctness on any justly intoned instrument

---

* [Including both the modern kingdom of the Netherlands or Holland, and the still more modern kingdom of Belgium. Josquin, 1450– 1532, was born in Hainault in the present Belgium.—*Translator*.]

[as the Harmonical]. The present system of tempered intonation certainly oblite-
rates somewhat of the more delicate distinctions, without, however, entirely
destroying them.

————————

Having thus concluded that part of our investigations which rests upon purely
scientific principles, it will be advisable to look back upon the road we have travelled
in order to review our gains, and examine the relation of our results to the views of
older theoreticians. We started from the acoustical phenomena of upper partial tones,
combinational tones and beats. These phenomena were long well known both to
musicians and acousticians, and the laws of their occurrence were, at least in their
essential features, correctly recognised and enunciated. We had only to pursue
these phenomena into further detail than had hitherto been done. We succeeded
in finding methods for observing upper partial tones, which rendered comparatively
¶ easy an observation previously very difficult to make. And with the help of this
method we endeavoured to shew that, with few exceptions, the tones of all musical
instruments were compounded of partial tones, and that, in especial, those qualities
of tone which are more particularly favourable for musical purposes, possess at
least a series of the lower partial tones in tolerable force, while the simple tones,
like those of stopped organ pipes, have a very unsatisfactory musical effect,
although even these tones when loudly sounded are accompanied in the ear itself
by some weak harmonic upper partials. On the other hand we found that, for the
better musical qualities of tone, the higher partial tones, from the Seventh onwards,
must be weak, as otherwise the quality, and every combination of tones would be
too piercing. In reference to the beats, we had to discover what became of them
when they grew quicker and quicker. We found that they then fell into that
roughness which is the peculiar character of dissonance. The transition can be
effected very gradually, and observed in all its stages, and hence it is apparent to
¶ the simplest natural observation that the essence of dissonance consists merely* in
very rapid beats. The nerves of hearing feel these rapid beats as *rough* and
unpleasant, because every intermittent excitement of any nervous apparatus affects
us more powerfully than one that lasts unaltered. With this there is possibly
associated a psychological cause. The individual pulses of tone in a dissonant
combination give us certainly the same impression of separate pulses as slow beats,
although we are unable to recognise them separately and count them ; hence they
form a *tangled* mass of tone, which cannot be analysed into its constituents. The
cause of the unpleasantness of dissonance we attribute to this *roughness* and
entanglement. The meaning of this distinction may be thus briefly stated: *Consonance is a continuous, dissonance an intermittent sensation of tone.* Two consonant tones flow on quietly side by side in an undisturbed stream ; dissonant
tones cut one another up into separate pulses of tone. This description of the
distinction at which we have arrived agrees precisely with Euclid's old definition,
¶ 'Consonance is the blending of a higher with a lower tone. Dissonance is
incapacity to mix, when two tones cannot blend, but appear rough to the ear.' †
    After this principle had been once established there was nothing further to do
but to inquire under what circumstances, and with what degree of strength, beats

————————

* [But see also Prof. Preyer, in App. XX.
sect. L. art. 7, infra.—*Translator.*]

† *Euclides*, ed. Meibomius, p. 8: Ἔστι δὲ
συμφωνία μὲν κρᾶσις δύο φθόγγων, ὀξυτέρου καὶ
βαρυτέρου. Διαφωνία δὲ τοὐναντίον δύο φθόγγων
ἀμιξία, μὴ οἵων τε κραθῆναι, ἀλλὰ τραχυνθῆναι
τὴν ἀκοήν. [In translating this passage in the
text, I have endeavoured to make the distinction of μίξις and κρᾶσις ; the former is taken to
be of the nature of a mechanical, and the
latter a chemical mixture. *Mixing* and *blending* seem to convey the notion. In συμφωνία
(which Euclid admitted only for the Octave,

Fifth, and Fourth) he felt that the tones
blended. But the διαφωνία (which he applies
to *all* other intervals, for he used Pythagorean
major and minor Thirds, which are really dissonant) he found to consist in their not even
mixing, not even forming a mechanical, much
less a chemical unit, so that he goes on to explain that this *non-mixing* of the two tones
consisted in *inability to blend*, and resulted in
producing a *roughness*, as contradistinguished
from a *blending* in the ear. The *tones* are
φθόγγοι, properly tones sung, but used even for
tones of the lyre.—*Translator.*]

would arise in the various combinations of tones through either the partial or the combinational tones. This investigation had hitherto been completely worked out by Scheibler for the combinational tones of two simple tones only. The law of beats being known, it became easy to extend it to compound tones. Every theoretical conclusion on this field can be immediately checked by a proper observation, when the analysis of a mass of tone is facilitated by the use of resonators. All these beats of partial and combinational tones, of which so much has been said in the last chapter, are not inventions of empty theoretical speculation, but rather facts of observation, and can be really heard without difficulty by any practised observer who performs his experiments correctly. The knowledge of the acoustic law facilitates our discovery of the phenomena in question. But all the assertions on which we depend for establishing a theory of consonance and dissonance, such as was given in the last chapters, are founded wholly and solely on a careful analysis of the sensations of hearing, an analysis which a practised ¶ ear could have executed without any theoretical assistance, although of course the task was immensely facilitated by the guidance of theory and the assistance of appropriate instruments of observation.

For these reasons the reader is particularly requested to observe that my hypothesis concerning the sympathetic vibration of Corti's organs inside the ear has no immediate connection whatever with the explanation of consonance and dissonance. That explanation depends solely upon observed facts, on the beats of partial tones and the beats of combinational tones. Yet I thought it right not to suppress my hypothesis (which must of course be regarded solely as an hypothesis), because it gathers all the various acoustical phenomena with which we are concerned into one sheaf, and gives a clear, intelligible, and evident explanation of the whole phenomena and their connection.

The last chapters have shewn, that a correct and careful analysis of a mass of sound under the guidance of the principles cited, leads to precisely the same dis- ¶ tinctions between consonant and dissonant intervals and chords, as have been established under the old theory of harmony. We have even shewn that these investigations give more particular information concerning individual intervals and chords than was possible with the general rules of the former theory, and that the correctness of these rules is corroborated both by observation on justly intoned instruments and the practice of the best composers.

Hence I do not hesitate to assert that the preceding investigations, founded upon a more exact analysis of the sensations of tone, and upon purely scientific, as distinct from esthetic principles, exhibit the true and sufficient cause of consonance and dissonance in music.

One circumstance may, perhaps, cause the musician to pause in accepting this assertion. We have found that from the most perfect consonance to the most decided dissonance there is a continuous series of degrees, of combinations of sound, which continually increase in roughness, so that there cannot be any sharp ¶ line drawn between consonance and dissonance, and the distinction would therefore seem to be merely arbitrary. Musicians, on the contrary, have been in the habit of drawing a sharp line between consonances and dissonances, allowing of no intermediate links, and Hauptmann advances this as a principal reason against any attempt at deducing the theory of consonance from the relations of rational numbers.*

As a matter of fact we have already remarked that the chords of the natural

---

* *Harmonik und Metrik*, p. 4. [At the same time, by accepting equal temperament they accept as consonant a series of tones which really form only *one* consonant interval (the Octave) and only *two* others even approximatively consonant (the Fifth and Fourth), while the commonest intervals on which harmony rests, the Thirds, with their inversions the Sixths, are not merely dissonant but, on the sustained tones of the voice for example, *grossly* dissonant. It is difficult for any ear brought up among these dissonances, to understand the real distinction between consonance and dissonance. Hence the absolute necessity of testing all the above assertions by a justly intoned instrument such as the Harmonical.—*Translator.*]

or subminor Seventh 4 : 7 [c' to $^7b'\flat$ on the Harmonical], and of the subminor Tenth 3 : 7 [g to $^7b'\flat$] in many qualities of tone sound at least as well as the minor Sixth 5 : 8 [e' to c''], and that the subminor Tenth really sounds better than the ordinary minor Tenth 5 : 12 [e' to g'']. But we have already noticed a circumstance of great importance for musical practice which gives the minor Sixth an advantage over the intervals formed with the number 7. The inversion of the minor Sixth gives a better interval, the major Third [e' c'' inverted gives c' e'], and its importance as a consonance in modern music is especially due to this very relation to the major Third ; it is essentially necessary, and justified, just because it is the inversion of the major Third. On the other hand the inversion or transposition of an interval formed with the number 7 leads to intervals worse than itself. Hence, as it is necessary, for the purposes of harmony, to have the power of transposing the parts at pleasure, we have a sufficient reason for drawing the ¶ line between the minor Sixth on the one hand, and the intervals characterised by 7 on the other. It is not, however, till we come to construct scales, which we shall have to consider in the next chapter, that we find decisive reasons for making this the boundary. The scales of modern music cannot possibly accept tones determined by the number 7.* But in musical harmony we can only deal with chords formed of notes in the scale. Intervals characterised by 5, as the *Thirds* and *Sixths*, occur in the scale, as well as others characterised by 9, as the *major Second* 8 : 9, but there are none characterised by 7, which should form the transition between them. Here, then, there is a real gap in the series of chords arranged according to the degree of their harmonious effect, and this gap serves to determine the boundary between consonance and dissonance.

The decision does not depend, then, on the nature of the intervals themselves but on the construction of the whole tonal system. This is corroborated by the fact that the boundary between consonant and dissonant intervals has not been ¶ always the same. It has been already mentioned that the Greeks always represented Thirds as dissonant, and although the original Pythagorean Third 64 : 81, determined by a series of Fifths, was not a consonance, yet even when the natural major Third 4 : 5 was afterwards included in the so-called syntono-diatonic mode of Didymus and Ptolemaeus, it was not recognised as a consonance. It has already been mentioned that in the middle ages, first the Thirds and then the Sixths were acknowledged as imperfect consonances, that the Thirds were long omitted from the final chord, and that it was not till later that the major, and quite recently the minor Third was admitted in this position. It is quite a mistake to suppose, with modern musical theorists, that this was merely whimsical and unnatural, or that the older composers allowed themselves to be fettered by blind faith in Greek authority. The last was certainly partly true for writers on musical theory down to the sixteenth century. But we must distinguish carefully between composers and theoreticians. Neither the Greeks, nor the great musical ¶ composers of the sixteenth and seventeenth centuries, were people to be blinded by a theory which their ears could upset. The reason for these deviations is to be looked for rather in the difference between the tonal systems in early and recent times, with which we shall become acquainted in the next part. It will there be seen that our modern system gained the form under which we know it through the influence of a general use of harmonic chords. It was only in this system that a complete regard was paid to all the requisitions of interwoven harmonies. Owing to its strict consistency, we were not only able to allow many licences in the use of the more imperfect consonances and of dissonances, which older systems had to avoid, but we were often required to insert the Thirds in final chords, as a mode of distinguishing with certainty between the major and minor mode, in cases where this distinction was formerly evaded.

* [Poole's scale $f\,g\,a_1$ ^7bb $c'\,d'\,e'_1\,f'$, and Bosanquet's and White's tempered imitation of $^7b'b$, properly 969 cents, as 974 cents, shew the feeling that exists for using the 7th harmonic, which is the only acoustical justification for the greatly harsher dominant Seventh.— *Translator.*]

But if the boundary between consonance and dissonance has really changed with a change of tonal system, it is manifest that the reason for assigning this boundary does not depend on the intervals and their individual musical effect, but on the whole construction of the tonal system.

The enigma which, about 2500 years ago, Pythagoras proposed to science, which investigates the reasons of things, ' Why is consonance determined by the ratios of small whole numbers ? ' has been solved by the discovery that the ear resolves all complex sounds into pendular oscillations, according to the laws of sympathetic vibration, and that it regards as harmonious only such excitements of the nerves as continue without disturbance. The resolution into partial tones, mathematically expressed, is effected by Fourier's law, which shews how any periodically variable magnitude, whatever be its nature, can be expressed by a sum of the simplest periodic magnitudes.* The length of the periods of the simply periodic terms of this sum must be exactly such, that either one or two ¶ or three or four, and so on, of their periods are equal to the period of the given magnitude. This, reduced to tones, means that the pitch numbers of the partial tones must be exactly once, twice, three times, four times, and so on, respectively, as great as that of the prime tone. These are the whole numbers which determine the ratios of the consonances. For, as we have seen, the condition of consonance is that two of the lower partial tones of the notes combined shall be of exactly the same pitch ; when they are not, disturbance arises from beats. Ultimately, then, the reason of the rational numerical relations of Pythagoras is to be found in the theorem of Fourier, and in one sense this theorem may be considered as the prime source of the theory of harmony.†

The relation of whole numbers to consonance became in ancient times, in the middle ages, and especially among Oriental nations, the foundation of extravagant and fanciful speculation. 'Everything is Number and Harmony,' was the characteristic principle of the Pythagorean doctrine. The same numerical ratios ¶ which exist between the seven tones of the diatonic scale, were thought to be found again in the distances of the celestial bodies from the central fire. Hence the harmony of the spheres, which was heard by Pythagoras alone among mortal men, as his disciples asserted. The numerical speculations of the Chinese in primitive times reach as far. In the book of Tso-kiu-ming, a friend of Confucius (B.C. 500), the five tones of the old Chinese scale were compared with the five elements of their natural philosophy—water, fire, wood, metal, and earth. The whole numbers 1, 2, 3 and 4 were described as the source of all perfection. At a later time the 12 Semitones of the Octave were connected with the 12 months in the year, and so on. Similar references of musical tones to the elements, the temperaments, and the constellations are found abundantly scattered among the musical writings of the Arabs. The harmony of the spheres plays a great part throughout the middle ages. According to Athanasius Kircher, not only the macrocosm, but the microcosm is musical. Even Keppler, a man of the deepest scientific spirit, could not ¶ keep himself free from imaginations of this kind. Nay, even in the most recent times, theorising friends of music may be found who will rather feast on arithmetical mysticism than endeavour to hear upper partial tones.

The celebrated mathematician Leonard Euler‡ tried, in a more serious and more scientific manner, to found the relations of consonances to whole numbers upon psychological considerations, and his theory may certainly be regarded as the one which found most favour with scientific investigators during the last century, although it perhaps did not entirely satisfy them. Euler § begins by explaining that we are pleased with everything in which we can detect a certain amount of

---

* Namely magnitudes which vary as sines and cosines.

† [The coincidences or non-coincidences of combinational tones, which are independent of Fourier's law, are also considered of impor-

tance by Prof. Preyer. See infra, App. XX. sect L. art. 7.—*Translator*.]

‡ *Tentamen novae theoriae Musicae*, Petropoli, 1739.

§ *Loc. cit.* chap. ii. § 7.

perfection. Now the perfection of anything is determined by the co-operation of all its parts towards the attainment of its end. Hence it follows that wherever perfection is to be found there must be order ; for order consists in the arrangement of all parts by a certain law from which we can discover why each part lies where it is, rather than in any other place. Now in any perfect object such a law of arrangement is determined by the end to be attained which governs all the parts. For this reason order pleases us more than disorder. Now order can be perceived in two ways : either we know the law whence the arrangement is deduced, and compare the deductions from this law with the arrangements observed ; or, we observe these arrangements and endeavour to determine the law from them. The latter is the case in music. A combination of tones will please us when we can discover the law of their arrangement. Hence it may well happen that one hearer finds it and that another does not, and that their judgments consequently ¶ differ.

The more easily we perceive the order which characterises the objects contemplated, the more simple and more perfect will they appear, and the more easily and joyfully shall we acknowledge them. But an order which costs trouble to discover, though it will indeed also please us, will associate with that pleasure a certain degree of weariness and sadness (*tristitia*).

Now in tones there are two things in which order is displayed, pitch and duration. Pitch is ordered by intervals, duration by rhythm. Force of tone might also be ordered, had we a measure for it. Now in rhythm two or three or four equally long notes of one part may correspond with one or two or three of another, in which the regularity of the arrangement is easily observed, especially when frequently repeated, and gives considerable pleasure. Similarly in intervals we should derive more pleasure from observing that two, three, or four vibrations of one tone coincided with one, two, or three of another, than we could possibly experience if ¶ the ratios of the time of vibration were incommensurable with one another, or at least could not be expressed except by very high numbers. Hence it follows that the combination of two tones pleases us the more, the smaller the two numbers by which the ratios of their periods of vibration can be expressed. Euler also remarked that we could better endure more complicated ratios of the periods of vibration, and consequently less perfect consonances, for higher than for deeper tones, because for the former the groups of vibrations which were arranged to occur in equal times, were repeated more frequently than in the latter, and we were consequently better able to recognise the regularity of éven a more involved arrangement.

Hereupon Euler develops an arithmetical rule for calculating the degree of harmoniousness of an interval or a chord from the ratios of the periods of the vibrations which characterise the intervals. The *Unison* belongs to the first degree, the *Octave* to the second, the *Twelfth* and *Double Octave* to the third, the ¶ *Fifth* to the fourth, the *Fourth* to the fifth, the major *Tenth* and *Eleventh* to the sixth, the *major Sixth* and *major Third* to the seventh, the *minor Sixth* and *minor Third* to the eighth, the *subminor Seventh* 4 : 7 to the ninth, and so on. To the ninth degree belongs also the *major triad*, both in its closest position and in the *position of the Sixth and Fourth*. The *major chord of the Sixth and Third* belongs, however, to the tenth degree. The *minor triad*, both in its closest and in its *position of the Sixth and Third*, also belongs to the ninth degree, but its *position of the Sixth and Fourth* to the tenth degree. In this arrangement the consequences of Euler's system agree tolerably well with our own results, except that in determining the relation of the major to the minor triad, the influence of combinational tones was not taken into account, but only the kinds of interval. Hence both triads in their close position appear to be equally harmonious, although again both the *major chord of the Sixth and Third*, and the *minor chord of the Sixth and Fourth* are inferior with him as with us.*

---

* The principle on which Euler calculated the degrees of harmoniousness for intervals

Euler has not confined these speculations to single consonances and chords, but has extended them to their results, to the construction of scales, and to modulations, and brought out many surprising specialities correctly. But without taking into account that Euler's system gives no explanation of the reason why a consonance when slightly out of tune sounds almost as well as one justly tuned, and much better than one greatly out of tune, although the numerical ratio for the former is generally much more complicated, it is very evident that the principal difficulty in Euler's theory is that it says nothing at all of the mode in which the mind contrives to perceive the numerical ratios of two combined tones. We must not forget that a man left to himself is scarcely aware that a tone depends upon vibrations. Moreover, immediate and conscious perception by the senses has no means of discovering that the numbers of vibrations performed in the same time are different, greater for high than for low tones, and that determinate intervals have determinate ratios of these numbers. There are certainly many perceptions of the ¶ senses in which a person is not precisely able to account for the way in which he has attained to his knowledge, as when from the resonance of a space he judges of its size and form, or when he reads the character of a man in his features. But in such cases a person has generally had a large experience in such relations, which helps him to form a judgment in analogous circumstances, without having the previous circumstances on which his judgment depends clearly present to his mind. But it is quite different with pitch numbers. A man that has never made physical experiments has never in the whole course of his life had the slightest opportunity of knowing anything about pitch numbers or their ratios. And almost every one who delights in music remains in this state of ignorance from birth to death.

Hence it would certainly be necessary to shew how the ratios of pitch numbers can be perceived by the senses. It has been my endeavour to do this, and hence the results of my investigation may be said, in one sense, to fill up the gap which Euler's left. But the physiological processes which make the difference sensible ¶ between consonance and dissonance, or, in Euler's language, orderly and disorderly relations of tone, ultimately bring to light an essential difference between our method of explanation and Euler's. According to the latter, the human mind perceives commensurable ratios of pitch numbers *as such* ; according to our method, it perceives *only the physical effect* of these ratios, namely the continuous or intermittent sensation of the auditory nerves.* The physicist knows, indeed, that the reason why the sensation of a consonance is continuous is that the ratios of its pitch numbers are commensurable, but when a man who is unacquainted with physics, hears a piece of music, nothing of the sort occurs to him,† nor does the physicist find a chord in any respect more harmonious because he is better acquainted with the cause of its harmoniousness.‡ It is quite different with the order of rhythm. That exactly two crotchets, or three in a triplet, or four quavers

and chords, is here annexed, because its consequences are very correct, if combinational tones are disregarded. When p is a prime number, the degree is $= p$. All other numbers are products of prime numbers. The number of the degree for a product of two factors a and b, for which separately the numbers of degree are a and β respectively $= a + \beta - 1$. To find the number of the degree of a chord, which can be expressed by $p : q : r : s$, &c., in smallest whole numbers, Euler finds the least common multiple of p, q, r, s, &c., and the number of its degree is that of the chord. Thus, for example :

The number of the degree of 2 is 2, and of 3 is 3,
$$\text{of } 4 = 2 \times 2, \text{ it is } 2 + 2 - 1 = 3,$$
$$\text{of } 12 = 4 \times 3, \text{ it is } 3 + 3 - 1 = 5,$$
$$\text{of } 60 = 12 \times 5, \text{ it is } 5 + 5 - 1 = 9.$$

That of the major triad 4 : 5 : 6 is that of 60,

because 60 is the least common multiple of ¶ 4, 5, 6, that is, the least number which all of them will divide without a remainder.

* [With possibly Prof. Preyer's addition, see App. XX. sect. L. art. 7.—*Translator.*]

† [In point of fact, as he always hears tempered tones, he never hears the exact commensurable ratios. Indeed, on account of the impossibility of tuning with perfect exactness, the exact ratios are probably never heard, except from the double siren and wave-siren.—*Translator.*]

‡ [Does a man breathe more easily and aerate his blood better because he knows the constitution of the atmosphere and its relation to his carbonised blood? Does a man feel a weight greater or less, because he knows the laws of gravitation? These are quite similar questions.—*Translator*

go to one minim is perceived by any attentive listener without the least instruction. But while the orderly relation (or commensurable ratio) of the vibrations of two combined tones, on the other hand, undoubtedly affects the ear in a certain way which distinguishes it from any disorderly relation (incommensurable ratio), this difference of consonance and dissonance depends on physical, not psychological grounds.

The considerations advanced by Rameau* and d'Alembert† on the one side, and Tartini ‡ on the other, concerning the cause of consonance agree better with our theory. The last founded his theory on the existence of combinational tones, the two first on that of upper partial tones. As we see, they had found the proper points of attack, but the acoustical knowledge of last century did not allow of their drawing sufficient consequences from them. According to d'Alembert, Tartini's book was so darkly and obscurely written that he, as well as other well-¶ instructed people, were unable to form a judgment upon it. D'Alembert's book, on the other hand, is an extremely clear and masterly performance, such as was to be expected from a sharp and exact thinker, who was at the same time one of the greatest physicists and mathematicians of his time. Rameau and d'Alembert lay down two facts as the foundation of their system. The first is that every resonant body audibly produces at the same time as the prime (*générateur*) its Twelfth and next higher Third, as upper partials (*harmoniques*). The second is that the resemblance between any tone and its Octave is generally apparent. The first fact is used to shew that the major chord is *the most natural* of all chords, and the second to establish the possibility of lowering the Fifth and the Third by one or two Octaves without altering the nature of the chord, and hence to obtain the major triad in all its different inversions and positions. The minor triad is then found by the condition that all three tones should have the same upper partial or harmonic, namely the Fifth of the chord (in fact C, $E\flat$, and G have all the same ¶ upper partial g'). Hence although the minor chord is not so perfect and natural as the major, it is nevertheless prescribed *by nature*.

In the middle of the eighteenth century, when much suffering arose from an artificial social condition, it may have been enough to shew that a thing was *natural*, in order at the same time to prove that it must *also* be beautiful and desirable. Of course no one who considers the great perfection and suitability of all organic arrangements in the human body, would, even at the present day, deny that when the existence of such natural relations have been proved as Rameau discovered between the tones of the major triad, they ought to be most carefully considered, at least as starting-points for further research. And Rameau had indeed quite correctly conjectured, as we can now perceive, that this fact was the proper basis of a theory of harmony. But that is by no means everything. For in nature we find not only beauty but ugliness, not only help but hurt. Hence the mere proof that anything is natural does not suffice to justify it esthetically. ¶ Moreover if Rameau had listened to the effects of striking rods, bells, and mem-branes, or blowing over hollow chambers, he might have heard many a perfectly dissonant chord. And yet such chords cannot but be considered equally *natural*. That all musical instruments exhibit harmonic upper partials depends upon the selection of qualities of tone which man has made to satisfy the requirements of his ear.

Again the resemblance of the Octave to its fundamental tone, which was one of Rameau's initial facts, is a musical phenomenon quite as much in need of explanation as consonance itself.

No one knew better than d'Alembert himself the gaps in this system. Hence

---

* [*Traité de l'harmonie réduite à des prin-cipes naturels*, 1721.—*Translator.*]

† *Éléments de Musique, suivant les prin-cipes de M. Rameau*, par M. d'Alembert. Lyon,

1762.

‡ [*Trattato di Musica secondo la vera scienza dell' armonia*. Padova, 1751.—*Trans-lator.*]

in the preface to his book he especially guards himself against the expression :
' Demonstration of the Principle of Harmony,' which Rameau had used. He
declares that so far as he himself is concerned, he meant only to give a well-
connected and consistent account of all the laws of the theory of harmony, by
deriving them from a single fundamental fact, the existence of upper partial tones
or harmonics, which he assumes as given, without further inquiry respecting its
source.   He consequently limits himself to proving the *naturalness* of the major
and minor triads.   In his book there is no mention of beats, and hence of the
real source of distinction between consonance and dissonance.   Of the laws of beats
very little indeed was known at that time, and combinational tones had only been just
brought under the notice of French savants, by Tartini (1751) and Romieu (1753).
They had been discovered a few years previously in Germany by Sorge (1745), but
the fact was probably little known.   Hence the materials were wanting for building up a more perfect theory.   ¶

Nevertheless this attempt of Rameau and d'Alembert is historically of great importance, in so far as the theory of consonance was thus for the first time shifted from
metaphysical to physical ground.   It is astonishing what these two thinkers effected
with the scanty materials at their command, and what a clear, precise, comprehensive
system the old vague and lumbering theory of music became under their hands.
The important progress which Rameau made in the specially musical portion of
the theory of harmony will be seen hereafter.

If, then, I have been myself able to present something more complete, I owe it
merely to the circumstance that I had at command a large mass of preliminary
physical results, which had accumulated in the century that has since elapsed.

PART III.

THE RELATIONSHIP OF MUSICAL TONES.

SCALES, AND TONALITY.

CHAPTER XIII.

GENERAL VIEW OF THE DIFFERENT PRINCIPLES OF MUSICAL STYLE IN THE DEVELOPMENT OF MUSIC.

UP to this point our investigation has been purely physical. We have analysed the sensations of hearing, and investigated the physical and physiological causes for the phenomena discovered,—partial tones, combinational tones, and beats. In the whole of this research we have dealt solely with natural phenomena, which present themselves mechanically, without any choice, to all living beings whose ears are constructed on the same anatomical plan as our own. In such a field, where necessity is paramount and nothing is arbitrary, science is rightfully called upon to establish constant laws of phenomena, and to demonstrate strictly a strict connection between cause and effect. As there is nothing arbitrary in the phenomena embraced by the theory, so also nothing arbitrary can be admitted into the laws which regulate the phenomena, or into the explanations given for their occurrence. As long as anything arbitrary remains in these laws and explanations, it is the duty of science (a duty which it is generally able to discharge) to exclude it, by continuing the investigations.

But in this third part of our inquiry into the theory of music we have to furnish a satisfactory foundation for the elementary rules of musical composition, and here we tread on new ground, which is no longer subject to physical laws alone, although the knowledge which we have gained of the nature of hearing, will still find numerous applications. We pass on to a problem which by its very nature belongs to the domain of esthetics. When we spoke previously, in the theory of consonance, of agreeable and disagreeable, we referred solely to the immediate impression made on the senses when an isolated combination of sounds strikes the ear, and paid no attention at all to artistic contrasts and means of expression; we thought only of sensuous pleasure, not of esthetic beauty. The two must be kept strictly apart, although the first is an important means for attaining the second.

The altered nature of the matters now to be treated betrays itself by a purely external characteristic. At every step we encounter historical and national differences of taste. Whether one combination is rougher or smoother than another, depends solely on the anatomical structure of the ear, and has nothing to do with psychological motives. But what degree of roughness a hearer is inclined to endure as a means of musical expression depends on taste and habit; hence the boundary between consonances and dissonances has been frequently changed. Similarly Scales, Modes, and their Modulations have undergone multifarious alterations, not merely among uncultivated or savage people, but even in those periods of the world's history and among those nations where the noblest flowers of human culture have expanded.

Hence it follows,—and the proposition is not even now sufficiently present to the minds of our musical theoreticians and historians—*that the system of Scales, Modes, and Harmonic Tissues does not rest solely upon inalterable natural laws, but is also, at least partly, the result of esthetical principles, which have already changed, and will still further change, with the progressive development of humanity*.

But it does not follow from this that the choice of those elements of musical art was perfectly arbitrary, and that they do not allow of being derived from some more general law. On the contrary the rules of any style of art form a well-connected system whenever that style has attained a full and perfect development. These rules of art were certainly never developed into a system by the artists them- ¶ selves with conscious intention and consistency. They are rather the result of tentative exploration or the play of imagination, as the artists think out or execute their plans, and by trial gradually discover what kind or manner of performance best pleases them. Yet science can endeavour to discover the motors, whether psychological or technical, which have been at work in this artistic process. Scientific *esthetics* have to deal with the psychological motor; scientific *physics* with the technical. When the artist's aim in the style he has adopted, and its principal direction, have once been rightly conceived, it can be more or less correctly determined why he was forced to follow this or that rule, or employ this or that technical means. In musical theory, namely where the peculiar physiological functions of the ear, while not immediately present to conscious self-examination, play an important part, a large and rich field is thrown open for scientific investigation to shew the necessary character of the technical rules for each individual direction in the development of our art. ¶

It does not rest with natural science to characterise the chief problem worked out by each school of art, and the elementary principle of its style. This must be gathered from the results of historical and esthetical inquiry.

The relation we have to treat may be illustrated by a comparison with architecture, which, like music, has pursued essentially different directions at different times. The *Greeks*, in their stone temples, imitated the original wooden constructions; that was the principle of their architectural style. The whole division and arrangement of their decorations clearly shew that it was their intention to imitate wooden constructions. The verticality of the supporting columns, the general horizontality of the supported beam, forced them to divide all the subordinate parts for the great majority of cases into vertical and horizontal lines. The purposes of Grecian worship, which performed its principal functions in the open air, were satisfied by erections, of this kind, in which the internal spaces were necessarily narrowly limited by the length of the stone or wooden beams which could be em- ¶ ployed. The old *Italians* (Etruscans), on the other hand, discovered the principle of the arch, composed of wedge-shaped stones. This discovery rendered it possible to cover in much more extensive buildings with arched roofs, than the Greeks could do with their wooden beams. Among these arched buildings the halls of justice (basil'icae) became important, as is well known, for the subsequent development of architecture. The arched roof made the circular arch the chief principle in division and decoration for *Roman* (*Byzantine*) art. The columns, pressed by heavy weights, were transformed into pillars, on which, after the style was fully developed, columns merely appeared in diminished forms, half sunk in the mass of the pillar, as simply decorative articulations, and as the downward continuation of the ribs of the arches which radiated towards the ceiling from the upper end of the pillar.

In the arch the wedge-shaped stones press against each other, but as they all uniformly press inwards, each one prevents the other from falling. The most powerful and most dangerous degree of pressure is exerted by the stones in the horizontal parts of the arch, where they have either no support or no obliquely placed support, and are prevented from falling solely by the greater thickness of their upper extremities. In very large arches the horizontal middle portion is consequently the most dangerous, and would be precipitated by the slightest yielding of the materials. As, then, medieval ecclesiastical structures assumed continually larger dimensions, the idea occurred of leaving out the middle horizontal part of the arch altogether, and of making the sides ascend with moderate obliquity until they met in a pointed arch. From thenceforward the pointed arch became the dominant principle. The building was divided into sections externally by the projecting buttresses. These, and the omnipresent pointed arch, made the outlines hard, and the churches became enormously high. But both characters suited the vigorous minds of the northern nations, and perhaps the very hardness of the forms, ¶ thoroughly subdued by that marvellous consistency which runs through the varied magnificence of form in a *gothic* cathedral, served to heighten the impression of immensity and power.

We see here, then, how the technical discoveries which were associated with the problems as they rose successively created three entirely distinct principles of style—the horizontal line, the circular arch, the pointed arch—and how at each new change in the main plan of construction, all the subordinate individualities, down to the smallest decorations, were altered accordingly ; and hence how the individual rules of construction can only be comprehended from the general principle of construction. Although the gothic style has developed the richest, the most consistent, the mightiest and most imposing of architectural forms, just as modern music among other musical styles, no one would certainly for a moment think of asserting that the pointed arch is nature's original form of all architectural beauty, and must consequently be introduced everywhere. And at the present day ¶ it is well known that it is an artistic absurdity to put gothic windows in a Greek building. Conversely any one can unfortunately convince himself on visiting most of our gothic cathedrals how detestably unsuitable to the whole effect are those numerous little chapels of the *renaissance* period built in the Greek or Roman style. Just as little as the gothic pointed arch, should our diatonic major scale be regarded as a *natural product*. At least such an expression is quite inapplicable, except in so far as both are necessary and *natural* consequences of the principle of style selected. And just as little as we should use gothic ornamentation in a Greek temple, should we venture upon improving compositions written in ecclesiastical modes, by providing their notes with marks of sharps and flats in accordance with the scheme of our major and minor harmonies. The feeling for historical artistic conception has certainly made little progress as yet among our musicians, even among those who are at the same time musical historians. They judge old music by the rules of modern harmony, and are inclined to consider every deviation from ¶ it as mere unskilfulness in the old composer, or even as barbarous want of taste.*

Hence before we proceed to the construction of scales and rules for a tissue of harmony, we must endeavour to characterise the principles of style, at least for the chief phases of the development of musical art. For present purposes we may divide these into three principal periods :—

1. The *Homophonic* or Unison Music of the ancients, to which also belongs the existing music of Oriental and Asiatic nations.

2. The *Polyphonic* Music of the middle ages, with several parts, but without regard to any independent musical significance of the harmonies, extending from the tenth to the seventeenth century, when it passes into

3. *Harmonic* or *Modern* Music, characterised by the independent significance

---

* Thus in R. G. Kiesewetter's historico-musical writings, which are otherwise so rich in facts industriously collected, there is evi-dently an exaggerated zeal to deny everything which will not fit into the modern major and minor modes.

attributed to the harmonies as such. Its sources date back from the sixteenth century.

1. HOMOPHONIC MUSIC.

One part music is the original form of music with all people. It still exists among the Chinese, Indians, Arabs, Turks, and modern Greeks, notwithstanding the greatly developed systems of music possessed by some of these nations.* That music in the time of highest Grecian culture, neglecting perhaps individual instrumental ornamentation, cadences, and interludes, was written in one part, or that the voices at most sang in Octaves, can now be considered as established. In the problems of Aristotle we find the question : 'Why is the consonance of the Octave alone sung? For this and no other consonance is played on the magădis.' This was a harp-shaped instrument [with a bridge dividing the strings at one-third their length]. In another place he remarks that the voices of boys and men form an Octave in singing.†

One part music, considered independently and unaccompanied by words, is too poor in forms and changes, to develop any of the greater and richer forms of art. Hence purely instrumental music at this stage is necessarily limited to short dances or marches. We really find no more among nations that have no harmonic music.‡ Performers on the flute § have certainly repeatedly gained the prize in the Pythian games, but it is possible to perform feats of execution in instrumental music in concise forms of composition, as, for example, in the variations of a short melody. That the principle of varying (μεταβολή) a melody with reference to dramatical expression (μίμησις), was known to the Greeks, follows also from Aristotle. He describes the matter very plainly, and remarks that choruses must simply repeat the melodies in the antistrophes, because it is easier for one than for several to introduce variations. But public competitors (ἀγωνισταί) and actors (ὑποκριταί) are able to grapple with these difficulties.**

---

* [See App. XX. sect. K. for some of these scales.—*Translator.*]

† Διὰ τί ἡ διὰ πασῶν συμφωνία ᾄδεται μόνη ; μαγαδίζουσι γὰρ ταύτην, ἄλλην δὲ οὐδεμίαν. *Prob.* xix. 18. [Translated in the text.] Διὰ τί ἥδιόν ἐστι τὸ σύμφωνον τοῦ ὁμοφώνου ; Ἢ καὶ τὸ μὲν ἀντίφωνον σύμφωνόν ἐστι διὰ πασῶν ; ἐκ παίδων γὰρ νέων καὶ ἀνδρῶν γίνεται τὸ ἀντίφωνον· οἳ διεστᾶσι τοῖς τόνοις, ὡς νήτη πρὸς τὴν ὑπάτην. *Prob.* xix. 39. ['Why is a consonant union of voices pleasanter than a single voice ? Is the singing of voice against voice, a consonant union of voices in Octaves ? This singing of voice against voice occurs when young boys and men sing together, and their tones differ as the highest from the lowest of the scale.'] Towards the end of the songs the instrumental accompaniment seems to have separated itself from the voice. Probably this is what is meant by the *krousis* in the passage τελευτώσαις δ' εἰς ταὐτὸν, ἐν καὶ κοινὸν τὸ ἔργον συμβαίνει γίνεσθαι, καθάπερ τοῖς ὑπὸ τὴν ᾠδὴν κρούουσι· καὶ γὰρ οὗτοι τὰ ἄλλα οὐ προσαυλοῦντες, ἐὰν εἰς ταὐτὸν καταστρέφωσιν, εὐφραίνουσι μᾶλλον τῷ τέλει ἢ λυποῦσι ταῖς πρὸ τοῦ τέλους διαφοραῖς, τῷ τὸ ἐκ διαφόρων τὸ κοινόν, ἥδιστον ἐκ τοῦ διὰ πασῶν γίνεσθαι. Arist. *Prob.* xix. 39. ['But when they end in the same, the matter is precisely similar to what occurs *when they play an accompaniment to a song.* For the accompanyists do not follow the rest, but when the singers return to the same, they please more in the end, than they displease in the differences before the end, by which means the common part in what is generally different, pleases more than anything but the Octave.'] See also Plutarch, *De Musica*, xix. xxviii. That the Greeks knew the effect of consonances and did not like it, appears by the following passage from Aristotle, *De Audibilibus*, ed. Bekker, p. 801 : 'For this reason we understand a single speaker better than many who are saying the same things at the same time. And so with strings. And much less, when flute and lyre are played at the same time, because the voices are confused by the others. And this is very plain for the consonances. For both tones are concealed, one by the other.' Διὸ καὶ μᾶλλον ἑνὸς ἀκούοντες συνίεμεν, ἢ πολλῶν ἅμα ταὐτὰ λεγόντων· καθάπερ καὶ ἐπὶ τῶν χορδῶν· καὶ πολὺ ἧττον, ὅταν προσαυλῇ τις ἅμα καὶ κιθαρίζῃ, διὰ τὸ συγχεῖσθαι τὰς φωνὰς ὑπὸ τῶν ἑτέρων. Οὐχ ἥκιστα δὲ τοῦτο ἐπὶ τῶν συμφωνιῶν φανερόν ἐστιν. Ἀμφοτέρους γὰρ ἀποκρύπτεσθαι συμβαίνει τοὺς ἤχους ὑπ' ἀλλήλων.

‡ [In Java long pieces of music in non-harmonic scales occur to accompany actions and develop the feeling of a plot. Many instruments play together, but there is no harmony. —*Translator.*]

§ The αὐλοί were perhaps more like our oboes.

** [Διὰ τί οἱ μὲν νόμοι οὐκ ἐν ἀντιστρόφοις ἐποιοῦντο · αἱ δὲ ἄλλαι ᾠδαὶ αἱ χορικαί ; Ἢ ὅτι οἱ μὲν νόμοι ἀγωνιστῶν ἦσαν, ὧν ἤδη μιμεῖσθαι δυναμένων καὶ διατείνεσθαι, ἡ ᾠδὴ ἐγίνετο μακρὰ καὶ πολυειδής ; καθάπερ οὖν καὶ τὰ μέλη τῇ μιμήσει ἠκολούθει ἀεὶ ἕτερα γινόμενα. Μᾶλλον γὰρ τῷ μέλει ἀνάγκη μιμεῖσθαι, ἢ τοῖς ῥήμασι. Δι' ὃ καὶ οἱ διθύραμβοι, ἐπειδὴ μιμητικοὶ ἐγένοντο, οὐκ ἔτι ἔχουσιν ἀντιστρόφους, πρότερον δὲ εἶχον. Αἴτιον δὲ, ὅτι τὸ παλαιὸν οἱ ἐλεύθεροι ἐχόρευον αὐτοί. Πολλοὺς οὖν ἀγωνιστικῶς ᾄδειν, χαλεπὸν ἦν. Ὥστε ἐναρμόνια μᾶλλον καὶ ἐνῇδον. Μεταβάλλειν γὰρ πολλὰς μεταβολὰς τῷ ἑνὶ ῥᾷον, ἢ τοῖς πολλοῖς, καὶ τῷ ἀγωνιστῇ, ἢ τοῖς τὸ ἦθος

Extensive works of art, in homophonic music, are only possible in connection with poetry, and this was also the way in which music was applied in classical antiquity. Not only were songs (odes) and religious hymns sung, but even tragedies and long epic poems were performed in some musical manner, and accompanied by the lyre. We are scarcely in a condition to form a conception of how this was done, because modern taste points in precisely the opposite direction, and demands from a great declaimer or public reader that he should produce a dramatic effect true to nature by the speaking voice alone, rating all approach to singing as one of the greatest of faults. Perhaps we have some echoes of the ancient spoken song in the singing tone of Italian declaimers, and the liturgical recitations (intoning) of the Roman Catholic priests. Indeed, attentive observations on ordinary conversation shews us that regular musical intervals involuntarily recur, although the singing tone of the voice is concealed under the noises ¶ which characterise the individual letters, and the pitch is not held firmly, but is frequently allowed to glide up and down. When simple sentences are spoken without being affected by feeling, a certain middle pitch is maintained, and it is only the emphatic words and the conclusions of sentences and clauses which are indicated by change of pitch.* The end of an affirmative sentence followed by a pause, is usually marked by the voice falling a Fourth from the middle pitch. An interrogative ending rises, often as much as a Fifth above the middle pitch. For example a bass voice would say :

Ich   bin   spa - tzie - ren   ge - gan - gen.
I   have been walk - ing   this morn - ing.

¶

Bist   du   spa - tzie - ren   ge - gan - gen?
Have   you   been walk - ing   this morn - ing?

Emphasised words are also rendered prominent by their being spoken about a Tone higher than the rest,† and so on. In solemn declamation the alterations of pitch are more numerous and complicated. Modern recitative has arisen from attempting to imitate these alterations of pitch by musical notes. Its inventor, Giacomo Peri, in the preface to his opera of *Eurydice*, published in 1600, distinctly says as much. An attempt was then made to restore the declamation of ancient tragedies by means of recitative. Ancient recitative certainly differed somewhat from modern recitative, by preserving the metre of the poems more exactly, and by

¶ φυλάττουσι. Δι' ὃ ἁπλούστερα ἐποιοῦντο αὐτοῖς τὰ μέλη. Ἡ δὲ ἀντίστροφος, ἁπλοῦν. Ἀριθμὸς γάρ ἐστι, καὶ ἑνὶ μετρεῖται. Τὸ δ' αὐτὸ αἴτιον καὶ διότι τὰ μὲν ἀπὸ τῆς σκηνῆς οὐκ ἀντίστροφα, τὰ δὲ τοῦ χοροῦ ἀντίστροφα. Ὁ μὲν γὰρ ὑποκριτὴς ἀγωνιστὴς [καὶ μιμητής·] ὁ δὲ χορὸς, ἧττον μιμεῖται. Arist. *Prob.* xix. 15. 'Why are themes (*nomoi*) not used in antistrophic singing, while all other choral singing is employed? Is it because themes belong to public performers who are already able to imitate and extend, and hence would make their song long and very figurate? For melodies, like words, follow imitation and change. It is more necessary for melody to imitate, than for words. Wherefore dithyrambic poets also when they became mimetic, disused their previous antistrophic singing. The reason is that formerly gentlemen (*eleutheroi*) used to sing the choruses themselves. It was difficult for many to sing like public performers. So they rather intoned

suitable melodies. For it is easier for one to make numerous variations than for many to do so, and for a public performer than for those who retained old usage. Hence the melodies were made simpler for them. Now antistrophic singing is simple, for it depends on number, and is measured by a unit. The same reason shews why the parts of the actors are not antistrophic, but those of the chorus are so. For the actor is a public performer [and a mime], but the chorus does not imitate so well.'—*Translator.*]

* [Prof. Helmholtz's observations on speaking must be read in reference to North German habits only.—*Translator.*]

† [By no means uniformly, even in North Germany. The habits of different nations here vary greatly. In Norway and in Sweden the voice is regularly raised on unemphatic syllables. In Scotland the emphasis is often marked by lowering the pitch.—*Translator.*]

having no accompanying harmonies.　Nevertheless our recitative, when well performed, will give us a better conception of the degree in which the expression of the words can be enhanced by musical recitation, than we can obtain from the monotonous repetition of the Roman liturgy, although the latter perhaps is more nearly related in kind to ancient recitation than the former.　The settlement of the Roman liturgy by Pope Gregory the Great (A.D. 590 to 604) reaches back to a time in which reminiscences of the ancient art, although faded and deformed, might have been in some degree handed down by tradition, especially if, as we are probably entitled to assume, Gregory really did little more than finally establish the Roman school of singing which had existed from the time of Pope Sylvester (A.D. 314 to 335).* The majority of these formulae for lessons, collects, &c., evidently imitate the cadence of ordinary speech.　They proceed at an equal height; particular, emphatic, or non-Latin words are somewhat altered in pitch; and for the punctuation certain concluding forms are prescribed, as the following ¶ for lessons, according to the customs of Münster.†

Sic can - ta com - ma, sic du - o punc-ta: sic ve - ro punc-tum.
Thus sing the com - ma, and thus the co - lon: and thus the full stop.

Sic sig - num in - ter - ro - ga - ti - o - nis?
Thus sing the mark of in - ter - ro - ga - tion?

These and similar final formulae were varied according to the solemnity of the feast, the subject treated, the rank of the priest that sang and that answered, and ¶ so on.† It is easy to see that they strove to imitate the natural cadences of ordinary speech, and to give them solemnity by eliminating their individual irregularities.　Of course in such fixed formula no regard can be paid to the grammatical sense of the clauses, which suffers much in various ways from the intoning. Similarly we may suppose that the ancient tragic poets prescribed the cadences of speech to their actors, and preserved them by a musical accompaniment.　And since ancient tragedy kept much further aloof from immediate external realism than our modern drama, as is shewn by the artificial rhythms, the unusual rolling words, the immovable strange masks, it could admit of a more singing tone for declamation than would, perhaps, please our modern ears.　Then we must remember that by emphasising or increasing the loudness of certain words, and by rapidity or slowness of speech, or pantomimic action, much life can be thrown into delivery of this kind, which would certainly be insufferably monotonous if not thus enlivened.

But in any case homophonic music, even when in olden time it had to ac- ¶ company extensive poems of the highest character, necessarily played an utterly subordinate part.　The musical turns must have entirely depended on the changing sense of the words, and could have had no independent artistic value or connection without them.　A peculiar melody for singing hexameters throughout an epic, or iambic trimeters throughout a tragedy, would have been insupportable.‡ Those

* [These are the dates of his reign. Hullah says the school was founded in A.D. 350.—*Translator.*]

† Antony, 'Lehrbuch des Gregorianischen Kirchengesanges' [*Manual of Gregorian Church-music*], Münster, 1829.　According to the information collected by Fétis (in his *Histoire générale de Musique*, Paris, 1869, vol. i. chap. vi.), it has become doubtful whether this system of declamation with prescribed cadences, is not rather to be deduced

from the Jewish ritual chants.　In the oldest manuscripts of the Old Testament, there are 25 different signs employed to denote cadences and melodic phrases of this kind.　The fact that the corresponding signs of the Greek Church are Egyptian demotic characters, hints at a still older Egyptian origin for this notation.

‡ [We must remember that the Greek and Latin so-called *accents* consisted solely in alterations of pitch, and hence to a certain

melodies (νόμοι) which were allotted to odes and tragic choruses, were certainly
freer and more independent.   For odes there were also well-known melodies (the
names of some of them are preserved) to which fresh poems were continually
composed.

In the great artistic works just mentioned, then, music must have been entirely
subordinate ; independently, it could only have formed short pieces.   Now this is
closely connected with the development of homophonic music as a musical system.
Among the nations who possess such music we always find certain degrees of pitch
selected for the melodies to move in.   These scales are very various in kind, partly,
it would seem, very arbitrary, so that many appear to us quite strange and incomprehensible, and yet the best gifted among those nations which possess them, as
the Greeks, Arabs, and Indians, have developed them in an extremely subtile and
varied manner.   [See App. XX. sect. K.]

¶   When speaking of these systems of tones, it becomes a question of essential
importance for our present purpose, to inquire whether they are based upon any
determinate reference of all the tones in the scale to one single principal and
fundamental tone, the *tonic* or *key-note*.   Modern music effects a purely musical
internal connection among all the tones in a composition, by making their relationship to one tone as perceptible as possible to the ear.   This predominance
of the tonic, as the link which connects all the tones of a piece, we may, with
Fétis, term the principle of *tonality*.   This learned musician has properly drawn
attention to the fact that tonality is developed in very different degrees and
manners in the melodies of different nations.   Thus in the songs of the modern
Greeks, and chants of the Greek Church, and the Gregorian tones of the Roman
Church, they are not developed in a manner which is easy to harmonise, whereas,
according to Fétis,* it is on the whole easy to add accompanying harmonies
to the old melodies of the northern nations of German, Celtic, and Sclavonic
¶ origin.

It is indeed remarkable that though the musical writings of the Greeks often
treat subtile points at great length, and give the most exact information about all
other peculiarities of the scale, they say nothing intelligible about a relation which
in our modern system stands first of all, and always makes itself most distinctly
sensible.   The only hints to be found concerning the existence of the tonic are
not in especial musical writings, but as before in the works of Aristotle, who
asks :—

' Why is it that if any one alters the tone on the middle string (μέση) after the
others have been tuned, and plays, every thing sounds amiss, not merely when he
comes to this middle tone, but throughout the whole melody ? but if he alters the
tone played by the forefinger † or any other, the difference is only perceived when
that string is struck? Is there a good reason for this?   All good melodies often
employ the tone of the middle string, and good composers often come upon it,
¶ and if they leave it recur to it again ; but this is not the case with any other
tone.'   Then he compares the tone of the middle string with conjunctions in
language, such as ' and ' [and ' then '], without which language could not exist, and
proceeds to say : ' In this way the tone of the middle string is a link between
tones, especially of the best tones, because its tone most frequently recurs.' ‡

extent determined a melody.   See Dionysius
of Halicarnassus, περὶ συνθέσεως ὀνομάτων,
chap. xi., where we also find that in his day
(first century before Christ) the musical composers transgressed at pleasure the rules of
both accent and quantity.   But if the written
accents in Greek, and the accents as determined by the rules of grammarians in Latin,
are carefully examined, it will be found that
every line in a Greek or Latin poem had its
own distinct melody, the art of the poet being
shewn by the great variability of pitch con-

joined with a constant quantity or rhythm.—
Translator.]
    * Fétis, *Biographie universelle des Musi-
ciens*, vol. i. p. 126.
    † [The forefinger is ὁ λιχανός, the note
played by it is ἡ λίχανος, accent and gender
both differing.--*Translator.*]
    ‡ Διὰ τί, ἐὰν μέν τις τὴν μέσην κινήσῃ ἡμῶν,
ἁρμόσας [δὲ] τὰς ἄλλας χορδὰς, κέχρηται τῷ
ὀργάνῳ, οὐ μόνον ὅταν κατὰ τὸν τῆς μέσης γένη-
ται φθόγγον, λυπεῖ, καὶ φαίνεται ἀνάρμοστον,
ἀλλὰ καὶ κατὰ τὴν ἄλλην μελῳδίαν · ἐὰν δὲ τὴν

And in another place we find the same question with a slightly different answer, 'Why do the other tones sound badly when the tone of the middle string is altered? but if the tone of the middle string remains, and one of the others is altered, the altered one alone is spoiled? Is it because that all are tuned and have a certain relation to the tone of the middle string, and the order of each is determined by that? The reason of the tuning and connection being removed, then, things no longer appear the same.'* In these sentences the esthetic significance of the tonic, under the name of 'the tone of the middle string,' is very accurately described. To this we may add that the Pythagoreans compared the tone of the middle string with the sun, and the other tones in the scale with the planets.† It appears as if it had been usual to begin with the tone of the middle string above mentioned, for we read in the 33rd problem of Aristotle: 'Why is it more agreeable to proceed from high pitch to low pitch, than from low pitch to high pitch? Can it be that we thus begin at the beginning? for the tone of the middle string is also ¶ the leader of the tetrachord and highest in pitch. The second way would be to begin at the end instead of at the beginning. Or can it be that tones of lower pitch sound nobler and more euphonious after tones of high pitch?'‡ This seems also to shew that it was not the custom to end with the tone of the middle string, which commenced, but with the tone of lowest pitch [produced by the uppermost string or *Hypatē*], of which last tone Aristotle, in his 4th problem, says that, as opposed to its neighbour, the tone of lowest pitch but one, [due to the string of highest position but one, or *Parhypatē*,] it is sung with complete relaxation of all the effort that is felt in the other.§ These words of Aristotle may certainly be applied to the national Doric scale of the Greeks, which, increased by Pythagoras to eight tones, was as follows:—

λίχανον, ἤ τινα ἄλλον φθόγγον, τότε φαίνεται διαφέρειν μόνον, ὅταν κἀκείνη τὶς χρῆται; *Ἡ εὐλόγως τοῦτο συμβαίνει; πάντα γὰρ τὰ χρηστὰ μέλη, πολλάκις τῇ μέσῃ χρῆται· καὶ πάντες οἱ ἀγαθοὶ ποιηταί, πυκνὰ πρὸς τὴν μέσην ἀπαντῶσι· κἂν ἀπέλθωσι ταχὺ ἐπανέρχονται· πρὸς δὲ ἄλλην οὕτως οὐδεμίαν. Καθάπερ ἐκ τῶν λόγων ἐνίων ἐξαιρεθέντων συνδέσμων, οὐκ ἔστιν ὁ λόγος Ἑλληνικός· (οἷον τὸ τὲ, καὶ τὸ τοὶ) καὶ ἔνιοι δὲ οὐθὲν λυποῦσι· διὰ τὸ τοῖς μὲν, ἀναγκαῖον εἶναι χρῆσθαι πολλάκις, ἢ οὐκ ἔσται λόγος Ἑλληνικός· τοῖς δὲ, μὴ· οὕτω καὶ τῶν φθόγγων ἡ μέση, ὥσπερ σύνδεσμός ἐστι, καὶ μάλιστα τῶν καλῶν, διὰ τὸ πλειστάκις ἐνυπάρχειν τὸν φθόγγον αὐτῆς. Arist. *Prob.* xix. 20. This passage has also been partly quoted by Ambrosch. [The names of Greek tones were those of the strings on the lyre by which they were played, just as if in English we were to call the tones *g, d', a', e''*, the tones of the fourth, third, second, and first strings respectively, because they are produced as the open notes of these strings on the violin, and contracted them to *fourth, third,* &c., only omitting the word *string*. As the violin when held sideways in playing, throws the *g* string uppermost, and the *e''* string the lowermost, we might in the same way call *g* the 'uppermost note,' ὑπάτη, although lowest in pitch, and *e''* the 'lowermost note,' νήτη, although highest in pitch. Then *d* might be called the *middle*, μέση, being really the key-note of the violin and one of the two middle strings. This illustrates the Greek names very closely, for the lyre was held with the string sounding the lowest note, uppermost. See the scale on the next page.—*Translator.*]

* Διὰ τί, ἐὰν μὲν ἡ μέση κινηθῇ, καὶ αἱ ἄλλαι χορδαὶ ἠχοῦσι φθεγγόμεναι· (one of my colleagues, Prof. Stark, conjectures that in place of φθεγγόμεναι, which makes nonsense, we should read φθειρόμεναι·) ἐὰν δὲ αὖ ἡ μὲν μένῃ, τῶν δ' ἄλλων τὶς κινηθῇ, κινηθεῖσα μόνη φθέγ- ¶ γεται; (for which Prof. Stark again proposes φθείρεται;) *Ἡ ὅτι τὸ ἡρμόσθαι ἐστὶν ἁπάσαις, [τὸ δὲ ἔχειν πως πρὸς τὴν μέσην ἁπάσαις], καὶ ἡ τάξις ἡ ἑκάστης, ἤδη δι' ἐκείνην; ἀρθέντος οὖν τοῦ αἰτίου τοῦ ἡρμόσθαι καὶ τοῦ συνέχοντος, οὐκ ἔτι ὁμοίως φαίνεται ὑπάρχειν. Arist. *Prob.* xix. 36.

† Nicomachus, *Harmonicē*, lib. i. p. 6, ed. Meibomii. [The following is Nicomachus's arrangement of the comparison, with his reasons:

Saturn *hypatē*, as being highest in position, ὕπατον γὰρ τὸ ἀνώτατον.

Jupiter *parhypatē*, as next highest to Saturn.

Mars *lichanos* or *hypermesē*, as between Jupiter and the Sun.

The Sun *mesē*, as lying in the middle, the fourth from either end, middlemost string and planet. ¶

Mercury *paramesē*, as lying between the Sun and Venus.

Venus *parancatē*, as lying just above the moon.

The Moon *neatē*, as being lowest of all in position and next the earth, καὶ γὰρ νέατον, τὸ κατώτατον.—*Translator.*]

‡ Διὰ τί εὐαρμοστότερον ἀπὸ τοῦ ὀξέος ἐπὶ τὸ βαρὺ, ἢ ἀπὸ τοῦ βαρέος ἐπὶ τὸ ὀξύ; Πότερον ὅτι τὸ ἀπὸ τῆς ἀρχῆς γίνεται ἄρχεσθαι; ἡ γὰρ μέση καὶ ἡγεμὼν ὀξυτάτη τοῦ τετραχόρδου. Τὸ δὲ οὐκ ἀπ' ἀρχῆς, ἀλλ' ἀπὸ τελευτῆς. *Ἡ ὅτι τὸ βαρὺ ἀπὸ τοῦ ὀξέος γενναιότερον, καὶ εὐφωνότερον; Arist. *Prob.* xix. 33.

§ Διὰ τί δὲ ταύτην [τὴν παρυπάτην] χαλεπῶς [ᾄδουσι], τὴν δὲ ὑπάτην ῥᾳδίως· καίτοι δίεσις ἑκατέρας; ἢ ὅτι μετ' ἀνέσεως ἡ ὑπάτη, καὶ ἅμα μετὰ τὴν σύστασιν ἐλαφρὸν τὸ ἄνω βάλλειν; Arist. *Prob.* xix, 4.

Tetrachord of lowest pitch	⎧ E Hypătē	. . . .	. [ὑπάτη] uppermost string]
	F Parhypătē	. . .	. [παρυπάτη] next to uppermost string]
	G Lichănos	. . .	. [λίχανος] forefinger string]
	⎩ A Mĕsē (tone of middle string)		. [μέση] middle string]
Tetrachord of highest pitch	⎧ B Paramĕsē	. . .	. [παραμέση] next to middle string]
	C Trĭte	. . . .	. [τρίτη] third string]
	D Paranētē	. . . .	. [παρανήτη] next to lowermost string]
	⎩ E Nētē	. . . .	. [νήτη] lowermost string]

In modern phraseology the last description cited from Aristotle implies that the Parhypatē was a kind of descending 'leading note' to the Hypatē. In the leading tone there is perceptible effort, which ceases on its falling into the fundamental tone.

¶ If, then, the tone of the middle string answers to the tonic, the Hypatē, which is its Fifth, will answer to the dominant. For our modern feeling it is far more necessary to close with the tonic than to begin with it, and hence we usually take the final tone of a piece to be its tonic without further inquiry. Modern music, however, usually introduces the tonic also in the first beat of the opening bar. The whole mass of tone is developed from the tonic and returns into it. Modern musicians cannot obtain complete repose at the end unless the series of tones converges into its connecting centre.

Ancient Greek music seems, then, to have deviated from ours by ending on the dominant instead of the tonic. And this is in full agreement with the intonation of speech. We have seen that the end of an affirmative sentence is likewise formed on the Fifth next below the principal tone.* This peculiarity has also been generally preserved in modern recitative, in which the singer usually ends on the dominant; the accompanying instruments then make this tone part of the chord of the dominant Seventh, leading to the tonic chord, and thus make a close ¶ on the tonic in accordance with our present musical feeling. Now since Greek music was cultivated by the recitation of epic hexameters and iambic trimeters, we should not be surprised if the above-mentioned peculiarities of chanting were so predominant in the melodies of odes that Aristotle could regard them as the rule.†

From the facts just adduced it follows (and this is what we are chiefly concerned with) that the Greeks, among whom our diatonic scale first arose, were not without a certain esthetic feeling for tonality, but that they had not developed it so decisively as in modern music. Indeed, it does not appear to have even entered into the technical rules for constructing melodies. Hence Aristotle, who treated music esthetically, is the only known writer who mentions it; musical writers proper do not speak of it at all. And unfortunately the indications furnished by Aristotle are so meagre, that doubt enough still exists. For example, he says nothing about the differences of the various musical modes in reference to their ¶ principal tone, so that the most important point of all from which we should wish to regard the construction of the musical scale, is almost entirely obscured.

The reference to a tonic is more distinctly made out in the scales of the old Christian ecclesiastical music. Originally the four so-called *authentic* scales were distinguished, as they had been laid down by Ambrose of Milan (elected Bishop A.D. 374, died A.D. 398). Not one of these agrees with any one of our scales. The four *plagal* scales afterwards added by Gregory, are no scales at all in our sense of the word. The four authentic scales of Ambrose ‡ are :

* [This would be entirely crossed by the ancient Greek system of pitch-accents, just as it now is by a similar system in Norwegian, where the pitch may rise for *both* affirmative *and* interrogative sentences. See p. 239*d'*, note ‡.—*Translator.*]

† Among the presumed ancient melodies which have been handed down to us, the frag-ment of the Homeric Ode to Dēmētēr, which has been published by B. Marcello, shews the above-mentioned peculiarity very distinctly.

‡ [Mr. Rockstro in his article 'Ambrosian Chant,' in Grove's *Dictionary of Music*, states that this attribution of four authentic scales to St. Ambrose has not been proved.—*Translator.*]

$$
\begin{aligned}
&\text{1)} \quad D \quad E \quad F \quad G \quad A \quad B \quad c \quad d \\
&\text{2)} \quad E \quad F \quad G \quad A \quad B \quad c \quad d \quad e \\
&\text{3)} \quad F \quad G \quad A \quad B \quad c \quad d \quad e \quad f \\
&\text{4)} \quad G \quad A \quad B \quad c \quad d \quad e \quad f \quad g
\end{aligned}
$$

Perhaps, however, the change of B into $B\flat$ was allowed from the first, and this would make the first scale agree with our descending scale of D minor, and the third scale would become our scale of F major. The old rule was that songs in the first scale should end in D, those in the second in E, those in the third in F, and those in the fourth in G. This marked out these tones as tonics in our sense of the word. But the rule was not strictly observed. The conclusion might fall on other tones of the scale, the so-called *confinal* tones, and at last the confusion became so great that no one was able to say exactly how the scale was to be recognised, all kinds of insufficient rules were formulated, and at last musicians clung to the mechanical expedient of fixing upon certain initial and concluding ¶ phrases, called *tropes*, as characterising the scale.

Hence although the rule of tonality had been already remarked in these medieval ecclesiastical scales, the rule was so unsettled and admitted so many exceptions, that the feeling of tonality must have been much less developed than in modern music.

The Indians also hit upon the conception of a tonic, although their music is likewise unisonal. They called the tonic *Ansa*.* Indian melodies as transcribed by English travellers, seem to be very like modern European melodies.† Fétis and Coussemaker ‡ have made the same remark respecting the few known remains of old German and Celtic melodies.

Although, therefore, homophonic music was [possibly] not entirely without a reference to some tonic, or predominant tone, such a tone was beyond all dispute much more weakly developed than in modern music, where a few consecutive chords suffice to establish the scale in which that portion of the piece is written. The ¶ cause of this seems to me traceable to the undeveloped condition and subordinate part which characterises homophonic music. Melodies which move up and down in a few tones which are easily comprehended, and are connected, not by some musical contrivance, but by the words of a poem, do not require the consistent application of any contrivance, to combine them. Even in modern recitative tonality is much less firmly established than in other forms of composition. The necessity for a steady connection of masses of tone by purely musical relations, does not dawn distinctly on our feeling, until we have to form into one artistic whole large masses of tone, which have their own independent significance without the cement of poetry.

---

* Jones, *On the Music of the Indians*, translated by Dalberg, pp. 36, 37. [Sir W. Jones's tract, with many others, is reprinted in Sourindro Mohun Tagore's *Hindu Music from various Authors*. This is what he says of the ans'a, p. 149 of Tagore: 'Since it appears from the Náráyan [a Sanscrit treatise on music], that 36 modes are in general use, and the rest very rarely applied to practice, I shall exhibit only the scales of the 6 Rágas [tunes] and 30 Ráginis [female personification of tunes in Hindu music] according to Sóma. . . . Three distinguished sounds in each mode [as Sir W. Jones translates *rág*] are called *graha*, *nyása*, *ans'a*, and the writer of the Náráyan defines them in the two following couplets. [I give the translation only.] " The note called *graha* is placed at the beginning, and that named *nyása* at the end, of a song ; that note, which displays the peculiar melody, and to which all the others are subordinate, that, which is always of the greatest use, is like a sovereign, though a mere *ans'a* or portion."

" By the word *vádi*," says the commentator, "he means the note, which announces and ascertains the Rága [tune], and which may be considered as the parent and origin of the *graha* and *nyása*." This clearly shews, I think ¶ [says Sir W. Jones], that the *ans'a* must be the tonic ; and we shall find that the two other notes are generally its third and fifth, or the mediant and dominant. In the poem entitled *Mágha* there is a musical simile, which may illustrate and confirm our idea. [I give the translation only.] " From the greatness, from the transcendent qualities of that hero, eager for conquest, other kings march in subordination to him, as other notes are subordinate to the *ans'a*."—*Translator*.]

† [The construction and time are very different. The scales are extremely variable. The results are very imperfectly represented by our present musical notation.—*Translator*.]

‡ *Histoire de l'Harmonie au moyen Age*, Paris, 1852, pp. 5-7.

2. Polyphonic Music.

The second stage of musical development is the polyphonic music of the middle ages. It is usual to cite as the first invented part-music, the so-called *organum* or *diaphony*, as originally described by the Flemish monk Hucbald at the beginning of the tenth century. In this, two voices are said to have proceeded in Fifths or Fourths, with occasional doublings of one or both in Octaves. This would produce intolerable music for modern ears. But according to O. Paul* the meaning is not that the two voices sang at the same time, but that there was a responsive repetition of a melody in a transposed condition, in which case Hucbald would have been the inventor of a principle which subsequently became so important in the fugue and sonata.

¶ The first undoubted form of part-music intentionally for several voices, was the so-called *discantus*, which became known at the end of the eleventh century in France and Flanders. The oldest specimens of this kind of music which have been preserved are of the following description. Two entirely different melodies —and to all appearance the more different the better—were adapted to one another by slight changes in rhythm or pitch, until they formed a tolerably consonant whole. At first, indeed, there seems to have been an inclination for coupling a liturgical formula with a rather 'slippery' song. The first of such examples could scarcely have been intended for more than musical tricks to amuse social meetings. It was a new and amusing discovery that two totally independent melodies might be sung together and yet sound well.

The principle of discant was fertile, and its nature was suitable for development at that period. Polyphonic music proper was its issue. Different voices, each proceeding independently and singing its own melody, had to be united in such a way as to produce either no dissonances, or merely transient ones which ¶ were readily resolved. Consonance was not the object in view, but its opposite, dissonance, was to be avoided. All interest was concentrated on the motion of the voices. To keep the various parts together, time had to be strictly observed, and hence the influence of discant developed a system of musical rhythm, which again contributed to infuse greater power and importance into melodic progression. There was no division of time in the Gregorian *Cantus firmus*. The rhythm of dance music was probably extremely simple. Moreover, melodic movement increased in richness and interest as the parts were multiplied. But the establishment of an artistic connection between the different voices, which, as we have seen, were at first perfectly free, required a new invention, and this, though it cropped up at first in a very humble form, has ended by obtaining predominant importance in the whole art of modern musical composition. This invention consisted in causing a musical phrase which had been sung by one voice to be repeated by another. Thus arose *canonic imitation*, which may be met with sporadically as ¶ early as in the twelfth century.† This subsequently developed into a highly artistic system, especially among Netherland composers, who, it must be owned, ended by often shewing more calculation than taste in their compositions.

But by this kind of polyphonic music—the repetition of the same melodic phrases in succession by different voices—it first became possible to compose musical pieces on an extensive plan, owing their connection not to any union with another fine art—poetry, but to purely musical contrivances. This kind of music also was especially suited to ecclesiastical songs, in which the chorus had to express the feelings of a whole congregation of worshippers, each with his own peculiar disposition. It was, however, not confined to ecclesiastical compositions, but was also applied to secular songs (madrigals). The sole form of harmonic music yet known, which could be adapted to artistic cultivation, was that founded on canonic

* *Geschichte des Claviers* [History of the Pianoforte], Leipzig, 1868, p. 49.
　† Coussemaker, *loc. cit.* Discant : *Custodi*

nos, Pl. xxvii. No. iv., translated in p. xxvii. No. xxix.

repetitions. If this had been rejected, nothing but homophonic music remained. Hence we find a number of songs set as strict canons or with canonical repetitions, although they were entirely unsuited for such a heavy form of composition. Even the oldest examples of instrumental compositions in several parts, the dance music of 1529,* are written in the form of madrigals and motets, a character of composition which, more freely treated, lasted down to the suites of S. Bach and Handel's times. Even in the first attempts at musical dramas in the sixteenth century, there was no other way of making the personages express their feelings musically, than by causing a chorus behind or upon the stage to sing over some madrigals in the fugue style. It is scarcely possible for us, from our present point of view, to conceive the condition of an art which was able to build up the most complicated constructions of voice parts in chorus, and was yet incapable of adding a simple accompaniment to the melody of a song or a duet, for the purpose of filling up the harmony. And yet when we read how Giacomo Peri's invention of recitative ¶ with a simple accompaniment of chorus was applauded and admired and what contentions arose as to the renown of the invention; what attention Viadana excited when he invented the addition of a *Basso continuo* for songs in one or two parts, as a dependent part serving only to fill up the harmony †; it is impossible to doubt that this art of accompanying a melody by chords (as any amateur can now do in the simplest manner possible) was completely unknown to musicians up to the end of the sixteenth century. It was not till the sixteenth century that composers became aware of the meaning possessed by chords as forming an harmonic tissue independently of the progression of parts.

To this condition of the art corresponded the condition of the tonal system. The old ecclesiastical scales were retained in their essentials, the first from D to d, the second from E to e, the third from F to f, and the fourth from G to g. Of these the scale from F to f was useless for harmonic purposes, because it contained the Tritone F—B, in place of the Fourth F—$B\flat$. Again, there was no reason for ¶ excluding the scales from C to c and A to a. And thus the ecclesiastical scales altered under the influence of polyphonic music. But as the old unsuitable names were retained notwithstanding the changes, there arose a terrible confusion in the meaning attached to modes. It was not till nearly the end of this period that a learned theoretician, Glarean, undertook in his *Dodecachordon* (Basle 1547) to put some order into the theory of modes. He distinguished twelve of them, six authentic and six plagal, and assigned them Greek names, which were, however, incorrectly transferred. However, his nomenclature for ecclesiastical modes has been generally followed ever since. The following are Glarean's six authentic ecclesiastical modes, keys, or scales, with the incorrect Greek names he assigned to them.

Ionic .	.	. $C\ D\ E\ F\ G\ A\ B\ c$
Doric .	.	. $D\ E\ F\ G\ A\ B\ c\ d$
Phrygian	.	. $E\ F\ G\ A\ B\ c\ d\ e$
Lydian	.	. $F\ G\ A\ B\ c\ d\ e\ f$
Mixolydian .	.	$G\ A\ B\ c\ d\ e\ f\ g$
Eolic .	.	. $A\ B\ c\ d\ e\ f\ g\ a$

¶

Ionic answers to our major, *Eolic* to our minor system. *Lydian* was scarcely ever used in polyphonic music owing to the false Fourth F—B, and when it was employed it was altered in many different ways.

Inability to judge of the musical significance of a connected tissue of harmonies again appears in the theory of the keys, by the rule, that the key of a polyphonic composition was determined by considering the separate voices independently. Glarean in certain compositions attributes different keys to the tenor and bass, the soprano and alto. Zarlino assumes the tenor as the chief part for determining the key.

* Winterfeld, *Johannes Gabrieli und sein Zeitalter*, vol. ii. p. 41.
† Winterfeld, *ibid.*, vol. ii. p. 19 and p. 59.

The practical consequences of this neglect of harmony are conspicuous in various ways in musical compositions. The composers confined themselves on the whole to the diatonic scale; 'accidentals,' or signs of alterations of tone, were seldom used. The Greeks had introduced the depression of the tone B to $B\flat$ in a peculiar tetrachord, that of the *synēmmenoi*, and this was retained. Besides this $f\sharp$, $c\sharp$ and $g\sharp$ were used, to introduce leading tones in the cadences. Hence modulation, as we understand it, from the key of one tonic to that of another with a different signature was almost entirely absent. Moreover, the chords used by preference down to the end of the fifteenth century, were formed of the Octave and Fifth without the Third, and such chords now sound poor and are avoided as much as possible. To medieval composers who only felt the want of the most perfect consonances, these chords appeared the most agreeable, and none others might be used at the close of a piece. The dissonances which occur are universally those which ¶ arise from suspended and passing tones; chords of the dominant Seventh, which, in modern harmony, play such an important part in marking the key, and in connecting and facilitating progressions, were quite unknown.

Great, then, as was the artistic advance in rhythm and the progression of parts, during this period, it did little more for harmony and the tonal system than to accumulate an undigested mass of experiments. Since the involved progression of the parts gave rise to chords in extremely varied transpositions and sequences, the musicians of this period could not but hear these chords and become acquainted with their effects, however little skill they shewed in making use of them. At any rate, the experience of this period prepared the way for harmonic music proper, and made it possible for musicians to produce it, when external circumstances forced on the discovery.

3. Harmonic Music.

¶   Modern harmonic music is characterised by the independent significance of its harmonies, for the expression and the artistic connection of a musical composition. The external inducements for this transformation of music were of various kinds. First there was the Protestant ecclesiastical chorus. It was a principle of Protestantism that the congregation itself should undertake the singing. But a congregation could not be expected to execute the artistic rhythmical labyrinths of Netherland polyphony. On the other hand, the founders of the new confession, with Luther at their head, were far too penetrated with the power and significance of music, to reduce it at once to an unadorned unison. Hence the composers of Protestant ecclesiastical music had to solve the problem of producing simply harmonised chorales, in which all the voices progressed at the same time. This excluded those canonic repetitions of the same melodic phrases in different parts, which had hitherto formed the chief unity of the whole piece. A new connecting principle had to be looked for in the sound of the tones themselves, and this was ¶ found in a stricter reference of all to one predominant tonic. The success of this problem was facilitated by the fact that the Protestant hymns were chiefly adapted to existing popular melodies, and the popular songs of the Germanic and Celtic races, as already remarked, betrayed a stricter feeling for tonality in the modern sense, than those of southern nations. Thus as early as in the sixteenth century, the system of the harmony of the ecclesiastical Ionic mode (our present major) developed itself with tolerable correctness, so that these chorales do not strike modern ears as strange, although they were still without many of our later contrivances for marking the key, as, for example, the chord of the dominant Seventh. On the other hand, it was much longer before the other ecclesiastical modes, in harmonising which much uncertainty still prevailed, were fused into the modern minor mode. The Protestant ecclesiastical hymns of that time produced great effects on the feelings of contemporaries—a fact emphasised on all sides in the liveliest language, so that no doubt can exist that the impression made by such music, was something as new as it was peculiarly powerful.

In the Roman Church also a desire arose for altering their music.   The divisions of polyphonic music scattered the sense of the words, and made them unintelligible to the unpractised public, and occasioned even a learned and cultivated hearer great difficulties in endeavouring to disentangle the knot of voices.   In consequence of the proceedings of the Council of Trent, and by an order of Pope Pius IV. (A.D. 1559–1565), Palestrina (A.D. 1524–1594) carried out this simplification and embellishment of ecclesiastical music, and the simple beauty of his compositions is said to have prevented the complete banishment of part music from the Roman liturgy. Palestrina, who wrote for choruses of singers practised in their art, did not entirely drop the more complicated progression of parts found in polyphonic music, but by appropriate sections and divisions he separated and connected both the mass of tones and the mass of voices, and generally distributed the latter into several distinct choirs.   The voices also are more or less frequently heard together in such progressions as were used in chorales, and in this case consonant chords greatly ¶ predominated.   By this means he made his pieces more comprehensible and intelligible, and in general extremely agreeable to the ear.   But the deviation of ecclesiastical modes from the new modes invented in modern times for the treatment of harmonies, is nowhere so remarkable as in the compositions of Palestrina and those of contemporary Italian composers of ecclesiastical music, among whom Giovanni Gabrieli, a Venetian, should be particularly named.   Palestrina was a pupil of Claude Goudimel (a Huguenot, slain at Lyons in the massacre of St. Bartholomew), who had harmonised French psalms in a way which, when the scale was major, was but very slightly different from modern habits.   These psalm melodies had been borrowed, or at least imitated from popular songs.   Hence Palestrina was certainly acquainted with this mode of treatment, through his teacher, but he had to deal with themes from the Gregorian *Cantus firmus* that moved in ecclesiastical tones, which he was forced to maintain strictly even in pieces where he himself invented or adapted the melodies.   Now these modes necessitated a ¶ totally different harmonic treatment, which sounds very strange to moderns.   As a specimen I will only cite the commencement of his eight-part *Stabat mater*.

Sta - bat   ma - ter   do - - - lo - ro - - sa

Here, at the commencement of a piece, just where we should require a steady characterisation of the key, we find a series of chords in the most varied keys, ¶ from *A* major to *F* major, apparently thrown together at haphazard, contrary to all our rules of modulation.   What person that was ignorant of ecclesiastical modes could guess the tonic of the piece from this commencement?   As such we find *D* at the end of the first strophe, and the sharpening of *C* to *C*♯ in the first chord also points to *D*.   The principal melody too, which is given to the tenor, shews from the commencement that *D* is the tonic.   But we do not get a minor chord of *D* till the eighth bar, whereas a modern composer would have been forced to introduce it in the first good place he could find in the first bar.

We see from these characters how greatly the nature of the whole system of ecclesiastical modes differed from our modern keys.   We cannot but assume that masters like Palestrina founded their method of harmonisation upon a correct feeling for the peculiar character of those modes, and that, as they could not fail to be acquainted with the contemporary advances in Protestant ecclesiastical music, their work was neither arbitrary nor unskilful.

What we miss in such examples as the one just adduced, is first, that the tonic chord does not play the same prominent part at the very commencement that is assigned to it in modern music. In the latter, the tonic chord has the same prominent and connecting significance among chords as the tonic or key-note among the tones of the scale. Next we miss altogether that feeling for the connection of consecutive chords which in modern times has led to the very general custom of giving them a common tone. This is evidently related to the fact that, as we shall see hereafter, it was not possible in the old ecclesiastical modes to produce chains of chords so closely connected with each other and with the tonic chord, as in the modern major and minor modes.

¶ Hence, although we recognise in Palestrina and Gabrieli a delicate artistic sensitiveness for the esthetic effect of separate chords of various kinds, and in so far a certain independent significance in their harmonies, yet we see that the means of establishing an internal connection in the tissue of chords had still to be discovered. This problem, however, required a reduction and transformation of the previous scales, to our major and minor. On the other hand, this reduction sacrificed the great variety of expression which depended on diversity of scale. The old scales partly form transitions between major and minor, and partly enhance the character of the minor, as in the ecclesiastical Phrygian mode [p. 245d]. This diversity being lost, it had to be replaced by new contrivances, such as the transposition of the scales for different tonics, and the modulational passage from one key to another.

¶ This transformation was completed during the seventeenth century. But the most active cause for the development of harmonic music is due to the commencement of opera. This had been occasioned by a revival of acquaintance with classical antiquity, and its avowed object was to rehabilitate ancient tragedy, which was known to have been recited musically. Here arose immediately the problem of allowing one or two voices to execute solos; but these again had to be harmonised so as to fit in between the choruses, which were treated in the polyphonic manner, the object being to make the solo parts stand prominently forward and keep the accompanying voices well under. These conditions first gave rise to Recitative, invented by Giacomo Peri and Caccini in 1600, and solo songs with airs, invented by Claudio Monteverde and Viadana. The new view taken of harmony shews itself in written music by the appearance of figured basses in the works of these composers. Every figured bass note represented a chord, so that the chords themselves were settled, but the progression of the parts of which they were constituted was left to the taste of the player. And thus what was merely secondary in polyphonic music, became principal, and conversely.

¶ Opera also necessitated the discovery of more powerful means of expression than were admissible in ecclesiastical music. Monteverde, who was extremely prolific in inventions, is the first composer who used chords of the dominant Seventh without preparation, for which he was severely blamed by his contemporary Artusi. Generally we find a bolder use of dissonances, which were employed independently, to express sharp contrasts of expression, and not, as before, as accidental results of the progression of parts.

Under these influences, even as early as in Monteverde's time, the Doric, Eolic, and Phrygian ecclesiastical modes [p. 254c, d] began to be transformed and fused into our modern minor mode. This was completed in the seventeenth century, and these modes were thus made more suitable for giving prominence to the tonic of the harmony, as will be more fully shewn hereafter.

We have already given an outline of the nature of the influence which these changes exerted on the constitution of the tonal system. The mode of connecting musical phrases hitherto in vogue—canonic repetitions of similar melodic figures —had necessarily to be abandoned as soon as a simple harmonic accompaniment

had to be subordinated to a melody.   Hence some new means of artistic connection
had to be discovered in the sound of the chords themselves.   This was effected,
first by making the harmonies refer their tones much more definitely to one predominant tonic than before, and secondly by giving fresh strength to the relations between the chords themselves and between all other chords and the tonic
chord.   In the course of our investigations we shall see that the distinctive peculiarities of the modern system of tones can be deduced from this principle, and
that the principle itself is very strictly carried out in our present music.   In
reality the mode in which the materials of music are now worked up for artistic
use, is in itself a wondrous work of art, at which the experience, ingenuity, and
esthetic feeling of European nations has laboured for between two and three thousand years, since the days of Terpander and Pythagoras.   But the complete formation of the essential features as we now see it, is scarcely two hundred years
old in the practice of musical composers, and theoretical expression was not given ¶
to the new principle till the time of Rameau at the beginning of last century.   In
the historical point of view, therefore, it is wholly the product of modern times,
limited nationally to the German, Roman, Celtic, and Sclavonic races.

  With this tonal system, which admits great wealth of form with strictly defined
artistic consistency, it has become possible to construct works of art, of much
greater extent, and much richer in forms and parts, much more energetic in
expression, than any producible in past ages ; and hence we are by no means
inclined to quarrel with modern musicians for esteeming it the best of all, and
devoting their attention to it exclusively.   But scientifically, when we proceed to
explain its construction and display its consistency we must not forget that our
modern system was not developed from a natural necessity, but from a freely
chosen principle of style ; that beside it, and before it, other tonal systems have
been developed from other principles, and that in each such system the highest
pitch of artistic beauty has been reached, by the successful solution of more ¶
limited problems.

  This reference to the history of music was necessitated by our inability in this
case to appeal to observation and experiment for establishing our explanations,
because, educated in a modern system of music, we cannot thoroughly throw ourselves back into the condition of our ancestors, who knew nothing about what we
have been familiar with from childhood, and who had to find it all out for themselves.   The only observations and experiments, therefore, to which we can appeal,
are those which mankind themselves have undertaken in the development of music.
If our theory of the modern tonal system is correct it must also suffice to furnish
the requisite explanation of the former less perfect stages of development.

  As the fundamental principle for the development of the European tonal
system, we shall assume *that the whole mass of tones and the connection of harmonies must stand in a close and always distinctly perceptible relationship to some
arbitrarily selected tonic, and that the mass of tone which forms the whole compo-* ¶
sition, must be developed from this tonic, and must finally return to it.   The
ancient world developed this principle in homophonic music, the modern world in
harmonic music.   But it is evident that this is merely an esthetical principle, not
a natural law.

  The correctness of this principle cannot be established *à priori.*   It must
be tested by its results.   The origin of such esthetical principles should not be
ascribed to a natural necessity.   They are the inventions of genius, as we previously
endeavoured to illustrate by a reference to the principles of architectural style.

CHAPTER XIV.

THE TONALITY OF HOMOPHONIC MUSIC.

MUSIC was forced first to select artistically, and then to shape for itself, the material on which it works. Painting and sculpture find the fundamental character of their materials, form and colour, in nature itself, which they strive to imitate. Poetry finds its material ready formed in the words of language. Architecture has, indeed, also to create its own forms; but they are partly forced upon it by technical and not by purely artistic considerations. Music alone finds an infinitely rich but totally shapeless plastic material in the tones of the human voice and artificial musical instruments, which must be shaped on purely artistic principles, unfettered by any reference to utility as in architecture, or to the imitation of ¶ nature as in the fine arts, or to the existing symbolical meaning of sounds as in poetry. There is a greater and more absolute freedom in the use of the material for music than for any other of the arts. But certainly it is more difficult to make a proper use of absolute freedom, than to advance where external irremovable landmarks limit the width of the path which the artist has to traverse. Hence also the cultivation of the tonal material of music has, as we have seen, proceeded much more slowly than the development of the other arts.

It is now our business to investigate this cultivation.

The first fact that we meet with in the music of all nations, so far as is yet known, is that *alterations of pitch in melodies take place by intervals, and not by continuous transitions.* The psychological reason of this fact would seem to be the same as that which led to rhythmic subdivision periodically repeated. All melodies are motions within extremes of pitch. The incorporeal material of tones is much more adapted for following the musician's intention in the most delicate ¶ and pliant manner for every species of motion, than any corporeal material however light. Graceful rapidity, grave procession, quiet advance, wild leaping, all these different characters of motion and a thousand others in the most varied combinations and degrees, can be represented by successions of tones. And as music expresses these motions, it gives an expression also to those mental conditions which naturally evoke similar motions, whether of the body and the voice, or of the thinking and feeling principle itself. Every motion is an expression of the power which produces it, and we instinctively measure the motive force by the amount of motion which it produces. This holds equally and perhaps more for the motions due to the exertion of power by the human will and human impulses, than for the mechanical motions of external nature. In this way melodic progression can become the expression of the most diverse conditions of human disposition, not precisely of human *feelings,*[*] but at least of that *state of sensitiveness* which is produced by feelings. In English the phrase *out of tune, unstrung,* and ¶ in German the word *stimmung,* literally *tuning,* are transferred from music to mental states. The words are meant to denote those peculiarities of mental condition which are capable of musical representation. I think we might appropriately define *gemüthsstimmung,* or *mental tune,* as representing that general character temporarily shewn by the motion of our conceptions, and correspondingly impressed on the motions of our body and voice. Our thoughts may move fast or slowly, may wander about restlessly and aimlessly in anxious excitement, or may keep a determinate aim distinctly and energetically in view; they may lounge about without care or effort in pleasant fancies, or, driven back by some sad memories, may return slowly and heavily from the spot with short weak steps. All this may be imitated and expressed by the melodic motion of the tones, and the listener may thus receive a more perfect and impressive image of the ' tune ' of

---

[*] Hanslick seems to me to have the advantage over other esthetic writers in this point, because music, unassisted by poetry, has no means of clearly characterising the object of feeling.

another person's mind, than by any other means, except perhaps by a very perfect dramatic representation of the way in which such a person really spoke and acted.

Aristotle also formed a similar conception of the effect of music. In his 29th problem he says : ' Why do rhythms and melodies, which are composed of sound, resemble the feelings ; while this is not the case for tastes, colours, or smells ? Can it be because they are motions, as actions are also motions ? Energy itself belongs to feeling and creates feeling. But tastes and colours do not act in the same way.' *
And at the end of the 27th problem he says : ' These motions, i.e. rhythms and melodies, are active, and action is the sign of feeling.' †

Not merely music but even other kinds of motions may produce similar effects. Water in motion, as in cascades or sea waves, has an effect in some respects similar to music. How long and how often can we sit and look at the waves rolling in to shore ! Their rhythmic motion, perpetually varied in detail, produces a peculiar feeling of pleasant repose or weariness, and the impression of a mighty ¶ orderly life, finely linked together. When the sea is quiet and smooth we can enjoy its colouring for a while, but this gives no such lasting pleasure as the rolling waves. Small undulations, on the other hand, on small surfaces of water, follow one another too rapidly, and disturb rather than please.

But the motion of tone surpasses all motion of corporeal masses in the delicacy and ease with which it can receive and imitate the most varied descriptions of expression. Hence it arrogates to itself by right the representation of states of mind, which the other arts can only indirectly touch by shewing the situations which caused the emotion, or by giving the resulting words, acts, or outward appearance of the body. The union of music to words is most important, because words can represent the cause of the frame of mind, the object to which it refers, and the feeling which lies at its root, while music expresses the kind of mental transition which is due to the feeling. When different hearers endeavour to describe the impression of instrumental music, they often adduce entirely different ¶ situations or feelings which they suppose to have been symbolised by the music. One who knows nothing of the matter is then very apt to ridicule such enthusiasts, and yet they may have been all more or less right, because music does not represent feelings and situations, but only frames of mind, which the hearer is unable to describe except by adducing such outward circumstances as he has himself noticed when experiencing the corresponding mental states. Now different feelings may occur under different circumstances and produce the same states of mind in different individuals, while the same feelings may give rise to different states of mind. Love is a feeling. But music cannot represent it directly as such. The mental states of a lover may, as we know, shew the extremest variety of change. Now music may perhaps express the dreamy longing for transcendent bliss which love

* Διὰ τί οἱ ῥυθμοὶ καὶ τὰ μέλη φωνὴ οὖσα, ἤθεσιν ἔοικεν· οἱ δὲ χυμοὶ οὐ, ἀλλ' οὐδὲ τὰ χρώματα καὶ αἱ ὀσμαί ; Ἢ ὅτι κινήσεις εἰσίν, ὥσπερ καὶ αἱ πράξεις ; ἤδη δὲ ἡ μὲν ἐνέργεια ἠθικὸν, καὶ ποιεῖ ἦθος· οἱ δὲ χυμοὶ καὶ τὰ χρώματα οὐ ποιοῦσιν ὁμοίως. Arist. *Prob.* xix. 29.

† [The above words conclude the problem, which it seems best to cite in full. Διὰ τί τὸ ἀκουστὸν μόνον ἦθος ἔχει τῶν αἰσθητῶν ; καὶ γὰρ ἐὰν ᾖ ἄνευ λόγου μέλος, ὅμως ἔχει ἦθος· ἀλλ' οὐ τὸ χρῶμα, οὐδὲ ἡ ὀσμὴ, οὐδὲ ὁ χυμὸς ἔχει. Ἢ ὅτι κίνησιν ἔχει μονονουχὶ, ἣν ὁ ψόφος ἡμᾶς κινεῖ ; Τοιαύτη μὲν γὰρ καὶ τοῖς ἄλλοις ὑπάρχει, κινεῖ γὰρ καὶ τὸ χρῶμα [καὶ] τὴν ὄψιν · ἀλλὰ τῆς ἑπομένης τῷ τοιούτῳ ψόφῳ αἰσθανόμεθα κινήσεως. Αὕτη δὲ ἔχει ὁμοιότητα, ἔν τε τοῖς ῥυθμοῖς καὶ ἐν τῇ τῶν φθόγγων τάξει τῶν ὀξέων καὶ βαρέων, οὐκ ἐν τῇ μίξει. Ἀλλ' ἡ συμφωνία οὐκ ἔχει ἦθος. Ἐν δὲ τοῖς ἄλλοις αἰσθητοῖς τοῦτο οὐκ ἔστιν. Αἱ δὲ κινήσεις αὗται, πρακτικαί εἰσιν. Αἱ δὲ πράξεις, ἤθους σημασία ἐστί. Arist. *Prob.* xix. 27. Which we may perhaps translate thus : ' Why is sound

the only sensation which excites the feelings ? Even melody without words has feeling. But this is not the case for colour, or smell, or ¶ taste. Is it because they have none of the motion which sound excites in us ? For the others excite motion ; thus colour moves the eye. But we feel the motion which follows sound. And this is alike, in rhythm, and alteration in pitch, but not in united sounds. Sounding notes together does not excite feeling. This is not the case for other sensations. Now these motions stimulate action, and this action is the sign of feeling.' Aristotle seems to have required motion to excite feeling, and in sounding two notes *together*, there was no motion of one towards the other. It is evident that he had not the slightest inkling of a progression of *harmonies*, and this utter blank in his mind is one of the strongest proofs that the Greeks had never tried harmony. Ἁρμονία had the modern meaning of *melody* ; μελῳδία was words set to music.—*Translator.*]

may excite. But precisely the same state of mind might arise from religious enthusiasm. Hence when a piece of music expresses this mental state it is not a contradiction for one hearer to find in it the longing of love, and another the longing of enthusiastic piety. In this sense Vischer's rather paradoxical statement that the mechanics of mental emotion are perhaps best studied in their musical expression, may be not altogether incorrect. We really possess no other means of expressing them so exactly and delicately.

As we have seen, then, melody has to express a motion, in such a manner that the hearer may easily, clearly, and certainly appreciate the character of that motion by *immediate perception*. This is only possible when the steps of this motion, their rapidity and their amount, are also exactly *measurable* by immediate sensible perception. Melodic motion is change of pitch in time. To measure it perfectly, the length of time elapsed, and the distance between the pitches, must be measur-
¶ able. This is possible for immediate audition only on condition that the alterations both in time and pitch should proceed by regular and determinate degrees. This is immediately clear for time, for even the scientific, as well as all other measurement of time, depends on the rhythmical recurrence of similar events, the revolution of the earth or moon, or the swings of a pendulum. Thus also the regular alternation of accentuated and unaccentuated sounds in music and poetry gives the measure of time for the composition. But whereas in poetry the construction of the verse serves only to reduce the external accidents of linguistic expression to artistic order; in music, rhythm, as the measure of time, belongs to the inmost nature of expression. Hence also a much more delicate and elaborate development of rhythm was required in music than in verse.

It was also necessary that the alteration of pitch should proceed by intervals, because motion is not measurable by immediate perception unless the amount of space to be measured is divided off into degrees. Even in scientific investigations
¶ we are unable to measure the velocity of continuous motion except by comparing the space described with the standard measure, as we compare time with the seconds pendulum.

It may be objected that architecture in its arabesques, which have been justly compared in many respects with musical figures, and which also shew a certain orderly arrangement, constantly employs curved lines and not lines broken into determinate lengths. But in the first place the art of arabesques really began with the Greek meander, which is composed of straight lines set at right angles to each other, following at exactly equal lengths, and cutting one another off in degrees. In the second place, the eye which contemplates arabesques can take in and compare all parts of the curved lines at once, and can glance to and fro, and return to its first contemplation. Hence, notwithstanding the continuous curvature of the lines, their paths are perfectly comprehensible, and it became possible to renounce the strict regularity of the Grecian arabesques in favour of
¶ the curvilinear freedom. But whilst freer forms are thus admitted for individual small decorations in architecture, the division of any great whole, whether it be a series of arabesques or a row of windows or columns, &c., throughout a building, is still tied down to the simple arithmetical law of repetition of similar parts at equal intervals.

The individual parts of a melody reach the ear in succession. We cannot perceive them all at once. We cannot observe backwards and forwards at pleasure. Hence for a clear and sure measurement of the change of pitch, no means was left but progression by determinate degrees. This series of degrees is laid down in the musical scale. When the wind howls and its pitch rises or falls in insensible gradations without any break, we have nothing to measure the variations of pitch, nothing by which we can compare the later with the earlier sounds, and comprehend the extent of the change. The whole phenomenon produces a confused, unpleasant impression. The musical scale is as it were the divided rod, by which we measure progression in pitch, as rhythm measures progression in time. Hence the analogy between the scale of tones and rhythm naturally occurred to musical theoreticians of ancient as well as modern times.

We consequently find the most complete agreement among all nations that use music at all, from the earliest to the latest times, as to the separation of certain determinate degrees of tone from the possible mass of continuous gradations of sound, all of which are audible, and these degrees form the scale in which the melody moves. But in selecting the particular degrees of pitch, deviations of national taste become immediately apparent. The number of scales used by different nations and at different times is by no means small.

Let us inquire, then, what motive there can be for selecting one tone rather than another in its neighbourhood for the step succeeding any given tone. We remember that in sounding two tones together such a relation was observed. We found that under such circumstances certain particular intervals, namely the consonances, were distinguished from all other intervals which were nearly the same, ¶ by the absence of beats. Now some of these intervals, the Octave, Fifth, and Fourth, are found in all the musical scales known.* Recent theoreticians that have been born and bred in the system of harmonic music, have consequently supposed that they could explain the origin of the scales, by the assumption that all melodies arise from thinking of a harmony to them, and that the scale itself, considered as the melody of the key, arose from resolving the fundamental chords of the key into their separate tones. This view is certainly correct for modern scales ; at least these have been modified to suit the requirements of the harmony. But scales existed long before there was any knowledge or experience of harmony at all. And when we see historically what a long period of time musicians required to learn how to accompany a melody by harmonies, and how awkward their first attempts were, we cannot feel a doubt that ancient composers had no feeling at all for harmonic accompaniment, just as even at the present day many of the more gifted Orientals are opposed to our own harmonic music. We must also not forget ¶ that many popular melodies, of older times or foreign origin, scarcely admit of any harmonic accompaniment at all, without injury to their character.

The same remark applies to Rameau's assumption of an ' understood ' fundamental bass in the construction of melodies or scales for a single voice. A modern composer ‧would certainly imagine to himself at once the fundamental bass to the melody he invents. But how could that be the case with musicians who had never heard any harmonic music, and had no idea how to compose any ? Granted that an artist's genius often unconsciously ' feels out ' many relations, we should be imputing too much to it‧ if we asserted that the artist could observe relations of tones which he had never or very rarely heard, and which were destined not to be discovered and employed till many centuries after his time.

It is clear that in the period of homophonic music, the scale could not have been constructed so as to suit the requirements of chordal connections unconsciously supplied. Yet a meaning may be assigned, in a somewhat altered form, ¶ to the views and hypotheses of musicians above mentioned, by supposing that the same physical and physiological relations of the tones, which become sensible when they are sounded together and determine the magnitude of the consonant intervals, might also have had an effect in the construction of the scale, although under somewhat different circumstances.

Let us begin with the Octave, in which the relationship to the fundamental tone is most remarkable. Let any melody be executed on any instrument which has a good musical quality of tone, such as a human voice ; the hearer must have heard not only the primes of the compound tones, but also their upper Octaves, and, less strongly, the remaining upper partials. When, then, a higher voice afterwards executes the same melody an Octave higher, *we hear again a part of what we heard before*, namely the evenly numbered partial tones (p. 49*d*) of the former

* [It will be seen in App. XX. sect. K. that the Fourth and Fifth are often materially inexact or designedly altered.—*Translator.*]

compound tones, and *at the same time we hear nothing that we had not previously heard.* Hence the repetition of a melody in the higher Octave is a real repetition of what has been previously heard, not of all of it, but of a part.* If we allow a low voice to be accompanied by a higher in the Octave above it, the only part music which the Greeks employed, we add nothing new, we merely reinforce the evenly numbered partials. In this sense, then, the compound tones of an Octave above are really repetitions of the tones of the lower Octaves, or at least of part of their constituents. Hence the first and chief division of our musical scale is that into a series of Octaves. In reference to both melody and harmony, we assume tones of different Octaves which bear the same name, to have the same value, and, in the sense intended, and up to a certain point, this assumption is correct. An accompaniment of Octaves gives perfect consonance, but it gives nothing additional ; it merely reinforces tones already present. Hence it is musically applicable for in-

¶ creasing the power of a melody which has to be brought out strongly, but it has none of the variety of polyphonic music, and therefore is felt to be *monotonous,* and it is consequently forbidden in polyphonic music.

What is true of the *Octave* is true in a less degree for the *Twelfth.* If a melody is repeated in the Twelfth we again hear only what we had already heard, but the repeated part of what we heard is much weaker, because only the third, sixth, ninth, &c., partial tone is repeated, whereas for repetition in the Octave, instead of the third partial, the much stronger second and weaker fourth partial is heard, and in place of the ninth, the eighth and tenth occur, &c. Hence repetition of a melody in the Twelfth is less complete than repetition in the Octave, because only a smaller part of what had been already heard is repeated. In place of this repetition in the Twelfth, we may substitute one an Octave lower, namely in the *Fifth.* Repetition in the Fifth is not a pure repetition, as that in the Twelfth is. Taking 2 for the pitch number of the prime tone, the partials are (197*c, d*)

¶

for the fundamental compound	.	2		4		6	8		10	12
for the Twelfth	. . . .					6				12
for the Fifth .	. . . .		3			6		9		12

When we strike the Twelfth we repeat the simple tones 6 and 12, which already existed in the fundamental compound tone. When we strike the Fifth, we continue to repeat the same simple tones, but we also add two others, 3 and 9. Hence for the repetition in the Fifth, only a part of the new sound is identical with a part of what had been heard, but it is nevertheless the most perfect repetition which can be executed at a smaller interval than an Octave. This is clearly the reason why unpractised singers, when they wish to join in the chorus to a song that does not suit the compass of their voice, often take a Fifth to it. This is also a very evident proof that the uncultivated ear regards repetition in the Fifth as natural. Such an accompaniment in the Fifth and Fourth is said to have been systematically developed in the early part of the middle ages. Even in modern music, repetition

¶ in the Fifth plays a prominent part next to repetition in the Octave. In normal fugues the theme, as is well known, is first repeated in the Fifth ; in the normal form of instrumental pieces, that of the Sonata, the theme in the first movement is transposed to the Fifth, returning in the second part to the fundamental tone. This kind of imperfect repetition of the impression in the Fifth induced the Greeks also to divide the interval of the Octave into two equivalent sections, namely two *Tetrachords.* Our major scale on being divided in this manner would be :—

$$c \quad d \quad e \quad f \quad \overbrace{g \quad a \quad b \quad c'} \quad \overbrace{d' \quad e' \quad f'}$$

I.　　　　　　II.　　　III.

* [Some considerations have been omitted, probably by design. The quality of tone of the voice which sings the Octave above is materially different. The evenly numbered partials of the lower tone are by no means so powerful as in the higher tone. The upper partials of the higher tone, which are still quite effective, would be inaudible in the lower tone.—*Translator.*]

The succession of tones in the second tetrachord is a repetition of that in the first, transposed a Fifth.* To pass into the Octave division, the successive tetrachords must be alternately separate and connected. They are said to be connected, or *conjunct*, when, as in II. and III., the last tone c of the lower becomes the first of the higher tetrachord ; and separate, or *disjunct*, when, as in I. and II., the last tone of the lower is different from the upper. In the second tetrachord g to c, every ascending series of tones necessarily leads to c′ as the final tone, and this c′ is also the Octave of the fundamental tone of the first tetrachord. Now this c′ is the Fourth of g, the fundamental tone of the second tetrachord. To make the succession of tones the same in both tetrachords, the lower tetrachord had to be increased by the tone f which answers to c′. The Fourth f, however, would have suggested itself in the same way as the Fifth, independently of this analogy of the tetrachords. The Fifth is a compound tone in which the second partial is the third partial of the fundamental compound tone ; the Fourth is a compound tone in which the third ¶ partial is the same as the second of the Octave. Hence the limits of the two analogous divisions of the Octave are settled, namely :—

$$c - f, \qquad g - c',$$

but the mode of filling up these gaps remains arbitrary, and different plans for doing so were adopted by the Greeks themselves at different periods, and others again by other nations. But the division of the scale into octaves, and the octave into two analogous tetrachords, occurs everywhere, almost without exception.

Boethius (*De Musica*, lib. i. cap. 20) informs us that according to Nicomachus the most ancient method of tuning the lyre down to the time of Orpheus, consisted of open tetrachords,

$$c - f - g - c',$$

with which certainly it was scarcely possible to construct a melody. But as it ¶ contained the chief degrees of the pitch of ordinary speech, a lyre of this kind might possibly have served to accompany declamation.

The relationship of the Fifth, and its inversion the Fourth, to the fundamental tone, is so close that it has been acknowledged in all known systems of music.† On the other hand, many variations occur in the choice of the intermediate tones which have to be inserted between the terminal tones of the tetrachord. The interval of a Third is by no means so clearly defined by easily appreciable partial tones, as to have forced itself from the first on the ear of unpractised musicians. We must remember that even if the fifth partial tone existed in the compound tones of the musical instruments employed, it would have had to contend with the much louder prime tone, and would also have been covered by the three adjacent and lower partials. As a matter of fact, the history of musical systems shews that there was much and long hesitation as to the tuning of the Thirds. And the doubt is even yet felt when Thirds are used in pure melody, unconnected with any har- ¶ monies. I must own that on observing isolated intervals of this kind, I cannot come to perfectly certain results, but I do so when I hear them in a well-constructed melody with distinct tonality. The natural major Thirds of 4 : 5 thus seem to me calmer and quieter than the sharper major Thirds of our equally tempered modern instruments, or with the still sharper major Thirds which result from the Pythagorean tuning with perfect Fifths. Both of the latter intervals have a strained effect.. Most of our modern musicians, accustomed to the major Thirds of the equal temperament, prefer them to the perfect major Thirds, when melody alone is concerned. But I have convinced myself that artists of the first rank, like Joachim, use the Thirds of 4 : 5 even in melody. For harmony there is no doubt at all.

---

* [This applies to the Pythagorean scale and hence to Greek music, and also to all tempered music. But in just intonation c to d is a major Tone, and d to e a minor Tone, whereas g to a is a minor Tone and a to b a major Tone. These distinctions were of course purposely omitted in the text.—*Translator.*]

† [But see App. XX. sect. K.—*Translator.*]

Every one chooses the natural major Thirds.   In Chapter XVI. I shall describe an instrument which will enable any one to perform experiments of this kind.*

Under these circumstances another principle for determining the small intervals of the scale was resorted to during the infancy of music, and seems to be still employed among the less civilised nations. This principle, which has subsequently had to yield to that of tonal relationship, consists in an endeavour to distinguish equal intervals by ear, and thus make the differences of pitch perceptibly uniform.

This attempt has never prevailed over the feeling of tonal relationship for the division of the Fourth, at least in artistically developed music. But in the division of smaller intervals we shall find it applied as an auxiliary in many of the less usual divisions of the Greek tetrachord and in the scales of Oriental nations. But arbitrary divisions which are independent of tonal relationship, disappeared
¶ everywhere in exact proportion to the higher development of the musical art.

We will now inquire what kind of a scale we should obtain by pursuing to its consequences the natural relationship of the tones. *We shall consider musical tones to be related in the first degree which have two identical partial tones ; and related in the second degree, when they are both related in the first degree to some third musical tone.* The louder the coincident in proportion to the non-coincident partials of compound tones related in the first degree, the *closer* is their relationship and the more easily will both singers and hearers feel the common character of both the tones. Hence it follows that the feeling for tonal relationship ought to differ with the qualities of tone : and I believe that this states a fact in nature, because flutes and the soft stops of organs, on which chords are somewhat colourless owing to an absence of upper partials and a consequent incomplete definition of dissonances, retain much of the same colourless character in melodies. This,
I think, depends upon the fact, that, for such qualities of tone, the recognition of
¶ the natural intervals of the Thirds and Sixths, and perhaps even of the Fourths and Fifths, does not result from the immediate sensation of the hearer, but at most from his recollection. When he knows that on other instruments and in singing he has been able by immediate sensation to recognise the Thirds and Sixths as naturally related tones, he acknowledges them as well-known intervals even when executed by a flute or on the soft stops of an organ. But the mere recollection of an impression cannot possibly have the same freshness and power as the immediate sensation itself.

Since the closeness of relationship depends on the loudness of the coincident upper partial tones, and those having a higher ordinal number are usually weaker than those having a lower one, the relationship of two tones is generally weaker, the greater the ordinal number of the coincident partials. These ordinal numbers, as the reader will recollect from the theory of consonant intervals, also give the ratio of the vibrational numbers of the corresponding notes.

¶   In the following table, the first horizontal line contains the ordinal numbers of the partial tones of the tonic *c*, and the first vertical column those of the corresponding tone in the scale. Where the corresponding vertical columns and horizontal lines intersect, the name of the tone of the scale is given for which this coincidence holds. Only such notes are admitted as are distant from the tonic by less than an Octave. Below each degree of the scale are placed the two ordinal numbers of the coincident partials, which will serve as a scale for measuring the closeness of the relationship.

---

* [Other experimental instruments will be described in App. XX. sect. F. The Harmonical gives only the just major Third 4 : 5. Its nearest approach to the Pythagorean 64 : 81, or 408 cents, is *Bb* : D_1 = 63 : 80, or 413 cents, not so harsh but quite near enough to shew its character. For the intervals used by violinists, see also App. XX. sect. G. arts. 6 and 7.— *Translator.*]

Partial Tones of the Tonic						
	1	2	3	4	5	6
1	c 1 : 1	c' 1 : 2				
2	C 2 : 1	c 2 : 2	g 2 : 3	c' 2 : 4		
3		F 3 : 2	c 3 : 3	f 3 : 4	a 3 : 5	c' 3 : 6
4		C 4 : 2	G 4 : 3	c 4 : 4	e 4 : 5	g 4 : 6
5			$E\flat$ 5 : 3	$A\flat$ 5 : 4	c 5 : 5	$e\flat$ 5 : 6
6			C 6 : 3	F 6 : 4	A 6 : 5	c 6 : 6

In this systematic comparison we find the following series of notes lying in the octave above the fundamental note c, and related to the tonic c in the first degree, arranged in the order of their relationship:

$$c \quad\quad c' \quad\quad g \quad\quad f \quad\quad a \quad\quad e \quad\quad e\flat$$
$$1:1 \quad 1:2 \quad 2:3 \quad 3:4 \quad 3:5 \quad 4:5 \quad 5:6$$

and the following series in the descending octave:

$$c \quad\quad C \quad\quad F \quad\quad G \quad\quad E\flat \quad\quad A\flat \quad\quad A$$
$$1:1 \quad 1:2 \quad 3:2 \quad 4:3 \quad 5:3 \quad 5:4 \quad 6:5$$

The series is discontinued when the resultant intervals become very close. Intervals adapted for practical use must not be too close to be easily taken and distinguished. What is the smallest interval admissible in a scale is a question which different nations have answered differently according to the different direction of their taste, and perhaps also according to the different delicacy of their ear.

It seems that in the first stages of the development of music many nations avoided the use of intervals of less than a Tone, and hence formed scales, which alternated in intervals from a Tone to a Tone and a half. According to examples collected by M. Fétis,[*] a scale of this kind is found not only among the Chinese but also among the other branches of the Mongol race, among the Malays of Java and Sumatra, the inhabitants of Hudson's Bay, the Papuas of New Guinea, the inhabitants of New Caledonia, and the Fullah negroes.[†] The five-stringed lyre (Kissar) of the inhabitants of North Africa and Abyssinia, which is represented in the bas-reliefs of the Assyrian palaces as an instrument played on by captives, was also, according to Villoteau,[‡] tuned by the scale of five degrees:

$$g - a - b - d' - e' \text{ §}$$

Traces of an ancient scale of this kind are clearly furnished by the five-stringed lyre or lute ($\kappa\iota\theta\acute{a}\rho a$) of the Greeks. At least Terpander (circa B.C. 700–650), who played a conspicuous part in the development of ancient Greek music, and who added a seventh string to the former Cithara of six strings, used a scale composed of a tetrachord and a trichord, having the compass of an Octave and tuned thus:—

$$e \smallsmile f - g - a - b \smallsmile - d' - e'\text{**}$$

* Histoire Générale de la Musique, Paris, 1869, vol. i.

† [See App. XX. sect. K. for pentatonic scales in Java, China, and Japan.—Translator.]

‡ Descriptions des Instruments de Musique des Orientaux; chap. xiii. in the Description de l'Egypte. État Moderne.

§ [This is probably only a rude approximation or a guess. See App. XX. sect. K. for observations on existing pentatonic scales actually heard.—Translator.]

** Nicomachus makes Philolaus say (edit. Meibomii, p. 17), 'From the Hypatē (e) to the

in which there is no c', and the upper tetrachord has no interval of a Semitone, although there is an interval of this kind in the lower.*

Olympos (circa B.C. 660–620), who introduced Asiatic flute music into Greece and adapted it to Greek tastes, transformed the Greek Doric scale into one of five tones, the *old enharmonic* scale

$$b \smile c \text{---} e \smile f \text{---} a\dagger$$

This seems to indicate that he brought a scale of five tones with him from Asia, and merely borrowed the use of the intervals of a Semitone from the Greeks. Among the more cultivated nations, the Chinese and the Celts of Scotland and Ireland still retain the scale of five notes without Semitones, although both have also become acquainted with the complete scale of seven notes.

¶ Among the Chinese, a certain prince Tsay-yu is said to have introduced the scale of seven notes amid great opposition from conservative musicians. The division of the Octave into twelve Semitones, and the transposition of scales have also been discovered by this intelligent and skilful nation. But the melodies transcribed by travellers mostly belong to the scale of five notes. The Gaels and Erse have likewise become acquainted with the diatonic scale of seven tones by means of psalmody, and in the present form of their popular melodies the missing tones are sometimes just touched as appoggiature or passing notes. These are, however, in many cases merely modern improvements, as may be seen on comparing the older forms of the melodies, and it is usually possible to omit the notes which do not belong to the scale of five tones without impairing the melody. This is not only true of the older melodies, but of more modern popular airs which were composed during the last two centuries, whether by learned or unlearned musicians. Hence the Gaels as well as the Chinese, notwithstanding their acquaintance with the modern tonal system, hold fast by the old.‡ And it cannot be denied that by ¶ avoiding the Semitones of the diatonic scale, Scotch airs receive a peculiarly bright and mobile character, although we cannot say as much for the Chinese. Both Gaels and Chinese make up for the small number of tones within the Octave by great compass of voice.§

The scale of five tones admits of a certain variety in its construction. Assume c as the tonic and add to it the nearest related notes in the ascending Octave, till you come to a Semitone. This gives

$$c - c' - g - f - a.$$

The next note e would form a Semitone with f. In the descending Octave we find in the same way

$$c - C - F - G - E\flat.$$

The great gaps in the scales between c and f in the first, and between G and c ¶ in the second are filled up by tones related in the second degree. Since the tones related to the Octave can only be repetitions of those directly related to the tonic,

---

Mesē (a) was a Fourth, from the Mesē (a) to the Nētē (e') a Fifth, from the Nētē (e') to the Tritē (b) a Fourth, from the Tritē (b) to the Hypatē (e) a Fifth.' This shews that c, not b, was the missing note.

* [The upper tetrachord was thus reduced to a trichord, while the lower remained a perfect tetrachord. If we take Pythagorean intonation the cents are e 204 g 204 a 204 b 294 d' 204 e'.—*Translator.*]

† [Taking Pythagorean intonation, the cents in the intervals are b 90 c 408 e 90 f 408 a 204 b. The account of the popular tuning of the Ko-to, the national Japanese instrument, furnished by the Japanese, but in European notes, at the International Health Exhi-

bition in London, 1884, gives many varieties of this scale, see App. XX. sect. K. Japan.—*Translator.*]

‡ Chinese Melodies, in Ambrosch's *Geschichte der Musik*, vol. i. pp. 30, 34, 35. Of Scotch melodies there is a fine collection with reference to the authorities and the older forms in G. F. Graham's *Songs of Scotland*, 3 vols. Edinburgh, 1859. The modern pianoforte accompaniment which has been added, is often ill enough suited to the character of the airs.

§ [Exclusive of the two drones there are only 9 tones on the bagpipe. For the whole of these observations see App. XX. sect. K.—*Translator.*]

the next tones to be considered are those related to the upper Fifth g, and lower Fifth F, and these are d (the Fifth above the upper Fifth g) and $B\flat$ (the Fifth below the lower Fifth F). We thus obtain the scales*

1) Ascending $c - d - \smile f - g - a - \smile c'$

                       $1 \quad \frac{9}{8} \quad\quad \frac{4}{3} \quad \frac{3}{2} \quad \frac{5}{3} \quad\quad 2$

2) Descending $C - \smile E\flat - F - G - \smile B\flat - c$

                       $1 \quad\quad \frac{6}{5} \quad \frac{4}{3} \quad \frac{3}{2} \quad\quad \frac{16}{9} \quad 2$

But in place of the tones more distantly related to the tonic in the first degree, both systems of tones related in the second degree might be used, and this would give a scale resulting from a simple progression by Fifths, as

3) $c - d - \smile f - g - \smile b\flat - c'$

                       $1 \quad \frac{9}{8} \quad\quad \frac{4}{3} \quad \frac{3}{2} \quad\quad \frac{16}{9} \quad 2$

¶

Then there are also some more irregular forms of this scale of five tones, in which the major Third e replaces the Fourth f, which is more nearly related to the tonic c. This transformation is probably due to the modern preference for the major mode, and it has made its appearance in very many Scotch melodies. The scale is then

4) $c - d - e - \smile g - a - \smile c'$

                       $1 \quad \frac{9}{8} \quad \frac{5}{4} \quad\quad \frac{3}{2} \quad \frac{5}{3} \quad\quad 2$

The examples of a similar exchange of the Fifth g for the minor Sixth $a\flat$ are doubtful. This would give the scale

5) $C - \smile E\flat - F - \smile A\flat - B\flat - c$

                       $1 \quad\quad \frac{6}{5} \quad \frac{4}{3} \quad\quad \frac{8}{5} \quad \frac{16}{9} \quad 2$

The scale   6) $c - \smile e\flat - f - g - a - \smile c$

                       $1 \quad\quad \frac{6}{5} \quad \frac{4}{3} \quad \frac{3}{2} \quad \frac{5}{3} \quad\quad 2$

¶

in which all the notes are related in the first degree, but for which the nearest notes to the tonic, either way, are a Tone and a Semitone distant from it, has not yet been discovered in actual use.

The above five forms of the scale of five tones can all be so transposed that they can be played on the black notes of a piano without touching the white ones.† This is the well-known simple rule for composing Scotch melodies.‡ Any one of

---

* [In the following investigation the Author all along assumes harmonic forms of the intervals, which are certainly modern. The cents in the five forms cited, as determined from the ratios given, are:

1) c 204 d 294 f 204 g 204 a 316 c'.
2) C 316 $E^1\flat$ 182 F 204 G 294 $B\flat$ 204 c.
3) c 204 d 294 f 204 g 294 $b\flat$ 204 c'.

4) c 204 d 182 e_1 316 g 182 a_1 316 c'.
5) C 316 $E^1\flat$ 182 F 316 $A^1\flat$ 204 $B^1\flat$182 c.
6) c 316 $e^1\flat$ 182 f 204 g 182 a_1 316 c.— *Translator.*]

† [In the following way—the numbers referring to the schemes in this page, and the corresponding cents, of course, belonging to equal temperament:

¶

1) $c\sharp$ 200 $d\sharp$ 300 $f\sharp$ 200 $g\sharp$ 200 $a\sharp$ 300 $c'\sharp$
2) $d\sharp$ 300 $f\sharp$ 200 $g\sharp$ 200 $a\sharp$ 300 $c'\sharp$ 200 $d'\sharp$
3) $g\sharp$ 200 $a\sharp$ 300 $c'\sharp$ 200 $d'\sharp$ 300 $f'\sharp$ 200 $g'\sharp$
4) $f\sharp$ 200 $g\sharp$ 200 $a\sharp$ 300 $c'\sharp$ 200 $d'\sharp$ 300 $f'\sharp$
5) $A\sharp$ 300 $c\sharp$ 200 $d\sharp$ 300 $f\sharp$ 200 $g\sharp$ 200 $a\sharp$

And this shews that all five are formed by a simple succession of tempered Fifths, for the five black notes arranged in order of Fifths are $f\sharp$ 700 $c\sharp$ 700 $g\sharp$ 700 $d\sharp$ 700 $a\sharp$. The piano being tuned in equal temperament gives very nearly perfect Fifths, and hence very well imitates a succession of five notes thus tuned. If, however, the Fifths are perfect, then every 200 and 300 cents in the above scheme becomes 204 and 294, differences which few ears will perceive in melody.

‡ [Mr. Colin Brown, Euing Lecturer on the Science, Theory, and History of Music,

Anderson's College, Glasgow, in *The Thistle*, ' a miscellany of Scottish song, with notes critical and historical ; the melodies arranged in their natural modes ; with an introduction, explaining the construction and characteristics of Scottish music, the Principles, Laws, and Origin of Melody ' (Glasgow, Dec. 1883), says, p. viii.: ' The pentatonic form of the scale is used in Scotland, but not to a greater extent than in the national music of some other countries. A general idea seems to prevail that Scottish music can be played upon the five black digitals of the pianoforte, representing

¶

the five black notes may then be used as a tonic, but the $B\flat$ or $A\sharp$ having no Fifth (F or $E\sharp$) among the black notes has a very doubtful effect as a tonic.

The following are examples of the use of these various scales of five tones:—

1. *The First Scale* without Third or Seventh. Chinese, after John Barrow.*

2. *To the Second Scale*, without Second or Sixth, belong most Scotch airs which have a minor character. In the modern forms of these airs one or other of the missing tones is often transiently touched. Here follows an older form of the air called *Cockle Shells* :†

what is popularly known as the Caledonian scale; but any one who will take the trouble to examine Scottish music will find that not more than a twentieth part of our old melodies are pentatonic, or constructed upon this form of the scale. In Dauney's work, where the Skene MSS. (the oldest collection extant) are noted, this statement is fully verified.' I have examined the first 36 airs as printed in *The Thistle*, and I found only one which was ¶ strictly pentatonic, p. 51, No. 8, *Lament for Ruaridh Mor*, Macleod of Macleod—Dunvegan 1626. But in nearly a quarter of the airs the Semitones were introduced by an unaccented note which looked to be modern, as in *Roy's Wife*, p. 10, and the *Banks and Braes o' Bonnie Doon*, p. 48, on the last of which Mr. Brown observes, p. 49: " With pentatonic theorists *Ye Banks and Braes* is a favourite example of this assumed peculiarity of Scottish music. But it can only be brought into the pentatonic scale by being played in an incomplete form.' The only places in which the Seventh $g\sharp$ occurs are the cadence $e'\ f\sharp\ g\sharp\ a'$ (which occurs twice, and is evidently out of character, and should be $e'\ f\sharp\ a'\ a'$), and the flourished ad libitum cadence $f''\sharp\ e''\ d''\ c''\sharp\ b''$ containing the Fourth d, (which should clearly be $f''\sharp\ e''\ c''\sharp\ b'$). And many of the others can be probably ' restored ' in a similar fashion.

Thus of *Roy's Wife* Mr. Brown himself says, p. 11, ' played as a dance tune it is pentatonic,' and gives the substitutes for his version, which are clearly the more ancient forms. Mr. Brown gives as the marks of Scotch music (pp. ix., x.) 1. *its modal character*, being constructed on the ancient seven modes ; 2. *its modulation* or change of mode, which is constant ; 3. *almost absence of transition* or change of key ; 4. *preponderance of minor forms of the scale* ; 5. *almost absence of sharp Sevenths in the minors* ; 6. *cadences* on to every note of the scale, and double cadences closing on an unaccented note, which are simple (repeating the cadential tone) or compound (the unaccented tone differing from the preceding).—*Translator.*]

* [Scale, tempered d 200 e 300 g 200 a 200 b 300, d'. That is, no $f\sharp$ and no $c\sharp$. All these scales are merely the best representatives in European notation of the sensations produced by the scales on European listeners. They cannot be received as correct representations of the notes actually played.—*Translator.*]

† Playford's *Dancing Master*, ed. 1721. The first edition appeared in 1657.—*Songs of Scotland*, vol. iii. p. 170. [Scale, d 300 f 200 g 200 a 300 c' 200 d', without e or $b\flat$.—*Translator.*]

3. *For the Third Scale*, without Third and Sixth. Gaelic. Probably an old bagpipe tune.*

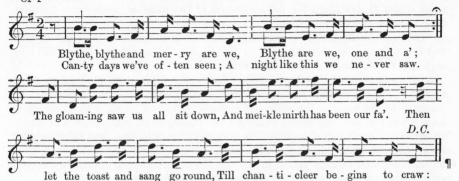

Blythe, blythe and mer - ry are we, Blythe are we, one and a';
Can-ty days we've of - ten seen; A night like this we ne - ver saw.

The gloam-ing saw us all sit down, And mei-kle mirth has been our fa'. Then
D.C.

let the toast and sang go round, Till chan - ti - cleer be - gins to craw:

4. *To the Fourth Scale*, without Fourth or Seventh, belong most Scotch airs which have the character of a major mode. Since dozens of Scotch tunes of this kind are to be found in every collection, and are perfectly well known, I give here a Chinese temple hymn, after Bitschurin,† as an example:

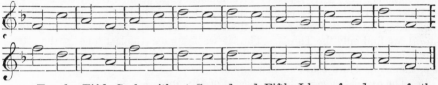

5. *For the Fifth Scale*, without Second and Fifth, I have found no perfectly pure examples. But there are melodies with either only the Fifth or else with a ¶ mere transient use of both Second and Fifth. In the latter case the *minor* Second is used, giving it the character of the ecclesiastical Phrygian tone, for example in the very beautiful air, *Auld Robin Gray*. I give an example with the tonic *f♯*, in which the Second (*g♯* or *g*) is altogether absent, and the fifth *c♯* is only once transiently touched, so that it might just as well have been omitted.

Will ye go, las - sie, go To the braes o' Bal -
Let us

- quhid - der, Where the blae - ber - ries grow, Mang the

Fine.

bon - nie bloom - in hea - ther; Where the deer and the
Sport the lang sim - mer

rae, Light - ly bound - ing the - ge - ther,
day, On the braes o' Bal - quhid - der?

---

* There is a Chinese tune of the same kind in Ambrosch, *loc. cit.* vol. i. p. 34, second piece. Another, with a single occurrence of the Sixth, *My Peggie is a young thing*, may be seen in *Songs of Scotland*, vol. iii. p. 10. [Scale, *e* 200 *f♯* 300 *a* 200 *b* 300 *d'* 200 *e'*, without *g* or *c*. On the bagpipe, see App. XX. sect. K.

Probably the scale of the bagpipe has been unaltered since its importation from the East, and it probably never could have played such a scale as it is here supposed capable of performing.—*Translator.*]

† Ambrosch, *loc. cit.* vol. i. p. 30. To the same class belongs the first piece on p. 35 after

We might also in this example assume b as the tonic, and regard the conclusions as formed upon the dominant and subdominant in the old-fashioned way.* In these scales of five tones the determination of the tonic is much more doubtful than in the scales of seven tones.

The rule usually given for the Gaelic and Chinese scales, to omit the Fourth and Seventh, applies therefore only to the fourth of the above scales, which corresponds to our major scale. True this scale is often used in the usual Scotch airs of the present day, and is probably due to the reaction of our modern tonal system. But the examples here adduced shew that every possible position may be assumed by the tonic in the scale of five tones, if indeed we allow these scales to have a tonic at all. In Scotch melodies the omissions in both major and minor scales are so contrived as to avoid the intervals of a Semitone, and substitute for them intervals of a Tone and a half. Among the Chinese airs, however, I have ¶ found one which belongs rather to the old Greek enharmonic system, to be considered presently, and it will be explained at the same time (p. 265c).

We now proceed to the construction of *scales with seven degrees*. The first form was developed in Greece under the influence of the tetrachordal divisions. The ancient Greek melodies had a small compass and few degrees, a peculiarity especially emphasised even by later authors, as Plutarch, but it is also found among most nations in the early stages of their musical cultivation. Hence the scale was at first formed within a less compass than an Octave, namely within the tetrachord. On looking within this compass for the tones nearest related to the limiting tonic ($\mu\acute{\epsilon}\sigma\eta$), we find only the Thirds. Thus if we assume e (the last tone in the tetrachord, $b-e$) as a tonic, its next related tone within the compass of that tetrachord is c, the major Third below e. This gives :—

1. The *ancient enharmonic tetrachord* of Olympos—

¶
$$b \smile c \ \text{———} \ e$$
$$\tfrac{3}{4} \quad \tfrac{4}{5} \qquad 1$$

Archytas was the first to settle that the tuning of $c : e$ must be $4 : 5$ in the enharmonic mode. The next most closely related tone to e would be the minor Third below it. Adding this we obtain :

2. The older *chromatic tetrachord* of the Greeks—

$$b \smile c \smile c\sharp \ \text{—} \smile e$$
$$\tfrac{3}{4} \quad \tfrac{4}{5} \quad \tfrac{5}{6} \qquad 1$$

The method of tuning the intervals here assigned agrees with the data of Eratosthenes (in the third century before Christ). The interval between c and $c\sharp$ in this case corresponds to the small ratio $\tfrac{25}{24}$ [$= 70$ cents], which is less than the Semitone $\tfrac{16}{15}$ [$= 112$ cents]. Next to it comes the much wider interval, $c\sharp - e$, corresponding to a minor Third. We should obtain a more even distribution of ¶ intervals, by measuring the minor Third upwards from the lowest tone of the tetrachord. This gives rise to

3. The *diatonic tetrachord*—

$$b \smile c \ \text{—} \ d \ \text{—} \ e$$
$$\tfrac{3}{4} \quad \tfrac{4}{5} \quad \tfrac{9}{10} \quad 1$$

This is the tuning assigned by Ptolemy for the diatonic tetrachord. Here we

Barrow and Amiot. [Scale, f 200 g 200 a 300 c' 200 d' 300 f', without bb or e.—*Translator*.]

* [Taking $f\sharp$ as the Tonic, the scale would be No. 5, without Second and Fifth, thus :

$$f\sharp \ 300 \ a \ 200 \ b \ 300 \ d \ 200 \ e \ 200 \ f\sharp$$

but taking b as the tonic the scale would be No. 2, without Second and Sixth, as

$$b \ 300 \ d \ 200 \ e \ 200 \ f\sharp \ 300 \ a \ 200 \ b$$

which is altogether different. Any reference

to tonic, dominant and subdominant, implies harmonic scales, which pentatonic scales could not have been originally. Mr. C. Brown gives this air (*Thistle*, p. 198) as here printed, but says it varies between his modes of the 3rd (Greek Doric, Ecclesiastical Phrygian) and 5th of the scale (Gr. Ionic, Eccl. Mixolydian). The spelling of the words has been corrected by his edition.—*Translator*.]

must observe that if e continued to be regarded as the tonic, d would have only a distant relation with it in the second degree through the auxiliary tone b. If two tetrachords had been connected, as was very early done, thus :

$$b \text{ —— } e \text{ —— } a$$

a closer connection in the second degree between d and e might have been obtained by tuning d as a fifth below a. Taking e as i, a will be $\frac{4}{3}$, and the Fifth below it is $d = \frac{8}{9}$. We thus obtain the tetrachord

4. $\qquad\qquad\qquad\qquad b \smile c - d - e$

$\qquad\qquad\qquad\qquad\quad \frac{3}{4} \quad\ \frac{4}{5} \quad\ \frac{8}{9} \quad\ \mathrm{I}$

which agrees with the tuning assigned by Didymus (in the first century before Christ). ¶

According to the old theory of Pythagoras, which will be examined presently, all the intervals of the diatonic scale should be tuned by means of intervals of a Fifth, giving :

5. $\qquad\qquad\qquad\qquad b \ \smile\ c - d - e$

$\qquad\qquad\qquad\qquad \frac{4}{3} \quad\ \frac{81}{64} \quad\ \frac{9}{8} \quad\ \mathrm{I}$

$\qquad\qquad\qquad\qquad\ \underbrace{\quad}\ \underbrace{\quad}\ \underbrace{\quad}$

$\qquad\qquad\qquad\qquad \frac{243}{256} \quad\ \frac{9}{8} \quad\ \frac{9}{8}$

The tetrachord thus obtained is the Greek *Doric*, which is considered as normal, and made the basis of all considerations on other scales. Accordingly those tones which formed the lower notes of the semitonic intervals of the scale, were, at least theoretically, considered as the immovable limiting tones of the tetrachord while the intermediate tones might change their position. Practically the intonation of even these fixed tones was a little changed, as Plutarch tells us, which may mean that in the *Lydian*, and *Phrygian* modes, &c., the tonic is not selected from one of ¶ these so-called fixed tones of the tetrachords. Thus we shall see further on, that when d is the tonic, the b in the natural intonation of such a scale does not form a perfect Fifth with e.

The tetrachord could, however, be differently completed by inserting tones which formed major or minor Thirds with either of the extreme tones.

Two minor Thirds give the *Phrygian* tetrachord—

6. $\qquad\qquad\qquad\qquad d - e \smile f - g$

$\qquad\qquad\qquad\qquad\quad \frac{3}{4} \quad\ \frac{5}{6} \quad\ \frac{9}{10} \quad\ \mathrm{I}$

If a major Third were taken upwards from the lower extreme tone, and a minor Third downwards from the upper extreme, we should obtain the *Lydian tetrachord*—

7. $\qquad\qquad\qquad\qquad c - d - e \smile f$

$\qquad\qquad\qquad\qquad\quad \frac{3}{4} \quad\ \frac{5}{6} \quad\ \frac{15}{16} \quad\ \mathrm{I}$

¶

8. Two major Thirds, as in $\ b \smile c - d\sharp \smile e$, would form a variety of the chromatic scale, which does not seem to have been used, or at any rate not to have been distinguished from the chromatic form.*

---

* [Adopting the notation explained later on in this chapter, these tetrachords may be accurately written as follows ; Nos. 1, 3, 4 and 7 may be played as they stand on the Harmonical, and Nos. 2, 6, 8 by transposition as shewn below, but No. 5 requires the six notes forming 5 perfect Fifths, and these do not occur on the Harmonical, but can be played sufficiently well on any tempered harmonium. Between the names of the notes are inserted the number of cents in the interval between them. By referring to the table called the Duodēnārium, App. XX. sect. E. art. 18, which employs the same notation, the exact position of the notes

may be seen, and the correctness of the transpositions verified.

1. Olympos . . $b_1\,112\,c'\,386\,e_1{'}$
2. Old Chromatic . $b_1\,112\,c'\,70\,c_2{'}\sharp\,316\,e_1{'}$
     (play . $g\,112\,a^{\mathsf{I}}b\,70\,a_1\,316\,c'$)
3. Diatonic . . $b_1\,112\,c'\,204\,d'\,182\,e_1{'}$
4. Didymus . . $b_1\,112\,c'\,182\,d_1{'}\,204e_1{'}$
5. Doric . . $b\,90\,c'\,204\,d'\,204\,e'$
     (not playable on the Harmonical)
6. Phrygian . . $d\,182\,e_1\,134\,f^{\mathsf{I}}\,182\,g$
     (play . $g\,182\,a_1\,134\,b^{\mathsf{I}}b\,182\,c'$)
7. Lydian . . $c\,182\,d_1\,204\,e_1\,112\,f$
8. Unused . . $b_1\,112\,c'\,274\,d_2{'}\sharp\,112\,e_:{'}$
     (play . $g\,112\,a^{\mathsf{I}}b\,274\,b_1\,112\,c'$)

These are all the normal subdivisions of the tetrachord that have been used. But other subdivisions occur which the Greeks themselves termed *irrational* (ἄλογα),* and we do not know with certainty how far they were practically used. One of them, the *soft diatonic mode*, makes use of the interval 6 : 7, which is at any rate very near to a natural consonance, being that between the Fifth and the subminor Seventh of the fundamental note, an interval occasionally used in harmonic music when unaccompanied singers take the minor Seventh of the chord of the dominant Seventh. The intervals † are :

9.

$$3 : 4$$

$$\underbrace{\frac{21}{20} \qquad \frac{10}{9} \qquad \frac{8}{7}}$$

$$6 : 7$$

By lowering the Lichanos the Parhypatē is also flattened. However, the small ¶ interval $\frac{21}{20}$ is very nearly the Pythagorean Semitone, which expressed approximately is $\frac{20}{19}$.

The equal diatonic mode of Ptolemy, which was divided thus : ‡

10.

$$3 : 4$$

$$\underbrace{\frac{12}{11} \qquad \frac{11}{10} \qquad \frac{10}{9}}$$

$$5 : 6$$

contained a perfect minor Third divided as evenly as possible.

There is a similar succession of tones, in an inverse order, in the modern Arabic scale as measured by the Syrian, Michael Meshāqah.§ In this case the Octave is divided into twenty-four Quartertones **; the tetrachord 10 has ten of them, its lowest interval four, and each of the upper intervals three. Under these circumstances the two upper intervals together form very nearly a minor Third, ¶ which, as in the *equal diatonic* scale of the Greeks, is divided into two equal intervals, without paying regard to any sensible relationship of the intermediate tone thus produced.

The closer the interval, the more easy and certain is its division into two intervals, by the mere feeling for difference of pitch. This is, in particular, possible for intervals which approach to the limits at which differences of pitch are

If the minor Thirds df and eg were taken as Pythagorean = 294 cents, tetrachord 6 would become d 204 e 90 f 204 g, which is more intelligible.

On referring to App. XX. sect. D. the ratios corresponding to each of these numbers of cents will be found.—*Translator.*]

* [That is, strictly, having a ratio not expressible by whole numbers.—*Translator.*]

† [The notes which would form tetrachord 9 ¶ might be written in the Translator's notation, descending from left to right,

$$^7b\flat \; 85 \; a_1 \; 182 \; g \; 231 \; {}^7f.$$

The three first notes could be played on the Harmonical. The interval 231 cents could be played on it downwards as c' 231 $^7b\flat$, but the whole tetrachord cannot be played on it. Here 85 cents represent 21 : 20, while the Pythagorean Semitone 256 : 243 is 90 cents. The difference is small but perceptible.—*Translator.*]

‡ [Using the notation ^{11}f for the 11th harmonic of c, so that ^{11} is equivalent to 33 : 32 or 53 cents, tetrachord 10 may be written downwards :

$$g \; 151 \; {}^{11}f \; 165 \; e_1 \; 182 \; d.$$

This is simply, in order, the 12, 11, 10, and 9th harmonic of c, and can be played on the horn or trumpet, and on the 5th octave of the Har-

monical, as d''' e_1''' $^{11}f'''$ g''', downwards as g''' $^{11}f'''$ e_1''' d'''. The division of the minor Third $g : e_1 = 316$ cents into 151 and 165 cents is of course only approximative. But it is a purely natural tetrachord of which g 204 f 112 e_1 182 d is a deformation.—*Translator.*]

§ *Journal of the American Oriental Society*, vol. i. p. 173, 1847.

** [If the Octave is divided equally into 24 quarters, each of which is half an equal Semitone or 50 cents, we can write it by using the additional sign q (a turned ♭, standing for q the initial of *quarter*) to represent an added Quartertone, ♯ being two Quartertones, and ♯q three Quartertones, thus ascending c cq $c♯$ c $♯q$ d or descending d $d\flat q$ $d\flat$ $d\flat cq$ c, using $d\flat$ as the equivalent of $c♯$. Then the principal scale of Meshāqah (see App. XX. sect. K.) is

$$a \; 200 \; b \; 150 \; c'q \; 150 \; d' \; 200 \; e' \; 150 \; f'q \; 150 \; g' \; 200 \; a'.$$

Hence the tetrachord $a : d'$, which represents 10 in the text, has one interval of 200 cents or 4 Quartertones, and two of 150 cents or 3 Quartertones. This interval of 3 Quartertones represents the trumpet intervals $^{11}f : g = 11 : 12$ = 151 cents, and $e : {}^{11}f = 10 : 11 = 165$ cents, and was introduced into Arabia by the lutist Zalzal, who died about 1000 years ago, and is much used in the East.—*Translator.*]

distinguishable by the ear. The distinctness with which the yet sensible difference can be felt then furnishes a measure of its magnitude. In this sense we have probably to explain the possibility of the later *enharmonic mode* of the Greeks, which, however, had already fallen into disuse in the time of Aristoxenus, and was perhaps hunted up again by later writers as an antiquarian curiosity. In this mode the Semitone of the ancient enharmonic mode already mentioned (No. 1, p. 262b) was again subdivided into two Quartertones, so that a tetrachord was produced like the chromatic one, but with closer intervals between the adjacent tones. The division of this *enharmonic tetrachord** was

11.

This Quartertone can only be considered as a transition in the melodic movement towards the lowest extreme of the tetrachord. A similar interval occurs in ¶ this way in existing Oriental music. A distinguished musician whom I requested to pay attention to it on a visit to Cairo, wrote to me as follows : ' This evening I have been listening attentively to the song on the minarets, to try to appreciate the Quartertones, which I had not supposed to exist, as I had thought that the Arabs sang *out of tune.* But to-day as I was with the dervishes I became certain that such Quartertones existed, and for the following reasons. Many passages in litanies of this kind end with a tone which was at first the Quartertone and then ended in the pure tone.† As the passage was frequently repeated, I was able to observe this every time, and I found the intonation invariable.' The Greek writers on music themselves say that it is difficult to distinguish the enharmonic Quartertones.‡

The later interpreters of Greek musical theory have mostly advanced the opinion that the above-mentioned differences, which the Greeks called *colourings* (χρόαι), were merely speculative and never came into practical use.§ They con- ¶ sider that these distinctions were too delicate to produce any esthetic effect except on an incredibly well cultivated ear. But it seems to me that this opinion could never have been entertained or advanced by modern theorists, if any of them had practically attempted to form these various tonal modes and to compare them by ear. On an harmonium which will shortly be described I am able to compare

---

* [It is not to be supposed that these two Quartertones, differing only by two cents (32 : 31 = 55 cents, 31 : 30 = 57 cents), were exactly produced. The lutist or lyrist would tune his Fourth *c* : *f* by ear (tolerably correctly), then a major Third *c* : *f* below *f* or *d'*♭ also by ear (and probably very incorrectly on account of the great difficulty of tuning a major Third), and then would by feeling divide the remaining interval in halves as well as he could. Using *c* : *c*q for the approximate Quartertone he would have *about, c* 56 *c*q 56 *d'*♭ 386 *f*, or something sufficiently like it. Meshāqah's *c* 50 *c*q 50 *c*♯ 400 *f* would doubtless have been near enough. Probably no two lyrists tuned alike. My experience of tuning by ear is quite against any approach to the accuracy which the figures in the text would imply.—*Translator.*]

† [Probably the effect was like that which I heard produced by Rája Rám Pál Singh on his Sitár. Here the tone of the note, played by pressing the string against a fret, was sharpened a quarter of a Tone by sliding the finger along the fret (thus deflecting the string and increasing the tension), and then it was allowed to glide on to the proper note by straightening the string without replucking it. I determined the amount of sharpening by observing the distance of deflection, and then, at leisure, measuring by my forks the

number of vibrations for the sharpened and normal note, which gave the interval as 48 cents. The effect was very peculiar, but can of course be easily imitated on the violin. On the classical Indian instrument, the Vina, the frets are very high, sometimes about an inch. Hence by pressing down the string behind the fret, the tension could be greatly increased, and as much as a Semitone could be easily added, so that the scale could be indefinitely altered without changing the frets, which were fixed ¶ with wax. On the Arabic Rabāb and the curious Chinese fiddles, which have no frets or finger-board, a note could be instantaneously sharpened in a similar manner by pressing more strongly.—*Translator.*]

‡ [And yet a quarter of a Tone is between 2 and 3 commas, and all the difficulties of tuning in just and tempered intonation arise from intervals of a single comma or less.—*Translator.*]

§ Even Bellerman is of this opinion (*Tonleiter der Griechen*, p. 27). Westphal, in his *Fragmenten der Griechischen Rhythmiker*, p. 209, has collected passages from Greek writers proving the real practical use of these intervals. According to Plutarch (*De Musica*, pp. 38 and 39), the later Greeks had even a preference for these surviving archaic intervals.

natural intonation with Pythagorean, and to play the diatonic mode at one time after the method of Didymus and at another after that of Ptolemy, and also to make other deviations. It is not at all difficult to distinguish the difference of a comma $\frac{81}{80}$ in the intonation of the different degrees of the scale, when well-known melodies are performed in different ' colourings,' and every musician with whom I have made the experiment has immediately heard the difference. Melodic passages with Pythagorean Thirds have a strained and restless effect, while the natural Thirds make them quiet and soft, although our ears are habituated to the Thirds of the equal temperament, which are nearer to the Pythagorean than to the natural intervals. Of course where delicacy in any artistic observations made with the senses, comes into consideration, moderns must look upon the Greeks in general as unsurpassed masters. And in this particular case they had very good reason and abundance of opportunity for cultivating their ear better than ours. From

¶ youth upwards we are accustomed to accommodate our ears to the inaccuracies of equal temperament, and the whole of the former variety of tonal modes, with their different expression, has reduced itself to such an easily apprehended difference as that between major and minor. But the varied gradations of expressions which moderns attain by harmony and modulation, had to be effected by the Greeks and other nations that use homophonic music, by a more delicate and varied gradation of the tonal modes. Can we be surprised, then, if their ear became much more finely cultivated for differences of this kind than it is possible for ours to be?

The Greek scale was soon extended to an octave. Pythagoras is said to have been the first to establish the eight complete degrees of the diatonic scale. At first two tetrachords were connected in such a way as to have a common tone, the μέση :

$$e \smile f - g - a \smile b\flat - c - d$$

¶ which produced a scale of seven degrees. Then this scale was changed into the following form :

$$e \smile f - g - a - b \smile - d - e$$

and thus made to consist of a tetrachord and a trichord, of which mention has already been made (p. 257d). Finally Lichaon of Samos, (according to Boethius,) or Pythagoras, (according to Nicomachus,) completed the trichord into a tetrachord, and thus established a scale consisting of two disjunct tetrachords.

The diatonic scale thus obtained could be continued either way at pleasure by adding higher and lower octaves, and it then produced a regularly alternating series of Tones and Semitones. But for each piece of music a portion only of this unlimited diatonic scale was employed, and the tonal systems were distinguished by the character of the portions selected.

These sectional scales might be produced in very different ways. The first

¶ practical object which necessarily forces itself on attention, as soon as an instrument with a limited number of strings, like the Greek lyre, is used for executing a piece of music, is, of course, that there should be a string for every musical tone required. This prescribes a certain series of tones which must be provided and tuned on the instrument. Now as a rule when a certain series of tones is thus prescribed as a scale for the tuning of a lyre, no question is raised as to whether a tonic is to be distinguished or not, or if so which it should be. A tolerable number of melodies may be found in which the lowest tone is the tonic : others in which an interval below the tonic is touched ; and others, again, in which the Fifth or Fourth above the Octave below the tonic is used. This is the kind of difference between the *authentic* and *plagal* scales of the middle ages. In the authentic scales the deepest tone of the scale, in the plagal its Fifth below or Fourth above, was the tonic ; thus : *—

* [See Mr. Rockstro's article, 'Modes Ecclesiastical,' vol. ii. p. 340, and Rev. T. Helmore's on 'Gregorian Modes,' vol. i. p. 625, in Grove's *Dictionary of Music.* What Prof. Helmholtz

First Authentic Ecclesiastical Scale, tonic d.

$$d — e — f — g — a — b — c — d$$

Fourth Plagal Scale, tonic g.

$$d — e — f — g — a — b — c — d$$

The scales were looked upon as composed of a Fifth and a Fourth, as the braces shew. In the authentic tone the Fifth lay below; in plagal, above. Now if we have nothing else before us but a scale of this kind, which marks out the accidental compass of a series of melodies, we can collect but little respecting the key. Such scales themselves may be fittingly termed *accidental*. They comprise, among others, the medieval plagal scales. On the other hand, those ¶ scales which, like the modern, are bounded at each extremity by the tonic, may be termed *essential*. Now practical needs clearly lead in the first place to accidental scales alone. When a lyre had to be tuned to accompany the human voice in unison, it was indispensably necessary that all the tones required should be present. There was no immediate practical need for marking the tonic of a song sung in unison, or even to become fully aware that it had a tonic at all. In modern music, where the structure of the harmony essentially depends on the tonic, the case is entirely different. Theoretical considerations on the structure of melody could alone lead to distinguishing one tone as tonic. It has been already mentioned in the preceding chapter, that Aristotle, as a writer on esthetics, has left a few notices indicating such a conception, but that the authors who have specially written on music say nothing about it.

In the best times of Greece, song was usually accompanied by an eight-stringed lyre, tuned so as to embrace an Octave of tones selected from the diatonic scale. ¶ These were the following :

1. Lydian . . . . . $c — d — e — f — g — a — b — c$

2. Phrygian . . . . . $d — e — f — g — a — b — c — d$

3. Doric . . . . . $e — f — g — a — b — c — d — e$

4. Hypolydian . . . $f — g — a — b — c — d — e — f$

5. Hypophrygian (Ionic) . . $g — a — b — c — d — e — f — g$

6. Hypodoric (Eolic or Locrian) . $a — b — c — d — e — f — g — a$

7. Mixolydian . . . . $b — c — d — e — f — g — a — b — (c)$

¶

Hence any one of the tones in the diatonic scale could be used as the initial or final extremity of such a tonal mode. The Lydian and Hypolydian scales contain Lydian, the Phrygian and Hypophrygian contain Phrygian, and the Doric and Hypodoric contain Doric tetrachords. In the Mixolydian two Lydian tetrachords seem to have been assumed, one of which was divided, as shewn by the braces in the above examples.*

calls the *tonic* was termed the *final*. What was the exact intonation of this music it is perhaps impossible to say. Perhaps we may assume it to have been Pythagorean, as

d 204 e 90 f 204 g 204 a 204 b 90 c 204 d.

Of course, modern musicians play them

on the piano and organ in equally tempered intonation, as their ancestors played them in meantone intonation. But either of the latter admit of being harmonised; not so the former, so that there is an essential difference. —*Translator.*]

* [By a reference to p. 263d, note, it will be

The scales or *tropes* of the best Greek period have hitherto been considered as essential, that is, the lowest tone or *hypatē* has been considered as the tonic. But I cannot find any definite ground for this assumption. What Aristotle says, as we have seen, makes the middle tone or *mesē*, function as the tonic, but yet it cannot be denied that other attributes of our tonic belong to the *hypatē*.* Whatever may have been the real state of the case, whether the *mesē* or *hypatē* be regarded as the tonic, whether the scales be considered as all authentic or all plagal, it is extremely probable that the Greeks, among whom we first find the complete diatonic scale, took the liberty of using every tone of this scale as a tonic, just as we have seen that every one of the five tones forming the scales of the Chinese and Gaels occasionally functions as a tonic. The same scales are also found, probably handed down immediately by ancient tradition, in the ancient Christian ecclesiastical music.

¶ Hence if we disregard the chromatic and enharmonic scales, and the apparently arbitrary scales of the Asiatics, none of which have shewn themselves capable of further development,† homophonic vocal music developed seven diatonic scales, which differ from one another in about the same way as our major and minor scales. These differences will be better appreciated by making them all begin with the same tonic *c*.‡

seen that this paragraph materially alters the intonation from what would result from a mere beginning of each mode with a different note of the Pythagorean or diatonic scale. I therefore repeat the scales as defined by this paragraph in the notation explained on pp. 276*a* to 277*a* and note *, and write between each pair of notes the number of cents in the interval between each pair of notes, which will be found useful in future comparisons. These scales ¶ should be traced out on the Duodenarium, App. XX. sect. E. art. 18. They cannot be played on the Harmonical.

1. Lydian, c 182 d_1 204 e_1 112 f 204 g 182 a_1 204 b_1 112 c

2. Phrygian, d 182 e_1 134 f^1 182 g 204 a 182 b_1 134 c^1 182 d

3. Doric, e 90 f 204 g 204 a 204 b 90 c 204 d 204 e

4. Hypolydian, f 204 g 182 a_1 204 b_1 112 c 182 d_1 204 e_1 112 f

5. Hypophrygian (Ionic), g 204 a 182 b_1 112 c 204 d 182 e_1 134 f^1 182 g

6. Hypodoric (Eolic or Locrian), a 204 b 90 c 204 d 204 e 90 f 204 g 204 a

7. Mixolydian, b_1 112 c 182 d_1 204 e_1 112 f 204 g 182 a_1 204 b_1.—*Translator.*]

* R. Westphal, in his *Geschichte der alten* ¶ *und mittelalterlichen Musik*, Breslau, 1864, which is unfortunately still incomplete, uses the previous citations from Aristotle, to frame an hypothesis on the tonic and final cadence of the above scales. But he applies the remarks of Aristotle only to the Doric, Phrygian, Lydian, Mixolydian and Locrian scales, and not to the Eolic and Ionic, which were also known at that time, although the ground for their exclusion is not apparent. In the first four of these he takes the *mesē* as tonic and the *hypatē* as the terminal tone. In those scales distinguished by the prefix *Hypo-*, the *hypatē* was both tonic and terminal; but in those having the prefix *Syntono-*, the *hypatē* was both the terminal and the Third of the tonic, and the same was the case perhaps for the *Boeotian* scale, which is only mentioned once. Hence it follows that the minor scale of *A* occurs as Doric with the terminal *c*, as Hypo-

doric with the terminal *a*, as Boeotian with the terminal *c*. Moreover the Mixolydian would be a minor scale of *E*, with a minor Second, and a terminal in *b*; the Locrian a minor scale of *D* with a major Sixth, and a terminal in *a*; the Phrygian, Hypophrygian or Iastic, and the Syntonoiastic, major scales of *G*, with a minor Seventh, the terminals being *d*, *g*, and *b* respectively. Finally the Lydian, Hypolydian and Syntonolydian would be major scales of F, with superfluous Fourth, and with the terminals *c*, *f*, and *a* respectively. But according to Westphal the normal major scale was entirely absent. If the Ionic were interpreted according to the words of Aristotle, it would yield a correct major scale. The tonic *F* with *B* (instead of *B*♭) as its Fourth, has a totally impossible appearance to modern musical feeling.

† [In India there is a highly developed system with a vast variety of scales.—*Translator.*]

‡ [Continuing to use the notation of p. 268*b*, note, these transposed scales may be written as follows. As the order is different from that in p. 267*c*, the numbers there used are added in ( ). The number of cents in each interval will complete the identification. I give only the ancient Greek names, and the names proposed by Prof. Helmholtz.

1. Lydian—Mode of the First (Major) (1), c 182 d_1 204 e_1 112 f 204 g 182 a_1 204 b_1 112 c.

2. Ionic or Hypophrygian — Mode of the Fourth (5), c 204 d 182 e_1 112 f 204 g 182 a_1 134 b^1♭ 182 c.

3. Phrygian—Mode of the minor Seventh (2), c 182 d_1 134 e^1♭ 182 f 204 g 182 a_1 134 b^1♭ 182 c.

4. Eolic—Mode of the minor Third (Minor) (6), c 204 d 90 eb 204 f 204 g 90 ab 204 bb 204 c.

5. Doric—Mode of the minor Sixth (3), c 90 db 204 eb 204 f 204 g 90 ab 204 bb 204 c.

6. Mixolydian—Mode of the minor Second (7), c 112 d^1♭ 182 eb 204 f 112 g^1♭ 204 a^1♭ 182 bb 204 c.

7. Syntonolydian—Mode of the Fifth, not in

Ancient Greek Names	Scales beginning with c	Glarean's Ecclesiastical Names	Proposed new Names*
			Mode of the :—
1. Lydian	$c-d\ -e\ -f\ -g\ -a\ -b\ -c'$	Ionic	First (major)
2. Ionic or Hypophrygian .	$c-d\ -e\ -f\ -g\ -a\ -bb-c'$	Mixolydian	Fourth
3. Phrygian . . .	$c-d\ -eb-f\ -g\ -a\ -bb-c'$	Doric	minor Seventh
4. Eolic	$c-d\ -eb-f\ -g\ -ab-bb-c'$	Eolic	minor Third (minor)
5. Doric . . .	$c-db-eb-f\ -g\ -ab-bb-c'$	Phrygian	minor Sixth
6. ⎰ Mixolydian . . .	$c-db-eb-f\ -gb-ab-bb-c'$	⎰ Lydian ⎰	minor Second
7. ⎱ Syntonolydian . .	$c-d\ -e\ -f\sharp-g\ -a\ -b\ -c'$		Fifth

To assist the reader I have added the names assigned to the ecclesiastical modes by Glarean, which were wrongly distributed among the scales owing to his confusing the older tonal modes with the later (transposed) minor Greek scales, ¶ but which are more known among musicians than the proper Greek names. But I shall not use Glarean's names without expressly mentioning that they refer to an ecclesiastical mode. It would be really better to forget them altogether. The old numerical notation of Ambrose was much more suitable, but as his figures have been altered again and do not suffice for all modes, I have ventured to propose a new nomenclature in the above table, which will save the reader the trouble of memorising the systems of Greek names, of which Glarean's are certainly wrong, and the others are also perhaps not quite correctly applied. The principle of the new nomenclature is this. By 'the mode of Fourth of C,' is meant a mode of which C is the tonic, but which has the same *signature* (or additional $\sharp$ and $\flat$ signs) as the major scale formed on the Fourth of the diatonic scale beginning with C; that is on F. The minor Seventh, minor Third, minor Sixth, and minor Second must always be understood as the intervals intended in this case.† If the major intervals were selected the tonic would not occur in their scales. Thus ¶ 'the mode of the minor Third of C' is the scale with the tonic C, having the signature of $E\flat$ major (that is $B\flat$, $E\flat$, $A\flat$), because $E\flat$ is the minor Third of C; this is therefore C minor, at least as it is played in the descending scale. I hope the reader will have no difficulty in understanding what is meant by this notation.‡

This was the tonal system in the best times of Greek art, up to the Macedonian empire. Airs were at first limited to a tetrachord, as is still often the case in the Roman Catholic liturgy. They were afterwards extended to an Octave. Longer scales were not necessary for singing, as the Greeks refused to employ the straining upper notes, and unmetallic deep notes of the human voice. Modern Greek songs, of which Weitzmann has made a collection,§ have also a surprisingly small compass. If Phrynis (victor in the Panathenaic competitions, B.C. 457) added a ninth string to his cithara, the chief advantage of the arrangement was to allow of passing from one kind of scale to another.

The later Greek scale, which first occurs in Euclid's works of the third century ¶ B.C., embraces two Octaves, thus arranged :

the former table under this name, but really the Hypolydian (4), c 204 d 182 e_1 204 $f_1\sharp$ 112 g 182 a_1 204 b_1 112 c
Refer to the Duodenarium, App. XX. sect. E. art. 18.—*Translator.*]

* [If we subtract each of the numbers in the names of the modes here proposed, from 9, (reckoning 1 as 8 its Octave,) we obtain the numbers on p. 267c, which shew the number of the note in the major scale determined by the signature, on which the special scale begins. Thus as 9 less 7 is 2, the mode of the minor Seventh is that numbered 2 on pp. 267c, 268c. If we call the major scale, when reduced to a harmonisable form, 1. *do*, 2. *re*, 3. *mi*, 4. *fa*, 5. *so*, 6. *la*, 7 *ti*, then these transformed modes

may be called with the Tonic Sol-faists the *do*, *re*, *mi*, &c., modes respectively.—*Translator.*]

† [The qualification *minor* will therefore be always used in this translation, and has been inserted in the above table.—*Translator.*]

‡ [In App. XX. sect. E. No. 10, I have endeavoured to deduce scales for harmonic use, from a general theory of harmony which determines the precise value of each tone as part of a chord, and I have given precise names for them, there exemplified. This harmonic deduction of scales is quite independent of the historical melodic deduction in the text. —*Translator.*]

§ *Geschichte der Griechischen Musik*, Berlin, 1855.

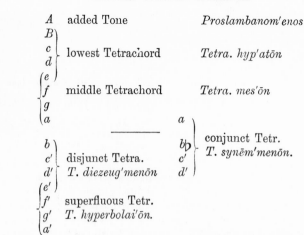

This scheme gives first the Hypodoric [Eolic, or Locrian] scale* for two Octaves, and then an added tetrachord which introduces a $b\flat$ in addition to the b, and thus, in modern language, allows of modulation from the principal scale into that of the subdominant.†

This scale, essentially of a minor character, was transposed, and thus a new series of scales were generated that correspond with the (descending) minor scales of modern music. To these were applied the old names of the tonal modes, by giving originally to each minor mode the name belonging to that tonal mode which was formed by the section of the minor scale which lay between the extreme tones of the Hypodoric ‡ scale. According to the Greek method of representing the notes, these extreme tones would have to be written $f \ldots f$. Their actual pitch was probably a Third lower. Thus the minor scale of D was called Lydian, because in this scale—

$$d-e- \ | \ f-g-a-b\flat-c--d-e-f \ | \ g-a-b\flat-c-d$$

the section of the scale lying between the extreme tones f and f belonged to the Lydian tonal mode. In this way the old names of the *tonal modes* altered their meaning into those of *tonal keys*. The following table shews the correspondence of these names :—

1) Hypo-doric	$= F$ minor		8) Phrygian	$= C$ minor	
2) Hypo-ionic	$= F\sharp$ minor		9) Eolic	$= C\sharp$ minor	
(deeper Hypo-phrygian)			(deeper Lydian)		
3) Hypo-phrygian	$= G$ minor		10) Lydian	$= D$ minor	
4) Hypo-eolic	$= G\sharp$ minor		11) Hyper-doric	$= E\flat$ minor	
(deeper Hypo-lydian)			(Mixo-lydian)		
5) Hypo-lydian	$= A$ minor		12) Hyper-ionic	$= E$ minor	
			(higher Mixo-lydian)		
6) Doric	$= B\flat$ minor		13) Hyper-phrygian	$= f$ minor	later additions
			(Hyper-mixo-lydian)		
7) Ionic	$= B$ minor		14) Hyper-eolic	$= f\sharp$ minor	
(deeper Phrygian)			15) Hyper-lydian	$= g$ minor	

Within each of these scales each of the previously mentioned tonal modes might be formed, by using the corresponding part of the scale. Besides this it was possible to pass into the conjunct tetrachord and thus modulate into the tonal key of the subdominant.

The experiments on transposition which formed the basis of these scales

---

* [See No. 6, of p. 267c, text, assuming Pythagorean intonation.—*Translator*.]

† Singularly enough this species of musical scale has been preserved in the Zillerthal in Tyrol, for the wood-harmonicon. This scale has two rows of bars. One forms a regular diatonic scale with the disjunct tetrachord. The other, which lies deep, has the conjunct tetrachord in its upper part.

‡ [This seems to be an error for Hypo-lydian, No. 4 of p. 267c, of which the extreme tones are f and f.—*Translator*.]

shewed that the Octave might be considered as composed approximatively of twelve Semitones. Even Aristoxenus knew that by taking a series of twelve Fifths we reached a tone that was at least very near to a higher Octave of the initial tone. Thus in the series

$$f-c-g-d-a-e-b-f\sharp-c\sharp-g\sharp-d\sharp-a\sharp-e\sharp$$

he identified $e\sharp$ with f, and by thus closing the series of tones he obtained a cycle of Fifths. Mathematicians denied the fact, and with reason, because if the Fifths are taken perfectly true, $e\sharp$ is a little sharper than f. For practical purposes, however, the error was quite insensible, and might be justly neglected in homophonic music in particular.*

This closes the development of the Greek tonal system. Complete as is our acquaintance with its outward form, we know but little of its real nature, because the examples of melodies which we possess are not only few in number, but very ¶ doubtful in origin.

Whatever may have been the nature of tonality in Greek scales, and however numerous may be the questions about it that are still unresolved, yet so far as the theory of the general historical development of tonal modes is concerned we learn all we want from the laws of the earliest Christian ecclesiastical music, which at its commencement touched upon the ancient construction as it died out. In the fourth century of our era, Bishop Ambrose, of Milan, established four scales for ecclesiastical song, which in the untransposed diatonic scale were:

First mode : $d-e-f-g-a-b-c-d$ mode of the minor Seventh.
Second mode : $e-f-g-a-b-c-d-e$ mode of the minor Sixth.
Third mode : $f-g-a-b-c-d-e-f$ mode of the Fifth (unmelodic).
Fourth mode : $g-a-b-c-d-e-f-g$ mode of the Fourth.

The variable character of the tone b, which was transmutable into $b\flat$ in the ¶ later Greek scales, remained, and produced the following scales :—

First : $d-e-f-g-a-b\flat-c-d$ mode of the minor Third.
Second : $e-f-g-a-b\flat-c-d-e$ { mode of the minor Second (unmelodic).
Third : $f-g-a-b\flat-c-d-e-f$ mode of the First (*major*).
Fourth : $g-a-b\flat-c-d-e-f-g$ mode of the minor Seventh.

There can be no doubt that these Ambrosian scales are to be regarded as essential (see p. 267b), for the old rule is that melodies in the first are to end in d, those in the second in e, those in the third in f, and those in the fourth in g, and this marks the initial tones of the scale as tonics. We may certainly assume that this arrangement was made by Ambrose for his choristers as a practical simplification of the old musical theory, which was overloaded with an inconsistent nomenclature. And this leads us to conclude that we were right in conjecturing that the ¶ similar older Greek scales could have been really used as different essential scales.

Pope Gregory the Great inserted between the Ambrosian essential scales the same number of accidental scales (p. 267a), called *plagal*, proceeding from the Fifth to the Fifth of the tonic. The Ambrosian scales were, then, called *authentic*

---

* It is by no means an unimportant fact, for our appreciation of the Greek scale, that a flute was found in the royal tombs at Thebes in Egypt (now in the Florentine Museum, No. 2688), which, according to M. Fétis, who examined it, gave an almost perfect scale of Semitones for about an Octave and a half; namely,

Series of primes, a $b\flat$ b c' $c'\sharp$ d'
First upper partial tones, a' $b'\flat$ b' c'' $c''\sharp$ d''
Second upper partial tones, e'' f'' $f''\sharp$ g'' $g''\sharp$ a''
Third upper partial tones, a'' $b''\flat$ b'' c''' $c'''\sharp$ d''

Representations of such flutes are found in the very oldest Egyptian monuments. They are very long, the holes are all near the end, and hence the arms must have been greatly stretched, giving the player a characteristic position. The Greeks can scarcely have been ignorant of this scale of Semitones. That it was not introduced into their theory till after the time of Alexander, clearly shows the preference they gave to the diatonic scale. [M. Fétis's deductions must be treated with much caution.—*Translator*.]

for distinction. The existence of these plagal ecclesiastical scales helped to increase the confusion which broke over the ecclesiastical scales towards the end of the middle ages, as composers began to neglect the rules which fixed the terminal tones, and this confusion assisted in favouring a freer development of the tonal system. This confusion also shewed, as we remarked in the last chapter (p. 243b), that no feeling for the thorough predominance of the tonic was much developed in the middle ages. But a step, at least, was made in advance of the Greeks, by recognising as a rule that the piece should close on the tonic, although this rule was not always observed.

Glarean endeavoured in 1547 to reduce the theory of the scales to order again, in his *Dodecachordon*. He shewed by an examination of the musical compositions of his contemporaries, that *six*, and not *four*, authentic scales should be distinguished, and adorned them with the Greek names in the table on p. 269a. Then he assumed
¶ six plagal scales, and hence on the whole distinguished twelve modes, whence the name of his book. Hence down to the sixteenth century essential and accidental scales were reckoned as parts of one series. Among Glarean's scales one is unmelodic, namely the mode of the Fifth, which he calls the Lydian. There are no examples of these to be found, as we know from a careful examination of medieval compositions made by Winterfeld,* and this confirms Plato's opinion of the Mixolydian and Hypolydian modes.

Hence there remain the following five melodic tonal modes applicable strictly for homophonic and polyphonic vocal music, namely :

	In our Nomenclature	Ancient Greek	Glarean's Names	Scale
1	Major Mode	Lydian	Ionic	$C-c$
2	Mode of the Fourth . .	Ionic	Mixolydian	$G-g$
3	Mode of the minor Seventh .	Phrygian	Doric	$D-d$
¶ 4	Mode of the minor Third .	Eolic	Eolic	$A-a$
5	Mode of the minor Sixth .	Doric	Phrygian	$E-e$

The *rational construction* of these scales when extended to the Octave or beyond the Octave results from the principle of tonal relationship already explained.† The limits of the extent to which tones related in the first degree should be used, are determined by the necessity of avoiding intervals too close to be distinguished with certainty. The larger gaps thus left have to be filled with the tones most nearly related in the second degree.

The Chinese and Gaels made the whole Tone $\frac{10}{9}$ [$=182$ cents] the smallest interval.‡ The Orientals, as we have seen, still retain Quartertones. The Greeks experimented with them, but soon gave them up and kept to the Semitone $\frac{16}{15}$ [$=112$ cents] as the smallest.

European nations have followed Greek habits, and retained the Semitone $\frac{16}{15}$ as
¶ the limit. The interval between $E\flat$ ($\frac{6}{5}$) [$=316$ cents] and E ($\frac{5}{4}$) [$=386$ cents], and between $A\flat$ ($\frac{8}{5}$) [$=814$ cents] and A ($\frac{5}{3}$) [$=884$ cents], in the natural scale is smaller, being $\frac{25}{24}$ [$=70$ cents], and we consequently avoid using both $E\flat$ and E, or both $A\flat$ and A in the same scale. We thus obtain the following two series of intervals between the most nearly related tones for ascending and descending scales :

Ascending : c — — e — f — g — a — — c'
  $\frac{5}{4}$ $\frac{16}{15}$ $\frac{9}{8}$ $\frac{10}{9}$ $\frac{6}{5}$

Descending : c — — $A\flat$ — G — F — $E\flat$ — — C
  $\frac{5}{4}$ $\frac{16}{15}$ $\frac{9}{8}$ $\frac{10}{9}$ $\frac{6}{5}$

* von Winterfeld's *Johannes Gabrieli und sein Zeitalter*, Berlin, 1834, vol. i. pp. 73 to 108.

† [The following is not an attempt to *restore* the Greek originals, which have already been treated, but to form harmonic scales on the same, and these are obtained by another process in App. XX. sect. E. art. 9.—*Translator*.]

‡ [I have found much smaller intervals in Chinese instruments. See App. XX. sect. K.—*Translator*.]

The numbers below the series shew the intervals between the two tones between which they are placed.*

It is at once seen that the intervals from and to the tonic are too large, and might be further divided. But as we have come to the limits of relationship in the first degree, we have to fill these gaps by tones related in the second degree.

The closest relationship in the second degree is necessarily furnished by the tones most nearly related to the tonic. Among these the Octave stands first. The tones related to the Octave of the tonic are of course the same as those related to the tonic itself; but by passing to the Octave of the tonic we obtain the descending in place of the ascending scale, and conversely.

Thus, ascending from c we found the following degrees of our major scale—

$$c \; -- \; e \; - f \; - g \; - a \; -- \; c'$$

¶

But taking the tones related to c', we obtain—

$$c \; -- \; e\flat \; - f \; - g \; - a\flat \; -- \; c'$$

Hence the second degree of relationship to the tonic gives an ascending minor scale. In this scale $e\flat$ is given as the major Sixth below c'. But it has also the weak relationship to c marked by $5 : 6$. Now we found that the sixth partial of a compound tone was clearly audible in many qualities of tone for which the seventh or eighth could not be heard; for example, on the pianoforte, the narrower organ pipes, and the mixture stops of the organ. Hence the relationship expressed by $5 : 6$ may often become evident as a natural relationship in the first degree. This, however, could scarcely be the case for the relationship $c - a\flat$ or $5 : 8$. Hence it is more natural to change e into $e\flat$ than a into $a\flat$ in the ascending scale. The latter, $a\flat$, can only be related to the tonic in the second degree. The three ascending scales in order of intelligibility are, therefore—†

¶

$$c \; -- \; e \; \; - f \; - g \; - a \; \; -- \; c'$$
$$c \; -- \; e\flat \; - f \; - g \; - a \; \; -- \; c'$$
$$c \; -- \; e\flat \; - f \; - g \; - a\flat \; -- \; c'$$

These distinctions based on a relationship in the second degree, through the medium of the Octave, are certainly very slight, but they make themselves felt in the well-known transformation of the ascending minor scale, to which these distinctions clearly refer.

Descending from c, instead of the relations in the first degree, given in

$$c \; -- \; A\flat \; - G \; - F \; - E\flat \; -- \; C$$

we may assume relations in the second degree, that is of the deeper C, and obtain

$$c \; -- \; A \; - G \; - F \; - E \; -- \; C$$

¶

In the latter, A is connected with the initial tone by the distant relationship in the first degree, $5 : 6$, and E only by a relationship in the second degree. Hence the third descending scale

$$c \; -- \; A \; - G \; - F \; - E\flat \; -- \; C$$

which we also found as an ascending scale. For descending scales we have therefore the following series.‡

* [With the subsequent notation and intervals expressed in cents:

c 386 e_1 112 f 204 g 182 a_1 316 c'
c 386 A^1b 112 G 204 F 182 E^1b 316 C
Translator.]

† [With the subsequent notation and intervals in cents:

c 386 e_1 112 f 204 g 182 a_1 316 c'
c 316 e^1b182 f 204 g 182 a_1 316 c'
c 316 e^1b182 f 204 g 112 a^1b386 c'
Translator.]

‡ [These are the same three scales as in the last note, read backwards.—*Translator.*]

$$c - - A\flat - G - F - E\flat - - C$$
$$c - - A - G - F - E\flat - - C$$
$$c - - A - G - F - E - - C$$

Generally, since all Octaves of the tonic, distant or near, higher or lower, are so closely related that they can be almost identified with it, all higher and lower Octaves of the individual degrees of the scale are almost as closely related to the tonic, as those of the next adjacent tonic of the same name.

Next to the relations of the Octave c' of c, follow those of g, the Fifth above, and F the Fifth below c. We must therefore proceed to study their effect in the construction of the scale. Let us begin with the relations of g, the Fifth above the tonic.*

ASCENDING SCALES.

¶

Related to c : $c - - e - f - g - a - - c'$
Related to g : $c\ d\ e\flat - - g - - b - c'$

Uniting the two, we have—

1) *The Major Scale* (Lydian mode of the ancient Greeks) :

$$c - d - e - f - g - a - b - c'$$
$$\text{I} \quad \tfrac{9}{8} \quad \tfrac{5}{4} \quad \tfrac{4}{3} \quad \tfrac{3}{2} \quad \tfrac{5}{3} \quad \tfrac{15}{8} \quad 2$$

The change of e into $e\flat$ is here facilitated by its second relationship to g. This gives—

2) *The Ascending Minor Scale :*

$$c - d - e\flat - f - g - a - b - c'$$
$$\text{I} \quad \tfrac{9}{8} \quad \tfrac{6}{5} \quad \tfrac{4}{3} \quad \tfrac{3}{2} \quad \tfrac{5}{3} \quad \tfrac{15}{8} \quad 2$$

DESCENDING SCALES.

¶

Related to c : $c - - - A\flat - G - F - E\flat - - - - C$
Related to g : $c\ B\flat - - - - G - - - - E\flat - D - C$

giving :—

3) *The Descending Minor Scale* (Hypodoric or Eolic mode of the ancient Greeks—our mode of the minor Third) :

$$c - B\flat - A\flat - G - F - E\flat - D - C$$
$$2 \quad \tfrac{9}{5} \quad \tfrac{8}{5} \quad \tfrac{3}{2} \quad \tfrac{4}{3} \quad \tfrac{6}{5} \quad \tfrac{9}{8} \quad \text{I}$$

or in the mixed scale, changing $A\flat$ into A,

* [In the complete notation, and with intervals in cents, these scales are:

ASCENDING SCALES.

Related to c: c 386 e_1 112 f 204 g 182 a_1 316 c'
Related to g: c 204 d 112 $e^1\flat$ 386 g 386 b_1 112 c'

¶ 1) *Major Scale :*
c 204 d 182 e_1 112 f 204 g 182 a_1 204 b_1 112 c
This is quite the Greek Lydian, see p. 268d', note ‡, No. 1. It is I C ma.ma.ma. of App. XX. sect. E. art. 9, I.

2) *The Ascending Minor Scale :*
c 204 d 112 $e^1\flat$ 182 f 204 g 182 a_1 204 b_1 112 c'
This is I C ma.mi.ma. (*ibid.* III.).

DESCENDING SCALES.

Related to c: c 386 $A^1\flat$ 112 G 204 F 182 $E^1\flat$ 316 C
Related to g: c 182 $B^1\flat$ 316 G 386 $E^1\flat$ 112 D 204 C

3) *The Descending Minor Scale :*
c 182 $B^1\flat$ 204 $A^1\flat$ 112 G 204 F 182 $E^1\flat$ 112 D 204 C
This is not quite the Greek Eolic, see p. 268d', note ‡, No. 4. It is I C mi.mi.mi. (*ibid.* VIII.).

4) *Mode of the minor Seventh :*
c 182 $B^1\flat$ 134 A_1 182 G 204 F 182 $E^1\flat$ 112 D 204 C

This is different from the Greek Phrygian, p. 268d', note ‡, No. 3, in the two last intervals. It is I C ma.mi.mi. (*ibid.* VII.).

ASCENDING SCALES.

Related to c: c 386 e_1 112 f 204 g 182 a_1 316 c'
Related to F: c 182 d_1 316 f 386 a_1 112 $b\flat$ 204 c'

5) *Mode of the Fourth :*
c 182 d_1 204 e_1 112 f 204 g 182 a_1 112 $b\flat$ 204 c'
This is not quite Greek Ionic or Hypophrygian, p. 268d', note ‡, No. 2. It is 5 F ma.ma.ma. (*ibid* I.).

6) *New form of mode of the minor Seventh :*
c 182 d_1 134 $e^1\flat$ 182 f 204 g 182 a_1 112 $b\flat$ 204 c'
This is 5 F ma.ma.mi. (*ibid.* V.).

DESCENDING SCALES.

Related to c: c 386 $A^1\flat$ 112 G 204 F 182 $E^1\flat$ 316 C
Related to F: c 204 $B\flat$ 112 A_1 386 F 386 $D^1\flat$ 112 C

7) *Mode of the minor Sixth :*
c 204 $B\flat$ 182 $A^1\flat$ 112 G 204 F 182 $E^1\flat$ 204 $D^1\flat$ 112 C
This is not quite the Greek Doric, p. 268d', note ‡, No. 5. It is 5 F mi.mi.mi. (*ibid.* VIII.).
—*Translator.*]

4) *Mode of the minor Seventh* (ancient Greek Phrygian):

$$c - B\flat - A - G - F - E\flat - D - C$$
$$2 \quad \tfrac{9}{5} \quad \tfrac{5}{3} \quad \tfrac{3}{2} \quad \tfrac{4}{3} \quad \tfrac{6}{5} \quad \tfrac{9}{8} \quad 1$$

On examining the relations of F, the Fifth below the tonic c, the following scales result:

ASCENDING SCALES.

Related to c: $c - - - - e - f - g - a - - - - c'$
Related to F: $c - d - - - - f - - - - a - b\flat - c'$

This gives—

5) *The mode of the Fourth* (ancient Greek Hypophrygian or Ionic):

$$c - d - e - f - g - a - b\flat - c'$$
$$1 \quad \tfrac{10}{9} \quad \tfrac{5}{4} \quad \tfrac{4}{3} \quad \tfrac{3}{2} \quad \tfrac{5}{3} \quad \tfrac{16}{9} \quad 2$$

By changing e into $e\flat$, we again obtain—   ¶

6) *The mode of the minor Seventh*, but with a different determination of the intercalary tones d and $b\flat$, from those in No. 4:

$$c - d - e\flat - f - g - a - b\flat - c'$$
$$1 \quad \tfrac{10}{9} \quad \tfrac{6}{5} \quad \tfrac{4}{3} \quad \tfrac{3}{2} \quad \tfrac{5}{3} \quad \tfrac{16}{9} \quad 2$$

DESCENDING SCALES.

Related to c: $c - - - - - A\flat - G - F - E\flat - - - - C$
Related to F: $c - B\flat - A - - - - - F - - - D\flat - C$

giving:—

7) *The mode of the minor Sixth* (ancient Greek Doric):

$$c - B\flat - A\flat - G - F - E\flat - D\flat - C$$
$$2 \quad \tfrac{16}{9} \quad \tfrac{8}{5} \quad \tfrac{3}{2} \quad \tfrac{4}{3} \quad \tfrac{6}{5} \quad \tfrac{16}{15} \quad 1$$ ¶

In this way the melodic tonal modes of the ancient Greeks and Christian Church have all been rediscovered by a consistent method of derivation. As long as homophonic vocal music is alone considered, all these tonal modes are equally justified in their construction.

The scales have been given above in the order in which they are most naturally deduced. But, as we have seen, each of the three scales

$$c - - e - f - g - a - - - c'$$
$$c - - e\flat - f - g - a - - - c'$$
$$c - - e\flat - f - g - a\flat - - - c'$$

can be played either upwards or downwards, although the first is best suited to ascending and the last to descending progression, and hence the gaps of any one of them may be filled up with either the relations of F or the relations of g, or even one gap with those of F and the other with those of g.   ¶

The pitch numbers of the tones directly related to the tonic are of course fixed * and unchangeable, because they are given by the condition that the tones should form consonances with the tonic, and are thus more strictly determined than by any more distant connection. On the other hand, the intercalary tones related in the second degree are by no means so precisely fixed.

Taking $c = 1$, we have for the Second—

1) the d derived from $g = \tfrac{9}{8}$, [$= 204$ cents]
2) the d derived from $f = \tfrac{10}{9} = \tfrac{80}{81} \times \tfrac{9}{8}$, [$= 182$ cents]
3) the $d\flat$ derived from $f = \tfrac{16}{15}$, [$= 112$ cents]

---

* Thus I cannot agree with Hauptmann, in allowing a Pythagorean a, the Fifth above d, in the ascending minor scale of c. d'Alembert introduces the same tone even in the major scale, by passing from g to b through the fun- damental bass d. But this would indicate a distinct modulation into G major, which is not required when the natural relations of the tones to the tonic are preserved. See Haupt- mann, *Harmonik und Metrik*, p. 60.

and for the Seventh—

    1) the b derived from $g = \frac{15}{8}$, [=1088 cents]
    2) the $b\flat$ derived from $g = \frac{9}{5}$, [=1018 cents]
    3) the $b\flat$ derived from $f = \frac{16}{9} = \frac{80}{81} \times \frac{9}{5}$, [=996 cents]

Hence while b and $d\flat$ are given with certainty, $b\flat$ and d are uncertain. Either of them may be distant from the tonic by the major Tone $\frac{9}{8}$ [=204 cents] or the minor Tone $\frac{10}{9}$. [=182 cents].

In order henceforth to mark this difference of intonation with certainty and without ambiguity, we will introduce a method of distinguishing the tones determined by a progression of Fifths, from those given by the relationship of a Third to the tonic. We have already seen that these two methods of determining the tones lead to somewhat different pitches, and hence in accurate theoretical re-
¶ searches both kinds of tones must be kept distinct, although in modern music they are practically confused.

The idea of this notation belongs to Hauptmann, but as the capital and small letter which he uses, and which I also, in consequence, employed in the first edition of this book, have a different meaning in our method of writing tones, I now introduce a slight modification of his notation.

Let C be the initial tone, and write * its Fifth G, the Fifth of this Fifth D, and so on. In the same way let the Fourth of C be F, the Fourth of this Fourth $B\flat$ and so on. In this way we have a series of Tones, here written with simple capitals, all distant from each other by a perfect Fifth or a perfect Fourth : †

$$B\flat \pm F \pm C \pm G \pm D \pm A \pm E, \&c.$$

The pitch of every tone in the whole series is, therefore, known when that of any one is known.

¶   The major Third of C, on the other hand, will be expressed by E_1, that of F by A_1, and so on. Hence the series of tones

$$B\flat_1 + D_1 - F + A_1 - C + E_1 - G + B_1 - D + F_1\sharp - A, \&c.,$$

is a series of alternate major and minor Thirds. It is therefore clear that the Tones

$$D_1 \pm A_1 \pm E_1 \pm B_1 \pm F_1\sharp, \&c.,$$

also form a series of perfect Fifths.

We have already found that the tone D_1, that is the minor Third below or major Sixth above F, is lower in pitch than the tone D, which would be reached by a series of Fifths from F, and that the difference of pitch is that known as a *comma*, the numerical value of which is $\frac{81}{80}$, or musically about the tenth part of a whole Tone.‡ Since, then, $D \pm A$ and $D_1 \pm A_1$ are both perfect Fifths, A must be also a
¶ comma higher than A_1, and so also every letter with an inferior number, as 1, 2, 3, &c., attached to it, will represent a tone which is 1, 2, 3, &c., commas *lower* in pitch than

---

    * *Die Natur der Harmonik und Metrik*, Leipzig, 1853, pp. 26 and following. I cannot but join with C. E. Naumann in expressing my regret that so many delicate musical apperceptions as this work contains, should have been needlessly buried under the abstruse terminology of Hegelian dialectics, and hence have been rendered inaccessible to any large circle of readers.

    † [Prof. Helmholtz uses (−) between the letters in all such cases. I have taken the liberty from this place onwards, whenever a line or combination of Thirds occurs to leave (−) only in the just minor Thirds of 316 cents, to use (|) in the Pythagorean minor Thirds of 294 cents, as Prof. Helmholtz does subse-

quently, and change (−) into (+) for the major Third of 386 cents. In the case of Fifths which consist of a major and a minor Third $702 = 386 + 316$ cents, the symbol is properly ± which I here also take the liberty to use. For other intervals I shall use (...) for (−), and generally give the precise interval in cents elsewhere. I trust that this change will be found suggestive as well as convenient, and may therefore not be considered presumptuous. —*Translator.*]

    ‡ [The comma 81 : 80 is just over $21\frac{1}{2}$ cents, for which I use 22 cents, see App. XX. sect. A. art. 4, and sect. D. Hence a major tone of 204 cents contains about $9\frac{1}{2}$ commas.— *Translator.*]

that represented by the same letter with *no* inferior number attached, as is easily seen by carrying on the series.

A major triad will therefore be written thus:

$$C + E_1 - G$$

and a minor triad

$$A_1 - C + E_1 \text{ or } C_1 - E\flat + G_1$$

Now if we lay it down as a rule that as every inferior figure, 1, 2, 3, &c., *depresses* its tone by the 1, 2, 3, &c., comma, every superior figure, 1, 2, 3, &c., shall *raise* its tone by the same 1, 2, 3, &c., commas, we may write the major triad as

$$c + e_1 - g \text{ or } c^1 + e - g^1$$

and the minor triad as

$$c - e^1\flat + g \text{ or } c_1 - e\flat + g_1,$$

or even

$$c^1 - e^2\flat - g^1 \text{ or } c_2 - e_1\flat - g_2.^*\qquad\qquad ¶$$

The three series of Tones directly related to C are consequently to be written thus:

$$C - - E_1 \ - F - G - A_1 \ - - c$$
$$C - - E^1\flat - F - G - A_1 \ - - c$$
$$C - - E^1\flat - F - G - A^1\flat - - c$$

and the intercalary tones are—

Between the tonic and Third, D, D_1, or $D^1\flat$.

Between the Sixth and Octave, B_1 and $B\flat$ or $B^1\flat$.

Consequently the melodic tonal modes of the ancient Greeks and old Christian Church are,†

---

* In the 1st [German] edition of this book, as in Hauptmann's, the small letters were supposed to be a comma lower than the capital letters, and a stroke above or below the letters was only occasionally used for raising or depressing the pitch by two commas. Hence a major triad was written $C - e - G$ or $\bar{c} - E - \bar{g}$; a minor triad $a - C - e$, or $\underline{A} - c - \underline{E}$, &c. The notation used here [in the 3rd and 4th German and the 1st English editions] and also in the French translation is due to Herr A. v. Oettingen, and is much more readily comprehended. [Herr v. Oettingen's notation of lines above and below, which was at Prof. Helmholtz's request retained in the 1st English edition of this translation, was found extremely inconvenient for the printer, and actually delayed the work three months in passing through the press. I have now for some years employed the very easy substitute here introduced. By referring to the table called the Duodēnārium, in App. XX. sect. E. art. 18, where this new notation is systematically carried out for 117 notes, the whole bearing of it will be better appreciated. Another notation which I had used formerly, and into which I translated Herr v. Oettingen's in the footnotes to the 1st edition of this translation, and employed in Table IV., there corresponding to my present Duodenarium, is consequently abandoned, and is now only mentioned to account for the difference in notation between the two editions of this translation. The spirit of Herr v. Oettingen's notation is therefore retained, while its use has been rendered typographically convenient.—*Translator.*]

† [This variation of the intercalary tones really amounts to a change of mode, so that

the names used in the text become ambiguous. ¶ This difficulty is overcome by the trichordal notation proposed in App. XX. sect. E. art. 9.

1) The major mode of C with D, has the 3 major chords $F + A_1 - C$, $C + E_1 - G$, $G + B_1 - D$ and is 1 C ma.ma.ma. But with D_1 in place of D, it has the 3 minor chords $D_1 - F + A_1$, $A_1 - C + E_1$, $E_1 - G + B_1$ (of which the two last belong also to the first form), and is therefore 3 A_1 mi.mi.mi. This is a related, but very different, mode.

2) The mode of the Fourth, as it stands in the first line, is not trichordal, but by using D and $B^1\flat$ it has the 3 chords $F + A_1 - C$, $C + E_1 - G$, $G - B^1\flat + D$, and is hence 1 C ma.ma.mi. If we take D_1 and $B\flat$ it has the 3 chords $B\flat + D_1 - F$, $F + A_1 - C$, $C + E_1 - G$, and is hence 5 F ma.ma.ma. With both D_1 and $B^1\flat$ it is again not trichordal.

3) The mode of the minor Seventh. If we ¶ take the upper line as it stands, this is also not trichordal. But if we use D and $B^1\flat$, it has the 3 chords $F + A_1 - C$, $C - E^1\flat + G$, $G - B^1\flat + D$, and is hence 1 C ma.mi.mi. If we take D_1 and $B\flat$, the 3 chords are $B\flat + D_1 - F$, $F + A_1 - C$, $C - E^1\flat + G$, and the scale is 5 F ma.ma.mi. With D_1 and $B^1\flat$ the scale is again not trichordal.

4) Mode of the minor Third. The first line as it stands is not trichordal. Taking D and $B^1\flat$ the 3 chords are $F - A^1\flat + C$, $C - E^1\flat + G$, $G - B^1\flat + D$, and the scale is 1 C mi.mi.mi. Taking D_1 and $B\flat$ the 3 chords are $B\flat + D_1 - F$, $F - A^1\flat + C$, $C - E^1\flat + G$, and the scale is 5 F ma.mi.mi. With D_1 and $B^1\flat$ again the scale is not trichordal.

5) Mode of the minor Sixth. The first line as it stands gives the 3 chords $B\flat - D^1\flat + F$, $F - A^1\flat + C$, $C - E^1\flat + G$, and the scale is

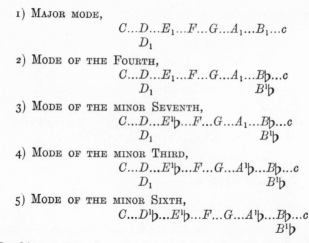

1) MAJOR MODE,
$$C...D...E_1...F...G...A_1...B_1...c$$
$$D_1$$

2) MODE OF THE FOURTH,
$$C...D...E_1...F...G...A_1...B\flat...c$$
$$D_1 \qquad\qquad B^1\flat$$

3) MODE OF THE MINOR SEVENTH,
$$C...D...E^1\flat...F...G...A_1...B\flat...c$$
$$D_1 \qquad\qquad B^1\flat$$

4) MODE OF THE MINOR THIRD,
$$C...D...E^1\flat...F...G...A^1\flat...B\flat...c$$
$$D_1 \qquad\qquad B^1\flat$$

¶ 5) MODE OF THE MINOR SIXTH,
$$C...D^1\flat...E^1\flat...F...G...A^1\flat...B\flat...c$$
$$B^1\flat$$

By this notation, then, the intonation is always exactly expressed, and the kind of consonance which each tone makes with the tonic, or the tones related to it is clearly shewn.

In the ancient Greek Pythagorean intonation these scales would have to be written:

Major mode— $C...D...E...F...G...A...B...C.$

and the others in a similar manner, all with letters of the same kind, belonging to the same series of Fifths.*

In the formulae here given for the diatonic tonal modes, the intonation of the Second and Seventh is partly undetermined. In these cases I have given D the
¶ preference over D_1, and $B\flat$ the preference over $B^1\flat$, because the relationship of the Fifth is closer than that of the Third; but $B\flat$ and D stand in the relation of the Fifth respectively to F, G, the tones nearest related to the tonic, while D_1 and $B^1\flat$ are only in the relation of the minor Third to F and G. But this reason is certainly not sufficient entirely to exclude the two last tones in homophonic vocal music. For if in a melodic phrase, the Second of the tonic came into close connection with tones related to F—for instance, if it fell between F and A_1, or followed them—an accurate singer would certainly find it more natural to use the D_1, which is directly related to F and A_1, than the D which is related to them only in the third degree. The slightly closer relationship of the latter to the tonic could scarcely give the decision in its favour in such a case.

This ambiguity in the intercalary tones cannot, I think, be considered as a fault in the tonal system, since in our modern minor mode, the Sixth and Seventh of the tonic are often altered, not merely by a comma, but by a whole Semitone,
¶ according to the direction of the melodic progression. We shall find, however, more decisive reasons for the use of D in place of D_1 in the next chapter, when we pass from homophonic music to the influence of harmonic music on the scales.

The account here given of the rational construction of scales and the corresponding intonation of intervals, deviates essentially from that given to the Greeks by Pythagoras, which has thence descended to the latest musical theories, and even now serves as the basis of our system of musical notation. Pythagoras constructed the whole diatonic scale from the following series of Fifths :—

$$F \pm C \pm G \pm D \pm A \pm E \pm B,$$

5 F mi.mi.mi. If we use $B^1\flat$ in place of $B\flat$, the 3 chords are $D^1\flat + F - A^1\flat$, $A^1\flat + C - E^1\flat$, $E^1\flat + G - B^1\flat$, and the scale is 3 $A^1\flat$ mi.mi.mi.

The modes formed by taking one intercalary tone or the other are therefore quite distinct, though purposely confused in the nomenclature of the text, apparently as an accommodation to the usual tempered notation.—*Translator.*]

* [In this case the intonation becomes altogether different. *Translator.*]

and calculated the intervals from it as they have been given above. In his diatonic scale there are but two kinds of small intervals, the *whole Tone* $\frac{9}{8}$, [= 204 cents] and the *Limma* $\frac{256}{243}$, [= 90 cents].*

In this series if C be taken as the tonic, A would be related to the tonic in the Third degree, E in the Fourth, and B in the Fifth. Such a relationship would be absolutely insensible to any ear that has no guide but direct sensation.

A series of Fifths may certainly be tuned on any instrument, and continued as far as we please; but neither singer nor hearer could possibly discover in passing from *c* to *e* that the latter is the fourth from the former in the series of Fifths. Even in a relation of the second degree through Fifths, as of *c* to *d*, it is doubtful whether a hearer can discover the relation of the tones. But in this case when we pass from one tone to the other we can imagine the insertion of ' a silent *g*,' so to speak, forming the Fourth below *c*, and the Fifth below *d*, and thus establish a connection, for the mind's ear at least, if not for the body's. This is probably the meaning to ¶ be attached to Rameau's and d'Alembert's explanation that a singer effects the passage from *c* to *d* by means of the fundamental bass *G*. If the singer does not hear the bass note *G* at the same time as *d*, he cannot possibly bring his *d* into consonance with that *G*; but the melodic progression may certainly be facilitated by conceiving the existence of such a tone. This is a well-known means for striking the more difficult intervals, and is often applied with advantage. But of course it completely fails when the transition has to be made between tones widely separated in the series of Fifths.†

* [The fact that the Greek scale was derived from the tetrachord, or divisions of the Fourth, and *not* the Fifth, leads me to suppose that the tuning was founded on the Fourth and not the Fifth. On proceeding *up*wards from *C* by Fourths, we get *C F B♭ E♭ A♭ D♭ G♭ C♭ F♭ B♭♭ E♭♭ A♭♭ D♭♭*, and on proceeding *down*wards we get *C G D A E*. Now these notes after *G♭* in the first series, are precisely those of Abdulqadir, written as ʼ*B₁* ʼ*E₁* ʼ*A₁* ʼ*D₁* *G₁* ʼ*C₁* on p. 282*b*, according to the notation explained on p. 281*b*, note *. Of course the Arabic lute, tuned in Fourths, naturally led to this. It is most convenient for modern habits of thought to consider the series as one of Fifths. But I wish to draw attention to the fact that in all probability it was historically a series of Fourths.—*Translator*.]

† [One of the practical results of the Tonic Solfa system of teaching to sing the diatonic major scale as marked on p. 274*b*, No. 1, in just intonation, (see App. XVIII.,) has been the discovery that it is not so easy to learn to strike the proper tone by a knowledge of the *interval* between two adjacent tones in a melodic passage, as by a knowledge of the *mental effect* produced by each tone of the scale in relation to the tonic. These mental effects are perhaps not very clearly characterised by the mere names given to them in the Tonic Solfa books, but the teacher soon makes his class understand them, and then finds them the most valuable instrument which he possesses for inspiring a feeling for just intonation. On these characters of each tone in the (just) diatonic scale, a system of manual signs has been formed, by which classes are constantly led. Particulars are given in ' *The Standard Course of Lessons and Exercises in the Tonic Solfa Method of Teaching Music*, with additional exercises, by John Curwen, new edition, re-written, A.D. 1872.' But it may be convenient to mention in this place the characters and manual signs there given (*ib.* p. iv.).

I. First step.
Do, Tonic, ' the STRONG or *firm* tone,' fist closed, horizontal, thumb down.
So, Fifth, ' the GRAND or *bright* tone,' the fingers extended and horizontal, hand with little finger below and thumb above, so that the palm of the hand is vertical. ¶
Mi, Major Third, ' the STEADY or *calm* tone,' fingers extended and horizontal, palm of hand horizontal and undermost.

II. Second step.
Re, Second, ' the ROUSING or *hopeful* tone,' fingers extended, hand forming half a right angle with ground pointing upwards, palm downwards.
Ti, Seventh, ' the PIERCING or *sensitive* tone,' only the forefinger extended and pointing up, the other fingers and thumb closed, hand forming half a right angle with ground, *back of hand* downwards.

III. Third step.
Fa, Fourth, ' the DESOLATE or *awe-inspiring* tone,' only the forefinger extended and pointing down, at half a right angle with the ground, the back of hand upwards.
La, major Sixth, ' the SAD or *weeping* tone,' ¶ fingers fully extended, whole hand pointing down with a weak fall, back of hand upwards.

It is thus seen that the order of teaching takes the tonic chord first, then the dominant, and lastly the subdominant. The doubtful Second thus comes early on. ' The teacher first sings the exercise to [the names of] consecutive figures, telling his pupils that he is about to introduce a new tone (that is one *not* DO, MI, or SO), and asking them to tell him on which figure it falls. When they have distinguished the new tone, he sings the exercise again—laa-ing it [this is calling each note *la*] —and asks them to tell him how that tone " makes them feel." Those who can *describe* the feeling hold up their hands, and the teacher asks one for a description. But others, who are not satisfied with words, may also perceive and feel. The teacher can tell by

Finally there is no perceptible reason in the series of Fifths why they should not be carried further, after the gaps in the diatonic scale have been supplied. Why do we not go on till we reach the chromatic scale of Semitones? To what purpose do we conclude our diatonic scale with the following singularly unequal arrangement of intervals—

$$1, \ 1, \ \tfrac{1}{2}, \ 1, \ 1, \ 1, \ \tfrac{1}{2}$$

The new tones introduced by continuing the series of Fifths would lead to no closer intervals than those which already exist. The old scale of five tones appears to have avoided Semitones as being too close. But when two such intervals already appear in the scale, why not introduce more?

¶ The *Arabic and Persian musical system*, so far as its nature is shewn in the writings of the older theorists, also knew no method of tuning but by Fifths. But this system, which seems to have developed its peculiarities in the Persian dynasty of the Sassanides (A.D. 226–651) before the Arabian conquest, shews an essential advance on the Pythagorean system of Fifths.

In order to judge of this system of music, which has been hitherto completely misunderstood, the following relation has to be known. By tuning four Fifths upwards from C

$$C \pm G \pm D \pm A \pm E$$

we come to a tone, E, which is $\tfrac{81}{80}$ or a comma higher than the natural major Third of C, which we write E_1. The former E forms the major Third in the Pythagorean scale. But if we tune eight Fifths downwards from C, thus—

$$C \pm F \pm B\flat \pm E\flat \pm A\flat \pm D\flat \pm G\flat \pm C\flat \pm F\flat$$

¶ we come to a tone, $F\flat$, which is almost exactly the same as the natural E_1. The interval of C to $F\flat$ is expressed by

$$\tfrac{8192}{6561} = \tfrac{5}{4} \times \tfrac{32768}{32805}, \text{ or nearly } \tfrac{5}{4} \times \tfrac{885 \cdot 6}{886 \cdot 6} \ [=384 \text{ cents}].$$

their eyes whether they have done so. He multiplies examples until *all* the class have their attention fully awakened to the effect of the new tone. This done he tells his pupils the Sol-fa name and the manual sign for the new tone, and guides them by the signs to Sol-fa the exercise and themselves produce the proper effect. The signs are better in this case than the notation, because with them the teacher can best command the attention of every eye and ear and voice, and at the first introduction of a tone, attention should be acute' (*ibid.* p. 15). This passage, the result of practice with hundreds of thousands of children, shews that a totally new ¶ principle of understanding the relation of the tones in a scale to the tonic has not only been introduced, but worked out on a large scale practically, and, as I myself know, successfully. See Prof. Helmholtz's own impression of the success, as long ago as 1864, in App. XVIII. Since that time great experience has been gained and many methods improved. But the object of introducing this notice here is to shew that *proper training* (such as the ancient Greeks certainly had) could produce the corresponding *feeling* for the effect of any tone in any scale anyhow divided, independently of the relation of consonances, and that this consideration may help to explain the persistence of many scales which are harmonically inexplicable. No doubt Pythagorean singers hit the degrees of their scale quite correctly, and no doubt the 'mental effects' of their A, E, B, were very different from those of the harmonisable A_1,

E_1, B_1. We can partially judge of them by the effects of equal temperament, which melodically cannot differ much, although they certainly differ sensibly, from those of Pythagorean intonation. And it must be remembered that singers actually learn to sing in equal temperament, in which all major Thirds are 14 cents too sharp, and then find just major Thirds intolerably flat! To this I would add the following anecdote quoted from Fétis (*Hist. Générale de la Musique*, vol. ii. p. 27) by Prof. Land (*Gamme Arabe*, p. 19 footnote), containing 'a fact,' as he says, ' which could not be believed, if it were not attested by the person whom it concerns. The celebrated organist M. Lemmens, who was born in a village of Campine [or Kempenland, a district in the Belgian province of Limbourg, 51°15′ N. lat., 5°20′ E. long.], studied music in early youth upon a clavecin (harpsichord), which had been long dreadfully out of tune, because no tuner existed in the district. Fortunately, an organ-builder was summoned to repair the organ at the abbey of Everbode near that village. By chance he called upon the young musician's father, and heard the boy play on his miserable instrument. Shocked at the multitude of false notes which struck his ear, he immediately determined to tune the clavecin. When he had done so, M. Lemmens experienced the most disagreeable sensations, and it was some time before he could habituate his ear to the correct intervals, having been so long misled by different relations.' Hence, false intervals may seem natural.—*Translator.*]

Hence the tone Fb is lower than the natural major Third E_1 [$=386$ cents] by the extremely small interval $\frac{887}{886}$ [$=2$ cents], which is about the eleventh part of a comma [$=22$ cents]. This interval between Fb and E_1 is practically scarcely perceptible, or at most only perceptible by the extremely slow beats produced by the chord $C...Fb...G$ [$=C$ 384 Fb 318 G] upon an instrument most exactly tuned. Practically, then, we may without hesitation assume that the two tones Fb and E_1 are identical, and of course that their Fifths are also identical, or

$$Fb = E_1, \quad Cb = B_1, \quad Gb = F_1\sharp, \quad \&c.^*$$

Now in the Arabic and Persian scale the Octave is divided into 16 intervals, but in our equal temperament it is divided into 6 whole Tones. Modern [European] interpreters of the Arabic and Persian system of music have hence been misled into the conclusion that each of the 17 degrees of the scale corresponded to about the third of a Tone in our music. In that case the intonation of the degrees in ¶ the Arabic and Persian scale would not be executable on our instruments. But in Kiesewetter's work on the Music of the Arabs,† which was written with the

---

* [On this substitution, which amounts to a temperament with perfect Fifths, and major Thirds too flat by a skhisma, or nearly the eleventh of a comma, and which I therefore call skhismic temperament, see Appendix XX. section A. art. 17. It is convenient to use a grave accent prefixed thus, 'E_1, to show flattening by a skhisma, and to read it as skhismic, thus, 'skhismic E one.' The above equations can therefore be made precise by writing $Fb =$'E_1, $Cb =$'B_1, $Gb =$'$F_1\sharp$, &c.—Translator.]

† R. G. Kiesewetter, Die Musik der Araber nach Originalquellen dargestellt, mit einem Vorworte von dem Freiherrn von Hammer-Purgstall. Leipzig, 1842, pp. 32, 33. The directions given in an anonymous manuscript of the 666th year of the Hegira, A.D. 1267, in the possession of Prof. Salisbury [of Yale Coll.], are essentially the same. See Journal of the American Oriental Society, vol. i. p. 174. [Since the publication of the 4th German edition of this work in 1877, the whole history of the Arabic scale has been reinvestigated from the original Arabic sources by Herr J. P. N. Land, D.D., Professor of Mental Philosophy at Leyden, an Oriental scholar and a musician, and the results were published first in Dutch as a paper in the Transactions of the Dutch Academy of Sciences, division Literature, 2nd series, vol. ix., and separately under the title of Over de Toonladders der Arabische Muziek (On the Scales of Arabic Music) in 1880, and secondly in French as a paper communicated to the International Congress of Orientalists at Leyden in 1882 and published in vol. ii. of their 'Transactions,' and also separately in 1884 as Recherches sur l'histoire de la Gamme Arabe. This paper supersedes in many respects the work of Kiesewetter and von Hammer-Purgstall, of whom the first was a musician but not an Orientalist, and the second an Orientalist but not a musician. Alfārābī's scale was produced by a succession of Fifths [or rather Fourths, see p. 42, note], but a century and a half previously Zalzal had introduced a new interval 22 : 27 = 355 cents, which Prof. Land terms a neutral Third. It is actually $\frac{12}{11} \times \frac{9}{8}$ or $151 + 204$ cents, that is, three quarters of a Tone sharper than a major tone, whereas the major Third is 182 cents or a minor Tone sharper, and the minor Third was only a diatonic Semitone 112 cents sharper. The interval 12 : 11 = 151 cents is the

well-known trumpet interval between the sharpened Fourth and Fifth, the 11th and 12th harmonics, as may be heard in the fifth Octave of the Harmonical $^{11}f''' : g'''$. This on the Arabic lute was necessarily accompanied by a similar interval on the next string, $498 + 355 = 853$ cents. These two notes eventually superseded the old Pythagorean minor Third of 294 cents and the Fourth above it of 792 cents; and seem entirely out of the reach of a succession of Fifths or Fourths. But it was the object of Abdulqadir and others to form a succession of Fifths (or rather Fourths) which would include these two intervals, at least approximately. ¶ This they accomplished within less than 30 cents by their 384 and 882 cents. It does not appear to have been Abdulqadir's object to approximate to the just major Third 386, and just major Sixth 884, but to get by means of Fifths or Fourths certain tones which would pass as Zalzal's. The list in the text (p. 282b) gives the seventeen tones thus produced with the intervals that they form with each other, and Prof. Helmholtz's names of the notes, completed by a grave accent. Here I re-arrange them in order of Fifths down or Fourths up, the approximate Thirds being added immediately to the right, and the numbers showing the interval in cents from C:

E	408		
A	906,	$Db =$'$C_1\sharp$	90
D	204,	$Gb =$'$F_1\sharp$	588
G	702,	$Cb =$'B_1	1086
C	0,	$Fb =$'E_1	384
F	498,	$Bbb =$'A_1	882
Bb	996,	$Ebb =$'D_1	180
Eb	294,	$Abb =$'G_1	678
Ab	792,	$Dbb =$'C_1	1176

Observe that the real major Third was the Pythagorean 408 cents, as the minor Third was the Pythagorean 294 cents. Also that 180 cents was within two cents of the minor tone 182 cents. But these approximations were probably not contemplated.

An English concertina, which has fourteen notes to the Octave, was tuned with thirteen consecutive Fifths from Gb to $C\sharp$, so that I was able to try the chords $ADbE$ $DGbA$, that is, A'$C_1\sharp$ E, D'$F_1\sharp$ A, where the major Thirds are two cents too flat, and compare them with the Pythagorean chords $AC\sharp E$,

assistance of the celebrated Orientalist von Hammer-Purgstall, there is given a translation of the directions for the division of the monochord laid down by Abdul Kadir, a celebrated Persian theorist of the fourteenth century of our era, that lived at the courts of Timur and Bajazet. These directions enable us to calculate the intonation of the Oriental scale with perfect certainty. These directions also agree in essentials both with those of the much older Farabi,* (who died in A.D. 950), and of his own contemporary, Mahmud Shirazi,† (who died in 1315), for dividing the fingerboard of lutes. According to the directions of Abdul Kadir all the tonal degrees of the Arabic scale are obtained by a series of 16 Fifths, and if we call the lowest degree C, and arrange them in order of pitch within the compass of an Octave, they will be the following, as expressed in our notation [with the addition of the grave accent explained in p. 281b, note *].

¶   1) C — 2) $D\flat$ — 3) ʻD_1 ‿ 4) D — 5) $E\flat$ — 6) ʻE_1‿
     7) E — 8) F — 9) $G\flat$ —10) ʻG_1‿11) G —12) $A\flat$—
     13) ʻA_1‿14) A —15) $B\flat$ —16) ʻB_1—17) ʻc_1 ‿18) c

where the line — between two tones indicates the interval of a Pythagorean Limma $\frac{256}{243}$ (which is nearly $\frac{20}{19}$ [=90 cents]), and the sign ‿ a Pythagorean comma [=531441 : 524288=24 cents]. The Limma is about $\frac{4}{5}$ and the Pythagorean comma a little more than $\frac{1}{5}$ and less than $\frac{2}{9}$ of the natural Semitone $\frac{16}{15}$ [=112 cents].

Abdul Kadir assigns the following intonation to the three first of the 12 principal tonal modes or *Makamat* :—

<table>
<tr><td>Arabic</td><td></td><td></td><td></td><td></td><td></td><td></td><td>Ancient Greek</td></tr>
<tr><td>1. Uschak</td><td>$C...D$</td><td>...E</td><td>...F...G</td><td>...A</td><td>...$B\flat$...c</td><td>Hypophrygian or Ionic.</td></tr>
<tr><td>2. Newa</td><td>$C...D$</td><td>...$E\flat$...F...G</td><td></td><td>...$A\flat$...$B\flat$...c</td><td></td><td>Hypodorian or Eolic.</td></tr>
<tr><td>3. Buselik</td><td>$C...D\flat$...$E\flat$...F...$G\flat$...$A\flat$...$B\flat$...c</td><td></td><td></td><td></td><td></td><td>Mixolydian, [all on p. 269a.]</td></tr>
</table>

These three are therefore completely identical with the ancient Greek scales in Pythagorean intonation.‡ Since the Arabic theoreticians divide these scales into the Fourth $C...F$ and the Fifth $F \pm c$, and since C, F and $B\flat$ are considered to be invariable tones, and the others to be variable, it is probable that F must be regarded as the tonic. In this case

     1. Uschak would be $= F$ major.
     2. Newa would be $=$ the mode of the minor Seventh of F.§
     3. Buselik would be $=$ the mode of the minor Sixth of F.

all three in Pythagorean intonation. The Persian school also considers the scales to be related.

$DF\sharp A$. The latter were offensive, the former ¶ indistinguishable from just. It seems remarkable, therefore, how with such a collection of notes the Arabs escaped harmonic music. But it will be seen on examining the scales formed from them (see especially p. 284d, note), that they were perfectly unadapted for harmony, which would have occasioned a perfect revolution in their musical systems.

There was certainly no attempt to divide the scale as Villoteau supposed into seventeen equal parts each of about 70·6 cents, for no such intervals occur, still less any third parts of a (tempered) tone of 66$\frac{2}{3}$ cents, which was a mere hallucination of Villoteau's.

This system of Abdulqadir prevailed from the thirteenth to the fifteenth century. The modern division into twenty-four Quartertones is noticed on p. 264b and note **.

The Arabs, however, had also entirely different scales for other instruments than

their classical lute, to which alone the above refers.—*Translator.*]

* J. G. L. Kosegarten, *Alii Ispahanensis Liber Cantilenarum*, pp. 76-86.

† Kiesewetter, *Die Musik der Araber nach Originalquellen darg.*, p. 33.

‡ [Not therefore according to the forms on p. 268d', note, but on the more recent Pythagorean imitation of those forms. They are respectively the representatives of scales 2, 4, and 6 of that note.—*Translator.*]

§ [In the German text, *Quartengeschlecht*, or the mode of the Fourth of F. The tones in the mode of the Fourth of F are those in the Pythagorean scale of $B\flat$, or, in order of Fifths, $E\flat \pm B\flat \pm F \pm C \pm G \pm D \pm A$, and the tones of the mode of the minor Seventh of F are those in the Pythagorean scale of $E\flat$, or, in order of Fifths, $A\flat \pm E\flat \pm B\flat \pm F \pm C \pm G \pm D$. The correction is therefore evident.—*Translator.*]

The next group consists of five tonal modes having just or natural intonation, namely :

4. Rast . . . . . $C \ldots {}^`D_1 \ldots {}^`E_1 \ldots F \ldots G \ldots {}^`A_1 \ldots B\flat \ldots c$
5. Husseini. . . . $C \ldots {}^`D_1 \ldots E\flat \ldots F \ldots {}^`G_1 \ldots A\flat \ldots B\flat \ldots c$
6. Hidschaf. . . . $C \ldots {}^`D_1 \ldots E\flat \ldots F \ldots {}^`G_1 \ldots {}^`A_1 \ldots B\flat \ldots c$
7. Rahewi . . . . $C \ldots {}^`D_1 \ldots {}^`E_1 \ldots F \ldots {}^`G_1 \ldots A\flat \ldots B\flat \ldots c$
8. Sengule . . . . $C \ldots D \ldots {}^`E_1 \ldots F \ldots {}^`G_1 \ldots {}^`A_1 \ldots B\flat \ldots c$

Rast may be regarded as the mode of the Fourth of C; *Hidschaf* as the mode of the Fourth of F; *Husseini* as the mode of the Fourth of $B\flat$; as such they would have perfectly natural intonation. In *Rahewi*, if we refer it to the tonic F, the minor third $A\flat$ is in Pythagorean, not natural, intonation. It might be regarded as the mode of the minor Seventh of F in which the major Seventh E_1 is used as the leading note in place of the minor Seventh, as in our own minor mode. The natural intonation of such a tonal mode cannot, indeed, be properly represented by the existing 17 tonal degrees. It becomes necessary to take either Pythagorean minor Thirds and natural major Thirds or conversely. *Husseini* may be regarded as the same tonal mode with *Rahewi*, having the same false minor Third, but a minor Seventh. Finally *Sengule* may be regarded as F major with a Pythagorean Sixth. *Rast* may be conceived in the same way; they are merely distinguished by the different values of the Seconds G or ${}^`G_1$.

The four last *Makamat* have each 8 tones, new intercalary tones being employed. Two of them resemble the modes *Rast* and *Sengule*, and between $B\flat$ and C there is an intercalary tone ${}^`c_1$ introduced ; named

9. Irak . . . $C \ldots {}^`D_1 \ldots {}^`E_1 \ldots F \ldots G \ldots {}^`A_1 \ldots B\flat \ldots {}^`c_1 \ldots c$
10. Iszfahan . . $C \ldots D \ldots {}^`E_1 \ldots F \ldots {}^`G_1 \ldots {}^`A_1 \ldots B\flat \ldots {}^`c_1 \ldots c$

The last transposed a Fourth gives

11. Büsürg . . $C \ldots D \ldots {}^`E_1 \ldots F \ldots {}^`G_1 \ldots G \ldots A \ldots {}^`B_1 \ldots c$

The last tonal mode is

12. Zirefkend . . $C \ldots {}^`D_1 \ldots E\flat \ldots F \ldots {}^`G_1 \ldots A\flat \ldots {}^`A_1 \ldots {}^`B_1 \ldots c$

which certainly, if rightly reported, is a very singular creation. It might be looked upon as a minor scale with a major Seventh, and both a major and minor Sixth, but then the Fifth ${}^`G_1$ is wrong. On the other hand, if F is taken as the tonic, it has no Fourth, for which certainly there is some analogy in the Mixolydian and Hypolydian scales. The instructions for scales of eight notes are very contradictory, to judge by the different authorities cited by Kiesewetter.

The following four are distinguished as the principal modes of the *Makamat* :—

1. Uschak = Pythagorean F major.
2. Rast = Natural mode of the Fourth of C, or natural F major with acute Sixth.
3. Husseini = Natural mode of the minor Seventh of F.
4. Hidschaf = Natural mode of the Fourth of F.

We find, then, a decided predominance of scales with a perfectly correct natural intonation, which has been attained by a skilful use of a continued series of Fifths. This makes the Arabic and Persian tonal system very noteworthy in the history of the development of music. Moreover, in some of these scales we find ascending leading notes, which are perfectly foreign to the Greek scales. Thus in Rahewi, E_1 is the leading note to F, although the minor Third $A\flat$ stands above F, while no Greek scale could have allowed this without at the same time changing E_1 into $E\flat$.

Similarly in Zirefkend the B_1 is used as a leading note to C, although the minor Third $E\flat$ is used above C.*

* [Prof. Land (*Gamme Arabe*, p. 38, note 3) says 'some of the descriptions of Prof. Helmholtz, borrowed from Kiesewetter, do not quite correspond with the original data.' It will be interesting therefore to give these scales as Prof. Land describes them with his (more exact) French orthography of the Arabic names and in his order. The notation is the Translator's, $'A_1$ being 24 cents flatter than A.

1. *'Ochaq.* Our F major commencing (as shewn by [) with the dominant, $FGAB\flat$ $[CDEF$. 'This commencement is the inevitable consequence of the progression by conjunct tetrachords which belongs to the lute. 'Ochaq is as it were the type of all these *maqāmāt*, the others of which differ at one time like the tropes or modes of the Greeks and of the middle ages, by the displacement of both the Semitones at once, and at other times like the Greek genera, by exchanges of intervals without disturbing the scheme of two conjunct tetrachords followed by a tone, with the exception of Nos. 7 and 8, which are more distinct from the model *maqāma*.'

2. *Nawā.* 'We may say that the scale is that of $E\flat$ major, beginning with the Sixth.' $E\flat FGAB\flat [CDE\flat.$

3. *Bousilik* or *Abou-silik*. 'The scale of $D\flat$ major beginning at the Seventh, $D\flat E\flat$ $F G\flat A\flat B\flat [C D\flat.$' The Pythagorean intonation of the three first scales renders them non-harmonic.

¶ 4. *Rāst.* 'The same as 'Ochaq except that the Third A and the Seventh E are depressed by a Pyth. comma, $FG'A_1B\flat [CD'E_1F,$ which makes them just rather than Pythagorean.' The subdominant $B\flat DF$ is non-harmonic.

5. *'Irāq.* 'Like Rāst, but with the second and the sixth above diminished by a Pyth. comma, which makes the second nearly the minor Second 10 : 9, and with grave supplementary Fifth.' $F'G_1 'A_1 B\flat 'C_1 [C'D_1 'E_1 F.$ This has the proper subdominant $B\flat'D_1F,$ but the double Fifth is quite non-harmonic.

6. *Içfahān.* 'Rāst enriched with a grave supplementary Fifth.' $FG'A_1B\flat'C_1[CD'E_1F.$ Here both the subdominant $B\flat DF$ and double Fifth render the scale non-harmonic.

7. *Zirafkend.* $C'D_1 E\flat F'G_1 A\flat 'A_1 'B_1 C.$ 'An artificial scale composed of fragments of those of $E\flat$ $(e\flat f 'g_1 a\flat c 'd_1 e\flat,$ Third and ¶ Seventh almost just) and of C $(c 'd_1 f 'a_1 'b_1 c,$ Second minor and Sixth nearly just) varied also with Pythagorean A or D and $'B_1.$' Of course entirely non-harmonic.

8. *Bouzourk.* 'C major with the Second, Third, and Seventh diminished by a Pyth. comma, and with a grave supplementary Fifth.' $C'D_1'E_1F'G_1GA'B_1C.$ Both subdominant and dominant are non-harmonic.

9. *Zenkouleh.* 'Differs from Rāst only in having the Second minor.' $F'G_1 'A_1 B\flat [C D'E_1F.$ Subdominant non-harmonic.

10. *Rāhawi.* 'F minor commencing with the Fifth, but with the Sixth and Seventh each increased by a Limma = 90 cents, and the Second diminished by a Pythagorean comma, very nearly our just ascending scale of F minor.' $F'G_1A\flat B\flat [CD_1'E_1F.$ The Pyth. scale of F minor is ˙$F G A\flat B\flat [C D\flat E\flat F.$ Here $D\flat = 90$ cents; $90 + 90 = 180 = 204 - 24$

cents $= 'D_1,$ 24 cents being the Pyth. comma. Similarly $E\flat = 294$ cents; $294 + 90 = 384 = 408 - 24$ cents $= 'E_1.$ Entirely non-harmonic.

11. *Hhosaïni.* 'Like Nawā, but with the Third and Seventh diminished by a Pyth. comma.' $E\flat F'G_1 A B\flat [C'D_1 E\flat.$ Entirely non-harmonic.

12. *Hhidjāzi.* '$B\flat$ major, beginning with the Second and with the Third, Sixth, and Seventh diminished and therefore nearly just.' $B\flat[C'D_1E\flat F'G_1'A_1B\flat.$ This is the only one of these scales which is practically harmonic.

If we restore the proper names of the notes in the series of Fifths or Fourths, (as in p. 281d') calculate the cents between each pair of notes and from the first to each note, and begin with the note indicated, we shall have a better idea of the real nature of these scales, thus :

1. *'Ochaq.* C 204 D 204 E 90 F 204 G 204
    o   204   408   498   702
A 90 $B\flat$ 204 C
906   996   1200

2. *Nawā.* C 204 D 90 $E\flat$ 204 F 204 G 204
    o   204   294   498   702
A 90 $B\flat$ 204 C
906   996   1200

3. *Bousilik.* C 90 $D\flat$ 204 $E\flat$ 204 F 90
    o   90   294   498
$G\flat$ 204 $A\flat$ 204 $B\flat$ 204 C
588   792   996   1200

4. *Rāst.* C 204 D 180 $F\flat$ 114 F 204 G 180
    o   204   384   498   702
$B\flat\flat$ 114 $B\flat$ 204 C
882   996   1200

5. *'Irāq.* C 180 $E\flat\flat$ 204 $F\flat$ 114 F 180
    o   180   384   498
$A\flat$ 204 $B\flat\flat$ 114 $B\flat$ 180 $D\flat\flat$ 24 $C.$
678   882   996   1176 1200

This double initial $D\flat\flat, C$ may be compared to our double second in just major scales, and possibly has to be explained in the same way as a real modulation.

6. *Içfahān.* C 180 $E\flat\flat$ 204 $F\flat$ 114 F 204
    o   180   384   498
G 180 $B\flat\flat$ 114 $B\flat$ 180 $D\flat\flat$ 24 C
702   882   996   1176 1200

7. *Zirafkend.* C 180 $E\flat\flat$ 114 $E\flat$ 204 F 180
    o   180   294   498
$A\flat\flat$ 114 $A\flat$ 90 $B\flat\flat$ 204 $C\flat$ 114 C
678   792   882   1086 1200

8. *Bouzourk.* C 180 $E\flat\flat$ 204 $F\flat$ 114 F 180
    o   180   384   498
$A\flat\flat$ 24 G 204 A 180 $C\flat$ 114 C
678   702   906   1086 1200

9. *Zenkouleh.* C 204 D 180 $F\flat$ 114 F 180
    o   204   384   498
$A\flat$ 204 $B\flat\flat$ 114 $B\flat$ 204 C
678   882   996   1200

10. *Rāhawi.* C 180 $E\flat\flat$ 204 $F\flat$ 114 F 180
    o   180   384   498
$A\flat\flat$ 114 $A\flat$ 204 $B\flat$ 204 C
678   792   996   1200

11. *Hhosaïni.* C 180 $E\flat\flat$ 114 $E\flat$ 204 F 180
    o   180   294   498
$A\flat\flat$ 228 A 90 $B\flat$ 204 C
678   906   996   1200

12. *Hhidjāzi.* C 180 $E\flat\flat$ 114 $E\flat$ 204 F 180
    o   180   294   498
$A\flat\flat$ 204 $B\flat\flat$ 114 $B\flat$ 204 C
678   882   996   1200

Of these I have been able to play 1, 2, 3 direct, and 4, 5, 10, 12 by transposition upon my Pythagorean concertina (p. 281d'). When 12 begins with $B\flat,$ or is played by transposition $a b d'b$ $d'e'g'b a'b a',$ it is indistinguishable from the

At a little later period a new musical system was developed in Persia with 12 Semitones to the Octave, analogous to the modern European system. Kiesewetter here hazards the very unlikely hypothesis that this scale was introduced into Persia by Christian missionaries. But it is clear that the system of 17 tonal degrees which had been previously in popular use, merely required the feeling for the finer distinctions to grow dull so that intervals which differed only by a [Pythagorean] comma should be confused, in order to generate the system of 12 Semitones.* No foreign influence was necessary here. Moreover, the Greek system of music had long been taught to the Arabs and Persians by Alfārābī. Again, the European theory of music had not made any essential advance in the fourteenth and fifteenth centuries, if we except the study of harmony, which never found favour with the Orientals. Hence the Europeans of those days could teach the Orientals nothing that they did not already know better themselves, except some imperfect rudiments of harmony which they did not want. There is much more reason, I think, for ¶ asking whether the imperfect fragments of the natural system which we find among the Alexandrine Greeks, do not depend on Persian traditions, and also, whether the Europeans in the time of the Crusades did not learn much music from the Orientals. It is very probable that they brought the lute-shaped instruments with fingerboards and the bowed instruments from the East. In the construction of tonal modes we might especially instance the use of the leading note, which we have here found existing in the East, and which at that time also began to figure in the Western music.

The use of the major Seventh of the scale as a leading note to the tonic marks a new conception, which admitted of being used for the further development of the tonal degrees of a scale, even within the domain of purely homophonic music. The tone B_1 in the major scale of C has the most distant relationship of all the tones to the tonic C, because as the major Third of the dominant G, it has a less close connection with it than its Fifth D. We may perhaps assume this to be the ¶ reason why, when a sixth tone was introduced into some Gaelic airs, the Seventh was usually omitted. But, on the other hand, the major Seventh B_1 developed a peculiar relation to the tonic, which in modern music is indicated by calling it the *leading note*. The major Seventh B_1 differs from the Octave c of the tonic by the smallest interval in the scale, namely a Semitone, and this proximity to the tonic allows the Seventh to be struck easily and pretty surely, even when starting from tones in the scale which are not at all related to B_1. The leap $F...B_1$ [$=45 : 32 = 590$ cents], for example, is difficult, because there is no relationship at all between the tones. But when a singer has to perform the passage $F...B_1...c$, he conceives the interval $F...c$, which he can easily execute, but does not force his voice up sufficiently high to reach c at first, and thus strikes B_1 on the way. Thus B_1 assumes the appearance of a preparation for c, and this view alone justifies it to the ears of a listener, by whom the transition into c is, therefore, expected. Hence it has been said that B_1 leads to c; or that B_1 is the *leading note* to c. In ¶

just scale $a b c_1' \sharp d' e' f_1' \sharp g_1' \sharp a'$. The three chords $d' g' b a'$, $a d' b e'$, $e' g' b b$ are perfectly good, and the passage $d' g' b a'$, $d' g' b a'$, $e' a' d'' b$, $e' a' b b'$, $d' b e' a'$ perfectly good, much better than on the piano. Yet it never occurred to Arabs to play in harmony.

'In face of these historical scales,' observes Prof. Land (*ibid.* p. 38), 'it is difficult to conceive how Kiesewetter could say that the 17 degrees of the complete scale were not treated like sharps and flats, but that each one had the same importance. On the contrary, the 17 degrees were like our 12 Semitones to the Octave, or, still better, like the 17 intervals of the so-called enharmonic scale, which distinguishes sharps and flats, without dividing the Semitones E to F, and B to C. To compose their

melodies the Orientals, as we do, selected from them several series of 7 [occasionally 8] tones, very slightly different from our diatonic scales.' But so materially different that any attempt to play harmonies upon them would result in frightful dissonance.—*Translator.*]

* [If we suppose the pairs of notes in ( ) to have been confused into one by neglecting the Pythagorean comma, then the series of notes on p. 282b becomes C Db ($'D_1$ D) Eb ($'E_1$ E) F Gb ($'G_1 G$) Ab ($'A_1 A$) Bb $'B_1$ ($'c_1 c$), whence the equally tempered scale C Db D Eb E F Gb G Ab A Bb B c immediately follows. In Meshāqah's scale of 24 Quartertones, p. 264c, that of 12 Semitones is also implicitly contained.—*Translator.*]

this sense it also becomes easy to sharpen B_1 somewhat, making it B, for example, to bring it near to c, and mark its reference to that tone more distinctly.

According to my own feeling, the leading effect of the tone B_1 is much more marked in such passages as $F...B_1...c$ or $F+A_1...B_1...c$, in which B_1 is not related to the preceding tones, than in such a passage as $G+B_1...c$ where it is. But as I have found nothing on this point in musical writings, I do not know whether musicians are likely to agree with me in this opinion. For the other Semitone of the scale, $E_1...F$, the E_1 does not seem to lead to F, if the tonality of the melody is well preserved, because in this case E_1 has its own independent relation to the tonic, and hence is musically quite determinate. The hearer, then, has no occasion to regard E_1 as a mere preparation for F. Similarly for the interval $G...A^1\flat$ [$=112$ cents] in the minor mode. The G is more nearly related to the tonic C than $A^1\flat$ is. On the other hand, Hauptmann is probably right in ¶ considering the interval $D...E^1\flat$ [$=112$ cents] in the minor mode, as one in which D leads to $E^1\flat$, because D has only a relationship of the second degree to the tonic C, although its relationship is certainly closer than that of B_1 to C.

But the relation of $D^1\flat$ in descending passages of the mode of the minor Sixth of C (the old Greek Doric) is perfectly similar to the effect of B_1 in the ascending scale of C major. It really forms a kind of descending leading note, and since in the best period of Greek music descending passages were felt as nobler and more harmonious than ascending ones,* this peculiarity of the Doric mode may have been of special importance and have been a reason for the preference given to this scale. The cadence with the chord of the extreme sharp Sixth [ratio $128:225$, cents 976]—

$$D^1\flat + F \quad ... \; G + B_1$$
$$C \quad -E^1\flat + G ... c$$

¶ is almost the only remnant of the ancient tonal modes. It is quite isolated and misunderstood. This is a (Greek) Doric cadence, in which $D^1\flat$ and B_1 are both used at the same time as leading notes to C.†

The relation of the second or *parhypatē* of the Greek Doric scale, to the lowest tone or *hypatē*, seems also to have been perfectly well felt by the Greeks themselves, to judge by Aristotle's remarks in the 3rd and 4th of his problems on harmony. I cannot abstain from adducing them here because they admirably and delicately characterise the relation. Aristotle inquires why the singer feels his voice more taxed in taking the *parhypatē* than in taking the *hypatē*, although they are separated by so small an interval. The *hypatē* is sung, he says, with a remission of effort. And then he adds that in order to reach an aim easily it is necessary that in addition to the motive which determines the will, the kind of volitional effort should be quite familiar and easy to the mind.‡ The effort felt in singing the lead-

---

* Aristotle, *Problems* xix. 33. [The pas-
¶ sage has already been cited at full, p. 241d',
note ‡.—*Translator.*]

† [This cadence is a *union* of the ancient Doric, beginning with c, rendered harmonisable as the mode of the minor Sixth, ($c...d'b...e'b$ $...f...g...a'b...b'b...c'$, p. 278d, note) with the modern minor, beginning with c, ($c...d...e'b...f$ $...g...a'b...b_1...c'$,) and will be more particularly considered in the next chapter, pp. 306d–308c. The intervals expressed in cents are $D^1\flat$ 386 F 204 G 386 B_1 and C 316 $E^1\flat$ 386 G 498 c.—*Translator.*]

‡ This periphrasis seems to me to render correctly the last clause in the following citation: Διὰ τί τὴν παρυπάτην ᾄδοντες μάλιστα ἀποῤῥήγνυνται, οὐχ ἧττον ἢ τὴν νήτην καὶ τὰ ἄνω, μετὰ δὲ διαστάσεως πλείονος; Ἢ ὅτι χαλεπώτατα ταύτην ᾄδουσι, καὶ αὕτη ἡ ἀρχή; τὸ δὲ χαλεπὸν, διὰ τὴν ἐπίτασιν [παὶ πίεσιν] τῆς φωνῆς; ἐν

τούτοις δὲ πόνος· πονοῦντα δὲ μᾶλλον διαφθείρεται· Διὰ τί δὲ ταύτην χαλεπῶς, τὴν δὲ ὑπάτην ῥᾳδίως· καίτοι δίεσις ἑκατέρας; Ἢ ὅτι μετ' ἀνέσεως ἡ ὑπάτη, καὶ ἅμα μετὰ τὴν σύστασιν ἐλαφρὸν τὸ ἄνω βάλλειν; Διὰ ταὐτὸ δὲ ἔοικε πρὸς μίαν ᾄδεσθαι τὰ πρὸς ταύτην ἢ παρανήτην ᾀδόμενα· δεῖ γὰρ μετὰ συννοίας καὶ καταστάσεως οἰκειοτάτης τῷ ἤθει πρὸς τὴν βούλησιν. Arist. *Probl.* xix. 3, 4. [The whole passage may perhaps be translated thus: 'Why do those who sing the *parhypatē* break down not less than those who sing the *nētē* and higher tones, though with a greater disagreement (διάστασις)? Is it because they sing this with the greatest difficulty, even when this is the beginning? Does not difficulty arise from straining [and forcing] the voice? This occasions effort, and things done with effort are most apt to fail. But why do they sing the *parhypatē* with difficulty, and yet take the *hypatē* easily, although there is only

ing note does not lie in the larynx, but in the difficulty we feel in fixing the voice upon it by mere volition while another tone is already in our mind, to which we desire to pass, and which by its proximity conducted us to the leading note. It is not till we reach the final tone that we feel ourselves at home and at rest, and this final tone is sung without any strain on the will.

Proximity in the scale then gives a new point of connection between two tones, which is not merely active in the case of the leading note, just considered, but also, as already mentioned, in interpolating tones between two others in the chromatic and enharmonic modes. Intervals of pitch are in this respect analogous to mea-surements of distance. When we have the means of determining one point (the tonic) with great exactness and certainty, we are able by its means to determine other points with certainty, when they are at a known small distance from it (the interval of a Semitone), although perhaps we could not have determined them with so much certainty independently. Thus the astronomer employs his fundamental ¶ fixed stars, of which the positions have been determined with the greatest possible accuracy, for accurately determining the positions of other stars in their neigh-bourhood.

We may also remark that the interval of a Semitone plays a peculiar part as the introduction (*appoggiatura*) to another note. As an appoggiatura in a melody any tone can be used, even when not in its scale, provided it makes the interval of a Semitone with the note in the scale which it introduces ; but a foreign tone which makes the interval of a whole Tone with that note in the scale, cannot be so used. The only justification of this use of the Semitone is certainly its existence as a well-known interval in the diatonic scale, which the voice can sing correctly and the ear can readily appreciate, even when the relations on which its magnitude depends are not clearly sensible in the passage where it is used. Hence also no arbitrarily chosen small interval can be thus employed. Although slight changes in the interval of the leading note may be introduced by practical musicians to give ¶ a stronger expression to its tendency towards the tonic, they must never go so far as to make those changes clearly felt.*

Hence the major Seventh in its character of leading note to the tonic acquires a new and closer relationship to it, unattainable by the minor Seventh. And in this way the note which is most distantly related to the tonic becomes peculiarly valuable in the scale. This circumstance has continually grown in importance in modern music, which aims at referring every tone to the tonic in the clearest pos-sible manner ; and hence, in ascending passages going to the tonic, a preference has been given to the major Seventh in all modern keys, even in those to which it did not properly belong. This transformation appears to have begun in Europe during the period of polyphonic music, but not in part songs only, for we find it also in the homophonic *Cantus firmus* of the Roman Catholic Church. It was blamed in an edict of Pope John XXII., in 1322, and in consequence the sharpening of the lead-ing note was omitted in writing, but was supplied by the singers, a practice which ¶ Winterfeld believes to have been followed by Protestant musical composers even down to the sixteenth and seventeenth centuries, because it had once come into use. And this makes it impossible to determine exactly what were the steps by which this change in the old tonal modes was effected.†

Even to the present day, according to A. v. Oettingen's report,‡ the Esthonians

---

a *diesis* (Semitone or Quartertone) between them ? Is it because the *hypatē* is sung with a remission of effort, and at the same time it is easy to go upwards after getting oneself together for the effort (σύστασις) ? For the same reason it is easy to sing what leads up to any note, or the *paranētē*. For the will requires not only conscious thought (σύννοια) but an inclination (κατάστασις) which is per-fectly familiar to the habit of mind (ἦθος).' The passage is very difficult, and there was

clearly a connection in the writer's mind be-tween διάστασις, σύστασις, and κατάστασις, which influenced his reasoning, but evaporates in translation.—*Translator.*]

* [See App. XX. sect. G. art. 6.—*Trans-lator.*]

† *Der evangelische Kirchengesang.* Leip-zig, 1843, vol. i. introduction.

‡ *Das Harmoniesystem in dualer Ent-wickelung.* Dorpat und Leipzig, 1866, p. 113.

struggle against singing the leading note in minor scales, although it may be clearly struck on the organ.

Among the ancient tonal modes, the Greek Lydian (major mode) and the unmelodic Hypolydian (mode of the Fifth, p. 269a, No. 7) had the major Seventh as the leading note to the tonic, and hence the first was developed into the principal tonal mode of modern music, the major mode. The Greek Ionic (mode of the Fourth) differed from it only in having a minor Seventh. On simply altering this into a major Seventh, this mode also became major. On giving a major Seventh to each of the other three, they gradually converged to our present minor mode during the seventeenth century. From the Greek Phrygian (mode of the minor Seventh) by changing $B\flat$ into B_1 we obtain

THE ASCENDING MINOR SCALE.

¶
$$C...D...E^1\flat...F...G...A_1...B_1...c$$

as we had already found from a simple consideration of the relationship of tones [p. 274b, No. 2]. The Greek Hypodoric or Eolic (mode of the minor Third), which answers to our descending minor scale, gives on changing $B^1\flat$ into B_1,

THE INSTRUMENTAL MINOR SCALE.

$$C ... D ... E^1\flat ... F ... G ... A^1\flat ... B_1 ... c$$

which is difficult for singers to execute, on account of the interval $A^1\flat ... B_1$ [= ratio 64 : 75, cents 274], but frequently occurs in modern music both ascending and descending.

The Greek Doric (mode of the minor Sixth) with a major instead of a minor Seventh, is still discoverable in the final cadence mentioned on p. 286b.

The general introduction of the leading tone represents, therefore, a continually increasing consistency in the development of a feeling for the predominance of the ¶ tonic in a scale. By this change, not only is the variety of character in the ancient tonal modes seriously injured, and the wealth of previous means of expression essentially diminished, but even the links of the chain of tones in the scale were disrupted or disturbed. We have seen that the most ancient theory made tonal systems consist of series of Fifths, and that each system had at first four and afterwards six intervals of a Fifth. The predominance of a tonic as the single focus of the system was not yet indicated, at least externally; it became apparent at most by a limitation of the number of Fifths to contain those tones only which occurred in the natural scale. All Greek tonal modes may be formed from the tones in the series of Fifths—

$$F \pm C \pm G \pm D \pm A \pm E \pm B.$$

Directly we proceed to the natural intonation of Thirds, the series of Fifths is interrupted by an imperfect Fifth, as in

¶
$$F \pm C \pm G \pm D ... A_1 \pm E_1 \pm B_1$$

where the Fifth $D...A_1$ [= 680 cents] is imperfect. And when finally the sharp leading note is introduced, as by the use of $G_2\sharp$ for G in A_1 minor, the series is entirely interrupted [$C : G_2\sharp = 16 : 25 = 772$ cents].

In the gradual development of the diatonic system, therefore, the various links of the chain which bound the tones together were sacrificed successively to the desire of connecting all the tones in a scale with one central tone, the tonic. And in exact proportion to the degree with which this was carried out, the conception of tonality consciously developed itself in the minds of musicians.

The further development of the European tonal system is due to the cultivation of harmony, which will occupy us in the next chapter.

But before leaving our present subject, some doubtful points have still to be considered. In the preceding chapter I have shewn that the melodic relationship of tones can be made to depend upon their upper partials, precisely in the same way as their consonance was shewn to be determined in Chapter X. Now this method of explanation may in a certain sense be considered to agree with the favourite assertion that '*melody is resolved harmony*,' on which musicians do not hesitate to form musical systems without staying to inquire how harmonies could have been resolved into melodies at times and places where harmonies had either never been heard, or were, after hearing, repudiated. According to our explanation, at least, the same physical peculiarities in the composition of musical tones, which determined consonances for tones struck simultaneously, would also determine melodic relations for tones struck in succession. The former then would not be the reason for the latter, as the above phrase suggests, but both would have a common cause in the natural formation of musical tones.

Again, in consonance we found other peculiar relations, due to combinational tones, which become effective even when simple tones, or tones with few and faint upper partials, are struck simultaneously. I have already shewn that combina- ¶ tional tones very imperfectly replace the effect of upper partial tones in a consonance, and that consequently a chord formed of simple tones is wanting in brightness and character, the distinctions between consonance and dissonance being only very imperfectly developed.

In melodic passages, however, combinational tones do not occur, and hence the question arises as to how far a melodic effect could be produced by a succession of simple tones. There is no doubt that we can recognise melodies which we have already heard, when they are executed on the stopped pipes of an organ, or are whistled with the mouth, or merely struck on a glass or wood or steel harmonicon, as a musical box, or are played on bells. But there is also no doubt that all these instruments, which generate simple tones, either alone or accompanied by weak and remote inharmonic secondary tones, are incapable of producing any effective melodic impression without an accompaniment of musical instruments proper. They may be often extremely effective for performing single parts when accom- ¶ panied by the organ, or the orchestra, or a pianoforte, but by themselves they produce very poor music indeed, which degenerates into absolute unpleasantness when the inharmonic secondary tones are somewhat too loud.

We are bound, however, to give some reason why any impression of melody at all can be produced by such instruments.

Now we must first remember that, as shewn at the end of Chapter VII., the actual construction of the ear favours the generation of weak harmonic upper partials within the ear itself, when powerful but objectively simple tones are sounded. Hence it is at most very weak objectively simple tones which can be regarded as also subjectively simple.

Next, there is an effect of memory to be brought into account. Supposing that I have been used to hear Fifths taken at all possible pitches, and have recognised them by aural sensation as having a very close melodic relationship, I should know the magnitude of this interval by experience for every tone in the scale, ¶ and should retain the knowledge thus acquired by the action of a man's memory of sensations, even of those for which he has no verbal expression.

When, then, I hear such an interval executed on tuning-forks, I am able to recognise it as an interval I have often heard, although its tones have either none, or only some faint remnants of those upper partials which formerly gave it a right to be considered as a favourite interval of close melodic relationship. And just in the same way I shall be able to recognise, as previously known, other melodic passages or whole melodies which are executed in simple tones, and even if I hear a melody for the first time in this way, whistled with the mouth or chimed by a clock, or struck on a glass harmonicon, I should be able to complete it by imagining how it would sound if executed on a real musical instrument, as the voice or a violin.

A practised musician is able to form a conception of a melody by merely reading the notes. If we give the prime tones of these notes on a glass harmonicon, we give a firmer basis to the conception by really exciting a large portion of the impression on the senses which the melody would have produced if sung. Simple tones, however, merely exhibit an outline of the melody. All that gives the melody its charm is absent. We know, indeed, the individual intervals which it contains, but we have no immediate impression on our senses which serves to distinguish those which are distantly from those which are closely related, or the related from the totally unrelated. Observe the great difference between merely whistling a melody or playing it on a violin ; between striking it on a glass harmonicon or on the piano ! The difference is somewhat of the same kind as that between viewing a single photograph of a landscape, and seeing two corresponding photographs of it through a stereoscope. The first enables me, by means of my memory, to form a conception of the relative distances of its parts, and this conception may be often very satisfactory. But the stereoscopic fusion of the two figures gives me the real impression on the senses which the relative distances of the parts of the landscape would have themselves produced, and which I am obliged in the case of a single image to supply by experience and memory. Hence the stereoscopic picture is more lively than the simple perspective view, exactly in the same way as immediate impressions on our senses are more lively than our recollections.

The case seems to be the same for melodies executed in simple tones. We recognise the melodies when we have heard them otherwise performed ; we can even, if we have sufficient musical imagination, picture to ourselves how they would sound if executed by other instruments, but they are decidedly without the immediate impression on the senses which gives music its charm.

CHAPTER XV.

THE CONSONANT CHORDS OF THE TONAL MODES.

POLYPHONY was the form in which music for several voices first attained a certain degree of artistic perfection. The peculiar characteristic of this style of music was that several voices were singing each its own independent melody at the same time, which might be a repetition of the melodies already sung by the other voices, or else quite a different one. Under these circumstances each voice had to obey the general law of tonality common to the construction of all melodies, and, moreover, every tone of a polyphonic passage had to be referred to the same tonic. Hence each voice had to commence separately on the tonic or some tone closely related to it, and to close in the same way. In practice each part of a polyphonic piece was made to begin with the tonic or its Octave. This fulfilled the law of tonality, but necessitated the closing of a polyphonic piece with a unison.

The reason why higher Octaves might accompany the tonic at the close, lies, as we saw in the last chapter, in the fact that higher Octaves are merely repetitions of portions of the fundamental tone. Hence by adding its Octave to the tonic at the close, we merely reinforce part of its compound tone ; no new compound or simple tone is added, and the union of all the tones contains only the constituents of the tonic itself.

The same is true for all the other partial tones which are contained in the tonic. The next step in the development of the final chord was to add the Twelfth of the tonic. Now the chord $c...c'\pm g'$ contains no element which is not also a constituent of the compound tone c when sounded alone, and consequently, being a mere representative of the single musical tone c, it is suitable for the termination of a piece of music having the tonic c.

Nay even the chord $c'\pm g'...c''$ might be so used, for when it is struck we hear, weakly indeed, but still sensibly, the combinational tone c, so that the whole mass of tone again contains nothing more than the constituents of the tone c. It must be owned, however, that this combination would answer to a rather unusual quality of tone, with a proportionably weak prime partial.

On the other hand, it was not possible to use the chords $c...c'...f'$ or $c'...f'\pm c''$

to end a piece having the tonic c, although these chords are consonances as well as the preceding, because f is not an element of the compound tone c, and hence the closing chord would contain something which was not the tonic at all. It is here probably that we have to look for the reason why some medieval theoreticians wished to reckon the Fourth among the dissonances. But perfect consonance was not sufficient to make an interval available for the final chord. There was a second condition which the theoreticians did not clearly understand. The tones of the final chord had to be constituents of the compound tone of the tonic. This was the only case in which those tones could be employed.

The Sixth of the tonic is as ill suited as the Fourth for use in the final chord. But the major Third can be used, because it occurs as the fifth partial tone of the tonic. Since the qualities of tone which are fit for music generally allow the fifth and sixth partial tone to be audible, but make the higher partials either entirely inaudible or at least very faint, and since, moreover, the seventh partial is dissonant ¶ with the fifth, sixth, and eighth, and is not used in the scale, the series of tones available for the closing chord terminates with the Third. Thus we actually find down to the beginning of the eighteenth century, that the final chord has either no Third, or only a major Third, even in tonal modes which contain only the minor and not the major Third of the tonic. To attain fulness, it was preferred to do violence to the scale by using the major Third in the closing chord. The minor Third of the tonic can never stand for a constituent of its compound tone. Hence it was originally as much forbidden as the Fourth and Sixth of the tonic. Before a minor chord could be used to close a piece of music the feeling for harmony had to be cultivated in a new direction.

The ear is the more satisfied with a closing major chord, the more closely the order of the tones used imitates the arrangement of the partial tones in a compound. Since in modern music the upper voice is most conspicuous, and hence has the principal melody, this voice must usually finish with the tonic. Bearing this in ¶ mind, we can use any of the following chords for the close (combinational tones are added as crotchets) :—

In the chords 1 and 2 all the notes coincide with partials of C, and they therefore most closely resemble the compound tone C itself. And then closer positions of the chord can be substituted, provided they resemble the first by ¶ having C for the fundamental tone as in 3, 4, 5. They still retain sufficient resemblance to the compound tone of the low C to be used in its place. Moreover, the combinational tones, written as crotchets in 3, 4 and 5, assist in the effect of making the deeper partials of the compound C, at least faintly, audible. But the first two positions always give the most satisfactory close. The tendency towards a deep final tone in harmonic music is very characteristic, and I believe that the above is its proper explanation. There is nothing of the kind in the construction of homophonic melodies. It is peculiar to the bass of part music.

Precisely in the same way that the tonic, when used as the bass of its major chord at the close, gives it a resemblance to its own compound tone, and is hence felt as the essential tone of the chord, all major chords sound best when the lowest tone of their closest triad position (No. 4, p. 219c) is made the bass. The other major chords in the scale are those on its Fourth and Fifth, and hence for the scale of C major, are $F + A_1 - C$ and $G + B_1 - D$. Hence if we make the harmony of a piece of music to consist of these major chords only, each having its fundamental tone in the bass, the effect is almost that of a compound tonic in different qualities of tone passing into its two nearest related compound tones, the Fourth and Fifth. This makes the harmonisation transparent and definite, but it would be too uniform for long pieces. Modern popular tunes, songs and dances, are however, as is well known, constructed in this manner. The people, and generally persons of small musical cultivation, can be pleased only by extremely simple and intelligible musical relations. Now the relations of the tones are generally much easier to feel with distinctness in harmonised than in homophonic music. In the latter the feeling of relationship of tone depends solely on the sameness of pitch of two partials in two consecutive musical tones. But when we hear the second compound tone we can at most remember the first, and hence we are driven to complete the comparison by an act of memory. The consonance, on ¶ the other hand, gives the relation by an immediate act of sensation ; we are no longer driven to have recourse to memory ; we hear beats, or there is a roughness in the combined sound, when the proper relations are not preserved. Again, when two chords having a common note occur in succession, our recognition of their relationship does not depend upon weak upper partials, but upon the comparison of two independent notes, having the same force as the other notes of the corresponding chord.

When, for example, I ascend from C to its Sixth A_1, I recognise their mutual relationship in an unaccompanied melody, by the fact that e'_1, the fifth partial of C, which is already very weak, is identical with the third partial of A_1. But if I accompany the A_1 with the chord $F + A_1 - c$, I hear the former c sound on powerfully in the chord, and know by immediate sensation that A_1 and C are consonant, and that both of them are constituents of the compound tone F.

When I pass melodically from C to B_1 or D, I am obliged to imagine a kind of ¶ mute G between them, in order to recognise their relationship, which is of the second degree. But if I audibly sustain the note G while the others are sounded, their common relationship becomes really sensible to my ear.

Habituation to the tonal relations so evidently displayed in harmonic music, has had an indisputable influence on modern musical taste. Unaccompanied songs no longer please us ; they seem poor and incomplete. But if merely the twanging of a guitar adds the fundamental chords of the key, and indicates the harmonic relations of the tones, we are satisfied. Again, we cannot fail to see that the clearer perception of tonal relationship in harmonic music has greatly increased the practicable variety in the relations of tones, by allowing those which are less marked to be freely used, and has also rendered possible the construction of long musical pieces which require powerful links to connect their parts into one whole.

The closest and simplest relation of the tones is reached in the major mode, ¶ when all the tones of a melody are treated as constituents of the compound tone of the tonic, or of the Fifth above or the Fifth below it. By this means all the relations of tones are reduced to the simplest and closest relation existing in any musical system—that of the Fifth.

The relation of the chord of the dominant G to that of the tonic C, is somewhat different from that of the chord of the subdominant F to the tonic chord. When we pass from $C + E_1 - G$ to $G + B_1 - d$ we use a compound tone, G, which is already contained in the first chord, and is consequently properly prepared, while at the same time such a step leads us to those degrees of the scale which are most distant from the tonic, and have only an indirect relationship with it. Hence this passage forms a distinct progression in the harmony, which is at once well assured and properly based. It is quite different with the passage from $C + E_1 - G$ to $F + A_1 - c$. The compound tone F is not prepared in the first chord, and it has therefore to be discovered and struck. Hence the justification of this passage as correct and closely related, is not complete until the step is actually made and it is felt that the chord of F contains no tones which are not directly related to the tonic C. In the passage from the chord of C to that of F, therefore, we miss that distinct and well-assured progression which marked the passage from the chord of C to that of G. But as a compensation, the progression from the chord of C to that of F has a softer and calmer kind of beauty, due, perhaps, to its keeping within tones directly related to the tonic C. Popular music, however, favours the other passage from the tonic to the Fifth above (hence called the *dominant* of the key), and many of the simpler popular songs and dances consist merely of an interchange of tonic and dominant chords. Hence also the common harmonicon (accordion, German concertina), which is arranged for them, gives the tonic chord on opening the bellows, and the dominant chord on closing them. The Fifth below the tonic is called the *subdominant* of the key. Its chord is seldom introduced at all into usual popular melodies, except, perhaps, once near the close, to restore the equilibrium of the harmony, which had chiefly inclined towards the dominant.

When a section of a piece of music terminates with the passage of the dominant into the tonic chord, musicians call the close a *complete cadence*. We thus return from the tones most distantly related to the tonic, to the tonic itself, and, as befits a close, make a distinct passage from the remotest parts of the scale to the centre of the system itself. If, on the other hand, we close by passing from the subdominant to the tonic chord, the result is called an *imperfect* or *plagal cadence*. The tones of the subdominant triad are all directly related to the tonic, so that we are already close upon the tonic before we pass over to it. Hence the imperfect cadence corresponds to a much quieter return of the music to the tonic chord, and the progression is much less distinct than before.

In the complete cadence the chord of the tonic follows that of the dominant, but to preserve the equilibrium of the system in relation to the subdominant, its chord is made to precede that of the dominant as in 1 or 2.

This succession really forms the *complete close*, by bringing all the tones of the whole scale together again, and thus in conclusion collecting and fixing every part of the key.

The major mode, as we have seen, permits the requisitions of tonality to be most easily and completely united with harmonic completeness. Every tone of its scale can be employed as a constituent of the musical tone of the tonic, the dominant, or the subdominant, because these fundamental tones of the mode are also fundamental tones of major chords. This is not equally the case in the other ancient tonal modes.

1. MAJOR MODE *

$$\underbrace{f \underbrace{+ a_1 - c}_{\text{major}} \overbrace{+ e_1 - g}^{\text{major}} \underbrace{+ b_1 - d}_{\text{major}}}$$

2. MODE OF THE FOURTH *

$$\underbrace{f \underbrace{+ a_1 - c}_{\text{major}} \overbrace{+ e_1 - g}^{\text{major}} \underbrace{- b^1\natural + d}_{\text{minor}}}$$

* [Of course when the modes are thus reduced to harmonic combinations, the tones of the old modes, as given in footnote to p. 268c, are all altered, and become those in footnote to p. 274c. See also App. XX. sect. E. arts. 9 and 10.—*Translator*.]

3. MODE OF THE
    MINOR SEVENTH *

$$\overbrace{f + a_1 - c}^{\text{major}} - \overbrace{e^1\flat + g}^{\text{minor}} - \underbrace{b^1\flat + d}_{\text{minor}}$$

4. MODE OF THE
    MINOR THIRD *
    (*minor mode*)

$$\underbrace{f - a^1\flat + c}_{\text{minor}} - \overbrace{e^1\flat + g}^{\text{minor}} - \underbrace{b^1\flat + d}_{\text{minor}}$$

5. MODE OF THE
    MINOR SIXTH *

$$\underbrace{b\flat - d^1\flat + f}_{\text{minor}} - \overbrace{a^1\flat + c}^{\text{minor}} - e^1\flat + g$$

¶ In the minor chords, the Third does not belong to the compound tone of its fundamental note, and hence cannot appear as a constituent of its quality; so that the
relation of all the parts of a minor chord to the fundamental note is not so immediate as that for the major chord, and this is a source of difficulty in the final
chord. For this reason we find almost all popular dance and song music written
in the major mode †; indeed, the minor mode forms a rare exception. The people
must have the clearest and simplest intelligibility in their music, and this can only
be furnished by the major mode. But there was nothing like this predominance of
the major key in homophonic music. For the same reason the harmonic accompaniment of chorales in major keys was developed with tolerable completeness as
early as the sixteenth century, so that many of them correspond with the cultivated
musical taste of the present day; but the harmonic treatment of the minor and
the other ecclesiastical modes was still in a very unsettled condition, and strikes
modern ears as very strange.

¶ In a major chord $c + e_1 - g$, we may regard both g and e_1 as constituents of the
compound tone of c, but neither c nor g as constituents of the compound tone of e_1,
and neither c nor e_1 as constituents of the compound tone of g.‡ Hence the major
chord $c + e_1 - g$ is completely unambiguous, and can be compared only with the
compound tone of c, and consequently c is the predominant tone in the chord, its
root, or, in Rameau's language, its *fundamental bass*; and neither of the other two
tones in the chord has the slightest claim to be so considered.

In the minor chord $c - e^1\flat + g$, the g is a constituent of the compound tones of
both c and $e^1\flat$. Neither $e^1\flat$ nor c occurs in either of the other two compound
tones c, g. Hence it is clear that g at least is a dependent tone. But, on the other
hand, this minor chord can be regarded either as a compound tone of c with an
added $e^1\flat$ or as a compound tone of $e^1\flat$ with an added c. Both views are entertained at different times, but the first usually prevails. If we regard the chord as
the compound tone of c, we find g for its third partial, while the foreign tone $e^1\flat$
only occupies the place of the weak fifth partial e_1. But if we regarded the chord
¶ as a compound tone of $e^1\flat$, although the weak fifth partial g would be properly represented, the stronger third partial, which ought to be $b^1\flat$, is replaced by the foreign
tone c. Hence in modern music we usually find the minor chord $c - e^1\flat + g$ treated
as if its root or fundamental bass were c, so that the chord appears as a somewhat altered and obscured compound tone of c. But the chord also occurs in the
position $e^1\flat + g \ldots c$ (or better still as $e^1\flat + g \ldots c^1$) even in the key of $B^1\flat$ major, as a
substitute for the chord of the subdominant $e^1\flat$. Rameau then calls it the chord of
the great Sixth [in English ' added Sixth '], and, more correctly than most modern
theoreticians, regards $e^1\flat$ as its fundamental bass.§

---

* [See p. 293, note.]   † [This remark does not apply to old English music.—*Translator.*]

‡ [Taking only six partials, we have for—

Compound Tones	Simple Partial Tones					
	1	2	3	4	5	6
C	C	c	g	c^1	$e^1{}_1$	g^1
E_1	E_1	e_1	b_1	$e^1{}_1$	$g^1{}_2\sharp$	$b^1{}_1$
G_1	G	g	d^1	g^1	$b^1{}_1$	d''
$E^1\flat$	$E^1\flat$	$e^1\flat$	$b^1\flat$	$e^{1}\flat$	g^1	$b''\flat$.

—*Translator.*]

§ [The scale of $B^1\flat$ major has the chords
$e^1\flat + g - b^1\flat + d - f^1 + a - c^1$; hence, regarding
the chord as made up of the notes of this
scale, it would be $c^1 \mid e^1\flat + g$, which is not a
minor chord at all, like $c - e^1\flat + g$, because it
has a Pythagorean in place of a just minor
Third. It was only tempered intonation which
confused the two cases. Attention will be

When it is important to guide the ear in selecting one or other of these two meanings of the minor chord, the root intended may be emphasised by giving it a low position or by throwing several voices upon it. The low position of the root allows such other tones as could be fitted into its compound tone, to be considered directly as its partials, whereas the low compound tone itself cannot be considered as the partial of another much higher tone. In the first half of last century, when the minor chord was first used as a close, composers endeavoured to give prominence to the tonic by increasing the loudness of the tonic note in comparison with its minor Third. Thus in Handel's oratorios, when he concludes with a minor chord, most of the conspicuous vocal and instrumental parts are concentrated on the tonic, while the minor Third is either touched by one voice alone, or merely by the accompanying pianoforte or organ. The cases are much rarer where in minor keys he gives only two voices to the tonic in the closing chord, and one to its Fifth and another to its Third, which is his rule in major modes. ¶

When the minor chord appears in its second subordinate signification, as $e^1\flat + g...c$ with the root $e^1\flat$, this fact is shewn partly by the position of $e^1\flat$ in the bass, and partly by its close relationship to the tonic $b^1\flat$. Modern music even makes this interpretation of the chord still clearer by adding $b^1\flat$ as the Fifth of $e^1\flat$, so that the chord becomes dissonant in the form $e^1\flat - g + b^1\flat...c^1$.*

The disinclination of older composers to close with a minor chord, may be explained partly by the obscuration of its consonance from false combinational tones, and partly because, as already mentioned, it does not give a mere quality of the tonic tone, but mixes foreign constituents with it. But in addition to the minor Third, which does not fit into the compound tone of the tonic, the combinational tones of a minor chord are equally foreign to it. As long as the feeling of tonality required a definite single compound tone for the connecting centre of the key, it was impossible to form a satisfactory close except by a reproduction of the pure compound tone of the tonic with no foreign admixture. It was not till a further ¶ development of musical feeling had given the chords of the mode an independent significance, that the minor chord, notwithstanding its possession of constituents foreign to the compound tone of the tonic, could be justified in its use as a close.

Hauptmann † gives a different reason for avoiding the minor chord at the close. He asserts that before the chord of the dominant Seventh came into use, there was no voice-part suitable for falling into the minor Third of the tonic. Thus if the final cadence consisted of the chords $G + B_1 - D$, $C - E^1\flat + G$, the D of the first chord was the only one which could pass melodiously in $E^1\flat$, but this would have appeared like the passage of the leading note D in the key of $E^1\flat$ major into its tonic $E^1\flat$, and hence have called up the feeling of $E^1\flat$ major in lieu of C minor. We may admit that this relation of the leading note would have drawn the hearer's special attention to the two tones in question, and to a certain extent disturbed his recognition of the key, but yet it is clear that even without the help of this chord of the dominant Seventh, there were several ways for the voices to pass through ¶ dissonances into the minor Third of the closing chord, if composers had felt any wish to do so. Thus in the plagal cadence

$$c - e^1\flat + g \ ...c'$$
$$F...f \ - a^1\flat + c'$$
$$C - e^1\flat + g \ ...c'$$

which is so often used on other occasions, the Fourth f could be made to descend to the minor Third $e^1\flat$ without any inconvenience. Indeed, we find that when

---

hereafter drawn to this important distinction, see p. 299a.—*Translator.*]

* [Transposing the c^1 the chord becomes $c^1 \mid e^1\flat + g - b^1\flat$, so that we have a major chord with the Sixth of its root added, that is, the subdominant of the key of $B^1\flat$ rendered dissonant by introducing c^1, the Second of the key, or the Sixth above the subdominant $c^1\flat$.

Observe that it is c^1 which is now introduced in the text, in place of c. If c is retained, thus $c - e^1\flat + g - b^1\flat$, the chord is one of those chords of the Seventh considered in Chapter XVI.—*Translator.*]

† *Harmonik und Metrik*, Leipzig, 1853, p. 216.

the chord of the dominant Seventh had actually come into use, and the Seventh F of the chord $G+B_1-D \mid F$ ought by every right to have descended into the minor Third $E^1\flat$ of the closing chord, musical pieces of the fifteenth century * avoid this progression, and make this Seventh F either ascend to the Fifth G, or descend to the major Third E_1 of the final chord, instead of to $E^1\flat$, its minor Third. This custom prevailed down to Bach's time.

In Chapter XIII. (p. 249a) we characterised modern harmonic music, as contrasted with medieval polyphony, by its development of a feeling for the independent significance of chords. In Palestrina, Gabrieli, and still more in Monteverde and the first composers of operas, we find the various degrees of harmoniousness in chords carefully used for the purposes of expression. But these masters are almost entirely without any feeling for the mutual relation of consecutive chords. These chords often follow one another by entirely unconnected leaps, and the only bond of union ¶ is the scale, to which all their notes belong.

The transformation which took place from the sixteenth to the beginning of the eighteenth century, may, I think, be characterised by the development of a feeling for the independent relationship of chords one to the other, and by the establishment of a central core, the *tonic chord*, round which were grouped the whole of the consonant chords that could be formed out of the notes of the scale. For these chords there was a repetition of the same effort which was formerly shewn in the construction of the scale, where interrelations of the tones were first grounded on a chain of intervals, and afterwards on a reference of each note to a central compound tone, the tonic.

Two chords which have one or more tones in common will here be termed *directly related*.

Chords which are directly related to the same chord will be here said to be *related to each other in the second degree*.

¶   Thus $c+e_1-g$ and $g+b_1-d$ are directly related, and so are $c+e_1-g$ and a_1-c+e_1; but $g+b_1-d$ and a_1-c+e_1 are related only in the second degree.

When two chords have two tones in common they are more closely related than when they have only one tone in common. Thus $c+e_1-g$ and a_1-c+e_1 are more closely related than $c+e_1-g$ and $g+b_1-d$.

The tonic chord of any tonal mode can of course only be one which more or less perfectly represents the compound tone of the tonic, that is, that major or minor chord of which the tonic is the root. The tonic note, as the connecting core of all the tones in a regularly constructed melody, must be heard on the first accented part of a bar, and also at the close, so that the melody starts from it and returns to it; the same is true for the tonic chord in a succession of chords. In both of these positions in the scale we require to hear the tonic note, accompanied not by any arbitrary chord, but only by the tonic chord, having the tonic note itself as its root. This was not the case even as late as the sixteenth century, as ¶ is seen by the example on p. 247c taken from Palestrina.

When the tonic chord is major, the domination of all the tones by the tonic note is readily reconciled with the domination of all the chords by the tonic chord, for as the piece begins and ends with the tonic chord, it also begins and ends at the same time with the pure unmixed compound tone of the tonic note. But when the tonic chord is minor, all these conditions cannot be so perfectly satisfied. We are obliged to sacrifice somewhat of the strictness of the tonality in order to admit the minor Third into the tonic chord at the beginning and end. At the commencement of the eighteenth century we find Sebastian Bach using minor chords at the end of his preludes, because these were merely introductory pieces, but not at the end of fugues and chorales, and at other complete closes. In Handel and even in the ecclesiastical pieces of Mozart, the close in a minor chord is used alternately

---

* See an example in Anton Brumel, in Forkel's *Geschichte der Musik*, vol. ii. p. 647. Another, with a plagal cadence by Josquin, will be found, *ibid.* p. 550, where the voices might have easily been led to the minor Third.

with the close in a chord without any Third, or with the major Third.   And the last composer cannot be accused of external imitation of old habits, for we find that in these usages they always observe the expression of the piece.   When at the close of a composition in the minor mode, a major chord is introduced, it has the effect of a sudden and unexpected brightening up of the sadness of the minor key, producing a cheering, enlightening, and reconciling effect after the sorrow, grief, or restlessness of the minor.   Thus a close in the major suits the prayer for the peace of the departed in the words, 'et lux perpetua luceat eis,' or the conclusion of the *Confutatis maledictis*, which runs thus :—

> Oro supplex et acclinis,
> Cor contritum quasi cinis ;
> Gere curam mei finis.

But such a closing major chord is certainly somewhat startling for our present ¶ musical feeling, even though its introduction may, at one time, add wondrous beauty and solemnity, or, at another, dart like a beam of hope into the gloom of deepest despair.   If the restlessness remains to the last, as in the *Dies irae* of Mozart's *Requiem*, the minor chord, in which an unresolved disturbance exists, forms a fitting close.   Mozart was wont to terminate ecclesiastical pieces of a less decided character with a chord that had no Third.   There are many similar examples in Handel.   Hence although both masters stood on the very same plat-form as modern musical feeling, and themselves gave, as it were, the finishing touch to the construction of the modern tonal system, they were not altogether strangers to the feeling which had prevented older musicians from using the minor Third of the tonic in the final chord.   They followed no strict rule, however, but acted according to the expression and character of the piece and the sense of the words with which they had to close.

Those tonal modes which furnish the greatest number of consonant chords ¶ related to one another or to the common chord, are best adapted for artistically connected harmonies.   Since all consonant chords, when reduced to their closest position and simplest form, are triads consisting of a major and a minor Third, all the consonant chords of any key can be found by simply arranging them in order of Thirds, as in the following tables.   The braces above and below connect the chords together.   The ordinary round braces, which are placed above, point out minor chords ; the square braces below indicate major chords.   The tonic chord is printed in capitals.

1) MAJOR MODE

$$d_1 - f + a_1 - C + E_1 - G + b_1 - d$$

2) MODE OF THE FOURTH

$$b\flat + d_1 - f + a_1 - C + E_1 - G - b^1\flat + d$$

3) MODE OF THE MINOR SEVENTH

$$b\flat + d_1 - f + a_1 - C - E^1\flat + G - b^1\flat + d$$

4) MODE OF THE MINOR THIRD

$$b\flat + d_1 - f - a^1\flat + C - E^1\flat + G - b^1\flat + d$$

5) MODE OF THE MINOR SIXTH

$$b\flat - d^1\flat + f - a^1\flat + C - E^1\flat + G - b^1\flat$$

In this arrangement I have introduced the different intonations of the Second and Seventh of the key, which we found in the construction of the scales for homophonic music.* But we observe that the chords directly related to the tonic chord contain every tone in the scale, excepting in the mode of the minor Sixth. The Second and Seventh of the tonic occur first in the chord of G, which is directly related to the tonic chord, and next in chords containing F, which are, however, not directly connected with the tonic chord. The supplementary tones of the scale which are related to the dominant thus acquire in harmonic music an important preponderance over those related to the subdominant. We must necessarily prefer direct to indirect relations for determining scalar degrees. Hence by confining ourselves to the chords which are directly related to the tonic chord, we obtain the following arrangement of the tonal modes :†—

¶   1) MAJOR MODE

$$f + a_1 - C + E_1 - G + b_1 - d$$

2) MODE OF THE FOURTH

$$f + a_1 - C + E_1 - G - b^1\flat + d$$

3) MODE OF THE MINOR SEVENTH

$$f + a_1 - C - E^1\flat + G - b^1\flat + d$$

4) MODE OF THE MINOR THIRD

¶

$$f - a^1\flat + C - E^1\flat + G - b^1\flat + d$$

5) MODE OF THE MINOR SIXTH

$$d^1\flat + f - a^1\flat + C - E^1\flat + G - b^1\flat.$$

A glance at this table shews that the *major mode* and *mode of the minor Third* (*minor mode*) possess the most complete and connected series of chords, so that these two are decidedly superior to the rest for harmonic purposes. This is also the reason which led to the preference given to them in modern music.

And in this way we obtain a final settlement of the proper intonation of the supplementary tones of the scale, at least for the first four modes. Hauptmann, with whom I agree, considers the tone D alone to be the essential constituent of both the major and minor modes of C. This D forms an imperfect (Pythagorean) ¶ minor Third with F, so that the chord $D \mid F + A_1$ must be considered as dissonant.‡ This chord thus intoned is in reality most decidedly dissonant to the ear. On the other hand, Hauptmann admits a major mode which reaches over to the subdominant, and uses D_1 in place of D. I consider this conception to be a very

---

* [These scales differ from those transcribed in pp. 293*d* and 294*a*, only in the addition of the secondary forms of intercalary tones, d_1, $b\flat$, or $b^1\flat$, which, in fact, imply modulations into adjacent modes, or else give a double and ambiguous character to each mode, as shewn on p. 277, footnote †, and by referring to the Duodenarium, App. XX. sect. E. art. 18, it will be seen that there is a real change of duodene, which always must happen when changes of a comma occur.—*Translator.*]

† [The first four are the same as in pp. 293*d* and 294*d*. The settlement in the text avoids

the *double modality* alluded to in the last note, and fixes the modes in the meanings of App. XX. sect. E. art. 9, as

(1) 1 C ma.ma.ma.
(2) 1 C ma.ma.mi.
(3) 1 C ma.mi.mi.
(4) 1 C mi.mi.mi.
(5) 3 $A^1\flat$ ma.ma.ma.

In the last scale it is more usual, however, to take $b\flat$ in place of $b^1\flat$, which makes the scale $b\flat - d^1\flat + f - a^1\flat + C - E^1\flat + G = 5F$ mi.mi.mi. But temperament obscures all these differences.—*Translator.*]

‡ [See p. 295*d*, note *.—*Translator.*]

happy expression of the real state of things. When the consonant chord $D_1 - F + A_1$ occurs in any composition it is impossible to return immediately, without any transitional tone, to the tonic chord $C + E_1 - G$. The result would be felt as an harmonic leap without adequate notice. Hence it is a correct expression of the state of affairs to look upon the use of this chord as the beginning of a modulation beyond the boundaries of the key of C major, that is, beyond the limits of direct relationship to its tonic chord. In the minor mode this would correspond to a modulation into the chord of $D^1\flat + F - A^1\flat$. Of course in the modern tempered intonation the consonant chord $D_1 - F + A_1$ is not distinguished from the dissonant $D \mid F + A_1$, and hence the feeling of musicians has not been sufficiently cultivated to make them appreciate this difference on which Hauptmann insists.*

As regards the other supplementary tone $b^1\flat$, which may occur in the chords $e^1\flat + g - b^1\flat$ and $g - b^1\flat + d'$, I have already shewn in the last chapter that even in homophonic music it is almost always replaced by b_1. Harmonic considerations likewise favour the use of b_1, independently of melodic progression. It has been already shewn that when the two tones of the scale which are but distantly related to the tonic, make their appearance as constituents of the dominant, they enter into close relation to the tonic. Now this can only be the case with the compound tones of the major chord $g + b_1 - d$, and not with those of the minor chord $g - b^1\flat + d$. Considered independently, the tones $b^1\flat$ and d are quite as closely related to c as the tones b_1 and d. But by regarding the two latter as constituents of the compound tone g, we connect them with c by the same closeness of relationship that g is itself connected with c. Hence, in all modern music, wherever $b^1\flat$ might occur as a constituent of the dominant chord of the key of c minor, or of some dissonant chord replacing the dominant chord, it is usual to change it into b_1 and otherwise to use either $b^1\flat$ or b_1 according to the melodic progression, but more frequently the latter, as I have already remarked when treating of the construction of minor scales. It is this systematic use of the major Seventh b_1 in place of the minor Seventh $b^1\flat$ of the key which now distinguishes the modern minor mode from the ancient Hypodoric,† or the mode of the minor Third. Here again some part of the consistency of the scale is sacrificed in order to bind the harmony closer together.

The chain of consonant chords in the mode of the minor Third is certainly impaired when that mode is transformed into our minor by the introduction of b_1. In place of the chain

$$\overbrace{f - a^1\flat + C - E^1\flat + G - b^1\flat + d}$$

our minor furnishes only

$$\overbrace{f - a^1\flat + C - E^1\flat + G + b_1 - d}$$

which has one triad less. But the composer is at liberty to alternate the two tones $b^1\flat$ and b_1.

The introduction of the leading note b_1 into the key of c minor generated a new difficulty in the complete closing cadence of this key. When the chord $g + b_1 - d$ is followed by the chord $c - e^1\flat + g$, the first being a perfectly harmonious major chord, and the latter an obscurely harmonious minor chord, the defect in the harmonious-

---

* [This was referred to in p. 294d, note §. See App. XX. sect. E. art. 26, example of the use of duodenals. It is a real, though temporary modulation into a new duodene, one Fifth lower. But for $D \mid F + A_1$ we might use $D - F^1 + A$, which is again a modulation into a new duodene, one Fifth higher. This should be traced on the Duodenarium.—*Translator.*]

† [Hypodoric, also called Eolic, p. 268d, footnote No. 6, but here the harmonic alteration of that mode is meant as in p. 274, footnote No. 3. This confusion is here regular and intentional.—*Translator.*]

ness of the latter is made much more evident by the contrast. But it is precisely in the final chord that perfect consonance is essential to satisfy the feeling of the hearer. Hence this close could not become satisfactory until the chord of the dominant Seventh had been invented, which changed the dominant consonance into a dissonance.

The preceding explanation shews that when we try to institute a close connection among all the chords peculiar to a mode similar to the close connection among the tones of the scale, (that is, when we require all the consonant triads in the harmonic tissue to be related to one of their number, the tonic triad, in a manner analogous to that in which the notes of the scale are related to one of their number, the tonic tone), there are only *two tonal modes*, the *major* and *minor*, which properly satisfy such conditions of related tones and related chords.

The *major mode* fulfils the two conditions of chordal relationship and tonal relationship in the most perfect manner. It has four triads which are immediately related to the tonic chord

$$f + \overbrace{a_1 - C + E_1} - \overbrace{G + b_1} - d$$

Its harmonisation can be so conducted, (indeed, in popular pieces which must be readily intelligible, it is so conducted,) that all tones appear as constituents of the three major chords of the system, those of the tonic, dominant, and subdominant. These major chords, when their roots lie low, appear to the ear as reinforcements of the compound tones of the tonic, dominant, and subdominant, which tones are themselves connected by the closest possible relationship of Fifths. Hence in this mode everything can be reduced to the closest musical relationship in existence. And since the tonic chord in this case represents the compound tone of the tonic immediately and completely, the two conditions—predominance of the tonic tone and of the tonic chord—go hand in hand, without the possibility of any contradiction, or the necessity of making any changes in the scale.

The major mode has, therefore, the character of possessing the most complete melodic and harmonic consistency, combined with the greatest simplicity and clearness in all its relations. Moreover, its predominant chords being major, are distinguished by full unobscured harmoniousness, when such positions are selected for them as do not introduce inappropriate combinational tones.

The major scale is purely diatonic, and possesses the ascending leading note of the major Seventh, whence it results that the tone most distantly related to the tonic is brought into closest melodic connection with it.

The three predominant major chords furnish tones sufficient to produce two minor chords, which are closely related to them, and can be employed to diversify the succession of major chords.

The *minor mode* is in many respects inferior to the major. The chain of chords for its modern form is—

$$f - \overbrace{a^1\flat + C} - \overbrace{E^1\flat + G} + \overbrace{b_1 - d}$$

Minor chords do not represent the compound tone of their root as well as the major chords ; their Third, indeed, does not form any part of this compound tone. The dominant chord alone* is major, and it contains the two supplementary tones of the scale. Hence when these appear as constituents of the dominant triad, and therefore of the compound tone of the dominant, they are connected with the tonic by the close relationship of Fifths. On the other hand, the tonic and subdominant triads do not simply represent the compound tones of the tonic and subdominant notes, but are accompanied by Thirds which cannot be reduced to the close relation-

---

* [That is, among the characteristic chords. The two minor chords, as is shewn in the text, contain the tones of one major chord, $a^1\flat + c - c^1\flat$.—*Translator*.]

ship of Fifths. The tones of the minor scale can therefore not be harmonised in such a way as to link them with the tonic note by so close a relationship as in the major mode.

The conditions of tonality cannot be so simply reconciled with the predominance of the tonic chord as in the major mode. When a piece concludes with a minor chord, we hear, in addition to the compound tone of the tonic note, a second compound tone which is not a constituent of the first. This accounts for the long hesitation of musical composers respecting the admissibility of a minor chord in the close.

The predominant minor chords have not the clearness and unobscured harmoniousness of the major chords, because they are accompanied by combinational tones which do not. fit into the chord.

The minor scale contains an interval $a^1\flat...b_1$, which exceeds a whole Tone in the diatonic scale,[*] and answers to the numerical ratio 75 : 64 [$=274$ cents]. ¶ To make the minor scale melodic it must have a different form in descending from what it has in ascending, as mentioned in the last chapter.

The minor mode, therefore, has no such simple, clear, intelligible consistency as the major mode ; it has arisen, as it were, from a compromise between the different conditions of the laws of tonality and the interlinking of harmonies. Hence it is also much more variable, much more inclined to modulations into other modes.

This assertion that the minor system is much less consistent than the major, will be combated by many modern musicians, just as they have contested the assertion already made by me, and by other physicists before me, that minor triads are generally inferior in harmoniousness to major triads. There are many eager assurances of the contrary in recent books on the theory of harmony.[†] But the history of music, the extremely slow and careful development of the minor system in the sixteenth and seventeenth centuries, the guarded use of the minor close by Handel, the partial avoidance of a minor close even by Mozart,—all these seem to ¶ leave no doubt that the artistic feeling of the great composers agreed with our conclusions.[‡] To this must be added the varied use of the major and minor Seventh, and the major and minor Sixth of the scale, the modulations rapidly introduced and rapidly changing, and finally, but very decisively, popular custom. Popular melodies can contain none but clear transparent relations. Look through collections of songs now preferred by those classes among the Western nations which have often an opportunity of hearing harmonic music, as students, soldiers, artisans. There are scarcely one or two per cent. in minor keys, and those are mostly old popular songs which have descended from the times of homophonic music. It is also characteristic that, as I have been assured by an experienced

---

* [The interval is so strange, when unaccompanied, that if it had to be taken merely as an interval, $a^1\flat$ 274b_1, a singer would probably fail. But the $a^1\flat$ is taken as the minor Third of f with ease, and the b_1 is taken as the leading note to c', with equal ease, so that the perfectly unmelodic and inharmonic interval $a^1\flat$ 274b_1 never comes into consideration at all. To get rid of it, the subdominant is often taken major, producing the chords of 1 C ma.mi.ma., App. XX. sect. E. art. 10, III., which makes the scale c204 d 112$e^1$$\flat$182$f$ 204 g 182 a_1,204 b_1,112 c', and this differs from the major only by having $e^1\flat$ in place of e_1. In many pianoforte instruction books this is given as the only form of the ascending minor. Mr. Curwen (*Standard Course*, p. 86) says that this major Sixth ‘ascending is very difficult to sing,’ and ‘has a hard and by no means pleasant effect,’ and points out that it leads singers to forget the key, and in such a phrase as $g\,a_1\,b_1\,c'\,d'\,e^{'1}\flat$, the pupils will sing c', instead of $e^{'1}\flat$; and even in singing such a passage as $g\,a_1\,b_1\,c'\,g$, instead of

falling upon the same note with which they began, will take e'_1, the major Third of c'. Hence the difficulty is not avoided but increased by introducing the ambiguity of the ¶ major key, into which this is a real modulation from g onwards.—*Translator*.]

† [Can this be due to temperament ? The sharp equally tempered major Third of 400 cents is worse of its kind than the flat equally tempered minor Third of 300 cents, which approaches close to 16 : 19 = 298 cents, an interval which many like, and which may be tried as c''' : $^{19}e'''\flat$ on the Harmonical.—*Translator*.]

‡ [These composers played in meantone temperament (App. XX. sect. A. art. 16), in which the minor Third of 310 cents was much rougher than the equally tempered one of 300 cents, having much slower beats. Possibly this difference in the modes of tempering the minor Third, may have led to the difference of opinion mentioned in the text.—*Translator*.]

teacher of singing, pupils of only moderate musical talent have much more difficulty in hitting the minor than the major Third.

But I am by no means of opinion that this character depreciates the minor system. The major mode is well suited for all frames of mind which are completely formed and clearly understood, for strong resolve, and for soft and gentle or even for sorrowing feelings, when the sorrow has passed into the condition of dreamy and yielding regret. But it is quite unsuited for indistinct, obscure, unformed frames of mind, or for the expression of the dismal, the dreary, the enigmatic, the mysterious, the rude, and whatever offends against artistic beauty ;—and it is precisely for these that we require the minor mode, with its veiled harmoniousness, its changeable scale, its ready modulation, and less intelligible basis of construction. The major mode would be an unsuitable form for such purposes, and hence the minor mode has its own proper artistic justification as a separate ¶ system.

The harmonic peculiarities of the modern keys are best seen by comparing them with the harmonisation of the other ancient tonal modes.

Major Mode.

Among the melodic tonal modes the *Lydian* of the Greeks (the ecclesiastical *Ionic* [p. 274, note, No. 1]), in agreement with our major, is the only one which has an ascending leading note in the form of a major Seventh. The four others had originally and naturally only minor Sevenths, which even in the later periods of the middle ages began to give place to major Sevenths, in order that the Seventh of the scale, which was in itself so loosely connected with the tonic, might be more closely united to it by becoming the leading note to the tonic at the close.

Mode of the Fourth.

¶   The *mode of the Fourth* (the Greek *Ionic*, and ecclesiastical *Mixolydian*) is principally distinguished from the major mode by its minor Seventh. By merely changing this into the major we obliterate the difference between them. Taking C as the tonic the chain of chords in the unaltered mode are as on p. 298b, No. 2,

$$f + \overbrace{a_1 - C + E_1} - \overbrace{G - b^1\flat + d}$$

If we attempt to form a complete cadence in this mode, as in the following examples 1 and 2, they will sound dull from want of the leading note, even when the dominant chord is extended to a chord of the Seventh $g - b^1\flat + d \,|\, f$, as in 2.

The second example, in which the leading note $b^1\flat$ lies uppermost, is even duller than the first example, in which that note $b^1\flat$ is more concealed. The $b^1\flat$ in these examples has a very uncertain sound. It is not closely enough related to the tonic, it is not part of the compound tone of the dominant note g, it is not sufficiently close in pitch to serve as a leading note to the tonic, and it has no tendency

---

* [The [C] is the duodenal of App. XX. sect. E. art. 26, shewing the exact pitch of all the notes. These examples have been trans-   posed to admit of their being played on the Harmonical.—*Translator*.]

to push on to the tonic. Hence when the older composers wished to distinguish pieces written in the mode of the Fourth from those in the major mode, by their close, they employed the imperfect or plagal cadence, as in example 3. And as such a cadence wants the decisive progression required for a close, the sluggishness previously caused by the absence of a leading tone ceases to be striking.*

In the course of a piece written in this tonal mode, the leading note b_1 may of course be used in ascending passages, provided the minor Seventh $b^1\flat$ is employed often enough in descending passages. But the effect of the mode is destroyed when an essential tone of the scale is changed at the close. Hence pieces in the mode of the Fourth sound like pieces in a major mode which have a decided inclination to modulate into the major mode of the subdominant.† For reasons already given, transition to the subdominant appears to be less active than transition to the dominant. This tonal mode has also no decided progression at the close, whereas major chords, of which the tonic is one, predominate in it owing to ¶ their greater harmoniousness. The mode of the Fourth is consequently as soft and harmonious as the major mode, but it wants the powerful forward impetus of major movement. This agrees with the character assigned to it by Winterfeld.‡ He describes the ecclesiastical *Ionic* (major) mode, as a scale which, ' strictly self-contained and founded on the clear and bright major triad—a naturally harmonious and satisfactory fusion of different tones,—also bears the stamp of bright and cheerful satisfaction.' On the other hand, the ecclesiastical *Mixolydian* (mode of the Fourth) is a scale ' in which every part by sound and movement hastens to the source of its fundamental tone ' (that is, to the major mode of its subdominant), ' and this gives it a yearning character in addition to the former cheerful satisfaction, not unlike to the Christian yearning for spiritual regeneration and redemption, and return of primitive innocency, though softened by the bliss of love and faith.'

Mode of the minor Seventh.   ¶

The *mode of the minor Seventh* (Greek *Phrygian* [p. 274, note, No. 4] ecclesiastical *Doric*) has a minor chord on c as the tonic, and originally another on g as the dominant, while it has a major chord on its subdominant f, and this last chord distinguishes the mode from the *mode of the minor Third* (Eolic [p. 294d, note, No. 3]) ; thus

$$f + a_1 - \overset{\frown}{C - E^1\flat} + \overset{\frown}{G - b^1\flat} + d$$

Both of these modes of the minor Seventh and minor Third may, without destroying their character, change the minor Seventh $b^1\flat$ into a leading note b_1, and our minor mode is a fusion of both. The ascending minor scale belongs to the mode of the minor Seventh, in which the leading note is used, and the ¶ descending to the mode of the minor Third. But when the mode of the minor Seventh admits the leading note, its chain of chords reduces to the three essential chords of the scale

$$f + a_1 - \overset{\frown}{C - E^1\flat} + G + b_1 - d$$

This tonal mode has all the character of a minor, but the transition to the chord of the subdominant has a brighter effect than in the normal minor, where the sub-

---

* [These can be played on the Harmonical. —*Translator.*]

† [This inclination seems to arise from the tempered confusion of $b^1\flat\, d$ with $bb\, d_1$, so that the scale $c\, d\, e_1\, f\, g\, a_1\, b^1\flat\, c'$ becomes confused

with its subdominant $f\, g\, a_1\, bb\, c'\, d'_1\, e'_1\, f'$.— *Translator.*]

‡ *Johannes Gabrieli und sein Zeitalter,* vol. i. p. 87.

dominant chord is also minor. On forming the complete cadence both dominant and subdominant chords are major, while the tonic remains minor. This has of course an unpleasant effect in the close, because it makes the final chord obscurer than either of the other two principal chords. Hence it is necessary to introduce strong dissonances into these two chords, to restore the balance. But if we follow the old composers and make the final chord major, we give the closing cadence of this mode an unmistakably major character. As in ecclesiastical modes it is always allowed to change A_1 into $A^1\flat$, which would change the subdominant chord of the mode of the minor Seventh into a minor chord,* we can protect the mode of the minor Seventh from confusion with the major mode in its final cadence, but then again it will entirely coincide with the old minor cadence.

Sebastian Bach introduces the major Sixth of the tonic, which is peculiar to this tonal mode, into other chords for the closing cadence, and thus avoids the
¶ major triad on the subdominant. He very usually employs the major Sixth as the Fifth of the chord of the Seventh on the Second of the scale,† as in the following examples. No. 1 is the conclusion of the chorale: *Was mein Gott will, das gescheh' allzeit*, in the St. Matthew Passion-Music. No. 2 is the conclusion of the hymn *Veni redemptor gentium*, at the end of the cantata: *Schwingt freudig Euch empor zu den erhabenen Sternen.* In both the tonic is b_1, the major Sixth $g_2\sharp.\ddagger$

Ex. 1. $[B_1.]$ $[E_1.]$ $[B_1.]$ $[F_1\sharp.]$ $[B_1.]$

* [In the original the scale was $g + b_1 - D - F^1 + A + C_1\sharp - e$ in order that it might run from D to d; and hence the statement was that it is allowable to change B into $B^1\flat$. But in order to keep to the same notes as were used previously, and to allow of the scale being played on the Harmonical, I have transposed it, and hence have had to make the same change here. The result is precisely the same, merely meaning that the Seventh
¶ might be taken minor.—*Translator*.]

† [In the scale $f + a_1 - c - e^1\flat + g + b_1 - d$, a_1 is the major Sixth of the tonic c and d the Second. The chord of the Seventh on the Second of the scale is therefore $d + f_1\sharp - a \mid c$,

hence if S. Bach makes this Fifth a agree with the major Sixth of the scale a_1, he is thinking in tempered music. When just intonation is restored, this occasions a restless modulation as shewn by the duodenals which I have introduced over the following examples.—*Translator*.]

‡ [The notes in the staff notation are the usual tempered scale, but the inserted duodenals convert them into just notes, on the principle of App. XX. sect. E. art. 26. The tonic is taken as B_1 in order to be within the duodene of C, and hence the subdominant is E_1 and the dominant $F_1\sharp$, giving the three duodenes: In Ex. 1 the $[B_1]$ indicates that the first two

Subdominant E_1			Tonic B_1			Dominant $F_1\sharp$		
D	$F_1\sharp$	$A_2\sharp$	A	$C_1\sharp$	$E_2\sharp$	E	$G_1\sharp$	$B_2\sharp$
G	B_1	$D_2\sharp$	D	$F_1\sharp$	$A_2\sharp$	A	$C_1\sharp$	$E_2\sharp$
C	E_1	$G_2\sharp$	G	B_1	$D_2\sharp$	D	$F_1\sharp$	$A_2\sharp$
F	A_1	$C_2\sharp$	C	E_1	$G_2\sharp$	G	B_1	$D_2\sharp$

chords are in the duodene of B_1. Then $[E_1]$ shews that the next two chords are in the duodene of E_1. The difference relates to the chords with A in the first case and A_1 in the second. But the next pair of chords return to the duodene of B_1, which remains till the last bar, when the notes are in the duodene of $F_1\sharp$. This is rendered necessary by the chord of the Seventh on $C_1\sharp$ the second of the scale, the Fifth of which is $G_1\sharp$ and not $G_2\sharp$, which is the Sixth of the scale of B_1. That is, it is $C_1\sharp + E_2\sharp$

$- G_1\sharp \mid B_1$. This is, however, only a temporary modulation, and the piece ends in the duodene of B_1. In Ex. 2 the modulations are only B_1, $F_1\sharp$ and B_1, that into $F_1\sharp$ being necessitated by the same chord as before. If these modulations were not taken, but the duodene of B_1 were persisted in throughout, frightful dissonances (much worse than the old 'wolves') would ensue from the imperfect Fifths E_1A and $C_1\sharp G_2\sharp$.—*Translator*.]

2. $[B_1.]$ $[F_1\sharp.]$ $[B_1.]$

There are many similar examples.   He evidently evades a regular close.

MINOR-MAJOR MODE.

Modern composers, when they wish to insert a tonal mode which lies between ¶ Major and Minor, to be used for a few phrases or cadences, have generally preferred giving the minor chord of the mode to the subdominant and not to the tonic. Hauptmann calls this the minor-major mode (*Moll-Durtonart*).* Its chain of chords is—

$$f - a^1\flat + C + \overbrace{E_1 - G + b_1} - d$$

This gives a leading note in the dominant chord, and a complete final cadence in the major chord of the tonic, while the minor relation of the subdominant chord remains undisturbed.   This minor-major mode is at all events much more suitable for harmonisation than the old mode of the minor Seventh.   But it is unsuitable for homophonic singing, unless in the ascending scale $a^1\flat$ is changed into a_1, because the voice would otherwise have to make the complicated step $a^1\flat...b_1[=274$ cents, see p. 301a, d].   The old modes were derived from homophonic singing, for ¶ which the mode of the minor Seventh is perfectly well fitted, as we know from its being still used as our ascending minor scale.†

MODE OF THE MINOR SIXTH.

While the *mode of the minor Seventh* oscillates indeterminately between major and minor without admitting of any consistent treatment, the *mode of the minor Sixth* (Greek *Doric*, [p. 274d', note No. 7] ecclesiastical *Phrygian*), with its minor Second, has a much more peculiar character, which distinguishes it altogether from all other modes.   This minor Second stands in the same melodic connection to the tonic as a leading note would do, but it requires a descending progression.   Hence for descending passages this mode possesses the same melodic advantages as the major mode does for ascending passages.   The minor Second has the more distant relationship with the tonic, due entirely to the subdominant.   The mode cannot ¶ form a dominant chord without exceeding its limits.   If we keep c as the tonic, the chain of chords is

$$b\flat - d^1\flat + f - a^1\flat + C - E^1\flat + G - b^1\flat ‡$$

In this case the chords $b\flat - d^1\flat + f$ and $d^1\flat + f - a^1\flat$ are not directly related to the tonic.   The tone $d^1\flat$ cannot enter into any consonant chord which is directly related to the tonic.   But since $d^1\flat$ is the characteristic minor Second of the mode, such chords cannot well be avoided, not even in the cadence.   Although, then,

---

* [It is 1 C mi.ma.ma. of App. XX. sect. E. art. 9.—*Translator*.]

† [After the introduction of the leading note to form a major dominant chord.—*Translator*.]

‡ [The notes have been transposed in order to keep the same tonic chord $C - E^1\flat + G$.

Observe that both Sevenths $b\flat$ and $b^1\flat$ are introduced. If $b^1\flat$ be omitted, the system of chords is that of 5 F mi.mi.mi.   On the Harmonical, on account of the absence of $b\flat$, it is necessary to use the system of chords $d_1 - f + a_1 - c + E_1 - G + B_1 - d$.—*Translator*.]

there is a close relationship between the consecutive links of the chain of chords, some of its indispensable terms are only distantly related to the tonic. Moreover, in the course of a piece in this mode, it will always be necessary to form the dominant chord $g + b_1 - d$* although it contains two tones foreign to the original mode, as otherwise we could not prevent the prevalence of the impression that f is the tonic and $f - a^1\flat + c$ the tonic chord. It follows, therefore, that the *mode of the minor Sixth* must be still less consistent in its harmonisation and still more loosely connected than the minor mode, although it admits of very consistent melodic treatment. It contains three essential minor chords, namely the tonic $c - e^1\flat + g$, the subdominant $f - a^1\flat + c$, and the chord which contains the two tones slightly related to the tonic $b\flat - d^1\flat + f$. It is exactly the reverse of the major mode, for whereas that mode proceeds towards the dominant, this mode proceeds towards the subdominant.

¶ Major: $f + a_1\ -C + E_1\ -G + b_1 - d$
   Mode of | | |
   minor Sixth : $b\flat - d^1\flat + f - a^1\flat + C - E^1\flat + G$

For harmonisation the difference of the two cases is, first, that the related tones introduced into the scale by the subdominant f, namely $b\flat$ and $d^1\flat$, are not partials of the compound tone of the subdominant, whereas tones b_1 and d, which are introduced by the dominant, are some of the partials of the tonic ; and, secondly, that the tonic chord always lies on the dominant side of the tonic tone. Hence in the harmonic connection, the tones $b\flat$ and $d^1\flat$ cannot be so closely united with either the tonic tone or the tonic chord, as is the case with the supplementary tones introduced by the dominant. This gives a kind of exaggerated minor character to the mode of the minor Sixth, when harmonised. Its tones and chords are certainly connected, but much less clearly and intelligibly than those of the minor
¶ system. The chords which can be brought together in this key, without obscuring reference to c as the tonic, are $b\flat$ minor and $d^1\flat$ major on the one hand and g major on the other, chords which in the major system could not be brought together without extraordinary modulational appliances.† The esthetical character of the mode of the minor Sixth corresponds with this fact. It is well suited for the expression of dark mystery, or of deepest depression, and an utter lapse into melancholy, in which it is impossible to collect one's thoughts. On the other hand, as its descending leading note gives it a certain amount of energy in descent, it is able to express earnest and majestic solemnity, to which the concurrence of those major chords which are so strangely connected gives a kind of peculiar magnificence and wondrous richness.

Notwithstanding that the mode of the minor Sixth has been rejected from modern musical theory, much more distinct traces of its existence have been left in musical practice than of any other ancient mode ; for the mode of the Fourth
¶ has been fused into the major, and the mode of the minor Seventh into the minor. Certainly a mode like that we have described is not suitable for frequent use ; it is not closely enough connected for long pieces, but its peculiar power of expression cannot be replaced by that of any other mode. Its occurrence is generally marked by its peculiar final cadence which starts from the minor Second in the root. In Handel the natural cadence of this system is used with great effect. Thus in the

* [The introduction of this chord shews that the composer is writing in the key of c, but has a prevailing tendency to modulate into the subdominant, from which $b\flat$, $d^1\flat$ are chosen. When $b^1\flat$ is used for $b\flat$, or b_1 for $b^1\flat$, the modulation into the subdominant does not take place. The major chord $e^1\flat + g - b^1\flat$ is entirely adventitious. If it is used in ascending, thus, c 112 $d^1\flat$ 204 $e^1\flat$ 182 f 204 g 112 $a^1\flat$ 204 $b^1\flat$ 182 c', the result is the scale of

3 $A^1\flat$ ma.ma.ma. of App. XX. sect. E. art. 9.—*Translator.*]

† [This, in fact, lengthens the original chain of chords into $b\flat - d^1\flat + f - a^1\flat + c - e^1\flat + g + b_1 - d$, and leads to the treatment of the mode as merely C minor, with a tendency to modulate into F minor. The C minor is, however, the modern minor C mi.mi.ma., and the F minor is F mi.mi.mi., which is much more gloomy.—*Translator.*]

Messiah, the magnificent fugue *And with his stripes we are healed*, which has the signature of *F* minor, but by its frequent use of the harmony of the dominant Seventh on *G*, shews that *C* is the real tonic, introduces the pure [ecclesiastical Phrygian] Doric cadence as follows : *

Similarly in *Samson*,† the chorus, *Hear, Jacob's God*, which, written in the ¶ Doric mode of *E*, finely characterises the earnest prayer of the anxious Israelites as contrasted with the noisy sacrificial songs of the Philistines in *G* major, which immediately follow. The cadence here also is purely Doric.‡

The chorus of Israelites which introduces the third part : *In Thunder come, O God,* ¶ *from heaven!* and is chiefly in *A* minor, has likely an intermediate Doric section.

Sebastian Bach also, in the chorales which he has harmonised, has left them in the mode of the minor Sixth, to which they melodically belonged, whenever the text requires a deeply sorrowful expression, as in the *De Profundis* or the *Aus tiefer Noth schrei' ich zu dir*, and again in Paul Gerhardt's song, *Wenn ich einmal soll scheiden, so scheide nicht von mir*. But he has harmonised the same melody arranged for other texts, as *Befiehl Du deine Wege*, and *O Haupt von Blut und Wunden*, &c., as major or minor, in which case the melody ends on the Third or Fifth of the key, instead of on the Doric tonic.

Fortlage § had already observed that Mozart had applied the Doric mode in

---

* [The cadence is produced by passing from the minor subdominant $Bb - D^1b + F$ to the major dominant, $C + E_1 - G$, in the key of *F* minor. This is the concluding cadence of the whole fugue, and for this reason apparently, the signature in Novello's edition is that of *C* minor, not of *F* minor, and the d^1b is marked as an accidental throughout. That is, Novello takes the key to be *C* minor with a constant tendency to modulate into the key of the subdominant, from which it borrows the chord $Bb - D^1b + F$. But the fugue begins with *F...f* in the bass, and the opening subject, in the treble, is c'', $a^{1'}b$, $d^{1''}b$, e_1', f', g', $a^{1'}b$, $b'b$, c'', which is clearly in the scale of *F* minor, with the chordal system $bb - d^1b + f - a^1b + c + e_1 - g$, of which it contains every note. In the text the [*F.*] is the duodenal and refers to the duodene of *F*, which contains all the tones in the passage. The whole fugue oscillates between the duodenes *C* and *F*.—*Translator*.]

† [Mr. H. Keatley Moore informs me that this chorus was taken by Handel from *Plorate filiae Israel* in Carissimi's *Jephthah*.—*Translator*.]

‡ [The duodene is that of A_1. The succes- ¶ sion of chords, each reduced to the simplest form, as referred to by the bracketed figures below the notes, is 1. $e_1 - g + b_1$, 2. $a_1 - c + e_1$, 3. $e_1 - g + b_1$, 4. $f + a_1 - c$, 5. $d_1 - f + a_1$, 6. $e_1 + g_2\sharp - b$, 7. $a_1 - c + e_1$, 8. $e_1 + g_2\sharp - b_1$. Hence, assuming the scale to have the chordal system $d_1 - f + a_1 - c + e_1 - g + b_1$, with $e_1 - g + b_1$ as the tonic chord, taken major as $e_1 + g_2\sharp - b_1$ in the close, we have the 'Doric cadence' between chords 5 and 6, which is then lengthened by introducing the remaining tones of the key in 7, the whole closing as in 8. It would be most probably received as in A_1 minor, closing on the dominant.—*Translator*.]

§ Examples from instrumental music are mentioned by Ekert in his *Habilitationsschrift, Die Principien der Modulation und musikalischen Idee.* Heidelberg, 1860, p. 12.

Pamina's air in the second act of *Il Flauto Magico* [No. 19]. One of the finest examples for the contrast between this and the major mode occurs in the same composer's *Don Giovanni*, in the Sestette of the second act [No. 21], where Ottavio and Donna Anna enter. Ottavio sings the comforting words—

> Tergi il ciglio, o vita mia,
> E dà calma al tuo dolore

in *D* major, which, however, is peculiarly coloured by a preponderating, although not uninterrupted, inclination to the subdominant, as in the mode of the Fourth. Then Anna, who is plunged in grief, begins in perfectly similar melodical phrases, and with a similar accompaniment, and after a short modulation through *D* minor, establishes herself in the mode of the minor Sixth for *C*, with the words—

¶

> Sol la morte, o mio tesoro,
> Il mio pianto può finir.

The contrast between gentle emotion and crushing grief is here represented with a most wonderfully beautiful effect, principally by the change of mode. The dying Commandant also, in the introduction to *Don Giovanni*, ends with a Doric cadence.* Similarly the *Agnus Dei* of Mozart's *Requiem*—although, of course, we are not quite certain how much of this was written by himself.

Among Beethoven's compositions we may notice the first movement of the Sonata, No. 90, in *E* minor, for the pianoforte, as an example of peculiar depression caused by repeated Doric cadences, whence the second (major) movement acquires a still softer expression.

Modern composers form a cadence which belongs to the mode of the minor Sixth, by means of the minor Second and the major Seventh, the so-called chord of the extreme sharp Sixth,† $f^1 + a ... d_1\sharp$, where both f^1 and $d_1\sharp$ have to move ¶ half a tone to reach the tonic *e* [p. 286b]. This chord cannot be deduced from the major and minor modes, and hence appears very enigmatical and inexplicable to many modern theoreticians. But it is easily explained as a remnant of the old mode of the minor Sixth, in which the major Seventh $d_1\sharp$, which belongs to the dominant chord $b + d_1\sharp - f\sharp$, is combined with the tones $f^1 + a$, which are taken from the subdominant side.‡

These examples may suffice to shew that there are still remnants of the mode of the minor Sixth in modern music. It would be easy to adduce more examples if they were looked for. The harmonic connection of the chords in this mode is not sufficiently firm and intelligible for the construction of long pieces. But in short pieces, chorales, or intermediate sections, and melodic phrases in larger musical works, it is so effective in its expression, that it should not be forgotten, especially as Handel, Bach, and Mozart have used it in such conspicuous places in their works.§

¶   * [No. 1 of the opera. Representing major chords by capitals and minor by small letters, the final chords of the vocal music are f, D^1b, G^1b, f, C, f, so that all the tones will lie in the scheme $g^1b + bb - d^1b + f - a^1b + c + e_1 - g$, or $c - e^1b + g$. The tonic is F.—*Translator.*]

† [Callcott (*Musical Grammar*, 1809, art. 441) calls it 'the chord of the *extreme sharp Sixth*,' and says that 'this harmony when accompanied simply by the Third, has been termed the *Italian Sixth*.' Of course he has no theory for it; the tone is 'accidentally sharpened.'—*Translator.*]

‡ [That is the chords of the scale are taken as $d - f^1 + a - c^1 + e - g^1 + b + d\sharp - f\sharp$, of which the two notes last are modern additions. See p. 286d, note †.—*Translator.*]

§ Herr A. von Oettingen, in his *Harmonie-system in dualer Entwickelung* (Dorpat and Leipzig, 1866), has carried out, in a most interesting manner, the complete analogy between the mode of the minor Sixth and the major mode, of which it is the direct conversion; and has shewn how this conversion leads to a peculiarly characteristic harmonisation of the mode of the minor Sixth. In this respect I wish emphatically to recommend this book to the attention of musicians. On the other hand, it seems to me that it is necessary to shew by musical practice, that the new principle, which is made the basis of that writer's theory of the mode of the minor Sixth, considered by him as the theoretically normal minor mode, really suffices for the construction of great musical pieces. The author, namely, considers the minor triad $c - e^1b + g$ as representing the tone g'' which is common to the three compound tones of which it is composed

Similar relations exist for the mode of the Fourth and of the minor Seventh, although these are less specifically different from the major and minor modes respectively. They are, however, capable of giving a peculiar expression to certain musical periods, although difficulties would arise in consistently carrying out these peculiarities through long pieces of music. The harmonic phrases which belong to these two last-named modes can of course also be executed within the usual major and minor systems. But perhaps it would facilitate the theoretical comprehension of certain modulations, if the conception of these modes and of their system of harmonisation were definitely laid down.

The only point, then, as historical development and physiological theory alike testify, for which modern music is superior to the ancient, is harmonisation. The development of modern music has been evoked by its theoretical principle, that the tonic chord should predominate among the series of chords by the same laws of relationship as the tonic note predominates among the notes of the scale. This ¶ principle did not become practically effective till the commencement of last century, when it was felt necessary to preserve the minor chord in the final cadence.

The physiological phenomenon which this esthetical principle brought into action, is the compound character of musical tones which are of themselves chords composed of partials, and consequently, conversely, the possibility under certain circumstances of replacing compound tones by chords. Hence in every chord the principal tone is that of which the whole chord may be considered to express its compound form. Practically this principle was acknowledged from the time that pieces of music were allowed to end in chords of several parts. Then it was immediately felt that the concluding tone of the bass might be accompanied by a higher Octave, Fifth, and, finally, major Third, but not by a Fourth, or minor Sixth, and for a long time also the minor Third was rejected; and we know that the first three intervals (the Octave, Fifth, and major Third) occur among the partials of the compound tone which lies in the bass, and that the others do not. ¶

The various values of the tones of a chord were first theoretically recognised by Rameau in his theory of the fundamental bass, although Rameau was not acquainted with the cause here assigned for these different values. That compound tone which represents a chord according to our view, constitutes its *Fundamental Bass*, *Radical Tone* or *Root*, as distinguished from its *bass*, that is, the tone which belongs to the lowest part. The major triad has the same *root* whatever be its inversion or position. In the chords $c + e_1 - g$, or $g...c + e_1$, the root is still c. The minor chord $d - f^1 + a$ has also as a rule only d as its root in all its inversions, but in the chord of the great [or added] Sixth $f^1 + a...d^1$ we may also consider f^1 as the root, and it is in this sense that it occurs in the cadence of c^1 major. Rameau's successors have partly given up this last distinction; but it is one in which Rameau's fine artistic feeling fully corresponded with the facts in nature. The minor chord really admits of this double interpretation, as we have already shewn (p. 294*d*).

The essential difference between the old and new tonal modes is this: the old ¶ have their minor chords on the dominant, the new on the subdominant side.

The reasons for the following construction have been already investigated.*

In the		The chord of the		
		Subdominant is	Tonic is	Dominant is
Old	Mode of the minor Third . .	Minor	Minor	Minor
	Mode of the minor Seventh .	Major	Minor	Minor
	Mode of the Fourth . . .	Major	Major	Minor
New	Major Mode	Major	Major	Major
	Minor-Major Mode . . .	Minor	Major	Major
	Minor Mode	Minor	Minor	Major

(being a higher Octave of g, of the Fifth of c, and of the major Third of e^1b), and hence calls it 'the phonic g tone,' whereas he considers $c + e_1 - g$ in the same way as we do, as the 'tonic c tone.'

* [It will be seen that this arrangement

CHAPTER XVI.

THE SYSTEM OF KEYS.

THERE is nothing in the nature of music itself to determine the pitch of the tonic of any composition. If different melodies and musical pieces have to be executed by musical instruments or singing voices of definite compass, the tonic must be chosen of a suitable pitch, differing when the melody rises far above the tonic and when it sinks much below it. In short, the pitch of the tonic must be chosen so as to bring the compass of the tones of the piece within the compass of the executants, vocal or instrumental. This inevitable practical necessity entails the condition of being able to give any required pitch to the tonic.

¶   Moreover, in the longer pieces of music it is necessary to be able to make a temporary change of tonic, that is, to *modulate*, in order to avoid uniformity and to utilise the musical effects resulting from changing and then returning to the original key. Just as consonances are made more prominent and effective by means of dissonances, the feeling for the predominant tonality and the satisfaction which arises from it, is heightened by previous deviations into adjacent keys. The variety in musical turns produced by modulational connection has become all the more necessary for modern music, because we have been obliged entirely to renounce, or at any rate materially to circumscribe, the old principle of altering expression by means of the various tonal modes. The Greeks had a free choice among seven different tonal modes, the middle ages among five or six, but we can choose between two only, major and minor. Those old tonal modes presented a series of different degrees of tonal character, out of which two only remain suitable for harmonic music. But the clearer and firmer construction of an harmonic piece gives modern
¶ composers greater freedom in modulational deviations from the original key, and places at their command new sources of musical wealth, which were scarcely accessible to the ancients.

Finally I must just touch on the question, so much discussed, whether each different key has a peculiar character of its own.

It is quite clear that, within the course of a single piece of music, modulational deviations into the more or less distantly related keys on the dominant or subdominant side produce very different effects. This, however, arises simply from the contrast they offer to the original principal key, and would be merely a *relative character*. But the question here mooted is, whether individual keys have an *absolute character* of their own, independently of their relation to any other key.

This is often asserted, but it is difficult to determine how much truth the assertion contains, or even what it precisely means, because probably a variety of different things are included under the term *character*, and perhaps the amount of
¶ effect due to the particular instrument employed has not been allowed for. If an instrument of fixed tones is completely and uniformly tuned according to the equal temperament, so that all Semitones throughout the scale have precisely the same magnitude, and if also the musical quality of all the tones is precisely the same, there seems to be no ground for understanding how each different key should have a different character. Musicians fully capable of forming a judgment have also admitted to me, that no difference in the character of the keys can be observed on the organ, for example. And Hauptmann,[*] I think, is right when he makes the same assertion for singing voices with or without an organ accompaniment. A great change in the pitch of the tonic can at most cause all the higher notes to be strained or all the lower ones obscured.

On the other hand, there is a decidedly different character in different keys on

does not include the mode of the minor Sixth. It was this tabulation which led me to the richordal theory developed in App. XX. sect. E.   art. 9, and thence to the general theory of duodenes in that section.—*Translator*.]
    [*] *Harmonik und Metrik*, p. 188.

pianofortes and bowed instruments. C major and the adjacent $D\flat$ major have different effects. That this difference is not caused by difference of absolute pitch, can be readily determined by comparing two different instruments tuned to different pitches. The $D\flat$ of the one instrument may be as high as the C of the other, and yet on both the C major retains its brighter and stronger character, and the $D\flat$ its soft and veiled harmonious effect. It is scarcely possible to think of any other reason than that the method of striking the short narrow black digitals of the piano must produce a somewhat different quality of tone, and that difference of character arises from the different distribution of the stronger and gentler quality of tone among the different degrees of the scale.* The difference made in the tuning of those Fifths which the tuner keeps to the last, and on which are crowded the whole of the errors in tuning the other Fifths in the circle of Fifths, may possibly be regular, and may contribute to this effect, but of this I have no personal experience. [See App. XX. sect. G. art. 17.]   ¶

In bowed instruments the more powerful quality of tone in the open strings is conspicuous, and there are also probably differences in the quality of tone of strings which are stopped at short and long lengths, and these may alter the character of the key according to the degree of the scale on which they fall. This assumption is confirmed by the inquiries I have made of musicians respecting the mode in which they recognise keys under certain conditions. The inequality of intonation will add to this effect. The Fifths of the open strings are perfect Fifths. But it is impossible that all the other Fifths should be perfect if in playing in different keys each note has the same sound throughout, as appears at least to be the intention of elementary instruction on the violin. In this way the scales of the various keys will differ in intonation, and this will necessarily have a much more important influence on the character of the melody. [See App. XX. sect. G. arts. 6 and 7.]

The differences in quality of tone of different notes on wind instruments are still more striking.   ¶

If this view is correct, the character of the keys would be very different on different instruments, and I believe this to be the case. But it is a matter to be decided by a musician with delicate ears, who directs his attention to the points here raised.

It is, however, not impossible that by a peculiarity of the human ear, already touched upon in p. 116a, certain common features may enter into the character of keys, independent of the difference of musical instruments, and dependent solely on the absolute pitch of the tonic. Since g'''' is a proper tone of the human ear, it sounds peculiarly shrill under ordinary circumstances, and somewhat of this shrillness is common to $f''''\sharp$ and $a''''\flat$. To a somewhat less extent those musical tones of which g'''' is an upper partial, as g''', c'''', and g'', have a brighter and more piercing tone than their neighbours. It is possible, then, that it is not indifferent for pieces in C major to have its high Fifth g'' and high tonic c''' thus distinguished in brightness from other tones, but these differences must in all cases be very slight, and for the present I must leave it undecided whether they have any weight at all. ¶

All or some of these reasons, then, made it necessary for musicians to have free command over the pitch of the tonic, and hence even the later Greeks transposed their scales on to all degrees of the chromatic scale. For singers these transpositions offer no difficulties. They can begin with any required pitch, and find in their vocal instrument all such of the corresponding degrees as lie within the extreme limits of their voice. But the matter becomes much more difficult for musical instruments, especially for such as only possess tones of certain definite degrees of pitch. The difficulty is not entirely removed even on bowed instruments. It is true that these can produce every required degree of pitch ; but players are unable to hit the pitch, as correctly as the ear desires, without acquiring a certain

---

* [Mr. H. Keatley Moore, Mus. B., thinks the difference is due to the different leverage of the black digitals. Although in well constructed digitals the black have as much action at the further end as the white ones, they gain this by a quicker motion, each arm of the lever being shorter, and short keys differing altogether from long ones in the feeling produced in the hand. See also App. XX. sect. N. No. 6.—*Translator.*]

mechanical use of their fingers, which can only result from an immense amount of practice.

The Greek system was not accompanied with great difficulties, even for instruments, so long as no deviations into remote keys were permitted, and hence but few marks of sharps and flats had to be used. Up to the beginning of the seventeenth century musicians were content with two signs of depression for the notes $B\flat$ and $E\flat$, and with the sign $\sharp$ for $F\sharp$, $C\sharp$, $G\sharp$, in order to have the leading tones for the tonics G, D, and A. They took care to avoid the enharmonically equivalent tones $A\sharp$ for $B\flat$, $D\sharp$ for $E\flat$, $G\flat$ for $F\sharp$, $D\flat$ for $C\sharp$, and $A\flat$ for $G\sharp$. By help of $B\flat$ for B* every tonal mode could be transposed to the key of its subdominant, and no other transposition was made.

In the Pythagorean system, which maintained its predominance over theory to the time of Zarlino in the sixteenth century, tuning proceeded by ascending
¶ Fifths, thus—

$$C \quad G \quad D \quad A \quad E \quad B \quad F\sharp \quad C\sharp \quad G\sharp \quad D\sharp \quad A\sharp \quad E\sharp \quad B\sharp$$

Now if we tune two Fifths upwards and an Octave downwards, we make a step having the ratio $\frac{3}{2} \times \frac{3}{2} \times \frac{1}{2} = \frac{9}{8}$, which is a major Second. This gives for the pitch of every second tone in the last list—

$$C \quad D \quad E \quad F\sharp \quad G\sharp \quad A\sharp \quad B\sharp$$
$$1 \quad \tfrac{9}{8} \quad (\tfrac{9}{8})^2 \quad (\tfrac{9}{8})^3 \quad (\tfrac{9}{8})^4 \quad (\tfrac{9}{8})^5 \quad (\tfrac{9}{8})^6$$

Now if we proceed *down*wards by Fifths from C we obtain the series—

$$C \quad F \quad B\flat \quad E\flat \quad A\flat \quad D\flat \quad G\flat \quad C\flat \quad F\flat \quad B\flat\flat \quad E\flat\flat \quad A\flat\flat \quad D\flat\flat.$$

If we descend two Fifths and rise an Octave, we may obtain the tones—

$$C \quad B\flat \quad A\flat \quad G\flat \quad F\flat \quad E\flat\flat \quad D\flat\flat$$
$$1 \quad \tfrac{8}{9} \quad (\tfrac{8}{9})^2 \quad (\tfrac{8}{9})^3 \quad (\tfrac{8}{9})^4 \quad (\tfrac{8}{9})^5 \quad (\tfrac{8}{9})^6$$

¶ Now the interval $(\tfrac{8}{9})^6 = \tfrac{262144}{531441} = \tfrac{1}{2} \times \tfrac{524288}{531441}$

or, approximately $(\tfrac{8}{9})^6 = \tfrac{1}{2} \times \tfrac{73}{74}$

$$(\tfrac{8}{9})^6 = 2 \times \tfrac{74}{73}.$$

Hence the tone $B\sharp$ is higher than the Octave of C by the small interval $\frac{74}{73}$ [$=24$ cents], and the tone $D\flat\flat$ is lower than the Octave below C by the same interval. If we ascend by perfect Fifths from C and $D\flat\flat$, we shall find the same constant difference between

$$C \quad G \quad D \quad A \quad E \quad B \quad F\sharp \quad C\sharp \quad G\sharp \quad D\sharp \quad A\sharp \quad E\sharp \quad B\sharp \text{ and}$$
$$D\flat\flat \quad A\flat\flat \quad E\flat\flat \quad B\flat\flat \quad F\flat \quad C\flat \quad G\flat \quad D\flat \quad A\flat \quad E\flat \quad B\flat \quad F \quad C.$$

The tones in the upper line are all higher than those in the lower by the small interval $\frac{74}{73}$ [$=24$ cents]. Our staff notation had its principles settled before the development of the modern musical system, and has consequently preserved these differences of pitch. But for practice on instruments with fixed tones the distinc-
¶ tion between degrees of tone which lie so near to each other, was inconvenient, and attempts were made to fuse them together. This led to many imperfect attempts, in which individual intervals were more or less altered in order to keep the rest

---

* [In the oldest printed book on music, (*Franchini Gafori Laudensis Musici professoris theoricum opus armonice discipline.* Neapolis M.CCCC.LXXX., for a sight of which I am indebted to Mr. Quaritch, who bought it at the sale of the *Syston Library*) $b\flat$ is in the printed text represented by a small Roman b, and $b\natural$ by a capital Roman B. But in a plate attached are given eight varieties of the written form of $b\natural$, by which it would seem to have been intended for b with a square instead of a round bottom, like $\natural$, which is almost indistinguishable from a mutilated Roman b. As it was clearly made in two parts ᘯ ᒋ, the second was often long, and then the resemblance to $\natural$ was great, and this was almost the cursive written form ᒋ of h in Germany. On the other hand it was often made with two strokes ‖ afterwards crossed, like $\natural$, and then it degenerated into $\natural$, which is apparently the precursor of our $\sharp$. In this case both $\natural$ and $\sharp$ and also h would have arisen from the same square-bottomed b, the French *bécarre*, and Prætorius $\natural$ *quadratum*, which, however, he identified with h, H in subsequent writing. The Italian names for b, $b\flat$ are *si minore, si maggiore*. Whether these refer to the musical intervals $a\ b\flat$, $a\ b\natural$, which Gafori printed a b, a B, or to these printed forms, it is difficult to say with certainty. The Germans accepted the forms b $\natural$, as b h, calling the latter ha. The meaning that

true, producing the so-called *unequal temperaments,* and finally to the system of *equal temperament,* in which the Octave was divided into 12 precisely equal degrees of tone.* We have seen that we can ascend from C by 12 perfect Fifths to $B\sharp$, which differs from c by about $\frac{1}{5}$ of a Semitone, namely by the interval $\frac{74}{73}$. In the same way we can descend by 12 perfect Fifths to $D\flat\flat$, which is as much lower than C, as $B\sharp$ is higher. If, then, we put $C=B\sharp=D\flat\flat$, and distribute this little deviation of $\frac{74}{73}$ equally among all the 12 Fifths of the circle, each Fifth will be erroneous by about $\frac{1}{60}$ of a Semitone [or $\frac{1}{11}$ of a comma or 2 cents], which is certainly a very small interval. By this means all varieties of tonal degrees within an Octave are reduced to 12, as on our modern keyed instruments.

The Fifth in the system of equal temperament, is, when expressed approximately in the smallest possible numbers,$= \frac{3}{2} \times \frac{885}{886}$. It is very seldom that any difficulty could result from its use in place of the perfect Fifth. The root struck with its tempered Fifth makes one beat in the time that the Fifth makes $442\frac{1}{2}$ complete ¶ vibrations. Now since a' makes 440 vibrations in a second, it follows that the tempered Fifth $d' \pm a'$ will produce exactly *one* beat in a second. In long-sustained tones this would, indeed, be perceptible, but by no means disturbing, and for quick passages it would have no time to occur. The beats are still less disturbing in lower positions, where they decrease in rapidity with the pitch numbers of the tones. In higher positions they certainly become more striking; $d''' \pm a'''$ gives four, and $a''' \pm e'''$ six beats in a second; but chords very seldom occur with such high notes in slow passages. The Fourths of the equal temperament are $\frac{4}{3} \times \frac{886}{885}[=498+2$ cents]. There is one beat for every $221\frac{1}{4}$ vibrations of the lower tone of the Fourth. Hence the Fourth $a . d'$ makes one beat in a second, the same as the Fifth $d' \pm a'$. The pure consonances retained in the Pythagorean system are therefore not injured to any extent worth notice by equal temperament. In melodic progression of tones the interval $\frac{885}{886}$ borders on the very limits of distinguishable differences of pitch, according to Preyer's experiments (see p. 147b). ¶ In the doubly accented Octave it would be easily distinguished. In the unaccented or lower Octaves it would not be felt at all.

The Thirds and Sixths of the equal temperament are nearer the perfect intervals than are the Pythagorean.†

Intervals	Perfect		Equally Tempered		Pythagorean	
	ratios	cents	ratios	cents	ratios	cents
Major Third . . .	$\frac{5}{4}$	386	$\frac{5}{4} \times \frac{127}{126}$	400	$\frac{5}{4} \times \frac{81}{80}$	408
Minor Sixth	$\frac{8}{5}$	814	$\frac{8}{5} \times \frac{126}{127}$	800	$\frac{8}{5} \times \frac{80}{81}$	792
Minor Third . . .	$\frac{6}{5}$	316	$\frac{6}{5} \times \frac{121}{122}$	300	$\frac{6}{5} \times \frac{80}{81}$	294
Major Sixth . . .	$\frac{5}{3}$	884	$\frac{5}{3} \times \frac{122}{121}$	900	$\frac{5}{3} \times \frac{81}{80}$	906
Semitone . . .	$\frac{16}{15}$	182	$\frac{18}{17}$ or $\frac{16}{25} \times \frac{147}{148}$	100	$\frac{256}{243}$ or $\frac{16}{15} \times \frac{80}{81}$	90

The dissonances occasioned by the upper partial tones are consequently some- ¶ what milder than those due to Pythagorean intervals, but the combinational tones

Gafori attached to b B (which in one plate he also gives in the same sense in black letter), is shewn by the following quotation which he makes from Guido's hexachord, and this also shews that he used Pythagorean intonation, meaning in our notation:

F fa ut — f
   tonus — 204 cents = Tone
G sol re ut — g
   tonus — 204 cents = Tone
a la mi re — a
   semitonium — 90 cents = Semitone
b fa — $b\flat$
   apothome — 114 cents = Apotome
B mi Trittonus — b Tritone 612 cents
   semitonium — 90 cents = Semitone
c sol fa ut — c' Fifth 702 cents

The Germans generally speak of $\flat$ and $\sharp$ as Be, Kreuz (cross). I do not remember ever having heard the $\natural$ named, but I find in Flügel's Dictionary *Bequadratum* (square b) and *Wiederherstellungszeichen* (sign of restitution, for which $\natural$ was not used till the seventeenth century). Germans never have occasion to use the word, because, instead of 'd flat, d natural, d sharp,' they say 'des, de, dis,' while $b\flat$, $b\natural$ are termed 'be, ha.' On older organ pipes $b\natural$ are constantly used for $b\flat$ b $\natural$, and some organ-builders still use them.—*Translator.*]

* [The general relations on which the schemes of temperament depend will be found in App. XX. sect. A.—*Translator.*]

† [The cents in the Table were of course inserted by me.—*Translator.*]

are much more disagreeable. For the Pythagorean Thirds $c' + e'$ and $e' - g'$ the combinational tones are nearly $C\sharp$ and $B_{\prime}$, both differing by a Semitone from the combinational tone C, which would result from the perfect intervals in both cases. For the Pythagorean minor chord $e' - g' + b'$ the combinational tones are $B_{\prime}$ and very nearly $G\sharp$. The first, $B_{\prime}$, is very suitable, better even than the combinational tone C which results from perfect intonation. But the second, $G\sharp$, belongs to the major and not to the minor chord of E. However, as in perfect intonation one of the two combinational tones C and G is false, the Pythagorean minor chord can hardly be considered inferior in this respect. But the combinational tones of the equally tempered Thirds lie between those of the perfect and Pythagorean Thirds, and are less than a Semitone different from those of just intonation. Hence they correspond to no possible modulation, no tone of the chromatic scale, no dissonance that could possibly be introduced by the progression of the melody; they ¶ simply sound out of tune and wrong.*

These bad combinational tones have always been to me the most annoying part of equally tempered harmonies. When moderately slow passages in Thirds at rather a high pitch are played, they form a horrible bass to them, which is all the more disagreeable for coming tolerably near to the correct bass, and hence sounding as if they were played on some other instrument which was dreadfully out of tune. They are heard most distinctly on the harmonium and violin. Here every professional and even every amateur musician observes them immediately, when their attention is properly directed. And when the ear has once become accustomed to note them, it can even discover them on the piano. In the Pythagorean intonation the combinational tones sound rather as if some one were intentionally playing dissonances. Which of these two evils is the worse, I will not venture to decide. In lower positions where the very low combinational tones can be scarcely, if at all, heard, the equally tempered Thirds have the advantage ¶ over the Greek, because they are not so rough, and produce fewer beats. In higher positions the latter advantage is perhaps destroyed by their combinational tones. However, the equally tempered system is capable of effecting everything that can be done by the Pythagorean, and with less expenditure of means.

C. E. Naumann,† who has lately defended the Pythagorean as opposed to the equally tempered system, grounds his reasons chiefly on the fact that the Semitones which separate the ascending leading tone from the tonic, and the descending minor Seventh from the Third of the chord on which it has to be resolved, are smaller in the Pythagorean (where they are about $\frac{21}{20}$; as appears in the Table on p. 313c) than in the equally tempered, where they are about $\frac{18}{17}$; while they are greatest of all in just intonation, viz. $\frac{16}{15}$. Now in the equally tempered scale there is only one tone between f and g, which is accepted at one time as $f\sharp$ to be a leading note to g, and at another as $g\flat$ to act as a Seventh resolving upon f; but in the Pythagorean there are two tones, $f\sharp$ and $g\flat$, of which the latter is the flatter.

¶   * [This may be seen more clearly by calculating the pitch numbers, assuming c' to be 264 as on p. 17. Then—

Notes	Just, Difference		Pythagorean, Difference		Tempered, Difference	
c'	264		264		264	
		$66 = C$		70·12		68·61
e'	330		334·12		332·61	
		$66 = C$		$61·88 = B_{\prime}$		62·94
g'	396		396		395·55	
		$99 = G$		105·19		102·81
b'	495		501·19		498·36	

The 'differences' give the pitch numbers of the combinational tones. Now we have by p. 17a, $C = 66$, $G = 99$, $B_{\prime} = 61·88$, but the others correspond to no precise tones. The nearest equally *tempered* intervals are B 62·3, C♯

= 69·93, and $G\sharp = 104·76$.—*Translator*.]

† *Ueber die verschiedenen Bestimmungen der Tonverhältnisse.* [On the various determinations of the ratios of tones.] Leipzig, 1858.

Hence the Semitone always approaches the tone on to which it would fall in regular resolution, and the height of the pitch determines the direction of resolution. But although the leading tone plays an important part in modulations, it is perfectly clear that we are not justified in changing its pitch at will in order to bring it nearer to the note on which it has to be resolved. There would otherwise be no limit to our making it come nearer and nearer to that tone, as in the ancient Greek enharmonic mode.* Suppose we replace the Pythagorean Semitone, which is about $\frac{4}{5}$ of the natural Semitone, by another still smaller, about $\frac{3}{5}$ of the natural one, say $\frac{16}{15} \times \frac{80}{81} \times \frac{80}{81}$; the result would be perfectly unnatural as a leading note.† We have already seen that the character of the leading note essentially depends upon its being that tone in the scale which is most distantly related to the tonic, and hence most uncertain and alterable [melodically]. Hence we are perfectly unjustified in deducing from such a tone the principle of construction for the whole scale. ¶

The principal fault of our present tempered intonation, therefore, does not lie in the Fifths; for their imperfection is really not worth speaking of, and is scarcely perceptible in chords. The fault rather lies in the Thirds, and this error is not due to forming the Thirds by means of a series of imperfect Fifths, but it is the old Pythagorean error of forming the Thirds by means of an ascending series of four Fifths. Perfect Fifths in this case give even a worse result than flat Fifths. The natural relation of the major Third to the tonic, both melodically and harmonically, depends on the ratio $\frac{5}{4}$ of the pitch numbers. Any other Third is only a more or less unsatisfactory substitute for the natural major Third. The only correct system of tones is that in which, as Hauptmann proposed, the system of tones generated by Fifths should be separated from those generated by major Thirds. Now as it is important for the solution of many theoretical questions to be able to make experiments on tones which really form with each other the natural intervals required by theory, to prevent the ear from being deceived by ¶ the imperfections of the equal temperament, I have endeavoured to have an instrument constructed by which I could modulate by perfect intervals into all keys.

If we were really obliged to produce in all its completeness the system of tones distinguished by Hauptmann, in order to obtain perfect intervals in all keys, it would certainly be scarcely possible to overcome the difficulties of the problem. Fortunately it is possible to introduce a great and essential simplification by means of the artifice originally invented by the Arabic and Persian musicians, and previously mentioned on p. 281a.

We have already seen that the tones of Hauptmann's system which are generated by Fifths, and are marked by letters without any subscribed or superscribed lines, as $c \pm g \pm d \pm a \pm$, &c., are one comma or $\frac{81}{80}$ [=22 cents] higher than the notes which bear the same names, when generated by major Thirds, and which are here distinguished by an inferior figure as $c_1 \pm g_1 \pm d_1 \pm a_1 \pm$, &c. We have further seen that if we descend from b by a series of 12 Fifths down to $c\flat$, the last tone, reduced to ¶ the proper Octave, is lower than b by about $\frac{74}{73}$ [=24 cents]. Hence we have—

$$b : b_1 = 81 : 80$$
$$b : c\flat = 74 : 73$$

Now these two intervals are very nearly alike; b_1 is rather higher than $c\flat$, but only in the proportion—

$$c\flat : b_1 = 32768 : 32805 \ [=2 \text{ cents}]$$

---

* [However justifiable such alterations may be in unaccompanied melody, they are destructive of harmony, and hence do not belong to harmonic music proper. Of all the older temperaments, the meantone is most harmonious, but this makes the leading tones still further from the tone on which they are resolved, than even in just intonation (117 in place of 112 cents). No diminution of the just Semitone can be made without injury to the major Thirds.—*Translator.*]

† [This would have $112 - 2 \times 22 = 68$ cents, which approaches very closely to the small Semitone $25 : 24 = 70$ cents, so that the effect can be judged from playing $b^1\flat \dots b_1$ on the Harmonical.—*Translator.*]

or, using the approximation obtained by continued fractions—

$$c\flat : b_1 = 885 : 886.$$

The interval between $c\flat$ and b_1 is consequently about the same as that between a perfect and an equally tempered Fifth.*

Now b_1 is the true major Third of g, and if we descend 8 Fifths from g we arrive at $c\flat$ thus :

$$g \pm c \pm f \pm b\flat \pm e\flat \pm a\flat \pm d\flat \pm g\flat \pm c\flat$$

Now, as $c\flat$ is flatter than b_1, if we diminished † all the Fifths by $\frac{1}{8}$ of the small interval $\frac{886}{885}$ we should arrive at b_1 instead of $c\flat$.

Now, since the interval $\frac{886}{885}$ is itself on the limits of sensible difference of pitch, the eighth part of this interval cannot be taken into account at all, and we may ¶ consequently identify the following tones of Hauptmann's system, by proceeding in a series of Fifths from $c\flat = b_1$, that is, the upper line with the lower, or—

$$f\flat \pm c\flat \pm g\flat \pm d\flat \pm a\flat \pm e\flat \pm b\flat$$
$$= e_1 \pm b_1 \pm f_1\sharp \pm c_1\sharp \pm g_1\sharp \pm d_1\sharp \pm a_1\sharp$$

Among musical instruments, the harmonium, on account of its uniformly sustained sound, the piercing character of its quality of tone, and its tolerably distinct combinational tones, is particularly sensitive to inaccuracies of intonation. And as its vibrators also admit of a delicate and durable tuning, it appeared to me peculiarly suitable for experiments on a more perfect system of tones. I therefore selected an harmonium of the larger kind,‡ with two manuals, and a set of vibrators for each, and had it so tuned that by using the tones of the two manuals I could play all the major chords from $F\flat$ major to $F\sharp$ major. The tones are thus dis-¶ tributed :

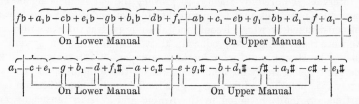

This instrument therefore furnishes 15 major chords and as many minor chords, with perfectly pure Thirds, but with Fifths too flat by $\frac{1}{8}$ of the interval by which an equally tempered Fifth is too flat.§ On the Lower Manual we have the

* [On account of the approximate character of the calculation, the extreme closeness of result is not well shewn. Taking the accurate ¶ numbers, the ratio
$b_1 \div c\flat = \frac{32805}{32768}$, giving cents $1 \cdot 953721$.
Perfect Fifth÷tempered Fifth$= \frac{3}{2} \div \sqrt[12]{2}^7$, giving cents $701 \cdot 955001 - 700 = 1 \cdot 955001$. Difference $\cdot 001280$ cents. Human ears, however much assisted by human contrivances, could never hear the difference.—*Translator.*]

† [Accidentally misprinted 'increased,' that is, 'zu gross,' and 'too sharp,' that is, 'zu hoch,' instead of 'zu klein,' and 'zu tief' in all the four German editions. This error evidently arose merely from forgetting for the moment that the Fifths were taken *down* and not *up*. Now 8 perfect Fifths down $= -8 \times 702$ cents $= -5616$ cents, which on adding 5 octaves $= 6000$ cents, gives 384 cents, and this is less than the major Thirds of 386 cents by 2 cents. Hence if we diminish each Fifth by $\frac{1}{4}$ cent, 8 diminished Fifths down $= -8 \times 701\frac{3}{4} = -5614$ cents, which, on adding 5 octaves or 6000 cents

gives 386 cents, and this is the correct major Third. But to tune Fifths of this kind, if possible, would be a work of immense labour even with tuning-forks, the most permanent of existing conveyors of pitch, and the most perfect apparatus known. Thus such a Fifth reckoned from c' 264 vib. gives g' $395 \cdot 944$ vib., while the perfect g' is $396 \cdot 000$ vib., difference $\cdot 056$ vib., which it is hopeless to tune exactly. Hence these Fifths can only be regarded as products of calculation which could not be realised. In App. XX. sect A. art. 18 I term the resulting temperament Helmholtzian, although, as will be seen in the following note §, Prof. Helmholtz himself did not attempt to realise it.—*Translator.*]

‡ Made by Messrs. J. & P. Schiedmayer, in Stuttgart.

§ The tuning of this instrument was easily managed. Herr Schiedmayer succeeded at the first attempt by the following direction. Starting from a on the lower manual, tune the Fifths $d \pm a$, $g \pm d$, $c \pm g$ perfectly just, and thus

complete scales of $C\flat$ major and G major and in the upper the complete scales of $E\flat$ major and B major inclusive complete. All the major scales exist from $C\flat$ major to B major, and they can all be played with perfect exactness in the natural intonation. But to modulate beyond B major on the one side and $C\flat$ major on the other, it is necessary to make a really enharmonic interchange between B_1 and $C\flat$, which perceptibly alters the pitch (by a comma $\frac{81}{80}$).* The minor modes on the lower manual are B_1 minor or $C\flat$ minor complete, on the upper manual $D_1\sharp$ minor or $E\flat$ minor.

For the minor keys this series of tones is not quite so satisfactory as for the major keys. The dominant of a minor key is the Fifth of a minor triad and the root of a major triad. But as the minor chord has to be written as $a_1 - c + e_1$, and the major chord as $f\flat + a_1\flat - c\flat$, the corresponding dominant must be written in the first chord with an *inferior* number, and in the second with a letter without any number attached ; that is, they must be tones of the kind which we have ¶ identified by means of the assumptions here made, as in the present case where e_1

obtain c, g, d. Then tune the major chords $c + e_1 - g$, $g + b_1 - d$, $d + f_1\sharp - a$, giving the three tones $e_1 b_1 f_1\sharp$, and finally the Fifth, $f_1\sharp \pm c_1\sharp$, to obtain $c_1\sharp$. Then putting $e_1 = f\flat$, $b_1 = c\flat$, $f_1\sharp = g\flat$, $c_1\sharp = d\flat$, tune the major chords $f\flat + a_1\flat - c\flat$, $c\flat + e_1\flat - g\flat$, $g\flat + b_1\flat - d\flat$ with pure Thirds giving no beats, thus obtaining $a_1\flat, e_1\flat, b_1\flat$, and finally the Fifth $b_1\flat \pm f_1$, giving f_1. This completes the tuning of the notes on the lower manual. For the upper manual first tune e as the perfect Fifth of the a in the lower manual, and then the three major chords $e + g_1\sharp - b$, $b + d_1\sharp - f\sharp$, $f\sharp + a_1\sharp - c\sharp$, and the Fifth $a_1\sharp \pm e_1\sharp$, giving $b, f\sharp, c\sharp$, and then $g_1\sharp$, $d_1\sharp$, $a_1\sharp$ and also $e_1\sharp$. Then put $g_1\sharp = ab$, $d_1\sharp = eb$, $a_1\sharp = bb$, $e_1\sharp = f$, and tune the Thirds in the major chords $ab + c_1 - eb$, $eb + g_1 - bb$, $bb + d_1 - f$, and the Fifth $d_1 \pm a_1$. This gives c_1, g_1, d_1, and a_1, and completes the whole tuning, which is much easier than for a series of equally tempered tones.

[The theoretical flattening of all the Fifths by $\frac{1}{8}$ of a skhisma is here neglected, as it would be impossible by ear only, and in all probability many other errors in tuning were committed, which could not be detected. The result is that the two manuals were tuned to the following tones, using capital letters to

represent the large or white digitals, and the small letters the small or black digitals. The Roman letters below shew the secondary meaning attached to the letters above them for the tuning of the notes marked with a * above them.

Upper Manual.

$\overset{*}{C_1}\; c\sharp\; \overset{*}{D_1}\; d_1\sharp\; E\; E_1\sharp\; f\sharp\; \overset{*}{G_1}\; g_1\sharp\; \overset{*}{A_1}\; a_1\sharp\; B$
$\qquad\quad eb \qquad\quad f \qquad\qquad\quad ab \qquad bb$

Lower Manual.

$C\; c_1\sharp\; D\; \overset{*}{e_1\flat}\; E_1\; F_1\; f_1\sharp\; G\; \overset{*}{a_1\flat}\; A\; \overset{*}{b_1\flat}\; B_1$
$\quad\; db \qquad\;\; F\flat \qquad gb \qquad\qquad\qquad\;\; C\flat$

To make it more clear how the 24 notes of ¶ this instrument represent 48 by neglecting the skhisma, I have below arranged the scale on the duodenary system (major Thirds in lines, Fifths in columns, App. XX. sect. E. art. 18), and given the proper number of cents for each note, using capitals for the notes actually tuned, and small letters for those obtained by substitution. The notes above the horizontal line were in the upper manual, those below it on the lower.

It is thus seen 1) that the notes in cols. I., II.,

I.		II.		III.		IV.		V.		VI.	
$C\sharp$	114	$E_1\sharp$	500	$g_2\sharp\sharp$	886	d^1b	112	f	498	A_1	884
$F\sharp$	612	$A_1\sharp$	998	$c_2\sharp\sharp$	184	g^1b	610	bb	996	D_1	182
B	1110	$D_1\sharp$	296	$f_2\sharp\sharp$	682	c^1b	1108	eb	294	G_1	680
E	408	$G_1\sharp$	794	$b_2\sharp$	1180	f^1b	406	ab	792	C_1	1178
A	906	$C_1\sharp$	92	$e_2\sharp$	478	b^1bb	904	db	90	F_1	476
D	204	$F_1\sharp$	590	$a_2\sharp$	976	e^1bb	202	gb	588	B_1b	974
G	702	B_1	1088	$d_2\sharp$	274	a^1bb	700	cb	1086	E_1b	272
C	0	E_1	386	$g_2\sharp$	772	d^1bb	1198	fb	384	A_1b	770

¶

III. are exactly 2 cents sharper than those in cols. IV., V., VI. 2) that only cols. I., II., VI. were tuned, and that IV., V., and III. without being tuned were assumed to be identical with them respectively. 3) That cols. I., II., III. form a series of Fifths down or Fourths up, of which only two, namely C to $E_1\sharp$ and E_1 to $g_2\sharp\sharp$, are defective, being both 700 cents down or 500 up, in place of 702 and 498 as all the others. 4) That the simplest way of tuning would be to take A to pitch, and then $A + C_1\sharp$, and $C_1\sharp + e_2\sharp$ as perfect major Thirds, and then from A, $C_1\sharp$, $e_2\sharp$ to tune the rest of the notes

in their columns by perfect Fifths and Fourths, naming the notes in col. III. for convenience as those in col. VI. Afterwards the identity of the first three with the last three columns would be assumed. All the properties and defects of this system of tuning can be immediately deduced from the above diagram.—*Translator.*]

* [For instead of the keys of $F\sharp$ and $F\flat$, the absence of $G\sharp$, and bbb, d_1b obliges us to use the keys of $F^1_1\sharp$ and $F^1\flat$, which are respectively a comma lower and higher.—*Translator.*]

is identified with $f\flat$. Hence the instrument furnishes the following eight perfectly just minor scales [where the letters in brackets indicate those which are not written in the account of the manuals in the text of p. $316c$] :—

$$1)\ a_1\ \text{or}\ b\flat\flat\flat\ \text{minor}: d_1 - f + a_1 - c + e_1 + [g_2\sharp - b_1\ ]$$
$$f\flat + a_1\flat - c\flat$$

$$2)\ e_1\ \text{or}\ f\flat\ \ \text{minor}: a_1 - c + e_1 - g + b_1 + [d_2\sharp - f_1\sharp]$$
$$c\flat + e_1\flat - g\flat$$

$$3)\ b_1\ \text{or}\ c\flat\ \ \text{minor}: e_1 - g + b_1 - d + f_1\sharp + [a_2\sharp - c_1\sharp]$$
$$g\flat + b_1\flat - d\flat$$

$$4)\ f_1\sharp\ \text{or}\ g\flat\ \ \text{minor}: b_1 - d + f_1\sharp - a + c_1\sharp + [e_2\sharp - g_1\sharp]$$
$$d\flat + f_1 - a\flat$$

$$5)\ c_1\sharp\ \text{or}\ d\flat\ \ \text{minor}: f_1\sharp - a + c_1\sharp - e + g_1\sharp + [b_2\sharp - d_1\sharp]$$
$$a\flat + c_1 - e\flat$$

$$6)\ g_1\sharp\ \text{or}\ a\flat\ \ \text{minor}: c_1\sharp - e + g_1\sharp - b + d_1\sharp + [f_2\sharp\sharp - a_1\sharp]$$
$$e\flat + g_1 - b\flat$$

$$7)\ d_1\sharp\ \text{or}\ e\flat\ \ \text{minor}: g_1\sharp - b + d_1\sharp - f\sharp + a_1\sharp + [c_2\sharp\sharp - e_1\sharp]$$
$$b\flat + d_1 - f$$

$$8)\ a_1\sharp\ \text{or}\ b\flat\ \ \text{minor}: d_1\sharp - f\sharp + a_1\sharp - c\sharp + e_1\sharp + [g_2\sharp\sharp - b_1\sharp]$$
$$f + a_1 - c$$

Of these, the six last tonics from $C\flat$ to $B\flat$ are also provided with major scales. Hence there are complete minor scales on all degrees of the scales of B_1 major and E_1 major; and complete minor and major scales on all degrees of the scale of B_1 major, with the exception of E_1.

After previous experiments on another harmonium, where I had at command only the two sets of tones of one octave common to two stops with one manual, I had expected that it would be scarcely observed if either the other minor keys had a somewhat too sharp Pythagorean Seventh, or if minor chords which are themselves rather obscurely harmonious, were executed in Pythagorean intonation. When isolated minor chords are struck the difference is, indeed, not much observed. But when long series of justly-intoned chords have been employed, and the ear has grown accustomed to their effect, it becomes so sensitive to any intermixture of chords in imperfect intonation, that the disturbance is very appreciable.*

The least disturbance is caused by taking the Pythagorean Seventh, because this leading tone is in modern compositions scarcely ever used but in the chord of the dominant Seventh, or other dissonances. In a pure major triad its effect is certainly very harsh. But in a discord it has a less disturbing effect, because by its sharpness it brings out the character of the leading note more distinctly. On the other hand, I have found minor chords with Pythagorean Thirds absolutely intolerable when coming between justly-intoned major and minor chords.* By allowing, then, a Pythagorean Seventh in the scale, or a Pythagorean major Third in the chord of the dominant Seventh, we may form the following minor scales :†—

$$9)\ d_1\ \ \ \text{minor}: g_1\ \ - b\flat + d_1\ \ - f\ \ + a_1\ \ ...\ c_1\sharp\ |\ e_1$$
$$10)\ g_1\ \ \ \text{minor}: c_1\ \ - e\flat + g_1\ \ - b\flat + d_1\ ...f_1\sharp\ |\ a_1$$
$$11)\ c_1\ \ \ \text{minor}: f_1\ \ - a\flat + c_1\ \ - e\flat + g_1\ ...\ b_1\ \ |\ d_1$$
$$12)\ f_1\ \ \ \text{minor}: b_1\flat - d\flat + f_1\ \ - a\flat + c_1\ ...\ e_1\ \ |\ g_1$$
$$13)\ b_1\flat\ \text{minor}: e_1\flat - g\flat + b_1\flat - d\flat + f_1\ ...\ a_1\ \ |\ c_1$$
$$14)\ e_1\flat\ \text{minor}: a_1\flat - c\flat + e_1\flat - g\flat + b_1\flat ...\ d_1\ \ |\ f_1$$

* [My own experience is that the minor chords even more than the major shew the vast superiority of the just intonation over the equal temperament; and that the occasional introduction of Pythagorean among justly-intoned chords, major or minor, is comparable only to the 'wolves' on the 'bad keys,' as $E\flat$, or E, of the old organ tuning. My instru-ments enable me to compare these effects readily, and both arise from similar, though not the same causes.—*Translator*.]

† [In which (...) represents the Pythagorean major Third of 408 cents and (|) the Pythagorean minor Third of 294 cents.—*Translator*.]

In the former series, Nos. 8 and 7, we had already $b\flat$ minor and $e\flat$ minor, which are a comma sharper than Nos. 13 and 14. Hence the series of minor keys is also completed by the fusion of their extremities through enharmonic interchange.

In most cases it is possible to transpose the music to be played on such instruments, so as to avoid the necessity of making these enharmonic interchanges, provided the modulations do not extend too far into different keys. But if it is not possible to avoid enharmonic interchanges, they must be introduced where two unrelated * chords follow each other. This is best done between dissonant chords. Naturally this enharmonic change is always necessary when a piece of music modulates through the whole circle of Fifths—from C major to $B\sharp$ major, for example. But Hauptmann is certainly right when he characterises such circular modulation as unnatural artificiality, which could only be rendered possible by the imperfections of our modern system of temperament. Such a process must cer- ¶ tainly destroy the hearer's feeling for the unity of the tonic. For although $B\sharp$ has very nearly the same pitch as C, or can be even improperly identified with it, the hearer can only restore his feeling for the former tonic by going back on the same path that he advanced. He cannot possibly retain his recollection of the absolute pitch of the first tonic C after his long modulations up to $B\sharp$, with such a degree of exactness as to be able to recognise that they are identical. For any fine artistic feeling $B\sharp$ must remain a tonic far removed from C on the dominant side ; or, more probably, after such distant modulations, the hearer's whole feeling for tonality will have become confused, and it will then be perfectly indifferent to him in what key the piece ends. Generally speaking, an immoderate use of striking modulations is a suitable and easy instrument in the hands of modern composers, to make their pieces piquant and highly coloured. But a man cannot live upon spice, and the consequence of restless modulation is almost always the obliteration of artistic connection. It must not be forgotten that modulations ¶ should be only a means of giving prominence to the tonic by contrasting it with another and then returning into it, or of attaining isolated and peculiar effects of expression.

Since harmoniums with two manuals have usually two sets of vibrators for each manual of which the above system of tuning only uses one, I have had the two others (an 8-foot and a 16-foot stop) tuned in the usual equal temperament, which renders it very easy to compare the effect of this tuning with just intonation, as I have merely to pull out or push in a stop to make the difference.†

As regards musical effect, the difference between the just and the equally-tempered, or the just and the Pythagorean intonations, is very remarkable. The justly-intoned chords, in favourable positions, notwithstanding the rather piercing quality of the tone of the vibrators, possess a full and as it were saturated harmoniousness ; they flow on, with a full stream, calm and smooth, without tremor or beat. Equally-tempered or Pythagorean chords sound beside them rough, dull, ¶ trembling, restless. The difference is so marked that every one, whether he is musically cultivated or not, observes it at once. Chords of the dominant Seventh in just intonation have nearly the same degree of roughness as a common major chord of the same pitch in tempered intonation. The difference between natural and tempered intonation is greatest and most unpleasant in the higher Octaves of the scale, because here the false combinational tones of the tempered intonation are more observable, and the number of beats for equal differences of pitch becomes larger, and hence the roughness greater.

A second circumstance of essential importance is, that the differences of effect between major and minor chords, between different inversions and positions of

---

* [That is, chords not having a common tone.—*Translator*.]

† Proposals for making the series of tones in this system of intonation more complete and at the same time greatly facilitating fingering by the use of a single manual, will be found in Appendix XVII.

chords of the same kind, and between consonances and dissonances are much more decided and conspicuous, than in the equal temperament. Hence modulations become much more expressive. Many fine distinctions are sensible, which otherwise almost disappear, as, for instance, those which depend on the different inversions and positions of chords, while, on the other hand, the intensity of the harsher dissonances is much increased by their contrast with perfect chords. The chord of the diminished Seventh, for example, which is so much used in modern music, borders upon the insupportable, when the other chords are tuned justly.*

Modern musicians who, with rare exceptions, have never heard any music executed except in equal temperament, mostly make light of the inexactness of tempered intonation. The errors of the Fifths are very small. There is no doubt of that. And it is usual to say that the Thirds are much less perfect consonances than the Fifths, and consequently also less sensitive to errors of intonation. The ¶ last assertion is also correct, so long as homophonic music is considered, in which the Thirds occur only as melodic intervals and not in harmonic combinations. In a consonant triad every tone is equally sensitive to false intonation, as theory and experience alike testify, and the bad effect of the tempered triads depends especially on the imperfect Thirds.†

There can be no question that the simplicity of tempered intonation is extremely advantageous for instrumental music, that any other intonation requires an extraordinarily greater complication in the mechanism of the instrument, and would materially increase the difficulties of manipulation, and that consequently the high development of modern instrumental music would not have been possible without tempered intonation. But it must not be imagined that the difference between tempered and just intonation is a mere mathematical subtilty without any practical value. That this difference is really very striking even to unmusical ears, is shewn immediately by actual experiments with properly tuned instruments. ¶ And that the early musicians, who were still accustomed to the perfect intervals of vocal music, which was then most carefully practised, felt the same, is immediately seen by a glance at the musical writings of the latter half of the seventeenth and the earlier part of the eighteenth centuries, at which time there was much discussion about the introduction of different kinds of temperament, and one new method after another was invented and rejected for escaping the difficulties, and the most ingenious forms of instrument were designed for practically executing the enharmonic differences of the tones. Praetorius ‡ mentions a universal cymbalum, which he saw at the house of the court-organist of the Emperor Rudolph II. in Prague, and which had 77 digitals in 4 octaves, or 19 to the octave, the black digitals being doubled, and others inserted between those for *e* and *f*, and between those for *b* and *c*.§ In the older directions for tuning, several tones are usually tuned by Fifths which beat slightly, and then others as perfect major Thirds. The intervals on which the errors accumulated were called *wolves*.

¶   * [This should be tried on the Harmonical as $b_1 - d \mid f - a^1 b$. Although it has two perfect minor Thirds and only one Pythagorean, it is a mere piece of noise, of a much worse kind than the noise of the equally-tempered imitation of the same chord in the best quality of tone. In just intonation the chord of the diminished Seventh can therefore be used only with mild qualities of tone. But the real intonation of this chord is 10 : 12 : 14 : 17, which can also be played on the Harmonical as $e_1'' 316$ $g'' 267$ $^7b'' b$ 336 $^{17}d''' b$.— *Translator*.]

† [A triad in which the major Third is perfect, but the Fifth and minor Third both too small by a quarter of a comma or $5\frac{1}{3}$ cents (as in the meantone temperament, in which I have a concertina tuned), has a very much better effect than the equally-tempered triad, where the Fifth is only one-eleventh of a comma or 2 cents too flat, and the major Third is seven-elevenths of a comma or 14 cents too sharp, and hence the minor Third is eight-elevenths of a comma or 16 cents too flat. The effect is much more strongly felt in playing passages than in playing isolated chords.—*Translator*.]

‡ *Syntagma musicum*, II., Chap. XI., p. 63.

§ [This was to make the meantone scale more complete, the scale being *C c♯ d♭ D d♯ e♭ E e♯ F f♯ g♭ G g♯ a♭ A a♯ b♭ B b♯ C*, where the capitals represent the white, and small letters the black digitals, all in meantone temperament, the effect of which would have been very good on the organ. For the intonation of these notes see App. XX. sect. A. art. 16.— *Translator*.]

Prætorius says : ' It is best for the wolf to remain in the wood with its abominable howling, and not disturb our *harmonicas concordantias*.' Rameau, too, who at a later period contributed greatly to the introduction of equal temperament, in 1726 * still defended a different style of tuning, in which the Thirds of the more usual keys were kept perfect at the expense of the Fifths and of the unusual keys. Thus he tuned up from C, in Fifths so much diminished, that the fourth Fifth, instead of being E, became the perfect Third of C, namely $E_1 = F\flat$. Then again four Fifths more to $A_1\flat$, the perfect Third of $F\flat$, instead of to $A\flat$. But then the four Fifths between this $A_1\flat$ and C had necessarily † to be made too large, because it is not $A_1\flat$ but $A\flat$ which is four perfect Fifths distant from C. This plan of tuning gives the perfect major Thirds, $C+E_1$, $G+B_1$, $D+F_1\sharp$, $E_1+G_2\sharp$, but when we proceed further from E on the dominant side, or from C on the sub-dominant side, we find Thirds which become worse and worse. The error in the Fifths is about three times that in the equal temperament. Even in 1762, this ¶ system could be characterised by d'Alembert as that commonly used in France, in opposition to the equal temperament which Rameau subsequently proposed. Marpurg ‡ has collected a long series of other systems of tuning. Since players found themselves compelled by the use of only 12 digitals to the octave, to put up with a series of false intervals, and to let their ears become accustomed to them, it was certainly better to make up their minds to give up their few perfect major Thirds still remaining in the scale, and to make all the major Thirds equally erroneous. It necessarily produces more disturbance to hear very falsely tuned Thirds amidst correct intervals, than to hear intervals which are all equally out of tune and are not contrasted with others in perfect intonation. Hence as long as it is necessary practically to limit the number of separate tones within the octave to 12, there can be no question at all as to the superiority of the equal temperament with its 12 equal Semitones, over all others, and, as a natural consequence, this has become the sole acknowledged method of tuning. It is only bowed instruments, with [including ¶ the *Tenor*] their four perfect Fifths $C \pm G \pm D \pm A \pm E$, which still deviate from it.

The equal temperament came into use in Germany before it was introduced into France. In the second volume of Matheson's *Critica Musica*, which appeared in 1752, he mentions *Neidhard* and *Werckmeister* as the inventors of this tempera-ment.§ Sebastian Bach had already used it for the clavichord (*clavier*), as we must conclude from Marpurg's report of Kirnberger's assertion, that when he was a pupil of the elder Bach he had been made to tune all the major Thirds too sharp. Sebastian's son, Emanuel, who was a celebrated pianist, and published in 1753 a

---

* *Nouveau Système de Musique*, Chap. XXIV.

† [That is, if only twelve digitals might be used, so that the temperament became unequal. But this style of tuning, which at first was the meantone temperament, where the Fifths are made a quarter of a comma too flat, should be carried out through twenty-six Fifths, requiring twenty-seven tones (namely, 7 natural, 7 sharp, 7 flat, 3 double sharp, and 3 double flat) to be really effective, and if any fewer are employed no attempt should be made to modulate into keys not provided with proper tones. It is a temperament with which I am practically familiar. It is harmonically far superior to the equally tempered, and is even endurable on the concertina, which used to be always so tuned, but having fourteen digitals, extends from $A\flat$ to $D\sharp$. The twelve digitals could play then only in $B\flat$, F, C, G, D, and A major, and in G, D, and A minor, which of course failed to satisfy the requirements of modulation. Hence players sought to identify $D\sharp$, $A\sharp$, $E\sharp$, $F\sharp\sharp$, $C\sharp\sharp$, $G\sharp\sharp$ with $E\flat$, $B\flat$, F, G, D, A on the one hand, and $D\flat$, $G\flat$, $C\flat$, $F\flat$, $B\flat\flat$, $E\flat\flat$, $A\flat\flat$ with $C\sharp$, $F\sharp$, B, E, A, D, G on the other. Hence came the wolves. And a system of tuning was blamed for not doing what it never professed to do. As long as twelve digitals only are insisted on, the equal temperament, by dividing the Octave into twelve equal Semitones, is a necessity. But with Mr. Bosanquet's fingerboard (App. ¶ XX. sect. F. No. 8) there is no longer any need to limit organs to 12 notes to the Octave. When, however, he played on such an organ before the Musical Association, great objection was taken to the flatness of the leading note, which was $5\frac{1}{3}$ cents flatter than just, as musicians are accustomed to one which is 12 cents too sharp.—*Translator*.]

‡ *Versuch über die musikalische Temperatur*, Breslau, 1776.

§ *Op. cit.* p. 162. The following works of these two authors are cited by Forkel : *Neidhard* (Royal Prussian Band-conductor), *Die beste und leichteste Temperatur des Monochordi* (the best and easiest temperament of the monochord), Jena, 1706 ; *Sectio canonis harmonici*, Königsberg, 1724. *Werckmeister* (organist at Quedlinburg, born 1645), *Musikalische Temperatur*, Frankfurt and Leipzig, 1691.

work of great authority in its day ' on the true art of playing the clavier,' requires this instrument to be always tuned in the equal temperament.*

The old attempts to introduce more than 12 degrees into the scale have led to nothing practical, because they did not start from any right principle. They always attached themselves to the Greek system of Pythagoras, and imagined only that it was necessary to make a difference between $c\sharp$ and $d\flat$, or between $f\sharp$ and $g\flat$, and so on. But that is not by any means sufficient, and is not even always correct. According to our system of notation we may identify $c_1\sharp$ with $d\flat$, but we must distinguish the $c\sharp$ found from the relation of Fifths, from the $c_1\sharp$ found from the relation of Thirds.† Hence the attempts to construct instruments with complex arrangements of manuals and digitals, have led to no result, which was at all commensurate with the trouble bestowed upon them, and the increased difficulties of fingering which they occasioned. The only instrument of the kind ¶ which is still used is the pedal harp *à double mouvement*, on which the intonation can be changed by the foot.

Not only habitual use, and the absence of any power to compare its effects with those of just intonation, but some other circumstances are favourable to equal temperament.

First, it should be observed that the disturbances due to beats in the tempered scale, are the less observable the swifter the motion and the shorter the duration of the single notes. When the note is so short that but very few beats can possibly occur while it lasts, the ear has no time to remark their presence. The beats produced by a tempered triad are the following :

1. Beats of the tempered Fifth. Suppose we take the number of vibrations of c' to be 264, the tempered Fifth $c' \pm g'$ would produce 9 beats in 10 seconds, partly by the upper partials, and partly by the combinational tones. These beats are always quite audible.

¶   2. Beats of the two first combinational tones of $c' + e'$ and $e' - g'$ in tempered intonation ; $5\frac{2}{3}$ in the second. These are plainly audible in all qualities of tones, if the tones themselves are not too weak.

3. Beats of the major Third $c' + e'$ alone, $10\frac{1}{2}$ in the second, which, however, are not plainly audible unless the qualities of tone employed have high upper partials.

4. Beats of the minor Third $e' - g'$, 18 in the second, mostly much weaker than those of the major Third, and also heard only in qualities of tone having high upper partials.

All these beats occur twice as fast when the chord lies an Octave higher, and half as fast when it occurs an Octave lower.

Of these beats, the first, arising from the tempered Fifths, have the least injurious influence on the harmoniousness of the chord. They are so slow that they can be heard only for very slow notes in the middle parts of the scale, and ¶ then they produce a slow undulation of the chord which may occasionally have a good effect. Beats of the second kind are most striking for the softer quality of tone. In an *Allegro*, of four crotchets in a bar, two bars occupy about three seconds. If, then, the tempered chord $c' + e' - g'$ is played on a crotchet in this bar, $2\frac{1}{8}$ of these beats will be heard, so that if the tone begins soft, it will swell, decrease, swell again, decrease and then finish. It would be certainly worse if this chord were played an Octave or two higher, so that $4\frac{1}{4}$ or $8\frac{1}{2}$ beats could be heard, because these could not fail to strike the ear as a marked roughness.

For the same reason the beats of the third and fourth kinds, arising from the Thirds, which are clearly audible on harsher qualities of tone (as on the harmonium), are also decidedly disturbing in the middle positions, even in quick time,

---

* [Equal temperament was not commercially established in England till 1841–1846. See App. XX. sect. N. No. 5. With regard to both Sebastian and Emanuel Bach's relation to it, see Bosanquet's *Musical Intervals and Temperament*, pp. 27–32.—*Translator.*]

† [That is, $c\sharp$ found from $c \pm g \pm d \pm a \pm e \pm b \pm f\sharp \pm c\sharp$, from $c_1\sharp$ found from first the major Third $c + e_1$ and then the Fifths $e_1 \pm b_1 \pm f_1\sharp \pm c_1\sharp$.— *Translator.*]

and essentially injure the calmness of the triad, because they are twice and thrice as fast as the others.   It is only in soft qualities of tone that they are but little observed, or, when observed, are so covered by stronger unbroken tones as to be very slightly marked.

Hence in rapid passages, with a soft quality and moderate intensity of tone, the evils of tempered intonation are but little apparent.   Now, almost all instrumental music is designed for rapid movement, and this forms its essential advantage over vocal music.   We might, indeed, raise the question whether instrumental music had not rather been forced into rapidity of movement by this very tempered intonation, which did not allow us to feel the full harmoniousness of slow chords to the same extent as is possible from well-trained singers, and that instruments had consequently been forced to renounce this branch of music.

Tempered intonation was first and especially developed on the pianoforte, and hence gradually transferred to other instruments.   Now, on the pianoforte circum- ¶ stances really favour the concealment of the imperfections due to the temperament. The tones of a pianoforte are very loud only at the moment of striking, and their loudness rapidly diminishes.   This, as I have already had occasion to mention, causes their combinational tones to be audible at the first moment only, and hence makes them very difficult to hear.   Beats from that source must therefore be left out of consideration.   The beats which depend on the upper partials have been eliminated in modern pianofortes (especially in the higher Octaves, where they would have done most harm), owing to the mode in which upper partials are greatly weakened and the quality of tone much softened by regulating the striking place, as I have explained in Chap. V. (p. 77b).   Hence on a pianoforte the deficiencies of the intonation are less marked than on any instrument with sustained tones, and yet are not quite absent.   When I go from my justly-intoned harmonium to a grand pianoforte, every note of the latter sounds false and disturbing, especially when I strike isolated successions of chords.   In rapid melodic figures ¶ and passages, and in arpeggio chords, the effect is less disagreeable.   Hence older musicians especially recommended the equal temperament for the pianoforte alone. Matheson, in doing so, acknowledges that for organs Silberman's unequal temperament, in which the usual keys were kept pure,* is more advantageous.   Emanuel Bach says that a correctly tuned pianoforte has the *most perfect intonation of all instruments*, which in the above sense is correct.   The great diffusion and convenience of pianofortes made it subsequently the chief instrument for the study of music and its intonation the pattern for that of all other instruments.

On the other hand, for the harsher stops on the organ, as the mixture and reed stops, the deficiencies of equal temperament are extremely striking.   It is considered inevitable that when the mixture stops are played with full chords an awful din (*höllenlärm*) must ensue, and organists have submitted to their fate.   Now this is mainly due to equal temperament, because if the Fifths and Thirds in the pipes for each digital of the mixture stops were not tuned justly, every single note would ¶ produce beats by itself.   But when the Fifths and Thirds between the notes belonging to the different digitals are tuned in equal temperament, every chord furnishes at once tempered and just Fifths and Thirds, and the result is a restless blurred confusion of sounds.   And yet it is precisely on the organ it would be so easy by a few stops to regulate the action for each key so as to produce harmonious chords.†

Whoever has heard the difference between justly-intoned and tempered chords, can feel no doubt that it would be the greatest possible gain for a large organ to omit half its stops, which are mostly mere toys, and double the number of tones

---

* [Probably this was the meantone temperament explained on p. 321, note †.—*Translator.*]

† From Zamminer's book (p. 140), I see that in Silliman's *American Journal of Science* for 1850 there is a description of an organ by Poole, which is tuned justly for all keys by means of stops.   [See App. XVIII. second paragraph, and App. XX. sect. F. No. 7, where Poole's new keyboard without stops is figured.]

in the Octave, so as to be able, by means of suitable stops, to play correctly in all keys.*

The case is the same for the harmonium as for these stops on the organ. Its powerful false combinational tones and its gritty trembling chords, both due to tempered intonation, are certainly the reason why many musicians pronounce this instrument to be out of tune, and dismiss it at once as too trying to the nerves.

Orchestral instruments can generally alter their pitch slightly. Bowed instruments are perfectly unfettered as to intonation ; wind instruments can be made a little sharper or flatter by blowing with more or less force. They are, indeed, all adapted for equal temperament, but good players have the means of indulging the ear to some extent. Hence, passages in Thirds for wind instruments, when executed by indifferent players, often sound desperately false (*verzweifelt falsch*), whereas good performers, with delicate ears, make them sound perfectly well.

¶ The bowed instruments are peculiar. From the first they have retained their tuning in perfect Fifths. The violins themselves have the perfect Fifths, $G \pm D \pm A \pm E$. The tenor and violoncello give the Fifth $C \pm G$ in addition. Now, every scale has its own peculiar fingering, and hence every pupil could be easily practised in playing each scale in its proper intonation, and then, of course, tones of the same name but in different keys could not be stopped in the same way, and even the major Third of the major scale of C, when the C of the tenor is taken as the tonic, must not be played on the E string of the violin, because this gives E and not E_1. Nevertheless, the modern school of violin-playing since the time of Spohr, aims especially at producing equally-tempered intonation, although this cannot be completely attained, owing to the perfect Fifths of the open strings. At any rate, the acknowledged intention of present violin-players is to produce only 12 degrees in the Octave. The sole exception which they allow is for double-stop passages, in which the notes have to be somewhat differently stopped from what

¶ they are when played alone. But this exception is decisive. In double-stop passages the individual player feels himself responsible for the harmoniousness of the interval, and it lies completely within his power to make it good or bad. Any violin-player will easily be able to verify the following fact. Tune the strings of the violin in the perfect Fifths $G \pm D \pm A \pm E$, and find where the finger must be pressed on the A string to produce the B, which will give a perfect Fourth $B...E$. Now, let him, without moving his finger, strike this same B together with the open D string. The interval $D...B$ would, according to the usual view, be a major Sixth, but it would be a Pythagorean one [of 906 cents]. In order to obtain the consonant Sixth $D...B_1$ [of 884 cents], the finger would have to be drawn back for about $1\frac{3}{5}$ Paris lines (nearly $\frac{3}{20}$ inch), a distance quite appreciable in stoppings, and sufficient to alter the pitch and the beauty of the consonance most perceptibly.

But it is clear that if individual players feel themselves obliged to distinguish

¶ the different values of the notes in the different consonances, there is no reason why the bad Thirds of the Pythagorean series of Fifths should be retained in quartett playing. Chords of several parts, executed by several performers in a quartett, often sound very ill, even when each single one of these performers can perform solo pieces very well and pleasantly ; and, on the other hand, when quartetts are played by finely-cultivated artists, it is impossible to detect any false consonances. To my mind the only assignable reason for these results is that practised violinists with a delicate sense of harmony, know how to stop the tones they want to hear, and hence do not submit to the rules of an imperfect school.

---

* [That is, as correctly as on the Author's justly-intoned harmonium, but that is far too deficient in power of modulation into minor keys, to make it worth while to construct it on a great organ. Nothing short of the 53 division of the Octave (p. 328c) would suffice, and this would necessitate the omission of more than *three-quarters* instead of only one-half of the stops. And then musicians would have to learn how to use a practically just scale, and how to adapt tempered music to it, both of which present considerable difficulties. It is, therefore, safe to say that nothing of the kind will be done.—*Translator.*]

That performers of the highest rank do really play in just intonation, has been directly proved by the very interesting and exact results of Delezenne.* This observer determined the individual notes of the major scale, as it was played by distinguished violinists and violoncellists, by means of an accurately gauged string, and found that these players produced correctly perfect Thirds and Sixths, and neither equally tempered nor Pythagorean Thirds or Sixths. I was fortunate enough to have an opportunity of making similar observations by means of my harmonium on Herr Joachim. He tuned his violin exactly with the $g \pm d \pm a \pm e$ of my instrument. I then requested him to play the scale, and immediately he had played the Third or Sixth, I gave the corresponding note on the harmonium. By means of beats it was easy to determine that this distinguished musician used b_1 and not b as the major Third to g, and e_1 not e as the Sixth.† But if the best players who are thoroughly acquainted with what they are playing are able to overcome the defects of their school and of the tempered system, it would ¶ certainly wonderfully smooth the path of performers of the second order, in their attempts to attain a perfect *ensemble*, if they had been accustomed from the first to play the scales by natural intervals. The greater trouble attending the first attempts would be amply repaid by the result when the ear has once become accustomed to hear perfect consonances. It is really much easier to apprehend the differences between notes of the same name in just intonation than people usually imagine, when the ear has once become accustomed to the effect of just consonances. A confusion between a_1 and a in a consonant chord on my harmonium strikes me with the same readiness and certainty as a confusion between A and $A\flat$ on a pianoforte.‡

I am, however, too little acquainted with the technicalities of violin-playing, to attempt making any proposals for a definite regulation of the tonal system of bowed instruments. This must be left to masters of this instrument who at the same time possess the powers of a composer. Such men will readily convince ¶ themselves by the testimony of their ears, that the facts here adduced are correct, and perceive that, far from being useless mathematical speculations, they are practical questions of very great importance.

The case is precisely similar for our present singers. For singing, intonation is perfectly free, whereas on bowed instruments, the five tones of the open strings at least have an unalterable pitch. In singing the pitch can be made most easily and perfectly to follow the wishes of a fine musical ear. Hence all music began with singing;§ and singing will always remain the true and natural school of all music. The only intervals which singers can strike with certainty and perfection, are such as they can comprehend with certainty and perfection, and what the singer easily and naturally sings the hearer will also easily and naturally understand.**

* *Recueil des Travaux de la Société des Sciences, de l'Agriculture, et des Arts de Lille, 1826 et premier semestre* 1827 ; *Mémoire sur les Valeurs numérique des Notes de la Gamme,* par M. Delezenne. [See especially pp. 55-6.] For observations on corresponding circumstances in singing, see Appendix XVIII.

† Messrs. Cornu and Mercadier have indeed published contradictory observations. (*Comptes Rendus de l'Acad. des Sc. de Paris,* 8 et 22 Février, 1869.) They let a musician play the Third of a major chord first in melodic succession, and then in harmonious consonance. In the latter case it was always 4 : 5. But in melodic succession the performer selected a somewhat sharper Third. I am bound to reply, that in melodic succession the major Third is not a very characteristically determined interval, and that all living musicians have been accustomed to sharp Thirds on the pianoforte. In the simple succession $c + e - g_1$

when isolated from the rest of the scale, I find it difficult to distinguish between the just ¶ and the Pythagorean major Third. But when I play on my harmonium the complete melody of some well-known air without harmonies the Pythagorean Third always feels to me strained, the perfect Third calm and soft. It is only in the leading note, perhaps, that the sharper Third is more expressive. [See App. XX. sect. G. arts. 6 and 7, for the results of later experiments by Messrs. Cornu and Mercadier. —*Translator.*]

‡ [In a consonant *chord* the difference is striking, melodically not so. An eminent teacher of singing could only by great attention tell the difference when I alternated a_1, a and d_1, d in the major scale of C.—*Translator.*]

§ [It must not be forgotten, however, that the voice was the only musical instrument at first known.—*Translator.*]

** [That this must also apply to non-har-

Down to the seventeenth century singers were practised by the monochord, for which Zarlino in the middle of the sixteenth century reintroduced the correct natural intonation. Singers were then practised with a degree of care of which we have at present no conception. We can even now see from the Italian music of the fifteenth and sixteenth centuries that they were calculated for most perfect intonation of the chords, and that their whole effect is destroyed as soon as this intonation is executed with insufficient precision.

But it is impossible not to acknowledge that at the present day few even of our opera singers are able to execute a little piece for several voices, when either totally unaccompanied, or at most accompanied by occasional chords, (as, for example, the trio for the three masks, *Protegga il giusto cielo*, from the finale to the first act of Mozart's *Don Giovanni*,) in a manner suited to give the hearer a full enjoyment of its perfect harmony. The chords almost always sound a little ¶ sharp or uncertain, so that they disturb a musical hearer. But where are our singers to learn just intonation and make their ears sensitive for perfect chords? They are from the first taught to sing to the equally-tempered pianoforte. If a major chord is struck as an accompaniment, they may sing a perfect consonance with its root, its Fifth, or its Third. This gives them about the fifth part of a Semitone for their voices to choose from without decidedly singing out of harmony, and even if they sing a little sharper than consonance with the sharp Third requires, or a little flatter than consonance with the flat Fifth requires, the harmoniousness of the chord will not be really much more damaged. The singer who practises to a tempered instrument has no principle at all for exactly and certainly determining the pitch of his voice.*

On the other hand, we often hear four musical amateurs who have practised much together, singing quartetts in perfectly just intonation. Indeed, my own experience leads me almost to affirm that quartetts are more frequently heard ¶ with just intonation when sung by young men who scarcely sing anything else, and often and regularly practise them, than when sung by instructed solo singers who are accustomed to the accompaniment of the pianoforte or the orchestra. But correct intonation in singing is so far above all others the first condition of beauty, that a song when sung in correct intonation even by a weak and unpractised voice always sounds agreeable, whereas the richest and most practised voice offends the hearer when it sings false, or sharpens.

The case is the same as for bowed instruments. The instruction of our present singers by means of tempered instruments is unsatisfactory, but those who possess good musical talents are ultimately able by their own practice to strike out the right path for themselves, and overcome the error of their original instruction. They even succeed the earlier, perhaps, the sooner they quit school, although, of course, I do not mean to deny that fluency in singing, and the disuse of all kinds of bad ways can only be acquired in school.

¶   It is clearly not necessary to temper the instruments to which the singer practises. A single key suffices for these exercises, and that can be correctly tuned. We do not require to use the same piano for the teaching to sing and for playing sonatas. Of course it would be better to practise the singer to a justly-intoned organ or harmonium in which by means of two manuals all keys may be used.† Sustained tones are preferable as an accompaniment because the singer himself can immediately hear the beats between the instrument and his voice when he alters the pitch slightly. Draw his attention to these beats, and he will

monic scales is evident from the fact that music has existed for thousands of years, but harmonic scales have been in use only a few centuries, and are far from being even yet universal.— *Translator*.]

* See Appendix XVIII.

† [Voices differ so much that the same pitch for the tonic, that is the same key, would not suit. Still the Harmonical, or, for modulating purposes, the just harmonium or just concertina, may prove of service. Otherwise special instruments must be used, as Mr. Colin Brown's Voice-Harmonium. The Tonic So'-faists teach without any accompaniment, not even that of the teacher's voice, but rapidly introduce part music.—*Translator*.]

then have a means of checking his own voice in the most decisive manner. This is very easy on my justly-intoned harmonium, as I know by experience. It is only after the singer has learned to hear every slight deviation from correctness announced by a striking incident, that it becomes possible for him to regulate the motions of his larynx and the tension of his vocal chords with sufficient delicacy to produce the tone which his ear demands. When we require a delicate use of the muscles of any part of the human body, as, in this case, of the larynx, there must be some sure means of ascertaining whether success has been attained. Now the presence or absence of beats gives such a means of detecting success or failure when a voice is accompanied by sustained chords in just intonation. But tempered chords which produce beats of their own are necessarily quite unsuited for such a purpose.

Finally, we cannot, I think, fail to recognise the influence of tempered intonation upon the style of composition. The first effect of this influence was favourable. ¶ It allowed composers as well as players to move freely and easily into all keys, and thus opened up a new wealth of modulation. On the other hand, we likewise cannot fail to recognise that the alteration of intonation also compelled composers to have recourse to some such wealth of modulation. For when the intonation of consonant chords ceased to be perfect, and the differences between their various inversions and positions were, as a consequence, nearly obliterated, it was necessary to use more powerful means, to have recourse to a frequent employment of harsh dissonances, and to endeavour by less usual modulations to replace the characteristic expression, which the harmonies proper to the key itself had ceased to possess. Hence in many modern compositions dissonant chords of the dominant Seventh form the majority, and consonant chords the minority, yet no one can doubt that this is the reverse of what ought to be the case ; and continual bold modulational leaps threaten entirely to destroy the feeling for tonality. These are unpleasant symptoms for the further development of art. The mechanism of instruments ¶ and attention to their convenience, threaten to lord it over the natural requirements of the ear, and to destroy once more the principle upon which modern musical art is founded, the steady predominance of the tonic tone and tonic chord. Among our great composers, Mozart and Beethoven were yet at the commencement of the reign of equal temperament. Mozart had still an opportunity of making extensive studies in the composition of song. He is master of the sweetest possible harmoniousness, where he desires it, but he is almost the last of such masters. Beethoven eagerly and boldly seized the wealth offered by instrumental music, and in his powerful hands it became the appropriate and ready tool for producing effects which none had hitherto attempted. But he used the human voice as a mere handmaid, and consequently she has also not lavished on him the highest magic of her beauty.

And after all, I do not know that it was so necessary to sacrifice correctness of intonation to the convenience of musical instruments. As soon as violinists have ¶ resolved to play every scale in just intonation, which can scarcely occasion any difficulty, the other orchestral instruments will have to suit themselves to the correcter intonation of the violins. Horns and trumpets have already naturally just intonation.*

Moreover, we must observe that when just intonation is made the groundwork of modulations, even comparatively simple modulational excursions will occasion enharmonic confusions (amounting to a comma) which do not appear as such in the tempered system.†

To me it seems necessary that the new tonic into which we modulate should

---

* [Referring to this passage, Mr. Blaikley says (*Proceedings of the Musical Association*, vol. iv. p. 56) : 'This must be taken as being particularly and not generally true, that is, though the ideal instrument has such characteristics, this ideal is not necessarily attained in practice.' See the whole of this paper, and the discussion on it, and also see supra, pp. 97*d*, 99*b*, notes.—*Translator*.]

† [See p. 324*d*, note *, and p. 340*c*, note *. —*Translator*.]

be related to the tonic in which we are playing; the nearer the relationship, the more striking the transition. Again, it is not advisable to remain long in a key which is not related to the principal tonic of the piece. With these principles the rules for modulation usually given coincide. The easiest and most usual transitions are into the key of the dominant or subdominant, these tones being, as is well known, the nearest relations of the first tonic. Hence if the original key is C, we can pass immediately into G major, and thus change the tones F and A_1 of the scale of C major into $F_1\sharp$ and A. Or we can pass into F major by exchanging B_1 and D for $B\flat$ and D_1. After this step has been made, the music will often pass into a key with a tonic related to C in the second degree only, as from G to D or from F to $B\flat$. By proceeding in this way we should come to keys as A and $E\flat$, of which the relation to the original tonic C would be very obscure and in which it would certainly not be advisable to remain long for fear of too much weakening
¶ the feeling for the original tonic.

Again, we may also modulate from the principal tonic C to its Thirds and Sixths, to E_1 and A_1 or $E^1\flat$ and $A^1\flat$. In tempered intonation these changes seem to be the same as from G and D to A and E, or from F and $B\flat$ to $E\flat$ and $A\flat$. But they differ in the pitch, as shewn by the different marks A and A_1, &c. In the tempered intonation it seems allowable to go by a Sixth from c to the key a, and then by Fifths back, to d, g, and then c again. But in reality we thus reach a different c from that with which we began. By such a transition, which is certainly not quite natural, we should be obliged to make an enharmonic exchange [alteration of pitch by a comma], and this would be best done while in the key of d, since both d and d_1 are related to c in the second degree. In the complicated modulations of modern composers such enharmonic changes will of course have to be often made. A cultivated taste will have to judge in each individual case how they are to be introduced, but it will be probably advisable to retain the rules
¶ already mentioned, and to choose the intonation of the new tonics introduced by modulation in such a manner as will keep them as closely related to the principal tonic as possible. Enharmonic changes are least observed when they are made immediately before or after strongly dissonant chords, as those of the diminished Seventh. Such enharmonic changes of pitch are already sometimes clearly and intentionally made by violinists, and where they are suitable even produce a very good effect.*

† If we desire to produce a scale in almost precisely just intonation, which will allow of an indefinite power of modulation without having recourse to enharmonic changes,‡ we can effect our purpose by the division of the Octave into 53 exactly equal parts, as was long ago proposed by Mercator [to represent Pythagorean intonation]. Mr. R. H. M. Bosanquet § has recently provided this temperament, as realised on an harmonium, with a symmetrically arranged fingerboard. When the Octave is divided into 53 equal intervals or degrees, 31 such
¶ degrees give an almost perfect Fifth, the error of which is only $\frac{1}{28}$ of the error of the Fifth of the usual equal temperament, and 17 of these degrees give a major Third, of which the error in defect is only $\frac{5}{7}$ of the above-named error of the Fifth in equal temperament.** The error of the Fifth in this system must be considered

* See examples in C. E. Naumann's *Bestimmungen der Tonverhältnisse*' (Determinations of the Tonal Ratios), Leipzig, 1858, pp. 48, sqq.

† [From here to the end of the chapter is an addition to the 4th German edition.—*Translator.*]

‡ [This is unfortunately not the case when translating equally tempered music, as shewn by the last example.—*Translator.*]

§ *An Elementary Treatise on Musical Intervals and Temperament*, London, Macmillan, 1875. The instrument described was exhibited in the Scientific Loan Exhibition at South Kensington [in May 1876, and is still

in Room Q of the Scientific Collections at the South Kensington Museum].

** On converting the ratio of the extent of the interval of a Fifth to that of an Octave (that is log. 1·5 : log. 2) into a continued fraction, we get the following approximations :

12	53	306 Fifths.
nearly = 7	31	179 Octaves.

And by a similar approximation

3	28	59 major Thirds.
nearly = 1	9	19 Octaves.

[As these approximations give no concep-

as quite inappreciable, that of the major Third is still more difficult to perceive than that of the equally tempered Fifth.* In these degrees the major scale will be

$$C \quad D \quad E_1 \quad F \quad G \quad A_1 \quad B_1 \quad C$$

degrees	0		9		17		22		31		39		48		53
differences		9		8		5		9		8		9		5	

These differences of 9, 8, 5 correspond to the major, minor, and half Tone of the just scale. Each separate degree of the scale corresponds nearly with the interval $77 : 76$ [$= 22 \cdot 6$ cents] and is therefore extremely little greater than the comma $81 : 80$ [$= 21 \cdot 5$ cents], which in the just scale gives the difference between a large or diatonic Semitone [$16 : 15 = 112$ cents] and a small Semitone or limma [$256 : 243 = 90$ cents]. The ear cannot distinguish this scale from the just,† and in its practical applications it admits of unlimited modulation in what is equal to exact intonation. The difference between our c_1 and c, or our c and c^1 would answer to sharpening by one degree. Mr. Bosanquet therefore employs the convenien' ¶ signs $\backslash c$ for c_1 and $/c$ for c^1, $\backslash\backslash c$ for c_2, &c. These signs, $\backslash$ and $/$, he also employs before notes on the staff, exactly as we employ ♯ and ♭. The fingerboard is arranged in a very comprehensible and symmetrical way to make the fingering of all scales and all chords the same in all keys.‡ A diagram of the keyboard will be found in App. XIX.§

Perhaps a justification is needed for our having in this whole theory of keys and modulations, identified the key of the Octave with that of its root, while

tion of the extreme closeness with which the 53 division, if accurately tuned, would approximate to just intonation, I annex the following table :

Note	Just cents	53 division cents	No. of degrees
C	0	0	0
C^1	$21 \cdot 506$	$22 \cdot 642$	1
D_1	$182 \cdot 404$	$181 \cdot 132$	8
D	$203 \cdot 910$	$203 \cdot 774$	9
$E^1 \flat$	$315 \cdot 641$	$316 \cdot 981$	14
E	$386 \cdot 314$	$384 \cdot 906$	17
^7F	$470 \cdot 781$	$475 \cdot 472$	21
F	$498 \cdot 045$	$498 \cdot 113$	22
G	$701 \cdot 955$	$701 \cdot 887$	31
$A^1 \flat$	$813 \cdot 687$	$815 \cdot 094$	36
A_1	$884 \cdot 359$	$883 \cdot 019$	39
$^7B \flat$	$968 \cdot 826$	$973 \cdot 585$	43
$B \flat$	$996 \cdot 091$	$996 \cdot 226$	44
$B^1 \flat$	$1017 \cdot 597$	$1018 \cdot 868$	45
B_1	$1088 \cdot 269$	$1086 \cdot 792$	48
C	$1200 \cdot 000$	$1200 \cdot 000$	53
$^{17}D \flat$	$1304 \cdot 955$	$1313 \cdot 208$	58

Hence for all tertian intervals the approximation is within 2 cents, often within 1 cent. The septimal comma being greater than 1 degree, the $^7B\flat$ is too sharp by 5 cents. Still the 53 div. chord $C : E_1 : G : {}^7B\flat$ is a great improvement on the just $C : E_1 : G : B\flat$. The 17th harmonic is 8 cents too sharp, but the chord of the diminished Seventh in the 53 div. $E_1 : G : {}^7B\flat : {}^{17}D\flat$ is much superior to the just form $E_1 : G : B\flat : D^1\flat$, though from its just surroundings inferior in effect to the equally tempered $E : G : B\flat : D\flat$. — *Translator.*]

* [Both, however, give rise to beats which are of great importance to the tuner. See App. XX. sect. G. art. 20.--*Translator.*]

† [Melodically; but harmonically, at least as the intervals were tuned on Mr. Bosanquet's instrument, there was a decidedly perceptible difference to an ear, accustomed as mine was, to listen to just intonation. --*Translator.*]

‡ [Prof. Helmholtz adds, ' after a plan invented by the American, Mr. H. W. Poole.' I ¶ have omitted this line because, although Mr. Poole's remarkable fingerboard (figured in App. XX. sect. F. No. 6) also allows of playing with the same fingering in all keys, it was not intended for the 53 division, and it bears no resemblance to that of Mr. Bosanquet, who has the exclusive merit of inventing and practically carrying out his extraordinary ' generalised keyboard,' which is suitable for all cycles (except the ordinary one of 12) that resolve the tones used into a series of tempered Fifths. See App. XX. sect. A. art. 20 sqq.—*Translator.*]

§ [In App. XX. sect. F. No. 8, I have added a further account of this invention, and (*ibid.* No. 9) a notice of another keyboard for a reed instrument called the *Harmon*, also using the 53 division, invented and executed by Mr. James Paul White, a tuner, of Springfield, Massachusetts, U.S. America.--*Translator.*]

we have distinguished the key of the Twelfth. In the usual school of musical theory, the meaning of the sound of the Octave is completely identified with that of its root, and is so treated. For us, on the other hand, the Octave is only the Tone most nearly and clearly related to the root, but its relationship is the same in kind as that of the Twelfth, or the next higher major Third (Seventeenth) to the root.

Now we have shewn in p. 273a that in the particular relation of the formation of scales, that is of the determination of the key, the higher Octave introduces the same series of directly related tones as does the lower, although in a somewhat different order of strength of relationship. Only throughout the formation of the lower Octave the tones of the major scale are favoured, and in the formation of the upper Octave those of the minor scale are preferred, but not to the exclusion of those of other scales.

¶ When we proceed beyond the limits of the first Octave, the relationships of tone depending on the six first partials give only the Tenth and Twelfth. The other steps of the scales have then to be filled up with tones related in the second degree, and, among these, the relations of the Octave must have the preference, and next those of the Twelfth. H nce in the second Octave we have necessarily a repetition of the scale of the first. By this means, in the formation of scales an equivalence of Octaves is established, without any necessity for assuming a specifically different relation of similarity between them and the root, as we had to do for the other consonances. In the formation of consonant intervals, the usual theory of music also considers the Octaves as equivalent to the roots. This is within certain limits correct, because the intervals usually considered as consonant, remain consonant when one of their tones is transposed by an Octave, or at least produce intervals which lie on the limits of consonance. But here the usual rule of the school really gave a very imperfect expression of the facts, since, as we ¶ have shewn in Chapters X., XI., and XII., the degree and sequence of the consonance are really materially altered by these changes, and composers who have outgrown the rules of the school, have also very clearly had regard to these alterations.

CHAPTER XVII.

OF DISCORDS.

WHEN voices move forward melodically in part music, the general rule is that they must form consonances with each other. For it is only as long as they are consonant, that there is an uninterrupted fusion of the corresponding auditory sensations. As soon as they are dissonant the individual parts mutually disturb each ¶ other, and each is a hindrance to the free motion of the other. To this esthetic reason must be added the purely physical consideration, that consonances cause an agreeable kind of gentle and uniform excitement to the ear which is distinguished by its greater variety from that produced by a single compound tone, whereas the sensation caused by intermittent dissonances is distressing and exhausting.

However, the rule that the various parts should make consonances with each other, is not without exception. The esthetic reason for this rule is not opposed to an occasional and temporary dissonance among the various parts, provided the motion of the parts is so contrived as to make the directions of the different voices perfectly easy to follow by the ear. Hence, in addition to the general laws of scale and key, to which the direction of every part is subject, there are particular rules for the progression of voices through discords. Again, dissonances cannot be entirely excluded because consonances are physically more agreeable. That which is physically agreeable is an important adjunct and support to esthetic beauty, but it is certainly not identical with it. On the contrary, in all arts we frequently employ its opposite, that which is physically disagreeable, partly to bring the beauty of the first into relief, by contrast, and partly to gain a more powerful means for the expression of passion. Dissonances are used for similar purposes in music. They are partly means of contrast, to give prominence to the impression made by consonances, and partly means of expression, not merely for peculiar and isolated emotional disturbances, but generally to heighten the impression of musical progress and impetuosity, because when the ear has been distressed by dissonances it longs to return to the calm current of pure consonances. It is for this last reason that dissonances are prominently employed immediately before the conclusion of a piece, where they were regularly introduced even by the old masters of medieval polyphony. But to effect this object in using them, the motion of the parts must be so conducted that the hearer can feel throughout that the parts are pressing forward through the dissonance to a following consonance, and, although ¶ this may be delayed or frustrated, the anticipation of its approach is the only motive which justifies the existence of the dissonances.

Since any relation of pitch which cannot be expressed in small numbers is dissonant, and it is only the number of the consonances which is limited, the number of possible dissonances would be infinite were it not that the individual parts composing a discord in music must necessarily obey the laws of melodic motion, that is, must lie within the scale. Consonances have an independent right to exist. Our modern scales have been formed upon them. But dissonances are allowable only as transitions between consonances. They have no independent right of existence, and the parts composing them are consequently obliged to move within the degree of the scales, by the same laws that were established in favour of the consonances.

On proceeding to a detailed consideration of the separate dissonant intervals, it should be remembered that in theoretical music the normal position of discords is ¶ taken to be that which arranges their tones as a series of Thirds. This, for example, is the rule for the chord of the dominant Seventh, which consists of the root, its Third, Fifth, and Seventh. The Fifth forms a Third with the Third, and the Seventh forms a Third with the Fifth. Hence we can consider a Fifth to be composed of two, and a Seventh of three Thirds. By inverting Thirds we obtain Sixths, by inverting Fifths we obtain Fourths, and by inverting Sevenths we obtain Seconds. In this way all the intervals in the scale are reproduced.

Using the present modification of Hauptmann's notation, it is easily seen how different intervals of the same name must differ from each other in magnitude. We have only to remember that c^1 is a comma higher than c, and c_1 two commas lower than c^1 and one comma lower than c, and that the comma is about the fifth part of a Semitone.

To obtain a general view of both the magnitude and roughness of the dissonant intervals, I have constructed fig. 61 (p. 333a), in which the curve of roughness is ¶ copied from fig. 60 A (p. 193b). The base line X Y signifies the interval of an Octave, upon which the individual consonant and dissonant intervals are set off from X, according to their magnitude on this scale.* On the lower side of the base are marked the twelve equal Semitones of the equally-tempered scale [each distant from the other by 100 cents], and on the upper side the consonant and dissonant intervals which occur in justly-intoned scales. The magnitude of the interval is always to be measured on the base line from X to the corresponding vertical line. The vertical lines corresponding to the consonances have been produced to the upper margins of the diagram, and those for the dissonances have been made shorter. The length of the verticals intercepted between the base and the curve of roughness shews the comparative degree of roughness probably possessed by the interval when played in a violin quality of tone.

* [That is, assuming X Y to represent the cents in an Octave or 1200, the distance from X of any line shewing the interval, gives the cents in that interval.—*Translator*.]

* [TABULAR EXPRESSION OF THE DIAGRAM, FIG. 61 OPPOSITE.

Intervals	No.	Helmholtz's Notation, as in Diagram	Ellis's Notation of Intervals reckoned from c	Ratio	Cents	Roughness
Unison	1	$c : c$	$c : c$	$1 : 1$	0	0
Minor Seconds	2	$\|c : c\sharp$	$\|c : c\sharp$	$1 : \sqrt[12]{2}$	‖ 100	‖ 76
	3	$b_1 : c'$	$c : d^1\flat$	$15 : 16$	112	70
Major Seconds	4	$d : e_1$	$c : d_1$	$9 : 10$	182	38
	5	$\|c : d$	$\|c : d$	$1 : \sqrt[12]{4}$	‖ 200	‖ 25
	6	$c : d$	$c : d$	$8 : 9$	204	32
	7	$b_1 : d^1\flat$	$c : e^2\flat\flat$	$225 : 256$	224	30
Minor Thirds	8	$a^1\flat : b_1$	$c : d_2\sharp$	$64 : 75$	274	24
	9	$d : f$	$c : e\flat$	$27 : 32$	294	26
	10	$\|c : d\sharp$	$\|c : d\sharp$	$1 : \sqrt[12]{8}$	‖ 300	‖ 24
	11	$*c : e^1\flat$	$*c : e^1\flat$	$*5 : 6$	* 316	* 20
Major Thirds	12	$*c : e_1$	$*c : e_1$	$*4 : 5$	* 386	* 8
	13	$\|c : e$	$\|c : e$	$1 : \sqrt[12]{16}$	‖ 400	‖ 18
	14	$b_1 : e^{1\prime}\flat$	$c : f^2\flat$	$25 : 32$	428	25
Fourths	15	$*c : f$	$*c : f$	$*3 : 4$	* 498	* 2
	16	$\|c : f$	$\|c : f$	$1 : \sqrt[12]{32}$	‖ 500	‖ 3
	17	$a_1 : d'$	$c : f^1$	$20 : 27$	520	27
Sharp Fourths or Flat Fifths	18	$b^1\flat : e'_1$	$c : f_2\sharp$	$18 : 25$	568	32
	19	$f : b_1$	$c : f_1\sharp$	$32 : 45$	590	20
	20	$\|c : f\sharp$	$\|c : f\sharp$	$1 : \sqrt[12]{64}$	‖ 600	‖ 18
	21	$b_1 : f'$	$c : g^1\flat$	$45 : 64$	610	28
	22	$e_1 : b^1\flat$	$c : g^2\flat$	$25 : 36$	632	35
Fifths	23	$d : a_1$	$c : g_1$	$27 : 40$	680	44
	24	$\|c : g$	$\|c : g$	$1 : \sqrt[12]{128}$	‖ 700	‖ 1
	25	$*c : g$	$*c : g$	$*2 : 3$	* 702	* 0
Minor Sixths	26	$e^1\flat : b_1$	$c : g_2\sharp$	$16 : 25$	772	39
	27	$\|c : g\sharp$	$\|c : g\sharp$	$1 : \sqrt[12]{256}$	‖ 800	‖ 22
	28	$*c : a^1\flat$	$*c : a^1\flat$	$*5 : 8$	* 814	* 20
Major Sixths	29	$*c : a_1$	$*c : a_1$	$*3 : 5$	* 884	* 3
	30	$\|c : a$	$\|c : a$	$1 : \sqrt[12]{512}$	‖ 900	‖ 22
	31	$f : d'$	$c : a$	$16 : 27$	906	24
	32	$b_1 : a^1\flat$	$c : b^2\flat\flat$	$75 : 128$	926	24
	33	$d^1\flat : b_1$	$c : a_2\sharp$	$128 : 225$	976	15
Minor Sevenths	34	$d : c'$	$c : b\flat$	$9 : 16$	996	23
	35	$\|c : b\flat$	$\|c : b\flat$	$1 : \sqrt[12]{1024}$	‖ 1000	‖ 24
	36	$e_1 : d'$	$c : b^1\flat$	$5 : 9$	1018	25
Major Sevenths	37	$c : b_1$	$c : b$	$8 : 15$	1088	42
	38	$\|c : b$	$\|c : b$	$1 : \sqrt[12]{2048}$	‖ 1100	‖ 48
Octave	39	$*c : c'$	$*c : c'$	$*1 : 2$	* 1200	* 0

FIG. 61.   (*See note * opposite.*)

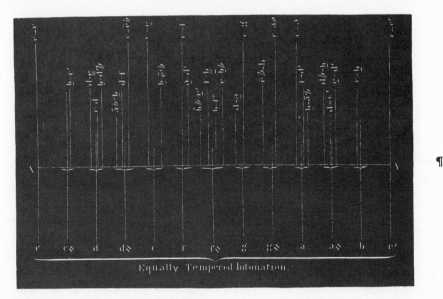

Equally Tempered Intonation.

The preceding tabular expression of the diagram will be often found convenient. The degree of roughness was determined by measuring the lengths of the verticals in the diagram in hundredths of an inch. The names of the notes are given in the notation of the text, using superior and inferior figures for the lines above and below in the diagram. The sign ‖ means 'equally tempered,' and * 'consonance.' The cross lines group the just intervals represented by a single tempered interval. The cents are cyclical, as in the Duodenarium, App. XX. sect. E. art. 18.

The intervals in the diagram are not noted as from *C* to another tone, but as between the two tones where they usually occur, except in the equal intonation below. In the Table both are given. The verticals for the dissonances were placed in two rows ‘in re-cutting the diagram for the 1st edition of this translation, merely for the purpose of clearness, to prevent the letters from coming too close to each other, but without attaching any meaning to the difference of row ; the other differences described in the text have been retained. The diagram also uses the lines above and below the letters employed in the 1st edition, (p. 277*c*, note *) and separates the letters by (—), (p. 276*d*, note †) as it was not considered advisable to re-engrave it. In the Table, however, the notation of the text is restored.

Table of Roughness.

The following is a comparative arrangement of these intervals in order of roughness, the consonances being marked *, and the tempered intervals ‖. The number in a parenthesis is that of the interval when it is contained in the preceding Table. The name given to each interval in App. XX. sect. D. is annexed, following by its *roughness*, marked ‘ro.’

Roughness
* 0—(25) just Fifth.

Roughness
‖ 1—(24) tempered Fifth representing (25) just Fifth, ro. 0, and (23) grave Fifth, ro. 44.
* 2—(15) just Fourth.
‖ 3—(16) tempered Fourth, representing (15) just Fourth, ro. 2, and (17) acute Fourth, ro. 27.
* 3—(29) just major Sixth.
* 8—(12) just major Third.
  15—(33) extreme sharp Sixth.
‖ 18—(13) tempered major Third, representing (12) just major Third, ro. 8, and (14) diminished Fourth, ro. 25,— and also the Pythagorean major Third, if required, ro. 19.
‖ 18—(20) tempered sharp Fourth or flat Fifth, representing (19) false Fourth or Tritone, ro. 20, (21) diminished Fifth, ro. 28, (18) superfluous Fourth, ro. 32, and (22) acute diminished Fifth, ro. 35.
  19—Pythagorean major Third $c : e = \frac{81}{64} =$ 408 cents. See p. 334, note ‡.
* 20—(28) just minor Sixth.
* 20—(11) just minor Third.
  20—(19) false Fourth or Tritone.
‖ 22—(30) tempered major Sixth, representing (29) just major Sixth, ro. 3, (31) Pythagorean major Sixth, ro. 24, and (32) diminished Seventh, ro. 24.
‖ 22—(27) tempered minor Sixth, representing (28) just minor Sixth, ro. 20, and (26) grave superfluous Fifth, ro. 39.
  23—(34) minor Seventh.
‖ 24—(35) tempered minor Seventh, representing (33) the extreme sharp Sixth, ro. 15 ; (34) the minor Seventh, ro. 23, and (36) the acute minor Seventh, ro. 25.

[*Continued on next page.*]

The various Thirds, Fifths, and Sevenths of the scale are found by arranging it in Thirds thus:—

A. Tones of the Major Scale.

$$b_1 - d \mid f + a_1 - c + e_1 - g + b_1 - d \mid f - a_1$$
$$\tfrac{6}{5} \quad \tfrac{32}{27} \quad \tfrac{5}{4} \quad \tfrac{6}{5} \quad \tfrac{5}{4} \quad \tfrac{6}{5} \quad \tfrac{5}{4} \quad \tfrac{6}{5} \quad \tfrac{32}{27} \quad \tfrac{5}{4}$$

B. Tones of the Minor Scale.

$$b_1 - d \mid f - a^1\flat + c - e^1\flat + g + b_1 - d \mid f - a^1\flat$$
$$\tfrac{6}{5} \quad \tfrac{32}{27} \quad \tfrac{6}{5} \quad \tfrac{5}{4} \quad \tfrac{6}{5} \quad \tfrac{5}{4} \quad \tfrac{5}{4_1} \quad \tfrac{6}{5} \quad \tfrac{32}{27} \quad \tfrac{6}{5}$$

For the minor scale I have assumed the usual form with the major Seventh, ¶ because scales with the minor Seventh yield the same intervals as the major scale.*

I. Thirds and Sixths.

The above schemes shew that in the justly-intoned major and minor scales, three kinds of Thirds occur, and their inversions give three kinds of Sixths. These are :

1) The *justly-intoned major Third* $\tfrac{5}{4}$, [12, cents 386, roughness 8],† and its inversion the *minor Sixth* $\tfrac{8}{5}$, [28, cents 814, roughness 20], both consonant.

2) The *justly-intoned minor Third* $\tfrac{6}{5}$, [11, cents 316, roughness 20], and its inversion the *major Sixth* $\tfrac{5}{3}$, [29, cents 884, roughness 3], also both consonant.

3) The *Pythagorean minor Third* $\tfrac{32}{27}$, [9, cents 294, roughness 26], between the extreme tones of the key, d and f. If we used d_1 in place of d, this interval would occur between b_1 and d_1. On comparing this dissonant minor Third $d \mid f$ ¶ with the consonant minor Third $d_1 - f$, we find that the former is a comma closer than the latter, since d is a comma sharper than d_1. The Pythagorean minor Third is somewhat less harmonious than the just minor Third, but the difference between them is not so great as that between the two corresponding major Thirds.‡ The difference of the two cases consists, first, in the major Third being a much more perfect consonance than the minor Third, and consequently much more liable to injury from defects of intonation ; and secondly in the nature

Roughness
|| 24—(10) tempered minor Third, representing (11) just minor Third, ro. 20 ; (8) acute augmented Tone, ro. 24, and (9) Pythagorean minor Third, ro. 26.

24—(31) Pythagorean major Sixth.
24— Pythagorean minor Sixth $c : a\flat = \tfrac{128}{81} = 792$ cents.
¶   24—(32) diminished Seventh.
24—(8) augmented Tone.
|| 25—(5) tempered major Second or whole Tone, representing (7) diminished minor Third, ro. 30, (6) major Tone, ro. 32, and (4) minor Tone, ro. 38.
25—(14) diminished Fourth.
25—(36) acute minor Seventh.
26—(9) Pythagorean minor Third.
27—(17) acute Fourth.
28—(21) diminished Fifth.
29—grave major Seventh $c : b_2 = \tfrac{50}{27} = 1067$.
30—(7) diminished minor Third.
32—(6) major Tone.
32—(18) superfluous Fourth.
35—(22) acute diminished Fifth.
38—(4) minor Tone.
39—(26) grave superfluous Fifth.
42—(37) just major Seventh.

Roughness
44—(23) grave Fifth.
|| 48—(38) tempered major Seventh, representing (37) just major Seventh, ro. 42.
56—great Limma $c : d^2\flat = \tfrac{27}{25} = 134$ cents.
70—(3) just minor Second, just or diatonic Semitone.
|| 76—(2) tempered Semitone, representing (3) just Semitone, ro. 70.— *Translator.*]

* [The remainder of this chapter should be followed step by step on the Harmonical, wherever it is possible, as is most frequently the case.—*Translator.*]

† [For immediate comparison I have, after each interval as it arises, inserted in square brackets the number of the interval, the number of cents it contains, and its degree of roughness as given in the Table on p. 332. —*Translator.*]

‡ [The roughness of the just major Third, $c + e$, is only 8, while that of the Pythagorean = $\tfrac{81}{64}$ (which is not given in the Table on p. 332, because it does not occur in the scale) is necessarily close to that of the tempered major Third, 18, and may probably be taken as 19, as will be seen by the curve in fig. 61, p. 333*a*. —*Translator.*]

of the two combinational tones. The just minor Third $d_1''' - f'''$ has $b\flat$ for its combinational tone, which completes it into the just major triad of $b\flat$. The Pythagorean minor Third $d''' \mid f'''$ has a_1 for its combinational tone, which completes it into the chord $d \mid f + a_1$, and this is not a perfectly correct minor chord. But as the incorrect Fifth a_1 lies among the deep combinational tones and is very weak, the difference is scarcely perceptible. Moreover, it is practically almost impossible to tune the interval so precisely as to insure the combinational tone a_1 in place of a. But for the Pythagorean major Third $c''.. e''$ the combinational tone is $c\sharp$, which is, of course, much more annoying than the rather imperfect Fifth a_1 when added to the chord $d \mid f$.*

The Pythagorean major Third does not occur in scales tuned according to the conditions of harmonic music. If we used the minor Seventh $b\flat$ in place of $b^1\flat$ for the minor scale, $b\flat \ldots d$ would be a Pythagorean major Third.†

The inversion of the Third $d \mid f$ is the *Pythagorean major Sixth* $f \ldots d'$, $\frac{27}{16}$, [31, ¶ cents 906, roughness 24], which is a comma wider than the just major Sixth, and is greatly inferior to it in harmoniousness, as is clearly seen in fig. 61 (p. 333a).

II. FIFTHS AND FOURTHS.

The Fifth is simply composed of two Thirds, and the different varieties of Fifths depend upon the nature of those Thirds.

4) The *just Fifth* $\frac{3}{2}$, [25, cents 702, roughness 0], consists of a just major and a just minor Third, or $\frac{3}{2} = \frac{5}{4} \times \frac{6}{5}$ [cents $702 = 386 + 316$]. Its inversion is the *just Fourth* $\frac{4}{3}$, [15, cents 498, roughness 2]. Both are consonant. Examples in the major scale, $f \pm c'$, $a_1 \pm e_1'$, $c \pm g$, $e_1 \pm b_1$, $g \pm d$.

5) The *grave* or *imperfect Fifth* $d \ldots a_1$ $\frac{40}{27}$, [23, cents 680, roughness 44], a comma [of 22 cents] less than the just Fifth, consists of a Pythagorean minor and a just major Third, $\frac{40}{27} = \frac{32}{27} \times \frac{5}{4}$ [cents $680 = 294 + 386$]. It sounds like a badly-tuned Fifth, and makes clearly sensible beats. In the Octave $c' \ldots c''$, the number of these beats in a second is 11. Its inversion, the *acute* or *imperfect Fourth*, $a_1 \ldots d'$, $\frac{27}{20}$, [17, cents 520, roughness 27], is also decidedly dissonant. The Fourth $A_1 \ldots d$ makes as many beats in a second as the Fifth $d \ldots a_1$, the d being the same in each, [see App. XX. sect. G. art. 16]. ¶

6) The *false* or diminished *Fifth*, $b_1 \ldots f'$, $\frac{64}{45}$, [21, cents 610, roughness 28], consists of a just and Pythagorean minor Third, $\frac{64}{45} = \frac{6}{5} \times \frac{32}{27}$, [cents $610 = 316 + 294$] and is, hence, as the composition shews, [92 cents or] about half a Tone closer than the just Fifth. It is a tolerably rough dissonance, nearly equal in roughness to a major Second [6, cents 204, roughness 32]. Its inversion is the *false Fourth* or *Tritone*, $f \ldots b_1$, $\frac{45}{32}$, [19, cents 590, roughness 20], consisting of three whole Tones, major $f \ldots g$, minor $g \ldots a_1$, and major $a_1 \ldots b_1$, $\frac{9}{8} \times \frac{10}{9} \times \frac{9}{8} = \frac{45}{32}$, [cents $590 = 204 + 182 + 204$]; it has very nearly the same degree of roughness as the last [or false Fifth], and is [20 cents or] about a comma closer. For the false Fifth $b_1 \ldots f$ ¶ is nearly the same as $c\flat \ldots f$, and if we diminish this interval by a comma we obtain $c\flat - f_1$, which is a false Fourth. Strictly speaking, as $c\flat$ is not precisely the same as b_1, the difference between the intervals is not precisely a comma, $\frac{81}{80}$, but about $\frac{89}{88}$, [or $\frac{10}{11}$ of a comma $= 20$ cents]. On keyed instruments they coincide.

7) The *superfluous* or *extreme sharp Fifth* of the minor scale, $e^1\flat \ldots b_1$, $\frac{25}{16}$, [26, cents 772, roughness 39], consists of two major Thirds, $e^1\flat + g$, and $g + b_1$, $\frac{25}{16} = \frac{5}{4} \times \frac{5}{4}$ [cents $772 = 386 + 386$]. It is seen to be [42 cents or] nearly two commas [44 cents] closer than the minor Sixth, [cents 814] by putting for b: the nearly identical $c\flat$, so that $e^1\flat \ldots b_1$ is nearly the same as $e^1\flat \ldots c\flat$, whereas the consonant minor Sixth is $e_1\flat \ldots c\flat$, where $e_1\flat$ is two commas flatter than $e^1\flat$. The superfluous Fifth [26, cents 772, roughness 39] is markedly rougher than the minor Sixth [28,

---

* [In just intonation, however, the difference between $d_1 - f$ and $d \mid f$ is very marked, as may be readily observed on the Harmonical.— *Translator.*]

† [The Pythagorean major Third of 408 cents does not occur on the Harmonical. The nearest interval $^7bb \ldots d_1$ of 413 cents is superior in effect.— *Translator.*]

cents 814, roughness 20], with which it coincides upon keyed instruments. Its inversion, the *diminished Fourth* $b_1...e^{1'}\flat$, $\frac{32}{25}$, [14, cents 428, roughness 25], is [42 cents or] about two commas higher than the just major Third, [12, cents 386, roughness 8], and considerably rougher, although the two intervals coincide on keyed instruments, [13, cents 400, common roughness 18].

Two just or two Pythagorean minor Thirds cannot occur consecutively in the natural series of Thirds of the just major and minor scales. In the modes of the minor Seventh and of the Fourth, we may find the intervals $a_1...e^1\flat$ and $e_1...b^1\flat$ $= \frac{36}{25}$, [22, cents 632, roughness 35], composed of two minor Thirds, $\frac{36}{25} = \frac{6}{5} \times \frac{6}{5}$ [cents 632 = 316 + 316]; these are a comma wider than the usual false Fifths $b_1...f'$ (or $a_1..e'\flat$ in the key of $b\flat$ major, and $e_1...b\flat$ in the key of f major), and are decidedly rougher than these, [21, cents 610, roughness 28].

¶

III. Sevenths and Seconds.

Any three successive Thirds give a Seventh. Beginning with the smallest we obtain the following different magnitudes :

8) The *diminished Seventh* of the minor scale $b_1...a^{1'}\flat$ [32, cents 926, roughness 24], $= (b_1 - d') + (d' + f') + (f' - a^{1'}\flat)$, or two just minor Thirds and one Pythagorean minor Third. Its numerical ratio is $\frac{128}{75} = \frac{6}{5} \times \frac{32}{27} \times \frac{6}{5}$, [cents 926 = 316 + 294 + 316,] which is [42 cents or] about two commas greater than the major Sixth [29, cents 884, roughness 3], as is seen by putting $b_1...a^{1'}\flat = c\flat .. a^{1'}\flat$. The interval $c\flat...a_1'\flat$, which is two commas flatter than the last, would be a just major Sixth. Its dissonance is harsh and rough, the same as that of the Pythagorean major Sixth $c...a$, [31, cents 906], which is [20 cents or] about a comma less. But its inversion, the *superfluous Second* $a^1\flat...b_1$ [8, cents 274, having the same roughness 24], is not much rougher than the just minor Third [11, cents 316, roughness 20; the tempered minor Third 10, cents 300, has exactly the same roughness 24]. Its numerical ratio $\frac{75}{64}$ [cents 274] is very nearly $\frac{7}{6}$ [cents 267] (since $\frac{75}{64} = \frac{7}{6} \times \frac{225}{224}$ [cents 274 = 267 + 7]). If this Second is extended to a Ninth, $\frac{75}{32}$, [having 1474 cents or] nearly $\frac{7}{3}$, [cents 1467] it becomes tolerably harmonious, as much so as the minor Tenth $\frac{12}{5}$, [cents 1516] which, however, is a very imperfect consonance, [see fig. 60 B, p. 193c].*

9) The *closer minor Seventh* of the scale $g...f'$, $b_1...a_1'$, or $d-c'$, $\frac{16}{9}$, [34, cents 996, roughness 23], consists of a just major, a just minor, and a Pythagorean minor Third, $g - f' = (g + b_1) + (b_1 - d') + (d'^1 - f')$, [or $\frac{16}{9} = \frac{5}{4} \times \frac{6}{5} \times \frac{32}{27}$ cents 996 = 386 + 316 + 294.] It is a comparatively mild dissonance, milder than the diminished Seventh [32, cents 926, roughness 24], and this is of importance for the effect of the chord of the dominant Seventh, in which the Seventh has this form. This closer minor Seventh is the interval of a Seventh in the scale nearest to the natural Seventh or seventh harmonic, $\frac{7}{4}$ [$=\frac{16}{9} \times \frac{63}{64}$, cents 969=996 − 27], although not so close as the extreme sharp Sixth [33, cents 976=996−20, roughness 15]. It has been already shewn that the natural Seventh belongs to harmonious combinations (pp. 195a, 217c). The inversion of this Seventh is the *major Second*, $c...d$, $a_1...b_1$, $f...g$, $\frac{9}{8}$, [6, cents 204, roughness 32], a powerful dissonance.

10) The *acute* or *wider minor Seventh*, $e_1...d'$, $a_1...g'$, $\frac{9}{5}$, [36, cents 1018, roughness 25], a comma greater than the last, is distinctly harsher than that interval, because it is nearer to the Octave; its roughness [25] is nearly the same as that of the diminished Seventh [24]. It consists of a just major and two just minor Thirds : $e_1...d' = (e_1 - g) + (g + b_1) + (b_1 - d')$, [or $\frac{9}{5} = \frac{6}{5} \times \frac{5}{4} \times \frac{6}{5}$, cents 1018 = 316 + 386 + 316.] The last-mentioned *closer minor Seventh* has its root on the

---

* [Compare on Harmonical $a^{1'}b...b_1''$ with $g'...b^{1''}b$. The $g'...^7b''b$ will be found much $g'...^7b''b$, and $a^1b...b_1'$ with $g'...^7b''b$ and with   the most harmonious.—*Translator.*]

dominant side of the scale, and its Seventh on the subdominant side, because it contains the Pythagorean minor Third $d \mid f$. The wider minor Seventh, on the other hand, has its Seventh on the dominant side. Its inversion, the *minor Tone*, $\frac{10}{9}$, $d...e_1$, $g...a_1$, [4, cents 182, roughness 38] is somewhat harsher than the major Tone [6, cents 204, roughness 32].

11) The *major Seventh* $f...e'_1$, $c...b_1$, $\frac{15}{8}$, [37, cents 1088, roughness 42], consists of two just major and one just minor Third: $c...b_1 = (c + e_1) + (e_1 - g) + (g + b_1)$ [or $\frac{15}{8} = \frac{5}{4} \times \frac{6}{5} \times \frac{5}{4}$, cents $1088 = 386 + 316 + 386$]. It is a harsh dissonance, about the same as the minor Tone [4, cents 182, roughness 38]. Its inversion, the *minor Second* or *Semitone* $b_1...c'$, $e_1...f$, $\frac{16}{15}$, [3, cents 112, roughness 70], is the harshest dissonance in the scale.*

In the mode of the Fourth or minor Seventh, we find a somewhat closer *major Seventh*, $b^1\flat...a'_1$, which is a comma closer than the usual major Seventh, and hence somewhat milder in effect.†   ¶

Finally we have to mention an interval peculiar to the Doric mode of the minor Sixth, namely:

12) The *superfluous* or *extreme sharp Sixth* $d^1\flat...b_1$, which arises in this mode from connecting the peculiar minor Second of the mode $d^1\flat$ with the leading note b_1 [see p. 286b].

The numerical ratio is $\frac{225}{128}$, [33, cents 976, roughness 15], so that it is [20 cents or] about a comma less than the closer minor Seventh of the chord of the dominant Seventh [cents 996], as is seen by putting $d^1\flat...b_1 = d^1\flat...c'\flat$; the interval $d\flat$. $c'\flat$ would be the closer minor Seventh, and $d^1\flat$ is a comma higher than $d\flat$. The superfluous Sixth may be conceived as composed of two just major Thirds and one just major Tone: $(d^1\flat + f) + (f...g) + (g + b_1) = d^1\flat ... b_1$, [or $\frac{225}{128} = \frac{5}{4} \times \frac{9}{8} \times \frac{5}{4}$, cents $976 = 386 + 204 + 386$]. Its harmoniousness is equal to that of the minor Sixth, because it is almost exactly the natural Seventh $\frac{7}{4}$,‡ since $\frac{225}{128} = \frac{7}{4} \times \frac{225}{224}$ [or $976 = 969 + 7$]. Taken alone it cannot be regarded as a dissonance, but ¶ it makes no other consonant combinations, and hence is unfit for use in consonant chords. When it is inverted into the *diminished Third* $\frac{256}{225}$ [cents 224], or nearly $\frac{8}{7}$ [cents 231], it is, as already observed, considerably damaged [7, cents 224, the roughness rises to 30], but it is improved by taking the upper tone b_1 an Octave higher, in which case it is [cents 2176 or] nearly $\frac{7}{2}$ [= cents 2169]. Its near agreement with the natural Seventh and its comparative harmoniousness, seem to have preserved this remarkable interval in certain cadences, although it is quite foreign to our present tonal system. It is characteristic that musicians forbid its inversion into the diminished Third (which lessens its harmoniousness), but allow its extension into the corresponding Thirteenth (which improves its harmoniousness). On keyed instruments this interval coincides with the minor Seventh [35, cents 1000, roughness 24].

Generally, a glance at fig. 61 (p. 333a) will shew to what an extraordinary extent different intervals are fused on keyed instruments.§   On the lower side of the base ¶ line X Y are marked the places of the tones of the equally tempered scale, and the small braces below the base line shew those different tonal degrees which are

---

* [That is in the just major scale; the Semitone of the tempered scale, 2, reaches 76 degrees of roughness.—*Translator.*]

† [Its numerical ratio is $\frac{50}{27} = \frac{15}{8} \times \frac{80}{81}$, cents $1088 - 22 = 1066$, so that it is the interval $c...b_2$, which by fig. 61 (p. 333a) should have a roughness of about 29, in place of 42, the roughness of $c...b_1$.—*Translator.*]

‡ [The diagram, fig. 61 (p. 333a), gives the roughness of the superfluous Sixth as 15, and that of the minor Sixth as 20; see p. 333c', d'. This would make the former more harmonious than the latter. This interval does not exist on the Harmonical. In meantone intonation, the extreme sharp Sixth has only 966 cents, and is therefore still closer to the subminor Seventh

$\frac{7}{4} = 969$ cents. As a matter of fact, on my meantone concertina I find f 966 $d'\sharp$ much smoother than f 1007 $e\flat$. The chord introducing this interval occurs in three forms. The 'Italian' $D^1\flat$ 386 F 590 B_1, and the 'German' $D^1\flat$ 386 F 316 $A^1\flat$ 274 B_1, are simply imitations of the true chord of the dominant Seventh $D^1\flat$ 386 F 316 $A^1\flat$ 267 $^7C^1\flat$. The 'French' form, (the only one considered in the text and on p. 286b,) $D^1\flat$ 386 F 204 G 386 B_1, is the harshest of all. The G seems to be merely an *anticipation* of the note of the chord C 316 $E^1\flat$ 386 G 498 c on which it resolves.—*Translator.*]

§ [This is shewn in detail on pp. 332–4, note. *Translator.*]

usually expressed by the corresponding tone of the tempered scale. The interval $b_1...a^1\flat$ [cents 926] is identified on the pianoforte with the major Sixth $c\flat...a\flat$ [cents 884, or 42 cents closer], while the interval $d^1\flat...b_1$ [cents 976, or 50 cents wider than the first] is made a (tempered) Semitone [cents 100] wider [being identified with 1000 cents], and yet the last is scarcely more different from the first, than the first from a major Sixth. The figure 61 shews also very clearly what an immense difference of harmoniousness ought to exist between the first and either of the two last of the following intervals $c...a_1$, $f...d'$, and $b_1...a^1\flat$, [29, 31, 32, respective cents 884, 906, 926, respective roughness 3, 24, 24], which are all expressed by the sufficiently harsh sound of the tempered interval $c...a$ [30, cents 900, roughness 22]. The justly-intoned harmonium with two rows of keys* allows all these intervals to be given accurately, by which the difference of their sound becomes extremely striking. In this evidently lies one of the ¶ greatest imperfections of tempered intonation.

DISSONANT TRIADS.

Dissonant triads with a single dissonant interval are obtained by taking two tones which are consonant to the root, but dissonant to each other. Thus:

1) *Fifth and Fourth :* $c...f...g$, [or $f \pm c \pm g$].
2) *Third and Fourth :* $c + e_1...f$ or $c - e^1\flat$ f, [or $f \pm c + e_1$ and $f \pm c - e^1\flat$].
3) *Fifth and Sixth :* $c \pm g$. a_1 or $c \pm g...a^1\flat$, [or $a - c \pm g$, and $a^1\flat + c \pm g$].
4) *Dissimilar Thirds and Sixths :* $c - e^1\flat...a_1$ or $c + e_1...a^1\flat$, [or $a_1 - c - e^1\flat$ and $a^1\flat + c + e_1$].†

In all these c is consonant with each of the other two tones. The first chord alone plays a great part in the older polyphonic music as a *chord of suspension.* ¶ The others we shall meet with again in the chords of the Seventh.

The chords named in the fourth series above ‡ admit of an inversion which makes them appear as triads with diminished or superfluous Fifths, namely :

$$a_1 - c - e^1\flat \text{ and } a^1\flat + c + e_1.$$

The first of these is composed of two just minor Thirds, [so that the Fifth $a_1...e^1\flat$, No. 22, ratio 25 : 36, cents 632, roughness 35, is the acute diminished Fifth,] and the second of two just major Thirds, [so that the Fifth $a^1\flat...e_1$, No. 26, ratio 25 : 16, cents 772, roughness 39, is the grave superfluous Fifth]. Both are dissonant on account of the altered Fifth, although the dissonance of the second has to be played as the consonance $g_2\sharp...e$ [minor Sixth 814 cents] upon keyed instruments. The first of these two chords can only appear in the mode of the minor Third, and the above would be heard in that of F.§ The second, on the other hand, belongs to F minor.**

¶   If we suppose this series of tones to be continued as

$$a^1\flat + c' + e_1'... a^1{}'\flat + c'' + e_1''$$

$$\tfrac{5}{4} \quad \tfrac{5}{4} \quad \tfrac{32}{25} \quad \tfrac{5}{4} \quad \tfrac{5}{4}$$

* [And, with the exception of the extreme sharp Sixth $d^1\flat...b_1$, on the Harmonical also. The extreme sharp Sixth $c...a_2\sharp$ may be sufficiently imitated as $c...'b\flat$.—*Translator.*]

† [These triads I propose to term *con-dissonant*, and the two last especially I call the minor and major trine. See App. XX. sect. E. art. 5.—*Translator.*]

‡ [From p. 338c, beginning with these words to the paragraph ending ‘as in concords,’ on p. 339b, is an insertion in the 4th German edition.—*Translator.*]

§ [It is evident that $a_1 - c - e^1\flat$ can only occur when the chain of chords contains $f + a_1$ $-c - e^1\flat + g$, that is in one of the forms

$b\flat + d_1 - f + a_1 - c - e^1\flat + g = 1$ F ma.ma.mi.
$b\flat - d^1\flat + f + a_1 - c - e^1\flat + g = 1$ F mi.ma.mi.
$f + a_1 - c - e^1\flat + g + b_1 - d = 1$ C ma.mi.ma.
$f + a_1 - c - e^1\flat + g - b^1\flat + d = 1$ C mi.mi.mi.

But not one of these belongs to the mode of the minor Third, which for F is 1 F mi.mi.mi., unless the second is taken to be such with a major tonic. The last, however, is the mode of the minor Seventh of C.—*Translator.*]

** [In the major dominant form $b\flat - d^1\flat + f - a^1\flat + c + e_1 - g$.—*Translator.*]

an interval glides in of $\frac{32}{25} = \frac{5}{4} \cdot \frac{128}{125} = \frac{5}{4} \cdot \frac{43}{42}$ approximatively [cents $428 = 386 + 42$], which is slightly (about 2 commas) greater than a just major Third. By small alterations of pitch other chords are formed which belong to other keys :

$$A^1\flat + c + e_1 \quad ...a^1\flat \text{ in } F \text{ minor}$$
$$\underset{\frac{5}{4}}{} \quad \underset{\frac{5}{4}}{} \quad \underset{\frac{32}{25}}{}$$

$$G_2\sharp ...c + e_1 + g_2\sharp \text{ in } A_1 \text{ minor}$$
$$\underset{\frac{32}{25}}{} \quad \underset{\frac{5}{4}}{} \quad \underset{\frac{5}{4}}{}$$

$$A^1\flat + c... f^2\flat + a^1\flat \text{ in } D^1\flat \text{ minor}$$
$$\underset{\frac{5}{4}}{} \quad \underset{\frac{32}{25}}{} \quad \underset{\frac{5}{4}}{}$$

The roots of these three minor keys
$$D^1\flat + F + A_1$$

form a similar chord, of which the roots are a Semitone higher than those of the ¶ preceding.* Since $A^1\flat$ is nearly the same as $G_2\sharp$, and $F^2\flat$ nearly the same as E_1, these transformations alter the pitch of one of the tones in the chord by about two commas, or, at least in the resolution of the chord, this tone is treated as a leading note just as if it were thus altered. Hence we obtain modulations which with a single step lead us to comparatively distant keys, and we can as easily resolve into the minor as into the major keys of the three roots named. This means of modulation is often employed by modern composers, (for example R. Wagner) in place of using the chord of the diminished Seventh, which is much rougher but was also applied for the same purpose. In just intonation these chords are not by any means so unpleasant as in the tempered intonation of the pianoforte. Generally it may be observed that when one is accustomed to play in just intonation, the ear becomes quite as sensitive to a pitch which is wrong by a comma in discords as in concords.

For modern music *triads with two dissonances*, formed by including the ¶ extremes of the key, are more important.

In the series of chords in any key, major and minor Thirds follow each other alternately, and any two adjacent Thirds produce a consonant triad. But the interval between the extreme tones d and f is a Pythagorean minor Third, and when these are connected as a chord with one of either of the two adjacent tones to make a new triad, it will be dissonant.

MAJOR : $c + e_1 \quad - g + b_1 - d \mid f + a_1 \quad - c + e_1 \quad - g$
$$\underset{\frac{5}{4}}{} \quad \underset{\frac{6}{5}}{} \quad \underset{\frac{5}{4}}{} \quad \underset{\frac{6}{5}}{} \quad \underset{\frac{32}{27}}{} \quad \underset{\frac{5}{4}}{} \quad \underset{\frac{6}{5}}{} \quad \underset{\frac{5}{4}}{} \quad \underset{\frac{6}{5}}{}$$

MINOR : $c - e^1\flat + g + b_1 - d \mid f - a^1\flat + c - e^1\flat + g$
$$\underset{\frac{6}{5}}{} \quad \underset{\frac{5}{4}}{} \quad \underset{\frac{5}{4}}{} \quad \underset{\frac{6}{5}}{} \quad \underset{\frac{32}{27}}{} \quad \underset{\frac{6}{5}}{} \quad \underset{\frac{5}{4}}{} \quad \underset{\frac{6}{5}}{} \quad \underset{\frac{5}{4}}{}$$

The major system gives two triads of this kind :

$$b_1 - d + f \qquad \text{and} \qquad d \mid f + a_1$$
$$\underset{\frac{6}{5}}{} \quad \underset{\frac{32}{27}}{} \qquad\qquad\qquad \underset{\frac{32}{27}}{} \quad \underset{\frac{5}{4}}{}$$

The minor scale also gives two :

$$b_1 - d \mid f \qquad \text{and} \qquad d \mid f - a^1\flat$$
$$\underset{\frac{6}{5}}{} \quad \underset{\frac{32}{27}}{} \qquad\qquad\qquad \underset{\frac{32}{27}}{} \quad \underset{\frac{6}{5}}{}$$

In the two triads $b_1 - d \mid f$ and $d \mid f - a^1\flat$, which combine a Pythagorean with a just minor Third, there are also second dissonances, namely the false Fifths $b_1 ... f$ and $d ... a^1\flat$, which make the chord more strongly dissonant than the Pythagorean minor Third $\frac{32}{27}$ alone could make them. They are hence called *diminished*

---

* [Only in the form $c + e_1 + g_2\sharp$. From what follows it is evident that the transformation could only take place in tempered intonation. The tones confounded are all 42 cents apart, and could not possibly be confounded in just intonation. Of course Wagner thought only in equal temperament, in which the tones are absolutely identical.—*Translator.*]

triads. The chord $d \mid f + a_1$, which in the usual musical notation is not distinguished from the minor triad $d_1 - f + a_1$, and may hence be called the *false minor triad*, is, as Hauptmann has correctly shewn, dissonant, and on justly-intoned instruments it is very decidedly dissonant. It sounds almost as rough as the chord $b_1 - d \mid f$. If in C major, without confounding d with d_1, we form the cadence 1 or 2

the chords $a_1'...d'' \mid f''$ and $f' + a_1'...d'' \mid f''$ are quite as dissonant in their effects as the following $b_1' - d'' \mid f'$ and $g' + b_1' - d'' \mid f''$. But on account of the incorrect intonation of our musical instruments we cannot produce the same effect ¶ without combining an inverted chord of the Seventh with the subdominant in the cadence, as $f + a_1 - c'...d'$. Hauptmann doubts whether in practice the false minor chord of the key of C major can be distinguished from the minor chord of D. I find that this is most distinctly and undoubtedly effected on my justly intoned harmonium, but allow that we cannot expect the correct intonation from singers. They will involuntarily pass into the minor chord, unless the progression of the parts which execute D, strongly emphasise its connection with the dominant G.*

These chords, and among them most decisively and distinctly the chord $b_1 - d \mid f$, have for musical composition the especially important advantage of combining those limiting tones of the key, which separate it from the nearest related keys, and are consequently extremely well suited for marking the key in which the harmony is moving at any given time. If the harmony passed into G major or G minor, f would have to be replaced by $f_1\sharp$. If it passed into F major, d would become d_1 and if into F minor d would become $d^1\flat$ and b_1 would in the ¶ same chords become $b^1\flat$. Thus—

$$
\begin{array}{llll}
\text{in } G \text{ major :} & b_1 & - d & + f_1\sharp & d & + f_1\sharp - a \\
\text{in } C \text{ major :} & b_1 & - d & \mid f & d & \mid f & + a_1 \\
\text{in } F \text{ major :} & b\flat & + d_1 & - f & d_1 & - f & + a_1 \\
\text{in } G \text{ minor :} & b^1\flat & + d & + f_1\sharp & d & + f_1\sharp - a \\
\text{in } C \text{ minor :} & b_1 & - d & \mid f & d & \mid f & - a^1\flat \\
\text{in } F \text{ minor :} & b\flat & - d^1\flat & + f & d^1\flat & + f & - a^1\flat \\
\end{array}
$$

This shews that the chords in the nearest related keys are all distinctly different, with the exception of $d \mid f + a_1$ and $d_1 - f + a_1$, the distinction between which in

---

* [The chord on the Second of the major scale is in fact the *crux* of the translation of tempered into just intonation. It is easy to ¶ play Ex. 1 and 2, and Ex. 3, here added, as

$$
\begin{array}{ll}
a' \quad d_1'' \quad f'' & f' \quad a_1' \quad d_1'' \quad f'' \\
b_1' \quad d'' \quad f' \text{ and } g' & b_1' \quad d'' \quad f'' \text{ and } g' \\
c'' \quad c'' \quad e_1'' & c' \quad c'' \quad c'' \quad e_1'' \\
\end{array}
\qquad
\begin{array}{l}
f' \quad a_1' \quad d_1'' \\
b_1' \quad d'' \\
f \quad b_1' \quad d'' \\
e' \quad c'' \quad c''
\end{array}
$$

and the effect is not bad. In the first the d_1'' might be held on to the second chord, as $b' \, d_1''$ f'', without materially increasing the harshness of the dissonance, but in the second this would give $g' \, b_1' \, d_1'' \, f''$, where the grave Fifth is very harsh. In the second case, then, there is least harshness in playing d'' in both chords. And in both cases there is most smoothness in playing them as just written. The effect is one on which I have repeatedly experimented, but I find that the small interval $d_1'' \, d''$ in the highest or lowest part, produces a strange effect, which in singing, and perhaps on the violin, seems to be overcome by a glide, if the other voices are strong enough to pull this voice out of its course, even though the words and parts are written so as to imply that this note is sustained. When the d'' is in the principal part in the melody, as in the third example, I find it best on the whole not to play as written, d_1'' d'', but to sustain d''. In some cases an attempt to avoid the dissonance, which is really harsh, would lead to such melodic phrases as $d \, d_1 \, d$, which would be simply impossible for an unaccompanied voice. If in the third example d'' were held throughout, and the accompanying voices sang the minor chord, we should get the succession $f' \, a' d''$, $g' \, b_1' \, d''$, $f \, b_1'$ d'', $e' \, c'' \, c''$, which amounts to a modulation into the minor of the dominant, instead of into the subdominant. Whether such is possible depends on the preceding chords. As f^1 does not occur on the Harmonical, I played Ex. 3 on my just concertina in A_1 major as d' $f_1\sharp \, b_1'$, $e_1' \, g_2\sharp \, b_1'$, $d_1' \, g_2\sharp \, b_1'$, $c_2\sharp \, a_1' \, a'$, and found that such chords produced the best effect of all for this isolated phrase.—*Translator.*]

singing might be doubtful. The rest are much more clearly distinguished from the chords in the nearest adjacent keys.   Nevertheless

$$b_1 - d \mid f \quad\text{and}\quad d \mid f - a^1\flat$$

$$\underbrace{}_{\frac{6}{5}} \underbrace{}_{\frac{32}{27}} \qquad\qquad \underbrace{}_{\frac{32}{27}} \underbrace{}_{\frac{6}{5}}$$

are easily confused with

$$b_1 \mid d_1 - f \quad\text{and}\quad d - f^1 \mid a^1\flat$$

$$\underbrace{}_{\frac{32}{27}} \underbrace{}_{\frac{6}{5}} \qquad\qquad \underbrace{}_{\frac{6}{5}} \underbrace{}_{\frac{32}{27}}$$

of which the former belongs to A_1 minor, and the latter to $E^1\flat$ major or minor, where A_1 minor is the minor key nearest related to C major, and $E^1\flat$ major is the major key nearest related to C minor.

Finally when we remember that the Pythagorean minor Third $\frac{32}{27}$ [cents 294] ¶ is nearer the superfluous second $\frac{75}{64}$ [cents 274] than to the normal minor Third [cents 316] ($\frac{32}{27} = \frac{6}{5} \times \frac{80}{81}$ [cents 294 = 316 − 22] and $\frac{32}{27} = \frac{75}{64} \times \frac{2048}{2025}$ [cents 294 = 274 + 20] or nearly $= \frac{75}{64} \times \frac{89}{88}$), it requires comparatively slight changes in intonation to convert the chord $b_1 - d \mid f$ into

$$b_1 - d \ldots e_2 \sharp \quad\text{and}\quad c^1\flat \ldots d_1 - f$$

$$\underbrace{}_{\frac{6}{5}} \underbrace{}_{\frac{75}{64}} \qquad\qquad \underbrace{}_{\frac{75}{64}} \underbrace{}_{\frac{6}{5}}$$

which belong to $F_1\sharp$ minor and $E\flat$ minor. Hence the diminished triad $b_1 - d \mid f$, by slight changes of intonation,* never exceeding $\frac{81}{80}$, can be referred to the keys of

$$C \text{ major, } C \text{ minor, } A_1 \text{ minor, } F_1\sharp \text{ minor, and } E\flat \text{ minor.}$$

Hence although the use of the diminished triad $b_1 - d \mid f$ excludes the keys ¶ most nearly related to C, it allows of a confusion with more distant keys, and hence also the characterisation of the key by these triads will not be complete without a fourth note, converting the triad into a tetrad. This leads us to the chords of the Seventh proper.

CHORDS OF THE SEVENTH.

A. *Formed of two Consonant Triads.*

Consonant tetrads, or chords in four parts, as shewn on p. 222b, cannot be constructed without using the Octave of one of the tones, but dissonant tetrads are easily constructed. The least dissonant of such chords are those in which only a single interval is dissonant, and the rest are consonant. These are most readily formed by uniting two consonant triads which have two tones in common. In this case the tones which are not in common to the two chords are dissonant to each ¶ other, and the rest are consonant, so that the dissonance is comparatively un-observed † amid the mass of consonances. Thus the triads

$$c + e_1 - g$$
$$e_1 - g + b_1$$

on being fused give the tetrad

$$c + e_1 - g + b_1$$

in which the major Seventh $c \ldots b_1$ is a dissonant interval and the other intervals are consonant, as the annexed scheme shews :—

* [Which are made spontaneously in equally tempered intonation, where all three chords are absolutely identical, but would otherwise require an entire sacrifice of the feeling of tonality. Follow these chords on the Duo-denarium, App. XX. sect. E. art. 18.—*Translator.*]

† [To my sensation the dissonant tones utterly destroy the consonance.—*Translator.*]

[In cents :
c 702 g, e_1 702 b_1
C 386 E_1 316 G 386 B_1
c 1088 b_1.]

This position of the chord of the Seventh, deduced from the closest positions of the triads, is regarded as fundamental or primary. The intervals between the individual tones appear as Thirds, and when chords of the Seventh are formed from the consonant triads of the scale, these Thirds will be alternately major and minor, because consonant triads always unite a major with a minor Third. Haupt-mann calls these chords of the Seventh which occur spontaneously in the natural ¶ series of Thirds of a key

$$f + a_1 - c + e_1 - g + b_1 - d$$

the *chords of the direct system* or simply *direct chords*. There are two kinds of these chords. In one a minor Third lies between two major Thirds, as in the tetrad $c + e_1 - g + b_1$ already cited, and similarly in $f + a_1 - c' + e_1'$ in C major, and $A^1\flat + c - e^1\flat + g$ in C minor. In the other a major Third lies between two minor Thirds, as in

[In cents :
a 702 e'_1, c' 702 g'
A_1 316 C 386 E_1 316 G
a_1 1018 g'.]

¶ and similarly in $e_1 - g + b_1 - d_1$ in C major and $f - a^1\flat + c - e^1\flat$ in C minor. In this second species the dissonance is a minor Seventh, $\frac{9}{5}$, [roughness 25, p. 332, Table, No. 36, cents 1018,] which is much milder than the major Seventh, $\frac{15}{8}$ [*ibid.* No. 37, cents 1088, roughness 42].

B. *Chords of the Seventh formed of Dissonant Triads.*

Other chords of the Seventh may be formed from the dissonant triads of the key, each united with one consonant triad, and also from the two dissonant triads themselves. By thus uniting the limiting tones of the series of chords in the key,

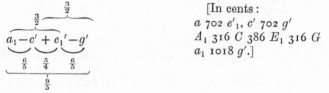

we obtain the following *Chords of the Seventh in the reverted system*, or *indirect* ¶ *tetrads :*

1)

[In cents :
1) g 702 d', b_1 610 f'
      G 386 B_1 316 D 294 F
      g 996 f']

2)

[2) d 680 a_1, f 702 c'
       D 294 F 386 A_1 316 C
       d 996 c']

3)

$$\overbrace{\underbrace{\frac{64}{45}}\; \overbrace{\frac{3}{2}}}$$

$$d \mid f - a^1\flat + c'$$

$$\underbrace{\frac{32}{27} \quad \frac{6}{5} \quad \frac{5}{4}}_{\frac{16}{9}}$$

[3] d 610 $a^1\flat$, f 702 c'
      D 294 F 316 $A^1\flat$ 386 C
      d 996 c']

4)

$$\overbrace{\underbrace{\frac{64}{45}}\; \overbrace{\frac{40}{27}}}$$

$$b_1 - d \mid f + a_1$$

$$\underbrace{\frac{6}{5} \quad \frac{32}{27} \quad \frac{5}{4}}_{\frac{16}{9}}$$

[4] b_1 610 f, d 680 a_1
      B_1 316 D 294 F 386 A_1
      b_1 996 a_1]

5)

$$\overbrace{\underbrace{\frac{64}{45}}\; \overbrace{\frac{64}{45}}}$$

$$b_1 - d \mid f - a^1\flat$$

$$\underbrace{\frac{6}{5} \quad \frac{32}{27} \quad \frac{6}{5}}_{\frac{128}{75}}$$

[5] b_1 610 f, d 610 $a^1\flat$ ¶
      B_1 316 D 294 F 316 $A^1\flat$
      b_1 926 $a^1\flat$]

The Sevenths of these chords all come pretty near to the natural Seventh $\frac{7}{4}$ [cents 969], and are all smaller than the Sevenths in the chords of the Seventh formed from two consonant triads [cents 1088 and 1018]. The principal dissonances in these chords are the false and imperfect Fifths $b_1 \ldots f$, $d \ldots a_1$, and $d \ldots a^1\flat$, that is, the intervals $\frac{64}{45}$ and $\frac{40}{27}$ [p. 332, Table, Nos. 21 and 23, cents 610 and 680, roughnesses 28 and 44]. Hence the first three of these chords of the Seventh, $g + b_1 - d \mid f$, $d \mid f + a_1 - c$, and $d \mid f - a^1\flat + c$, each of which contains only one of these imperfect Fifths, are less harshly dissonant than the two last, each of ¶ which contains two of them. Such of these chords as contain a major triad, namely—

$$g + b_1 - d \mid f \quad \text{and} \quad d \mid f + a_1 - c$$

are about equal in dissonance to the milder chords of the Seventh in the direct system, which contains the larger and rougher kind of Sevenths, and, at the same time, only perfect Fifths, viz. :

$$a_1 - c + e_1 - g \quad \text{and} \quad e_1 - g + b_1 - d$$

The chord of the dominant Seventh $g + b_1 - d' \mid f'$ can be even rendered much milder by lowering its f' to f_1'. The interval $g \ldots f_1'$ corresponds to the ratio $\frac{1280}{729}$ [cents 974], which is very nearly equal to $\frac{7}{4}$ [cents 969], being $= \frac{7}{4} \times$ ¶ $\frac{5120}{5103}$ [cents 969 + 5], or approximately $\frac{7}{4} \times \frac{301}{300}$. Hence the chord $g + b_1 - d \mid f_1$ is on the verge of consonance.*

But the chord of the Seventh which contains a false Fifth and a minor triad, namely No. 3 above, or

$$d \mid f - a^1\flat + c$$

is about as rough as the tetrads of the direct system containing a major Seventh, namely—

$$f + a_1 - c + e_1 \quad \text{and} \quad c + e_1 - g - b_1.$$

---

* [That is, allowing $g\, b_1 d'f$ or $c\, e_1 g'' b b$ to be consonant. In the 53 division the player uses 44 degrees = 996 cents for $g \ldots f$, and 43 degrees = 974 cents for $g \ldots {}^{\prime}f$, and the latter is found a satisfactory imitation—especially by ears unaccustomed to the true interval, because it is so much superior to the former of 44 degrees = 996 cents.—*Translator.*]

It is curious that the first of these three tetrads contains exactly the same intervals as the chord of the dominant Seventh itself, $g + b_1 - d \mid f$, only in inverse order, thus—

$$d \mid f - a^1\flat + c' \qquad\qquad g + b_1 - d' \mid f'$$
$$\underbrace{}_{\frac{32}{27}} \underbrace{}_{\frac{6}{5}} \underbrace{}_{\frac{5}{4}} \qquad\qquad \underbrace{}_{\frac{5}{4}} \underbrace{}_{\frac{6}{5}} \underbrace{}_{\frac{32}{27}}$$

In the first the consonant portion is a minor triad, and this makes it decidedly harsher than the second where the consonant portion is a major triad.

Here also the difference of harshness depends on the nature of the combinational tones, of which those generated by the closer intervals are most distinctly heard. These are

¶   for $\underbrace{g' + b_1' - d'' \mid f''}_{G \quad G \quad A_1}$ and for $\underbrace{d'' \mid f'' - a^{1''}\flat + c'''}_{A_1 \quad d^1\flat \quad a^1\flat}$

Hence one combinational tone in the first chord, and two in the second, are unsuitable to the chord.

The harshest chords of the Seventh are those which each contain two false Fifths, namely, No. 4 or $b_1 - d \mid f' + a_1'$ and No. 5 or $b_1 - d' \mid f' - a^{1'}\flat$. But the first of these can be made much milder by a slight change in its intonation. Thus $b_1 - d \dots f_1' \dots a'$ contains tones which all belong to the compound tone of $G_{\prime\prime}$, and these sound tolerably well together.*

The chords of the reverted system play an important part in modulations, by serving to mark the key precisely. The most decisive in its action is the chord of the Seventh on the dominant of the key, that is the chord $g + b_1 - d \mid f$ for the tonic C. We saw (p. 341b) that the diminished triad $b_1 - d \mid f$ could be adapted by slight changes in its intonation to the keys of

¶ C major, C minor, A_1 minor, $F_1\sharp$ minor, and $E^1\flat$ minor.

Of these only the two first contain the tone G, so that the chord $g + b_1 - d \mid f$ can belong to no tonic but C.

The imperfect minor triad [or chord of the added Sixth] $d \mid f + a_1$, which, when the intonation is correct, belongs only to the key of C major, admits of being confused [and is in equal temperament always identified] with $d_1 - f + a_1$, which

---

* [This is only to be taken as an approximative statement, grounded on the assumption that the interval g to f_1' is correctly $\frac{7}{4}$, in which case the primes of the tones b_1, d, f_1', a' are the 5th, 6th, 7th, 9th partials of G. This chord in its true formation is used on Mr. Poole's *double* or *dichordal* scale $F\ G\ A\ {}^7B\flat\ c\ d\ e_1 f$, the two chords being $F : A_1 : c = 4 : 5 : 6$, and $C : E_1 : G : {}^7B\flat : d = 4 : 5 : 6 : 7 : 9$.
¶ In the text it is, in point of fact, proposed to use $B_1\flat$ in the chord $C\ E_1\ G\ B_1\flat\ d$, as an imitation of Mr. Poole's natural chord, which would be still closer than $C\ E\ G\ A_2\sharp\ d$, with the extreme sharp Sixth in place of the natural Seventh. In fact, $C : {}^7B\flat = 969$ cents, $C : B_1\flat = 974$ cents, and $C : A_2\sharp = 976$ cents.

To test the effect of septimal intonation I had an instrument tuned to give the chords—

$B\flat\ d_1\ f, F\ A_1\ C, C\ E_1\ G\ {}^7B\flat\ d, G\ B_1\ d, D\ F^1\ A^1$

perfectly (of which the second, third, and fourth occur on the Harmonical). The effect of the third of these chords far surpasses my expectations, and it is beyond comparison better than the usual chord of the Ninth with $B\flat\ d$ in place of ${}^7B\flat\ d$ (for which on the Harmonical $g\ b_1\ d\ f\ a_1$ can be played). The chord of the subminor Seventh and its inversions

$C\ E_1\ G\ {}^7B\flat, E_1\ G\ {}^7B\flat\ c, G^{\cdot}B\flat\ c\ c_1, {}^7B\flat\ c\ e_1\ g$

are all decidedly superior to the chord of the dominant Seventh, with $B\flat$ in place of ${}^7B\flat$, and its inversions (which on the Harmonical must be tried as $g\ b_1\ d\ f$ and its inversions). The septimal minor triad $G\ {}^7B\flat\ d$ is far superior to the Pythagorean minor triad $D\ F\ A$ (not on the Harmonical), or the false minor triad $D\ F\ A_1$, and is not far inferior to the true minor triad $D_1\ F\ A_1$ or $D\ F^1\ A$ (on the Harmonical compare $g\ {}^7b\flat\ d'$ with $g\ b^1\flat\ d$). The septimal diminished triad $E_1\ G\ {}^7B\flat$ approaches consonance much more nearly than the usual diminished triad $E_1\ G\ B\flat$ (play $b_1\ d\ f$ on the Harmonical). Though Poole's ascending scale makes too great a gap between ${}^7B\flat$ and c, yet by using ${}^7B\flat\ D$ as alternative tones with $B\flat\ D_1$, ascending with the sharper and descending with the flatter forms, we obtain the perfectly melodious scales of

$F\ G\ A_1\ B\flat\ c\ d\ e_1\ f$ and $f\ e_1\ d_1\ c\ {}^7B\flat\ A_1\ G\ F$

(of which the first, being the ordinary scale of F major, does not exist on the Harmonical, which has no $B\flat$, but the second can be played upon it). These facts shew the acoustic possibility of a septimal theory of harmony, which would include the tertian, or ordinary harmony of just intonation.—*Translator*.]

belongs to the keys of A_1 minor, F major, and $B\flat$ major. This confusion is not entirely obviated by adding the tone c, and the consequence is the chord of the Seventh $d \mid f + a_1 - c$ is usually employed only in alternation with the chord of the dominant Seventh in the cadence, where it distinguishes C major from C minor.* But the addition of the tone b_1 to the triad $d \mid f + a_1$ [as $b_1 - d \mid f + a_1$] is characteristic, because this last can at most be confused with $b_1 \mid d_1 - f + a_1$, which belongs to A_1 minor. The chord $b_1 - d \mid f + a_1$, however, sounds comparatively harsh in every position for which a_1 is not the highest note, and hence its application is very limited. It is often united with the chord of the dominant Seventh as a chord of the Ninth, thus $g + b_1 - d' \mid f' + a_1'$, in which g and a_1' must remain the extreme tones. More upon this hereafter.

In the key of C minor, the triad $d \mid f - a^1\flat$ would, in just intonation, be characteristic, but yet it is easily confused with other chords. Thus

¶

$d \mid f - a^1\flat$ [in cents d 294 f 316 $a^1\flat$] belongs to C minor

$\underbrace{\quad}_{\frac{32}{27}}\ \underbrace{\quad}_{\frac{6}{5}}$

$d_1 - f \mid a\flat$ [in cents d 316 f 294 $a\flat$] to $E\flat$ major and $E\flat$ minor

$\underbrace{\quad}_{\frac{6}{5}}\ \underbrace{\quad}_{\frac{32}{27}}$

$d - f^1 \ldots g_1\sharp$ [in cents d 316 f^1 274 $g_1\sharp$] to A minor

$\underbrace{\quad}_{\frac{6}{5}}\ \underbrace{\quad}_{\frac{75}{64}}$

$d^1 \ldots e_1\sharp - g\sharp$ [in cents d^1 274 $e_1\sharp$ 316 $g\sharp$] to $F\sharp$ minor.

$\underbrace{\quad}_{\frac{75}{64}}\ \underbrace{\quad}_{\frac{6}{5}}$

The addition of the tone C in the first chord of the Seventh above, thus $d \mid f - a^1\flat + c$, would decisively exclude the key $F\sharp$ minor only, and the addition of the tone b_1 (which in tempered intonation is confused with b or $c^1\flat$) would also readily be adapted to all the above keys. Thus altered it becomes the chord $b_1 -$ ¶ $d \mid f - a^1\flat$, and is called *the chord of the diminished Seventh*, which on keyed instruments appears as a series of minor Thirds. In reality a Pythagorean minor Third or else an acute augmented Second separates the normal minor Thirds, thus:

$$b_1 - d \mid f - a^1\flat \quad b_1 - d \mid f - a^1\flat \ldots b_1$$

$$\underbrace{\ }_{\frac{6}{5}}\ \underbrace{\ }_{\frac{32}{27}}\ \underbrace{\ }_{\frac{6}{5}}\ \underbrace{\ }_{\frac{75}{64}}\ \underbrace{\ }_{\frac{6}{5}}\ \underbrace{\ }_{\frac{32}{27}}\ \underbrace{\ }_{\frac{6}{5}}\ \underbrace{\ }_{\frac{75}{64}}$$

[In cents: 316 294 316   274 316 294 316   274]

Since the three intervals $\frac{6}{5}$, $\frac{32}{27}$, and $\frac{75}{64}$ [cents 316, 294, 274] differ but very slightly [by 20, 22, 42 cents respectively], they are readily confused,† and we obtain the following, nearly identical, series of tones:

$b_1 - d \mid f - a^1\flat \ldots b_1$ [in cents b_1 316 d 294 f 316 $a^1\flat$ 274 b_1] in C minor

$\underbrace{\ }_{\frac{6}{5}}\underbrace{\ }_{\frac{32}{27}}\underbrace{\ }_{\frac{6}{5}}\underbrace{\ }_{\frac{75}{64}}$

¶

$b \mid d - f^1 \ldots g_1\sharp - b$ [in cents b 294 d 316 f^1 274 $g_1\sharp$ 316 b] in A minor

$\underbrace{\ }_{\frac{32}{27}}\underbrace{\ }_{\frac{6}{5}}\underbrace{\ }_{\frac{75}{64}}\underbrace{\ }_{\frac{6}{5}}$

$b - d^1 \ldots e_1\sharp - g\sharp \mid b$ [in cents b 316 d^1 274 $e_1\sharp$ 316 $g\sharp$ 294 b] in $F\sharp$ minor

$\underbrace{\ }_{\frac{6}{5}}\underbrace{\ }_{\frac{75}{64}}\underbrace{\ }_{\frac{6}{5}}\underbrace{\ }_{\frac{32}{27}}$

$c^1\flat \ldots d_1 - f \mid a\flat - c^1\flat$ [in cents $c^1\flat$ 274 d_1 316 f 294 $a\flat$ 316 $c^1\flat$] in $E\flat$ minor.

$\underbrace{\ }_{\frac{75}{74}}\underbrace{\ }_{\frac{6}{5}}\underbrace{\ }_{\frac{32}{27}}\underbrace{\ }_{\frac{6}{5}}$

* [This arises entirely from temperament, which identifies the two chords $d \mid f + a_1 - c$, and $d_1 - f + a_1 - c$. Listen to the difference on the Harmonical.—*Translator.*]

† [It is quite impossible to confuse them in the just intonation of any harmonic intervals, but they are absolutely identified in equally tempered intonation as 300 cents, and hence in all written music they are treated as identical. The four following forms of the chord (of which only the first can be played on the Harmonical) are struck with absolutely the same digitals on a pianoforte. Trace them on the Duodenarium, App. XX. sect. E. art. 18.— *Translator.*]

These chords of the diminished Seventh do not form so sharp a contrast with the consonances in the minor mode, as the corresponding chord does in the major mode, although if the intonation is just the dissonance is always extremely harsh and cutting.* When they are followed by the triad of the tonic, the two chords together contain all the tones of the key, and hence completely characterise it. The chief use of the chord of the diminished Seventh is due to its variability, which readily leads the harmony into new keys. By merely subjoining the minor chords of $F\sharp$, A, C or $E\flat$ the new key will be completely established. It is readily seen that this series of keys itself forms a chord of the diminished Seventh, the tones of which lie a Semitone higher than those of the given chord. This gives a simple means of recollecting them.†

The comprehension of the whole of a key by these chords is of special importance in the cadence at the end of a composition or of one of its principal sections. ¶ For this purpose we have also to determine what fundamental primary tones can be represented by these chords of the Seventh.

It is clear that a single musical tone can never be more than imperfectly represented by the tones of a dissonant chord. But as a general rule some of these tones can be accepted as the constituents of a musical tone. This gives rise to a practically important difference between the different tones of such a chord. Those tones which can be considered as the elements of a compound tone, form a compact, well-defined mass of tone. Any one or two other tones in the chord, which do not belong to this mass of tone, have the appearance of unconnected tones, accidentally intruding. The latter are called by musicians the *dissonances* or the *dissonant notes* of the chord. Considered independently, of course, either tone in a dissonant interval is equally dissonant in respect to the other, and if there were only two tones it would be absurd to call one of them only *the* dissonant tone. In the Seventh $c \ldots b_1$, c is dissonant in respect to b_1, and b_1 in respect to c. In the ¶ chord $c + e_1 - g + b_1$ the notes $c + e_1 - g$ form a single mass of tone corresponding to the compound tone of c, and b_1 is an unconnected tone sounding at the same time. Hence the three tones $c + e_1 - g$ have an independent steadiness and compact-

---

* [As the ratios $4 : 5 : 6 : 7$ are the justification of the chord of the dominant Seventh $4 : 5 : 6 : 7\frac{1}{5}$, so the ratios $10 : 12 : 14 : 17$ are the justification of the chord of the diminished Seventh $10 : 12 : 14\frac{2}{9} : 17\frac{1}{15}$ taking the ratios of No. 5, p. 343*b*, and commencing with 10. That is, $e''316 \ g''267 \ ^7b''\flat \ 336 \ ^{17}d'''\flat$, which can be played on the Harmonical, is the just chord of the diminished Seventh, for which the form of ordinary just intonation is $e''316 \ g''294 \ b''\flat 316 \ d'\flat$, which must be played as $g'316 \ b_1'294 \ f''316 \ a'''\flat$ on the Harmonical, an intensely harsh chord, for which is played in equal temperament $g'300 \ b'300 \ f''300 \ a''\flat$. ¶ Observe that the diminished Seventh $10 : 17$ has 919 cents, the diminished Seventh of ordinary just intonation $10 : 17\frac{1}{15}$ has 926 cents, 7 too sharp; while in equally tempered intonation it is only 900 cents or 19 too flat. And the tempered major Sixth is represented by the same interval of 900 cents, which is 16 cents too sharp. It is remarkable that any sense of interval or tonality survives these confusions. Of course the introduction of the 17th harmonic into the scale is a sheer impossibility. The chord $10 : 12 : 14\frac{2}{9} : 17\frac{1}{15}$ is simple noise. The chord $10 : 12 : 14 : 17$ which I have tried on Appunn's tonometer in its inversions, is a comparatively smooth discord superior to the tempered form. But the chord is really due to tempered intonation only. For further notes on this chord see App. XX. sect. E. art. 23, and sect. F. towards end of No. 7.—*Translator.*]

† [It is correctly stated in the text that the four keys into which a slight alteration of the pitches of the notes in the chord of the diminished Seventh will make it fit, are $F\sharp$, A, C, $E\flat$. These notes, however, do not form a chord with the same intervals, but $F\sharp \ 294 \ A \ 294 \ C \ 294 \ E\flat$, that is a succession of Pythagorean minor Thirds, the result of which is simply hideous. It is only in equally tempered intonation in which the four forms above given of the chord of the diminished Seventh agree absolutely in sound, though they differ in writing, because signs originally intended for other temperaments (as the Pythagorean, meantone, or other which distinguished $C\sharp$ and $D\flat$, but did not distinguish the comma) have continued in use, with confounded meanings. This is precisely the same as in ordinary English spelling, where combinations of letters originally representing very different sounds, are now confused, as I have demonstrated historically in my *Early English Pronunciation*. In equally tempered intonation the roots $f\sharp \ 300 \ a \ 300 \ c \ 300 \ eb$ do also form a chord of the diminished Seventh. But this does not end the confusion, for the key of $f\sharp$ may be taken as that of gb, of a as that of bbb, c as that of $b\sharp$, eb as that of $d\sharp$, and these four roots, gb, $bb \ b, b\sharp$, $d\sharp$, being played with the same digitals represent the same chord, but the four keys are now totally unrelated. What then becomes of the feeling of tonality? and how are we to feel the right amid this mass of wrong, as Sir George Macfarren says we can, and as I must therefore suppose he himself has succeeded in doing?—*Translator.*]

ness of their own. But the unsupported solitary Seventh b_1 has to stand against the preponderance of the other tones, and it could not do so either when executed by a singer, or heard by a listener, unless the melodic progression were kept very simple and readily intelligible. Consequently particular rules have to be observed for the progression of the part which produces this note, whereas the introduction of c, which is sufficiently justified by the chord itself, is perfectly free and unfettered. Musicians indicate this practical difference in the laws of progression of parts by terming b_1 alone the dissonant note of this chord; and although the expression is not a very happy one, we can have no hesitation in retaining it, after its real meaning has been thus explained.

We now proceed to examine each of the previous chords of the Seventh with a view to determine what compound musical tone they represent, and which are their dissonant tones.

1. The *chord of the dominant Seventh*, $g + b_1 - d \mid f$, contains three tones ¶ belonging to the compound tone of G, namely g, b_1, and d, and the Seventh f is the dissonant tone. But we must observe that the minor Seventh $g..f$ [or $\frac{16}{9} = \frac{64}{63} \times \frac{7}{4}$, or cents $996 = 969 + 27$] approaches so near to the ratio $\frac{7}{4}$ [cents 969] which would be almost exactly represented by $g...f_1$, [cents 974], that f may in any case pass as the seventh partial tone of the compound G.* Singers probably often exchange the f of the chord of the dominant Seventh for f_1, † partly because it usually passes into e_1, partly because they thus diminish the harshness of the dissonance. This can be easily done when the pitch of f is not determined in the preceding chord by some near relationship. Thus if the consonant chord $g + b_1 - d$ had already been struck and then f were added, it would readily fall into f_1, [that is ^7f] because f is to itself unrelated to g, b_1, or d.‡ Hence, although the chord of the dominant Seventh is dissonant, its dissonant tone so nearly corresponds to the corresponding partial tone in the compound tone of the dominant, that the whole chord may be very well regarded as a representative of that compound. For this reason, doubtless, the ¶ Seventh of this chord has been set free from many obligations in the progression of parts to which dissonant Sevenths are otherwise subjected. Thus it is allowed to be introduced freely without preparation, which is not the case for the other Sevenths. In modern compositions (as R. Wagner's) the chord of the dominant Seventh not unfrequently occurs as the concluding chord of a subordinate section of a piece of music.

The chord of the dominant Seventh consequently plays the second most important part in modern music, standing next to the tonic. It exactly defines the key, more exactly than the simple triad $g + b_1 - d$, or than the diminished triad $b_1 - d \mid f$. As a dissonant chord it urgently requires to be resolved on to the tonic chord, which the simple dominant triad does not. And finally its harmoniousness is so extremely little obscured, that it is the softest of all dissonant chords.§ Hence we could scarcely do without it in modern music. This chord appears to have been discovered in the beginning of the seventeenth century by Monteverde.   ¶

2. The *chord of the Seventh upon the Second of a major scale*, $d \mid f + a_1 - c$, has three tones, f, a_1, c, which belong to the compound tone of F. When the intonation is just, d is dissonant with each of the three tones of this chord, and hence must

---

* [It has, however, a very different effect on the ear.—*Translator.*]

† [Here f_1 must be considered as the representative of ^7f. Singers would not naturally take such a strange artificial approximation as f_1, unless led by an instrument. Unaccompanied singers could only choose between f and ^7f, and singers of unaccompanied melodies are said often to choose ^7f when descending to e. What is the custom in unaccompanied choirs, which have not been trained to give f, has, so far as I know, not been recorded.—*Translator.*]

‡ [And ^7f is, but f_1 again is not. It will be seen by the Duodenarium (App. XX. sect. E. art. 18) (which should be constantly consulted

on such points) that f_1 is very remote indeed from g.—*Translator.*]

§ [As we hear it only in tempered music as a rule, with the harsh major Third, which makes the major triad almost dissonant, the addition of the dominant Seventh increases the harshness surprisingly little. But in just intonation $g\ b_1\ d\ f$ is markedly harsher than $g\ b_1\ d\ ^7f$, as I have often had occasion to observe in Appunn's tonometer, where $g\ b_1\ d$ can be left sounding, and f suddenly transformed to ^7f and conversely. On the Harmonical we must compare $g\ b_1\ d\ f$ with $c\ e_1\ g\ ^7bb$, and that in all their inversions and positions.—*Translator.*]

be regarded as the dissonant note. This would make the fundamental position of this chord to be that which Rameau assigned, making f the root, thus : $f + a_1 - c \ldots d$, which is a position of the Sixth and Fifth, and the chord is called by Rameau the *chord of the great Sixth* [*grande Sixte*, in English 'added Sixth']. This is the position in which the chord usually appears in the final cadence of C major. Its meaning and its relation to the key is more certain than that of the *false minor chord* $d \mid f + a_1$, mentioned on p. 340a, which as executed by a singer or heard by a listener is readily apt to be confused with $d_1 - f + a_1$ in the key of A_1 minor. By changing $d \mid f + a_1$ into $d_1 - f + a_1$ we obtain a minor chord, to which there will be a great attraction when the relation of d to g is not made very distinct. But if we were to change d into d_1 in the chord $d \mid f + a_1 - c$, thus producing $d_1 - f + a_1 - c$, although d_1 would be consonant with f and a_1 it would not be so with c; on the contrary, the dissonance $d_1 \ldots c'$ [p. 332, No. 36, cents 1018, rough-
¶ ness 25] is much harsher than $d \ldots c'$, [*ibid.*, No. 34, cents 996, roughness 23, much the same as the other], and, after all, it would be only the tone a_1 which would enter into the compound tone of d_1, so that, notwithstanding this change, f, which contains three tones of the chord in its own compound tone, would predominate over d_1, which has only two. In accordance with this view, I find the chord $f + a_1 - c \ldots d$ when used on the justly-intoned harmonium, as subdominant of C major, produces a better effect than $f + a_1 - c \ldots d_1$.

3. The *corresponding chord of the Seventh on the Second of the minor scale*, $d \mid f - a^{1}\flat + c$, has only one tone, c, which can be regarded as a constituent of the compound tone of either f or $a^{1}\flat$. But since c is the third partial of f and only the fifth partial of $a^{1}\flat$, f as a rule predominates, and the chord must be regarded as a subdominant chord $f - a^{1}\flat + c$ with the addition of dissonant d. There is still less inducement to change d into d_1 in this case than in the last.

4. The *chord of the Seventh on the Seventh of the major scale*, $b_1 - d \mid f + a_1$,
¶ contains two tones, b_1 and d, belonging to the dominant g, and two others, f, and a_1, belonging to the subdominant f. Hence the chord splits into two equally important halves. But we must observe that the two tones f and a_1 approach very closely to the two next partial tones of the compound tone of G. The partials of this compound tone from the fourth onwards may be written—

$$g + b_1 - d \ldots f_1 \ldots g \ldots a$$
$$4 \quad 5 \quad 6 \quad 7^* \quad 8 \quad 9$$

Hence the chord of the Ninth $g + b_1 - d \mid f + a_1$ may represent the compound tone of the dominant g, provided that the similarity be kept clear by the position of the tones, g being the lowest and a_1 the highest; it is also best not to let f [standing for ^7f] fall too low. Since a is the ninth partial tone of the compound g, which is very weak in all usual qualities of tone, and is often inaudible, and since there is the interval of a comma between a and a_1, and also between f_1 and f [but ^7f and f
¶ differ by 27 cents], care must be taken to render the resemblance of the chord of the Ninth to the compound tone of g, as strong as possible, by adopting the device of keeping a_1 uppermost, and then the use of f, a_1, for f_1, a, [meaning $^7f a$] will not be very striking. In this case f and a_1 must be considered as the dissonant notes of the chord of the Ninth $g + b_1 - d \mid f + a_1$, because although they are very nearly the same, they are not quite the same, as the partial tones of G. No preparation is necessary for the introduction of a_1 into the chord, for the same reasons that f is allowed to be introduced into the chord of the dominant Seventh, $g + b_1 - d \mid f$ without preparation. Lastly, some of the tones of the pentad chord of the Ninth may be omitted, to reduce it to four parts; for example, its Fifth, as in $g + b_1 \ldots f + a_1$, or its root, as in $b_1 - d \mid f + a_1$. If only the order of the tones is preserved as much as possible, and especially the a_1 kept uppermost, the chord will always be recognised as a representative of the compound tone of G.

* [That is, supposing f_1 to be used for   so that the above chord represents $g\ b_1\ d\ ^7f\ g\ a$.
^7f, as already explained, see p. 347d, note †,   —*Translator*.]

This seems to me the simple reason why musicians find it desirable to make a_1 the highest tone in the chord $b_1 - d \mid f + a_1$. Hauptmann, indeed, gives this as a rule without exception, and assigns rather an artificial reason for it. The ambiguity of the chord will thus be obviated as far as possible, and it receives a clearly intelligible relation to the dominant of the key of C major, whereas in other positions of the same chord there would be too great a chance of confusing it with the subdominant of A_1 minor.[*] When the intonation is just, the chord $g + b_1 - d...f_1...a$,[†] which consists (very nearly indeed) of the partial tones of the compound tone of g, sounds very soft, and but slightly dissonant; the chord of the Ninth in the key of C major, $g + b_1 - d' \mid f' + a_1'$, and the chord of the Seventh in the position $b_1 - d' \mid f' + a_1'$, sound somewhat rougher, on account of the Pythagorean Third $d' \mid f'$, and the imperfect Fifth $d'...a_1'$, but they are not very harsh. If, however, a_1' is taken in a lower position, they become very rough indeed.

The chord of the Seventh $b_1 - d \mid f + a_1$ and the following triad $c + e_1 - g$, as ¶ already observed, contain all the tones in the key of C major, and hence this chordal succession is extremely well adapted for a brief and complete characterisation of the key.

5. The *chord of the diminished Seventh*, $b_1 - d \mid f - a^1\flat$, and the minor chord $c - e^1\flat + g$, have the last mentioned property for the minor key of C, and for this reason as well as for its great variability (p. 345d) it is largely, perhaps far too largely (p. 320d), employed in modern music, especially for modulations. It contains no note which belongs to the compound tone of any other note in the chord, but the three tones $b_1 - d \mid f$ may be regarded as belonging to the compound tone of g, so that it also presents the appearance of a chord of the Ninth in the form $g + b_1 - d \mid f - a^1\flat$. It therefore imperfectly represents the compound tone of the dominant, with an intruded tone $a^1\flat$, and f and $a^1\flat$ may therefore be regarded as its dissonant tones. But the connection of the three tones $b_1 - d \mid f$ with the compound tone of g is not so distinctly marked as to make it necessary to subordinate the pro- ¶ gression of the tones f and $a^1\flat$ to that of b_1 and d. At least the chord is allowed to commence without preparation, and it is resolved by the motion of all its tones to those tones of the scale which make the smallest intervals with them, for its elements are not sufficiently well connected with one another to allow of wide steps in its resolution.

6. The *chords of the major Seventh in the direct system of the key*, as $f + a_1 - c + e_1$ and $c + e_1 - g + b_1$ in C major, and $a^1\flat + c - e^1\flat + g$ in C minor, as already remarked, mainly represent a major chord with the major Seventh as dissonant tone. The major Seventh forms rather a rough dissonance, and is decidedly opposed to the triad below it, into which it will not fit at all.

7. The *chords of the minor Seventh in the direct system of the key*, as $a_1 - c + e_1 - g$ and $e_1 - g + b_1 - d$, give greatest prominence to the compound tone of their Thirds, to which their bass seems to be subjoined. Thus $c + e_1 - g...a_1$ is the compound tone of c with an added a_1, and $g + b_1 - d...e_1$ is the compound tone ¶ of g with an added e_1. But since $c + e_1 - g$ and $g + b_1 - d$, being the principal triads of the key, are constantly recurring, this addition of a_1 and e_1 respectively gives by contrast great prominence to these tones; moreover, the a_1 and e_1 in these chords of the Seventh are not so isolated as the d in $d \mid f + a_1 - c$, where d has no true Fifth in the chord. The a_1 in $a_1 - c + e_1 - g$ has the Fifth e_1, and even the Seventh $g\sharp$ which belongs to its compound tone; and in the same way the b^1 and

---

* [The rootless chord of the Ninth on the dominant of C major is $b_1 - d \mid f + a_1$, and the subdominant of A_1 minor is $b \mid d_1 - f + a_1$, which would not be confused with the former in just intonation, but in equal temperament is identical with it.—*Translator*.]

† [This is the form in which the Author was obliged to play it on his instrument, which had f_1, see p. 317c, note, but not ${}^7f$. On the Harmonical play $c + e_1 - g...^7b\flat...\dot{d}$ and com-

pare its effect with that of the next three chords as given in the text.—*Translator*.]

‡ [The tone e_1 of course represents the third partial of a_1. Does the Author mean that the acute minor Seventh g represents the seventh partial ${}^7g_1$ for which it is 49 cents, or about a quarter of a Tone too sharp? The usual minor Seventh g_1 has been allowed to do so, although 27 cents too sharp. Perhaps the expression 'even the Seventh' (*allenfalls auch*

d of $e_1 - g + b_1 - d$ may be considered to belong to the compound tone of e_1. Hence the tone a_1 in the first and e_1 in the second are not necessarily subject to the laws of the resolution of dissonant notes.

Writers on harmony are accustomed to consider the normal position of all these chords to be that of the chord of the Seventh, and to call the lowest tone its root. Perhaps it would be more natural to consider $c + e_1 - g \ldots a_1$ as the principal position of the chord $a_1 - c + e_1 - g$ and c as its root. But such a chord is a compound tone of c with an inclination to a_1, and in modulations this intrusion of the tone of a_1 is utilised for proceeding to those chords related to a_1 which are not related to the chord $c + e_1 - g$, for example to $d_1 - f + a_1$. In the same way we can proceed from $g + b_1 - d \ldots e_1$ to $a_1 - c + e_1$, which would be a jump from $g + b_1 - d$. For modulation, therefore, the a_1 and e_1 are essential parts of these chords respectively, and in this practical light they might be called the fundamental tones of their ¶ respective chords.

8. The *chord of the Seventh on the tonic of the minor key*, $c - e^1\flat + g + b_1$, is seldom used, because b_1 in the minor key belongs essentially to ascending motion, and a resolved Seventh habitually descends. Hence it would be always better to form the chord $c - e^1\flat + g - b^1\flat$, which is similar to the chords considered in No. 7.

CHAPTER XVIII.

LAWS OF THE PROGRESSION OF PARTS.

Up to this point we have considered only the relations of the tones in a piece of ¶ music with its tonic, and of its chords with its tonic chord. On these relations depends the connection of the parts of a mass of tone into one coherent whole. But besides this the succession of the tones and chords must be regulated by natural relations. The mass of sound thus becomes more intimately bound up together, and, as a general rule, we must aim at producing such a connection, although, exceptionally, peculiar expression may necessitate the selection of a more violent and less obvious plan of progression. In the development of the scale we saw that the connection of all the notes by means of their relation to the tonic, if originally perceived at all, was at most but very dimly seen, and was apparently replaced by the chain of Fifths; at any rate, the latter alone was sufficiently developed to be recognised in the Pythagorean construction of tonal systems. But by the side of our strongly developed feeling for the tonic in modern harmonic music, the necessity for a linked connection of individual tones and chords is still recognised, although the chain of Fifths, which originally connected the tones of the scale, as

¶
$$f \pm c \pm g \pm \quad d \quad \pm a \pm e \pm b,$$

has been interrupted by the introduction of perfect major Thirds, and now appears as
$$f \pm c \pm g \pm d \ldots d_1 \pm a_1 \pm e_1 \pm b_1.$$

The musical connection between two consecutive notes may be effected:
1. *By the relation of their compound tones.*
This is either:

a.) *direct*, when the two consecutive tones form a perfectly consonant interval, in which case, as we have previously seen, one of the clearly perceptible partial tones of the first note is identical with one of the second. The pitch of the follow-

---

die Septime) is intended to shew that this view is rather too loose. In equal temperament, indeed, the dissonant chord is $\| a - c + e$ $- g$, and the chord of a is $\| a + c\sharp - e - g$. But this is mere confusion.—*Translator.*]

ing compound tone is then clearly determined for the ear. This is the best and surest kind of connection. The closest relationship of this kind exists when the voice jumps a whole Octave; but this is not usual in melodies, except with the bass, as the alteration of pitch is felt to be too sudden for the upper part. Next to this comes the jump of a Fifth or Fourth, both of which are very definite and clear. After these follow the steps of a major Sixth or major Third, both of which can be readily taken, but some uncertainty begins to arise in the case of minor Sixths and minor Thirds. Esthetically it should be remarked, that of all the melodic steps just mentioned, the major Sixth and major Third have, I might almost say, the highest degree of thorough beauty. This possibly depends upon their position at the limit of clearly intelligible intervals. The steps of a Fifth or Fourth are too clear, and hence are, as it were, drily intelligible; the steps of minor Thirds, and especially minor Sixths, begin to sound indeterminate. The major Thirds and major Sixths seem to hold the right balance between darkness ¶ and light. The major Sixth and major Third seem also to stand in the same relation to the other intervals harmonically.

b.) *indirect*, of the second degree only. This occurs in the regular progression of the scale, proceeding by Tones or Semitones. For example :

$$\underbrace{c \dots d}_{G} \qquad \underbrace{d \dots e_1}_{G} \qquad \underbrace{e_1 \quad f}_{C}$$

The whole major Tone $c \dots d$ proceeds from the Fourth to the Fifth of the auxiliary tone G, which Rameau supposed to be subjoined as the fundamental bass of the above melodic progression. The minor Tone $d \dots e_1$ proceeds from the Fifth to the major Sixth of the auxiliary tone G, and the Semitone $e_1 \dots f$ from the major Third to the Fourth of the auxiliary tone C. But in order that these auxiliary tones may readily occur to both singer and hearer, they must be among the ¶ principal tones of the key. Thus the step $a_1 \dots b_1$ in the major scale of C causes the singers a little trouble, although it is only the interval of a major Tone, and could be easily referred to the auxiliary tone E_1. But the sound of e_1 is not so firm and ready in the mind, as the sound of C and its Fifth G and Fourth F. Hence the Hexachord of Guido of Arezzo, which was the normal scale for singers throughout the middle ages, ended at the Sixth.* This Hexachord was sung with different pitches of the first note, but always formed the same melody :

	Ut	Re	Mi	Fa	Sol	La
either	G	A	B	C	D	E
or	C	D	E	F	G	A
or	F	G	A	B♭	C	D

So that the interval $Mi \dots Fa$ always marked the Semitone.†

For the same reason Rameau preferred, in the minor scale, to refer the steps ¶ $d \dots e^1♭$ and $e^1♭ \dots f$ to G and C as auxiliary tones, rather than to $B♭$, the Seventh of

---

* For the same reason d'Alembert explains the limits of the old Greek heptachord, by means of two connected tetrachords—

$$b \dots c \dots d \dots e \dots f \dots g \dots a$$

in which the step $a \dots b$ is avoided. But this explanation would only suit a key in which c was the tonic, and this was probably not the case for the ancient Greek scale.

† [Prof. Helmholtz leaves the intonation unmarked. Guido d'Arezzo, the presumed inventor of the Hexachord, is said to have introduced it about 1024 A.D., that is long before meantone temperament existed. Hence we must assume Pythagorean intonation (see p. 313d). Yet in later times the Hexachord was certainly used for training singers in meantone temperament. It could not have been used for just intonation, because the melody $c\ d\ e\ f$ is assumed to be identical with $g\ a\ b\ c$ in the same scale, whereas in just intonation c 204 d 182 e_1 112 f and g 182 a_1 204b_1 112 c' are different. For an excellent account of the Hexachord see Mr. Rockstro's article 'Hexachord,' in Grove's *Dictionary*. To shew, however, how intonations are mixed up, it may be observed that he illustrates the use of the Hexachord in 'Real Fugue' by an example of Palestrina, who lived in the sixteenth century, and is often credited with just intonation, but who being junior to Salinas and Zarlino must have used meantone temperament.—*Translator*.]

the descending scale,* which had not a sufficiently close relationship to the tonic, and hence was not well enough impressed on the singer's mind for such a purpose. Taking g and c, the Octaves of G and C as the auxiliary tones, the motion in $d...e^1\flat$ is from the Fourth below g to the major Third below it, and in $e^1\flat...f$ from the major Sixth below c to the Fifth below it. On the other hand, it is impossible to reduce the step $a^1\flat...b_1$ [cents 274] in the minor scale to any relationship of the second degree. [See p. 301c, note *.] It is therefore also decidedly unmelodic and had to be entirely avoided in the old homophonic music, just as the steps of the false Fifths and Fourths, as $b_1...f$ [cents 610], or $f'...b_1'$ [cents 590]. Hence the alterations in the ascending and descending minor scales already mentioned.

In modern harmonic music many of these difficulties have disappeared, or become less sensible, because correct harmonisation can exhibit the connections which are absent in the melodic progression of an unaccompanied voice. Hence ¶ also it is much easier to take a part at sight in a harmony, written in pianoforte score, which shews its relations, than to sing it from an unconnected part. The former shews how the tone to be sung is connected with the whole harmony, the latter gives only its connection with the adjacent tones.†

2. Tones may be connected by *their approximation in pitch.*

This relation has been considered previously with reference to the leading note. The same holds good for the intercalated tones in chromatic passages. For example, if in C major, we replace $C...D$ by $C...C\sharp...D$, this $C\sharp$ has no relation either of the first or second degree with the tonic C, and also no harmonic or modulational significance. It is nothing but a step intercalated between two tones, which has no relation to the scale, and only serves to render its discontinuous progression more like the gliding motion of natural speech, or weeping or howling. The Greeks carried this subdivision still further than we do at present, by splitting up a Semitone into two parts in their enharmonic system (p. 265a). Notwithstand- ¶ ing the strangeness of the tone to be struck, chromatic progression in Semitones can be executed with sufficient certainty to allow it to be used in modulational transitions for the purpose of suddenly reaching very distant keys.

Italian melodies are especially rich in such intercalated tones. Investigations of the laws under which they occur will be found in two essays of Sig. A. Basevi.‡ The rule is without exception that tones foreign to the scale can be introduced only when they differ by a Semitone § from the note of the scale on to which they resolve, while any tones belonging to the scale itself can be freely introduced although out of harmony with the accompaniment, and even requiring the step of a whole Tone for their resolution.

---

* [The Author writes $B\flat$, and calls it 'the Seventh of the descending scale' of C minor, which, however, is $B^1\flat$, and this answers for the first interval $d...e^1\flat$, owing to $b^1\flat + d$, and ¶ $e^1\flat \pm b^1\flat$; but it does not answer to the second interval $e^1\flat...f$ as $f...b^1\flat$ is dissonant, and it would not do to use $b\flat$, although $b\flat \pm f$ is consonant, because $e^1\flat...b\flat$ is dissonant. But $a^1\flat$ would do, as we see from $a^1\flat \pm e^1\flat$, $f - a^1\flat$. Hence if the text gives Rameau's notes, he must have been misled by temperament.—*Translator.*]

† [Hence any means of shewing the relation of each tone to the tonic of the moment, as in the Tonic Solfa system, materially facilitates sight-singing, as perhaps the use of the duodenal (App. XX. sect. E. art. 26) when thoroughly understood might also do.—*Translator.*]

‡ *Introduction à un nouveau Système d'Harmonie*, traduit par L. Delâtre; Florence, 1855. *Studj sull' Armonia*, Firenze, 1865.

§ [Of course those who laid down the rule thought only of a tempered Semitone. But in Pythagorean temperament there were two Semi-

tones, the small 90 and the large 114 cents, and the rule was to make the Semitone closest to the note to which it led, thus c 114 $c\sharp$ 90 d, d 114 $d\flat$ 90 d. And this notation was retained even in meantone temperament, where the relations were reversed, as c 76 $c\sharp$ 117 d, d 76 $d\flat$ 117 c; but practically this made no difference except to the singer, as the player had only one Semitone at command. This writing is still continued in equal temperament, although the two Semitones are now equalised as 100 cents, thus c 100 $c\sharp$ 100 d, and d 100 $d\flat$ 100 c. But in just intonation we have Semitones of various dimensions, c 114 $c\sharp$ 90 d, c 112 $d^1\flat$ 92 d, c 92 $c_1\sharp$ 112 d, c 90 $d\flat$ 114 d, c 70 $c_2\sharp$ 134 d; which of these is the player to play, or the singer to sing (a question of importance when each part is sustained by many unaccompanied voices)? Practically the player will take the most handy interval, and the singers must arrange in rehearsal, but would possibly take c 92 $c_1\sharp$ 112 d, d 92 $d^1\flat$ 112 c, as these are the intervals used in modulation from f to $f_1\sharp$ to the dominant, and b_1 to $b\flat$ to the subdominant.—*Translator.*]

In the same way steps of a whole Tone may be made, provided the notes lie in the scale, when they serve merely to connect two other tones which belong to chords. These are the so-called *passing* or *changing notes*. Thus if while the triad of C major is sustained, a voice sings the passage $c...d...e_1...f...g$, the two notes d and f do not suit the chord, and have no relation to the harmony, but are simply justified by the melodic progression of the single voice. It is usual to place these passing notes on the unaccented parts of the bar, and to give them a short duration. Thus in the above example c, e_1, g would fall on accented parts of the bar. Then d is the passing note between c and e_1 and f between e_1 and g. It is essential for their intelligibility that they should make steps of Semitones or whole Tones. They thus produce a simple melodic progression, which flows on freely, without giving any prominence to the dissonances produced.

Even in the essentially dissonant chords the rule is, that dissonant tones which intrude isolatedly on the mass of the other tones must proceed in a melodic progression, ¶ which can be easily understood and easily performed. And since the feeling for the natural relations of such an isolated tone is almost overpowered by the simultaneous sound of the other tones which force themselves much more strongly on the attention, both singer and hearer are thrown upon the gradual diatonic progression as the only means of clearly fixing the melodic relations of a dissonant note of this description. Hence it is generally necessary that a dissonant note should enter and leave the chord by degrees of the scale.

Chords must be considered essentially dissonant, in which the dissonant notes do not enter as passing notes over a sustained chord, but are either accompanied by an especial chord, differing from the preceding and following chords, or else are rendered so prominent by their duration or accentuation, that they cannot possibly escape the attention of the hearer. It has been already remarked that these chords are not used for their own sakes, but principally as a means of increasing the feeling of onward progression in the composition. Hence it follows for the motion ¶ of the dissonant note, that when it enters and leaves the chord, it will either ascend on each occasion or descend on each. If we allowed it to reverse its motion in the second half, and thus return to its original position, there would seem to have been no motive for the dissonance. It would in that case have been better to leave the note at rest in its consonant position. A motion which returns to its origin and creates a dissonance by the way, had better be avoided; it has no object.

Secondly it may be laid down as a rule, that the motion of the dissonant note should not be such as to make the chord consonant without any change in the other notes. For a dissonance which disappears of itself provided we wait for the next step, gives no impetus to the progress of the harmony. It sounds poor and unjustified. This is the principal reason why chords of the Seventh which have to be resolved by the motion of the Seventh, can only permit the Seventh to descend. For if the Seventh ascended in the scale, it would pass into the octave of the lowest tone, and the dissonance of the chord would disappear. When Bach, ¶ Mozart, and others use such progressions for chords of the dominant Seventh, the Seventh has the effect of a passing note, and must be so treated. In that case it has no effect on the progression of the harmony.

The pitch of a single dissonant note in a chord of many parts is determined with greatest certainty, when it has been previously heard as a consonance in the preceding chord, and is merely sustained while the new chord is introduced. Thus if we take the following succession of chords:

$$G ... d ... g + b_1$$
$$c + e_1 - g + b_1$$

the b_1 in the first chord is determined by its consonance with G. It simply remains while the tones c and e_1 are introduced in place of G and d, and thus becomes a dissonance in the chord of the Seventh $c + e_1 - g + b_1$. In this case the dissonance is said to be *prepared*. This was the only way in which dissonances might be introduced down to the end of the sixteenth century.   Prepared dissonances produce a peculiarly powerful effect : a part of the preceding chord lingers on, and has to be forced from its position by the following chord.   In this way, an effort to advance against opposing obstacles which only slowly yield, is very effectively expressed. And for the same reason the newly introduced chord ($c + e_1 - g$ in the last example) must enter on a strongly accented part of the bar ; as it would otherwise not sufficiently express exertion.   The resolution of the prepared dissonance, on the contrary, naturally falls on an unaccented part of the bar.   Nothing sounds worse than dissonances played or sung in a dragging or uncertain manner.   In that case they appear to be simply out of tune.   They are, as a rule, only justified by expressing energy and vigorous progress.

Such prepared dissonances, termed *suspensions*, may occur in many other chords besides those of the Seventh.   For example :

¶

$$\text{Preparation : } \quad G \ldots c \; + e_1$$
$$\text{Suspension : } \quad G \ldots c \; \ldots d$$
$$\text{Resolution : } \quad G + B_1 - d$$

The tone c is the prepared dissonance ; in the second chord, which must fall on an accented part of the bar, d the Fifth of G is introduced and generates the dissonance $c \ldots d$, and then c must give way, and according to our second rule, must go further from d, which results in the resolution $G + B_1 - d$. The chords might also be played in the inverse order, and then d would be a prepared dissonance which was forced away by c. But this is not so good, because descending motion is better suited than ascending motion to an extruded note.   Heightened pitch always gives us involuntarily the impression of greater effort, because we have continually to exert our voice in order to reach high tones.   The dissonant note on descending seems to yield suitably to superior force, but on ascending it as it were rises by its ¶ own exertion.   But circumstances may render the latter course suitable, and its occurrence is not unfrequent.

In the other case, especially frequent for chords of the Seventh, when the dissonant note is not prepared but is struck simultaneously with the chord to which it is dissonant, the significance of the dissonance is different. Since these unprepared Sevenths must usually enter by the descent of the preceding note, they may be always considered as descending from the Octave of the root of their chord, by supposing a consonant major or minor chord having the same root as the chord of the Seventh to be inserted between that and the preceding chord.   In this case the entrance of the Seventh merely indicates that this consonant chord begins to break up immediately and that the melodic progression gives a new direction to the harmony.   This new direction, leading to the chord of resolution, must be emphasised, and hence the dissonance necessarily falls on the preceding unaccented part of a bar.

¶   The introduction of an isolated dissonant note into a chord of several parts cannot generally be used as the expression of exertion, but this character will attach to the introduction of a chord as against a single note, supposing that this single note is not too powerful.   Hence it lies in the nature of things that the first kind of introduction takes place on unaccented and the last on accented parts of a bar.

These rules for the introduction of dissonances may be often neglected for the chords of the Seventh in the reverted system, in which the Fourth and Second of the scale occur, and notes from the subdominant side are mixed up with notes from the dominant side.   These chords may also be introduced to enhance the dynamical impression of the advancing harmony, for they have the effect of keeping the extent of the key perpetually before the feeling of the hearer, and this object justifies their existence.

Of several voices which are leaving the chord of the tonic C, it is quite easy for some to pass on to notes of the dominant chord $g + b_1 - d$. and for others to proceed to the notes of the subdominant chords $f + a_1 - c$ or $f - a^1\flat + c$, as each voice will be able to strike the new note with perfect certainty, on account of the close relationship between the chords. When, however, the dissonant chord has been thus formed and sounded, the dissonant notes will have the feeling for their more distant relations obscured by the strangeness of the other parts of the chord, and must generally proceed according to the rule of resolution of dissonances. Thus the singer who sounds f in the chord $g + b_1 - d \mid f$, would vainly endeavour to picture to himself the sound of the a_1 which is related to f with sufficient clearness to leap up or down to it with certainty; but he is easily able to execute the small step of half a Tone, by which f descends to e_1 in the chord $c + e_1 - g$. But the note g itself, on the other hand, having its own compound tone approximatively indicated by the chord of the Seventh, has no difficulty in passing by a leap to its related notes, as c for example, or b_1 to g.

In the chords $b_1 - d \mid f + a_1$ and $b_1 - d \mid f - a^1\flat$, in which neither dominant nor ¶ subdominant prevails, it would not be advisable to let any note proceed by a leap.

And it would also not be advisable to pass by a leap into the chords of the reverted system from any other chord but the tonic, because that chord alone is related to both dominant and subdominant chords at the same time.

It is not possible to pass to chords of the Seventh in the direct system, from another chord related to both extremities of the chord of the Seventh, and hence in this case the dissonance must be introduced in accordance with the strict rules.

Musicians are divided in opinion as to the proper treatment of the subdominant chord with an added Sixth, $f + a_1 - c \ldots d$ in C major. The rule of Rameau is probably correct (p. 347d), making d the dissonant note, to be resolved by rising to e_1. This is also decidedly the most harmonious kind of resolution. Modern theorists, on the other hand, regard this chord as a chord of the Seventh on d, and take c as the dissonant note to be resolved by descent; whereas when c remains, d is quite free and may therefore even descend. ¶

Chordal Sequences.

Just as the older homophonic music required the notes of a melody to be linked together, modern music endeavours to link together the series of chords occurring in a tissue of harmony, and it thus obtains much greater freedom in the melodic succession of individual notes, because the natural relationship of the notes is much more decisively and emphatically marked in harmonic music than in homophonic melody. This desire for linking the chords together was but slightly developed in the sixteenth century. The great Italian masters of this period allow the chords of the key to succeed each other in leaps which are often surprising, and which we should at present admit only in exceptional cases. But during the seventeenth century the feeling for this peculiarity of harmony also was developed, so that we find Rameau laying down distinct rules on the subject in the beginning of the eighteenth century. In reference to his conception of fundamental bass, ¶ Rameau worded his rule thus : ' *The fundamental bass may, as a general rule, proceed only in perfect Fifths or Thirds, upwards or downwards.*' According to our view the fundamental bass of a chord is that compound tone which is either exclusively or principally represented by the notes of the chord. In this sense Rameau's rule coincides with that for the melodic progression of a single note to its nearest related notes. The compound tone of a chord, like the voice of a melody, may only proceed to its nearest related notes. It is much more difficult to assign a meaning to progression by relationship in the second degree for chords than for separate notes, and similarly for progression in small diatonic degrees without relationship. Hence Rameau's rule for the progression of the fundamental bass is on the whole stricter than the rules for the melodic progression of a single voice.

Thus if we take the chord $c + e_1 - g$, which belongs to the compound tone of C, we may pass by Fifths to $g + b_1 - d$, the compound tone of G, or to $f + a_1 - c$ the compound tone of F. Both of these chords are directly related to the first $c + e_1 - g$, because each has one note in common with it, g and c respectively.

But we can also allow the compound tone to proceed in Thirds, and then we obtain minor chords, that is, provided we keep to the same scale. The transition from the compound tone of C to that of E_1 is expressed by the sequence of chords $c + e_1 - g$ and $e_1 - g + b_1$, which are related by having two notes, e_1, g, in common. The sequence $c + e_1 - g$ and $a_1 - c + e_1$ from the compound tone of C to that of A_1 is of the same kind. The latter is even more natural than the former, because the chord $a_1 - c + e_1$ represents imperfectly the compound tone of A_1 into which that of C intrudes, so that the compound tone of C, which was clearly given in the preceding chord, persists with two of its tones, c, e_1, in the second chord, a relation which did not exist in the former case.

But if we prefer to leave the key of C major, we can pass to perfect compound ¶ tones in Thirds, as from $c + e_1 - g$ to $e_1 + g_2\sharp - b_1$ or $a_1 + c_2\sharp - e_1$, as is very usual in modulations.

Rameau will not allow a simple diatonic progression of the fundamental bass of consonant triads, except where major and minor chords alternate, as from $g + b_1 - d$ to $a_1 - c + e_1$, that is from the compound tone of G to that of A_1, but calls this a ' licence.' In reality this progression is readily explicable from our point of view, by considering $a_1 - c + e_1$ as a compound tone of C with an intrusive a_1. The transition is then one of the usual close relationship, from the compound tone of G to that of C, and the a_1 appears as a mere appendage to the latter. Every minor chord represents two compound tones in an imperfect manner. Rameau first formulated this ambiguity (*double emploi*) for the minor chord with added Seventh, which, in the form $d_1 - f + a_1 - c$, may represent the compound tone of D_1, and in the form $f + a_1 - c ... d$ that of F, or in Rameau's language its fundamental bass might be D_1 or F.* In this chord of the Seventh the ambiguity is ¶ more marked, because it contains the compound tone of F more completely ; but the ambiguity belongs in a less marked degree to the simple chord also.

With the false cadence in the major key

$$g + b_1 - d \qquad \text{to} \qquad a_1 - c + e_1$$

must be associated the corresponding cadence in the minor key,

$$g + b_1 - d \qquad \text{to} \qquad a^1\flat - c + e^1\flat$$

where the chord $a^1\flat - c + e^1\flat$ replaces the normal resolution $c - e^1\flat + g$. But here there is only a single note of the compound tone of C remaining, and the false cadence therefore becomes much more striking. It will be rendered milder by adding the Seventh f to the G chord, because f is related to $a^1\flat$.

When two chords having only a relationship of the second degree, are placed in juxtaposition, we usually feel the transition to be very abrupt. But if the chord ¶ which connects them is one of the principal chords of the key, and has consequently been frequently heard, the effect is not so striking. Thus in the final cadence it is not unusual to see the succession $f + a_1 - c$ and $g + b_1 - d$, the two chords being related through the tonic chord $c + e_1 - g$, thus :

$$\overbrace{f + a_1 - c} \qquad \overbrace{g + b_1 - d}$$
$$\underbrace{c - e_1 - g}$$

Generally we must remember that all these rules of progression are subject to many exceptions, partly because expression may require exceptional abruptness of transition, and partly because the hearer's recollection of previous chords may

---

* [Of course Rameau, writing in tempered notation, did not distinguish d_1 and d, so that the actual notes in the two chords $d_1 - f + a_1 - c$ and $f + a_1 - c ... d$ were to him identical. See pp. 340*a*, 345*a*, 348*a*.—*Translator*.]

sufficiently strengthen a naturally weak relationship. It is clearly an entirely false position which teachers of harmony have assumed, in declaring this or that to be ' forbidden' in music. In point of fact nothing musical is absolutely forbidden, and all rules for the progression of parts are actually violated in the most effective pieces of the greatest composers. It would have been much better to proceed from the principle that certain transitions, which are disallowed, produce striking and unusual effects upon the hearer, and consequently are unsuitable except for the expression of what is unusual. Generally speaking, the object of the rules laid down by theorists is to keep up a well-connected flow of melody and harmony, and make its course readily intelligible. If that is what we aim at, we had best observe their restrictions. But it cannot be denied that a too anxious avoidance of what is unusual places us in danger of becoming trivial and dull, while, on the other hand, inconsiderate and frequent infringement of rules makes compositions eccentric and unconnected. ¶

When disconnected triads would come together it is frequently advantageous to transform them into chords of the Seventh, and thus create a bond between them. In place of the preceding sequence of two triads

$$f + a_1 - c \quad \text{to} \quad g + b_1 - d$$

we can use a sequence of chords of the Seventh which represent the same compound tones

$$f + a_1 - c \; ...d \quad \text{to} \quad g + b_1 - d \mid f.$$

In this case two of the four notes remain unchanged; in the chord of F, the d belongs to the compound tone of the dominant, and in the chord of G the f to that of the subdominant.

Hence chords of the Seventh come to play an important part in modern music for the purpose of effecting well-connected and yet rapid transitions from chord to ¶ chord, and urging them forward by the action of dissonances. In this way particularly, transitions to the compound tone of the subdominant are easily effected.

Thus, for example, beginning with the triad $g + b_1 - d$ we can not merely pass to the chord of C, or $c + e_1 - g$, but, letting g remain as a Seventh, to the chord of the Seventh $a_1 - c + e_1 - g$, which unites the two chords $c + e_1 - g$ and $a_1 - c + e_1$, and then immediately pass to $d_1 - f + a_1$, which is related to the latter chord, so that two steps bring us to the other extremity of the system of C major. This transition also gives the best progression for the Seventh (g in the example), because it has been prepared in the previous chord, and is resolved by descent (to f) in the succeeding chord. If we tried the same transition backwards, we should have to obtain the Seventh g by progression from a_1 in the chord of $d_1 - f + a_1$, and then be compelled to introduce the c of the chord of the Seventh abruptly, because we should have a prohibited succession of Fifths ($d_1 \pm a_1$ and $c \pm g$) if we tried to descend from d_1. We must rather obtain c by a leap from f, ¶ because a_1 in the first triad must furnish both the a_1 and g of the chord of the Seventh. Thus the transition to the dominant is by no means easy, fluent, and natural; it is much more embarrassed than the passage to the subdominant. Consequently the regular and usual progression of the chord of the Seventh is for its Seventh to descend to the triad whose Fifth is the root of the chord of the Seventh. Supposing we denote the root of the chord of the Seventh by I, its Third by III, &c., a falling Seventh will lead us to either of these chords :

$$\text{I} - \text{III} - \text{V} - \text{VII} \quad \text{and} \quad \text{I} - \text{III} - \text{V} - \text{VII}$$
$$\text{I} \ - \ \text{IV} - \text{VI} \qquad\qquad \text{I} - \text{III} \ - \ \text{VI}$$

Of these two transitions, the first, which leads to a chord of which IV is the root, is the liveliest, because it introduces a chord with two new tones. The other, which leads to a triad having VI for its root, introduces only one new tone. Hence the first is regarded as the principal method of resolving chords of the Seventh. For example :

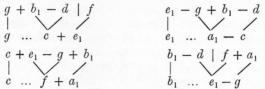

The descent of the tone VII introduces the tone VI. In the first case this is the Third of the new triad, and in the second its root.* But it may be its Fifth :

¶

$$\text{I — III — V — VII}$$
$$\text{II — IV — VI}$$

This, however, could only occur naturally in the two chords :

$$b_1 - d \mid f + a_1 \quad \text{and} \quad b_1 - d \mid f - a^1\flat$$
$$c + e_1 - g \qquad\qquad c - e^1\flat + g$$

because the two chords of the Seventh represent the compound tone of G, and the tonic chord establishes the bond of union between its two sections. In other cases our scheme gives so-called false cadences :

¶

$$g + b_1 - d \mid f \quad \text{and} \quad g + b_1 - d \mid f$$
$$a_1 - c + e_1 \qquad\qquad a^1\flat + c - e^1\flat$$

which are justified (the first as most natural) by the fact that either $c + e_1$ or $c - e^1\flat$ belongs to the chord of the normal resolution. Rameau therefore justly observes that this kind of resolution is only permissible when the IV of the second chord is the normal Fourth of the I in the chord of the Seventh.

This exhausts the resolutions by the descent of the Seventh. Those in which it remains unchanged take place according to the schemes :

$$\text{I — III — V — VII} \dagger \quad \text{and} \quad \text{I — III — V — VII} \ddagger$$
$$\text{II — IV — VII} \qquad\qquad \text{II — — V — VII}$$

In the first the Seventh becomes the root, in the second the Third of the new chord. If it were the Fifth, the new chord would coincide with part of the chord
¶ of the Seventh :

$$\text{I — III — V — VII}$$
$$\text{VII — III — V — VII.§}$$

* [As examples of the second method have been omitted in the text, take

$$g + b_1 - d \mid f \qquad e_1 - g + b_1 - d$$
$$g + b_1 \ldots e_1 \qquad e_1 - g \ldots c$$
—Translator.]

† [Examples :

$$g + b_1 - d \mid f \quad c + e_1 - g + b_1 \quad e_1 - g + b_1 - d$$
$$a_1 - c \ldots f \qquad d \mid f \ldots b_1 \qquad f + a_1 \ldots d$$

Here in the first example we obtain the major triad $f + a_1 - c$; in the second the diminished triad $b_1 - d \mid f$, itself a dissonance ; and in the third the imperfect minor $d \mid f + a_1$.—Translator.]

‡ [Examples :

$$g + b_1 - d \mid f, \; c + e_1 - g + b_1$$
$$a_1 \qquad d \; f, \quad d \ldots g + b_1$$
—Translator.]

§ [Examples :

$$g + b_1 - d \mid f \qquad c + e_1 - g + b_1$$
$$f \ldots b_1 - d \mid f \qquad b_1 \ldots e_1 - g + b_1$$
—Translator.

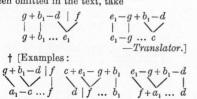

In these connections the resolution is towards the dominant side. The transition is most decisive in the first, where the Seventh becomes the root. These resolutions are on the whole less usual, because we pass more easily and frequently from chords on the dominant side into chords of the Seventh of the direct system. In the chords of the reverted system these transitions occur more frequently, because their Sevenths may enter by ascent, and hence we avoid the sequences of Fifths, which greatly embarrass the transitions from a triad to a chord of the Seventh on its dominant side.

As to the transitions from one chord of the Seventh to another, or to a dissonant triad of the direct system which may be regarded as a mutilated chord of the Seventh, all these matters are sufficiently developed in the ordinary manuals of Thorough Bass, and offer no difficulties that would justify us in dwelling upon them here.

On the other hand, we have to say a few words on certain rules respecting the ¶ progression of the individual parts in polyphonic compositions. Originally, as we have already remarked, all these parts were of equal importance, and had usually to repeat the same melodic figures in succession. The harmony was a secondary consideration, the melodic progression of the individual voices was the principal matter. Hence it was necessary to take care that each voice should stand out clear and distinct from all the others. The relation between the importance of harmony and melody has certainly altered essentially in modern music ; the former has attained a much higher independent significance. But, after all, perfection of harmony must arise from the simultaneous performance of several voices, each of which has its own beautiful and clear melodic progression, and each of which therefore moves in a direction that the hearer has no difficulty in understanding.

On this rests the prohibition of *consecutive Fifths and Octaves*. The meaning of this prohibition has given rise to much disputation. The meaning of prohibiting consecutive Octaves has been made clear by musical practice. In poly- ¶ phonic music two voices which lie one or two Octaves apart, are forbidden to move forward in such a way that after their next step they should be also one or two Octaves apart. But precisely in the same way, two voices in a polyphonic piece are forbidden to go on in unison for several notes, while for complete musical compositions it is not forbidden that two voices, or even all the voices, should proceed in Unisons or Octaves, for the purpose of strengthening the melodic progression. It is clear that the reason of this rule must lie in the limiting the wealth of the progression of parts by Unisons and Octaves. This is allowable when it is intentionally introduced for a whole melodic phrase, but it is not suited for a few notes in the course of a piece, where it can only give the impression of reducing the richness of the harmony by an unskilful accident. The accompaniment of a lower part by a voice singing an Octave higher, merely strengthens part of the compound tone of the lower voice, and hence where variety in the progression of parts is important, does not essentially differ from a Unison.   ¶

Now in this respect the nearest to an Octave are the Twelfth, and its lower octave, the Fifth. Hence, then, consecutive Twelfths and consecutive Fifths partake of the same imperfection as consecutive Octaves. But the case is somewhat worse. It is possible to accompany a whole melody in Octaves when desirable, without committing any error, but this cannot be done for Fifths and Twelfths without changing the key. It is impossible to proceed by a single diatonic step from the tonic as root with an accompaniment of Fifths, without departing from the key. In C major, we ascend from the Fifth $c \pm g$ to the Fifth $d \pm a$, but a does not belong to the scale, which requires the deeper a_1 ; we descend to $b_1 \pm f_1 \sharp$, and there is no $f_1 \sharp$ in the scale at all. The other upward steps from d exclusive to a_1 can of course be accompanied by perfect Fifths in the scale, as $e_1 \pm b_1$, $f \pm c'$, $g \pm d'$, $a_1 \pm e_1'$. It is therefore impossible to use the Twelfth consistently for increasing the richness of the tone. But again, when the intervals of a Twelfth or Fifth are continued for a few steps in melodic progression, they have simply the effect of strengthening the root. For the Twelfth this arises from its directly corresponding to one of the upper partial tones of the root. For the Fifth $c \pm g$, the c and g are the two first upper partials of the combinational tone C, which necessarily accompanies the Fifth. Hence an accompaniment in Fifths above, when it occurs isolatedly in the midst of a polyphonic piece, is not only open to the charge of monotony, but cannot be consistently carried out. It should therefore be always avoided.

But that consecutive Fifths merely infringe the laws of artistic composition, and are not disagreeable to the natural ear, is evident from the simple fact that all the tones of our voice, and those of most instruments, are accompanied by Twelfths, and that our whole tonal system reposes upon that fact. When the Fifths are introduced as merely mechanical constituents of the compound tone, they are therefore fully justified. So in the mixture stops of organs. In these stops the ¶ pipes which give the prime tones of the compound, are always accompanied by others which give its harmonics, as the Octaves, Twelfths variously repeated, and even the higher major Thirds. By this means the performer is able to compose a tone of a much more penetrating, piercing quality, than it would be possible to produce by the simple organ pipes with their relatively weak upper partial tones. It is only by such means that an organ is able to dominate over the singing of a large congregation. Almost all musicians have blamed an accompaniment of Fifths, or even Thirds, but fortunately have not been able to effect anything against the practice of organ-builders. In fact the mixture stops of an organ merely reproduce the masses of tone which would have been created by bowed instruments, trombones, and trumpets, if they had executed the same music. It would be quite different if we collected independent parts, from each of which we should have to expect an independent melodic progression in the tones of the scale. Such independent parts cannot possibly move with the precision of a ¶ machine; they would soon betray their independence by slight mistakes, and we should be led to subject them to the laws of the scale, which, as we have seen, render a consistent accompaniment in Fifths impossible.

The prohibition of Fifths and Octaves extends also, but with less strictness, to the next adjacent consonant intervals, when two of them are so placed as to form a connected group of upper partials in a compound tone. Thus transitions like

$$d \dots g + b_1 \quad \text{to} \quad c \dots f + a_1,$$

are ruled by musical theorists to be inferior to transitions like

$$b_1 - d' \dots g' \quad \text{to} \quad a_1 - c' \dots f'.$$

For d, g, b_1 are the third, fourth, and fifth partial tones of the compound $G_{\prime}$, but b_1, d', g' could only be regarded as its fifth, sixth, and eighth. Hence the first ¶ position of the chord expresses a single compound tone much more decidedly than the second, which is often allowed to be continued through long passages, when of course the nature of the Thirds and Fourths varies.

The prohibition of consecutive Fifths was perhaps historically a reaction against the first imperfect attempts at polyphonic music, which were confined to an accompaniment in Fourths or Fifths, and then, like all reactions, it was carried too far, in a barren mechanical period, till absolute purity from consecutive Fifths became one of the principal characteristics of good musical composition. Modern harmonists agree in allowing that other beauties in the progression of parts are not to be rejected because they introduce consecutive Fifths, although it is advisable to avoid them, when there is no need to make such a sacrifice.

There is also another point in the prohibition of Fifths to which Hauptmann has drawn attention. We are not tempted to use consecutive Fifths when we pass from one consonant triad to another which is nearly related to it, because other progressions lie nearer at hand. Thus we pass from the triad of C major to the four related triads in the following manner, the fundamental bass proceeding by Thirds or Fifths :

$$c + e_1 - g \qquad c + e_1 - g \qquad c + e_1 - g \qquad \text{and } c + e_1 - g$$
$$\text{to } c + e_1 \ldots a_1, \quad \text{to } c \ldots f + a_1, \quad \text{to } B_1 \ldots e_1 - g, \quad \text{to } B_1 \ldots d \ldots g.$$

But when the fundamental bass proceeds in Seconds, and hence does not pass to a directly related chord, the nearest position of the new chord is certainly one which produces consecutive Fifths.   For example :

$$g + b_1 - d' \qquad \text{or } g + b_1 - d'$$
$$\text{to } a - c' + e_1', \qquad \text{to } f + a_1 - c$$

In such cases, then, we must have recourse to other transitions by larger intervals, as :

$$g + b_1 - d' \qquad \text{or } g + b_1 - d'$$
$$\text{to } e_1 \ldots a_1 - c', \qquad \text{to } a_1 - c \ldots f$$

which avoid consecutive Fifths.

Hence when the chords are closely connected by near relationship and small distance in the scale, consecutive Fifths do not present themselves.   When they occur, therefore, they are always signs of abrupt chordal transition, and it is then better to assimilate the progression of parts to that which spontaneously arises in the case of related chords.

This consideration respecting consecutive Fifths, which was emphasised by Hauptmann, appears to give the law greater importance.   That it is not the only motive for the prohibition of consecutive Fifths appears from the fact, that the forbidden sequence

$$g + b_1 - d' \quad \text{to} \quad f + a_1 - c'$$

is allowed, when the chords are in the position

$$b_1 - d' \ldots g \quad \text{to} \quad a_1 - c' \ldots f',$$

although the step in the fundamental bass is the same.

The prohibition of so-called *hidden Fifths* and *Octaves* has been added on to the prohibition of consecutive Fifths and Octaves, at least for the two extreme voices of a composition in several parts.   This prohibition forbids the lowest and uppermost voice in a piece to proceed by direct motion [that is, both parts ascending or both parts descending] into the consonance of an Octave or Fifth (including Twelfth).   They should rather come into such a consonance by contrary motion (one descending and the other ascending).   In duets this would also hold for the unison.   The meaning of this law must certainly be, that whenever the extreme voices unite to form the partial tones of a compound, they ought to have reached a state of relative rest.   It must be conceded that the equilibrium will be more perfect when the extreme parts of the whole mass of tone approach their junction from opposite sides, than when the centre of gravity, so to speak, of the sonorous mass is displaced by the parallel motion of the extreme voices, and these voices catch one another up with different velocities.   But where the motion proceeds in the same direction, and no relative rest is intended, the hidden Fifths are also not avoided, as in the usual formulae :

in which the $g \pm d$ is reached by passages involving hidden Fifths.

Another rule in the progression of parts, prohibiting *false relations*, must have had its origin in the requirements of the singer.   But what the singer finds a

difficulty in hitting, must naturally also appear an unusual and forced skip to the hearer. By *false relations* is meant the case when two tones in consecutive chords, which belong to different voices, form false Octaves or false Fifths. For example, if one voice in the first chord sings b_1 and another voice in the next chord sings $b\flat$, or the first has c and the second $c_2\sharp$, there are false Octave relations. False Fifth relations are forbidden for the extreme voices only. Thus in the first chord the bass has b_1, in the second the soprano has f, or conversely, where $b_1 \ldots f$ is a false Fifth. The meaning of this rule is, probably, that the singer would find it difficult to hit the new tone which is not in the scale, if he had just heard the next nearest tone of the scale given by another singer. Similarly, when he has to take the false Fifth of a tone which is prominent in present harmony as lowest or highest. There is therefore a certain sense in the prohibition, but numerous exceptions have arisen, as the ear of modern musicians,
¶ singers and hearers, has become accustomed to bolder combinations and livelier progressions. All these rules were essentially intended for the old ecclesiastical music, where a quiet, gentle, well-contrived, and well-adjusted stream of sound was aimed at, without any intentional effort or disturbance of the smoothest equilibrium. Where music has to express effort and excitement, these rules become meaningless. Hidden Fifths and Octaves and even false relations of Fifths are found in abundance in the chorales of Sebastian Bach, who is otherwise so strict in his harmonies, but it must be admitted that the motion of his voices is much more powerfully expressed than in the old Italian ecclesiastical music.

¶

CHAPTER XIX.

ESTHETICAL RELATIONS.

LET us review the results of the preceding investigation.

Compound tones of a certain class are preferred for all kinds of music, melodic or harmonic ; and are almost exclusively employed for the more delicate and artistic development of music : these are the compound tones which have harmonic upper partial tones, that is compound tones in which the higher partial tones have vibrational numbers which are integral multiples of the vibrational number of the lowest partial tone, or prime. For a good musical effect we require a certain moderate degree of force in the five or six lowest partial tones, and a low degree of force in the higher partial tones.

This class of compound tones with harmonic upper partials is objectively distinguished by including all sonorous motions which are generated by a mechanical
¶ process that continues to act uniformly, and which consequently produce a uniform and sustained sensation. In the first rank among them stand the compound tones of the human voice, man's first musical instrument in time and value. The compound tones of all wind and bowed instruments belong to this class.

Among the bodies which are made to emit tones by striking, some, as strings, have also harmonic upper partials, and these can be used for artistic music.

The greater number of the rest, as membranes, rods, plates, &c., have inharmonic upper partial tones, and only such of them as have not very strong secondary tones of this kind can be singly and occasionally employed in connection with musical instruments proper.

Although sonorous bodies excited by blows may continue to sound for some time, their tones do not proceed with uniform force, but diminish more or less slowly and die away. Constant power over the intensity of tone, therefore, which is indispensable for expressive performance, can only be attained on instruments of the first kind, which can be maintained in a state of excitement, and which produce only harmonic upper partial tones. On the other hand, bodies excited by blows have a peculiar value for clearly defining the rhythm.

A second reason for preferring compound tones with harmonic upper partials is subjective and conditioned by the construction of our ear. In the ear even every simple tone, if sufficiently intense, excites feeble sensations of harmonic upper partials, and each combination of several simple tones generates combinational tones, as I have explained at the end of Chap. VII. (p. $157d–159c$). A single compound tone with irrational partials, when sounded with sufficient force, thus produces the sensation of dissonance, and simple tones acquire in the ear itself something of the nature of composition out of harmonic upper partial tones.

We are justified in assuming that historically all music was developed from song. Afterwards the power of producing similar melodic effects was attained by means of other instruments, which had a quality of tone compounded in a manner resembling that of the human voice. The reason why, even when constructive ¶ art was most advanced, the choice of musical instruments was necessarily limited to those which produced compound tones with harmonic upper partials, is clear from the above conditions.

This invariable and peculiar selection of instruments makes us perfectly certain that harmonic upper partials have from all time played an essential part in musical constructions, not merely for harmony, as the second part of this book shews, but also for melody.

Again, we can at any moment convince ourselves of the essential importance of upper partial tones to melody, by the absence of all expression in melodies executed with objectively simple tones, as, for example, those of wide-stopped organ pipes, for which the harmonic upper partials are formed only subjectively and weakly in the ear.

A necessity was always felt for music of all kinds to proceed by certain definite degrees of pitch; but the choice of these degrees was long unsettled. To distin- ¶ guish small differences of pitch and intonate them with certainty, requires a greater amount of technical musical power and cultivation of ear, than when the intervals are larger. Hence among almost all uncivilised people we find the Semitones neglected, and only the larger intervals retained. For some of the more cultivated nations, as the Chinese and Gaels, a scale of this kind has become established.*

It might perhaps have seemed most simple to make all such degrees of pitch of equal amount, that is, equally well distinguishable by our sensations. Such a graduation is possible for all our sensations, as Fechner has shewn in his investigations on psychophysical laws. We find such graduations used for the divisions of musical rhythm, and the astronomers use them in reference to the intensity of light in determining stellar magnitudes. Even in the field of musical pitch, the modern equally tempered chromatic scale presents us with a similar graduation. But although in certain of the less usual Greek scales and in modern Oriental music, cases occur where some particular small intervals have been divided on the ¶ principle of equal graduations, yet there seems at no time or place to have been a system of music in which melodies constantly moved in equal degrees of pitch, but smaller and larger intervals have always been mixed in the musical scales in a way that must appear entirely arbitrary and irregular until the relationship of compound tones is taken into consideration.*

On the contrary, in all known musical systems the intervals of Octave and Fifth have been decisively emphasised. Their difference is the Fourth, and the difference between this and the Fifth, is the Pythagorean major Tone 8 : 9, by which (but not by the Fourth or Fifth) the Octave might be approximatively divided.

The sole remnants that I can find in modern music of the endeavour sometimes made in homophonic music to introduce degrees depending on equality of interval and not on relationship of tone, are the chromatic intercalated notes, and

* [See, however, App. XX. sect. K.—*Translator.*]

the leading note of the key when similarly used. But this is always a Semitone
(p. 352c), an interval well known in the series of related tones, which, owing to
its smallness, is easily measured by the sensation of its difference, even in places
where its tonal relationship is not immediately sensible.

The decisive importance acquired by the Octave and Fifth in all musical scales
from the earliest times shews, that the construction of scales must have been
originally influenced by another principle, which finally became the sole regulator
of every artistic form of a complete scale. This is the principle which we have
termed *tonal relationship*.

Relationship in the first degree between two compound tones consists in their
each having a partial tone of the same pitch.

In singing, the similarity of two musical tones which stand in the relation of
Octave or Fifth to one another, must have been very soon observed. As already
¶ remarked, this gives also the Fourth, which has itself a sufficiently perceptible
natural relationship to have been remarked independently. To discover the tonal
similarity of the major Third and major Sixth, required a finer cultivation of the
musical ear, and perhaps also peculiar beauty of voice. Even yet we are easily led
by the familiar sharp major Thirds of equal temperament, to endure any major
Thirds which are somewhat too sharp, provided they occur melodically and are
not sounded together. On the other hand, we must not forget that the rules of
Archytas and Abdul Kadir,* both of which were applicable to homophonic music
only, gave a preference to the natural major Third, although its introduction
obliged both musicians to renounce a musical system so theoretically consistent
and invested with such high authority as that of Pythagoras.

Hence the principle of tonal relationship did not at all times exclusively determine the construction of the scale, and does not even yet determine it exclusively
among all nations. This principle must, therefore, be regarded to some extent as
¶ a *freely selected principle of style*, as I have endeavoured to shew in Chapter XIII.
But, on the other hand, the art of music in Europe was historically developed
from that principle, and on this fact depends the main proof that it was really as
important as we have assumed it to be. The preference first given to the *diatonic
scale*, and finally the exclusive use of that scale, introduced the principle of tonal
relationship in all its integrity into the musical scale. Within the diatonic scale
various methods of execution were possible, and these generated the ancient modes,
which had equal claims to attention in homophonic song, and hence stood on a
level.

But the principle of tonal relationship penetrated far deeper in its harmonic
than it did in its melodic form. In melodic sequence the identity of two partial
tones is a matter of memory, but when the notes are sounded together the immediate sensible impression of the beats, or else of the undisturbed flow of sound
forces itself on the hearer's attention. The liveliness of melodic and harmonic
¶ impressions differs in the same way as a recollected image differs from the actual
impression made by the original. As an immediate consequence arose that far
superior sensibility for the correctness of the intervals which is seen in the harmonic union of tones, and which admitted of being developed into the finest
physical methods of measurement.

It must also be remembered that relationship in the second degree can in
harmonic music be reduced to audible relationships of the first degree, by a proper
selection of the fundamental bass, and that generally more distant relationships
can easily be made clearly audible. By this means, notwithstanding the variety
of progression, a much clearer connection of all parts with their origin, the tonic,
can be maintained and rendered objectively sensible to the hearer. It cannot be
doubted that these are the essential foundations of the great breadth and wealth of
expression which modern compositions can attain without losing their artistic unity.

---

* [For Archytas of Tarentum, about b.c.   note †.—*Translator*.]
400, see p. 262c, and for Abdulqadir, see p. 281,

We then saw that the requirements of harmonic music reacted in a peculiar manner on the construction of scales ; that properly speaking only one of the old tonal modes (our major mode) could be retained unaltered,* and that the rest after undergoing peculiar modifications were fused into our minor mode, which, though most like the ancient mode of the minor Third, can at one time resemble the mode of the minor Sixth, and at another time that of the minor Seventh, but does not perfectly correspond with any one of these.

This process of the development of the elements of our modern musical system lasted down to the middle of the last century. It was not until composers ventured to put a minor chord at the close of compositions written in the minor mode, that the musical feeling of European musicians and hearers can be admitted to have become perfectly and surely habituated to the new system. The minor chord was allowed to be a real, although obscured, chord of its tonic.

Whether this admission of the minor chord expressed a feeling for another ¶ mode of unifying its three tones, as A. von Oettingen † has assumed,—relying on the fact that the three tones $c-e^1 \flat + g$ have a common upper partial g'',— must be left to future experience to decide, should it be found practicable to construct long and well-connected musical compositions in *Oettingen's phonic system* (this is the name which he gives to the minor system which he has theoretically developed, and which is essentially different from the historical minor mode). At any rate, the minor mode has historically developed itself as a compromise between different kinds of claims. Thus it is only major triads which can perfectly indicate the compound tone of the tonic ; minor chords contain in their Third an element which, although nearly related to the tonic and its Fifth, does not thoroughly fuse with them, and hence in their final cadence they do not so thoroughly agree with the principle of tonality which had ruled the previous development of music. I have endeavoured to make it probable that the peculiar esthetic expression of the minor mode proceeded partly from this cause and partly ¶ from the heterogenous combinational tones of the minor chord.

In the last part of my book, I have endeavoured to shew that the construction of scales and of harmonic tissue is a product of artistic invention, and by no means furnished by the natural formation or natural function of our ear, as it has been hitherto most generally asserted. Of course the laws of the natural function of our ear play a great and influential part in this result ; these laws are, as it were, the building stones with which the edifice of our musical system has been

* [But see supra, p. 274, note *, scale 1.— *Translator.*]

† *The System of Harmony Dually Developed*, Dorpat and Leipzig, 1866. Herr v. Oettingen, as already observed, p. 308, note §, regards the minor chord as representing the harmonic undertones of its Fifth, and hence as standing in place of a part of its compound tone. He calls it the 'phonic' chord, as opposed to the 'tonic' major chord which stands in place of the upper partials of its root. He proceeds to deduce the formation of the minor system from the relations of the harmonic undertones in a manner precisely analogous to that by which I have deduced the major system from the relations of the upper partial tones. The tonal mode thus constructed is, however, in our language the mode of the minor Sixth (p. 274, note *, scale 7), and the usual minor, a mixed mode. Latterly Dr. Hugo Riemann has given in his adhesion to this view, and in his lately published *Musical Syntaxis* has attempted to examine and establish the consequences of this system by examples from acknowledged composers. The application of this critical method appears to me very commendable, and to be the indispensable condition to advancing in the theory of composition. For the rest this author justifies (p. 54) the assertion I have made in the text by remarking : ' I am sorry to say that I am unable to adduce a single example from the whole of our musical literature, of the carrying out of (v. Oettingen's) pure minor mode harmony even in the simplest manner.' I have not been able to convince ¶ myself of the correctness of the fact adduced on p. 6, that the undertones of a tone strongly struck on the piano sound when the corresponding dampers are raised. Perhaps the author has been deceived by the circumstance that with very resonant instruments (especially older ones) any strong shake, and therefore probably a violent blow on the digitals, will cause some one or several of the deeper strings to sound its note. [The undertones have always each an upper partial tone of the pitch of the note struck ; the striking of this note must then sympathetically excite those upper partials of the undertones, and thus reinforce the prime of the note struck, just as striking the undertone sympathetically excites the higher tone itself. Can this have deceived Dr. Riemann ?—*Translator.*]

erected, and the necessity of accurately understanding the nature of these materials in order to understand the construction of the edifice itself, has been clearly shewn by the course of our investigations upon this very subject. But just as people with differently directed tastes can erect extremely different kinds of buildings with the same stones, so also the history of music shews us that the same properties of the human ear could serve as the foundation of very different musical systems. Consequently it seems to me that we cannot doubt, that not merely the composition of perfect musical works of art, but even the construction of our system of scales, keys, chords, in short of all that is usually comprehended in a treatise on Thorough Bass, is the work of artistic invention, and hence must be subject to the laws of artistic beauty. In point of fact, mankind has been at work on the diatonic system for more than 2500 years since the days of Terpander and Pythagoras, and in many cases we are still able to determine that the pro- ¶ gressive changes made in the tonal system have been due to the most distin- guished composers themselves, partly through their own independent inventions, and partly through the sanction which they gave to the inventions of others, by employing them artistically.

The esthetic analysis of complete musical works of art, and the comprehension of the reasons of their beauty, encounter apparently invincible obstacles at almost every point. But in the field of elementary musical art we have now gained so much insight into its internal connection that we are able to bring the results of our investigations to bear on the views which have been formed and in modern times nearly universally accepted respecting the cause and character of artistic beauty in general. It is, in fact, not difficult to discover a close connection and agreement between them; nay, there are probably fewer examples more suitable than the theory of musical scales and harmony, to illustrate the darkest and most difficult points of general esthetics. Hence I feel that I should not be justified in ¶ passing over these considerations, more especially as they are closely connected with the theory of sensual perception, and hence with physiology in general.

No doubt is now entertained that beauty is subject to laws and rules dependent on the nature of human intelligence. The difficulty consists in the fact that these laws and rules, on whose fulfilment beauty depends and by which it must be judged, are not consciously present to the mind, either of the artist who creates the work, or the observer who contemplates it. Art works with design, but the work of art ought to have the appearance of being undesigned, and must be judged on that ground. Art creates as imagination pictures, regularly without conscious law, designedly without conscious aim. A work, known and acknowledged as the pro- duct of mere intelligence, will never be accepted as a work of art, however perfect be its adaptation to its end. Whenever we see that conscious reflection has acted in the arrangement of the whole, we find it poor.

¶   Man fühlt die Absicht, und man wird verstimmt.
                     (We feel the purpose, and it jars upon us.)

And yet we require every work of art to be reasonable, and we shew this by subjecting it to a critical examination, and by seeking to enhance our enjoyment and our interest in it by tracing out the suitability, connection, and equilibrium of all its separate parts. The more we succeed in making the harmony and beauty of all its peculiarities clear and distinct, the richer we find it, and we even regard as the principal characteristic of a great work of art that deeper thought, reiterated observation, and continued reflection shew us more and more clearly the reason- ableness of all its individual parts. Our endeavour to comprehend the beauty of such a work by critical examination, in which we partly succeed, shews that we assume a certain adaptation to reason in works of art, which may possibly rise to a conscious understanding, although such understanding is neither necessary for the invention nor for the enjoyment of the beautiful. For what is esthetically beau- tiful is recognised by the immediate judgment of a cultivated taste, which declares it pleasing or displeasing, without any comparison whatever with law or conception.

But that we do not accept delight in the beautiful as something individual, but rather hold it to be in regular accordance with the nature of mind in general, appears by our expecting and requiring from every other healthy human intellect the same homage that we ourselves pay to what we call beautiful. At most we allow that national or individual peculiarities of taste incline to this or that artistic ideal, and are most easily moved by it, precisely in the same way that a certain amount of education and practice in the contemplation of fine works of art is undeniably necessary for penetration into their deeper meaning.

The principal difficulty in pursuing this object, is to understand how regularity can be apprehended by intuition without being consciously felt to exist. And this unconsciousness of regularity is not a mere accident in the effect of the beautiful on our mind, which may indifferently exist or not; it is, on the contrary, most ¶ clearly, prominently, and essentially important. For through apprehending everywhere traces of regularity, connection, and order, without being able to grasp the law and plan of the whole, there arises in our mind a feeling that the work of art which we are contemplating is the product of a design which far exceeds anything we can conceive at the moment, and which hence partakes of the character of the illimitable. Remembering the poet's words:

> Du gleichst dem Geist, den du begreifst,
> (Thou'rt like the spirit thou conceivest),

we feel that those intellectual powers which were at work in the artist, are far above our conscious mental action, and that were it even possible at all, infinite time, meditation, and labour would have been necessary to attain by conscious thought that degree of order, connection, and equilibrium of all parts and all internal relations, which the artist has accomplished under the sole guidance of tact and ¶ taste, and which we have in turn to appreciate and comprehend by our own tact and taste, long before we begin a critical analysis of the work.

It is clear that all high appreciation of the artist and his work reposes essentially on this feeling. In the first we honour a genius, a spark of divine creative fire, which far transcends the limits of our intelligent and conscious forecast. And yet the artist is a man as we are, in whom work the same mental powers as in ourselves, only in their own peculiar direction, purer, brighter, steadier; and by the greater or less readiness and completeness with which we grasp the artist's language we measure our own share of those powers which produced the wonder.

Herein is manifestly the cause of that moral elevation and feeling of ecstatic satisfaction which is called forth by thorough absorption in genuine and lofty works of art. We learn from them to feel that even in the obscure depths of a healthy and harmoniously developed human mind, which are at least for the present inaccessible to analysis by conscious thought, there slumbers a germ of order that ¶ is capable of rich intellectual cultivation, and we learn to recognise and admire in the work of art, though draughted in unimportant material, the picture of a similar arrangement of the universe, governed by law and reason in all its parts. The contemplation of a real work of art awakens our confidence in the originally healthy nature of the human mind, when uncribbed, unharassed, unobscured, and unfalsified.

But for all this it is an essential condition that the whole extent of the regularity and design of a work of art should *not* be apprehended consciously. It is precisely from that part of its regular subjection to reason, which escapes our conscious apprehension, that a work of art exalts and delights us, and that the chief effects of the artistically beautiful proceed, *not* from the part which we are able fully to analyse.

If we now apply these considerations to the system of musical tones and harmony, we see of course that these are objects belonging to an entirely subordinate and elementary domain, but nevertheless they, too, are slowly matured inventions of the artistic taste of musicians, and consequently they, too, must be governed by the general rules of artistic beauty. Precisely because we are here still treading the lower walks of art, and are not dealing with the expression of deep psychological problems, we are able to discover a comparatively simple and transparent solution of that fundamental enigma of esthetics.

The whole of the last part of this book has explained how musicians gradually discovered the relationships between tones and chords, and how the invention of harmonic music rendered these relationships closer, and clearer, and richer. We have been able to deduce the whole system of rules which constitute Thorough Bass, from an endeavour to introduce a clearly sensible connection into the series of tones which form a piece of music.

¶ A feeling for the melodic relationship of consecutive tones, was first developed, commencing with Octave and Fifth and advancing to the Third. We have taken pains to prove that this feeling of relationship was founded on the perception of identical partial tones in the corresponding compound tones. Now these partial tones are of course present in the sensations excited in our auditory apparatus, and yet they are not generally the subject of conscious perception as independent sensations. The conscious perception of everyday life is limited to the apprehension of the tone compounded of these partials, as a whole, just as we apprehend the taste of a very compound dish as a whole, without clearly feeling how much of it is due to the salt, or the pepper, or other spices and condiments. A critical examination of our auditory sensations as such was required before we could discover the existence of upper partial tones. Hence the real reason of the melodic relationship of two tones (with the exception of a few more or less clearly expressed conjectures, as, for example, by Rameau and d'Alembert) remained so long undiscovered, or at least was not in any respect clearly and definitely formulated. I believe that I have ¶ been able to furnish the required explanation, and hence clearly to exhibit the whole connection of the phenomena. The esthetic problem is thus referred to the common property of all sensual perceptions, namely, the apprehension of compound aggregates of sensations as sensible symbols of simple external objects, without analysing them. In our usual observations on external nature our attention is so thoroughly engaged by external objects that we are entirely unpractised in taking for the subjects of conscious observation, any properties of our sensations themselves, which we do not already know as the sensible expression of some individual external object or event.

After musicians had long been content with the melodic relationship of tones, they began in the middle ages to make use of harmonic relationship as shewn in consonance. The effects of various combinations of tones also depend partly on the identity or difference of two of their different partial tones, but they likewise partly depend on their combinational tones. Whereas, however, in melodic ¶ relationship the equality of the upper partial tones can only be perceived by *remembering* the preceding compound tone, in harmonic relationship it is determined by *immediate sensation*, by the presence or absence of beats. Hence in harmonic combinations of tone, tonal relationship is felt with that greater liveliness due to a present sensation as compared with the recollection of a past sensation. The wealth of clearly perceptible relations grows with the number of tones combined. Beats are easy to recognise as such when they occur slowly; but those which characterise dissonances are, almost without exception, very rapid, and are partly covered by sustained tones which do not beat, so that a careful comparison of slower and quicker beats is necessary to gain the conviction that the essence of dissonance consists precisely in rapid beats. Slow beats do not create the feeling of dissonance, which does not arise till the rapidity of the beats confuses the ear and makes it unable to distinguish them. In this case also the ear feels the difference between the undisturbed combination of sound in the case of two consonant tones, and the disturbed rough combination resulting from a dissonance. But, as a general rule, the hearer is then perfectly unconscious of the cause to which the disturbance and roughness are due.

The development of harmony gave rise to a much richer opening out of musical art than was previously possible, because the far clearer characterisation of related combinations of tones by means of chords and chordal sequences, allowed of the use of much more distant relationships than were previously available, by modulating into different keys. In this way the means of expression greatly increased as well as the rapidity of the melodic and harmonic transitions which could now be introduced without destroying the musical connection.

As the independent significance of chords came to be appreciated in the fifteenth and sixteenth centuries, a feeling arose for the relationship of chords to one another and to the tonic chord, in accordance with the same law which had long ago unconsciously regulated the relationship of compound tones. The relationship of compound tones depended on the identity of two or more partial tones, that of chords on the identity of two or more notes. For the musician, of course, the law of the relationship of chords and keys is much more intelligible than that of compound tones. He readily hears the identical tones, or sees them in the notes before him. But the unprejudiced and uninstructed hearer is as little conscious of the reason of the connection of a clear and agreeable series of fluent chords, as he is of the reason of a well-connected melody. He is startled by a false cadence and feels its unexpectedness, but is not at all necessarily conscious of the reason of its unexpectedness.

Then, again, we have seen that the reason why a chord in music appears to be the chord of a determinate root, depends as before upon the analysis of a compound tone into its partial tones, that is, as before upon those elements of a sensation which cannot readily become subjects of conscious perception. This relation between chords is of great importance, both in the relation of the tonic chord to the tonic tone, and in the sequence of chords.

The recognition of these resemblances between compound tones and between chords, reminds us of other exactly analogous circumstances which we must have often experienced. We recognise the resemblance between the faces of two near relations, without being at all able to say in what the resemblance consists, especially when age and sex are different, and the coarser outlines of the features consequently present striking differences. And yet notwithstanding these differences—notwithstanding that we are unable to fix upon a single point in the two countenances which is absolutely alike—the resemblance is often so extraordinarily striking and convincing, that we have not a moment's doubt about it. Precisely the same thing occurs in recognising the relationship between two compound tones.

Again, we are often able to assert with perfect certainty, that a passage not previously heard is due to a particular author or composer whose other works we know. Occasionally, but by no means always, individual mannerisms in verbal or musical phrases determine our judgment, but as a rule we are mostly unable to fix upon the exact points of resemblance between the new piece and the known works of the author or composer.

The analogy of these different cases may be even carried farther. When a father and daughter are strikingly alike in some well-marked feature, as the nose or forehead, we observe it at once, and think no more about it. But if the resemblance is so enigmatically concealed that we cannot detect it, we are fascinated, and cannot help continuing to compare their countenances. And if a painter drew two such heads having, say, a somewhat different expression of character combined with a predominant and striking, though indefinable, resemblance, we should undoubtedly value it as one of the principal beauties of his painting. Our admiration would certainly not be due merely to his technical skill; we should rather look upon his painting as evidencing an unusually delicate feeling for the significance of the human countenance, and find in this the artistic justification of his work.

Now the case is similar for musical intervals. The resemblance of an Octave to its root is so great and striking that the dullest ear perceives it ; the Octave seems to be almost a pure repetition of the root, as it, in fact, merely repeats a part of the compound tone of its root, without adding anything new. Hence the esthetical effect of an Octave is that of a perfectly simple, but little attractive interval. The most attractive of the intervals, melodically and harmonically, are clearly the Thirds and Sixths,—the intervals which lie at the very boundary of those that the ear can grasp. The major Third and the major Sixth cannot be properly appreciated unless the first five partial tones are audible. These are present in good musical qualities of tone. The minor Third and the minor Sixth are for the most part justifiable only as inversions of the former intervals. The more complicated ¶ intervals in the scale cease to have any direct or easily intelligible relationship. They have no longer the charm of the Thirds.

Moreover, it is by no means a merely external indifferent regularity which the employment of diatonic scales, founded on the relationship of compound tones, has introduced into the tonal material of music, as, for instance, rhythm introduced some such external arrangement into the words of poetry. I have shewn, on the contrary, in Chapter XIV., that this construction of the scale furnished a means of measuring the intervals of their tones, so that the equality of two intervals lying in different sections of the scale would be recognised by immediate sensation. Thus the melodic step of a Fifth is always characterised by having the second partial tone of the second note identical with the third of the first. This produces a definiteness and certainty in the measurement of intervals for our sensation, such as might be looked for in vain in the system of colours, otherwise so similar, or in the estimation of mere differences of intensity in our various sensual ¶ perceptions.

Upon this reposes also the characteristic resemblance between the relations of the musical scale and of space, a resemblance which appears to me of vital importance for the peculiar effects of music. It is an essential character of space that at every position within it like bodies can be placed, and like motions can occur. Everything that is possible to happen in one part of space is equally possible in every other part of space and is perceived by us in precisely the same way. This is the case also with the musical scale. Every melodic phrase, every chord, which can be executed at any pitch, can be also executed at any other pitch in such a way that we immediately perceive the characteristic marks of their similarity. On the other hand, also, different voices, executing the same or different melodic phrases, can move at the same time within the compass of the scale, like two bodies in space, and, provided they are consonant in the accented parts of bars, without creating any musical disturbances. Such a close analogy consequently exists in ¶ all essential relations between the musical scale and space, that even alteration of pitch has a readily recognised and unmistakable resemblance to motion in space, and is often metaphorically termed the ascending or descending *motion* or *progression* of a part. Hence, again, it becomes possible for motion in music to imitate the peculiar characteristics of motive forces in space, that is, to form an image of the various impulses and forces which lie at the root of motion. And on this, as I believe, essentially depends the power of music to picture emotion.

It is not my intention to deny that music in its initial state and simplest forms may have been originally an artistic imitation of the instinctive modulations of the voice that correspond to various conditions of the feelings. But I cannot think that this is opposed to the above explanation ; for a great part of the natural means of vocal expression may be reduced to such facts as the following : its rhythm and accentuation are an immediate expression of the rapidity or force of the corresponding psychical motives—all effort drives the voice up—a desire to make a pleasant impression on another mind leads to selecting a softer, pleasanter quality of tone—and so forth. An endeavour to imitate the involuntary modulations of the voice and make its recitation richer and more expressive, may therefore very possibly have led our ancestors to the discovery of the first means of musical expression, just as the imitation of weeping, shouting, or sobbing, and other musical delineations may play a part in even cultivated music, (as in operas), although such modifications of the voice are not confined to the action of free mental motives, but embrace really mechanical and even involuntary muscular contractions. But it is quite clear that every completely developed melody goes far beyond an imitation of nature, even if we include the cases of the most varied alteration of voice under the influence of passion. Nay, the very fact that music introduces progression by fixed degrees both in rhythm and in the scale, renders even an approximatively correct representation of nature simply impossible, for most of the passionate affections of the voice are characterised by a gliding transition in pitch. The imitation of nature is thus rendered as imperfect as the imitation of ¶ a picture by embroidery on a canvas with separate little squares for each shade of colour. Music, too, departed still further from nature when it introduced the greater compass, the mobility, and the strange qualities of tone belonging to musical instruments, by which the field of attainable musical effects has become so much wider than it was or could be when the human voice alone was employed.

Hence though it is probably correct to say that mankind, in historical development, first learned the means of musical expression from the human voice, it can hardly be denied that these same means of expressing melodic progression act, in artistically developed music, without the slightest reference to the application made of them in the modulations of the human voice, and have a more general significance than any that can be attributed to innate instinctive cries. That this is the case appears above all in the modern development of instrumental music, which possesses an effective power and artistic justification that need not be gainsaid, although we may not yet be able to explain it in all its details.　　¶

Here I close my work. It appears to me that I have carried it as far as the physiological properties of the sensation of hearing exercise a direct influence on the construction of a musical system, that is, as far as the work especially belongs to natural philosophy. For even if I could not avoid mixing up esthetic problems with physical, the former were comparatively simple, and the latter much more complicated. This relation would necessarily become inverted if I attempted to proceed further into the esthetics of music, and to enter on the theory of rhythm, forms of composition, and means of musical expression. In all these fields the properties of sensual perception would of course have an influence at times, but only in a very subordinate degree. The real difficulty would lie in the development of the psychical motives which here assert themselves. Certainly this is the point ¶ where the more interesting part of musical esthetics begins, the aim being to explain the wonders of great works of art, and to learn the utterances and actions of the various affections of the mind. But, however alluring such an aim may be, I prefer leaving others to carry out such investigations, in which I should feel myself too much of an amateur, while I myself remain on the safe ground of natural philosophy, in which I am at home.

APPENDICES.

APPENDIX I.

¶   ON AN ELECTRO-MAGNETIC DRIVING MACHINE FOR THE SIREN.

(See p. 13a.)

I HAVE lately had a small electro-magnetic machine constructed with a constant velocity of rotation, and it has proved of great service in driving the siren. A rotating electro-magnet, in which the direction of current is changed every semi-rotation, moves between two fixed magnetic poles. The current in this electro-magnet is interrupted, as soon as the velocity begins to exceed the desired amount, by the centrifugal force of a mass of metal fastened to its axis of rotation. Two spiral springs whose elasticity is opposed to the centrifugal force, may be tightened or loosened, and thus made stronger or weaker at pleasure. By this means the velocity can be maintained at any required rate. A figure and description of this machine were given by Herr S. Exner, in the 'Proceedings' (*Sitzungsberichte*) of the Vienna Academy : 'Math. Naturw. Cl.' vol. lviii. part 2, 8 Oct. 1868.

    The machine was improved in 1875 by separating from it the centrifugal
¶ regulator, and letting it only open and close the weak current for a relay. It is the relay which makes or breaks the strong current that drives the electro-magnetic machine.

    The siren is connected with the machine by a thin driving band, and then it does not require to be blown. Instead of blowing, I placed on the disc a small turbine constructed of stiff paper, which drove the air through the openings whenever they coincided with those in the chest. This arrangement gave me extremely constant tones on the siren, rivalling those on the best constructed organ pipes. Latterly I have given the siren straight holes, so that the strength of the wind has no longer any influence on its speed, and then I blow through the box. [See App. XX. sect. B. No. 2.]

APPENDIX II.

ON THE SIZE AND CONSTRUCTION OF RESONATORS.

(See pp. 44b and 166d, note *.)

SPHERICAL RESONATORS with a short funnel-shaped neck for insertion into the ear as shewn in fig. 16 a (p. 43b), are most efficient. Their advantage consists partly in their other proper tones being very distant indeed from their prime tones, and hence being very slightly reinforced, and partly in the spherical form giving the most powerful resonance. But the walls of the sphere must be firm and smooth, to oppose the necessary resistance to the powerful vibrations of air which take place within them, and to impede the motion of the air as little as possible by friction. At first I employed any spherical glass vessels that came to hand, as the receivers of retorts, and inserted into one of their openings a glass tube which had been adapted to my ear. Afterwards Herr R. Koenig, maker of acoustical instruments, Paris [now of 27 Quai d'Anjou], constructed a series of these glass spheres properly tuned, and afterwards had them made of brass in the form shewn in fig. 16 a, p. 43*b*. This is the most appropriate form for resonators. When the openings are relatively very narrow, their pitch can be determined by the formula which I have developed, namely *

$$n = \frac{a \sqrt[4]{\sigma}}{\sqrt[4]{\pi^5} . \sqrt{(2\,S)}},$$

where a is the velocity of sound, σ the area of the circular opening, and S the volume of the cavity. Or if we assume as its value

$$a = 332 \cdot 260 \text{ metres,}$$

which corresponds with a temperature of zero centigrade, the above formula gives

$$n = 56174 \; \sqrt[4]{\left(\frac{\sigma}{S^2} \right)} \qquad \P$$

Herr Sondhauss [Pogg. vol. lxxxi. pp. 347-373] had discovered the same formula experimentally, but used 52400 for the numerical coefficient, which agrees with the experimental results better when the openings are not very small. When the diameter of the opening is smaller than one-tenth of the diameter of the sphere, the formula deduced from theory agrees well with Wertheim's experiments. For resonators which have the diameter of their opening between a fourth and a fifth of the diameter of the sphere, I have experimentally determined the coefficient as 47000. The second opening of the resonator may be regarded as closed because it is brought firmly against the ear. If the cavity is spherical with radius R, while r is that of the opening, the above formula becomes

$$n = a \sqrt{\left(\frac{3r}{8\,\pi^3 . R^3} \right)}$$

Here follows a list of the measurements of my glass resonators. ¶

Pitch	Diameter of the Sphere in millimetres [and inches]	Diameter of the Orifice in millimetres [and inches]	Volumes of the Interior in cubic centimetres [and cubic inches]	Remarks
1) *g*	154 [6·06]	35·5 [1·40]	1773 [108·19]	⎫
2) *b*b	131 [5·16]	28·5 [1·12]	1092 [66·64]	⎪
3) *c′*	130 [5·12]	30·2 [1·19]	1053 [64·26]	⎬ Neck
4) *e′*	115 [4·53]	30 [1·18]	546 [33·32]	⎰ funnel shaped
5) *g′*	79 [3·11]	18·5 [·73]	235 [14·341]	⎪
6) *b′*b	76 [2·99]	22 [·87]	214 [13·06]	⎭
7) *c″*	70 [2·76]	20·5 [·81]	162 [9·89]	
8) *b′*b	53·5 [2·11]	8 [·31]	74 [4·52]	Neck cylindrical
9) *b″*b	46 [1·81]	15 [·59]	49 [2·99]	⎰ Neck cylindrical, ⎱ mouth at side
10) *d‴*	43 [1·69]	15 [·59]	37 [2·26]	Neck cylindrical ¶

Smaller spheres did not answer well. In order to tune the resonators, Herr Koenig has also made them of two short cylinders of which one runs into the other, each having a pierced lid at its external end. One opening serves for a connection with the ear or a sympathetic flame, the other is free.† The measurements given in App. IV., p. 377*d*, will serve for the manufacture of such tubes, as the second opening is of no consequence because it is firmly inserted into the ear.

Since metal cubes are troublesome to manufacture and hence proportionately dear, we can use double cones of tin or pasteboard, the vertices of which have been removed. The cone next the ear is more sharply pointed so that its end readily fits the ear.

Conical resonators of thin sheet zinc, such as Herr G. Appunn ‡ of Hanau manu-

* 'Theory of Vibrations of Air in Tubes with Open Ends,' in *Journal für reine u. angew. Mathematik*, vol. lvii., equation 30a, and following.

† Poggendorff's *Annals*, vol. cxlvi. p. 189.

‡ [Deceased, now Anton Appunn.]

factures,* are easily made, and are useful for most purposes. But they reinforce all the partials of their fundamental tone at the same time. Their length is about the same as that of open organ pipes of the same pitch.

Resonators with a very narrow opening generally produce a much more considerable reinforcement of the tone, but then there must be a much more precise agreement between the pitch of the tone to be heard, and the proper tone of the resonator. It is just as in microscopes; the greater the magnifying power, the smaller the field of view. Reducing the size of the orifice also deepens the pitch of the resonator, and this gives an easy means of tuning it to any required pitch. But, for the above reason, the opening must not be reduced too much.

I should also mention Herr Koenig's plan of transferring vibrations of air to gas flames, and thus making them visible. Flames of this sort act well when connected with resonators, which are then best made of a spherical form, and should have two equal openings. To one of the openings the small gas-chamber is fixed. This chamber is a small flat box, about big enough to contain two ¶ shilling pieces laid flat on one another. It is cut out of a plate of wood, and closed on the side next the resonator with a very thin membrane of india-rubber, which, while it completely separates the air of the resonator from the gas in the chamber, allows the vibrations of the air to be freely communicated to the gas. Through the plate of wood two narrow pipes enter the chamber, one introducing the inflammable gas and the other conducting it away. This last ends in a very fine point at which the gas is lighted. The vibrations of the air in the resonator being communicated to the gas cause the flame to leap up and down. These oscillations of the flame are so rapid and regular that, when viewed directly, the flame appears to be quite steady. Its altered condition, however, betrays itself by an altered form and colour. Thus to recognise the beats of two tones reinforced by the resonator, it is enough to look at the flame and observe how it alternates between its forms of rest and of oscillation. But to see the separate oscillations the flames should be viewed in a rotating glass, in which the flame at rest appears to be drawn out into a long uniform ribbon, while the oscillating flame ¶ appears as a series of separate images of flames. It is thus possible to allow a large number of persons at once to determine whether or not a given tone is reinforced by the resonator.†

An extremely sensitive means of making the vibrations of the air in a resonator visible, is a flat film of glycerined soap and water which is placed over its opening.

[Mr. Bosanquet finding that for observations on beats all these resonators imperfectly plug one ear and leave the other open, has invented another kind, for which see App. XX. sect. L. art. 4, *b*.]

APPENDIX III.

ON THE MOTION OF PLUCKED STRINGS.

(See p. 52*b*.)

LET *x* be the distance of a point in the string from one of its extremities, and *l* the length of the string, so that for one extremity $x=0$, and for the other $x=l$. It is sufficient to investigate the case for which the motion takes place in one plane passing through the position of rest. Let *y* be the distance of the point *x* from its position of rest at the time *t*. And let *μ* be the weight of the unit

---

* [There is a set in the Science Collections at South Kensington Museum.—*Translator.*]

† [All these instruments and appliances can be obtained of Herr Koenig, by whom they were exhibited in London at the International Exhibition of 1872. Large drawings of the appearances of the flames just described, as viewed in the rotating mirror while two octaves were sung to the French vowels, were also exhibited. See Koenig's paper on the subject, with plates, in *Philosophical Magazine*, 1873, vol. xlv. pp. 1–18, 105–14.—On the principles of the use of revolving mirrors, first experimentally used by Sir C. Wheatstone, see Donkin's *Acoustics*, 1870, p. 142.—*Translator.*]

of length, and T the tension of the string. The differential equation of motion is then

$$\mu \cdot \frac{d^2y}{dt^2} = T \cdot \frac{d^2y}{dx^2} \quad \dots\dots\dots\dots\dots\dots\dots \quad (1)$$

Then, since the extremities of the string are assumed to be at rest, we must have

$$y = 0 \text{ when } x = 0, \text{ and also when } x = l \quad \dots\dots\dots\dots \quad (1a)$$

The most general integral of the equation (1) which fulfils the conditions (1a) and corresponds to a periodic motion of the string, is

$$\left. \begin{aligned}
y = {} & A_1 \cdot \sin \frac{\pi x}{l} \cdot \cos 2\pi nt + A_2 \cdot \sin \frac{2\pi x}{l} \cdot \cos 4\pi nt \\
& + A_3 \cdot \sin \frac{3\pi x}{l} \cdot \cos 6\pi nt + \&c. \\
& + B_1 \cdot \sin \frac{\pi x}{l} \cdot \sin 2\pi nt + B_2 \cdot \sin \frac{2\pi x}{l} \cdot \sin 4\pi nt \\
& + B_3 \cdot \sin \frac{3\pi x}{l} \cdot \sin 6\pi nt + \&c.
\end{aligned} \right\} \dots\dots \quad (1b) \quad ¶$$

where $n^2 = \dfrac{T}{4\mu l^2}$ and A_1, A_2, A_3, &c., and B_1, B_2, B_3, &c., are any constant co-efficients, which can be determined when the form and velocity of the string are known for any determinate time t.

For $t = 0$, the form of the string will be

$$y = A_1 \cdot \sin \frac{\pi x}{l} + A_2 \cdot \frac{\sin 2\pi x}{l} + A_3 \cdot \sin \frac{3\pi x}{l} + \&c. \quad \dots\dots \quad (1c)$$

and the velocity of the string will be

$$\frac{dy}{dt} = 2\pi n \left(B_1 \cdot \sin \frac{\pi x}{l} + 2 B_2 \cdot \sin \frac{2\pi x}{l} + 3 B_3 \cdot \sin \frac{3\pi x}{l} + \&c. \right) \dots \quad (1d) \quad ¶$$

Now suppose the string to have been drawn aside by a sharp point, and that the point was withdrawn at the time $t = 0$, so that the vibration commenced at that moment, then for $t = 0$ there was no velocity, that is $\dfrac{dy}{dt}$ was $= 0$ for all values of x. This can only be the case when in equation (1d), $0 = B_1 = B_2 = B_3 = \&c.$

The coefficients A_1, A_2, A_3, &c., depend on the shape of the string at the time $t = 0$. At the moment the sharp point quitted it the string must have assumed the position of fig. 18 A (p. 54a), that is, it must have formed two straight lines proceeding from the sharp point to the fixed extremities of the string. Supposing the position of the sharp point at that moment to be determined by $x = a$ and $y = b$, then for the time $t = 0$, the value of y, was

$$y = \frac{bx}{a} \text{ if } a > x > 0 \quad \dots\dots\dots\dots\dots\dots \quad (2) \quad ¶$$

$$\text{and } y = b \cdot \frac{l-x}{l-a} \text{ if } l > x > a \quad \dots\dots\dots\dots \quad (2a)$$

and the values of y in (1c) and (2), or else (2a) respectively, must be identical.

To find the coefficient A_m, the well-known method is to multiply both sides of the equation (1c) by $\sin \dfrac{m\pi x}{l} \cdot dx$, and to integrate between the limits $x = 0$ and $x = l$. In this case equation (1c) reduces to

$$\int_0^l y \cdot \sin \frac{m\pi x}{l} \cdot dx = A_m \cdot \int_0^l \sin^2 \frac{m\pi x}{l} \cdot dx \dots\dots\dots\dots\dots \quad (2b)$$

in which y must be replaced by its values in (2) and (2a). Performing the integrations indicated in (2b) we find

$$A_m = \frac{2bl^2}{m^2\pi^2 a(l-a)} \cdot \sin \frac{m\pi a}{l} \quad \dots\dots\dots\dots\dots \quad (3)$$

Hence A_m will $= 0$, and consequently the mth tone of the string will disappear, when $\sin \dfrac{m\pi a}{l} = 0$, that is, when $a = \dfrac{l}{m}$ or $= \dfrac{2l}{m}$, or $= \dfrac{3l}{m}$, &c. Hence if we suppose the string to be divided into m equal parts, and to be plucked in one of these divisions, the mth tone disappears, and this is the tone whose nodes fall upon these points.

Every node for an mth tone is also a node for the $2m$th, $3m$th, $4m$th, &c., tone, and hence all these tones also disappear.

The integral of equation (1) may also, as is well known, be exhibited in the following form :—

$$y = \phi(x - at) + \psi(x + at) \quad\text{......................} \quad (4)$$

where $a^2 = \dfrac{T}{\mu}$, and ϕ, ψ are arbitrary functions. The function $\phi(x - at)$ denotes any form of the string which advances in the direction of positive x with the
¶ velocity a, but without any other change; and the function $\psi(x + at)$ denotes a similar form proceeding with the same velocity in the direction of negative x. For any given value of the time t we must suppose both functions to be given from $x = -\infty$ to $x = +\infty$, and then the motion of the string is determined.

The determination of the motion of a plucked string will result in this second form of solution, from determining the functions ϕ and ψ, so that
1) for the values $x = 0$ and $x = l$, the value of y for any value of t will be constantly $= 0$. This will be the case, if for any value of t,

$$\phi(-at) = -\psi(+at) \quad\text{.........................} \quad (4a)$$

and
$$\phi(l - at) = -\psi(l + at) \quad\text{........................} \quad (4b)$$

If in the first equation we put $at = -v$, and in the second $l + at = -v$, we obtain

$$\phi(v) = -\psi(-v)$$

¶ and
$$\phi(2l + v) = -\psi(-v)$$

so that
$$\phi(2l + v) = \phi(v) \quad\text{.................................} \quad (5)$$

Hence the function ϕ is periodic, for its value becomes the same when its argument is increased by $2l$. The same results for ψ.

2) For $t = 0$, we must have $\dfrac{dy}{dt} = 0$ between the limits $x = 0$ and $x = l$. Hence writing $\psi'(v)$ for $\dfrac{d\psi(v)}{dv}$, and putting $\dfrac{dy}{dt} = 0$ in equation (4), we obtain

$$\phi'(x) = \psi'(x)$$

And integrating this with respect to x, we have

$$\phi(x) = \psi(x) + C$$

¶ Now since neither y nor $\dfrac{dy}{dt}$ are altered by adding the same constant to ϕ and subtracting it from ψ, the constant C is perfectly arbitrary, and we may consequently assume it to be $= 0$, and hence write

$$\phi(x) = \psi(x) \quad\text{...............................} \quad (5a)$$

3) Since finally at the time $t = 0$, and within the limits $x = 0$, $x = l$, the magnitude

$$y, \text{ which is} = \phi(x) + \psi(x) = 2\phi(x),$$

must have the value shewn in fig. 18 A (p. 54a), the ordinates of this figure immediately give the value of $2\phi(x)$ and of $2\psi(x)$, by means of equation (5) :—

between $x = 0$ and $x = l$
,,　　$x = 2l$,, $x = 3l$
,,　　$x = 4l$,, $x = 5l$

and so forth.

But since from (4a, 4b, 5) it follows that $\phi(-v) = -\phi(v)$, and $\phi(l-v) = -\phi(l+v)$, the value of $2\phi(x)$ is given by the triangle in fig. 18 G (p. 54*b*),

between $x = -l$ and $x = 0$
„ $x = -3l$ „ $x = -2l$
and in the same way between $x = l$ „ $x = 2l$
„ $x = 3l$ „ $x = 4l$

and so forth.

By this means the functions ϕ and ψ are completely determined, and on supposing that the two wave-lines proceed in opposite directions with the velocity a, we obtain the forms of the string given in fig. 18, p. 54*a*, *b*, which represent the changes of the string for every twelfth part of the periodic time of its vibration.

[See Donkin's *Acoustics*, Chaps. V. and VI.]

¶

APPENDIX IV.

ON THE PRODUCTION OF SIMPLE TONES BY RESONANCE.

(See pp. 55*a* and 69*c*.)

I HAVE given the theory of tubes and hollow spaces filled with air, so far as it can be at present mathematically expressed, in my paper, entitled 'The Theory of Aerial Vibrations in Tubes with Open Ends' (*Theorie der Luftschwingungen in Röhren mit offenen Enden*), in Crelle's *Journal für Mathematik*, vol. lvii. A comparison of the upper partial tones of tuning-forks and the corresponding resonance tubes, will be found in my paper, 'On Combinational Tones' (*Ueber Combinationstöne*), in Poggendorff's *Annalen*, vol. xcix. pp. 509 and 510.*

I add here the dimensions of the resonance tubes mentioned on p. 54*a*, which were made for me by Herr Fessel, in Cologne, in connection with the tuning-forks ¶ kept in motion by electricity as described in Appendix VIII. These were cylindrical tubes of pasteboard, with terminal surfaces of zinc plate, one entirely closed, the other provided with a circular opening. These tubes therefore had only *one* opening, not two like the resonators which were intended for insertion in the ear. A resonance tube of this kind can have its tone flattened by diminishing its opening. To sharpen the tone, when necessary, I threw in a little wax, and placed the closed end of the tube on a warm stove or hob, until the wax was melted, and uniformly distributed over the surface. It was then allowed to cool in the same position. To try whether the tone of a tube is a little sharper or flatter than that of the fork, cover its opening slightly while the excited fork is held before it. If the covering strengthens the resonance the tube was too sharp. But if the resonance begins to decrease decidedly as soon as any part of the opening is covered, the tube was too flat. The dimensions in millimetres [and inches] are as follows :—

No.	Pitch	Length of Tube		Diameter of Tube		Diameter of Opening	
1	Bb	425	[16·73]	138	[5·43]	31·5	[1·24]
2	bb	210	[8·27]	82	[3·23]	23·5	[·93]
3	f'	117	[4·61]	65	[2·56]	16	[·63]
4	$b'b$	88	[3·46]	55	[2·17]	14·3	[·56]
5	d''	58	[2·28]	55	[2·17]	14	[·55]
6	f''	53	[2·09]	44	[1·73]	12·5	[·49]
7	$a''b$	50	[1·97]	39	[1·54]	11·2	[·44]
8	$b''b$	40	[1·57]	39	[1·54]	11·5	[·45]
9	d'''	35	[1·38]	30·5	[1·20]	10·3	[·41]
10	f'''	26	[1·02]	26	[1·02]	8·5	[·34]

¶

The theory of the sympathetic resonance of strings is best developed by means

* The harmonic upper partials of the air vibrating in the neighbourhood of a tuning-fork, there mentioned, have also been observed with an interference apparatus by Herr Stefan (*Proceedings of the Vienna Academy*, vol. lxi. part 2, pp. 491-8) and by Herr Quincke (Poggendorff's *Annals*, vol. xxviii.).

of the experiments mentioned on p. 55c. Retain the notations of Appendix III. and assume that the end of the string for which $x = 0$, is connected with the stem of the tuning-fork, and must move in the same way, and that its motion is given by the equation

$$y = A \, . \, \sin mt, \quad \text{for } x = 0 \quad \dots\dots\dots\dots\dots\dots \quad (6)$$

Suppose the other end of the string to rest on the bridge which stands on the sounding board. The following forces act upon the bridge :—

1) The pressure of the string, which will increase or diminish according to the angle under which the extremity of the string is directed against the bridge. The tangent of this angle between the variable direction of the string and its position of rest is $\dfrac{dy}{dx}$, and hence we can put the variable pressure of the string $= - \, T \, . \, \dfrac{dy}{dx}$, for $x = l$, supposing the bridge to lie on the side of negative y.

2) The elastic force of the sounding board, which acts to bring the bridge back into its position of rest, may be put $= -f^2 y$.

3) The sounding board, which moves with the bridge, is resisted by the air, to which it imparts some of its motion. The resistance of the air may be considered to be approximatively proportional to the velocity of its motion, and hence be put $= - \, g^2 \dfrac{dy}{dt}$.

Then putting M for the mass of the bridge, we obtain the following equation for the motion of the bridge, and hence for that of the extremity of the string which rests upon it :

$$M \, . \, \frac{d^2 y}{dt^2} = - \, T \, . \, \frac{dy}{dx} - f^2 y - g^2 \, . \, \frac{dy}{dt}, \quad \text{for } x = l \, \dots\dots\dots \quad (6a)$$

For the motion of the other points in the string, we have, as in Appendix III., the condition

$$\mu \, . \, \frac{d^2 y}{dt^2} = T \, . \, \frac{d^2 y}{dx^2} \dots\dots\dots\dots\dots\dots\dots\dots\dots\dots \quad (1)$$

Since part of every motion of the string must be constantly given off to the air in the resonance chamber, the motion would gradually die away if it were not kept up by some continuous cause. Hence we may neglect the variable initial conditions of the motion, and proceed at once to determine the periodic motion, which finally remains constant under the influence of the periodic agitation of the tuning-fork. It is manifest that the period of the motion of the string must be the same as the period of the motion of the fork. Hence the required integral of (1) must be of the form

$$\left. \begin{aligned} y = D \, . \, \cos px \, . \, \sin mt + E \, . \, \cos px \, . \, \cos mt \\ + F \, . \, \sin px \, . \, \sin mt + G \, . \, \sin px \, . \, \cos mt \end{aligned} \right\} \dots\dots\dots\dots \quad (7)$$

And to satisfy equation (1) we must then have

$$\mu m^2 = T p^2 \dots\dots\dots\dots\dots\dots\dots\dots\dots\dots \quad (7a)$$

From the equation (7) we have, when $x = 0$,

$$y = D \, . \, \sin mt + E \, . \, \cos mt,$$

and on comparing this with equation (6) we find

$$D = A, \text{ and } E = 0 \dots\dots\dots\dots\dots\dots\dots\dots \quad (8)$$

The two other coefficients of the equation (7), namely F and G, must be determined by means of equation (6a). On substituting in (6a) the values of y from (7), the equation (6a) splits into two, as we must put the sum of the terms multiplied by $\sin mt$ separately $= 0$, and also the sum of those multiplied by $\cos mt$ separately $= 0$. These two equations are :

$$\left. \begin{aligned} F \, . \, [(f^2 - Mm^2) \, . \, \sin pl + pT \, . \, \cos pl] - Gmg^2 \, . \, \sin pl \\ = -A \, . \, [(f^2 - Mm^2) \, . \, \cos pl - pT \, . \, \sin pl] \\ Fmg^2 \, . \, \sin pl + G \, . \, [(f^2 - Mm^2) \, . \, \sin pl + pT \, . \, \cos pl] \\ = -Ag^2 m \, . \, \cos pl \end{aligned} \right\} \dots\dots \quad (8a)$$

Assume for abbreviation

$$\left.\begin{array}{c} \dfrac{pT}{f^2 - Mm^2} = \tan k \\ (f^2 - Mm^2)^2 + p^2 T^2 = C^2 \end{array}\right\} \dotfill \text{(8b)}$$

Then the values of F and G will be as follows:

$$\left.\begin{array}{l} F = -\dfrac{A}{2} \cdot \dfrac{C^2 \cdot \sin 2\,(pl + k) + g^4 m^2 \cdot \sin 2pl}{C^2 \cdot \sin^2\,(pl + k) + g^4 m^2 \cdot \sin^2 pl} \\ G = -A\ \dfrac{Cmg^2 \cdot \sin k}{C^2 \cdot \sin^2\,(pl + k) + g^4 m^2 \cdot \sin^2 pl} \end{array}\right\} \dotfill \text{(8c)}$$

Putting the amplitude of the vibration of the extremity of the string which rests upon the bridge $= V$, equation (7) becomes

$$V^2 = [F \cdot \sin pl + A \cdot \cos pl]^2 + G^2 \cdot \sin^2 pl, \qquad\qquad ¶$$

and on putting in this equation the values of F and G from (8c) we find

$$V = \dfrac{AC \cdot \sin k}{\sqrt{[C^2 \cdot \sin^2\,(pl + k) + g^4 m^2 \cdot \sin\,^2 pl]}} \dotfill \text{(9)}$$

The numerator in this expression is independent of the length of the string. Any alteration of its length therefore affects the denominator only. Under the radical sign is the sum of two squares, which can never $= 0$, because m, g, p, T, and hence k, can never $= 0$. The coefficient of the resistance of air, g, must certainly be considered as infinitesimal. Hence the denominator is a maximum, and V is a minimum, when $\sin (pl + k) = 0$, or when

$$pl = v\pi - k \dotfill \text{(9a)}$$

where v is any whole number. The maximum value of V is　　　　　　　　　¶

$$V_M = \frac{AC}{g^2 m}.$$

Hence, other circumstances being the same, this maximum value increases as g, the coefficient of the resistance of the air, decreases, and as C increases. To ascertain the circumstances on which the magnitude of C depends, put for p^2 in the second of the equations (8b), which defines the meaning of C, its value from (7a), and also put $n^2 = \dfrac{f^2}{M}$; this gives

$$C^2 = M^2 \cdot (n^2 - m^2)^2 + T\mu m^2.$$

Now n is the number of vibrations which the bridge would perform in 2π seconds, under the influence of the elastic sounding board alone, without the string and the resistance of the air; and m is the same number of vibrations for the tuning-fork. Hence the maximum value of V can now be written　　　¶

$$V_M = \frac{A}{g^2} \sqrt{\left[M^2 \cdot \left(1 - \frac{n^2}{m^2}\right)^2 + T\mu \right]}$$

in which everything is reduced to the weights M, T, μ and the magnitude of the interval $1 - \dfrac{n}{m}$.

If $m > n$, which is usually the case, it is advantageous to make the weight of the bridge M, rather large. Hence I have had it constructed of a plate of copper. If M is very large, k will be very small in consequence of (8b), and then the equation (9a) shews that the various tones of greatest resonance approach all the more nearly to those which correspond with the series of simple whole numbers. The heavier the bridge the sharper the boundaries of the tones of the string.

Observe that the rules here given for the influence of the bridge hold only for the assumption that the string is excited by a tuning-fork, and not, so far as this investigation extends, for other cases.

APPENDIX V.

ON THE VIBRATIONAL FORMS OF PIANOFORTE STRINGS.

(See pp. 74c to 80b.)

WHEN a stretched string is struck by a perfectly hard and narrow metal point, which is immediately withdrawn, the blow conveys a certain velocity to the point struck, while the rest of the string receives no motion. Let the moment of impact correspond to $t = 0$; the motion of the string can then be determined on the condition that at that moment the string as a whole was in its position of equilibrium, and that it was only the point struck that had any velocity. Hence in equations

¶ (1c) and (1d) of Appendix III. (p. 375b) put both $y = 0$ and $\dfrac{dy}{dt} = 0$ for $t = 0$, at all

points except that which is struck, for which suppose the co-ordinate to be a.

Hence it follows that in those equations

$$0 = A_1 = A_2 = A_3 = \&c.,$$

and the values of B are determined by an integration similar to that in (2b), p. 375d, giving

$$2\pi n m B_m \cdot \int \sin^2 \frac{m\pi x}{l} \cdot dx = \int_0^l \frac{dy}{dt} \cdot \sin \frac{m\pi x}{l} \cdot dx,$$

and
$$\pi n m l B_m = c \cdot \sin \frac{m\pi a}{l},$$

where c is the product of the velocity imparted to the struck portion of the string and of its infinitesimal length. Consequently

¶
$$y = \frac{c}{\pi n l} \cdot \left(\sin \frac{\pi a}{l} \cdot \sin \frac{\pi x}{l} \cdot \sin 2\pi n t \right.$$

$$\left. + \frac{1}{2} \cdot \sin \frac{2\pi a}{l} \cdot \sin \frac{\pi x}{l} \cdot \sin 4\pi n t + \frac{1}{3} \cdot \sin \frac{3\pi a}{l} \cdot \sin \frac{3\pi x}{l} \cdot \sin 6\pi n t + \&c. \right)$$

and
$$B_m = \frac{c}{\pi n l m} \cdot \sin \frac{m\pi a}{l} \dots\dots\dots\dots\dots\dots\dots\dots \quad (10)$$

The mth partial tone of the string, therefore, disappears in this case also when it is struck in a node of this string. Also the upper partial tones are stronger in comparison with the prime tone, than when the string is plucked, because the value of A_m in equation (3), p. 375d, has m^2 as a divisor, whereas the value of B_m in (10) has only m as a divisor. This is immediately confirmed by experiment, on striking the strings with the sharp edge of a metal ruler.

For a pianoforte, the discontinuity in the motion of the string is diminished by covering the hammer with an elastic pad. This sensibly diminishes the force of

¶ the higher upper partials, because the motion is no longer conveyed to a single point, but is imparted to a sensible length of string, and this too, not in an indivisible moment of time, as would be the case for a blow with a perfectly hard body. On the contrary, the elastic pad yields to the blow at first, and then recovers itself, so that while the hammer is in contact with the string, the motion is capable of extending over a considerable length. An exact analysis of the motion of a string excited by the hammer of a pianoforte would be rather complicated. But observing that the string moves but very slightly from its position of rest, and that the elastic pad of the hammer is very yielding and admits of much compression, we may simplify the mathematical theory, by assuming the pressure exerted by the hammer during the blow which it gives to the string to be as great as it would be if the string were a perfectly fixed and perfectly unyielding body. We are then able to assume the pressure of the hammer to be

$$P = A \sin mt,$$

for such moments of time that $0 < t < \dfrac{\pi}{m}$. This last magnitude $\dfrac{\pi}{m}$ is the length of time during which the hammer is in contact with the string. The magnitude of m increases therefore as the elastic power of the hammer increases and its weight decreases.

We have first to determine the motion of the string during the interval of time that the hammer is in contact with it, that is, from $t = 0$ to $t = \frac{\pi}{m}$. During this time, the hammer divides the string into two sections, and the motion of each section has to be separately determined. At the place of impact let x be written x_0. When $x < x_0$, distinguish the values of y by writing them y_1, and when $x > x_0$, by writing them y^1. At the point struck the pressure of the string against the hammer must be equal to the pressure P, which the hammer exerts against the string. The pressure of the string is to be calculated as in Appendix IV., equation (6a) (p. 378b), and we consequently obtain the equation

$$P = A \cdot \sin mt = T \cdot \left(\frac{dy_1}{dx} - \frac{dy^1}{dx} \right) \quad \dots \dots \dots \dots \dots \dots \dots \dots \dots \quad (11) \qquad ¶$$

Waves proceed towards both ends of the string from the place struck. Hence y_1 must have the form

$$y_1 = \phi \, (x - x_0 + at)$$

for values of t, for which $0 < t < \frac{\pi}{m}$, and $x_0 > x > x_0 - at$, and y^1 must have the form

$$y^1 = \phi \, (x_0 - x + at)$$

for the same values of t, and for values of x for which $x_0 < x < x_0 + at$. Using ϕ' for the differential coefficient of the function ϕ, equation (11) gives

$$P = A \cdot \sin mt = 2 \, T \cdot \phi' \, (at) \quad \dots \dots \dots \dots \dots \dots \dots \dots \dots \dots \quad (11a)$$

Integrating with respect to t we find   ¶

$$C - \frac{A}{m} \cdot \cos mt = \frac{2T}{a} \cdot \phi \, (at),$$

and then, determining the constant C, so that $y_1 = 0$ when $x = x_0 + at$, and $y^1 = 0$ when $x = x_0 - at$, we have

$$y_1 = \frac{aA}{2mT} \cdot \left\{ 1 - \cos \left[\frac{m}{a} (x - x_0) + mt \right] \right\},$$

$$y^1 = \frac{aA}{2mT} \cdot \left\{ 1 - \cos \left[\frac{m}{a} (x_0 - x) + mt \right] \right\}.$$

This determines the motion of the string for the time t, when $0 < t < \frac{\pi}{m}$, and on the supposition that the two waves proceeding from the place of impact have not reached one of the ends of the string. If the latter had been the case, they ¶ would have been reflected there.

When at has become greater than $\frac{\pi}{m}$, the pressure P will be $= 0$, and hence it follows from equation (11a) that from thenceforward

$$\phi' \, (at) = 0, \text{ and hence } \phi = \text{constant, when } at > \frac{\pi}{m}.$$

Hence both y_1 and y^1 remain $= \frac{aA}{mT}$ for all those parts of the string over which the waves have already advanced, until portions of the waves reflected from the extremities reach those parts of the string on their return.

To introduce the influence of the extremities of the string properly into calculation, suppose the string to be infinitely long, and that at all points distant from x_0 by multiples of $2l$, similar blows are given to it, so that from all these places waves proceed similar to those which proceed from x_0. Moreover suppose that in all those places for which $x = -x_0 \pm 2\nu l$, a blow be applied equal to that given to x_0 and at the same time, but in the opposite direction, so that from all these latter points waves will proceed of an identical form, but with a negative height. Those points of the infinite string which correspond with the extremities of the finite string will then be agitated by positive and negative waves of equal magnitude at the same time, and will hence be completely at rest, and consequently all the conditions of the real finite string will be fulfilled by the state of this section of the infinite string.

From the moment that the hammer quits the string, the motion of the string may be regarded as two systems of waves, one advancing (or in the direction of positive x), and the other retreating (or in the direction of negative x). Of these systems of waves we have as yet found only certain isolated portions, namely those which correspond to the sections of the string lying nearest the point struck. We have now to complete these waves properly and obtain a connected advancing ¶ and retreating system.

Advancing in the direction of the positive x on the string, we have $y = 0$ until we come to a positive retreating wave, and then it rises to $\dfrac{aA}{mT}$, which is its value in the positive striking points. If we proceed beyond the striking point, and over the wave thence proceeding, we again find values of y which $= 0$, and sink to $-\dfrac{aA}{mT}$ as soon as the first negative retreating wave has been passed over. This is the value of y in the first negative striking point. To connect the positive and negative retreating waves properly with each other, we must suppose the values of y_1 to be increased between every positive striking point and the next following negative striking point, by the magnitude $+\dfrac{aA}{mT}$, so that the height of the wave retains this value, which it had at x_0, until the corresponding negative wave begins. ¶ Here then the height of the wave becomes $\dfrac{aA}{2mT} - y_1$ and sinks to zero. Similarly, suppose that $-\dfrac{aA}{mT}$ is added to the height of the wave in advancing waves between any negative striking point and the next following positive striking point. The retreating waves will thus be everywhere positive, and the advancing waves everywhere negative. and the waves at the same time are so constituted that their continued motion will generate that kind of motion which we have found to exist in the string after the hammer quits it.

We have now to express this system of waves as the sum of simple waves. The length of the wave is $2l$, because the points of simultaneous impact lie at intervals of $2l$. Let us take the positive retreating waves at the time $t = \dfrac{\pi}{m}$ then

¶

$$1)\ y_1 = 0,\ \text{from}\ x = 0\ \text{to}\ x = x_0 - \frac{a\pi}{m}$$

$$2)\ y_1 = \frac{aA}{2mT} \cdot \left\{ 1 + \cos\left[\frac{m}{a}(x - x_0)\right] \right\}$$

$$\text{from}\ x = x_0 - \frac{a\pi}{m}\ \text{to}\ x = x_0$$

$$3)\ y_1 = \frac{aA}{mT},\ \text{from}\ x = x_0\ \text{to}\ x = 2l - x_0 - \frac{a\pi}{m}$$

$$4)\ y_1 = \frac{aA}{2mT} \cdot \left\{ 1 - \cos\left[\frac{m}{a}(2l - x_0 - x)\right] \right\}$$

$$\text{from}\ x = 2l - x_0 - \frac{a\pi}{m}\ \text{to}\ x = 2l - x_0$$

$$5)\ y_1 = 0,\ \text{from}\ x = 2l - x_0\ \text{to}\ x = 2l.$$

Hence if we assume

$$y_1 = A_0 + A_1 . \cos \frac{\pi}{l} (x + c) + A_2 . \cos \frac{2\pi}{l} (x + c) + A_3 . \cos \frac{3\pi}{l} (x + c) + \&c.$$

$$+ B_1 . \sin \frac{\pi}{l} (x + c) + B_2 . \sin \frac{2\pi}{l} (x + c) + B_3 . \sin \frac{3\pi}{l} (x + c) + \&c. \quad (12)$$

we shall have

$$\int_0^{2l} y_1 . \cos \frac{n\pi}{l} (x + c) . dx = A_n l,$$

$$\int_0^{2l} y_1 . \sin \frac{n\pi}{l} (x + c) . dx = B_n l.$$

If we put $c = \dfrac{a\pi}{2m}$, every B becomes $= 0$, because y_1 has the same values for $\dfrac{a\pi}{2m} + \xi$ and $\dfrac{a\pi}{2m} - \xi$, and the limits of the integration may be selected at pleasure ¶ provided only that their difference is $2l$. But on the other hand

$$A_n = -\frac{2aAml^2}{Tn\pi . (n^2\pi^2a^2 - m^2l^2)} . \sin \left(\frac{n\pi}{l} . x_0 \right) . \cos \left(\frac{n\pi}{l} . \frac{a\pi}{2m} \right) \quad \ldots\ldots (12a)$$

This equation gives the amplitude A_n of the several partial tones of the compound tone of the string which has been struck. When the point of impact is a node of the nth partial, the factor $\sin \left(\dfrac{n\pi}{l} . x_0 \right)$ will $= 0$, and hence all those partial tones disappear which have a node at the point of impact. The table on p. 79c was calculated from this equation.*

To determine the complete motion of the string we must further substitute $x + at$ for x in the equation (12) for y_1. The corresponding expression for y^1 then becomes

$$y^1 = -A_0 - A_1 . \cos \frac{\pi}{l} (x + at - c) - A_2 . \cos \frac{2\pi}{l} (x - at - c) - \&c.$$ ¶

and finally

$$y = y_1 + y^1 = 2A_1 . \cos \frac{\pi}{l} x . \cos \frac{\pi}{l} (at + c) + 2A_2 . \cos \frac{2\pi}{l} x . \cos \frac{2\pi}{l} (at + c) + \&c.$$

which completely solves the problem.

If m be infinite, that is, if the hammer be perfectly hard, the expression for A_n in (12a) becomes identical with that of B_m in equation (10), p. 380c. It must be remembered that m in (10) is identical with n in (12a) (and that a in (10) is then identical with x_0 in (12a), while a in (12a) has a different meaning).

If m is not infinite, as n increases the coefficients A_n decrease as $\dfrac{1}{n^3}$, but if m be infinite they decrease as $\dfrac{1}{n}$; for plucked strings they decrease as $\dfrac{1}{n^2}$. This corresponds to the theorems proved by Stokes (*Cambridge Transactions*, vol. viii. pp. 533 to 584) concerning the effect of the discontinuity of a function, when ¶ developed in Fourier's series, upon the magnitude of the terms with high ordinal numbers. Thus, if y is the function to be developed in a series

$$y = A_0 + A_1 . \sin (mx + c_1) + A_2 . \sin (2mx + c_2) + \&c.$$

the coefficient of A_n when n is very great,

1) is of the order $\dfrac{1}{n}$ when y itself suddenly alters;

---

* [It is shewn in the notes on p. 76d' and p. 77c, that when the blow is made with an ordinary pianoforte hammer, the partial tone, corresponding to the node struck, does not wholly vanish. The subject is resumed in App. XX. sect. N. No. 2, where the result of later experiments is given. In the meantime it must be borne in mind that though the force of the corresponding partial is materially weakened, it is not absolutely extinguished. It may therefore be necessary to re-open the mathematical investigation without having recourse to the facilitation due to the fundamental (but certainly only approximative) assumption on p. 380d, giving $P = A \sin mt$.— *Translator*.]

2) is of the order $\dfrac{1}{n^2}$ when the first differential coefficient $\dfrac{dy}{dx}$ suddenly alters;

3) is of the order $\dfrac{1}{n^3}$ when the second differential coefficient $\dfrac{d^2y}{dx^2}$ suddenly alters;

4) is at most of the order e^{-n} when the function itself and all its differential coefficients are continuous. [See note, p. 35d.]

Hence follow the laws of musical tones so often mentioned in the text, that their upper partial tones generally increase in power, with the greater discontinuity of the corresponding motion of the resonant body.

¶ [See Donkin's *Acoustics*, pp. 119–126, where, on p. 124, equation (14) corresponds to equation (12a) above.]

APPENDIX VI.

ANALYSIS OF THE MOTION OF VIOLIN STRINGS.

(See p. 83a.)

ASSUME the lens of the vibration microscope to make horizontal vibrations, then vibrational curves will be observed like those represented in fig. 23, p. 82b, c. Call the vertical ordinate y and the horizontal x; then y is directly proportional to the displacement of the vibrating point, and x to that of the vibrating lens. The latter performs a simple pendular vibration. If the number of its vibrations be n ¶ and the time t, we have generally

$$x = A \cdot \sin\left(2\pi nt + c\right)$$

where A and c are constant.

Now if y also makes n vibrations, x and y are both periodic and have the same periodic time. Hence, at the end of each period, x and y have the same values as before, and the observed point is at exactly the same place as at the beginning of the period. This holds for every point in the curve and for every fresh repetition of the vibratory motion, so that the curve appears stationary.

Suppose a vibrational form of the kind depicted in figs. 5, 6, 7, 8, 9, pp. 20 and 21, in which the horizontal abscissa is directly proportional to the time, to be wrapped round a cylinder, of which the circumference is equal to a single period of those curves, so that the time t is now to be measured along the circumference of the cylinder. Call x the distance of a point from a plane drawn through the axis of the cylinder. Then in this case also

¶ $$x = A \cdot \sin\left(2\pi nt + c\right),$$

where $A \cdot \sin c$ is the value of x for $t = 0$, and A is the radius of the cylinder. Hence, if the curve drawn upon the cylinder be viewed by an eye at an infinite distance in the line $x = 0$, $y = 0$, the curve has exactly the same appearance as in the vibration microscope.

If x and y have not exactly the same period; if, for example, y makes n vibrations while x makes $n + \Delta n$, where Δn is a very small number, the expression for x may be written

$$x = A \cdot \sin\left[2\pi nt + (c + 2\pi t\,\Delta n)\right].$$

In this case, then, c which was formerly constant, increases slowly. But c represents the angle between the plane $x = 0$ and the point in the drawing for which $t = 0$. In this case, then, the imaginary cylinder round which the drawing is supposed to be wrapped, revolves slowly.

Since a magnitude which is periodic after the period π, may be also considered as periodic after the periods 2π, or 3π, or $\nu\pi$, where ν is any whole number, these remarks apply also for the case when the period of y is an aliquot part of the period of x, or conversely, or both are aliquot parts of the same third period, that is, for the case when the tones of the tuning-fork and of the observed body stand in any consonant ratio; the only limitation is that the common period of vibration must not exceed the time required for a luminous impression to become extinct in the eye. [See Donkin's *Acoustics*, pp. 36-44.]

From the observed curves, fig. 23 B, C, p. 82b, and fig. 24 A, B, p. 83b, it follows that all points of the string ascend and descend alternately, that the ascent is made with a constant velocity, and also the descent with a constant velocity, which is however different from the velocity of ascent. When the bow is drawn across a node of one of the upper partials of the string, the motion takes place in all nodes of the same tone precisely in the manner described. For other points of the string, little crumples are perceptible in the vibrational figure, but they do not prevent us from clearly recognising the principal motion. ¶

FIG. 62.

A

If in fig. 62 we reckon the time from the abscissa of the point a, so that for a, $t = 0$, and further for the point β put $t = \tau$, and for the point γ put $t = T$, so that the last represents the length of a whole period; then

$$\left.\begin{array}{l} y = ft + h, \text{ from } t = 0 \text{ to } t = \tau; \\ y = g\,(T - t) + h, \text{ from } t = \tau \text{ to } t = T, \end{array}\right\} \quad\dots\dots\dots\dots\dots\dots\dots\dots\quad (1)$$

whence for $t = \tau$, it results that

$$f\tau = g\,(T - \tau).\qquad\qquad\qquad ¶$$

Now suppose y to be developed in one of Fourier's series:

$$y = A_1 . \sin \frac{2\pi t}{T} + A_2 . \sin \frac{4\pi t}{T} + A_3 . \sin \frac{6\pi t}{T} \&c.$$

$$+ B_1 . \cos \frac{2\pi t}{T} + B_2 . \cos \frac{4\pi t}{T} + B_3 . \cos \frac{6\pi t}{T} \&c.$$

then it results from integration that

$$A_n . \int_0^T \sin^2 \frac{2n\pi t}{T} . dt = \int_0^T y . \sin \frac{2n\pi t}{T} . dt$$

$$B_n . \int_0^T \cos^2 \frac{2n\pi t}{T} . dt = \int_0^T y . \cos \frac{2n\pi t}{T} . dt, \qquad ¶$$

and this gives the following values for A_n and B_n:

$$A_n = \frac{(g + f) . T}{2n^2\pi^2} . \sin \frac{2n\pi\tau}{T}$$

$$B_n = -\frac{(g + f) . T}{2n^2\pi^2} . \left\{ 1 - \cos \frac{2n\pi\tau}{T} \right\}$$

and y may then be written in the form

$$y = \frac{(g + f) . T}{\pi^2} . \sum_{n=1}^{n=\infty} \left\{ \frac{1}{n^2} . \sin \frac{\pi n\tau}{T} . \sin \frac{2\pi n}{T}\left(t - \frac{\tau}{2}\right) \right\} \quad\dots\dots\dots\quad (2)$$

In equation (2), y denotes merely the distance of any determinate point of the string from its position of rest. If x denotes the distance of this point from the beginning of the string, and L the length of the string, then the general form of y, as in equation (1b) of App. III., p. 575b, is

$$y = \sum_{n=1}^{n=\infty} \left\{ C_n . \sin \frac{n\pi x}{L} . \sin \frac{2\pi n}{T} \left(t - \frac{\tau}{2} \right) \right\}$$

$$+ \sum_{n=1}^{n=\infty} \left\{ D_n . \sin \frac{2\pi x}{L} . \cos \frac{2\pi n}{T} \left(t - \frac{\tau}{2} \right) \right\} \quad \ldots\ldots\ldots\ldots\ldots (3)$$

or   By comparing equations (2) and (3) we find immediately that all D's vanish,

$$D_n = 0, \qquad \text{and}$$

$$C_n . \sin \frac{n\pi x}{L} = \frac{g+f}{\pi^2} . \frac{T}{n^2} . \sin \frac{n\pi\tau}{T} \quad \ldots\ldots\ldots\ldots\ldots\ldots (3a)$$

¶ Here $g + f$ and τ are independent of x, but not of n. On taking the equations for $n = 1$ and $n = 2$, and then, dividing one by the other, there results

$$\frac{C_2}{C_1} . \cos \frac{\pi x}{L} = \frac{1}{4} . \cos \frac{\pi\tau}{T}.$$

From which it follows that for $x = \dfrac{L}{2}$, τ is also $= \dfrac{T}{2}$, as observation shews.

But if $x = 0$, then according to observations τ is also $= 0$. Hence

$$C_2 = \frac{1}{4} . C_1, \quad \text{and} \frac{x}{L} = \frac{\tau}{T} \quad \ldots\ldots\ldots\ldots\ldots (3b)$$

so that $g + f$ is independent of x. Let v be the amplitude of the vibration of the point x in the string, then

¶

$$f\tau = g (T - \tau) = 2v,$$

$$g + f = \frac{2v}{\tau} + \frac{2v}{T-\tau} = \frac{2vT}{\tau(T-\tau)} = \frac{2vL^2}{Tx(L-x)}.$$

And since $g + f$ is independent of x, we must have

$$v = 4V . \frac{x(L-x)}{L^2}$$

where V is the amplitude in the middle of the string. From equation (3b) it follows that the sections $\alpha\beta$ and $\beta\gamma$ of the vibrational figure, fig. 62, p. 385b, must be proportional to the corresponding parts of the string on both sides of the observed point. Hence we have finally

¶

$$y = \frac{8V}{\pi^2} \sum_{n=1}^{n=\infty} \left\{ \frac{1}{n^2} . \sin \frac{n\pi x}{L} . \sin \frac{2\pi n}{T} \left(t - \frac{\tau}{2} \right) \right\} \quad \ldots\ldots (3c)$$

for the complete expression of the motion of the string.

If we put $t - \dfrac{\tau}{2} = 0$, y will $= 0$ for all values of x, and hence all parts of the string pass through their position of rest simultaneously. From that time the velocity f of the point x is

$$f = \frac{2v}{\tau} = \frac{8V(L-x)}{LT}.$$

But this velocity lasts only during the time τ, where $\tau = \dfrac{Tx}{L}$. Hence after the time t, and

as long as $t < \dfrac{Tx}{L}$, we have $y = ft = \dfrac{8V}{LT} . (L - x) t \ldots\ldots\ldots (4)$

and hence $y < \dfrac{8V}{L^2} . x (L - x)$.

From that point y returns with the velocity

$$g = \frac{2v}{T-\tau} = \frac{8Vx}{LT}.$$

And hence after the time $t = \tau + t_1$,

$$y = \frac{8V}{L^2}.x(L-x) - \frac{8Vx}{Lt}.t_1.$$

And since

$$L - x = \frac{T-\tau}{T}.L$$

we find

$$y = \frac{8Vx}{LT}.\left\{ T - (\tau + t_1) \right\}$$

$$= \frac{8Vx}{LT}.(T-t) \quad \text{.................... ....} \quad (4a) \quad ¶$$

The deflection on one part of the string is therefore given by the equation (4), and on the other part by the equation (4a). Both equations show that the form of the string is a straight line, which in (4) passes through $x = L$, and in (4a) through $x = 0$. These are the two extremities of the string. The point where these straight lines intersect is given by the condition

$$y = \frac{8V}{LT}.(L-x)\,t = \frac{8V}{LT}.x(T-t).$$

Whence

$$(L-x)\,t = x(T-t)$$

and

$$Lt = xT.$$

Hence the abscissa x of the point of intersection increases in proportion to the time. This point of intersection, which is at the same time the point of the string most remote from the position of rest, passes, therefore, with a constant velocity from one end of the string to the other, and during this passage describes a ¶ parabolic arc, for which

$$y = v = \frac{8V}{L^2}.x(L-x).$$

Hence the motion of the string may be briefly thus described. In fig. 63 the foot d of the ordinate of its highest point moves backwards and forwards with a constant velocity on the horizontal line ab,

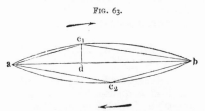

¶

while the highest point of the string describes in succession the two parabolic arcs ac_1b and bc_2a, and the string itself is always stretched in the two lines ac_1 and bc_1 or ac_2 and bc_2. [See Donkin's *Acoustics*, pp. 131–138.]

The small crumples of the vibrational figures which are so frequently observed, fig. 25, p. 84b, probably arise from the damping and disappearance of those tones which have nodes at the point bowed or in its immediate neighbourhood, and are consequently either unexcited or but slightly excited by the bow. When the bow is drawn across the string in a node of the mth partial tone situate near to the bridge, the vibrations of this mth, and further of the $2m$th, $3m$th, &c., tone have no influence on the motion of the point in the string touched by the bow, and they may consequently disappear, without changing the effect of the bow upon the string, and this really explains the crumples observed in the vibrational figure. [See also App. XX. sect. N. No. 5, on Prof. Mayer's Harmonic Curves.] I have not been able to determine by observation what happens when the bow is drawn across the string between two nodal points.

APPENDIX VII.

ON THE THEORY OF PIPES.

A. *Influence of Resonance in Reed Pipes.*

(See p. 102*b*.)

THE laws of resonance for cylindrical tubes have been developed mathematically in my paper on the 'Theory of Aerial Vibrations in Tubes with Open Ends' (*Theorie der Luftschwingungen in Röhren mit offenen Enden*, 'Journal für reine und angewandte Mathematik,' vol. lvii.). The example treated in § 7 of that paper is applicable to reed pipes, where the motion at the bottom of the pipe is assumed ¶ to be given. Let Vdt be the volume of air which enters the reed pipe in the infinitesimal time dt, then as this magnitude is periodical we can express V in one of Fourier's series, thus:—

$$V = C_0 + C_1 . \cos(2\pi nt + T_1) + C_2 . \cos(4\pi nt + T_2) + \&c. \dots\dots\dots \quad (1)$$

The resonance must be determined separately for each term, because the vibrations corresponding to each partial tone are superposed without modification. If we assume l to be the length of the tube, S its section, $l + a$ the reduced length of the tube (where in cylindrical tubes the difference a is equal to the radius multiplied by $\frac{\pi}{4}$ [but see p. 91*d*, note †]), k the magnitude $\frac{2\pi}{\lambda}$ (where λ is the length of the wave), and put the potential of the wave in free space, for the tone having the vibrational number νn,

¶
$$= \frac{M_\nu}{r} . \cos(\nu k r - 2\pi \nu n t + c),$$

where r is the distance from the middle of the opening; then the equations (15) and (12b) of the paper referred to, give

$$M_\nu = \frac{C_\nu}{\sqrt{(4\pi^2 \cos^2 \nu k (l + a) + \nu^4 k^4 S^2 \sin^2 \nu k l)}}.$$

Since the magnitude $k^2 S$ must always be considered as very small to make our theory applicable, this equation, for cases in which $l + a$ is not an uneven multiple of the length of a quarter of a wave, becomes approximatively,

$$M_\nu = \frac{C_\nu}{2\pi . \cos \nu k(l + a)}.$$

Hence the resonance is weakest when the reduced length of the tube is an even multiple of the length of a quarter of a wave, and becomes stronger as it ¶ approaches an uneven multiple of that length. When it absolutely reaches such a multiple the complete formula gives

$$M_\nu = \frac{C_\nu}{\nu^2 k^2 S}.$$

Hence the maximum of resonance increases as the length of the wave of the tone in question increases and the transverse section decreases. The smaller the transverse section, the more sharply defined is the limit of the pitch of the tone which is strongly reinforced by resonance; while when the transverse section is large, the reinforcement of resonance extends over a much greater length of the scale.

For hollow bodies of other shapes, with narrow mouths, similar equations may be obtained by means of the propositions given in § 10 of the paper cited.

Since the condition of powerful resonance is that $\cos \nu k (l + a) = 0$, cylindrical tubes (clarinet) reinforce only the prime and other unevenly numbered partial tones [but see note p. 99*c*].

In the interior of conical tubes we may assume the potential of the motion of the air for the tube n to be

$$V = \frac{A}{r} \cdot \sin(kr + c) \cos 2\pi nt,$$

where r is the distance from the vertex of the cone. If a vibrator is introduced at a distance a from the vertex, and l be the length of the tube, so that for the open end $r = l + a$, we may assume as an approximatively correct limiting condition for the free end, that the pressure there vanishes. This is the case when

$$\frac{dV}{dt} = -2\pi n \cdot \frac{A}{l+a} \cdot \sin[k(l+a)+c] \cdot \sin 2\pi nt = 0, \quad \text{and hence}$$

$$\sin[k(l+a)+c] = 0.$$

Hence we may assume

$$c = -k(l+a) \qquad \qquad \P$$

and

$$V = \frac{A}{r} \cdot \sin k(r-l-a) \cdot \cos 2\pi nt.$$

The most powerful resonance, then, in this case, as well as in cylindrical tubes, belongs to those tones which have a minimum velocity at the place where the vibrator is placed. For as during the development of velocity in the mouthpiece in equation (1) the coefficients C_ν have a determinate value which depends only on the motion of the vibrator and the pulses of air which it occasions, the coefficient A of the last equation must increase, as the velocity produced by the corresponding train of waves in the mouthpiece of the tube decreases. The velocity in the other parts of the tube will then increase. Now the velocity of a particle of air is

$$\frac{dV}{dr} = \frac{A}{r^2} \cdot \cos 2\pi nt \cdot [kr \cdot \cos k(r-l-a) - \sin k(r-l-a)].$$

Hence for maximum resonance we have the condition that for $r = a$ $\qquad$ $\P$

either $\quad kr = \tan k(r-l-a)$

or $\quad ka = -\tan kl.$

If, then, the magnitude a, that is, the distance of the vibrator from the vertex of the cone, be very small, ka and also $\tan kl$ will be very small, and $kl - \nu\pi$ must also be very small, if ν is any whole number. Hence we may develop the tangents according to powers of their arc, and retaining only the first term of this development we have

$$ka = \nu\pi - kl, \quad \text{and} \quad k(a+l) = \nu\pi$$

or, putting $\quad k = \dfrac{2\pi}{\lambda}$, we have

$$a + l = \nu \cdot \frac{\lambda}{2}.$$

$\qquad$ $\P$

This shews that conical tubes reinforce all those tones for which the whole length of the cone, reckoned up to its imaginary vertex, is a multiple of half the length of the wave, on the assumption that the distance of the vibrator from this imaginary vertex is infinitesimal in comparison with the length of the wave. Hence if the prime of the compound tone is reinforced by the tube, all the partial tones, both the evenly and unevenly numbered, will be reinforced up to a pitch where the wave-lengths of the higher partial tones cease to be very large in comparison with the distance a.*

Cæteris paribus, the number and magnitude of the terms of the series (1), which represents the exciting aerial motion will be the greater, the more perfectly the entering stream of air is interrupted. Free reeds must therefore fit their

* [The remainder of this Appendix VII. to p. 396b, is an addition to the 4th German edition. The additions on pp. 396-7 are principally from the 1st English edition.—*Translator.*]

frames very exactly, in order to produce a powerful tone. Striking reeds, which effect a more perfect stoppage, are superior in this respect. According to the information obtained by Mr. A. J. Ellis [see p. 95d', note †], organ-builders have really been more inclined in recent times to use striking reeds. But the vibrating laminæ are very slightly curved, so that they do not strike the frame all at once, but roll themselves gradually out upon it.

B. *Theory of the Blowing of Pipes.*

When longitudinal waves have once been excited in the mass of air in a tube, they may be reflected backwards and forwards many times between the ends of the tube, and form constant, periodically returning vibrations, before they die away.   At the closed end of a stopped pipe, the reflexion of every train of waves is tolerably complete, but at the open ends a perceptible fraction of the wave always passes into the open air, and hence the reflected wave has not the same ¶ intensity as the incident wave possessed.   Indeed the intensity of the waves reflected backwards and forwards in the tube continually diminishes, and finally dies off, if the lost work is not replaced at every backwards and forwards reflexion by some other kind of action.   What has to be replaced, however, is usually only a small part of the whole *vis viva* of the undulatory motion, that is, just as much as is lost by reflexion at the open ends.   If the inner radius at the open end of a cylindrical tube be R, the fraction of the amplitude which is lost at the open end for a tone having the wave-length λ, is, according to theory,

$$\frac{4\pi^2 R^2}{\lambda^2},$$

where R is small in comparison with λ.   In the pipes examined by Zamminer, the wave-length λ varied between $84R$ and $15\cdot6R$.   In the first case the loss would be $\frac{1}{200}$th, and in the latter about $\frac{1}{8}$th of the amplitude.

Now, this loss of *vis viva* can be replaced in various ways.   Supposing that ¶ the small volume dV, which was under the pressure p_0, were forced over into a space filled with air under the pressure p, the required work would be $(p - p_0)dV$. Hence if during the vibrations of sound, at those places and times where the air is condensed, either a small quantity of air is regularly forced in, or the pressure of the compressed air is increased by heating, this mass of air generates by its expansion a greater quantity of *vis viva* than was lost by its resistance to the condensation at the time the loss occurred.   The first of the two causes is effective in reed pipes, the second in the tubes of the Pyrophone [see App. XX. sect. N. No. 4], in which, together with the air which streams back into the tube, an increased quantity of gas is poured in from the gas tube, and this on burning increases the pressure during the time of re-expansion.

The conditions which must be fulfilled to make reed pipes speak were given by me in the ' Transactions of the Association for Natural History and Medicine ' (*Verhandlungen des naturhistorisch-medicinischen Vereins*) at Heidelberg (26 July 1861), and I take the liberty of reprinting this short explanation here with a few ¶ improvements.

I. *The Blowing of Reed Pipes.*

By a reed pipe I mean any kind of wind instrument in which the path of the stream of wind is alternately opened and closed by means of a vibrating elastic body.   The first work which made the mechanics of reed pipes accessible was that of W. Weber. But he experimented chiefly with metal reeds, which on account of their great mass and elasticity, could not be powerfully moved by the air unless the tone given by the pipe was not materially different from the proper tone of the reed independently of the pipe.   Hence pipes with metal reeds are usually capable of producing only a single tone, namely that one among those theoretically possible which is closest to the proper tone of the reed.

The case is different for reeds of light material which offers but little resistance, such as the cane reeds of the clarinet, oboe, bassoon, and the muscular reeds of the human lips in trumpets, trombones, and horns. Reeds of vulcanised india-rubber, placed similarly to the vocal chords in the larynx, are also well adapted for experiments; but, to make them speak easily and well, care must be taken to place them obliquely to the current of air (p. 97b).

The action of reeds differs essentially according as the passage which they close is opened when the reed moves against the wind towards the windchest, or moves with the wind towards the pipe. I shall say that the first *strike inwards*, and the second *strike outwards*. The reeds of the clarinet, oboe, bassoon, and organ all strike inwards. The human lips in brass instruments, on the other hand, are reeds striking outwards. The india-rubber reeds that I employ may be arranged to strike either inwards or outwards.

The laws for the pitch of reed pipes are completely found, when we have determined the motion of the reed as influenced by the alternating pressure of the air in the pipe and air chamber [see fig. 29, B, p p, on p. 96b]; remembering that the effluent air cannot attain its maximum velocity until the passage closed by the reed has been opened as widely as possible.

1) *Reeds with cylindrical pipe without air chamber.* The reed is regarded as ¶ a body which returns to its position of equilibrium by elastic forces, and is again brought out of that position by the pressure of the air in the pipe, which changes periodically with the sine of the time. The equations of motion* show that the instant of greatest pressure within the pipe must fall between a maximum displacement of the reed outwards, which precedes it, and a maximum displacement of the same inwards, which follows it. If the vibrational period be divided into 360°, like the circumference of a circle, the angle ϵ, by which the maximum pressure precedes the passage of the reed through its position of equilibrium, is given by the equation

$$\tan \epsilon = \frac{L^2 - \lambda^2}{\beta^2 L \lambda},$$

where L is the length of the wave of the reed in the air without the pipe, λ the wave-length of the tone which is actually produced, and β^2 a constant which is greater for reeds of light material and greater friction than for heavy and perfectly elastic materials. The angle ϵ must be taken between $-90°$ and $+90°$. ¶

In the same way we must determine the time at which the greatest pressure within the pipe separates from the greatest velocity. The latter must coincide with the position of the reed for which the opening is greatest. The calculation of this magnitude results from my investigations on the motion of air within an open cylindrical tube (*Journal für reine und angewandte Mathematik*, lvii.). The maximum of the velocity in the direction of the opening precedes the maximum of pressure by an angle δ (considering the vibrational period as the circumference of a circle) which is given by the equation

$$\tan \delta = \frac{-\lambda^2}{4\pi S} \sin \frac{4\pi(l+a)}{\lambda},$$

where S is the transverse section, l the length of the tube, and a a constant depending on the form of the opening, being 45° [but see note † and ‡ p. 91] for cylindrical tubes, of which the section has the radius ρ. The angle δ in this case is again to be taken between $-90°$ and $+90°$.

Now as air can only enter the pipe when the reed leaves the passage open, it ¶ follows that for reeds which *strike inwards* the maximum velocity of the air directed outwards must coincide with the maximum displacement of the reed inwards. Hence we must have

$$-\epsilon = \delta + \frac{1}{2}\pi,$$

and both δ and ϵ must be negative.

For reeds which *strike outwards*, on the other hand, the maximum effluence of the air must coincide with the maximum displacement of the reed outwards, and we must have

$$\frac{1}{2}\pi = \delta + \epsilon,$$

and both δ and ϵ must be positive.

* To be treated as in Seebeck's theory of sympathetic resonance, *Repertorium der Physik*, vol. viii. pp. 60–64. Also see equation 4c in the following App. IX. But the ϵ there is the complement of the ϵ here, and wave-lengths are here used instead of pitch numbers, as there.

Both cases are included in the equation

$$\tan \epsilon = \cot \delta$$

or

$$\sin \frac{4\pi(l+a)}{\lambda} = \frac{4\pi}{\lambda} S\beta^2 \cdot \frac{L}{\lambda^2 - L^2} \quad \dots\dots\dots\dots\dots \quad (\mathrm{I})$$

in which the reeds must strike inwards or outwards respectively, according as the quantities on each side of equation (1) are positive or negative.

Since S and β^2 are very small quantities, $\sin\{4\pi(l+a)\div\lambda\}$ cannot have any sensible value unless $\lambda^2 - L^2$ is very small, that is, unless the pitch of the pipe nearly coincides with that of the reed sounded separately from the pipe. This is generally the case with *metal reeds*. The value of λ is determined from equation (1).

On the contrary when the difference of the two tones $\lambda - L$ is great, then ¶ $\sin\{4\pi(l+a)\div\lambda\}$ must be very small, and hence approximatively

$$l + a = \nu \cdot \frac{\lambda}{4},$$

where ν is any whole number.

The alteration of pressure within the tube is proportional to $\sin\{2\pi(l+a)\div\lambda\}$, and hence is a minimum when

$$l + a = 2\nu \cdot \frac{\lambda}{4},$$

and a maximum when

$$l + a = (2\nu + 1) \cdot \frac{\lambda}{4}.$$

In the first case the force of the pressure of the air does not suffice to move the ¶ reed. In the second case it suffices when the reeds are not too heavy and have not too great a power of resistance. Hence the tones speak well for which approximatively

$$l + a = (2\nu + 1) \cdot \frac{\lambda}{4},$$

that is, for which the column of air in the pipe vibrates as in a stopped pipe. At the same time we see that these tones are almost independent of the proper tone of the reed.

Of this kind are the tones of the clarinet. Membranous reeds of india-rubber which strike inwards, attached to glass tubes 16 feet long, also speak easily and allow various upper partials to be produced which agree well with equation (1). The reeds which strike outwards must be tuned very low in order to give the pure tones of the tube. Hence the human lips are well adapted for this purpose, as the bundles of elastic fibres of which they are composed, are loaded with a ¶ large quantity of watery inelastic tissue [see footnote, p. 97d]. Cylindrical glass tubes may easily be blown as trumpets, and give the tones of a stopped pipe. Of these the upper tones, for which the difference $L^2 - \lambda^2$ is large, can be produced with firmness and correct intonation, but the lower tones, on the other hand, being not quite independent of the value of L, that is, of the tension and density of the lips, are uncertain and variable.

2) *Reeds with conical pipes without air chamber.* There is a remarkable difference between cylindrical and conical pipes. The motion of the air in the interior of the latter may be determined on the same principles that I have used for cylindrical pipes.

Put the potential of the motion of the air, inside the pipe, equal to

$$\frac{A}{r} \left\{ \sin\left[\frac{2\pi}{\lambda}(R - r + a)\right] \cdot \cos 2\pi nt \right\} + \frac{2\pi S}{\lambda^2} \cdot \cos \frac{2\pi}{\lambda}(R - r) \cdot \sin 2\pi nt,$$

where r is the distance from the vertex of the cone, R the value of r at the opening [or base] of the cone, S its section, a the difference between the true and reduced length, n the pitch number. Considering $a \div \lambda$ to be very small, and putting $R - r = l$, this gives

$$\tan \delta = \frac{-\lambda^2}{2\pi S} \cdot \sin \frac{2\pi(l + a)}{\lambda} \cdot \left[\cos \frac{2\pi(l + a)}{\lambda} + \frac{\lambda}{2\pi r} \sin \frac{2\pi(l + a)}{\lambda} \right]$$

in which l is to be referred to the place of the reed.

Here also we have to put

$$\cot \delta = \tan \epsilon.$$

We are at present chiefly interested in those tones of pipe which differ much from the tone of the reed, for which therefore $L^2 - l^2$ is great, $\tan \epsilon$ is also consequently very great, and $\tan \delta$ very small. For these then we must either have approximatively

$$\sin [2\pi(l + a) \div \lambda] = 0,$$

which gives no tone at all, because the alteration of pressure in the interior of the ¶ pipe is too weak, or else

$$\tan [2\pi(l + a) \div \lambda] = -2\pi r \div \lambda \dots\dots\dots\dots\dots \quad (2)$$

This is the equation for the powerful upper tones of the pipes.

Below is the series of tones calculated from equation (2) for a conical pipe of zinc of the following dimensions:

Length $l = 129 \cdot 7$ centimetres [$= 51 \cdot 06$ inch].
Diameter of the openings $5 \cdot 5$ and $0 \cdot 7$ ctm. [$= 2 \cdot 17$ and $\cdot 28$ inch].
Reduced length $l + a$, calculated $124 \cdot 77$ ctm. [$= 49 \cdot 12$ inch].

Approximate Tone	Wave-lengths calculated	Length of the corresponding open pipes stopped	
	centimetres	centimetres	centimetres
1. $B-$	$283 \cdot 61 =$	$\frac{2}{1} \times 141 \cdot 80$	$= \frac{4}{1} \times 70 \cdot 90$
2. $b-$	$139 \cdot 83 =$	$\frac{2}{2} \times 139 \cdot 84$	$= \frac{4}{3} \times 104 \cdot 88$
3. $f' \sharp$	$91 \cdot 81 =$	$\frac{2}{3} \times 137 \cdot 71$	$= \frac{4}{5} \times 114 \cdot 76$
4. $b' +$	$67 \cdot 94 =$	$\frac{2}{4} \times 135 \cdot 88$	$= \frac{4}{7} \times 118 \cdot 89$
5. $d'' \sharp +$	$53 \cdot 76 =$	$\frac{2}{5} \times 134 \cdot 39$	$= \frac{4}{9} \times 120 \cdot 95$
6. $g'' -$	$44 \cdot 40 =$	$\frac{2}{6} \times 133 \cdot 21$	$= \frac{4}{11} \times 122 \cdot 11$
7. $b'' b-$	$37 \cdot 79 =$	$\frac{2}{7} \times 132 \cdot 26$	$= \frac{4}{13} \times 122 \cdot 82$
8. $c''' -$	$32 \cdot 87 =$	$\frac{2}{8} \times 131 \cdot 50$	$= \frac{4}{15} \times 123 \cdot 28$
9. $d''' -$	$29 \cdot 22 =$	$\frac{2}{9} \times 131 \cdot 47$	$= \frac{4}{17} \times 124 \cdot 17$

The tones from the 2nd to the 9th could be observed, and were found to agree perfectly with the calculation. It appears from the last two columns that the higher tones were almost exactly those of a stopped pipe, the length of which is ¶ equal to the reduced length of the pipe $124 \cdot 77$ ctm., and that the deeper tones approach nearer to those of an open pipe, the length of which was that from the vertex to the foot of the cone. The reduced length of this would be $R + a = 142 \cdot 6$ ctm. [$= 56 \cdot 15$ inch]. The tones of brass instruments are usually assumed to be the same as those of an open pipe, but the higher tones of these instruments are relatively too sharp* for the lower ones, in the present case by more than half

* [The text has 'flat,' but this is against the figures. As it will appear that the notes assigned to the pitches in the first column are only roughly approximative, it is best to calculate out the intervals in cents, and assuming that the pitch varies inversely as the wave-length, we have in cents—

For tones . . .	1	2	3	4	5	6	7	8	9
Reed pipe . . .	0	1221	1953	2474	2879	3210	3489	3731	3935
Harmonics . . .	0	1200	1902	2400	2786	3102	3369	3600	3804
Difference . . .	0	21	51	74	93	108	120	131	131

a Tone.   In trumpets and horns this error is perhaps to some extent compensated by the cups of the mouthpiece.   In trombones the slides assist.*

Whereas trumpets, trombones, and horns belong to the reed pipes of this class with conical pipes, and deep reeds which strike outwards, oboes and bassoons have high reeds which strike inwards.   When strongly blown they also give the higher Octave and then the Twelfth, like an open pipe.   The calculation from equation (2) for the oboe agrees very well with Zamminer's measurements. [Zam. *ibid.* p. 306.]

II. *The Blowing of Flue Pipes.*

¶ In my memoir on 'The Discontinuous Motions of Fluids' (*Monthly Proceedings of the Academy of Sciences at Berlin*, April 23, 1868), I have described the mechanical peculiarities of such motions, and deduced from the theory how they are brought about by means of the blade-shaped current of air at the mouth of an organ pipe which is blown, as described on p. 92*a* to p. 93*a*.   The bounding surfaces of this current which cuts through and across the mass of air that runs into and out of the mouth of the pipe, are to be considered as *vortical surfaces*, that is, surfaces which are faced with a continuous stratum of vortical filaments or thread-like eddies.   Such surfaces have a very unstable equilibrium.   An infinitely extended plane surface uniformly covered with parallel straight vortical filaments might indeed continue stable ; but where the least flexure occurs at any time, the surface curls itself round in ever narrowing spiral coils, which continually involve more and more distant parts of the surface in their vortex.

shewing that the tones of the reed pipe are always too *sharp*, and not too *flat*, as the German text states, the sharpness being a comma for tone 2, a Quartertone for tone 3, $\frac{3}{4}$ Tone for 4, nearly a Semitone for 5, more than a Semitone for the rest, the last two being equally too sharp by about $1\frac{2}{3}$ Semitone. ¶ The misprint of $d'''\sharp$ for d''' in the German text

made the last tone appear to be a whole Tone too sharp.   In determining the notes in column 1, the Author has probably assumed the velocity of sound at 342 metres (which gives 1122 feet, or the velocity at about 60° F., see note p. 90*d*), and divided it by the wave-lengths reduced to metres.   This would give—

For tones . .	1	2	3	4	5	6	7	8	9
the pitch nos. .	120·6	244·6	372·5	503·4	636·2	770·3	905·0	1040·5	1170·4.
While pitch nos. .	124·6	249·2	373·5	498·4	628·0	781·6	940·0	1056·0	1184·2
belong to equally tempered notes	B	b	f'♯	b'	d''♯	g''	b''b	c'''	d'''

Whence it appears that g'' and $b''b$ are much, and c''', d''' somewhat, too sharp, so that the $d'''\sharp$ of the German text is a manifest error. This rough mode of comparing by vibrational

numbers, does not however convey a proper conception.   If we calculate the cents from *C*66, we find—

For tones . .	1	2	3	4	5	6	7	8	9
the cents . .	1044	2268	2996	3517	3923	4254	4533	4774	4978.
But cents . .	1100	2300	3000	3500	3900	4300	4600	4800	5000
belong to equally tempered notes	B	b	f'♯	b'	d''♯	g''	b''b	c'''	d'''
Differences .	− 56	− 32	− 4	+ 17	+ 23	− 46	− 67	− 26	− 22

And this shews how very rough are the approximations to the pitch which are made in the text by means of equally tempered notes.— *Translator.*]

* ['The conical tube examined by Prof. Helmholtz,' says Mr. Blaikley (see note, p. 97*d*), 'was not a perfect but a truncated cone, and any such would have its series of intervals intermediate between 1, 2, 3, 4, 5, &c., and 1, 3, 5, 7, 9, &c.; that is, a perfect cone, or one truncated to an infinitely small extent, would have the first, and an infinitely long cone (= a stopped cylindrical tube) would have the second.   Such a cone as Helmholtz describes is *not* a representative of the brass instrument family, for if cylindrical

tubing were added at the small end, the series with this added tube would not even be so near the theoretical 1, 2, 3, 4, 5, &c., as on the original cone.   There are brass instruments in which the series, so far from getting sharp on the higher tones, gets flat, 1, − 2, − 3, − 4, − 5, &c. Technically such instruments are said to be "sharp at the bottom."   In short, trumpets and trombones, &c., are not conical in the ordinary sense of the word, but have in most cases a cylindrical tube expanding into a bell by a line of increasing curvature, so that the boundaries are approximately hyperbolic.' MS. communication.   See also note †, p. 99*d*. —*Translator.*]

This inclination of the dividing surfaces of masses of air when moved discontinuously, to resolve themselves into vortices, can also be clearly seen on cylindrical streams of air, driven from cylindrical pipes and mixed with a little smoke to make them visible. In perfectly quiet air and under favourable conditions, they may reach a length of one foot to three feet. The least noise however makes them shrink up, as the vortices then commence close to their origin. Professor Tyndall has also observed and described a great number of similar phenomena of this kind, in burning gas jets.*

This resolution into vortices takes place in the blade of air at the mouth of the pipe, where it strikes against the lip. From this place on it is resolved into vortices, and thus mixes with the surrounding oscillating air of the pipe, and accordingly as it streams inwards or outwards, it reinforces its inward or outward velocity, and hence acts as an accelerating force with a periodically alternating direction, which turns from one side to the other with great rapidity. Such a blade of air follows the transversal oscillations of the surrounding mass of air ¶ without sensible resistance. During the phase of entrance of air, the blade is also directed inwards, and thus on its part reinforces the *vis viva* of the inward currents. Conversely, for the outward current.†

If we suppose the accelerating force of the current of air to be represented by one of Fourier's series, the amplitude of any term of the order m will in general diminish as $1 \div m$ (see p. 35d). In fact we require only to use the expression given in App. III. p. 375, equations (1b) and (3) for the displacement y of a plucked string, in order to find the value of $\frac{dy}{dx}$, for the time $t = 0$. We thus find the series which expresses the periodical alternation between a greater and smaller value of y, as shewn in fig. 19, p. 54c.

From my memoir 'On the Vibrations of Air in Tubes with Open Ends,' already cited (p. 388a), it follows that throughout the tube a positive component of pressure coincides with the maximum velocity in the direction of the opening, and when multiplied by such velocity this component has the value.   ¶

$$\frac{a\, A^2\, k^2\, S}{2\pi},$$

where

     a is the velocity of sound,
     A the maximum velocity at the end of the tube,
     S the transverse section of the cylindrical portion of the tube,
     $k = 2\pi \div \lambda$, λ being the length of the wave.

If, then, two trains of waves start from any transverse section in the directions of the two ends, and have the same velocity at that section, the above component of pressure must be directed in opposite ways in the two trains of waves. This holds for the place of blowing even when it is quite close to the end of the tube, so that one train of waves is infinitesimally short. Under these circumstances the acceleration produced by the air blown in, must correspond to twice that difference of pressure. Since A is the velocity at the opening, twice this difference of pres- ¶ sure for the mth tone, will be

$$\frac{a\, A_m\, 2\pi\, S m^2}{\lambda^2}.$$

This would be the only difference of pressure if the tone blown exactly corresponded to the proper tone of the tube. But it may be shewn that this cannot be made to agree with the mechanism of blowing, and there is always a length β which must be intercalated between the two trains of waves in order to reduce

---

* J. Tyndall *On Sound*, Lect. VI., also, in *Philosophical Magazine*, series iv. vol. xxxiii. pp. 92–99, and 375–391.

† The formation of this blade of air has been described by Messrs. Schneebeli (Pogg., *Ann.* cliii. p. 301), Sonreck (*ib.* clviii. p. 129), and Hermann Smith (*English Mechanics*' *Journal*, January, 1867; *Nature*, vol. viii. pp. 25, 45, 383, vol. x. pp. 161, 481, vol. xi. p. 325 [24 June 1875, p. 145; 27 April 1876, p. 511]). Herr Schneebeli also gave a mechanical explanation of the principal features of the process. [See the Translator's addition at the end of this Appendix, p. 396b.]

them to an accordant series of constant vibrations. In this case there is another additional difference of pressure equal to

$$-a\,A_m \sin \frac{2\pi m\beta}{\lambda}.$$

For the smaller numbers expressing the order, the sine may be replaced by the arc, and this latter term considered as the greater. Consequently the lower partials of the musical tone produced, allow the coefficient $m\,A_m$ to increase as $1 \div m$, that is A_m as $1 \div m^2$, and the higher partials allow A_m to increase as $1 \div m^3$. The velocities of the partial vibrations in more distant parts of the external free air contain the factor k once more than the velocities in the tubes (see equations 12g and 12h in my memoir). These will consequently increase as $1 \div m$ for lower values of m, which is also the case for the velocities of violin strings, but for higher values of m they decrease as $1 \div m^2$. The greater S is, the ¶ sooner will this more considerable decrease of the higher partials occur. It is for this reason especially that organ-builders compare the tones of narrow metal organ pipes with those of the violin and violoncello.

The circumstances which affect the blowing of pipes and the value of the magnitude β, require a more extended investigation, which I hope to be soon able to give elsewhere.

ADDITIONS BY TRANSLATOR.

[It may be convenient for those not conversant with mathematics to reproduce the account of the phenomena, which was given by me in pp. 708-711 of the 1st English edition. Herr Schneebeli had an experimental pipe constructed in the usual way; with glass back and a movable lip and slit or windway, through which was driven air impregnated with smoke, as is frequently done to make it visible. When ¶ he so placed the lip and slit that the stream of air passed *entirely outside* of the pipe, no sound occurred; but if he blew gently upon this sheet of air, at right angles, the pipe sounded, and the tone continued until he blew through the other end of the pipe; nevertheless under these circumstances it was very rare indeed to find that any smoke penetrated into the pipe. If the sheet of smoked air passed *entirely inside* of the pipe, there was also no sound; but then on blowing through the open end, so as to force some of it out, sound was produced, and it was stopped by blowing against the slit. This case was therefore the converse of the last. He concludes: 'That the stream of air which issues from the slit forms a species of air-reed (Luft-Lamelle, aerial lamina), and that this plays ¶ in the generation of vibrations in the mass of air, a part analogous to that of reeds in reed pipes.' He states that the vibrational nature of motion of the air between the slit and the lip can be shewn by attaching a piece of silk-paper to the edge of the lip or the split, and pressing a point against it. He further proposes a theory founded on Helmholtz's hydrodynamical investigations in the *Berichte der Berliner Akademie*, 1868, 23 April, Crelle 60, and states it to this effect: When the split is in the normal position the air-reed strikes the lip, a portion of the stream enters, and produces a compression as in reed pipes; the reaction of this compression affects the air-reed and bends it outwards; on the pressure ceasing the air-reed returns to its original position and the process begins afresh. Mr. Hermann Smith states that the air from the bellows is *not* directed 'against the

edge of the lip,' and that, if it were so directed, the pipe would not speak. He also states that the sharpness of the lip is immaterial to mere speaking, and that a pipe that speaks well may have the edge of its lip half an inch thick. (Compare suprà, p. 60c, where the wind which excites the sound in the bottle is blown *across* its mouth and the edges of the opening are rounded, not sharp.)

The source of tone, according to both Mr. Hermann Smith and Herr Schneebeli, is the formation of what the former calls an 'aeroplastic reed,' and also simply an 'air-reed,' and the latter a 'Luft-Lamelle' (Prof. Helmholtz's 'Luft-Blatt,' air-blade, or blattförmige Schicht, 'blade-shaped stratum or sheet,' as used suprà, p. 394a), which is produced outside of the pipe, and bends partly within it. For the formation of this reed both agree that it is essential for the exciting air to pass the lip, certainly not to enter the pipe. The existence of the reed is shewn by Mr. Hermann Smith by interposing a thin lamina, a shaving, or crisp tissue paper, which is caught by the air and vibrates as a reed[*], and by Herr Schneebeli by the smoke mixed with air which enables the experimenter to see its motion directly, and also by a piece of silk-paper.

Herr Schneebeli supposes this air-reed to act by producing condensation, but Mr. Hermann Smith's theory of its origin seems to be as follows. The air driven rapidly and closely from the slit past the mouth of the pipe, in a flat stream, just and only just avoiding the edge of the lip, creates a vacuum, precisely as in the tubes for ether spray or perfume spray in common use, or in ordinary chimney-pots. The air in the pipe under the action of the atmospheric pressure at

[*] Press a piece of crisp, but very light, thin paper firmly against the outside of the windway by means of a card or piece of wood. Let the paper project upwards till it nearly covers the mouth, but is quite clear of both lip and ears. The paper then resembles a free reed. On now blowing in the usual way through the slit, the paper will commence its vibrations, which Mr. Hermann Smith considers to correspond to those of his air-reed. The effect is easily observed.

the upper open end immediately descends to supply this vacuum, and by so doing not only bends the flat exciting stream of air outwards, but also of course produces a rarefaction in the tube, which by extending from the mouth upwards necessarily weakens the force of the outward rush of wind. The external (not the exciting) air, taking advantage of this relaxation of force, enters the tube at the lip, causing a condensation in the lower part of the pipe, and the resulting wave of condensation before it has proceeded half-way meets the former wave of rarefaction, which continued to proceed from the further end of the tube, and thus forms a node. The node is consequently always nearer the mouth than the end of the pipe. The ratio of the length of the segment further from the mouth to that nearer to it varies from 4 : 3 to 7 : 6 according to the diameter or scale of the pipe and the strength of wind (see p. 91*b*, and footnotes †, ‡). In the meantime the exciting stream of air rights itself, passes over the vertical, bends inwards, and a small portion of it enters the pipe with the external air, to be cast out again by the returning wave of rarefaction, and by this time the exciting stream of air has been converted into a vibrating air-reed. That the first (but momentary) effect of the upward rush of the exciting stream of air is to abstract air from the pipe Mr. Hermann Smith considers to be demonstrated by inserting on the 'languid,' just within the mouth, some filaments of cotton, fluff, or down, which, in larger pipes, are shot out with energy. He supposes the air-reed afterwards to abstract and admit air in constant succession, thus producing the necessary stimulus for the sound heard, which would on this theory depend, among other conditions, upon the force of blast, its inclination to the mouth, the size and form of the mouth and ears, the interposition of obstructions between the exciting air and outer air (as the shading bar of the *Gamba*), the capacity and length of the pipe, and whether a node has to be formed in the pipe in the mode explained for open pipes, or in the mode used for stopped pipes, which are acknowledged to speak more readily than the former. The external air of course passes continually, but intermittently, through the mouth of the pipe.

The law of vibration of the air-reed as stated by Mr. Hermann Smith, but unobserved and possibly unobtained by Herr Schneebeli, is: '*As its arcs of vibration are less, its speed is greater*,' or '*the times (of vibration) vary with the amplitude*,' being different from the usual law of a vibrating reed in which the time is independent of the amplitude of vibration. If by extraneous influence the pitch of the pipe is flattened, as by partly shading the mouth from the external air, Mr. Hermann Smith states that the path of the air-reed is lengthened, and conversely. When an organ pipe speaks the tones of the air-reed and pipe are distinct and may be separated or combined; and when a pipe is said 'to fly off to its Octave,' he says that the air-

reed leaps back to its Octave speed, compelling the pipe to follow, and that this can be made visible. The natural pitch of the air-reed is also, he states, far higher than the pitch of the tones of the pipe.

Herr Sonreck's account, although thoroughly independent, agrees with Mr. Hermann Smith's so closely as regards the origin of the motion of air in the pipe that it seems unnecessary to cite it. He calls the blade of air merely the *Anblasestrom*, i.e. 'blow-current' or 'blast.'

But so far as I have hitherto found, the first person who drew attention to the mode in which flue pipes were made to speak, was M. Aristide Cavaillé-Coll, the celebrated French organ-builder, who in a paper presented to the French Academy of Sciences on February 24, ¶ 1840 (which was never printed, owing apparently to the death of M. Savart, one of the referees), 'demonstrated,' as he states in his paper printed in the *Comptes Rendus* for 1860 (even then anterior to all the other writers), vol. l. p. 176, 'more clearly than had hitherto been done, the real function of the originator of the sound in the mouth of flue pipes, which originator he assimilated to a free aerial reed (*anche libre aérienne*).' He also investigated the mode of blowing the flute, by the mouth or a mouthpiece. And lastly he treated of the analogy between 'the transversal vibrations of vibrating laminæ of air, and solid vibrating laminæ,' which he supposed to be governed by the same laws, and from this examination he deduced 'positive data for determining the height of mouths of flue pipes ¶ in relation to their intonation and the elastic force of air which excites them.' He informs me that he intends to publish this paper.

In conclusion, it should be observed that this blade or flat current of air acts like a metal reed only so far as it oscillates backwards and forwards, and not in other respects. It does not shut off and open out a stream of air alternately. It is moved inwards by the outward air, and outwards by the inclosed air, whereas the metal reed is moved only one way by the current of air, and the other way by its own elasticity. There is therefore simply an analogy and not a substantive similarity, so that the terms aero-plastic reed, air-reed, free aerial reed, suggesting a different operation to what actually ensues, might be disused with advantage. The action seems really to be one of alternate rarefaction and condensation. But there are numerous little points ¶ which require very careful study—the shape of the upper lip; straight (as usual) or arched (as in Renatus Harris's flue pipes), the height of the opening of the mouth, the exact direction of the blade of air in relation to that of the upper lip, the presence and shape of the ears, and the general arts of the 'voicer' whereby he makes a pipe 'speak' satisfactorily. All of these matters influence both pitch and quality of tone, and though they are daily practised, their theory is as good as unknown.

For observations on the action of reeds see App. XX. sect. N. No. 8.]

APPENDIX VIII.

PRACTICAL DIRECTIONS FOR PERFORMING THE EXPERIMENTS ON THE
COMPOSITION OF VOWELS.

(See p. 122*d*.)

To make the forks vibrate powerfully, it is necessary that the ratios of their pitch numbers should agree with the simple arithmetical ratios to the utmost nicety. After the forks have been tuned by the maker by ear and to the piano as accurately as is possible in this way, the necessary greater exactness is obtained by the electrical current itself. First the interrupting fork, fig. 33, p. 122*c*, is connected with the fork that gives the prime tone, and the movable clamp on the former is
¶ arranged so as to make the unison perfect. This gives a maximum intensity to the prime tone, easily recognised by both eye and ear. The vibrations of this lowest fork are so powerful that the excursions of the extremities of the prongs under favourable circumstances amount to 2 or 3 millimetres (from ·08 to ·12 inch). It should also be observed that if the unison has not been perfectly attained, a few beats of the fork are heard when the electric current is first brought to bear upon it, although these disappear when the apparatus is in full action. [This is accounted for in Appendix IX.]

After perfect unison has been accomplished between the interrupting fork and that of the prime tone, the other forks are successively brought into electrical connection, with their resonance chambers wide open, and they are tuned until they reach a maximum intensity when excited by the current. The tuning is first performed with the file. The forks are sharpened, as is well known, by taking off some metal from their extremities, and flattened by reducing the thickness of the root of the prongs. Both must be done with the greatest possible uniformity to each prong. To discover whether the fork is too sharp or too flat, stick a little
¶ piece of wax at the ends of its prongs (which flattens the fork) and observe whether the tone becomes louder or weaker. If louder, the fork was too sharp ; if weaker, it was too flat. Since alterations of temperature, and, perhaps, other causes exert a slight influence on the pitch of the forks, I have preferred to make the higher forks a little too sharp by filing, and to bring them into exact tune by attaching small quantities of wax to the extremities of the prongs. The quantity of wax is easily altered at pleasure, and by this means slight accidental variations of pitch can be readily corrected.

It is not necessary to tune the resonance tubes so accurately ; if, when blown across, they give the same pitch as the forks, they are sufficiently well tuned. If they are too flat, some melted wax may be poured in to sharpen them. If they are too sharp, the opening must be reduced.

It cost me some trouble to get rid of the noise of the spark at the point of interruption. At first I inserted a large condenser of tin foil such as is used in induction machines. But this merely reduced the spark to a certain size. No
¶ good effect followed from increasing the size of the condenser. The layers of the condenser are separated by thin varnished paper ; one is connected with the interrupting fork, and the other with the cup of quicksilver into which its end dips. After many vain attempts, I at last found that, by inserting a very long and thin wire between the two extremities of the conduction at the point of interruption, the noise of the spark was almost entirely destroyed, without injuring the action of the current on the forks. The wire thus inserted must have an amount of resistance far greater than that occasioned by the coils in all the electro-magnets taken together. When this is the case no sensible portion of the current will go through this wire. It is not till the conduction is broken and the thin wire forms the only connection for the extra current of the electro-magnets, that the current discharges itself through the wire. But to prevent the thin wire itself from generating any secondary current, it must not be coiled round a cylinder, but must be stretched up and down on a board in such a way that two adjacent pieces of the wire should be traversed by two currents, proceeding in opposite directions. For this purpose I screwed two hard india-rubber combs at the two ends of a board

(one foot long), and passed a thin plated copper wire, such as is used for spinning over with silken threads, backwards and forwards (90 times) between the teeth of these combs. By this means a great length (90 feet) of this wire was brought well insulated into a comparatively small space, in such a way as not to produce any sensible secondary current. For when on breaking the primary current a secondary current would be formed in the wire, this would have a direction in the circuit formed by the electro-magnets and the thin wire, opposite to the secondary current in the electro-magnets, and the latter would be entirely or partly prevented from discharging itself through the thin wire.

For moving the forks I used two or three cells of a Grove's battery. The electro-magnets were placed in two rows beside one another. The whole arrangement is shewn in a diagram in fig. 64, below. The figures 1 to 8 shew the resonant chambers of the tuning-forks; the dotted lines which lead to m_1 and m_8 are the threads which remove the cover from the opening of the resonance chambers; a_1 to a_8 are the electro-magnets which set in motion the tuning-forks ¶

FIG. 64.

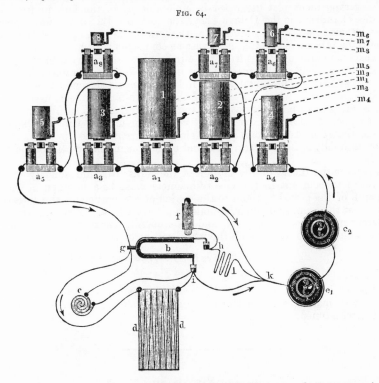

between their legs; b is the interrupting fork, and f its own electro-magnet. The ¶ relative position of the two last has been somewhat changed in order to make the connection of the direction of the currents more intelligible. The cells of the galvanic battery are marked e_1 and e_2; the great resistance-wire dd; the condenser c, but its plates which are rolled in a spiral are seen only in section.

The electric current passes from e_2, through all the electro-magnets in order, up to the handle of the interrupting fork g. It is sometimes more advantageous to arrange this part of the conduction so that it should be separated into two parallel branches, and that the three highest forks, which are the most difficult to set in motion, should be inserted into one branch, and the five lower forks into the other, thus allowing a stronger stream to pass through the former than the latter.

The remainder of the conduction from g to the second pole of the battery e_1 contains the interrupting apparatus, which is here so arranged that each vibration of the fork makes the current twice; once when the upper prong dips into the cup of mercury h, and once when the lower prong dips into the cup i. When the conduction is closed at h, it passes from g through the fork to h, and then

through the electro-magnet f to k and e_1. Between h and k it is generally necessary to insert a lateral branch h l k, having moderate resisting power, to weaken
the current in the electro-magnet f, sufficiently to prevent the fork b from making
violent vibrations. The zigzag at l shews this branch.

When the prongs of the fork move apart, the conduction will be broken at h,
and after a short interval again completed at i, so that the current now passes
from g through the lower prong of the fork to i, and thence by k to the battery at
e_1. But at the moment the conduction is broken either at h or at i, powerful
secondary currents are formed by induction in the 8 electro-magnets of the tuning-
forks, which would emit luminous and noisy sparks at the points of interruption,
if the rush of electricity were not partly stored for the moment in the condenser c,
and partly discharged through the very great resistance dd.

This resistance dd, as is seen by the figure, forms a perpetual connection between g and the battery, but it conducts so badly that no sensible part of the cur-
¶ rent can pass through it, except at the moment when, on breaking the conduction,
the great electro-motive force of the secondary currents is generated.

The arrangement just described is preferable when the fork in front of the
resonance chamber ı is the Octave above the fork b. But if the fork opposite ı
makes the same number of vibrations as the fork b, the wire i k must be removed,
and both the other wires ending in i must be connected with h.

To exclude particular forks from the circuit, short secondary connections of the
coils of wire of their electro-magnets are introduced. The arrangement is shewn
in fig. 32, p. 121b. The metal knobs h h are connected conductively with the
clamping screw g in which the wire of the electro-magnet terminates. If the
lever i is moved down, it presses with some friction on the nearer knob h, and
forms so good a secondary conducting connection for the wire of the electro-
magnets, that the greater part of the electric current passes by h h, and only
an infinitesimally small part travels round by the much longer path of the electro-
magnets.

As regards the theory of the motion of the forks, it is immediately seen that
¶ the force of the current in the electro-magnets must be a function of the time.
The length of the period is equal to the period of a vibration of the interrupting
fork b. Let the number of interruptions in a second be n. Then the strength of
the current in the electro-magnets, and consequently the magnitude of the force
exerted by the electro-magnets on the forks, will be of the form :

$$A_0 + A_1 \,.\, \cos{(2\pi nt + c_1)} + A_2 \,.\, \cos{(4\pi nt + c_2)} \\ + A_3 \,.\, \cos{(6\pi nt + c_3)} + \&c.$$

The general term of this series $A_m \,.\, \cos{(2\pi mnt + c_m)}$ is adapted for setting in
motion a fork making mn vibrations in a second, but would have little effect on
forks otherwise tuned.

¶

APPENDIX IX.

ON THE PHASES OF WAVES CAUSED BY RESONANCE.

(See p. 124b.)

LET a tuning-fork be brought near the opening of a resonance chamber, and suppose the ear of the listener to be at a great distance off in comparison with the
dimensions of the opening. In the *Journal für reine und angewandte Mathematik*, vol. lvii. pp. 1–72, in my paper on the ‘Theory of Aerial Vibrations in Tubes
with Open Ends,’ I have shewn that if a sonorous point exists at B in a space partly
bounded by firm walls, and partly unbounded, the motion of sound at another
point A, in the same space, will be identical in intensity and phase with that
which would have existed at B if A had been the sonorous point. Let B be the
position of the tuning-fork (or more properly of the end of one of its prongs), and
A that of the ear. The motion of the air which begins when the tuning-fork is
placed near the opening of the resonance chamber, is not easily determined, but

I have been able (in pp. 47 and 48 of the paper quoted above) to determine the motion when the tuning-fork is at a great distance. Let us suppose, then, that the fork is removed to the position of the ear A, and we shall then have to determine the motion of the sound at the point B near the opening. This motion is composed of two parts ; the first, having its potential denoted by Φ in the paper cited, corresponds with the motion which would also exist if the opening to the resonance chamber were closed, and in the above case is too small to be sensible ; the second, there marked Ψ, has, in open space and at some distance from the opening, the following value, using the notations explained in the above paper (p. 28, equation 12 h),

$$\Psi = -\frac{AQ}{2\pi\rho} \cdot \cos\left(k\rho - 2\pi nt\right) \quad\text{......................}\quad (1)$$

where Q is the sectional area of the resonance tube, ρ the distance from the middle point of the opening, n the pitch number, $\dfrac{2\pi}{k}$ the length of the wave. The motion ¶ at an infinitesimal distance r from the sonorous point A is given by the equation

$$\Phi = H \cdot \frac{\cos\left(2\pi nt - c\right)}{r} \quad\text{.... ....}\quad (2)$$

and if r_1 be the distance of the imaginary sonorous point A from the middle of the opening of the resonance tube, we find from equations (16c) and (13a) of the paper cited :

$$-\tan\left(kr_1 + c\right) = \tan \tau_2 = -\frac{k^2 \cdot Q \cdot \sin kl \cdot \cos ka}{2\pi \cdot \cos k(l + a)} \quad\text{..........}\quad (2a)$$

(l length of tube, and a a constant depending on the form of its opening), and finally by the same equations, the magnitude there called I is :

$$I = K \cdot \frac{2k \cdot \sin kl}{r_1} \pm AQ \cdot \frac{k^2 \cdot \sin kl}{2\pi \cdot \sin \tau_2},$$

¶

whence it follows that $\qquad A = \pm H \cdot \dfrac{4\pi \cdot \sin \tau_2}{k\,Q\,r_1} \quad\text{..........................}\quad (3)$

The sign $\pm$ is to be so determined that the consonants A and H have the same sign, and in that case τ_2 must lie between o and π.

In this case the strength of the resonance A is expressed in terms of the intensity of the sonorous point H, the section of the resonance tube Q, the distance r_1 of the sonorous point from the opening of the tube, and the magnitude τ_2. The difference of phase between the points A and B is shewn by equations (1), (2), and (2a) to be

$$\pi - k\rho + c = \pi - k\rho - kr_1 - \tau_2.$$

But the magnitude $k\rho$ at all such distances of the point B from the middle of the opening as we can use, may be regarded as infinitesimally small, so that when we weaken the tone by withdrawing the fork further from the opening of the tube, ¶ we do not sensibly change the phase of the aerial vibration. But if we change the pitch of the tube, the expression for the phase will be altered only through a change in τ_2, which by equation (2a) depends on kl, and to this change there always corresponds a change in the strength of the resonance, since $\sin \tau_2$ appears as a factor in the expression for that resonance in equation (3). The resonance is strongest when $\sin \tau_2 = 1$, or $\tau_2 = \frac{1}{4}\pi$. Calling this maximum resonance A_1, we have

$$A_1 = \frac{4\pi H}{kQr_1},$$

and for other pitches of the tube, supposing its sectional area Q to remain unchanged,

$$\sin \tau_2 = \frac{A}{A_1}.$$

Whether τ_2 is to be taken smaller or greater than a right angle, depends upon whether the value of $\tan \tau_2$ from equation (2a) is positive or negative. But since k, Q and $\cos ka$ are always positive, the value of $\tan \tau_2$ depends on the factor $\dfrac{\sin kl}{\cos k(l+a)}$. The maximum resonance corresponds to $\cos k(l+a) = 0$; the minimum to $\sin kl = 0$. Hence $\tau_2 < \frac{1}{2}\pi$ when by lengthening the tube the resonance is brought towards its minimum; but $\tau_2 > \frac{1}{2}\pi$ when the resonance is brought towards its maximum. In actual application the tube is always near its position of maximum resonance, and hence $\tau_2 < \frac{1}{2}\pi$ when the tube is too flat in pitch, and $\tau_2 > \frac{1}{2}\pi$ when the tube is too sharp in pitch.

If we put the tube out of tune to such an extent that $A^2 = \frac{1}{2} \cdot A_1{}^2$, the phase of vibration alters by $\frac{1}{4}\pi$. Hence we are always able to estimate the amount of alteration of phase by the alteration of strength of resonance.

¶ A similar law holds for the phases of the vibrating forks as compared with those of the exciting current. To simplify the treatment, I will consider the case of a single vibrating heavy point, which is constantly restored to its position of rest by an elastic force. When the heavy point is moved to the distance x from its position of rest, let $-a^2x$ be the elastic force. Suppose moreover that there act, first a periodic force, similar to that generated by the electrical currents in our experiments, which may be represented by $A \cdot \sin nt$, and secondly a force which damps the vibrations and is proportional to the velocity, so that we may write it $-b^2 \dfrac{dx}{dt}$. A force of the latter kind arises in our experiments partly from friction and resistance of the air, and also partly from the currents induced by the tuning-forks set in motion, and this latter part has most effect in damping the vibrations. If m is the mass of the heavy vibrating point, we have, therefore,

$$m \cdot \frac{d^2x}{dt^2} = -a^2x - b^2 \frac{dx}{dt} + A \cdot \sin nt \quad \ldots\ldots\ldots\ldots\ldots \quad (4)$$

¶ The complete integral of this equation is

$$x = \frac{A \cdot \sin \varepsilon}{b^2 n} \cdot \sin(nt - \varepsilon) + Be^{-\frac{b^2 t}{2m}} \cdot \sin \left\{ \frac{t}{m} \cdot \sqrt{(a^2 m - \tfrac{1}{4}b^4)} + c \right\} \ldots \quad (4a)$$

where

$$\tan \varepsilon = + \frac{b^2 n}{a^2 - mn^2} \quad \ldots\ldots\ldots\ldots\ldots\ldots\ldots\ldots \quad (4b)$$

The term having the coefficient B in (4a) is sensible only at the beginning of the motion; on account of the factor $e^{-\frac{b^2 t}{2m}}$ it decreases with the increase of the time t, and ultimately vanishes. But its existence at the beginning of the motion occasions those transient beats mentioned in App. VIII., p. 398b, when n is slightly different from

$$\frac{1}{m} \sqrt{(a^2 m - \tfrac{1}{4}b^2)}.$$

¶ The term with the coefficient A in equation (4a), on the other hand, corresponds to the sustained vibration of the heavy point. The *vis viva* i^2 of this motion is equal to the maximum value of $\frac{1}{2} m \cdot \left(\dfrac{dx}{dt} \right)^2$, or to

$$i^2 = \frac{mA^2 \cdot \sin^2 \varepsilon}{2b^4} \quad \ldots\ldots\ldots\ldots\ldots\ldots\ldots\ldots \quad (5)$$

When the pitch of the exciting tone, that is, n, can be altered, i^2 will reach its maximum (which we will call I^2), when

$$\sin^2 \varepsilon = 1, \quad \text{or} \quad \tan \varepsilon = \pm \infty,$$

giving

$$I^2 = \frac{mA^2}{2b^4}.$$

Hence we may also write

$$i^2 = I^2 \cdot \sin^2 \varepsilon \quad \ldots\ldots\ldots\ldots\ldots\ldots\ldots\ldots \quad (5a)$$

The same magnitude ϵ therefore determines in equation (4a) the difference of the phases between the periodically changing displacements x of the heavy point and the changing values of the force, and, in equation (5a), the strength of the resonance.

The condition $\tan \epsilon = \pm \infty$ is by (4b) fulfilled when $a^2 = m\,n^2$.

Hence if N be the value of n which answers to the maximum of the sympathetic vibration, we shall have

$$N^2 = \frac{a^2}{m} \quad\dots\dots\dots\dots\dots\dots\dots\dots\dots\dots (5b)$$

This tone of *strongest resonance* is the same as the tone which the heavy point would occasion, if it were set in vibration solely by the influence of the elastic force, without friction and without external excitement. Somewhat different from this is the *proper tone of the body*, which it produces under the influence of friction and resistance of the air. The pitch of this proper tone ν is given in ¶ the second term of the equation (4a)

$$\nu = \frac{1}{m} . \ \sqrt{} \ (a^2 m - \tfrac{1}{4}b^2).$$

Not until $b = 0$, that is, not until the friction and resistance of the air vanish,

will
$$\nu^2 = \frac{a^2}{m} = N^2.$$

Now in all practical cases where we have to observe the phenomenon of sympathetic vibration, b is infinitesimal, so that the difference between the tone of greatest resonance and the proper tone of the vibrating body may be disregarded, as in the text. Introducing the magnitude N the equation (4b) becomes

$$\tan \epsilon = \frac{b^2 n}{m(N^2 - n^2)} \quad\dots\dots\dots\dots\dots\dots\dots\dots (4c) \quad ¶$$

On account of the question raised on p. 150b as to the behaviour of the basilar membrane of the ear for noises, we are interested further in the integral of an equation in which $A \sin nt$ of equation (4) (p. 402b) is replaced by an arbitrary function of the time ψ_t. Of course, if this function vanishes for very great positive and negative values of the time, it could be transformed, by means of Fourier's integral, into a sum (integral) of terms such as $A \sin (nt + c)$, and then for each one of these terms, the solution just found might be applied, and finally the sum of all these solutions might be taken. But this form of solution becomes incomprehensible, because it exhibits a continuous series of tones each of which exists from $t = -\infty$ to $t = +\infty$. Hence we must proceed differently.

The differential equation to be integrated is

$$m \frac{d^2 x}{dt^2} + b^2 \frac{dx}{dt} + a^2 x = \psi \quad\dots\dots\dots\dots\dots\dots (5) \quad ¶$$

in which x is the required, and ψ the given function of the time, ψ being assumed to be finite for all values of t.

Assume

$$y + xi = A \int_{-\infty}^{t} \psi_s . e^{\kappa(t-s)} . ds \quad\dots\dots\dots\dots\dots\dots (6)$$

where κ represents one of the roots of the equation

$$m \kappa^2 + b^2 \kappa + a^2 = 0 \quad\dots\dots\dots\dots\dots\dots\dots (6a)$$

that is
$$\kappa = -\frac{b^2}{2m} \pm \sqrt{} \left(\frac{a^2}{m} - \frac{b^4}{4m^2} \right) \quad\dots\dots\dots\dots\dots\dots (6b)$$

which we will represent by

$$\kappa = -a + \beta i$$

assuming the coefficient of damping to be small enough for the root, which we represent by β, to be possible.

Hence, if ψ is a continuous function,

$$\frac{d}{dt}(y + x\mathrm{i}) = A\kappa \int_{-\infty}^{t} \psi_s \cdot e^{\kappa(t-s)} \cdot ds + A\,\psi_t \quad\quad\quad\quad (6c)$$

$$\frac{d^2}{dt^2}(y + x\mathrm{i}) = A\kappa^2 \int_{-\infty}^{t} \psi_s \cdot e^{\kappa(t-s)} \cdot ds + A\kappa\,\psi_t + A\frac{d\psi}{dt} \quad\quad (6d)$$

Then, multiplying (6) by a^2, (6c) by b^2, (6d) by m, and adding, and taking account of (6a), we obtain the following equation between the imaginary parts of the respective expressions,

¶
$$m\frac{d^2x}{dt^2} + b^2\frac{dx}{dt} + a^2x = m\,A\beta\,\psi_t.$$

Then assuming
$$A = \frac{1}{\beta m}$$

equation (6) gives a value for x which satisfies the differential equation in s, and is finite for all values of the time, namely

$$x = \frac{1}{\beta m}\int_{-\infty}^{t} \psi_s \cdot e^{-a(t-s)} \cdot \sin\beta(t-s) \cdot ds.$$

That is, x is shewn to be a sum of superposed expiring oscillations, of which the initial time is s, and the initial amplitude $\dfrac{\psi_s}{\beta m}$, and every moment preceding the point of time t contributes to the result. But this contribution vanishes for those parts of the motion which were excited for a long time before the moment con-
¶ sidered, that is, for those for which the exponent $a\,(t-s)$ is a large number, and the motion therefore depends at every instant only on those forces ψ which have acted a short time previously.

If the action of the force ψ takes place only during a limited time from t_0 to t_1, then x of the equation (6d) will not be $= 0$ except up to the time t_0, after which it will differ from nothing, and after t_1 the motion will be that of simply expiring vibrations. Also the magnitude of x will depend upon how often large positive values of ψ occur at the same time as large positive values of $\sin\beta t$, and negative with negative. The value of x will be comparatively the greatest when ψ and $\sin\beta t$ change their sign nearly at the same time.

If ψ_t has had a constant value p from $t = t_0$ to $t = t_1$, then

$$x = H\,e^{-a(t-t_1)}\sin\left[\beta(t-t_1) + \varepsilon - \eta\right]$$
on putting
$$k\cos\eta = -a, \quad k\sin\eta = \beta$$

¶
$$H\cos\varepsilon = \frac{p}{\beta mk}\left[1 - e^{-a(t_1-t_0)}\cos\beta(t_1-t_0)\right]$$

$$H\sin\varepsilon = \frac{p}{\beta mk}\,e^{-a(t_1-t_0)} \cdot \sin\beta(t_1-t_0).$$

If we suppose k to have the positive value of $\sqrt{(a^2 + \beta^2)}$, η will be an obtuse angle. If we give H the sign of the pressure p, then the angle ε, which lies between $+\frac{1}{2}\pi$ and $-\frac{1}{2}\pi$, will have the same sign as $\sin\beta(t_1-t_0)$. In this case the expression for x represents expiring vibrations, of which the initial amplitude (putting $\tau =$ the length of the action $= t_1 - t_0$) has the value

$$H = \frac{p}{pmk}\sqrt{(1 - 2\,e^{-a\tau}\cos\beta\tau + e^{-2a\tau})}.$$

This is a maximum for different values of τ, when $\cos(\beta\tau + \eta) = \cos\eta \cdot e^{-a\tau}$, or, for small values of a and τ, when $\beta\tau$ approximately contains an uneven number of half periods of vibration of the proper tone; and on the other hand H is a minimum for an even number of such vibrations.

After long continued action of the force p, however, the exponential functions vanish, and H receives the constant value

$$H_\infty = \frac{p}{\beta mk}.$$

On the other hand, for very small values of τ the initial maxima for $\beta\tau = \pi$, may attain the value

$$H_0 = \frac{p}{\beta mk}\left(1 + e^{-a\tau}\right).$$

If the pressure p changes its sign whenever $\cos \beta\tau$ does so, the amplitude H after n such changes of sign will be ¶

$$H = \frac{p}{\beta mk}\left(1 + e^{-a\tau}\right).\left(1 + e^{-a\tau} + e^{-2a\tau} + \ldots + e^{-na\tau}\right)$$

or

$$H = \frac{p}{\beta mk} \cdot \frac{1 + e^{-a\tau}}{1 - e^{-a\tau}} \cdot \left(1 - e^{-(n+1)a\tau}\right).$$

This expression shews the reinforcement, increasing with every change of sign, which ensues upon the coincidence of the period of change of pressure with the period of the proper tone. The denominator $(1 - e^{-a\tau})$ gives the amount of damping during half the periodic time of vibration. Finally, when this is very small H will be very large, and at last, after an infinite number of repetitions,

$$H = \frac{H_0}{1 - e^{-a\tau}}.$$

¶

APPENDIX X.

RELATION BETWEEN THE STRENGTH OF SYMPATHETIC RESONANCE AND THE LENGTH OF TIME REQUIRED FOR THE TONE TO DIE AWAY.

(See pp. 113a and 142d.)

RETAIN the notation of App. IX., for the motion of a heavy point, reduced to its position of rest by an elastic force. When such a point is agitated by an external periodic force, its motion is given by equation (4a), p. 402c. If we assume A, the intensity of this force, to vanish, equation (4a) reduces to ¶

$$x = B \cdot e^{-\frac{b^2 t}{2m}} \cdot \sin(\nu t + c)$$

where

$$\nu = \frac{1}{m} \cdot \sqrt{(a^2 m - \tfrac{1}{4} b^2)}.$$

On account of the factor which contains t in the exponent, the value of x continually diminishes. As in the text, measure t by the number of vibrations of the tone of strongest resonance, and for this purpose put

$$T = \frac{N}{2n} \cdot t$$

$$\beta = \frac{\pi b^2}{N m} = \pi \left(\frac{N}{n} - \frac{n}{N}\right) \cdot \tan \varepsilon \quad \ldots\ldots\ldots\ldots\ldots\ldots \quad (6)$$

Let L be the *vis viva* of the vibrations at the time $t = 0$, and l at the time $t = t$, then

$$L = B^2 \, \nu^2$$

$$l = B^2 \, \nu^2 \cdot e^{-2\beta T}$$

so that

$$\frac{l}{L} = e^{-2\beta T}$$

and

$$T = \frac{1}{2\beta} \cdot \log \mathrm{nat} \, \frac{L}{l} \ \dots\dots\dots\dots\dots\dots\dots \ (6a)$$

In the table on p. 143*a*, it is assumed that $L = 10 \ l$, and the value of T is calculated on this assumption, as follows, after finding the value of β. In equation (6), $\sin^2 \varepsilon$ is put $= \frac{1}{10}$, corresponding with the condition that the strength of the ¶ tone of the sympathetically resonant body should be $\frac{1}{10}$ of the maximum strength it can attain; and the ratio $N : n$ is calculated from the numerical ratios corresponding to the intervals mentioned in the first column of that table.

Equation (4b) in App. IX., p. 402*c*, may be written

$$\tan \varepsilon = \frac{b^2}{mN \cdot \left(\dfrac{N}{n} - \dfrac{n}{N} \right)} = \frac{\beta}{\pi \cdot \left(\dfrac{N}{n} - \dfrac{n}{N} \right)}$$

In this equation N, giving the pitch of strongest resonance; b^2, the strength of the friction; and m, the mass may be different for various fibres of Corti. Hence in applications to the ear, we must consider b^2 and m to be functions of N. Now since the degree of roughness of the closer dissonant intervals remains tolerably constant for constant intervals throughout the scale, the magnitude represented by $\tan \varepsilon$ must assume approximatively the same values for equal values of $\frac{N}{n}$, and ¶ hence the magnitude $\frac{b^2}{mN} = \frac{\beta}{\pi}$ must be tolerably independent of the values of N. No very exact result can be obtained. Hence in the calculations which will follow hereafter β is assumed to be independent of N.

APPENDIX XI.

VIBRATION OF THE MEMBRANA BASILARIS IN THE COCHLEA.

(See p. 146*b*.)

¶ THE mechanical problem here attempted is to examine whether a connected membrane with properties similar to those of the membrana basilāris in the cochlĕa, could vibrate as Herr Hensen has supposed this particular membrane to do; that is, in such a way that every bundle of nerves in the membrane could vibrate sympathetically with a tone corresponding to its length and tension, without being sensibly set in motion by the adjacent fibres. For this investigation we may disregard the spiral expansion of the basilar membrane, and assume it to be stretched between the legs of an angle, of the magnitude 2η. Let the axis of x bisect this angle, and the axis of y be drawn at right angles to it through the vertex of the angle. Let the tension of the membrane parallel to the axis of x be $= P$, and that parallel to the axis of y be $= Q$, both measured by the forces which when exerted on the sides of a unit square, parallel to x and y respectively, would balance the tension of the membrane. Let μ be the mass of this unit square, t the time, and z the displacement of a point in the membrane from its position of equilibrium. Moreover let Z be an external force, acting on the membrane in the direction of positive z, and setting it in vibration. The equation of the motion of the membrane, deduced without material difficulty from Hamilton's principle by Kirchhoff's process, is then

$$Z + P \cdot \frac{d^2z}{dx^2} + Q \cdot \frac{d^2z}{dy^2} = \mu \cdot \frac{d^2z}{dt^2} \qquad \text{.................} \qquad (1)$$

The limiting conditions are

1) that $z = 0$ along the legs of the angle, that is, $z = 0$, when $y = \pm x \cdot \tan \eta$.
2) that $z = 0$, when $x = y = 0$, that is, at the vertex of the angle, and finally
3) that z is finite, when x is infinite.

The further development of the problem will shew how these two last limiting equations, which suffice for our purpose, may be replaced by certain determinate curves acting as fixed boundaries between the legs of the angle (p. 411c, d).

By putting $x = \xi \cdot \sqrt{P}$ and $y = v \cdot \sqrt{Q}$, the equation (1) may be reduced to the ¶ better known form

$$Z + \frac{d^2z}{d\xi^2} + \frac{d^2z}{dv^2} = \mu \cdot \frac{d^2z}{dt^2} \qquad \text{.........................} \qquad (1a)$$

which is the equation of motion for a membrane stretched uniformly in all directions, ξ and v being the rectangular co-ordinates on its surface.

For this notation the limiting conditions become

1) $z = 0$ for $v = \pm \xi \cdot \sqrt{\dfrac{P}{Q}} \cdot \tan \eta$,

2) $z = 0$ for $\xi = v = 0$,
3) z finite, for $\xi = \infty$.

The transformed problem consequently differs from the original merely in having a uniformly stretched membrane, and a different amount of angle, which we will represent by 2ε.   ¶

Since in the applications which we have to make of the result, P will be very small in comparison with Q, the angle ε for the transformed membrane will also be very small, and upon this circumstance mainly depend the analytical difficulties of the problem.

After these preliminary remarks, we proceed to the analytical treatment of the equations (1) and (1a) by introducing polar co-ordinates, assuming

$$\left.\begin{aligned} x &= \xi \cdot \sqrt{P} = r \cdot \sqrt{P} \cdot \cos \omega \\ y &= v \cdot \sqrt{Q} = r \cdot \sqrt{Q} \cdot \sin \omega \end{aligned}\right\} \qquad \text{.....................} \qquad (1b)$$

The equations (1) and (1a) then take the form

$$\frac{d^2z}{dr^2} + \frac{1}{r} \cdot \frac{dz}{dr} + \frac{1}{r^2} \cdot \frac{d^2z}{d\omega^2} = \mu \cdot \frac{d^2z}{dt^2} - Z \qquad \text{.................} \qquad (1c)$$

The limiting conditions are now, that   ¶

1) $z = 0$, when $\omega = \pm \varepsilon$, and hence $\tan \varepsilon = \sqrt{\dfrac{P}{Q}} \cdot \tan \eta$,

2) $z = 0$ for $r = 0$,
3) z is finite, when r is infinite.

As regards the nature of the force Z, we shall assume that it consists of two parts; the *first* depending on the friction, which we may put $= -v \cdot \dfrac{dz}{dt}$, where v is a positive real constant; the *second*, depending on a periodically variable pressure exerted by the surrounding medium on the membrane, uniformly over its whole surface.   Consequently we put

$$Z = -v \cdot \frac{dz}{dt} + A \cdot \cos nt,$$

and obtain as the equation of motion

$$\frac{d^2z}{dr^2} + \frac{1}{r}\cdot\frac{dz}{dr} + \frac{1}{r^2}\cdot\frac{d^2z}{d\omega^2} = \mu\cdot\frac{d^2z}{dt^2} + \nu\cdot\frac{dz}{dt} - A\cdot\cos nt \dots\dots\dots \quad (2)$$

Of the various motions which the membrane could execute under these circumstances, we are interested solely in those which are maintained by the continuous periodical action of the force, and which must themselves have the same period. Let us consequently assume

$$z = \zeta\cdot e^{int}, \text{ where } i = \sqrt{(-1)} \dots\dots\dots\dots\dots \quad (2a)$$

and determine ζ by the equation

¶
$$\frac{d^2\zeta}{dr^2} + \frac{1}{r}\cdot\frac{d\zeta}{dr} + \frac{1}{r^2}\cdot\frac{d^2\zeta}{d\omega^2} + (\mu n^2 - i n\nu)\cdot\zeta = -A \dots\dots\dots \quad (2b)$$

In this case the real part of the value of z will satisfy equation (2) and correspond to a uniformly sustained oscillation of the membrane.

Having thus eliminated the variable t from the differential equation, we proceed to do the same for ω by means of the first limiting equation, after transforming both ζ and the constant A into a series of cosines of uneven multiples of the angle $\frac{\pi\,\omega}{2\varepsilon} = h\omega$. It is well known that between the limits $h\omega = +\frac{1}{2}\pi$ and $-\frac{1}{2}\pi$

$$A = \frac{4A}{\pi}\cdot\left(\cos h\omega - \frac{1}{3}\cdot\cos 3h\omega + \frac{1}{5}\cdot\cos 5h\omega + \dots\right) \dots\dots \quad (3)$$

If in the same way we put

¶
$$\zeta = s_1\cdot\cos h\omega - \frac{1}{3}\cdot s_3\cdot\cos 3h\omega + \frac{1}{5}\cdot s_5\cdot\cos 5h\omega + \dots\dots \quad (3a)$$

then for each coefficient s_m we must have

$$\frac{d^2s_m}{dr^2} + \frac{1}{r}\cdot\frac{ds_m}{dr} + \left(\mu n^2 - i n\nu - \frac{m^2 h^2}{r^2}\right)\cdot s_m = -\frac{4A}{\pi}\Bigg\} \dots \quad (3b)$$

And since the first of our limiting conditions is satisfied by the equation (3a), whenever the series converges, there remain only the conditions that

1) $s_m = 0$ for $r = 0$,
2) s_m finite, for $r = \infty$.

It is easily seen that every s_m is perfectly determined by these conditions. For if there were two different functions which satisfied the equation (3b) and the two limiting conditions, then their difference, which we will call δ, would satisfy the conditions

¶
$$\frac{d^2\delta}{dr^2} + \frac{1}{r}\cdot\frac{d\delta}{dr} + \left(\mu n^2 - i n\nu - \frac{m^2 h^2}{r^2}\right)\cdot\delta = 0 \dots\dots\dots\dots \quad (3c)$$

and hence be a Bessel's function, and at the same time we should have

1) $\delta = 0$ for $r = 0$,
2) δ finite for $r = \infty$.

But these two conditions cannot be satisfied at the same time by a Bessel's function, when ν has a value differing ever so little from 0. It is only when $\nu = 0$, that is, when there is no friction, that the determination is insufficient. In that case oscillations once induced may continue for ever, even when there is no force to give fresh impulses.

Particular integrals of the equation (3b) may be easily developed in the form of series, resembling the series for the related Bessel's functions which satisfy equation (3c). One of these series proceeds according to integral powers of r and is always convergent. But when the angle ε is very small, the number of terms in this series which are necessary to determine s is very large, and hence the series cannot be used for determining the progress of the function.—A second series which proceeds according to negative powers of r and gives a second particular integral is semi-convergent, and will not become an algebraical function, unless h is an uneven number. But in the latter case the first mentioned series will be infinite in its separate terms.

It is therefore preferable for our present purpose to obtain the expression for s in the form of definite integrals.

Let ϕ and ψ denote the following pair of integrals :—

$$\left.\begin{aligned}\psi &= \int_0^{\frac{1}{2}\pi} e^{-i\kappa r \cdot \sin t} \cdot \sin mht \cdot dt \\ \phi &= \int_1^{\infty-mh-1} u \cdot e^{-\frac{1}{2}\cdot i\kappa r \cdot \left(u+\frac{1}{u}\right)} \cdot du\end{aligned}\right\} \quad\dots\dots\dots\dots\dots\dots\dots \text{(4)}$$

where

$$\kappa = \sqrt{(\mu n^2 - i\, n\, \nu)} \dots\dots\dots\dots\dots\dots \text{(4a)} \quad ¶$$

and the sign of the root is so chosen that the possible part of $i\kappa$ is positive.

Then

$$s_m = \frac{4A}{\pi\kappa^2} \cdot (mh \cdot \psi + mh \cdot \phi \cdot \cos\tfrac{1}{2}\, m\, h\, \pi - 1) \dots\dots\dots\dots\dots \text{(4b)}$$

which is the required expression for s_m.

To shew that the expression in (4b) really satisfies the equation (3b), substitute this value for s_m in that equation, and in differentiating under the integral signs of ψ and ϕ, use partial integration to eliminate the factors $\cos t$ and $\left(u-\frac{1}{u}\right)$ which appear under the integral signs.

For $r = 0$ we find

$$\psi = \int_0^{\frac{1}{2}\pi} \sin mht \cdot dt = \frac{1}{mh} \cdot \left(1 - \cos\frac{mh\pi}{2}\right) \qquad ¶$$

$$\phi = \int_1^\infty \frac{du}{u^{mh+1}} = \frac{1}{mh}$$

and hence $s_m = 0$.

For $r = \infty$, we have $\phi = \psi = 0$, and hence

$$s_m = -\frac{4A}{\pi\kappa^2}.$$

Hence the function s_m also satisfies the two limiting conditions, which have been already shewn to be sufficient to determine it.

The equation (4b) may be used to determine the value of s_m when P, the tension of the membrane in the direction x, is infinitesimal. In this case, as (1b) shews, r must be the infinite; as also h, of which the value is   ¶

$$h = \frac{\pi \cdot \sqrt{Q}}{2 \cdot \sqrt{P} \cdot \tan\eta}.$$

Hence putting

$$r = h\rho$$

ρ will be the finite, namely

$$\rho = \frac{2x \cdot \tan\eta}{\pi \cdot \sqrt{Q}}.$$

It is easily seen that under these circumstances $m\, h\, \phi$ will $= 0$. For we may write

$$m\, h\, \phi = \int_1^\infty mh \cdot e^{-mh \cdot \log u - (l-i\lambda)\cdot\frac{h\rho}{2}\cdot\left(u+\frac{1}{u}\right)} \cdot \frac{du}{u} \dots\dots\dots\dots \text{(5)}$$

where I have put

$$i\kappa = l - i\lambda,$$

and l according to the above supposition will be positive. Since within the whole extent of the integration $u > 1$ and hence $\log u > 0$, the possible part of the exponent will be negative throughout the same extent, and will contain the infinite factor h. Consequently every part of the integral vanishes, and hence also the whole value $h\phi$.

On the other hand the integral ψ or

$$\psi = \int_0^{i\pi} e^{-(l-i\lambda).h\rho.\sin t} . \sin mht . dt$$

will have the possible part of the exponent negative and infinite for all those parts of the integral for which t is not infinitesimal, so that these will all $= 0$. But this is not the case for those parts of the integral for which t vanishes.

¶ Hence for an infinite h we may replace the above equations for ψ by the following:

$$\psi = \int_0^\infty e^{-(l-i\lambda).h\rho t} . \sin mht . dt$$

In this last form the integration may be effected and gives

$$\psi = \left. \frac{m}{h . [(l-i\lambda)^2 . \rho^2 + m^2]} \right\} \quad \dots\dots\dots\dots\dots\dots \quad (5a)$$

and

$$s_m = \frac{4A\rho^2}{\pi . (m^2 - \rho^2\kappa^2)}$$

or, by (4a),

$$s_m = \left. \frac{4A\rho^2}{\pi . (m^2 - \rho^2\mu n^2 + i\rho^2 n\nu)} \right\} \quad \dots\dots\dots\dots\dots\dots \quad (5b)$$

¶ Or if, in order to get rid of the auxiliary magnitude ρ, we represent by $\frac{1}{2}\beta$ the value of y on the limits of membrane, we have

$$\tfrac{1}{2}\beta = x \tan \eta,$$

and hence

$$\rho = \frac{\beta}{\pi . \sqrt{Q}},$$

so that [using S_m for the modulus of s_m],*

$$S_m = \frac{4A}{\pi . \sqrt{\left[\left(\dfrac{m^2\pi^2 Q}{\beta^2} - \mu n^2\right)^2 + n^2 \nu^2\right]}} \quad \dots\dots\dots\dots\dots \quad (5d)$$

This value is quite independent of the magnitude of the angle through which the membrane is stretched. In place of the distance ρ or x from the vertex, we have only β the breadth of the membrane at the point in question. Hence this expression will still hold when the angle is $= 0$, and the membrane vibrates like a string between two parallel lines, thus forming m vibrating segments which are ¶ separated by lines of nodes parallel to the edges.

The same expression also results for a string, if z is regarded from the first in equation (1) as only a function of y in a line, and supposed to be independent of x, but the limiting condition is retained that when $y = \pm\beta$, then $z = 0$. Hence the motion of the membrane is the same as that of a series of juxtaposed but unconnected strings.

The value of $\frac{1}{m} . S_m$ in (5d) gives us the amplitude of the corresponding form of vibration having the pitch number $\frac{n}{2\pi}$, and having m vibrating transverse divisions of the membrane. The maximum of S_m will occur when

$$m^2\pi^2 Q - \beta^2 \mu n^2 = 0 \dots\dots\dots\dots\dots\dots \quad (6)$$

---

* [In the 3rd German edition S_m is used without the explanation here inserted; in the 4th German edition by an error of the press s_m is henceforth used for S_m; consequently the reading of the 3rd edition has been retained.—*Translator.*]

The value of this maximum, which we call Σ_m is

$$\Sigma_m = \frac{4A}{\pi\, n\, \nu}.$$

The smaller the coefficient of friction ν, the larger will be this maximum at the point in question.

If we call b the value of β which satisfies the equation (6), we may write the equation (5d) thus

$$S_m = \frac{\Sigma_m}{\sqrt{\left[\, \mathrm{I} + \frac{m^4\, \pi^4\, Q^2}{n^2\, \nu^2} \cdot \left(\frac{\mathrm{I}}{\beta^2} - \frac{\mathrm{I}}{b^2}\right)^2\,\right]}}$$

When ν is infinitesimal, and the condition of the maximum is not fulfilled in equation (6), the denominator of this expression becomes infinite, and hence S_m infinitesimal. The amplitude of the vibrations $\frac{\mathrm{I}}{m} \cdot S_m$, will become finite, for ¶ those values of β only which are so nearly $= b$, that $b - \beta$ is of the same order as ν. Under these circumstances, therefore, each simple tone sets in vibration only some narrow strips of the membrane in the direction of x, of which the first has one, the second three, the third four, &c., vibrating segments, and in which $\frac{\beta}{m}$, that is the length of the vibrating segments, has always the same value.

The greater the coefficient of friction ν, the greater in general will be the extent of the vibrations of every tone over the membrane.

The present mathematical analysis shews that every superinduced tone must also excite all those transverse fibres of the membrane on which it can exist as a proper tone with the formation of nodes. Hence it would follow, that if the membrane of the labyrinth were of completely uniform structure, as the membrane here assumed, every excitement of a bundle of transverse fibres by the respective fundamental tone must be accompanied by weaker excitements of the unevenly ¶ numbered harmonic undertones, the intensity of which would, however, be multiplied by the factors $\frac{\mathrm{I}}{9}$, $\frac{\mathrm{I}}{25}$, and generally $\frac{\mathrm{I}}{m^2}$. Although this hypothesis has been advanced by Dr. Hugo Riemann in his *Musikalische Logik*, there is nothing of the kind observable in the ear. I think, however, that this cannot necessarily be urged as an objection against the present theory, because the appendages of the basilar membrane probably greatly impede the formation of tones with nodes.

The solution can also be extended without difficulty to the case where the membrane in the field of ξ, v is bounded by two circular arcs, with their centre at the vertex of the angle. To this case correspond as boundaries in the real case, that is, in the field of x, y, two elliptic limiting arcs, which when P vanishes become straight lines. It is only necessary to add to the value of s_m in (4b), a complete integral of the equation (3c), which can be expressed by Bessel's functions with two arbitrary constants, and to determine these constants in such a manner ¶ as to make $s_m = \mathrm{o}$ on the limiting curves selected. When ν is small this change in the limits has no essential effect on the motion of the membrane, except when the maximum of vibration itself falls in the neighbourhood of the limiting curves.

APPENDIX XII.

THEORY OF COMBINATIONAL TONES.

(See pp. 152, note †, and 158, note *.)

It is well known that the principle of the undisturbed superposition of oscillatory motions, holds only on the supposition that the motions are small,—so small, indeed, that the moving forces excited by the mutual displacements of the particles of the oscillating medium should be sensibly proportional to these displacements. Now it may be shewn that *combinational tones must arise whenever the vibrations are so large that the square of the displacements has a sensible influence on the motions.* It will suffice for the present to select, as the simplest example, the motion of a single heavy point under the influence of a system of waves, and develop the corresponding result. The motions of the air and other elastic media may be treated in a perfectly similar manner.

Suppose that a heavy point having the mass m is able to oscillate in the direction of the axis of x. And let the force which restores it to its position of equilibrium be

$$k = ax + bx^2.$$

Suppose two systems of sonorous waves to act upon it, with the respective forces

$$f \cdot \sin pt, \quad \text{and} \quad g \cdot \sin (qt + c)$$

¶ then its equation of motion is

$$-m \cdot \frac{d^2x}{dt^2} = ax + bx^2 + f \cdot \sin pt + g \cdot \sin (qt + c).$$

This equation may be integrated by a series, putting

$$x = \varepsilon x_1 + \varepsilon^2 x_2 + \varepsilon^3 x_3 + \dots$$
$$f = \varepsilon f_1$$
$$g = \varepsilon g_1,$$

and then equating the terms multiplied by like powers of ε, separately to zero. This gives

1) $ax_1 + m \cdot \dfrac{d^2x_1}{dt^2} = -f_1 \cdot \sin pt - g_1 \cdot \sin (qt + c),$

¶ 2) $ax_2 + m \cdot \dfrac{d^2x_2}{dt^2} = -b x_1{}^2,$

3) $ax_3 + m \cdot \dfrac{d^2x_3}{dt^2} = -2b x_1 x_2,$ and so on.

From the first equation we obtain

$$x_1 = A \cdot \sin\left(t \cdot \sqrt{\frac{a}{m}} + b\right) + u \cdot \sin pt + v \cdot \sin (qt + c)$$

where $\qquad u = \dfrac{f_1}{mp^2 - a} \quad \text{and} \quad v = \dfrac{g_1}{mq^2 - a}.$

This is the well-known result for infinitesimal vibrations, shewing that the body which vibrates sympathetically produces only its proper tone $\sqrt{\frac{a}{m}}$, together ¶ with those communicated to it, p and q. Since the proper tone in this case rapidly disappears, we may put $A = 0$. And then equation (2) gives

$$x_2 = -\frac{b}{2a} \cdot (u^2 + v^2)$$

$$-\frac{u^2}{2(4mp^2 - a)} \cdot \cos 2p\,t - \frac{u^2}{2(4mq^2 - a)} \cdot \cos 2(qt + c)$$

$$+\frac{uv}{m(p-q)^2 - a} \cdot \cos \left[(p-q)t - c\right] - \frac{uv}{m(p+q)^2 - a} \cdot \cos \left[(p+q)t + c\right].$$

The second term of the series for x [involving x_2], contains, then, a constant, and also the tones $2p$, $2q$, $(p-q)$, and $(p+q)$. If the proper tone $\sqrt{\frac{a}{m}}$ of the body which vibrates sympathetically is deeper than $(p-q)$, as may be certainly assumed in most cases for the drumskin of the ear in connection with the auditory ossicles, and if the intensities u and v are nearly the same, the tone $(p - q)$ will have the greatest intensity of all the tones in the terms of x_2; it corresponds with the well-known deep combinational tone. The tone $(p + q)$ will be much weaker, and the tones $2p$ and $2q$ will be heard with difficulty as weak harmonic upper partial tones of the generating tones.

The third term x_3 [of the series for x] contains the tones $3p$, $3q$, $2p+q$, $2p-q$, $p+2q$, $p-2q$, p and q. Of these $2p-q$ or $2q-p$ is a combinational tone of the second order according to Hallstroem's nomenclature (p. 154d). Similarly the fourth term x_4 [of the series for x] gives combinational tones of the third order; and so on.

If, then, we assume that in the vibrations of the tympanic membrane and its appendages, the square of the displacements has an effect on the vibrations, the preceding mechanical developments give a complete explanation of the origin of combinational tones. Thus the present new theory explains the origin of the tones $(p+q)$, as well as of the tones $(p-q)$, and shews us, why when the intensities ¶ u and v of the generating tones increase, the intensity of the combinational tones, which is proportional to uv, increases in a more rapid ratio.

The previous assumption respecting the magnitude of the force called into action, namely

$$k = ax + bx^2$$

implies that when x changes its sign, k changes not merely its sign, but also its absolute value. Hence this assumption can hold only for an elastic body which is unsymmetrically related to positive and negative displacements. It is only in such that the square of the displacement can affect the motion, and combinational tones of the first order arise. Now among the vibrating parts of the human ear, the drumskin is especially distinguished by its want of symmetry, because it is forcibly bent inwards to a considerable extent by the handle of the hammer, and I venture therefore to conjecture that this peculiar form of the tympanic membrane conditions the generation of combinational tones.

[See especially App. XX. sect. L. art. 5.]   ¶

APPENDIX XIII.

DESCRIPTION OF THE MECHANISM EMPLOYED FOR OPENING THE SEVERAL SERIES
OF HOLES IN THE POLYPHONIC SIREN.

(See p. 162, note *.)

FIG. 65 (p. 414a, b) shews the vertical section of the upper box of the double siren, in order to display its internal construction. E is the wind pipe which is prolonged into the interior of the box, and firmly fixed in the cross beam AA of the support of the apparatus. The prolongation of the wind pipe into the box B has conical surfaces at its upper and lower ends, on which slide corresponding hollow ¶ surfaces in the bottom and top surfaces of the box, so that this box can revolve freely about the wind pipe as an axis. At a may be seen a section of the toothed wheel fastened to the cover of the box. At β is the driving wheel which is turned by the handle γ; and δ is a pointer which is directed to the graduation on the edge of the disc $\epsilon\epsilon$.

D is the upper extremity of the axis of the movable discs, of which only the upper one CC is here shewn. The axis turns on fine points in conical cups. The upper cup is introduced into the lower end of the screw η, which can be more or less tightened by the milled screw head introduced above, so that any required degree of ease and steadiness in the motion of the axis may be attained.

Inside the box are seen the sections of four pierced rings $\kappa\lambda$, $\lambda\mu$, $\mu\nu$, and νo, which fit on to one another with oblique, tile-shaped edges, and thus mutually hold each other steady. Each of these rings lies beneath a series of holes in the cover, and contains precisely the same number of holes as the corresponding series of the cover and of the rotating disc. By means of studs, shewn at i i in fig. 56 (p. 162), these four rings can be slightly displaced, so as either to make the holes of the ring coincide with the holes of the box, and thus give free passage to the air and produce the corresponding tone; or else to close the holes of the cover by the

FIG. 65.

¶

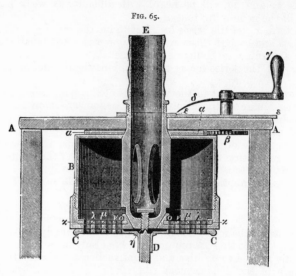

interspaces between the holes of the ring, and thus cut off their corresponding tone altogether.

In this way it is possible to sound the various tones of this siren in succession or simultaneously, and hence obtain separate or combined tones at pleasure.

¶

APPENDIX XIV.

VARIATION IN THE PITCH OF SIMPLE TONES THAT BEAT TOGETHER.

(See p. 165*b* and note *.)

LET v be the velocity of a particle, which vibrates under the influence of two tones, so that

$$v = A \cdot \sin mt + B \cdot \sin (nt + c)$$

where m differs very slightly from n, and $A > B$. We may then put

$$nt + c = mt - [(m - n) \, t - c], \text{ and}$$

¶ $v = \{A + B \cdot \cos [(m - n) \, t - c]\} \cdot \sin mt - B \cdot \sin [(m - n) \, t - c] \cdot \cos mt.$

Assume

$$A + B \cdot \cos [(m - n) \, t - c] = C \cdot \cos \varepsilon,$$

and
$$B \cdot \sin [(m - n) \, t - c] = C \cdot \sin \varepsilon,$$

so that

$$v = C \cdot \sin (mt - \varepsilon),$$

in which C and ε are functions of the time t, which will alter slowly, if, as we have assumed, $m - n$ is small in comparison with m.

The intensity C^2 of this oscillation is determined by

$$C^2 = A^2 + 2 \, AB \cdot \cos [(m - n) \, t - c] + B^2,$$

and it will be a maximum,

$$C^2 = (A + B)^2, \quad \text{when } \cos [(m - n) \, t - c] = + 1,$$

and a minimum,

$$C^2 = (A - B)^2, \quad \text{when } \cos\,[(m - n)\,t - c] = -1.$$

The variable phase ε of the motion is determined by

$$\tan \varepsilon = \frac{B \,.\, \sin\,[(m - n)\,t - c]}{A + B \,.\, \cos\,[(m - n)\,t - c]}$$

As $A > B$, this tangent never becomes infinite, and hence ε remains included between the limits $+\frac{1}{2}\pi$ and $-\frac{1}{2}\pi$, to which it alternately approaches. As long as ε increases, $mt - \varepsilon$ increases more slowly than mt; as long as ε diminishes, $mt - \varepsilon$ increases faster than mt; hence in the first case the tone flattens and in the second it sharpens.

The pitch number of the variable tone, multiplied by $2\,\pi$, is under these circumstances equal to ¶

$$m - \frac{d\varepsilon}{dt} = \frac{m\,A^2 + (m + n) \,.\, A\,B \,.\, \cos\,[(m - n)\,t - c] + n\,B^2}{A^2 + \quad 2 \,.\, A\,B \,.\, \cos\,[(m - n)\,t - c] + \quad B^2}$$

The limits for the pitch number therefore correspond

$$\text{to } \cos\,[(m - n)t - c] \text{ becoming } + 1 \text{ or } -1,$$

and hence also to a maximum or minimum strength of tone.

1) When the strength of tone is a maximum, the pitch number varies as

$$\frac{mA + nB}{A + B} = m - \frac{(m - n)\,B}{A + B} = n + \frac{(m - n)\,A}{A + B}.$$

2) When the strength of tone is a minimum, the pitch number varies as

$$\frac{mA - nB}{A - B} = m + \frac{(m - n)\,B}{A - B} = n + \frac{(m - n)\,A}{A - B}. \qquad ¶$$

Hence in the first case, [or during the maximum strength,] the pitch of the variable tone lies between the pitches of the two separate tones. But during the minimum strength, if the stronger tone is also the sharper, the pitch of the variable tone is sharper than that of either of the single tones; and if the stronger tone is the flatter, the pitch of the variable tone is flatter than that of either of the single tones.

These differences are well heard with two stopped pipes; and also with two tuning-forks when first the higher and then the lower is placed nearer to the resonance chamber.

[See Mr. Sedley Taylor's paper on this subject, *Philosophical Magazine*, July 1872, pp. 56–64, where he gives several figures illustrating the variability of the pitch, and deduces the above results (1) and (2) from the figures only.]

¶

APPENDIX XV.

CALCULATION OF THE INTENSITY OF THE BEATS OF DIFFERENT INTERVALS.

(See pp. 187a and note * and 193, note *.)

WE shall again employ the formulae for the strength of the sympathetic vibration developed in Appendix IX., equations (4a) and (4b), p. 402c, and (5) and (5a), p. 402d. For the tone of strongest resonance in one of Corti's elementary organs, let n be its number of vibrations in 2π seconds, n_1 and n_2 the corresponding numbers of vibrations in 2π seconds for two tones heard, and B', B'' the greatest velocities of the vibrations which they superinduce in those Corti's organs which have the same pitch, and B_1, B_2 the greatest velocities which both attain in their representation of the number of vibrations n. Then by equation (5a) of Appendix IX., p. 402d, we have

$$B_1 = B' \sin \varepsilon_1, \quad \text{and} \quad B_2 = B'' . \sin \varepsilon_2$$

where
$$\pi . \tan \varepsilon_1 = \frac{\beta}{\dfrac{n}{n_1} - \dfrac{n_1}{n}}, \quad \text{and} \quad \pi . \tan \varepsilon_2 = \frac{\beta}{\dfrac{n}{n_2} - \dfrac{n_2}{n}},$$

and β is a magnitude which may be regarded as independent of n. Hence the intensity of the vibrations of the organ for the number of vibrations n, when both tones n_1 and n_2 affect it simultaneously, fluctuates between the values

$$(B_1 + B_2)^2 \quad \text{and} \quad (B_1 - B_2)^2.$$

The difference of these two magnitudes, which measures the strength of the ¶ beats, is

$$4 B_1 B_2 = 4 B' B'' . \sin \varepsilon_1 . \sin \varepsilon_2 \quad \dots\dots\dots\dots \quad (7)$$

Hence for equal differences in the amount of pitch, the strength of the beats is dependent on the product $B' B''$. For the mth partial tone of the compound tone of a violin, we may, by Appendix VI., p. 597c, put $B'^2 = \dfrac{A'^2}{m^2}$, and hence if the m_1th and m_2th partial tone of two compound tones of a violin, beat, we may put the intensity of their beats for equal differences of interval $= \dfrac{A'^2}{m_1 . m_2}$.

This is the expression from which the numbers in the last column of the table on p. 187b have been calculated. [They are therefore 100 times the reciprocals of the products of the two numbers which give the ratio of the pitch numbers in the corresponding line of its third column.]

For the calculation of the degree of roughness of the various intervals, mentioned in pp. 193, 332, and 333, the following abbreviations of notation are in ¶ troduced:

$$n_1 + n_2 = 2 N.$$
$$n_1 \quad = N (1 + \delta).$$
$$n_2 \quad = N (1 - \delta).$$
$$n \quad = N (1 + \nu).$$

So that
$$\pi . \tan \varepsilon_1 = \frac{\beta}{\dfrac{1+\nu}{1+\delta} - \dfrac{1+\delta}{1+\nu}} \quad \text{and} \quad \pi . \tan \varepsilon_2 = \frac{\beta}{\dfrac{1+\nu}{1-\delta} - \dfrac{1-\delta}{1+\nu}}$$

Since powerful sympathetic resonance ensues only when ν and δ are very small, we may assume that, approximatively,

$$\tan \varepsilon_1 = \frac{\beta}{2\pi (\nu - \delta)}, \quad \text{and} \quad \tan \varepsilon_2 = \frac{\beta}{2\pi (\nu + \delta)}.$$

¶ Putting these values for ε_1 and ε_2, in equation (7) we have

$$4 B_1 B_2 = 4 B' B'' . \frac{\beta^2}{\sqrt{[\beta^2 + 4\pi^2 . (\nu - \delta)^2]} . \sqrt{[\beta^2 + 4\pi^2 . (\nu + \delta)^2]}} \quad \dots \quad (7a)$$

If then we consider ν, that is, the pitch of the Corti's organ which vibrates sympathetically, to be variable, $4 B_1 B_2$ will reach its maximum when $\nu = 0$, and hence $n = N = \frac{1}{2} (n_1 + n_2)$, and if we call the value of this maximum s we have

$$s = 4 B' B'' . \frac{\beta^2}{\beta^2 + 4\pi^2 \delta^2} \quad \dots\dots\dots\dots\dots \quad (7b)$$

In calculating the degree of roughness arising from sounding two tones together which differ from each other by the interval 2δ, I have thought it sufficient to consider this maximum value, which exists in those Corti's organs which are most favourably situated. Undoubtedly other beats of a weaker kind will be excited in the neighbouring fibres, but their intensity rapidly diminishes. It might therefore appear to be a more exact process to integrate the value of $4\,B_1\,B_2$ in equation (7a) with respect to ν, in order to obtain the sum of the beats in all Corti's organs. This would require an at least approximate knowledge of the density of Corti's organs for different values of ν, that is, for different parts of the scale, and of that we know nothing. In sensation, the highest degree of roughness is certainly more important than the distribution of a less degree of roughness over many sensitive organs. Hence I have preferred to take only the maximum of the vibrations from (7b) into account.

Finally we have to consider that very slow beats cause no roughness, and that when the intensity of the beats remains unaltered, and their number increases, the roughness reaches a maximum and then decreases. To express this, the value of s must be also multiplied by a factor, which vanishes when the number of beats is small, attains a maximum for about 30 beats in a second, and then diminishes, and again vanishes when the number of such beats is infinite. Suppose then that ¶ the roughness r_p, due to the pth partial tone, is expressed by

$$r_p = \frac{4\,\theta^2 \cdot \delta^2 \cdot p^2}{(\theta^2 + p^2 \cdot \delta^2)^2} \cdot s_p.$$

The factor of s_p reaches its maximum value $= 1$, when $p\delta = \theta$; and becomes $= 0$, when δ, that is, half the interval between the two tones in the scale, is $= 0$ or ∞. Since δ may be indifferently positive or negative, the expression can only involve even powers of δ. The above is the simplest expression which satisfies the conditions, but it is of course to a certain extent arbitrary.

For θ we must put half the extent of the interval which at the pitch of the lower beating tone causes 30 beats to be made in a second.

Since we have taken c' with 264 vibrations in a second, as the lower tone, θ has been put $= \frac{15}{264}$. Hence we have finally

$$r_p = 16\,B'\,B'' \cdot \frac{\beta^2\,b^2\,\delta^2\,p^2}{(\beta^2 + 4\pi^2\,\delta^2)(\theta^2 + p^2\,c^2)}.$$

¶

And from this formula I have calculated the roughness of the intervals, shewn graphically in the diagrams, fig. 60, A and B, p. 193b, c, and fig. 61, p. 333a. The roughnesses due to the separate partial tones have been drawn separately and superimposed on one another in the drawing.

Although the theory leaves much to be desired in the matter of exactness, it at least serves to shew that the theoretical view we have proposed is really capable of explaining such a distribution of dissonances and consonances as actually occurs in nature.

Professor Alfred M. Mayer, of Hoboken, New Jersey,* has instituted experiments on the duration of sensations of sound, and the number of audible beats. Between a vibrating tuning-fork and its resonator he interposed a revolving disc with openings of the same shape as that of the resonator, so that the sound was heard loudly when an opening in the disc came in front of that in the resonator, and faintly when the latter was covered. His results agree essentially with the assumptions I have made on pp. 143 to 145, and 183 to 185, but are more complete as they have been pursued throughout the whole scale. His conclusions are as follows :— ¶

---

* *Silliman's Journal*, ser. iii. vol. viii. October 1874, *Philosophical Magazine*, May 1875, vol. ii. [From the following table, p. 418a, it is seen that the interval of a minor Third as the limit of continuity applies only to the Octave c to c'. For g'''', supposing the law connecting D and N, given on p. 418d, to hold for such a high pitch, the interval of continuity would have the ratio 3072 : 3072 + 225, or 122 cents, and the interval of maximum dissonance would be 49 cents, so that the interval of one Semitone is near the limit of continuity ; hence it is not surprising that no beats were heard in the case referred to on p. 173c, note *. But for c'''' the interval of continuity would have the ratio 2048 : 2249, giving 162 cents, and the interval of maximum dissonance would be 67 cents, and hence the beats of $b'''\ c''''$ should be quite conspicuous, agreeing with observations. In reference to p. 144b, Prof. Mayer observes that the law abruptly breaks down for vibrations below 40 in a second, and thinks that this abrupt breaking down 'can only be explained by the highly probable supposition that co-vibrating bodies in the ear, tuned to vibrations below 40 per second, do not exist, and therefore . . . the inner ear . . . can only vibrate *en masse*,' and also that such oscillations cannot last $\frac{1}{30}$ sec. *Translator.*]

Note	Pitch number	Number of the beats for which the interruptions vanish	[Cents in the corresponding interval*	Number of beats for the greatest dissonance	[Cents in the corresponding interval†
C	64	16	386	6·4	165
c	128	26	308	10·4	135
c'	256	47	292	18·8	123
g'	384	60	251	24·0	105
c''	512	78	245	31·2	102
e''	640	90	228	36·0	95
g''	768	109	230	43·6	95
c'''	1024	135	214]	54·0	89]

APPENDIX XVI.

ON BEATS OF COMBINATIONAL TONES, AND ON COMBINATIONAL TONES IN THE SIREN AND HARMONIUM.

(See pp. 199a and note *, also 155c to 158a.)

LET a, b, c, d, e, f, g, h be whole numbers. Let an and $bn + \delta$ be the pitch numbers of the primes of two compound tones sounded simultaneously, where δ is supposed to be very small in comparison with n, and a and b are the smallest whole numbers by which the ratio $a : b$ can be expressed. The pitch numbers of any pair of partial tones of these two compound tones will be

$$acn \text{ and } bdn + d\delta.$$

These will beat with each other $d\delta$ times in a second, if

$$ac = bd \text{ or } \frac{a}{b} = \frac{d}{c}.$$

And since the ratio $\frac{a}{b}$ is expressed in its lowest terms, the smallest values of d and c are

$$d = a \text{ and } c = b,$$

and their other values are

$$d = ha \text{ and } c = hb.$$

Now c and d represent the ordinal numbers of the partial tones which beat together. Hence the lowest partials of this kind will be the bth partial of the compound an, and the ath partial of the compound $bn + \delta$. The resulting number of beats is $a\,\delta$.

In the same way the $2b$th partial of the first compound, and the $2a$th of the second give $2a\,\delta$ beats, and so on.

The first differential combinational tone of the two partials acn and $bdn + d\delta$ is

$$\pm[(bd - ac)n + d\delta]$$

where the $+$ or $-$ sign has to be taken so that the whole expression is positive.

* [The interval is found as the ratio of the pitch number to the same increased by the number in the next column to it; thus for C it is $64 : 64 + 16 = 4 : 5$, and for g' it is $384 : 384 + 60 = 96 : 111$, and from these I have calculated the cents as in p. 701 of the 1st edition.

If N be the pitch number, and $D =$ duration of residual sensations or the reciprocal of the numbers of vibrations producing a continuous sound, 16, 26, &c., as in the next pre-ceding column, then Prof. Mayer finds—

$$D = \left(\frac{53248}{N + 23} + 24\right) \cdot \frac{1}{10000} \text{ seconds.} - \textit{Translator.}]$$

† [The interval is found as the ratio of the pitch number to the same increased by the last mentioned number of beats, thus $64 : 64 + 6·4$. Prof. Mayer draws attention to the fact that his beats were all tones of the same pitch, whereas the beats of imperfect consonances are tones of variable pitch.—*Translator.*]

Two other partials fan and $gbn + g\delta$ give the differential combinational tone

$$\pm [(gb - af)n + g\delta].$$

When both sound together they produce $(g \mp d)\,\delta$ beats, if

$$bd - ac = \pm (gb - af)$$

or

$$\frac{a}{b} = \frac{g \mp d}{f \mp c}.$$

As before, it follows that the least value of $g \mp d$ is $= a$, and the other (greater) values are $= ha$, so that the smallest number of beats is $a\delta$.

To find the lowest values of the partials which must be present in order to beat with the first differential tones, we will take the lower signs for c and d, and we thus obtain:

$$g = d = \tfrac{1}{2}a, \text{ or } g = \tfrac{1}{2}(a + 1), \text{ and } d = \tfrac{1}{2}(a - 1) \qquad \P$$
$$f = c = \tfrac{1}{2}b, \text{ or } f = \tfrac{1}{2}(b + 1), \text{ and } c = \tfrac{1}{2}(b - 1),$$

according as a and b are even or odd. If b is the larger number, $\tfrac{1}{2}b$ or $\tfrac{1}{2}(b + 1)$ is the number of partials which any compound must have in order to produce beats when the two tones composing the interval are sounded. If the combinational tones are neglected, about double the number, that is b, are required.

When simple tones are sounded together, the beats arise from the combinational tones of higher orders. The general expression for a differential tone of a higher order arising from two tones with the vibrational numbers n and m is $\pm (an - bm)$, and this tone is then of the $(a + b - 1)$th order. Let the pitch number of a combinational tone of the $(c + d - 1)$ order arising from the tones an and $(bn + \delta)$ be

$$\pm [(bd - ca)\,.\,n + d\delta],$$

and of another of the $(f + g - 1)$th order be $\P$

$$\pm [(gb - fa)\,.\,n + g\delta],$$

then both produce $(g \mp d)\,.\,\delta$ beats, when

$$bd - ac = \pm (bg - af)$$

or

$$\frac{a}{b} = \frac{g \mp d}{f \mp c}.$$

The lowest number of beats is therefore again $a\delta$, and the lowest values of c, d, f, g, are found as in the former case, so that the ordinal numbers of combinational tones need not exceed $\tfrac{1}{2}(a + b - 2)$, if a and b are both odd, or $\tfrac{1}{2}(a + b - 1)$, if only one of them is odd, the other being even.

To what has been said in Chap. VII., pp. 154–159, I will add the following $\P$ remarks on the origin of combinational tones.

Combinational tones must always arise when the displacement of the vibrating particles from their position of rest is so large that the force of restitution is no longer simply proportional to the displacement. The mathematical theory of this case for a heavy vibrating point is given in App. XII., pp. 411d to 415b. The same holds for aerial vibrations of finite magnitude. The principles of the theory are given in my essay on the 'Theory of Aerial Vibrations in Tubes with Open Ends,' in Crelle's *Journal für Mathematik*, vol. lvii. p. 14. I will here draw attention to a third case, where combinational tones may also arise from infinitely small vibrations. This has already been mentioned in pp. 156d–157d. It occurs with sirens and harmoniums. We have here two openings, periodically altering in size, and with a greater pressure of air on one side than on the other. Since we are dealing only with very small differences of pressure, we may assume, that the mass of the escaping air is jointly proportional to the size of the opening ω, and to the difference of pressure p, so that

$$q = c\omega p$$

where c is some constant. If we now assume for ω the simplest periodic function which expresses an alternate shutting and opening, namely

$$\omega = A \cdot (1 - \sin 2\pi n t),$$

and consider p to be constant, that is, suppose ω to be so small and the influx of air so copious, that the periodical loss through the opening does not essentially alter the pressure, q will be of the form

$$q = B \cdot (1 - \sin 2\pi n t)$$

where $$B = cAp.$$

In this case the velocity of the motion of sound at any place of the space filled with air, must have a similar form, so that only a tone with the vibrational number n can arise. But if there is a second greater opening of variable size, through
¶ which there is sufficient escape of air to render the pressure p periodically variable, instead of being constant, as the air passes out through the other opening, that is, if p is of the form

$$p = P \cdot (1 - \sin 2\pi m t),$$

then q will have the form

$$q = cAP \cdot (1 - \sin 2\pi n t) \cdot (1 - \sin 2\pi m t)$$
$$= cAP \cdot [1 - \sin 2\pi n t - \sin 2\pi m t - \tfrac{1}{2}\cos 2\pi(m+n)t + \tfrac{1}{2}\cos 2\pi(m-n)t].$$

Hence, in addition to the two primary tones n and m, there will be also the tones $m+n$ and $m-n$, that is, the two combinational tones of the first order.

In reality the equations will always be much more complicated than those here selected for shewing the process in its simplest form. The tone n will influence the pressure p, as well as the tone m; even the combinational tones will alter p;
¶ and finally the magnitude of the opening may not be expressible by such a simple periodic function as we have selected for ω. This will occasion not merely the tones m, n, and $m+n$, $m-n$, to be produced, but also their upper partials, and the combinational tones of those upper partials, as may also be observed in experiments. The complete theory of such a case becomes extraordinarily complicated, and hence the above account of a very simple case may suffice to shew the nature of the process.

I will mention another experiment which may be similarly explained. The lower box of my double siren vibrates strongly in sympathy with the fork a' when it is held before the lower opening, and the holes are all covered, but not when the holes are open. On putting the disc of the siren in rotation so that the holes are alternately open and covered, the resonance of the tuning-fork varies periodically. If n is the vibrational number of the fork, and m the number of times that a single hole in the box is opened, the strength of the resonance will be a periodic function of the time, and consequently in its simplest case equal to $1 - \sin 2\pi m t$.
¶ Hence the vibrational motion of the air will be of the form

$$(1 - \sin 2\pi m t) \cdot \sin 2\pi n t = \sin 2\pi n t + \tfrac{1}{2}\cos 2\pi(m+n)t - \tfrac{1}{2}\cos 2\pi(m-n)t,$$

and consequently we hear the tones $m+n$, and $m-n$ or $n-m$. If the siren is rotated slowly, m will be very small, and these tones being all nearly the same, will beat. On rotating the disc rapidly, the ear separates them.*

* [For the whole subject of beats and combinational tones the reader is referred to the recent discussions in Appendix XX. sect. L.— *Translator.*]

APPENDIX XVII

PLAN FOR JUSTLY INTONED INSTRUMENTS WITH A SINGLE MANUAL.

(See p. 319c, and note †.)

To arrange an organ or harmonium with twenty-four tones to the octave in such a way as to play in just intonation in all keys, the tones of the instrument must be separated into four pairs of groups, thus

1 a)	f	a_1	c♯		1 b)	f_1	a	c_1♯
2 a)	c	e_1	a_1♭		2 b)	c_1	e	a♭
3 a)	g	b_1	e_1♭		3 b)	g_1	b	e♭
4 a)	d	f_1♯	b_1♭		4 b)	d_1	f♯	b♭

Each of these groups must have a separate portvent from the bellows, and valves must be introduced in such a way that the wind may be driven at pleasure either to the right or left group of any horizontal series. This would not be difficult on the organ. On the harmonium the digitals would have to be placed in a different order from the tongues, and consequently it would, as on the organ, be necessary to have a more complicated arrangement for conducting the effect of pressing down a digital to the valve.

Hence four valves are to be arranged by stops or pedals in a different way for every key. The following is a table of the arrangement of the stops for the four horizontal series of the tones named :—

Major keys	Series				Minor keys
	1	2	3	4	
Cb*	b	a	a	a	$(E_1$b$)$
Gb*	b	b	a	a	$(B_1$b$)$
Db*	b	b	b	a	(F_1)
Ab*	b	b	b	b	(C_1)
Eb*	a	b	b	b	(G_1)
Bb*	a	a	b	b	(D_1)
F	a	a	a	b	A_1
C	a	a	a	a	E_1
G	b	a	a	a	B_1* or Cb
D	b	b	a	a	F_1♯ * or Gb
A	b	b	b	a	C_1♯ * or Db
E	b	b	b	b	G_1♯ * or Ab
B	a	b	b	b	D_1♯ * or Eb
	a	a	b	b	A_1♯ * or Bb

The minor keys which have their names in parentheses, namely E_1♭, B_1♭, F_1, C_1, G_1, D_1, have a true minor Seventh, but too high a leading tone. [Their dominant chord has an impossible Pythagorean major Third.] For the six keys marked with (*), the arrangement of the stops is the same both for major and minor.†

In order to have a complete series of tonics, each with a perfect major and minor form, it would be necessary to cut off a_1♭, e_1♭, b_1♭, f_1, c_1, g_1, from the other notes, and to allow them to be replaced when needed by g♯, d♯, a♯, e♯, b♯, and f♯♯, by means of a fifth stop. We should thus have 30 tones to the octave. By drawing out this stop we should have the following system of keys :—

† [The series in the first six lines is the same as in the six last.—*Translator.*]

Major keys	Series marked with accented letters to shew that they are affected by the fifth stop				Minor keys
	1	2	3	4	
F	a'	a'	a'	b'	F
C	a'	a'	a'	a'	C
G	b'	a'	a'	a'	G
D	b'	b'	a'	a'	D
A	b'	b'	b'	a'	A
E	b'	b'	b'	b'	$E*$
B	a'	b'	b'	b'	$D_1\sharp$
$F\sharp$	a'	a'	b'	b'	$A_1\sharp$
$C\sharp$	a'	a'	a'	b'	$E_1\sharp$
$G\sharp$	a'	a'	a'	a'	$B_1\sharp$
$D\sharp$	b'	a'	a'	a'	$F_1\sharp\sharp$
$A\sharp$	b'	b'	a'	a'	$C_1\sharp\sharp$
$E\sharp$	b'	b'	b'	a'	$G_1\sharp\sharp$

¶ To have a complete series of minor keys, 28 instead of 30 tones to the octave would be enough. They would suffice for the 12 minor keys of A_1, E_1, B_1, $F_1\sharp$ or $G\flat$, $C_1\sharp$ or $D\flat$, $G_1\sharp$ or $A\flat$, $D_1\sharp$ or $E\flat$, B, F, C, G and D, and for 17 major keys from $C\flat$ major to $G\sharp$ major.†

APPENDIX XVIII.

JUST INTONATION IN SINGING.

(See p. 326*b*.)

¶ SINCE the publication of the first edition of this book, I have had an opportunity of seeing the *Enharmonic Organ*, constructed by General Perronet Thompson,‡ which allows of performance in 21 major and minor scales with different tonics harmonically connected. This instrument is much more complicated than my harmonium. It contains 40 pipes to the octave, and has three distinct manuals, with, on the whole, 65 digitals to the octave, as the same note has to be sometimes struck on two or all of the manuals. This instrument allows of the performance of much more extensive modulations than my harmonium, without requiring any enharmonic interchanges. It is even possible to execute tolerably quick passages and ornamentations upon it, notwithstanding its apparently involved fingering.

* [The E minor has the leading note, but not the minor Seventh. The other minor keys have both.—*Translator.*]

† [As Prof. Helmholtz has retained this Appendix in his 4th German edition it is given in the translation. But the scheme explained has never been tried. The plan for 24 notes is impracticable because of the defective state of the minor keys, and imperfect modulating power. It could only be used as an experimental instrument, and for that the double keyboard as explained on p. 316*b* suffices. The valve arrangement for 30 notes would be complicated, and even if it could be used would still have a very imperfect modulating power. The 53 division of the octave introduced by Mr. Bosanquet, and subsequently by Mr. Paul White, with fingerboards which have been actually used, as explained in App. XIX., and also App. XX. sect. F. Nos. 8 and 9, are so much superior in manipulation, musical effect, and power of modulation, that it is unnecessary to seek further. In App. XX. sect. F. the other principal methods that have been actually tried, are also explained.—*Translator.*]

‡ ['On the Principles and Practice of *Just Intonation*, with a view to the Abolition of Temperament, and embodying the results of the Tonic *Sol-fa* Associations, as illustrated on the *Enharmonic Organ* . . . presenting the power of performing correctly in 21 keys (with the minors to the extent of involving not more than 5 flats), and a correction for changes of temperature. . . . Calculated for taking the place of the choir organ in a cathedral, and learned by the blind in six lessons. With an Appendix tracing the identity of design with the Enharmonic of the Ancients.' By T. Perronet Thompson, F.R.S. Ninth edition, 1866. The exact compass of this organ will be explained in App. XX. sect. F. No. 6. General Thompson was born at Hull, in 1783, and died at Blackheath, 6 September 1869. He had been four years in the navy before joining the army, and was prominent during the Corn Law Abolition agitation. He was many years editor of the *Westminster Review*, and was first returned to Parliament for Hull in 1835.—*Translator.*]

The organ was erected in the Sunday School Chapel, 10 Jewin Street, Aldersgate, London,* and was built by Messrs. Robson, 101 St. Martin's Lane, London. It contains only one stop of the usual *principal* work, has Venetian shutters forming a swell throughout, and is provided with a peculiar mechanism for correcting the influence of temperature on the intonation.

Mr. H. W. Poole has lately transformed his organ† so as to get rid of stops for changing the intonation, and has constructed a peculiar arrangement of the digitals, which enables him to play in all keys with the same fingering. His scale contains not merely the just Fifths and Thirds in the series of major chords, but also the natural or subminor Sevenths for the tones of both series. There are 78 pipes to the octave, and $F'b$ has been identified with E_1, &c., as upon my harmonium.‡

Successions of chords on General Thompson's instrument are extraordinarily harmonious, and, perhaps, on account of their softer quality of tone, even more surprising in their agreeable character than on my harmonium.§ I had an opportunity, at the same time, of hearing a female singer, who had often sung to it, perform a piece to the accompaniment of the enharmonic organ, and her singing gave ¶ me a peculiarly satisfactory feeling of perfect certainty in intonation, which is usually absent when a pianoforte accompanies. There was also a violinist** present who had not been much accustomed to play with the organ, and accompanied well-known airs by ear. He hit off the intonation exactly as long as the key remained unchanged, and it was only in some rapid modulations that he was not able to follow it perfectly.

In London I had also an opportunity of comparing the intonation of this instrument with the natural intonation of singers who had learned to sing without any instrumental accompaniment at all, and are accustomed to follow their ear alone. This was the Society of Tonic Sol-faists, who are spread in great numbers (there were 150,000 in 1862††) over the large cities of England, and whose great

---

* [Shortly before his death General Perronet Thompson presented this organ to Mr. John Curwen, mentioned in note ††, below. The General's executors had it reconstructed in a schoolroom at Plaistow, Essex. It was afterwards exhibited at the Scientific Loan Exhibition at South Kensington Museum in 1876, and has remained there ever since, at the top of the staircase leading to room Q of the Science Collections.—*Translator.*]

† *Silliman's American Journal of Science and Arts*, vol. xliv., July 1867. [In its original form the instrument, with an ordinary keyboard and pedals, was termed the *Euharmonic Organ*, and is described in *Silliman's Journal*, vol. ix. p. 209, for May 1850. The new fingerboard is figured and described hereafter, App. XX. sect. F. No. 7.—*Translator.*]

‡ [The text is in error. There are 100 not 78 pipes to the octave, and E_1 is not identified with $F'b$.—*Translator.*]

§ [' On organs of many stops, one or more ought certainly to be tuned with mathematically correct intonation, on account of their wonderful effect, to be employed (of course without using any others at the same time) as the music of the spheres (*als Gesang der Sphären*). It is impossible to form any notion of the effect of a chord in mathematically just intonation, without having heard it. I have such a one to compare with the others. Every one who hears it expresses his delight and surprise at a correctness of intonation that it does one good to hear (*Jeder, der ihn hört, spricht sein frohes Erstaunen über diese wohlthuende Reinheit aus*).'—Scheibler, *Ueber mathematische Stimmung, Temperaturen und Orgelstimmung nach Vibrations-Differenzen oder Stössen*, 1838. I have given the original words of the last German sentence, as it was impossible to do justice to its homely force in

any translation. Every one who has heard just intonation will understand it.—*Translator.*]

** [A blind man, who had therefore no ¶ notes to guide him. I had the pleasure of taking Professor Helmholtz to hear the organ on this occasion (20 April 1864), and can corroborate his statements. Unfortunately the proper blind organist was not present. It is to this lady that General Thompson dedicates his little book, already recited, in these words : ' To Miss E. S. Northcote, Organist of St. Anne and St. Agnes, St. Martin's-le-Grand. In commemoration of the talent by which, after six lessons, she was able to perform in public on the enharmonic organ with 40 sounds to the octave; thereby settling the question of the practicability of just intonation on keyed instruments, and realising the visions of Guido and Mersenne, and the harmonists of classical antiquity.'—*Translator.*]

†† [The 20 years which have elapsed since Prof. Helmholtz's first acquaintance with the ¶ Tonic Sol-fa movement have made a struggling system, slowly elaborated by a Congregationalist minister in connection with his ministry, into a great national system of teaching singing. And as the system had the cordial approval of Prof. Helmholtz (see note p. 427*d*), I feel justified in adding a short account of its origin, progress, and present condition. In 1812 the two Miss Glovers, daughters of a clergyman of Norwich, then young women, now both dead at a very advanced age, invented and introduced into the schools under their superintendence a new sol-fa system, based upon the 'movable *doh*,' that is, the use of *doh* as the name of the keynote, whatever that might be. This was little known beyond the town where it was used, but was published about 1827 as a *Scheme for*

progress is of much importance for the theory of music. The Tonic Sol-faists represent the tones of the major scale by the syllables *Do, Re, Mi, Fa, So, La, Ti,*

rendering Psalmody Congregational, and passed through three editions. About 1841 John Curwen, then an unmarried Congregationalist minister (born 14 November 1816 at Heckmondwike, Yorks), visited the school, and at once saw that Miss Glover's scheme gave him the instrument he desired for his own work. In 1845 he married, and he and his wife struggled—it was a real and severe struggle, against small means—to make this system known and active. In the course of working it out various improvements suggested themselves, and the *Tonic Sol-fa* system, as he termed it, is not precisely the same as Miss
¶ Glover's; it is essentially John Curwen's. Thus Miss Glover's scheme (as she says in a MS. preface in 1862 to the 2nd edition of the description of her Harmonicon, in the Science Collections, South Kensington Museum) was founded on temperament; Curwen's on just intonation; and the alterations that this change involved were many and laborious. Here Curwen was, I believe, much assisted by the personal friendship of General Perronet Thompson, whose works he constantly quoted in the first book he issued, *Singing for Schools and Congregations,* 1843–8. A remarkable power of methodising, systematising, and teaching, of making friends and co-workers, and of utilising suggestions carried everything before it—at last. But the work was long, and the opposition strong. There was only an 'Association' when Prof. Helmholtz made the
¶ acquaintance of the Tonic Sol-fa system. But the Association grew to be a 'College,' which held its first 'summer term' on 10 July 1876, having been 'incorporated' on 26 June 1875, and there were in 1884, 1420 'Shareholders' in this College, which opened its 'Buildings' (at Forest Gate, London, E.) on 5 July 1879. John Curwen lived long enough to see the opening and to preside at the unveiling of Miss Glover's portrait in it, never having neglected to own his obligations to her initiative. On a stone at the entrance of the present College building he placed this inscription: 'This stone was laid by John Curwen, May 14, 1879, in memory of Miss Sarah Glover, on whose "Scheme for rendering Psalmody Congregational" the Tonic Sol-fa method was founded.' John Curwen died 26 May 1880, of weakness
¶ of the heart. His eldest son, John Spencer Curwen, Associate of the Royal Academy of Music, has been since that date annually elected as President of the College. The work of the College is chiefly examinational, carrying on classes by post in the various branches of music, and granting certificates shewing various degrees of attainment, on the authority of duly appointed examiners. From 1858 to 1884 the numbers of these certificates granted have been: Junior, 52,000; Elementary, 167,000; Intermediate, 44,000; Matriculation, 3350; Advanced, 520; Musical Theory (including Harmony, Composition, Form, Expression, Acoustics, &c.), 8200; total 275,070, as the Secretary informs me. During the summer there is always a term for the special *vivâ voce* instruction of teachers. Of course large classes are constantly going on everywhere. I quote the following from a letter dated 15 October

1884, written by John Spencer Curwen to the Editor of the *Times* :—

At the most modest estimate, during the 30 years our system has been at work, we have taught at least the elements of music to four million persons. There are now, in the elementary schools of the United Kingdom, about one million children learning to sing at sight upon our system. The Tonic Sol-fa College has 28 different kinds and grades of musical examinations, and these were passed last year by 18,716 persons. Every examination includes individual tests in singing at sight. We have between 4000 and 5000 teachers at work, and at the present time they have under instruction some 200,000 adults, in addition to the children already mentioned. I lately inquired of 16 of our most active professional teachers how many pupils, adults and children, they were instructing per week in their classes. The number proved to be 61,051. We have a well-organised movement. During the last four years I have attended 166 meetings in the length and breadth of the kingdom, my travels extending over 13,000 miles, and ranging from Plymouth to Londonderry, from Inverness to Norwich. These meetings, at which demonstrations of musical education are invariably given, have been attended by at least 100,000 people. I have further travelled in France, Germany, Austria, and Switzerland, studying the condition of popular musical instruction in schools, singing clubs, &c., so that we may bring our practice up to the best continental models. The quantity of music printed in the Tonic Sol-fa notation is enormous, and is increasing very rapidly. Two-thirds of our pupils, having been grounded in our notation, go on to learn the ordinary staff notation, and prove themselves excellent readers of that notation.

With regard to teaching music in schools, the following is compiled from the papers issued by the Educational Department in 1884, England and Wales C. 3941, and Scotland C. 3942. They refer to 27,330 subordinate educational departments for England and Wales. Of these, 21,743 teach music by ear only; 1429 by the staff or ordinary notation; 3871 by Tonic Sol-fa; 32 by both systems; and 2161 in some other way. For Scotland there are 3403 subordinate departments, of which 1238 teach by ear only; 8 by the late Dr. Hullah's modification of Wilhelm's method, 1746 by Tonic Sol-fa; 117 by old notation with movable *doh* (for which many teachers have a strong predilection), and 7 by more than one system. There are 94 departments in England, and 277 in Scotland making no returns. These returns shew that Tonic Sol-fa is the national system of teaching music *by note* in the primary schools of England and Scotland at the present day.

John Curwen having started his system from purely philanthropic motives, gladly placed his notation at the disposal of all who liked to use it. A strong proof of the success of his system is furnished by the fact that all the principal London publishers have availed themselves of this permission. Gounod's *Redemption* and Mackenzie's *Rose of Sharon* are among the latest additions to the Tonic Sol-fa répertoire. It is estimated that at the present time there are 40,000 pages of music printed and published in this latter notation. But the educational works on music and the system are the private property of the firm of

Do, where *Do* is always the tonic [vowels as in Italian]. Their vocal music is not written in ordinary musical notation, but is printed with common types, the initial letters of the above words representing the pitch.*

When the tonic is changed in modulations, the notation is also changed. The *new* tonic is now called *Do*, and the change is pointed out in the notation by giving two different marks to the note on which it occurs, one belonging to the old, and one to the new key. This notation, therefore, gives the very first place to representing the relation of every note to the tonic, while the absolute pitch in which the piece has to be performed is marked at the commencement only. Since the intervals of the natural major scale are transferred to each new tonic as it arises in the course of modulation, all keys are performed without tempering the intervals. That in the modulation from *C* major to *G* major, the *Mi* (or b_1) of the second scale answers precisely to the *Ti* of the first is not indicated in the notation, and is only taught in the further course of instruction. Hence the pupil has no inducement to confuse a with a_1.†   ¶

John Curwen & Sons, and are of such a remarkable character that a gold medal was awarded for them at the International Health and Education Exhibition at South Kensington in 1884. It would indeed be difficult to find so much information on music and the method of teaching it (in both notations), so succinctly and plainly given, and at so cheap a rate, as in the late John Curwen's *Teacher's Manual, Standard Course, Musical Theory, How to Observe Harmony, How to Read Music*, not to mention the very large number of books and music intended for immediate class use. John Curwen's especial desire was to teach 'the *thing* MUSIC,' as he words it, and the peculiar means which he elaborated for this purpose, he valued only because it proved effectual for that purpose.

As one who was personally acquainted with John Curwen and his work for a quarter of a century, I may be permitted to give this testimony, and to refer all those who would learn the history of this successful musical educationalist to the *Memorials of John Curwen*, compiled by his son, J. Spencer Curwen, 1882.--*Translator.*]

* [Great care has also been bestowed on the representation of rhythm, and exercises in rhythm form an important part of the *Standard Course* and the practice of Tonic Sol-fa teachers. --*Translator.*]

† [In a footnote to this passage Prof. Helmholtz gives a list of the Tonic Sol-fa works, which is superseded by the note I have inserted above, and at the end of it he says:] In France singing is taught by the *Galin-Paris-Chevé* system, on similar principles and with the help of a similar notation. [This statement is misleading. Neither principles nor notation are alike. In 1818 P. Galin, 'Instituteur à l'Ecole des Sourds-Muets de Bordeaux,' published his *Exposition d'une Nouvelle Méthode pour l'Enseignement de la Musique.* It follows from p. 162 of his book (3rd ed. 1862, reprinted by Émile Chevé) that Galin adopted as his normal intonation Huyghens's cycle of 31 divisions of the octave, which closely represents the meantone temperament (see App. XX. sect. A. art. 22, ii.), although Galin did not seem to be acquainted with it under that name, and seems to announce as his own discovery (*ibid.* p. 80, and especially p. 107) what was in fact Huyghens's more than 120 years previously : viz. that $\frac{1}{5}$ of a whole Tone $=\frac{1}{5}$ major Semitone $=\frac{1}{2}$ of a

minor Semitone, but the curious thing is that he considers the resulting flat Fifth of 696·773 cents to be *correct*, and the Fifth with 701·955 cents from the ratio of 2 : 3 to be *wrong*. This is enough to shew how widely Galin's *principles* differed from Curwen's. The notation of intervals which Galin used was Rousseau's numerical expression of the major scale as 1 2 3 4 5 6 7, indicating a rising Octave by overdotting and a falling by underdotting, but calling the figures *ut ré mi fa sol la si.* Here the only resemblance is the movable *ut* (=*doh*), as distinguished from the usual French custom of making *ut*=*C*. In marking sharps and flats and time the difference is greater, but need not be pursued. We should observe, however, that on this system, as Galin expressly states (p. 81), *g♯* is flatter than *a♭*. ¶ Galin was born 16 Dec. 1786, and died 30 Aug. 1822. His pupils, and especially Aimé and Nanine Paris (the latter of whom married Émile Chevé, a surgeon), continued to teach his system, and supplied it with text-books. The principal one is *Méthode élémentaire par Mme. Chevé* (Nanine Paris). *La partie théorique de cet ouvrage est rédigée par Émile Chevé.* In this theoretical part, p. 292, I find that Chevé imagined Galin to have called his single division half of a minor Second, whereas he says, as above, that it was half of a minor Semitone, which is totally different. The consequence is that Chevé makes Galin's scale a division of the octave into 29 divisions, instead of 31, and hence he obtained a sharp Fifth of 703·46 cents, a very sharp major Third of 413·8 cents, much sharper than the Pythagorean (App. XX. sect. A. art. 23, vi.). If he ¶ could have tuned an harmonium to this major scale and played the major chords, he would have been scared at the result. He makes *g♯* much sharper than *a♭*; his *ab* was indeed flatter and *g♯* sharper than on the Pythagorean system. It is evident that his pupils when they sang in chorus *could* not have used his theoretical scale. Hence his principles were entirely different from Curwen's. The notation remained the same as with Galin, sharps and flats being denoted by acute and grave accents drawn through the stems of the figures, but their meaning was altogether different. Also he retained the movable *ut* of Galin, and on p. 327 he made out a general table of the relation of modulations, which resembles my Duodenarium, App. XX. sect E. art. 18. M. Aimé Paris also introduced a plan for

It is impossible not to acknowledge that this method of notation has the great advantage to the singer of giving prominence to what is of the greatest importance to him, namely, the relation of each tone to the tonic. It is only a few persons, unusually gifted, who are able to fix in their mind, and re-discover absolute pitches, when other tones are sounded at the same time. But the ordinary notation* gives directly nothing but absolute pitch, and that too only for tempered intonation. Any one who has frequently sung at sight is aware how much easier it is to do so from a pianoforte vocal score, in which the harmony is shewn, than from the separate voice part. In the first case it is easy to see whether the note to be sung is the root, Third, Fifth, or dissonance of the chord which occurs, and it is then comparatively easy to find one's way; † in the second case the only resource of the singer is to go up and down by intervals as well as he can, and trust to the accompanying instruments and the other voices to force his own to the right pitch.

¶ Now the instruction conveyed to a singer who is familiar with musical theory by an examination of the pianoforte vocal score, is conveyed by the notation itself of the Tonic Sol-faist even to the uninstructed. I have convinced myself that by using this notation, it is much easier to sing from the separate part than in ordinary musical notation, and I had an opportunity when in one of the primary schools in London, of hearing more than forty children of between eight and

marking modulations which has a great resemblance to John Curwen's 'bridge tone,' but both plans were absolutely independent. The 'langue des durées' of Aimé Paris was, however, avowedly adapted to the Tonic Sol-fa system by John Curwen. Both M. and Mme. É. Chevé are dead, and after some time their son Amand Chevé revived the system, which has had great success in France, and gained many prizes in choral competitions, shewing that Émile Chevé's theoretical scale could not have been adopted. From a correspondence I ¶ had with M. Amand Chevé I found he did not hold with his father's 29 division of the octave, but adopted the 53 division (not however as representing just intonation with a major Third of 17 degrees or 384·9 cents, but) as representing the Pythagorean intonation with a major Third of 18 degrees or 4c7·5 cents (App. XX. sect. A. art. 22, iii.). As this would be frightful in part singing, it is probable that his pupils, although strictly taught to make g♯ sharper than ab (indeed to make the intervals g to ab and g♯ to a identical, each containing 4 degrees or 90·6 cents, with an interval of 1 degree or 22·6 cents between them), in choral singing insensibly use the equal temperament which Galin and Émile Chevé for different reasons inveighed against. At any rate the Galin-Paris-Chevé system, ¶ clever and successful as it is, is after all and was from the first a *tempered* system, and in its Chevé form a (theoretically) very badly tempered system, and hence not in the slightest degree similar in principle to the Tonic Sol-fa, which as taught by John Curwen was always a system of just intonation. Another immense difference must be noted. Curwen founds everything upon the major chord *do mi so* at all pitches, then proceeds to its dominant *so ti re*, and finally to its subdominant *fa la do*, in every case drawing attention to the *character* of the notes in the scale. The Chevé system began by teaching the *melody*, *ut re mi fa sol*, and not advancing till this melody was thoroughly impressed on the mind of the pupil for any *ut*, taken backwards or forwards, or stopping at any note and beginning again at that note. Afterwards the system took the melody *ut¹ si la sol*, and treated

it in the same way. Finally the two were united as *ut re mi fa sol la si ut¹*. On these melodies all is founded, and the pupil is told to take any other intervals *by imagining the intermediate notes, without uttering them*, thus (the notes in roman letters being merely imagined), *ut re mi fa sol la si ut*. This is developed in Mme. Chevé's *Science et Art de l'intonation, théorie et pratique, systéme des points d'appui*, 1868. On the title-page she says: 'Les grands ressorts de notre méthode, pour l'étude de l'intonation, consistent en ceci: 1° Chercher les sons *un à un* et les émettre aussi *un à un*, en les détachant les uns des autres. Hors de là point de succès possible. 2° Se servir de deux rapports que l'on connaît, pour trouver un troisième rapport qu'on ignore; c'est-à-dire, aller du connu à l'inconnu; ce qui conduit à penser par degrés conjoints, en chantant par degrés disjoints,' May 1868. The two systems of Chevé and Curwen are therefore distinct in principle, value of the signs, form of the signs, notation of rhythm, and mode of teaching. They are alike in being taught without an instrument, but for very different reasons; in the Tonic Sol-fa to allow just intonation to become the pupil's guide; in the Chevé to allow of taking g♯ sharper than ab, and to make e to f the same as g♯ to a, but different from ab to a. They are also alike in having a movable *do* or *ut*, a very ancient device. And also alike in their nomenclature of lengths, 'langue des durées,' which was an original invention of Aimé Paris. As to priority of invention, Miss Glover taught her system in 1812, Galin published his in 1818. Both used tempered systems.—*Translator.*]

* [Usually called 'the Staff Notation' or 'the Old Notation' by the Tonic Sol-faists by way of distinction.—*Translator.*]

† [After a pupil has thoroughly acquired music on the Sol-fa notation, it becomes part of his duty to learn the other, and a course of instruction has been prepared for this purpose by Mr. Curwen, which when properly mastered (a comparatively easy task) puts the pupil in a condition to sing at sight from the old notation as readily as from the new.—*Translator.*]

twelve years of age, that performed singing exercises in a manner that astonished me ('mich in Erstaunen setzten') by the certainty with which they read the notes, and by the accuracy of their intonation.* Every year the London schools of Sol-faists are accustomed to give a concert of two to three thousand children's voices in the Crystal Palace at Sydenham, which, I have been assured by persons who understand music, makes the best impression on the audience by the harmoniousness and exactness of its execution.†

The Tonic Sol-faists, then, sing by natural, and not by tempered intervals. When their choirs are accompanied by a tempered organ, there are marked differences and disturbances, whereas they are in perfect unison with General Thompson's Enharmonic Organ. Many expressions used are very characteristic. A young girl had to sing a solo in *F* minor, and took it home to study it at her pianoforte. When she returned she said that the *A♭* and *D♭* on her piano were all wrong. These are the Third and Sixth of the key in which the deviation of tempered from just intonation is most marked. Another girl was so charmed with the Enharmonic Organ that she remained practising for three hours in succession, ¶ declaring that it was pleasant to be able to play *real notes*. Generally in a large number of cases, young people who have learned to sing by the Sol-fa method, find out by themselves, without any instruction, how to use the complicated manuals of the Enharmonic Organ, and always select the proper intervals.

Singers find that it is easier to sing to the accompaniment of this organ, and also that they do not hear the instrument while they are singing, because it is in perfect harmony with their voice and makes no beats.

* [On 20 April 1864, after we had heard Gen. Perronet Thompson's organ, I had the pleasure of taking Prof. Helmholtz to hear the singing of the children in the British and Foreign School here alluded to, which was situate behind the chapel in Tottenham Court Road. The master of the school, Mr. Gardiner, was a very good Tonic Sol-fa teacher, but the children were those who ordinarily attended (about forty were then present) and had received only ordinary instruction. After hearing them sing a few tunes in parts, from the Tonic Sol-fa notation, Prof. Helmholtz himself 'pointed' out an air on the 'modulator' or scale drawn out large on a chart, from which the pupils learn to sing (that is, by means of a pointer shewed the Tonic Sol-fa names of the tones the children were to sing), and the class followed in unison at sight. Then, on the suggestion of Mr. Gardiner, the class was divided into two sections, and Prof. Helmholtz pointed a piece in two parts, one with each hand, while the class took them at sight. Of course the piece was simple, but the dissonance of a Semitone was purposely introduced in one place between the parts, and Prof. Helmholtz was delighted at the firmness and correctness with which the children took it. I recollect his saying to me afterwards, 'We could not do that in Germany!' meaning, as he subsequently explained, that there was no German system of teaching to sing which could produce such results on such materials. The following is an extract of a letter from Prof. Helmholtz to Mr. Curwen printed on p. 159 of the *Memorials*, dated 21 April 1864, the day after his visit to the class : 'We were really surprised by the readiness and surety [certainty] with which the children succeeded in reading music that they did not know before, and in following a series of notes which were indicated to them on their modulatory board [modulator]. I think that what I saw shewed the complete success of your system, and I was peculiarly interested by it, because during my researches in musical acoustics I came from theoretical reasons to the conviction that this was the natural way of learning music, but I did not know that it had been carried out in England with such beautiful results.'—*Translator.*]

† [I am informed by the Secretary of the Tonic Sol-fa College that the first Crystal Palace Festival of the Tonic Sol-faists was held on 2 September 1857, with a choir of ¶ about 3200 children and 300 adults. These concerts have been continued year by year to the present time. For many years two concerts were given, one juvenile and one adult, the singers varying in number from 3500 to 5000. Some of these performances were so popular that a repetition was given a few weeks later. The plan of testing the great choirs in sight singing was first tried at the Festival on 14 August 1867, at which I was present, when an anthem specially written for the occasion by Mr. (now Sir) G. A. Macfarren (Professor of Music at the University of Cambridge, and Principal of the Royal Academy of Music) was sung by a choir of 4500 voices. Of the performance of this anthem Mr. Macfarren wrote a short time after in the *Cornhill Magazine* thus : 'A piece of music which had been composed for the occasion, and had not until then been ¶ seen by human eyes save those of the writer and the printers, was handed forth to the members of the chorus there present, and then, before an audience furnished at the same time with copies to test the accuracy of the performance, forty-five hundred singers sang it at first sight in a manner to fulfil the highest requirements of the severest judges.' Mr. Macfarren was himself present, and publicly expressed his own satisfaction.

Sight-singing tests have been given almost every year since, and always with the same success. They have become a common part of public concerts intended as 'demonstrations,' and are regarded by Tonic Sol-faists as no more extraordinary than reading the words at sight would be considered.—*Translator.*]

I have myself observed, that singers accustomed to a pianoforte accompaniment, when they sang a simple melody to my justly intoned harmonium, sang natural Thirds and Sixths, not tempered, nor yet Pythagorean. I accompanied the commencement of the melody, and then paused while the singer took the Third or Sixth of the key. After he had struck it, I touched on the instrument the natural, or the Pythagorean, or the tempered interval. The first was always in unison with the singer, the others gave shrill beats.

After this experience, I think that no doubt can remain, if ever any doubt existed, that *the intervals which have been theoretically determined in the preceding pages, and there called natural, are really natural for uncorrupted ears ; that moreover the deviations of tempered intonation are really perceptible and unpleasant to uncorrupted ears ; and lastly that, notwithstanding the delicate distinctions in particular intervals, correct singing by natural intervals is much easier than singing in tempered intonation.* The complicated calculation of intervals which the natural scale necessitates, and which undoubtedly much increases the ¶ manual difficulty of performance on instruments with fixed tones, does not exist for either singer or violinist, if the latter only lets himself be guided by his ear. For in the natural progression of correctly modulated music they have always and only to proceed by the intervals of the natural diatonic scale. It is only the theoretician who finds the calculation complicated, when at the end of numerous such progressions he sums up the result, and compares it with the starting-point.

That the natural system can be carried out by singers, is proved by the English Tonic Sol-faists. That it can also be carried out on bowed instruments, and is really carried out by distinguished players, I have no doubt at all after the experiments of Delezenne already mentioned (p. 325, note *), and what I myself heard when I was listening to the violinist who accompanied the Enharmonic Organ. Among the other orchestral instruments, the brass instruments naturally play in just intonation, and can only be forced to the tempered system by being blown out of tune.* The wooden instruments could have their tones slightly changed so as ¶ to bring them into tune with the rest. Hence I do not think that the difficulties of the natural system are invincible. On the contrary, I think that many of our best musical performances owe their beauty to an unconscious introduction of the natural system, and that we should oftener enjoy their charms if that system were taught pedagogically, and made the foundation of all instruction in music, in place of the tempered intonation which endeavours to prevent the human voice and bowed instruments from developing their full harmoniousness, for the sake of not interfering with the convenience of performers on the pianoforte and the organ.

Musicians have contested, in a very dogmatic manner, the correctness of the propositions here advanced. I do not doubt for a moment that many of these antagonists of mine really perform very good music, because their ear forces them to play better than they intended, better than would really be the case if they actually carried out the regulations of the school, and played exactly in Pythagorean or tempered intonation. On the other hand, it is generally possible to convince oneself from their very writings, that these writers have never taken the trouble to make a methodical comparison of just and tempered intonation. I can ¶ only once more invite them to hear, before uttering judgments, founded on an imperfect school-theory, concerning matters which are not within their own personal experience. Those who have no time for such observations, should at any rate glance over the literature of the period during which equal temperament was introduced. When the organ took the lead among musical instruments it was not yet tempered. And the pianoforte is doubtless a very useful instrument for making the acquaintance of musical literature, or for domestic amusement, or for accompanying singers. But for artistic purposes its importance is not such as to require its mechanism to be made the basis of the whole system of music.†

* [On this sentence Mr. Blaikley observes (*Proceedings of the Musical Association*, vol. iv. p. 56) : 'It seemed to me worth attention that this must be taken as being particularly and not generally true : that is, that though the ideal brass instrument has such characteristics, this ideal is not necessarily attained to in practice.' See pp. 99 and 100.—*Translator.*]

† [This last paragraph, from 'Musicians have contested' to 'the whole system of music,' is an addition to the 4th German edition. The remainder of this Appendix, which concludes the work in the 3rd German edition, was occupied with a description of the musical notation which I employed in my footnotes and Appendix to the 1st English edition ; but

APPENDIX XIX.

PLAN OF MR. BOSANQUET'S MANUAL.

(See p. 328c.)

THE accompanying figure 66 [taken by permission from p. 23 of Mr. R. H. M. Bosanquet's *Elementary Treatise on Musical Intervals and Temperament,* 1876]

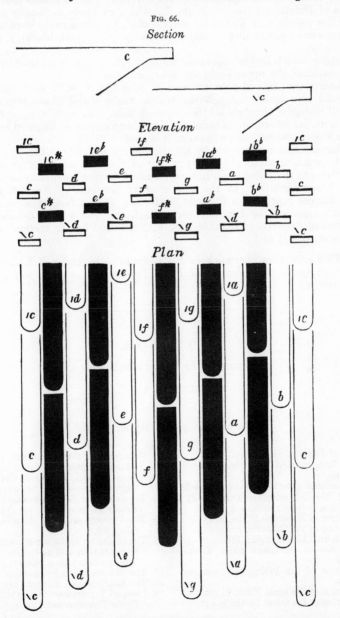

FIG. 66.

Section

Elevation

Plan

shews the arrangement of a part of this manual for 53 equal divisions of the octave.   The upper division gives a longitudinal section of two digitals standing one above the other. All the digitals are of the same form and differ only in colour. The middle part of the figure presents an elevation of the front ends of these digitals. In the lower part, there is a view as seen from above [plan]. Proceeding from one of the tones, as c, upwards and backwards, leaping over an intermediate digital in the way, we pass to d and e [on white digitals], and then continuing by major Seconds we pass to $f\sharp$, $g\sharp$, $a\sharp$ [on black digitals], and finally $b\sharp$ or $/c$ [on a white digital again]. The sign $/$ means, as has been explained in the text (p. 329b), sharpening by a comma $\frac{77}{76}$ [or one of the 53 degrees $= 22\cdot642$ cents] and is very nearly equivalent to our superior [1] [which means sharpening by a comma of Didymus $\frac{81}{80} = 21\cdot5$ cents, for which 22 cents is usually employed]. Between the members of this series are inserted, on the digitals leapt over, those of another series proceeding by major Seconds, $d\flat$, $e\flat$ [both black], f, g, a, b, $c\sharp$ [all five white].

¶ The series which lie just above one another differ from each other by a comma of the same kind, the upper being the sharper.

In playing the scale of c major [*] as c, d, $\diagdown e$, f, g, $\diagdown a$, $\diagdown b$, c', observe that a horizontal line drawn through the points where d and g are printed in the figure will just pass through all the required keys. At $\diagdown e$, $\diagdown a$, $\diagdown b$, we thus come on a deeper intermediate series.[†] Every major scale is fingered in precisely the same way no matter with what note it begins.

The harmonium constructed by Mr. Bosanquet distributes the 53 tones over 84 digitals, some of those at the upper part of the manual being identical with some of those at the lower part, in order to avoid having frequently to jump from upper to lower digitals. In the system of 53 divisions $///b = \diagdown\diagdown c$, since five smallest degrees represent a diatonic Semitone. [For a further account of Mr. Bosanquet's notation see App. XX. sect. A. art. 27. For a more detailed plan of his generalised fingerboard see *ibid.* sect. F. No. 8, and for his methods of tuning see *ibid.* sect. G. art 16.]

¶ ** [*This concludes the work in the German. Appendix XX. has been entirely written by the Translator, and Prof. Helmholtz is in no respect responsible for its contents.*]

APPENDIX XX.

ADDITIONS BY THE TRANSLATOR.

SECTION A.

ON TEMPERAMENT.

(See notes pp. 30, 281, 315, and 329.)

---

[*] [In the German edition a cross was placed on the digitals of the plan which were played in this key, but the German copy could not be used here because it followed German musical notation. The text has been altered accordingly.—*Translator.*]

[†] [A horizontal line through b in the figure will pass through $c\ d\ e\ f\ g\ a\ b\ c$, and thus give the Pythagorean major scale. *Translator.*]

Art. 1.—The object of temperament (literally 'tuning'), is to render possible the expression of an indefinite number of intervals by means of a limited number ¶ of tones without distressing the ear too much by the imperfections of the consonances. The general practice has been from the earliest invention of the keyboard of the organ to the present day to make 12 notes in the Octave suffice. This number has been in a very few instances increased to 14, 16, 19, and even to 31 and 53, but such instruments have never come into general use.

Art. 2.—The system which tuners at the present day intend to follow, though none of them absolutely succeed in so doing (see infrà, sect. G.), is to produce 12 notes reckoned from any tone exclusive to its Octave inclusive, such that the Octave should be just and the interval between any two consecutive notes, that is, the ratio of their pitch numbers, should be always the same. This is known as *Equal Temperament* (see suprà, pp. 320b to 327c). The interval between any two notes is an *Equal Semitone*, and its ratio is $1 : \sqrt[12]{2} = 1 : 1.0594631$, or very nearly 84 : 89. If we further supposed that 99 other notes were introduced so as to make 100 equal intervals between each pair of equal notes, these intervals would be those here termed *Cents*, having the common ratio $1 : \sqrt[1200]{2} = 1 : 1.0005778$, or very ¶ nearly 1730 : 1731. As the human ear is, except in very rare cases, insensible to the interval of a cent, we need not divide further, except occasionally for purely theoretical purposes, to avoid errors of accumulation, as in this section, when even the thousandth part of a cent may have to be dealt with. In practice the errors of tuning would soon far exceed the errors arising from systematically neglecting amounts of less than half a cent. The mode of finding cents from the ratios of pitch numbers, wave lengths, or vibrating lengths, is given in sect. C., and the values of most of the usually recognised intervals are represented in cents in sect. D. From these we take, up to 3 places of decimals,

One just Fifth	= 701.955 cents
" " major Third	= 386.314 "
" " Comma	= 21.506 "
" " Skhisma	= 1.954 "

Art. 3.—No recurrence of notes formed by taking intervals of Fifths, major ¶ Thirds and Octaves is possible because no powers of the numbers $\frac{3}{2}$, $\frac{5}{4}$, 2, or of any combination of them, however often repeated, can produce a power of any single one of them. When we only proceed to 3 places of decimals of cents (then of course using multiples for powers), there would be a recurrence, but so remote that it would be practically at an infinite distance, and would after all only arise from our not having carried the decimals far enough. The nearest approximations of any practical value are given in arts. 4, 6, 9, 11, 12.

Art. 4.—Four Fifths up and two Octaves down, together with a major Third down, give the Comma (of Didymus ratio, 80 : 81, which is always intended when no qualification is added), that is, in cents,

$$4 \times 701.955 - 2 \times 1200 - 386.314 = 21.506.$$

Art. 5.—Consequently if we used Fifths diminished by the small but sensible interval of a quarter of a Comma, that is 701.955 − ¼ × 21.506 or 696.578 cents, four of these Fifths would be precisely two Octaves more than an exact major Third. These (for a reason given in art. 16) were called meantone Fifths, and were long in use.

Art. 6.—Eight Fifths up and also a major Third up, with five Octaves down, give the Skhisma, that is, in cents,

$$8 \times 701\cdot955 + 386\cdot314 - 5 \times 1200 = 1\cdot954.$$

From this we can deduce two different usages.

Art. 7.—If we employed Fifths diminished by the insensible interval of $\frac{1}{8} \times 1\cdot954$, or $701\cdot955 - \cdot24425 = 701\cdot71075$, eight of these Fifths added to a just major Third would give exactly five Octaves. These are the Fifths proposed suprà, p. 316a, and may hence be called Helmholtz's.

Art. 8.—But if we diminished the major Third by a Skhisma, giving $386\cdot314 - 1\cdot954 = 384\cdot360$ cents, which may be called a Skhismic major Third, then this major Third added to five Octaves will give exactly eight Fifths. This is the relation which Prof. Helmholtz pointed out (as existing, but not designed) in medieval Arabic scales, suprà, p. 281a, and will be called Skhismic.

Art. 9.—Twelve Fifths up and seven Octaves down give the sum of a Comma ¶ and a Skhisma, known as the Pythagorean Comma, that is, in cents,

$$12 \times 701\cdot955 - 7 \times 1200 = 23\cdot460 = 21\cdot506 + 1\cdot954.$$

Art. 10.—Hence if we used Fifths diminished by the scarcely perceptible interval of $\frac{1}{12} \times 23\cdot460 = 1\cdot9550$, or $701\cdot955 - 1\cdot955 = 700$ cents, twelve of these Fifths, known as equal Fifths, would give exactly seven Octaves. These are the Fifths now in general use. The amount subtracted, $1\cdot95500$, is very nearly the Skhisma, which, more fully calculated, has $1\cdot95372$ cents. The difference between these two intervals is far beyond all powers of appreciation by any acoustical contrivance. The Skhisma will therefore be considered as the twelfth part of a Pythagorean Comma, and also as the error of an equal Fifth. See p. 316d.

Art. 11.—One Octave up and three major Thirds down give the difference between two Commas and a Skhisma, known as 'the Great Diësis,' that is, in cents,

$$1200 - 3 \times 386\cdot314 = 41\cdot059 = 2 \times 21\cdot506 - 1\cdot954.$$

¶   Art. 12.—Fifty-three Fifths up and thirty-one Octaves down give what may be called a Mercatorial, because on it depends the advantage arising from the use of Mercator's 53 division of the Octave. It is less than two Skhismas by about one-third of a cent, that is, in cents,

$$53 \times 701\cdot955 - 31 \times 1200 = 3\cdot615 = 2 \times 1\cdot954 - \cdot293 \ \ldots\ldots\ldots \ (5)$$

Consequently, as $\frac{1}{53} \times 3\cdot615 = \cdot068$, if we used Fifths which were too flat by this imperceptible interval, or had $701\cdot955 - \cdot068 = 701\cdot887$ cents (which may be called Mercator's Fifths, from their inventor), we should have precisely 53 Mercator's Fifths = 31 Octaves $\ldots\ldots\ldots\ldots\ldots\ldots\ldots\ldots\ldots\ldots\ldots\ldots\ldots\ldots\ldots\ldots\ldots\ldots\ldots\ldots\ldots\ldots$ (6)

On these relations depend all systems of temperament which are worth consideration.

Art. 13.—Let us suppose that, measured in cents, in any system of temperament, V represents the Fifth adopted, T the major Third adopted, and K and S the Comma and Skhisma adopted, so that $K+S$ will correspond to the Pythagorean ¶ Comma, and $2K-S$ to the Great Diësis. Then as the four first relations must hold for these tempered intervals, and an Octave has 1200 cents, we must have

from $\ldots\ldots\ldots\ldots\ldots$art. 9 ; $12V - 8400 = K+S$, whence $V = 700 + \frac{1}{12}(K+S)$
from $\ldots\ldots\ldots\ldots\ldots$art. 11 ; $1200 - 3T = 2K-S$, whence $T = 400 - \frac{1}{3}(2K-S)$

And deducing the values of K and S from these equations,

$K = 4V - T - 2400$, which is the relation in art. 4,
$S = 8V + T - 6000$, which is the relation in art. 6.

So that there are only 2 independent equations connecting the four intervals V, T, K, S. Hence, on assuming values for any two of them we may find corresponding values for the other two. But no results are of any European interest unless V and T both approximate very closely to the just values $701\cdot955$ and $386\cdot314$ cents.

Art. 14.—There are two quite different kinds of temperament, the Linear and the Cyclic. The Linear contains an endless series of notes which never recur in pitch. The Cyclic contains also an endless series of notes which, however, do recur in pitch, although usually under different names. Hence in Cyclic temperaments all the intervals are made up of aliquot parts of an Octave, or 1200 cents, which is not the case in Linear temperaments. In both of them the main object is to substitute a series of tempered Fifths for the several series of Fifths and major Thirds introduced, suprà, p. 276a, and exhibited at full in the Duodenarium, infrà, sect. E. art. 18. The advantage of the Cyclic over the Linear temperaments consists chiefly in a power of endless modulation—a very questionable advantage when harmoniousness is sacrificed to it.

LINEAR TEMPERAMENTS.

Art. 15.—*The Pythagorean or Ancient Greek Temperament.*

Assume $V=701\cdot955$ and $K=0$.

Then from art. 16, $S=12V-8400=23\cdot460$ ¶

and $T=400+\tfrac{1}{3}.S=407\cdot820$

$=386\cdot314+21\cdot506$

The major Third is a whole Comma too sharp, and hence this system is quite unfit for harmony. It was the theoretical Greek scale, and is still much used by violinists. See Cornu and Mercadier's experiments, infrà, sect. G. art. 6 and 7. The following are the 27 tones which this temperament would require for ordinary modulations, with the cents in the intervals from the lowest note, the logarithms of those interval ratios, and the pitch numbers to c′ 264.

Pythagorean Intonation.

No.	Note	Cents	Log	Pitch	No.	Note	Cents	Log	Pitch
1	c′	0	0	264·0	15	f′♯	611·7	15346	375·9
2	b♯	23·5	00589	267·6	16	a′b b	678·5	17021	390·7
3	d′b	90·2	02263	278·1	17	g′	702·0	17609	396·0
4	c′♯	113·7	02852	281·9	18	f′♯ ♯	725·4	18198	401·4
5	e′b b	180·5	04527	293·0	19	a′b	792·2	19873	417·2
6	d′	203·9	05115	297·0	20	g′♯	815·6	20461	422·9
7	c′♯ ♯	227·4	05704	301·1	21	b′b b	882·4	22136	439·5
8	e′b	294·1	07379	312·9	22	a′	905·9	22724	445·5
9	d′♯	317·6	07967	317·2	23	g′♯ ♯	929·3	23313	451·6
10	f′b	384·4	09642	329·6	24	b′b	996·1	24988	469·3
11	e′	407·8	10231	334·1	25	a′♯	1019·6	25576	475·7
12	f′	498·0	12494	352·0	26	c′b	1086·3	27251	494·4
13	e′♯	521·5	13082	356·8	27	b′	1109·8	27840	501·2
14	g′b	588·3	14757	370·8	1′	c″	1200·0	30103	528·0

These can be all exhibited and calculated as a series of 26 perfect Fifths up, namely,

ab♭ e♭♭ b♭♭, f♭ c♭ g♭ d♭ a♭ e♭ b♭, f c g d a e b,
f♯ c♯ g♯ d♯ a♯ e♯ b♯, f♯♯ c♯♯ g♯♯   ¶

The 17 notes of the medieval Arabic scale (suprà, p. 281c′) are those in the first line of this series, adding d♭♭ at the beginning, and omitting b at the end, with all the ♯ and ♯♯ notes in the Table.

Art. 16.—*The Meantone Temperament.*

The major Thirds are assumed to be perfect, and the Comma is left out of account. Hence,

$T=386\cdot314$, $K=0$. Whence by art. 13
$V=696\cdot578$ as in art. 5, and $S=-41\cdot059$.

Consequently the Second of the scale, which is always two Fifths less an Octave, will have 193·157 cents or be half a Comma or 10·753 cents flatter than the just major Tone of 203·910 cents too flat, and hence by the same amount sharper than the just minor Tone of 182·404 cents. From this *mean* value of the *Tone* the temperament receives its name. This was the temperament which prevailed all over the Continent and in England for centuries, and for this, and the Pythagorean, our musical staff notation was invented, with a distinct difference of meaning between sharps and flats, although that difference was different in each of the two cases. For the history of its invention see infrà, sect. N. No. 3. This temperament disappeared from pianofortes in England between 1840 and 1846. (See infrà, sect. N. No. 4.) But at the Great Exhibition of 1851 all English organs were thus tuned. If carried out to 27 notes bearing the same names as in art. 15, but having the different values in the following table, it would probably have still remained in use. Handel in his Foundling Hospital organ had 16 notes, tuned from db to $a\sharp$ in the series of Fifths in art. 15. Father Smith on Durham Cathedral and the Temple organ had 14 notes from ab to $d\sharp$, and the modern English concertina has the same compass and uses the same temperament, and the same number of notes. The only objection to this temperament was that the organ-builders, with rare exceptions, such as those just mentioned (see also 320c and note §), used only 12 notes to the Octave, $eb\ bb\ f\ c\ g\ d\ a\ e\ b f\sharp\ c\sharp\ g\sharp$. The consequence was that in place of the chords $ab\ c\ eb$, $f\ ab$ c, &c., organists had to play $g\sharp\ c\ eb$, $f\ g\sharp\ c$, where $g\sharp$ was a Great Diĕsis (41·059 cents) too flat, and the horrible effect was familiarly compared to the howling of 'wolves.' Similarly for $b\ d\sharp\ f\sharp$ it was necessary to use $b\ eb$ $f\sharp$, eb being a Great Diĕsis too sharp, with similar excruciating effects. In modern music it is quite customary to use keys requiring more than two flats and three sharps, and hence this temperament was first styled ' unequal ' (whereas the *organ*, not the *temperament* was—not unequal, but—*defective*) and then abandoned. But with the 27 notes here given there would have been nothing to offend the ears of Handel and Mozart. At the present day, ears accustomed to the sharp leading note of the equal temperament (where $b:c'$ has 100 cents) are shocked at the flat leading note of the meantone temperament when $b:c'$ has 117·1 cents. But played with 27 or 36 digitals on Mr. Bosanquet's generalised keyboard (Appendix XIX. and also XX., sect. F. No. 8) it is the only temperament suitable to the organ. In my examination of 50 temperaments (*Proc. Royal Soc.*, vol. xiii. p. 404) I found that this was decidedly the best for harmonic purposes. For simple melody perhaps the Pythagorean is preferred by violinists, but that was always absolutely impossible for harmony.

Meantone Intonation.

No.	Notes	Cents	Logs	Handel	Smart	Helmholtz	Durham
1	c'	0	0	252·7	259·1	264·0	283·6
2	$c'\sharp$	76·1	01908	264·1	272·0	275·9	296·3
3	$d'b$	117·1	02938	270·4	277·3	282·5	303·4
4	$c'\sharp\sharp$	152·1	03816	275·9	282·9	288·2	309·6.
5	d'	193·2	04846	282·5	289·7	295·2	317·1
6	$e'bb$	234·2	05876	289·3	296·7	302·2	324·7
7	$d'\sharp$	269·2	06753	295·2	302·7	308·4	331·3
8	$e'b$	310·3	07783	302·8	310·0	315·8	339·2
9	e'	386·3	09691	315·9	323·9	330·0	354·5
10	$f'b$	427·4	10721	328·5	331·7	337·9	363·0
11	$e'\sharp$	462·4	11599	330·1	338·4	344·8	370·4
12	f'	503·4	12629	338·0	346·6	353·1	379·2
13	$f'\sharp$	579·5	14537	353·2	362·1	369·0	396·3
14	$g'b$	620·5	15567	361·7	370·8	377·8	405·8
15	$f'\sharp\sharp$	655·5	16444	369·0	378·4	385·5	414·1
16	g'	696·6	17474	377·9	387·5	394·8	424·0
17	$a'bb$	737·6	18504	387·0	396·8	404·2	434·2
18	$g'\sharp$	772·6	19382	394·9	404·9	412·5	443·1
19	$a'b$	813·7	20412	404·3	414·6	422·4	453·7
20	$g'\sharp\sharp$	848·7	21290	412·6	423·0	431·0	463·0
21	a'	889·7	22320	422·5	433·2	441·4	474·1
22	$b'bb$	930·8	23350	432·6	443·6	452·0	485·5
23	$a'\sharp$	965·8	24228	441·5	452·7	461·2	495·4
24	$b'b$	1006·8	25258	452·1	463·5	472·3	507·3
25	b'	1082·9	27165	472·4	484·3	493·5	530·1
26	$c'b$	1124·0	28195	483·7	495·8	505·3	542·8
27	$b'\sharp$	1158·9	29073	493·6	508·4	515·6	553·9
1'	c''	1200·0	30103	505·4	518·2	528·0	567·2

On account of the great historical interest attaching to this temperament, I give the whole 27 notes, shewing their value in cents and logarithms, whence the pitch numbers can be calculated out for any pitch, and I have actually calculated them out for 4 pitches. That headed ' Handel ' has A 422·5, the pitch of Handel's own fork, the common pitch of Europe for two centuries (see infrà, sect. H.). The piano of the London Philharmonic Society was tuned to A 423·7 or very nearly this pitch when that Society was founded in 1813. But about 1828 Sir George Smart, the conductor of that Society's concerts, raised the pitch to A 433·2, as I have determined from his own fork, and column ' Smart ' gives the notes for this pitch. As Sir George considered the fork C 518 to correspond to his A 433·2 (it is only ·2 vib. too flat), he manifestly used meantone temperament even so late as this, for the equal C to A 433·2 would have been much flatter, namely C 515·1. This was a very curious anticipation of the French pitch of 1859. The next pitch which takes Helmholtz's *c'* 264 for comparison, gives *a'* 441·4, and Father Smith's pitch for the Hampton Court Palace organ, determined from an unaltered pipe, 1690, was *a'* 441·7. Hence this was a regular meantone pitch. The last column shews the extraordinarily high pitch used by Father Smith for the Durham Cathedral organ, determined from an original *g'♯* pipe, found with all the others by the organist Dr. Armes and measured by me. Now omitting Smart's pitch, which does not belong to the old organ period, a curious relation will be found to connect the other three. Handel's *c'♯* is Helmholtz's *c'*, and this pitch was therefore a 'small Semitone' of 76·1 cents sharper, and Handel's *d'* 282·5 is practically the Durham *c'* 283·6, which was therefore a Tone of 193·2 cents sharper than Handel's pitch.

Art. 17.—*The Skhismic Temperament.*

The condition is that the Fifths should be perfect and the Skhisma should be disregarded. This gives

$$V = 701\text{·}955, \quad S = 0, \text{ and hence by art. 13, } K = 23\text{·}460,$$
$$T = 400 - \tfrac{2}{3}K = 384\text{·}360 = 386\text{·}314 - 1\text{·}954 \text{ as in art. 8.} \qquad ¶$$

That is the major Third is too flat by a Skhisma, whence the name of the temperament (see suprà, p. 281*a*). The effect of this flat major Third is very good indeed. On looking at the Table in art. 15 we see that *c'* : *f'b* is such a major Third, and looking to the lists of Fifths given below the Table, we see that *f*b is the eighth Fifth below *c*, which follows from art. 13 whenever *S* = 0. Hence generally the notes in the top line will be major Thirds above those in the bottom line respectively—

abb	*ebb*	*bbb*	*fb*	*cb*	*gb*	*db*	*ab*	*eb*	*bb*	*f*	*c*	*g*	*d*	*a*	*e*	*b*	*f♯*	*c♯*
eb	*bb*	*f*	*c*	*g*	*d*	*a*	*e*	*b*	*f♯*	*c♯*	*g♯*	*d♯*	*a♯*	*e♯*	*b♯*	*f♯♯*	*c♯♯*	*g♯♯*

Having an English concertina (which has 14 notes) tuned in perfect Fifths from *g*b to *c*♯ in the series in art. 15, I have been able to verify this result for six of the major Thirds, and to determine that although *a c♯ e*, *e g♯ b*, are horrible chords, *a d♭ e*, *e a♭ b* are quite smooth and pleasant. The major Third *c'* : *f'*b of art. 15 beats 16 times in 10 seconds, which is scarcely perceptible and far from disagreeable. But it is evident that if this system of tuning were adopted, a different musical notation would be necessary, and a convenient typographical modification of Mr. Bosanquet's will be explained on p. 438*d*.

Art. 18.—*The Helmholtzian Temperament.* ¶

In this case the major Thirds are taken perfect, and the Skhisma is disregarded. Then by art. 13,

$$K = 20\text{·}534 = 21\text{·}506 - 1\text{·}072$$
$$\text{and } V = 701\text{·}711 = 701\text{·}955 - \tfrac{1}{8} \times 1\text{·}954 \text{ as in art. 7.}$$

The Comma and the Fifth are therefore imperceptibly flattened. In this case also the major Third would be the eighth Fifth down. And the same reason for altering the notation would hold as for art. 17.

Art. 19.—In their endeavours to avoid the 'wolves' of meantone temperament musicians invented numerous really *unequal* temperaments, which it would be uncharitable to resuscitate. There is, however, a really practicable unequal temperament which I call *Unequally Just*, but it cannot well be explained till the Duodenarium has been developed. (See infrà, sect. E. art. 25.) I proceed, therefore, to the consideration of the

CYCLIC TEMPERAMENTS. ¶

Art. 20.— The Octave of 1200 cents being divided into different sets of aliquot parts called *degrees*, certain numbers of those degrees may approach to the value of the just Fifth 701·955 and major Third 386·314, and from these the whole scale may be constructed, each interval being more or less well represented by a certain number of degrees. There would then be this advantage, that the number of *values* of notes (whatever happened to their *names*) would be strictly limited by the number of degrees in the Octave, and hence values would recur, and the whole scale could always be expressed by a limited number of cyclic Fifths.

Art. 21.—The equations for finding such cycles may be immediately derived from those in art. 13, thus :

Let *m* be the number of degrees, and *n* = 1200 ÷ *m* be the number of cents in one degree, so that 1200 = *mn*. Put *V* = *nv*, *T* = *nt*, *K* = *nk*, *S* = *ns* in the four equations art. 13, and divide out by *n*. Then 12*v* − 7*m* = *k* + *s*, *m* − 3*t* = 2*k* − *s*, whence *k* = 4*v* − *t* − 2*m*, *s* = 8*v* + *t* − 5*m*. On assuming values for *m*, beginning at 12, and putting first *k* = 0, and then *s* = 1, 2, 3, &c., or − 1, − 2, − 3, &c., and next

$s=0$, and then $k=1$, 2, 3, or $-1,-2,-3$, &c., we get the corresponding values of v and t, whence the scale. Most such scales would be useless. For practical tuning m should be small, and $v:m$, $t:m$ should be nearly the ratios of the numbers of cents in the just Fifth and major Third to 1200 (or of the logarithms of their interval ratios to $\log 2 = \cdot30103$). Now the approximate values of

$$\frac{701\cdot955}{1200} \text{ are } 1, \frac{1}{2}, \frac{3}{5}, \frac{7}{12}, \frac{31}{53}, \text{ &c.},$$

and of $\dfrac{386\cdot314}{1200}$ are $\dfrac{1}{3}=\dfrac{4}{12}$, $\dfrac{9}{28}=\dfrac{10-\frac{1}{28}}{31}$, $\dfrac{19}{59}=\dfrac{17\frac{4}{59}}{53}$, &c.

Next take the Pythagorean major Third from art. 15. The approximate values of

$$\frac{407\cdot810}{1200} \text{ are } \frac{1}{2}, \frac{1}{3}, \frac{17}{50}=\frac{18\frac{1}{59}}{53}, \text{ &c.}$$

Finally, take the meantone Fifth from art. 16. The approximate values of

$$\frac{696\cdot578}{1200} \text{ are } 1, \frac{1}{2}, \frac{4}{5}, \frac{5}{7}, \frac{9}{12}, \frac{14}{19}, \frac{23}{31}, \text{ &c.}$$

Art. 22.—These numbers suggest cycles of 12, 31, and 53 degrees.

i. The cycle of 12 with a Fifth of 7 and a major Third of 4 degrees would imitate Pythagorean intonation well and just intonation indifferently. It is the equal temperament of to-day. Here $m=12$, $v=7$, $t=4$, $k=s=0$. In cents, one degree = 100, $V=700$, $T=400$, $K=S=0$. See art. 25.

ii. The cycle of 31, with a Fifth of 23 and a major Third of 10 degrees would imitate meantone temperament very closely. It is the Harmonic Cycle of Huyghens. Here $m=31$, $v=18$, $t=10$, $k=0$, $s=-1$. In cents, one degree = 38·710, $V=696\cdot773$, $T=387\cdot097$, $K=0$, $S=-38\cdot710$.

iii. The cycle of 53, with a Fifth of 31 and a major Third of 18 degrees, is an extremely close approximation to Pythagorean temperament. It is Mercator's cycle. Here $m=53$, $v=31$, $t=18$, $k=0$, $s=1$. In cents, one degree = 22·642, $V=701\cdot886$, $T=407\cdot547$, $K=0$, $S=22\cdot642$.

iv. The cycle of 53, with the same Fifth of 31, but a major Third of 17 degrees would give a sufficiently close approximation to just intonation. (See p. 328d.) As it seems to have been first supplied with a fingerboard by Mr. Bosanquet, it is properly called Bosanquet's cycle, but, as will be seen (infrà, sect. F. No. 9), Mr. J. Paul White has also invented a keyboard for it. Long previously to either Gen. T. Perronet Thompson used it extensively in his works on the *Enharmonic Guitar* and *Just Intonation* as a convenient approximation to just intonation, and from his works it was introduced into the Tonic Sol-fa books for the same purpose. Here $m=53$, $v=31$, $t=17$, $k=1$, $s=0$. In cents, one degree $=22\cdot642$, $V=701\cdot886$, $T=384\cdot905$. Observe that in art. 21, Skhismic $V=701\cdot955$ is only the imperceptible interval ·069 cents sharper, and Skhismic $T=384\cdot360$ is only the imperceptible interval ·565 cents flatter. Hence Skhismic intonation and Bosanquet's cycle are audibly interchangeable within the limits of a few keys. It is only when the modulation beyond 53 degrees is required that the cyclic intonation has the advantage. See art. 27.

Art. 23.—Besides these the following temperaments have been at least proposed. They chiefly depend upon assigning imaginary or arbitrary evaluations of the ratio of a Tone to a Semitone, the Comma being neglected, and the Octave supposed to consist of 5 Tones and 2 Semitones, so that if Tone : Semitone$=p:q$, the number of degrees of the cycle will be $5p+2q$, and $v=3p+q$, $t=2p$, $k=0$, $2k-s$, the Diësis $=-s$ variable. This applies to all temperaments where $K=0$, thus (art. 22) in i., $p:q=2:1$; in ii., $p:q=5:3$; in iii., $p:q=9:4$.

We thus obtain, among others,

v. Woolhouse's cycle of 19, $p:q=3:2$, $v=11$, $t=6$, $k=0$, $s=-1$, $2k-s=1$. In cents, one degree = 63·16, $V=694\cdot76$, $T=378\cdot96$.

vi. Chevé's cycle of 29, $p:q=5:2$, $v=17$, $t=10$, $k=0$, $s=1$, $2k-s=-1$. In cents, one degree $=41\cdot380$, $V=703\cdot460$, $T=413\cdot80$.

vii. Sauveur's cycle of 43 mérides, $p:q=7:4$, $v=25$, $t=14$, $k=0$, $s=-1$, $2k-s=1$. In cents, one degree = 27·907, $V=697\cdot674$, $T=390\cdot698$.

viii. The Musician's cycle of 55, in 1755, according to Sauveur and Estève, $p:q=9:5$, $v=32$, $t=18$, $k=0$, $s=-1$, $2k-s=1$. In cents, one degree = 21·818, $V=698\cdot176$, $T=392\cdot724$.

ix. Henfling's cycle of 50, in 1710, $p:q=8:5$, $v=29$, $t=16$, $k=0$, $s=-2$. In cents, one degree = 24, $V=696$, $T=384$, $K=0$, $S=-48$.

Both Fifth and Third are much too flat in v. and too sharp in vi. Both vii. and viii. were decent approximations, and convenient on paper. It would not have been worth while to produce them on instruments.

Art. 24.—But paper cycles are sometimes extremely useful for the purposes of calculation, as in the following cases.

x. Cycle of 30103 jots (de Morgan's name), $v = 17609$, $t = 9691$, $k = 539$, $s = 48$, $2k-s = 1030$. In cents, one degree = ·039863, $V = 701·950$, $T = 386·3135$, $K = 21·4862$, $S = 1·91343$.

This is therefore an exceedingly accurate representation of just intonation. It is derived from $10000 \times \log 2$, using 5 place logarithms, and the only reason for the differences from just intonation is that it is taken strictly to the nearest integer. Thus $10000 \times \log \frac{3}{2} = 17609·13$, and this would have made V correct. Mr. John Curwen used this in his *Musical Statics*, to avoid logarithms.

xi. Cycle of 3010 degrees, $v = 1761$, $t = 969$, $k = 55$, $s = 7$, $2k-s = 103$. In cents, one degree = ·39871, $V = 702·060$, $T = 386·3135$, $K = 21·9269$, $S = 2·7907$. This is derived from $1000 \times \log 2$ to 4 places, and is consequently not quite so accurate as the last.

xii. Cycle of 301 degrees, $v = 176$, $t = 97$, $k = 5$, $s = 0$, $2k-s = 10$. In cents, one degree = $3·9866$, $V = 701·661$, $T = 386·711$, $K = 19·93355$, $S = 0$.

This was the cycle used by Sauveur (*Mém. de l'Académie*, 1701, p. 310) as a finer division than was given by his cycle of 43 mérides (see vii.). As $301 = 7 \times 43$, he called each degree a heptaméride, which he made = ·03987 of an (equal) Semitone. He also gives a rule for finding the number of heptamérides in any interval under $6 : 7 = 267$ cents, which is the equivalent of my rule for finding cents (infrà, sect. C., I. 4, note), only my rule extends to 498 cents. Sauveur's rule is: multiply 875 by the differences of the interval numbers and divide by their sum. For $6 : 7$, this gives 67 heptamérides, and as $\log \frac{7}{6} = ·067$ to three places, this is correct. It is the earliest instance I have met with of the bimodular method of finding logarithms. Sauveur's 875 is an augmented bimodulus for 869, for the same reason as I selected 3477 in place of 3462, p. 447c'.

I have here taken the values of v and t as they ought to be, but judging by vii. Sauveur took $v = 175$ and $t = 98$, and hence got the results there given, which agree better with meantone intonation.

Observe that the Helmholtzian $V = 701·711$ ¶ is only ·050 cents, or imperceptibly larger, and the cyclic T is only ·397 cents, also imperceptibly larger than just. Hence the Helmholtzian intonation and Sauveur's cycle of 301 are interchangeable within 301 degrees.

xiii. Cycle of 1200, or the Centesimal Cycle. If we refrain from using decimals of cents, we really use a cycle where one degree = 1 cent, $v = V = 702$, $t = T = 386$, $k = K = 22$, $s = S = 2$, $2k-s = 42$. These are quite imperceptibly different from the just for a single key, but when modulation is extended, the relative value of distant notes to the starting note will be slightly altered. See infrà, art. 28. In the body of this work and after this section 'cyclic cents,' as they may be called, will be used, unless accumulated fractions of a cent become sensible. But in the investigation of this section it was necessary to shew differences much more minute than a single cent. ¶

Art. 25.—The only cycles requiring further attention, then, are i., the Equal, and iv., Bosanquet's. The method of tuning Equal Temperament is given infrà, sect. G. art. 10. The pitch of the notes used is very variable. Six principal pitches are tabulated below.

Equal Intonation.

No.	Notes	Cents	Logs	Italian Military i.	French Normal ii.	Scheibler's Stuttgard iii.	Society of Arts iv.	English Band v.	Schnitger, 1688 vi.
I	c'	0	0	255·9	258·6	261·6	264·0	268·75	290·9
2	$c'\sharp$	100	02509	271·1	274·0	277·2	279·7	284·7	308·2
3	d'	200	05017	287·3	290·3	293·7	296·3	301·7	326·4
4	$d'\sharp$	300	07526	304·3	307·6	311·1	314·0	319·6	345·9
5	e'	400	10034	322·4	325·9	329·6	332·6	338·6	366·5
6	f'	500	12543	341·6	345·3	349·2	352·4	358·8	388·3
7	$f'\sharp$	600	15052	361·9	365·8	370·0	373·4	380·1	411·4
8	g'	700	17560	383·4	387·6	392·0	395·5	402·7	435·8
9	$g'\sharp$	800	20069	406·2	410·6	415·3	419·1	426·6	461·8
10	a'	900	22577	430·4	435·0	440·0	444·0	452·0	489·2
11	$a'\sharp$	1000	25086	456·0	460·9	466·2	470·4	478·9	518·3
12	b'	1100	27594	483·1	488·3	493·9	498·4	507·4	549·1
I'	c''	1200	30103	511·8	517·2	523·2	528·0	537·5	581·8

¶

i. The pitch officially adopted for Italian military bands in August 1884. The standard was a $B\flat 456$, because $B\flat$ can be produced on the brass instruments without using the valves. It is really the nearest approach in whole numbers to the old arithmetical pitch of $c''512$.

ii. French diapason normal intentionally $a'435$, giving equal $c''517·3$, is practically the same as Smart's pitch, which is the lowest that has been used for equal temperament in England, and was contemporary with its introduction there. On 19 March 1885 this was also officially adopted as the pitch of Belgian military bands, which had hitherto used $A451·7$, or say $A452$, as given in col. v.

iii. Scheibler's pitch, proposed at Stuttgardt, often called the German pitch, having $A440$.

iv. The pitch proposed by the Society of Arts, known in Germany as English pitch, having $c''528$, and $a'444$. Unfortunately the original fork tuned for the Society of Arts by Griesbach proved to be $c''534\cdot46$, equivalent to equal $a'449\cdot4$, and commercial copies vary from $c''533\cdot3$ to $c''535\cdot5$.

Helmholtz's just $c''528$ $a'440$, may be considered as represented by iii. and iv. jointly.

v. The highest usual English pitch, known as 'band pitch,' or 'Kneller Hall pitch,' to which military brass instruments are tuned in England, adopted as the pitch of musical instruments at the International Inventions and Music Exhibition, London, 1885.

vi. The original pitch of the St. Jacobi organ at Hamburg built by Schnitger 1688, the oldest as well as the sharpest example of equal temperament I have found. On these pitches, see the abstract of my *History of Musical Pitch*, infrà, sect. H.

Art. 26.—The notation of music (see art. 16) was adapted to either the Pythagorean or Meantone intonation, in which there was a Dièsis or interval between a sharp and a flat, not to the equal where the Dièsis disappears and sharp coalesces with flat. In the table only the sharps are noted as usual, but they ¶ imply flats. If we arrange the notes in three lines, in the same order of Fifths as in art. 15, but continue them to 36 tones, we shall have

$$g\flat\flat \quad a\flat\flat \quad a\flat\flat \quad e\flat\flat \quad b\flat\flat \quad f\flat \quad c\flat \quad g\flat \quad d\flat \quad a\flat \quad e\flat \quad b\flat$$
$$F \quad C \quad G \quad D \quad A \quad E \quad B \quad F\sharp \quad C\sharp \quad G\sharp \quad D\sharp \quad A\sharp$$
$$e\sharp \quad b\sharp \quad f\sharp\sharp \quad c\sharp\sharp \quad g\sharp\sharp \quad d\sharp\sharp \quad a\sharp\sharp \quad e\sharp\sharp \quad b\sharp\sharp \quad f\sharp\sharp\sharp \quad c\sharp\sharp\sharp \quad g\sharp\sharp\sharp$$

The middle line indicates the ordinary notes, but those in the upper and lower line are (at least from $a\flat\flat$ to $g\sharp\sharp$) occasionally met with in modulations. Now the three notes in each column have the same meaning precisely in equal temperament. They are absolutely identical. But in the Pythagorean and Meantone temperaments they have three different meanings, as shewn in the tables of arts. 15 and 16. This confusion arises from equal temperament being cyclic. If we begin at F and proceed by Fifths to $A\sharp$ we have exhausted all our 12 values. The Fifth above $A\sharp$ is played by F, but it would be considered 'bad spelling' to write it so, for $b\flat$ to f is called a Fifth, and $a\sharp$ to f a diminished Sixth, although they ¶ make the same interval precisely. This arises from history. It has been proposed to alter the notation, but the objections to changing are so great that the matter is mentioned here chiefly to explain how the apparently absurd synonymity of equal temperament arose, and also because this synonymity is a cardinal point in Mr. Bosanquet's notation.

Art. 27.—In Mr. Bosanquet's cycle, art. 22, iv., there are 53 notes to be supplied with names, and moreover after the 53 notes have been exhausted the values recur, but, if the old notation is to be in any way preserved, the old names do not recur. Hence there will be here also another and a different kind of synonymity, which will affect the position of the notes. In the following tables I have first arranged the notes by Fifths and then by regular ascent. The large letters may be considered regular. They are each supposed to have all the synonyms of equal temperament already explained, and hence are written only as the line of capitals in art. 26. They are divided into 'sets' of 12, each set being distinguished by a superior or inferior number, because each is one degree (22·642 cents, and hence very nearly a real Comma of 21·506 cents) sharper or flatter respectively. Under ¶ these capital letters is written the number of the note in the cycle according to Mr. Bosanquet's arrangement, dictated by practical convenience in performance. Now there are 12 notes in each line, and hence after 4 lines and 5 Fifths, indicated by ||, we have exhausted all the 53 values. The names of the letters are, however, continued on the same plan, but they are now synonymous with those at the beginning of the series, and hence the name of the 54th Fifth, or F^1, numbered 27, is written in small letters under the first note $F_3\sharp$, which is also numbered 27, and the series after F_1 in capitals is written after f^1 in small letters. These small letters have the same value as the large ones above them. This synonymity forms the chief difficulty of the instrument when modulations oblige the player to proceed beyond the first 53 notes. For this reason, partly, Mr. Bosanquet has in practice extended his keyboard to contain 7 sets or 84 notes. The inferior and superior numbers are my typographical contrivances, not Mr. Bosanquet's. He uses sloping lines like those suprà, p. 220c, last bar, ascending for my superiors, descending for my inferiors, and repeated twice, three times, &c., for my 2, 3, &c. These are very convenient in musical notation, and, on account of using the equal temperament synonyms, are the only alterations required to adapt ordinary music for performance in this cycle.

Mr. Bosanquet's Notes—in Fifths.

$F_3\sharp\sharp$	$C_3\sharp\sharp$	$G_3\sharp\sharp$	$D_3\sharp\sharp$	$A_3\sharp\sharp$	F_3	C_3	G_3	D_3	A_3	E_3	B_3
27	5	36	14	45	23	1	32	10	41	19	50
f^1	c^1	g^1	d^1	a^1	e^1	b^1	$f^2\sharp\sharp$	$c^2\sharp\sharp$	$g^2\sharp\sharp$	$d^2\sharp\sharp$	$a^2\sharp\sharp$

$F_2\sharp\sharp$	$C_2\sharp\sharp$	$G_2\sharp\sharp$	$D_2\sharp\sharp$	$A_2\sharp\sharp$	F_2	C_2	G_2	D_2	A_2	E_2	B_2
28	6	37	15	46	24	2	33	11	42	20	51
f^2	c^2	g^2	d^2	a^2	e^2	b^2	$f^3\sharp\sharp$	$c^3\sharp\sharp$	$g^3\sharp\sharp$	$d^3\sharp\sharp$	$a^3\sharp\sharp$

$F_1\sharp\sharp$	$C_1\sharp\sharp$	$G_1\sharp\sharp$	$D_1\sharp\sharp$	$A_1\sharp\sharp$	F_1	C_1	G_1	D_1	A_1	E_1	B_1
29	7	38	16	47	25	3	34	12	43	21	52
f^3	c^3	g^3	d^3	a^3	e^3	b^3	$f^4\sharp\sharp$	$c^4\sharp\sharp$	$g^4\sharp\sharp$	$d^4\sharp\sharp$	$a^4\sharp\sharp$

$F\sharp$	$C\sharp$	$G\sharp$	$D\sharp$	$A\sharp$	F	C	G	D	A	E	B
30	8	39	17	48	26	4	35	13	44	22	53
f^4	c^4	g^4	d^4	a^4	e^4	b^4	$f^5\sharp\sharp$	$c^5\sharp\sharp$	$g^5\sharp\sharp$	$d^5\sharp\sharp$	$a^5\sharp\sharp$

$F^1\sharp$	$C^1\sharp$	$G^1\sharp$	$D^1\sharp$	$A^1\sharp$	F^1	C^1	G^1	D^1	A^1	E^1	B^1
31	9	40	18	49	27	5	36	14	45	23	1
f^5	c^5	g^5	d^5	a^5	e^5	b^5	$f^6\sharp\sharp$	$c^6\sharp\sharp$	$g^6\sharp\sharp$	$d^6\sharp\sharp$	$a^6\sharp\sharp$

$F^2\sharp$	$C^2\sharp$	$G^2\sharp$	$D^2\sharp$	$A^2\sharp$	F^2	C^2	G^2	D^2	A^2	E^2	B^2
32	10	41	19	50	28	6	37	15	46	24	2
f^6	c^6	g^6	d^6	a^6	e^6	b^6	$f^7\sharp\sharp$	$c^7\sharp\sharp$	$g^7\sharp\sharp$	$d^7\sharp\sharp$	$a^7\sharp\sharp$

Mr. Bosanquet's Cycle of 53.

Cyclic Nos.	Names and Synonyms	Cents	Logs	Pitch Numbers	Cyclic Nos.	Names and Synonyms	Cents	Logs	Pitch Numbers
1	C_3 b^1 $a^6\sharp\sharp$	1132·1	28399	507·69	28	$F_2\sharp\sharp$ f^2 e^6	543·4	13632	361·34
2	C_2 b^2 $a^7\sharp\sharp$	1154·7	28967	514·37	29	$F_1\sharp\sharp$ f^3	566·0	14200	366·10
3	C_1 b^3	1177·4	29535	521·14	30	$F\sharp$ f^4	588·7	14768	370·92
4	C b^4	1200=0	0	264·00	31	$F^1\sharp$ f^5	611·3	15335	375·80
5	$C_3\sharp\sharp$ c^1 b^5	22·6	00568	267·48	32	G_3 $f^2\sharp\sharp$ f^6	634·0	15903	380·75
6	$C_2\sharp\sharp$ c^2 b^6	45·3	01136	271·00	33	G_2 $f^3\sharp\sharp$	656·6	16471	385·76
7	$C_1\sharp\sharp$ c^3	67·9	01704	274·56	34	G_1 $f^4\sharp\sharp$	679·2	17039	390·84
8	$C\sharp$ c^4	90·6	02272	278·18	35	G $f^5\sharp\sharp$	701·9	17607	395·98
9	$C^1\sharp$ c^5	113·2	02840	281·84	36	$G_3\sharp\sharp g^1$ $f^6\sharp\sharp$	724·5	18175	401·19
10	D_3 . $c^2\sharp\sharp$ c^6	135·8	03408	285·55	37	$G_2\sharp\sharp$ g^2 $f^7\sharp\sharp$	747·2	18744	406·48
11	D_2 $c^3\sharp\sharp$	158·5	03976	289·31	38	$G_1\sharp\sharp$ g^3	769·8	19311	411·83
12	D_1 $c^4\sharp\sharp$	181·1	04544	293·12	39	$G\sharp$ g^4	792·5	19879	417·25
13	D $c^5\sharp\sharp$	203·8	05112	297·00	40	$G^1\sharp$ g^5	815·1	20447	422·74
14	$D_3\sharp\sharp$ d^1 $c^6\sharp\sharp$	226·4	05680	300·89	41	A_3 $g^2\sharp\sharp$ g^6	837·7	21015	428·31
15	$D_2\sharp\sharp$ d^2 $c^7\sharp\sharp$	249·1	06248	304·85	42	A_2 $g^3\sharp\sharp$	860·4	21583	433·95
16	$D_1\sharp\sharp$ d^3	271·7	06816	308·86	43	A_1 $g^4\sharp\sharp$	883·0	22151	439·66
17	$D\sharp$ d^4	294·3	07384	312·93	44	A $g^5\sharp\sharp$	905·7	22719	445·45
18	$D^1\sharp$ d^5	317·0	07952	317·05	45	$A_3\sharp\sharp a^1$ $g^6\sharp\sharp$	928·3	23287	451·31
19	E_3 $d^2\sharp\sharp$ d^6	339·6	08520	321·22	46	$A_2\sharp\sharp a^2$ $g^7\sharp\sharp$	950·9	23855	457·25
20	E_2 $d^3\sharp\sharp$	362·3	09088	325·45	47	$A_1\sharp\sharp a^3$	973·6	24423	463·27
21	E_1 $d^4\sharp\sharp$	384·9	09656	329·73	48	$A\sharp$ a^4	996·2	24991	469·37
22	E $d^5\sharp\sharp$	407·5	10224	334·07	49	$A^1\sharp a^5$	1018·9	25559	475·55
23	F_3 e^1 $d^6\sharp\sharp$	430·2	10792	338·47	50	B_3 $a^2\sharp\sharp a^6$	1041·5	26127	481·81
24	F_2 e^2 $d^7\sharp\sharp$	452·8	11360	342·93	51	B_2 $a^3\sharp\sharp$	1064·2	26695	488·15
25	F_1 e^3	475·5	11928	347·44	52	B_1 $a^4\sharp\sharp$	1086·8	27263	494·58
26	F e^4	498·1	12496	352·01	53	B $a^5\sharp\sharp$	1109·4	27831	501·09
27	$F_3\sharp\sharp f^1$ e^5	520·8	13064	356·65	1'	B^1 $a^6\sharp\sharp c_3$	1132·1	28399	507·09

Art. 28.—The cycle of 1200 is especially used for indicating the relations of just notes and the mode in which the notes of tempered and inharmonic scales generally fit in among these just notes. The series of 117 just notes to the Octave is developed in sect. E. infrà, and the value of each note is given (*ibid.* art. 18) by the numbers of the corresponding note in the cycles of 53 and 1200. The number in the cycle of 53, by means of the table in art. 27, gives all the information required as to these substitutes, including Mr. Bosanquet's names, which are different from those assigned to just intonation on the principles of Chap. XIV. pp. 276*b* to 277*a*. To shew the difference between cyclic and just cents the following table is added, in which all the names of the 117 just notes in the

Duodenarium of sect. E. p. 463, are placed in alphabetical order for easy reference, with their cyclic and just cents, logarithms, and pitch. It must be remembered that the inferior and superior numbers in this table refer to differences of a *Comma* of 22 cyclic cents, or one of 21·5 just cents, and in that of the cycle of 53 on p. 439, to a *degree* of 22·6 cents, hence the same name has distinctly different meanings. Thus Bosanquet's No. 5, or $C_3\sharp = c^1$, has 22·6 cents, and just c^1 has 21·5 cents, which agrees well with 22·6, but just $C_3\sharp$ has 49·2 cents and agrees more nearly with Bosanquet's No. 6, or $C_2\sharp$ with 45·3 cents.

Expression of Just Intonation in the Cycle of 1200.

Note	Cyclic Cents	Just Cents	Logs	Pitch	Note	Cyclic Cents	Just Cents	Logs	Pitch
A	906	905·9	22724	445·5	Db	90	90·2	02263	278·1
A^1	928	927·4	23264	451·1	D^1b	112	111·7	02803	281·6
A_1	884	884·4	22185	440·0	D^2b	134	132·5	03342	285·0
A_2	862	862·9	21645	434·6	D^3b	156	154·7	03882	288·7
$A_1\sharp$	998	998·0	25037	469·9	D^2bb	20	19·6	00490	267·0
$A_2\sharp$	976	976·5	24497	464·1	D^3bb	42	41·1	01030	270·3
$A_3\sharp$	954	955·1	23958	458·3	D^1bb	64	62·6	01569	273·7
$A_3\sharp\sharp$	1068	1068·7	26809	480·4	D^1bbb	1150	1148·9	28821	512·6
$A_4\sharp\sharp$	1046	1047·2	26270	483·4	E	408	407·8	10231	334·1
Ab	792	792·0	19873	417·2	E_1	386	386·3	09691	330·0
A^1b	814	813·7	20412	422·4	E_2	364	364·8	09152	325·9
A^2b	836	835·2	20952	427·7	$E_1\sharp$	500	500·0	12542	352·4
A^2bb	722	721·5	18100	400·5	$E_2\sharp$	478	478·5	12003	348·0
A^3bb	744	743·0	18639	405·5	$E_3\sharp$	456	457·0	11464	343·8
A^4bb	766	764·5	19179	410·6	$E_4\sharp$	434	435·5	10924	339·5
A^4bbb	652	650·8	16327	384·5	$E_3\sharp\sharp$	570	570·7	14316	367·7
B	1110	1109·8	27840	501·2	$E_4\sharp\sharp$	548	549·2	13776	362·6
B_1	1088	1088·3	27300	495·0	Eb	294	294·1	07379	312·9
B_2	1066	1066·8	26761	488·9	E^1b	316	315·6	07918	316·8
$B_2\sharp$	1180	1180·4	29613	522·1	E^2b	338	337·1	08458	320·8
$B_3\sharp$	1158	1158·9	29073	515·6	E^2bb	224	223·5	05606	300·4
$B_4\sharp$	1136	1137·4	28534	509·3	E^3bb	246	245·0	06145	304·1
$B_3\sharp\sharp$	72	72·5	01822	275·3	E^4bb	268	266·5	06688	307·9
$B_4\sharp\sharp$	50	50·4	01282	271·8	E^3bbb	132	131·3	03293	284·8
Bb	996	996·1	24988	469·3	E^1bbb	154	152·8	03833	288·4
B^1b	1018	1017·6	25527	475·2	F	498	498·0	12493	352·0
B^2b	1040	1039·1	26067	481·1	F^1	520	519·6	13033	356·4
B^1bb	904	903·9	22675	445·0	F^2	542	541·1	13573	360·9
B^2bb	926	925·4	23215	450·6	F_1	476	476·5	11954	347·7
B^3bb	948	946·9	23754	456·2	$F\sharp$	612	611·7	15346	375·9
B^1bb	970	968·4	24294	461·9	$F_1\sharp$	590	590·2	14806	371·3
B^3bbb	834	833·2	20902	427·2	$F_2\sharp$	568	568·7	14267	366·7
B^1bbb	856	854·7	21442	432·5	$F_2\sharp\sharp$	682	682·4	17119	391·6
C	0	0	0	264·0	$F_3\sharp\sharp$	660	660·9	16579	386·7
C^1	22	21·5	00540	267·3	$F_4\sharp\sharp$	638	639·4	16040	381·9
C_1	1178	1178·5	29564	521·5	$F_4\sharp\sharp\sharp$	752	753·1	18892	407·9
$C\sharp$	114	113·7	02852	281·9	F^1b	406	405·9	10181	333·7
$C_1\sharp$	92	92·2	02312	278·4	F^2b	428	427·4	10721	337·9
$C_2\sharp$	70	70·7	01770	275·0	F^3b	450	448·9	11261	342·1
$C_3\sharp$	48	49·2	01233	271·6	F^3bb	336	335·2	08409	320·4
$C_2\sharp\sharp$	184	184·4	04625	293·7	F^1bb	358	356·7	08948	324·4
$C_3\sharp\sharp$	162	162·9	04085	290·0	G	702	702·0	17609	396·0
$C_4\sharp\sharp$	140	141·3	03546	286·6	G^1	724	723·5	18149	401·0
$C_4\sharp\sharp\sharp$	254	255·0	06398	305·9	G_1	680	680·4	17070	391·1
C^1b	1108	1107·8	27791	500·6	$G_1\sharp$	794	794·1	19922	417·7
C^2b	1130	429·3	28330	506·9	$G_2\sharp$	772	772·6	19382	412·5
C^3b	1152	1150·0	28870	513·2	$G_3\sharp$	750	751·1	18843	407·2
C^3bb	1038	1037·1	26018	480·7	$G_2\sharp\sharp$	886	886·3	22234	440·5
C^1bb	1060	1058·7	26557	486·6	$G_3\sharp\sharp$	864	864·8	21694	435·1
C^1bbb	946	945·0	23705	455·7	$G_4\sharp\sharp$	842	843·3	21155	429·7
D	204	203·9	05115	297·0	$G_4\sharp\sharp\sharp$	956	957·0	24007	458·9
D^1	226	225·4	05655	300·7	G^1b	610	609·8	15297	375·5
D_1	182	182·4	04576	293·3	G^2b	632	631·3	15836	380·2
$D_1\sharp$	296	296·1	07427	313·2	G^3b	654	652·8	16376	384·9
$D_2\sharp$	274	274·6	06888	309·4	G^2bb	518	517·6	12984	356·0
$D_3\sharp$	252	253·1	06349	305·6	G^3bb	540	539·1	13524	360·4
$D_3\sharp\sharp$	366	366·8	09201	326·3	G^1bb	562	560·6	14063	365·0
$D_4\sharp\sharp$	344	345·3	08661	322·3	G^1bbb	448	446·9	11210	341·8
$D_4\sharp\sharp\sharp$	458	458·9	11513	344·1					

If it is wished to introduce the series of natural harmonic Sevenths as in Poole (infrà, sect. F. No. 7), X being any note, $'X$ will have the number of X in Bosanquet's cycle diminished by 1, the cyclic cents of X diminished by 27, the just cents of X by 27·3, the log of X by ·00584, and the pitch number of X by $\frac{1}{64}$ of its value. Thus A has Bosanquet's number 44, cyclic cents 906, just cents 905·9, log ·22724,

and pitch number 445·5. Then for $'A$, Bosanquet's number is 43 (having 883 cents, log ·22151, pitch number 439·7), the cyclic cents are 906−27=879, the just cents 905·9−27·3 =878·6, log = ·22724 − ·00584 = ·22140, the pitch number = 445·5 − $\frac{1}{64}$ × 445·5 = 445·5 − 6·96 =438·5, shewing that Bosanquet's substitute is a trifle too sharp.

Art. 29.—Those who require more information are referred to my paper on 'Temperament,' *Proc. R. S.*, vol. xiii. pp. 404–422 (where the subject is treated more in detail and in an entirely different manner), and to the memoirs and essays of Salinas, Zarlino, Huyghens, Sauveur, Henflins, R. Smith, Marpurg, Estève, Cavallo, Romieu, Lambert, T. Young, Robison, Farey, Delezenne, Woolhouse, De Morgan, Drobisch, Naumann, there cited. Also to Mr. Bosanquet's treatise on *Musical Intervals and Temperament*, 1876, to his papers on 'Temperament' in the *Proc. of the Musical Association*, first year, pp. 4–17, 112–54, and his article on 'Temperament' in Stainer and Barrett's *Dictionary of Musical Terms*. Also to Mr. Lecky's article on 'Temperament' in Grove's *Dictionary of Music*.

SECTION B.

ON THE DETERMINATION OF PITCH NUMBERS.

(See notes pp. 11, 56, 168, 176.)

The determination of the pitch number of any note heard is a very difficult problem to solve. The following methods have been used.

1. *The String.* Supposing that a heavy string of uniform density, perfect elasticity, of no thickness, but capable of bearing a considerable strain, could have its vibrating length determined with perfect accuracy—none of which conditions can be more than roughly fulfilled—then the pitch numbers of its parts would be inversely proportional to their lengths. The formula has been worked out by Euler and Bernouilli, and amounts to this.

Let L be the vibrating length of a suspended string in English inches, l the same in French millimetres, W the stretching weight in any unit, w the weight of the vibrating length of the string in the same unit, V the pitch number. Then

$$2 \log \dot{V} = 1{\cdot}98485 + \log W - (\log w + \log L)$$
$$= 3{\cdot}38968 + \log W - (\log w + \log l).$$

This was used for careful measures by Euler, Dr. Robert Smith, Marpurg, Fischer, and De Prony. Probably on account of the necessary thickness of the string the results could not be trusted within 5 vib.

The work is also extremely difficult, and depends ultimately on determining a unison between two tones of very different qualities. General T. Perronet Thompson used such an instrument for tuning his organ (described in *Just Intonation*, 7th ed. p. 69). As his string was No. 20 (1·165 mm. in diam.) no reliance could be placed on the perfect exactness of his results. Mr. J. H. Griesbach in 1860 tuned a string 5·17 mm. in diam. till one quarter of its length was in unison with a given note, and then counted the vibrations of the whole string automatically. The instrument is in the South Kensington Museum, and was described in the *Journal of the Society of Arts*, 6 April 1860, p. 353. The results were 3 to 6 vibrations wrong.

Delezenne (*Mém. de la Soc. des Sciences à*

Lille, 1854, p. 1) made the best use of the string. He stretched 700 millimetres of wire on a violoncello body, and tuned it to Marloye's 128 vib. (which was probably accurate, as Marloye's 256 vib. certainly was), and then by a movable bridge cut off the length, which when bowed was in unison with the fork. This fork had been adjusted by sliding weights to the pitch of a note heard. Then measuring this length in millimetres, he divided 128 × 700 = 89,600 by it, to find the vib. This assumed that lengths were inversely as the vib. or pitch numbers. But he found that the same fork was in unison with 203·8 mm. of a string ·6154 mm. thick, and 198·9 mm. of a wire ·1280 mm. thick. The former gave 439·6, the latter 450·5 vib. The thick wire therefore gave a pitch 42 cents flatter than the thin. Hence he confined his observations to the thinnest wire that would bear the strain.

In conjunction with Mr. Hipkins of Broadwoods' and the foreman tuner, Mr. Hartan, I made some experiments on 2 April 1885 with the monochord at Broadwoods', having a string of ·98 mm. in thickness. Nineteen tuning-forks about 20 vib. apart, with pitches from 223·77 to 578·40 vib. accurately known, were brought into unison with lengths of the monochord limited by a movable bridge touching, but sliding easily under the string, which was pressed down on it by a knife-edge. The intervals between the lowest and each of the other forks, and the longest and each of the other lengths (assuming them to be inversely as the number of vibrations), were calculated in cents. The former were generally sharper by the following cents, the minus sign shewing when they were flatter : o 7 6 13 3 9 − ½ 2 23 22 9 1 11 8 9 5 5 − ⅓ − 4. These irregular differences demonstrate that such a monochord gives very uncertain results, even when the unisons are estimated by very sensitive ears.

2. *The Siren.* This has been described in the text, p. 12. But only the most carefully constructed sirens with bellows of constant pressure, as that described in App. I., or the ' Soufflerie de précision ' of M. Cavaillé-Coll, worked by well prac-
¶ tised operators, can give good results. Here also a unison between tones of very different qualities, one of which is fixed and the other variable, has to be determined. The best work that I know with the siren was that done by M. Lissajous (who used M. Cavaillé-Coll's bellows, as the latter tells me) in determining the pitch of the ' Diapason Normal ' at Paris, which was meant to give 435 vib. at 15° C.= 59° F., and actually gave 435·45 vib.

3. The *Optical Method* of Professor Herbert McLeod, F.R.S., and Lieut. R. G. Clarke, R.E. (*Proceedings of Royal Society*, Jan. 1879, vol. xxviii. p. 291, and *Philosophical Transactions*, vol. clxxi. pp. 1–14, plates 1 to 3), consisted in viewing white lines on a rotating cylinder through the shadow of a vibrating fork. The machine is difficult to manipulate, but in the hands of its inventors gave extremely accurate results.

4. The *Electrographic Method*, invented by Prof. A. Mayer, of Stevens Institute, Hoboken, New Jersey, U.S., consisted in causing a tuning-fork by means of a copper-foil point to scribe its vibrations on the camphor-smoked paper cover of a brass rotating cylinder, and marking seconds by passing a strong induction spark through the fork, scribing point, and paper cover at the passage of a seconds pen-
¶ dulum through a spot of mercury, and then counting the sinuosities at leisure. The weight of the scribing point had to be allowed for, but the results were very accurate.

For another means of determining the frequency of a fork used as an interrupter of electricity, see Mr. R. T. Glazebrook's paper in *Philosophical Magazine*, Aug. 1884, vol. xviii. pp. 98–105.

5. *The Clock.* Dr. Koenig (in Wiedemann's, late Puggendorff's *Annals*, 1880, pp. 394–417) describes a large tuning-fork having the pitch number 64, which was made to act on a clock at a constant temperature of 20° C.=68° F., functioning as a pendulum, so that every single vibration was registered for many hours, and he was thus enabled to determine one standard pitch with extreme accuracy. He also found from this that his old well-known forks of nominally 256 double vib. had been tuned at too high a temperature, and at 20° C. gave 256·1774, and at 15° C.=59° F., 256·28 d. vib. He also determined the pitch of the Diapason Normal at 15° C. as 435·45 d. vib. A rise in temperature of 1° F. flattens tuning-forks by 1 vib. in
¶ from 16,000 (Koenig) or 20,000 (Scheibler) or 22,000 (Mayer) double vibrations in a second (see my ' Notes of Observations on Musical Beats,' *Proc. R. S.*, 28 May 1880, vol. xxx. p. 523), and flattens harmonium reeds by about 1 in 10,000 vib. (See my paper ' On the Influence of Temperature on the Musical Pitch of Harmonium Reeds,' *Proc. R. S.*, Jan. 1881, p. 413.)

Lord Rayleigh (*Philosophical Transactions*, 1883, Part I., pp. 316–321) describes another method of determining the frequency of a standard fork by means of a clock. ' The observer looking over a plate carried by the upper prong of the fork [of intentionally 32 vib.] obtained 32 views per second, *i.e.* 64 views of the pendulum, in one complete vibration. The immediate subject of observation is a silvered bead attached to the bottom of the pendulum, upon which as it passes the position of equilibrium the light of a paraffin lamp is concentrated. Close in front of the pendulum is placed a screen perforated by a somewhat narrow vertical slit. If the period of the pendulum were a precise multiple of that of the fork, the flash of light, which to ordinary observers would be visible at each passage, would either be visible, or be obscured, in a permanent manner. If, as in practice, the coincidence be not perfect, the flashes appear and disappear in a regular cycle, whose period is the time in which the fork gains (or loses) one complete vibration. This period can be determined with any degree of precision by a sufficient prolongation of the observations.'

6. *Harmonium Reeds.*

Herr Georg Appunn, of Hanau, invented tonometers of 65, 33, and 57 reeds, serving for many important experiments (see for example suprà, p. 176, footnote †). Copies of all three are at the South Kensington Museum, and of the two first at the Museum of King's College, London.   These reeds were tuned to make 4 beats in a second with each other, so that the 65th reed made with the first, which was an Octave flatter, $4 \times 64 = 256$ beats, and consequently, according to the theory of beats (see text, Chap. VIII.), the pitch number of the lowest note should be 256. Unfortunately, the condensed air in which the beats took place accelerated the beats by 76 in 10,000, as I determined by long-continued observations and experiments (described generally in my first paper already cited, *Proc. R. S.*, vol. xxx. pp. 527–532), and consequently the results had to be lessened by that amount, making the pitch number of the lowest reed about 254.   These reeds, too, were not sufficiently permanent in pitch.   Hence this instrument, though otherwise very useful, failed ¶ in determining pitch with sufficient accuracy.

Lord Rayleigh (*Proc. Mus. Assn.* vol. v., 1878-9, p. 15) discovered a way of determining the pitch of two low harmonium reeds, the lowest *C* and *D* on his harmonium.   Keeping the wind for 10 minutes or 600 seconds as constant as possible and using resonators tuned by partially covering with the finger to about the 9th and 10th partials of the low *C*, two observers counted the beats, one between the 9th partial of *C* and the 8th of *D*, and the other between the 10th partial of *C* and the 9th of *D*.   They thus found

$$9\ C - \ 8\ D = 2392 \div 600$$
$$9\ D - 10\ C = 2341 \div 600$$

Whence

$$C = (9 \times 2392 + 8 \times 2341) \div 600$$
$$= (21528 + 18728) \div 600 = 67 \cdot 09$$

and similarly

$$D = (10 \times 2392 + 9 \times 2341) \div 600 = 74 \cdot 98$$

As these notes make an interval of 192·5 cents with each other, Lord Rayleigh had evidently (as he suggested) altered the interval to about a meantone 193·2 cents.   His object was to determine the pitch of a fork of Koenig's, supposed to vibrate 64 times in a second.   Now as Koenig's 256 is really 256·28 at 59° F., this 64 should be 64·07 at the same temperature.   Lord Rayleigh, to take account of the effect of the simultaneous beating of the two reeds, sounded both of them at the same time with the fork, and on different days obtained the following results (temperature not named) :—

Harmonium	67·09	Tuning-fork		64·06
,,	67·04	,,	,,	64·07
,,	67·17	,,	,,	64·17 ¶
,,	67·19	,,	,,	63·98

Which were wonderfully accurate.

7. *Tuning-forks.*

All the above methods have one important fault.   The measuring instruments are not easily portable and not readily applicable to all kinds of sustained tones, while the two first required trained ears to discriminate unisons between tones of different qualities, with great accuracy, that is to say, at least 1 vib. in 10 sec.   All these faults are obviated by the Tuning-fork Tonometer invented by J. Heinrich Scheibler (*b.* 1777, *d.* 1837), a silk manufacturer of Crefeld in Rhenish Prussia. The simplest process of making such a Tonometer, although not the one used by Scheibler (see his pamphlet cited in note ‡, p. 199*d*), is as follows.   I quote principally from my ' Notes on Musical Beats,' already cited.

Obtain a set of about 70 good forks with parallel prongs, and of a tolerably large size ; tune the lowest to about the *c'* (or *b* for English high pitch) and tune the rest roughly each about four beats in a second sharper than the preceding.   Then fit them with wooden collars or handles, and allow them to rest for three months, if possible in the same temperature at which they will be counted, and never alter their pitch again by filing, but count the beats between each set most carefully, at a temperature which remains as uniform as possible.   It may be necessary to use a high temperature ; thus Scheibler's was from 15° R. to 18° R. = 65·75 to 72°·5 F., which I reckon at 69° F. as a mean ; and Koenig now works at 20° C. = 68° F.   Count on one day the beats between forks 1 and 2, 3 and 4, 5 and 6, &c., and on another between forks 2 and 3, 4 and 5,

&c., so that the same fork is not used for ¶ two counts on the same day.   Excite by striking with a soft ball of fine flannel wound round the end of a piece of whalebone, as a bow is not convenient unless the forks are tightly fixed.   Each blow or bowing heats, and hence flattens, and this tells if the experiments on any one fork are long continued.   Count each set of beats for 40 seconds if possible, and many times over, registering the temperature and the beats, and taking the mean.   Having counted all, observe those forks which are near the Octave of the lowest fork.   Find two such, beating with the Octave (that is, the second partial) of the lowest fork less than they beat with each other.   Then the sum of all the beats from the lowest fork to the lower of the two forks, added to the beats of the Octave (that is, the second partial tone of the lower fork) with that fork, is the pitch of the lowest fork. Hence the pitch of all the forks is known. The extra high forks are for verifying by the Octaves of several low forks, and for the purpose of subsequently measuring. From such a tonometer any other can be made, and the value of each fork at another temperature calculated.

Scheibler made a 52-fork tonometer with infinite trouble, on another plan, described in his book and counted it with marvellous accuracy. This tonometer, which I have made many efforts to find, has absolutely disappeared and his family knows nothing of it. But he left behind him a 56-fork tonometer, believed to proceed from 220 to 440 vibrations, by steps of 4 vibrations, and through the kindness of Herr Amels, an old friend of ¶ the Scheibler family, who obtained it from Scheibler's grandson, I had the use of it for a year. I had to count it as well as I could, just as if it had been a set of forks such as I have described, and I found it was not what was thought, but that only 32 sets of beats were 4 in a second, and the other 23 sets varied from 38 to 42 in 10 seconds. I found also that the extremes were probably of the same pitch as in the original 52-fork tonometer. After then counting it as well as I could, and obtaining 219·27 vibrations in place of 219·67, at 69° F., I distributed the error of 4 beats in 10 seconds, as 2 in 100 seconds, among 20 of the 23 sets which were not exactly 4 beats in 10 seconds, leaving the first 3 sets, which I had repeatedly counted and felt sure of, unaltered. Then I reduced all the values from 69° to 59° F. Finally ¶ to verify my result I measured by beats with Scheibler's forks as thus determined; first 5 large forks of various pitches, which I had had made for me in Paris, and then 4 forks of Koenig's belonging to Professor McLeod. Professor McLeod himself kindly measured all of them, also, by his machine, and Professor Mayer also obligingly measured the first 5 forks by his electrographic method, both with the greatest care and precaution. The three sets of measurements agreed to less than 1 beat in 10 seconds, and more often less than 1 beat in 20 seconds, when reduced to the same temperature. Thus the value of the tonometrical measurement by beats only, and the possibility of counting a tonometer sufficiently, was fully established. Koenig's measurements of his own forks reduced to 59° F., and of the actual ¶ Diapason Normal at the Conservatoire, Paris, intended to be used at the same temperature, also agree with mine within less than the same limits.

The pitch of all sufficiently sustained tones can thus be determined mechanically. We find two forks, whose pitch is known, with each of which the new tone beats more slowly than the forks beat with each other, and we verify the count, by seeing that the sum of the beats with both forks is the number of beats of the forks with one another. The pitch number is then that of the lower fork increased by the number of beats made with it in a second, and that of the upper fork diminished by the number of beats made with it in a second. The following notes are the result of much experience.

Tuning-forks are comparatively simple in quality of tone but always possess an audible second partial or Octave, and sometimes higher partials still, capable of being so reinforced by resonance jars properly tuned to them, that beats can be separately obtained from them and counted. This, as we have seen, is a matter of great importance in the construction of a tuning-fork tonometer. When the tone is very compound, as in the case of bass reeds (especially those of Appunn's tonometer, furnished with a bellows giving, when properly managed, a perfectly steady blast for an indefinite length of time), beats can be obtained and counted from the 20th to the 30th and even the 40th partial, without any reinforcement by a resonance jar. (See p. 176d'.)

Taking tuning-forks first, I find it advantageous to hold the beating forks over one or two resonance jars, tuned, by pouring in water, to the pitch of the partial to be observed, whether it be the prime of both or the prime of one and the second partial (or Octave) of the other. There may be small differences, but I have not found any difference appreciable by my methods of observation in the number of beats in a second, whether the resonance jar is the same or different for the two forks, and whether it is exactly or very indifferently tuned to each fork, but a tolerably accurate tuning much improves the tone and length of the beat. In that case the resonance jar practically quenches all other partial tones, and the beats are distinctly heard as loudnesses separated by silences. If no jar is used, the other partials are heard. In the case of the Octave, the low prime becomes a drone and fills up the silences. In the case of beating primes, the Octaves, which are beating twice as fast, tend to confuse the ear. Sometimes the second partial of a fork is so much stronger than the prime, that when the fork is applied to a sounding-board, only the Octave is heard, which is often inconvenient to the fork tuner. This is entirely avoided by the resonance jar. Beats being a case of interference, the amplitude of the beating partials should be equalised as much as possible. With two forks of very different size and power, it is easy to regulate the amplitude by holding the louder fork further from the jar. Otherwise the beats become blurred and indistinct. For powerful reeds or organ pipes, beating with forks, it is best to go to a considerable distance from the reed or pipe and hold the fork close to the ear or over a jar. I find 30 or 40 feet necessary for organs; in Durham Cathedral, where the pressure of wind was strong and my forks weak, I found 60 or 70 feet distance much better. As I was not able latterly to go to a distance from Appunn's reed tonometer, having to pump it myself, I found it impossible to count the primes of the upper reeds by the Octaves of my forks, which were completely drowned by the reeds.

I find beats of all kinds most easy to count when about 3 or 4 in a second. They can be counted well from 2 to 5 in a second. Above 5 they are too rapid for accuracy; below 2, and certainly below 1, they are too slow, so that it is extremely difficult to tell from what part of the swell of sound the beat should be reckoned. Partly from this reason, perhaps, I have found great variety in counting successive sets of such slow beats. I never use beats of less than one in a second, if I can avoid it. When the beats are slow it is difficult to discover by ear which of the two beating tones is the sharper; and even fine ears are often deceived. It is easy to discover, however, by putting one of the forks under the arm for a minute. This heats and flattens it by 2 or 3 beats in 10 seconds. Hence if the beats with the heated fork are slower, that fork was sharper, because it has been brought nearer the other; if faster, it was flatter and has been brought further away. Count for 10, 20, or 40 seconds, according to the fork. Up to 20 or 30 beats in 10 seconds it is easy to count in ones, but from 30 to 50 it is best to count in twos, as one-ee, two-ee, &c., beginning with *one*, and hence throwing off one at the end. When counting for 20 seconds I always count in twos, and for 40 seconds in fours, as one-ee-ah-tee, two-ee-ah-tee, &c., because I have to divide the result by 20 or 40; and this division is avoided by the count itself. As my counting was never for more than 40 seconds, small errors of the clock or pendulum were imperceptible. I generally used a marine or pocket chronometer when making the principal count for 40 seconds, but for merely determining pitch from a fork of known pitch, 10 seconds of time, and any ordinary seconds watch suffice. For Prof. McLeod's observations, which lasted 5 minutes or more, extreme accuracy in rating an astronomical clock was necessary. Suppose a watch to gain 5 minutes or 300 seconds in 24 mean hours, which is an extreme case, so that 86,700 watch seconds = 86,400 mean seconds, then 10 watch seconds = 9·9654 mean seconds, and 40 watch seconds = 39·8616 mean seconds. Hence no perceptible error will arise from identifying watch and mean seconds. But 300 watch seconds = 298·962 mean seconds, and the error would have to be allowed for. Scheibler used a metronome corrected daily by an astronomical clock, and graduated. On this a movable weight enabled him to make one swing of the pendulum take place in the same time as 4 beats were heard, and then from the graduation he read off the rate. But counting by the seconds hand of a watch is much easier and altogether more convenient, while it is probably as accurate. Owing to difficulties in beginning and ending the count, I find the possible error per second to be 2 divided by the number of seconds through which the count extends; and that it is best to take a mean of 5 to 10 counts for each set of beats. As most persons, including myself, begin to count from *one* and not from *nought*, it must be remembered that the last number uttered on counting the last beat, is one in excess of the real number of beats. Thus if in counting for 10 seconds we end with 37, the number of beats was 36 in 10 seconds. If in counting by twos, one-ee, two-ee, &c., the last was 19, then we have counted only 18 pairs, and hence there were also 36 beats in 10 seconds. If we end with 19-ee, there were 18½ pairs or 37 beats in a second. It is best to count the same set of beats in one, twos and fours to realise these corrections, which are extremely important. Temperature must never be neglected.

Forks should not be touched with the unprotected hand; they otherwise easily flatten by 2 beats in 10 seconds. Interpose folds of paper. I use two folds of brown paper stitched between two pieces of wash-leather. Large forks are generally on resonance boxes and need not be touched, otherwise the same precautions should be used, as they are very sensitive, and retain the heat longer than small forks. Scheibler's forks are fitted with wooden handles. In tuning, the file heats and flattens; the result of tuning, therefore, can seldom be known for a day or two, when the forks have cooled and 'settled,' as they will be sure to 'jump up.' I find it best to leave off filing when the forks are two or three tenths of a vibration too flat. In sharpening there is, therefore, great danger of doing too much, as the fork remains apparently at the ¶ same pitch, the flattening by heat balancing the sharpening by filing. Hence all copies should be compared some days after, by means of a third fork about four vibrations flatter or sharper than each, to avoid the slow beats of approximate unisons. The filing also seems to interfere with the molecular arrangement of the forks.

The thermometer should be always consulted when beats are taken. But if the beats are between two forks, of which the pitch of one at a given temperature is known, and both forks may be assumed to be altered in the same ratio by heat, then the temperature need not be observed; but the unknown fork may be presumed to be as many vibrations sharper (or flatter) than the measured fork at the temperature at which the latter was measured, as beats in a second were observed ¶ to take place. This is because the alteration is very small, and would be quite inappreciable for the few vibrations between them. But for tonometrical purposes an allowance must be made.

When forks are counted without a resonance jar, they should not be applied to a sounding-board, or held one to one ear and one to the other, but should both be held about six inches from the same ear, and their strengths should be equalised by holding the weaker fork closer to the ear than the stronger.

When the forks are screwed on and off a sounding-board or resonance box, there is great danger of wrenching the prongs, unless they are held below the bend, but I have constantly seen this precaution neglected. A ¶ wrench immediately affects the pitch and duration of sound of a fork, and renders it comparatively worthless. Such cases have come within my observation. To prevent wrenching when filing forks, only one prong should be inserted in the vice. And for evenness file the same quantity off the inside of each prong, counting the number of strokes with the file, near the tips for sharpening, and near (not at) the bend for flattening.

The next enemy to be guarded against is rust. Forks should be kept dry, and occasionally oiled with gun-lock oil. Rust towards the tip affects the fork much less than rust at the bend. My observations and experiments shew that errors from rust can scarcely exceed a flattening of 1 vibration in 250, and are generally very much less. But as the amount is uncertain, rust spoils a fork for accurate tonometrical purposes. With care, however, the pitch of a tuning-fork remains remarkably permanent. Scheibler's had evidently not altered in more than forty years.

When the pitch to be ascertained is of a very short sounding tone, as that of a glass or wood harmonicon, or very high, it requires an extremely delicate ear indeed to be able to determine between which two forks, or the octaves of which two forks the tone lies. In this case I have been fortunate in having the kind assistance of Mr. A. J. Hipkins of Broadwoods', without whose accurate appreciation of differences of pitch I should have frequently been altogether at a loss.

¶ Dr. Koenig (27 Quai d'Anjou, Paris) makes tuning-fork tonometers of beautiful workmanship, but they are necessarily very expensive. The largest, proceeding from 64 to 21845 double vibrations at 20° F. with sliding weights, costs 1200*l.* The medium has 67 forks from 256 to 512 double vibrations at intervals of four beats each, with f and a' which fall between pairs of forks, mounted on resonance boxes, and costs 120*l.* A smaller set without resonators costs 60*l.* A small set of 13 forks in a case, giving the equally tempered Octave c' to c'' for a' 435 double vibrations, costs only 7*l.* 4*s.*, and for the same price another set 4 double vibrations lower, for tuning by beats to French pitch, can be obtained. This apparently high

price arises from the great difficulty of tuning such a succession of forks with perfect accuracy to particular pitches. This accuracy is, however, not necessary, provided the count be accurate. Any tuning-fork maker would make a set of forks such as has been described. The count must be made by the investigator himself, and he should verify by a set of Koenig's forks of c', e', g', c'', such as may be found in many places, remembering that all the older sets when reduced for temperature give at 15° C. = 59° F., c' 256·3, e' 320·3, g' 384·4, c'' 512·6. Prof. McLeod's determinations by his machine at this temperature, as the mean of many measurements, were c' 256·310, another copy 256·306, e' 320·372, g' 384·437, c'' 512·351. Koenig considered them correct at 26·2° C. = 79·16° F.

For my own use after returning Scheibler's forks to Mr. Amels, I had 105 forks constructed proceeding from 223·77 to 586·12 vib., which, after being tuned partly by Scheibler's maker in Crefeld by the forks already described, and partly constructed to differences of about 4 beats by the late Mr. Valantine of Sheffield, were all very carefully counted again by me with Scheibler's own forks, the means of many determinations up to two places of decimals being assumed as correct. With this perfectly unique set of forks I have, since that time, made all the determinations of pitch mentioned in this book.

SECTION C.

ON THE CALCULATION OF CENTS FROM INTERVAL RATIOS.

(See notes pp. 13, 41, and 70.)

1. When the 'interval numbers,' that is the pitch numbers of two notes, have been found (or the 'interval ratio,' that is ratio of those numbers given theoretically by means of pitch numbers, or of numbers in proportion to them, or of lengths of strings assumed to be perfect, or of wave-lengths), it is necessary, in order to have a proper conception of the interval itself by comparison with a piano or other instrument tuned in intentionally equal temperament, to determine the number of *cents*, or hundredths of an equal Semitone, in that interval. Such cents have been extensively used in the notes, and occasionally introduced into the text, of this translation, see pp. 41*d*, 50*a*, 56*a*, &c., and supra, sect. A. art. 2, p. 431*b*.

I. *First Method, without either Tables or Logarithms.*

2. If the greater number of the ratio be more than twice the smaller, divide the greater (or else multiply the less) by 2 until the greater number is not more than twice the smaller. This is equivalent to lowering the higher or raising the lower tone by so many Octaves. Hence for each division or multiplication by 2 add 1200 to the result.

Ex. To find the cents in 47th harmonic. Interval ratio 1 : 47. Multiplying the smaller number 5 times by 2, the result is 32 : 47, for which 'reduced' interval ratio we have to determine the cents as under and then add 5 × 1200 = 6000 to the result.

3. If the reduced interval ratio be such that 3 times the larger number is greater than 4 times the smaller, but twice the larger number is less than 3 times the smaller number, then multiply the larger number by 3, and the smaller by 4, for a new interval ratio, and add 498 cents to the result. ¶

Ex. For 32 : 47, then 3 × 47 = 141 is greater than 4 × 32 = 128, but 2 × 47 = 94 is less than 3 × 32 = 96. Hence we use the interval ratio 128 : 141 and add 498 cents to the result. If however as in 32 : 49, twice the larger number or 2 × 49 = 98, is not less than 3 times the smaller or 3 × 32 = 96, we use this interval ratio 96 : 98 or its equivalent 48 : 49 and add 702 cents to the result. In the first case the given interval having the ratio 32 : 47 lies between a Fourth and a Fifth, in the second case it is greater than a Fifth, but in both cases the reduced interval ratio 128 : 141 or 48 : 49 is less than a Fourth. The object of this reduction, which is seldom necessary, is to have to deal with ratios less than a Fourth.

4. Multiply 3477 by the difference of the numbers of the reduced interval ratio, and divide the result by their sum to the nearest whole number, and if the quotient is more than 450 add 1. To the result add the numbers of cents from arts. 1 and 2. The result is correct to 1 cent.

Ex. 1. Interval ratio 128 : 141

$$
\begin{array}{r}
3477 \\
13 \text{ difference} \\
\hline
10431 \\
3477 \\
\hline
\end{array}
$$

Sum
269)45201(168
        269   498 from art. 3
        ——   6000 from art. 2
        1830   ——
        1614   6666 cents result

        2161 or 5 Octaves
        2152   6 Semitones
        ——   and ⅔ Semitone
in the interval of 47th harmonic from the fundamental.

Ex. 2. Interval ratio 48 : 49, difference = 1
Sum
97)3477(36
   291   702
   ——
   567   738 cents in the interval
   588   32 : 49, or about 7⅓
   ——   Semitones.

*** The number 3477 depends on the principles explained in my paper 'On an ¶ Improved Bimodular Method of Computing Logarithms,' *Proc. R. S.*, Feb. 1881, vol. xxxi. p. 382. Cents are in fact a system of logs in which cent log 2 = 1200, and its bimodulus is 2 × cent log 2 ÷ nat log 2 = 2400 ÷ ·69315 = 3462·4. But if this number had been selected there would have been constant additive corrections from the first. Hence an augmented bimodulus 3477 has been selected = 9 × 386⅓, or 9 times the cents in the ratio 4 : 5. The result is that the rule is exact for intervals of a major Third. For less intervals it gives too great a result, but never by more than ·6 cent, which may be neglected. Between a major Third and a Fourth it gives too small a result, but only after 450 by about 1 cent. For small numbers, few calculations, and in the absence of tables this method is very convenient. For Sauveur's previous use of ¶ an augmented bimodulus, see suprà, p. 437*b*, sect. A. art. 24, xii.

II. *Second Method, with Tables but without Logarithms.*

5. The reduction for Octaves is always supposed to be made as in art. 2, and hence only intervals of less than an Octave will be considered.

If the interval numbers (art. 1) contain decimals they must be multiplied by 10 till the decimals disappear. Thus 264 : 478·5 is taken at 2640 : 4785, so that the rule applies to whole numbers only.

6. Rule. Annex 0000 to the larger number and divide by the smaller to the nearest whole number. If the quotient is less 11290, take the nearest 'quotient'

in the Principal Table below. The corresponding nearest whole number of 'cents' is on the same line.

Ex. 48 : 49.

$$12 \, | \, \overline{490000} \quad$$ Nearest quotient in Table 10210 for cents 36 as in art. 4, ex. 2.

Since 48 = 4 × 12 use short division.

$$4 \, | \, \overline{40833}$$

$$\overline{10208}$$

7. If the quotient exceeds 11290, look for the next less quotient in Auxiliary Table I., multiply and divide by the numbers in the col. of multipliers and divisors, and thus reduce the quotient to one in the table. Then proceed as before and add the number of cents on the line with the next least quotient in Auxiliary Table I.

¶

Ex. Ratio 32 : 47 as in art 3. Since 32 = 4 × 8 proceed thus :

$$4 \, | \, \overline{470000}$$

$$8 \, | \, \overline{117500}$$

$$\overline{14688}$$

$$3 \quad$$ Next less quotient 13333, mult. 3, div. 4, cents 498.

$$4 \, | \, \overline{44064}$$

$$11016 \quad$$ Nearest 11019 cents 168
$$\text{add } 498$$

as before, 666 cents for the 47th harmonic.

Actually 11016 lies exactly half-way between 11013 and 11019, and in that case the rule is to take the larger number. On account of the approximative nature of the calculation the error of 1 cent in the final result is always possible, and is here disregarded. It is avoided by the following methods.

¶

III. *Third Method, by Five Place Logarithms.*

8. This is by far the best, most exact, and at the same time easiest method, but as many musicians are not familiar with logarithms, and it is important that they should be able to reduce interval ratios to cents, the two preceding methods have been inserted.

9. After reducing for Octaves as in art. 2, subtract the log of the smaller number from that of the larger. If the resulting log is less than ·05268, the number of cents is given opposite to the nearest log in the Principal Table. The decimal point, zeroes after it, and characteristic are omitted in practice but are used here for completeness.

Ex. Interval ratio 48 : 49.
log 49 = 1·69020
log 48 = 1·68124

diff.   ·00896
¶   nearest log   ·00903, cents 36.

The difference between any two logarithms in the table is 25 or 26, hence the nearest is that which differs by less than 13. In this case it differs only by 7.

10. If the log exceeds ·05268, find the nearest log, in Aux. Table II., subtract it, take the cents from the Principal Table for this diff. as in art. 9, and add the cents opposite the next least log in Aux. Table II.

Ex. Interval ratio 32 : 47.
log 47 = 1·67210
log 32 = 1·50515

1st diff.   ·16695
Next least, Aux. T. II.   ·15051, cents 600

2nd diff.   ·01644
Nearest log   ·01656, cents 66

result 666, as before for the 47th harmonic.

11. If it is desired to find the number of cents to the nearest tenth of a cent,

take the next least number in the Principal Table, find the difference, and add the
tenths of cents from Aux. Table III.   Thus—

Ex. in art. 10.

$$
\begin{array}{llll}
\text{2nd diff.} & \cdot01644 & & \\
\text{Next least} & \cdot01631 & \text{cents } 65 & \\
\text{3rd diff.} & 13 & \text{,,} & \cdot5
\end{array}
$$

cents 65·5 result.

IV. *Fourth Method, by Seven Place Logarithms.*

12. As a general rule the approximation in art. 10 is amply sufficient, and is
generally used here.   But occasionally it is advisable to proceed to three places of
decimals of cents, as in the whole of sect. A. on ' Temperament.'   The process is
then conducted by Aux. Table IV., the method of using which will appear by the ¶
following examples :—

Ex. 1. Interval ratio 32 : 47.

$$
\begin{array}{ll}
\log 47 = 1\cdot672\ 0979 \\
\log 32 = 1\cdot505\ 1500
\end{array}
$$

difference		·166 9479
600	cents	·150 5150
		·016 4329
60	,,	·015 0515
		·001 3814
5	,,	·001 2543
		·000 1271
·5	,,	·000 1254
·007	,,	·000 0017

665·507 cents result.

Ex. 2. Interval ratio 264 : 478·5.

$$
\begin{array}{ll}
\log 478\cdot5 = 2\cdot679\ 8819 \\
\log 264\quad = 2\cdot421\ 6039
\end{array}
$$

difference		·258 2780
1000	cents	·250 8583
		·007 4197
20	,,	·005 0172
		·002 4025
9	,,	·002 2577
		·000 1448
·5	,,	·000 1254
		·000 0184
·07	,,	·000 0176
		·000 0018
·007	,,	·000 0018

¶

1029·5707 cents result.

V. *Method of finding the Interval Ratio from the Cents.*

13. *Without Logarithms.* If the cents are less than 210, the ratio is that of
10000 to the quotient opposite the cents in the Principal Table.   If the cents are
greater than 210, subtract the next least number of cents in Aux. Table I., and
multiply and divide the quotient opposite the diff. of cents in the Principal Table
by the corresponding divisor and multiplier respectively (observe this inversion,
multiply by divisor, and divide by multiplier) in Aux. Table I.

Thus, given 666 cents,
subtr. next least 498 in Aux. Table I.; take 3 as
                       —— div. and 4 as mult.
                        168

quo. to 168 cts. 11019 in Principal Table.

$$
\begin{array}{l}
4 \\
3\,|\,44076 \\
\end{array}
$$

14692, ratio 1·4692.

This is the correct ratio for 666 cents, it
is larger than 1·4688 obtained from 32 : 47 ¶
in art. 7, because the cents were in excess,
but the difference is quite unimportant.

14. *By Five Place Logarithms.*

Given 665·5 cents as results from art. 11—

Cents	Logs	
600	·15051	Aux. Table II.
65	·01631	Principal Table
·5	·00013	Aux. Table III.

Sum ·16695 = log 1·4687, which is now ·0001 smaller than in
                      art. 7, and is the correct value of 665·5 cents.

15. By Seven Place Logarithms.

From Aux. Table IV. Cents 665·507, art. 12.

which is the more correct value of 47÷32, obtained by carrying the division one step further than in art. 7.

600	cents	·150 5150
60	,,	·015 0515
5	,,	·001 2543
·5	,,	·000 1254
·007	,,	·000 0018

·166 9480 = log 1·46875,

Principal Table for the Calculation of Cents.

Cents	Quotients	Logs	Cents	Quotients	Logs	Cents	Quotients	Logs	Cents	Quotients	Logs
1	10006	00025	54	10317	01355	107	10638	02684	160	10968	04014
2	12	50	55	23	380	108	44	709	161	10974	04039
3	17	75	56	29	405	109	50	734	162	81	064
4	23	00100	57	35	430	110	56	760	163	87	089
5	29	125	58	41	455	111	10662	02785	164	94	114
6	35	151	59	47	480	112	69	810	165	11000	139
7	40	176	60	53	505	113	75	835	166	06	164
8	46	201	61	10359	01530	114	81	860	167	13	189
9	52	226	62	65	555	115	87	885	168	19	214
10	58	251	63	71	580	116	93	910	169	26	240
11	10064	00276	64	77	605	117	99	935	170	32	265
12	69	301	65	83	631	118	10705	960	171	11038	04290
13	75	326	66	89	656	119	11	985	172	45	315
14	81	351	67	95	681	120	18	03010	173	51	340
15	87	376	68	10401	706	121	10724	03035	174	57	365
16	93	401	69	07	731	122	30	061	175	64	390
17	99	427	70	13	756	123	37	086	176	70	415
18	10105	452	71	10419	01781	124	43	111	177	76	440
19	10	477	72	25	806	125	49	136	178	83	465
20	16	502	73	31	831	126	55	161	179	89	490
21	10122	00527	74	37	856	127	61	186	180	96	516
22	28	552	75	43	881	128	67	211	181	11102	04541
23	34	577	76	49	907	129	74	236	182	09	566
24	39	602	77	55	932	130	80	261	183	15	591
25	45	627	78	61	957	131	10786	03286	184	22	616
26	51	652	79	67	982	132	92	311	185	28	641
27	57	677	80	73	02007	133	99	337	186	34	666
28	63	702	81	10479	02032	134	10805	362	187	41	691
29	69	728	82	85	057	135	11	387	188	47	716
30	75	753	83	91	082	136	17	412	189	53	741
31	10181	00778	84	97	107	137	23	437	190	60	766
32	87	803	85	10503	132	138	30	462	191	11166	04791
33	93	828	86	09	157	139	36	487	192	73	816
34	98	853	87	15	183	140	43	512	193	79	842
35	10204	878	88	21	208	141	10849	03537	194	86	867
36	10	903	89	28	233	142	55	562	195	92	892
37	16	928	90	34	258	143	61	587	196	99	917
38	22	953	91	10540	02283	144	67	612	197	11205	942
39	28	978	92	46	308	145	74	637	198	12	967
40	33	01003	93	52	333	146	80	663	199	18	992
41	10240	01029	94	58	358	147	86	688	200	25	05017
42	46	054	95	64	383	148	93	713	201	11231	05042
43	52	079	96	70	408	149	99	738	202	37	067
44	57	104	97	76	433	150	10906	763	203	44	092
45	63	128	98	82	458	151	10912	03788	204	50	118
46	69	154	99	88	484	152	18	813	205	57	143
47	75	179	100	95	509	153	24	838	206	64	168
48	81	204	101	10601	02534	154	30	863	207	70	193
49	87	229	102	07	559	155	37	888	208	77	218
50	93	254	103	13	584	156	43	913	209	83	243
51	10299	01279	104	19	609	157	49	939	210	90	268
52	10305	305	105	25	634	158	56	964			
53	11	330	106	32	659	159	62	989			

16. Hence given a note of any pitch and the interval in cents between that and another note, it is easy to determine the pitch of this second note.

Ex. Required the reduced 47th harmonic to A 453·9, the concert organ pitch of Mr. H. Willis, to which is tuned the great organ at the Albert Hall.

$$\log 453 \cdot 9 = 2 \cdot 65696$$
$$\text{cents } 665 \cdot 5, \text{ art. } 14, \text{ give } \log = \cdot 16695$$
$$\overline{\log 666 \cdot 7 = 2 \cdot 82391}$$

Hence 666·7 is the pitch number of the note required. Thus it is possible, for any given pitch of the tuning note, to calculate the pitch of the notes for any given temperament, and hence, as will be shewn, to tune in that temperament.

Auxiliary Table I.

Cents	Quotients	Multipliers and Divisors
204	11250	× 8 ÷ 9
316	12000	× 5 ÷ 6
386	12500	× 4 ÷ 5
498	13333	× 3 ÷ 4
702	15000	× 2 ÷ 3
884	16667	× 6 ÷ 10
1018	18000	× 5 ÷ 9
1200	20000	× 1 ÷ 2

Aux. Table II.   Aux. Table III.

Cents	Logs	Cents	Logs
100	02509	·1	·00003
200	05017	·2	05
300	07526	·3	08
400	10034	·4	10
500	12543	·5	13
600	15051	·6	15
700	17560	·7	18
800	20069	·8	20
900	22577	·9	23
1000	25086	1·0	25
1100	27594		
1200	30103		

Auxiliary Table IV.

Cents	Logs	Cents	Logs
100	025 0858	·1	000 0251
200	050 1717	·2	0502
300	075 2575	·3	0753
400	100 3433	·4	1003
500	125 4292	·5	1254
600	150 5150	·6	1505
700	175 6008	·7	1756
800	200 6867	·8	2007
900	225 7725	·9	2258
1000	250 8583		
1100	275 9442		
1200	301 0300		
10	002 5086	·01	000 0025
20	05 0172	·02	050
30	07 5258	·03	075
40	10 0343	·04	100
50	12 5429	·05	125
60	15 0515	·06	151
70	17 5601	·07	176
80	20 0687	·08	201
90	22 5773	·09	226
1	000 2509	·001	000 0003
2	5017	·002	5
3	7526	·003	8
4	001 0034	·004	000 0010
5	2543	·005	13
6	5052	·006	15
7	7560	·007	18
8	002 0069	·008	20
9	2577	·009	23

SECTION D.

MUSICAL INTERVALS, NOT EXCEEDING AN OCTAVE, ARRANGED IN ORDER OF WIDTH.

(See notes pp. 13, 187, 213, 264, and 333.)

1. An interval was defined suprà, p. 13d, note ‡. The *width* of an interval is measured by the number of cents it contains. Beside the usual diatonic intervals, a large number of others occasionally occur, which it is convenient to have arranged

according to their widths as measured in cents.   Many of these are furnished in the following table.

2. The cents used are cyclic cents, as defined suprà, sect. A. art. 24, xiii. p. 437*b'*, that is, those intervals found by taking a certain number of Fifths and major Thirds up and down and reducing the result to the same octave, are assumed to have cyclic Fifths and major Thirds of 702 and 386 cents respectively. But as the actual Fifths and major Thirds have 701·955 and 386·314 cents respectively, a slight error in excess is made in every Fifth up and every major Third down, and in defect in every Fifth down and major Third up, which, when a great many are supposed to be taken for theoretical purposes, may reach to a sensible ¶ amount. These errors are of no consequence for ordinary purposes, but a means of correcting them is given in sect. E., p. 463*d*, and here it has been thought better to add the result to three places of decimals in many cases, and this is put in the last column, preceded by the letters ' ex.,' meaning ' more exact cents.'

3. Other intervals are given to the nearest whole number of cents, determined as in sect. C., which therefore belong to the cycle of 1200, and hence are rightly called cyclic. Here also is added the result to three places of decimals, when it appeared advisable for theoretical purposes. For ordinary purposes cyclic cents always suffice.

4. The *interval ratios*, being of historical

interest, are always given, although they are of no assistance to the eye in recognising the width of an interval. In these ratios the smaller number is always placed first. In the case of tempered intervals an approximate ratio, with † prefixed, is given in the second column, and the ' ex.' or more exact ' ratio' is given in the last column.

5. The (five place) logarithm of each interval ratio, considered as a fraction of which the larger number is the numerator, is added in each case, to enable those who understand logarithms to deal with them immediately in calculating pitch numbers, &c. The logarithms always give the exact intervals. The decimal point is omitted.

6. In the last column is given a variety of information. The name of the interval when any usual name exists, or the instrument on which it is found. The Greek and Arabic intervals were theoretical, and given in terms of lengths of string. As we see from sect. B. No. 1, p. 441*d*, there is every reason to suppose that the real intervals tuned differed from them materially. Some further information is occasionally added.

7. If the interval is found in the Duodenarium (sect. E., p. 463), then a mode of obtaining it by Fifths up and down with major Thirds up and down is annexed. Here

$$Vu = \text{one Fifth up} = 702 \text{ cents}, \quad 2Vu = 2 \times 702 \text{ cents, &c.}$$
$$Vd = \text{one Fifth down} = 498 \text{ cents up}, \quad 2Vd = 2 \times 498 \text{ cents, &c.}$$
$$Tu = \text{one major Third up} = 386 \text{ cents}, \quad 2Tu = 2 \times 386 \text{ cents, &c.}$$
$$Td = \text{one major Third down} = 814 \text{ cents}, \quad 2Td = 2 \times 814 \text{ cents, &c.}$$

When such additions are made, 1200 or multiples of 1200 must be subtracted till the result is less than 1200. That result will be the

number of cyclic cents in the first column. Of course, if we take the value to three places of decimals,

$$Vu = 701\text{·}955, \quad Vd = 498\text{·}045, \quad Tu = 386\text{·}314, \quad Td = 813\text{·}686,$$

and then the result will be correct to at least two places of decimals.

8. If we put $Vu = \frac{3}{2}$, $2Vu = \frac{3^2}{2^2}$, &c. $Vd = \frac{2}{3}$, $2Vd = \frac{2^2}{3^2}$, &c. $Tu = \frac{5}{4}$, $2Tu = \frac{5^2}{4^2}$, &c. $Td = \frac{4}{5}$, $2Td = \frac{4^2}{5^2}$, &c. and multiply instead of adding, and finally multiply or divide by 2 until the ¶ result lies between 1 and 2, these formulæ give the exact ratio. Thus $2Vd + 2Tu =$ cyclically $2 \times 498 + 2 \times 386 = 996 + 772 = 1768 - 1200 = 568$ cents as in the table. Or to three places of decimals, $2 \times 498\text{·}045 + 2 \times 386\text{·}314 = 996\text{·}090 + 772\text{·}628 = 1768\text{·}718 - 1200 = 568\text{·}718$, which is correct. Or again, $\frac{2^2}{3^2} \times \frac{5^2}{4^2} = \frac{4 \times 25}{9 \times 16} = \frac{25}{36} \times 2 = \frac{25}{18}$, the correct result. See Table I. under 568.

9. In Table II. are given all the unevenly numbered harmonics up to the 63rd in order of occurrence. The first column gives the number of the harmonic, in which those marked * will be found on the Harmonical as harmonics of both *C* 66 and *C* 132, and those marked † as harmonics of *C* 132 only. In the second column are the pitch numbers of all the harmonics of *C* 66. In the third column ' log ' are the logarithms of the harmonics of 1, pre-

ceded by a *plus* sign +, so that if to each of these be added the logarithm of the pitch number of the fundamental note, the result is the logarithm of the pitch number of the harmonic. Thus log $66 = 1\text{·}81954$, which added to $1\text{·}36173$, the log opposite 23rd harmonic, gives $3\text{·}18127 = \log 1518$, the pitch number (in the table) of the 23rd harmonic of *C* 66. The column is divided into octaves by cross lines, at the beginning of which, preceded by a *minus* sign −, is the number to be *subtracted* from the log given in order to find the log of the harmonic reduced to one octave as given in Table I. Thus for 23rd harmonic $1\text{·}36173 - 1\text{·}20412 = \text{·}15761$, which is the log opposite 628 cents in Table I. In the fourth column is given the cyclic cents in the ratio of the fundamental note to the harmonic reduced to the same octave, the same as given in Table I., where will be found the more exact number of cents. But to each octave is prefixed the number of cents, followed by a *plus* sign +, which have to be added in order to find the unreduced interval. Thus for 23rd harmonic it is $4800 + 628 = 5428$ cents. Finally in the last column there is given the *nearest* equally tempered tone, supposing the fundamental note is *C*, and the number of cents to be added to or subtracted from that note in order to produce the harmonic. Thus the 23rd harmonic is sharper than $\|f'''\sharp$ by 28 cents. These comparisons are readily made from the column of cyclic cents, and can be easily applied to any fundamental note. Thus the 23rd harmonic of B,b must be 4 Octaves and 6 Semitones and 28 cents sharper than B,b, and hence must be $e''' + 28$. The marking of the differences of the harmonics from equally tempered notes is convenient for repeating the experiments in sect. N. No. 2.

10. Table III. is constructed to shew how often each principal interval, not exceeding a Tritone, is contained in the Octave. The first column gives the cyclic cents in the interval for easy reference to Table I. The second column contains the names of the intervals. The third contains, up to one place of decimals, the number of times that the interval is contained in the Octave, found by dividing log 2 by the logarithm of the interval as given in Table I. This is therefore not always the same as the number arrived at by dividing 1200 by the number of *cyclic* cents, but only by the number of precise cents, as given in Table I. Thus, taking the skhisma of 2 cyclic and 1·953 ex. cents, $1200 \div 2 = 600$, too small, but $1200 \div 1·953 = 614·4$ as in Table III.

I. *Table of Intervals not exceeding one Octave.*

Cyclic Cents	Interval Ratios †Approximative	Logs	Name, &c.
0	1 : 1	0	Fundamental note of the open string, assumed as C 66
1	1730 : 1731	00025	Cent, hundredth of an equal Semitone, nearest approximate ratio, ex. 1 : 1·0005755
2	32768 : 32805	00049	Skhisma, $8Vu + Tu = C : B,\sharp$, ex. 1·953
7	255 : 256	00170	Ex. 6·775, the ratio $= \frac{17}{16} \cdot \frac{15}{16}$, and the result is the 17th harmonic of D^1b, a diatonic Semitone above C
18	95 : 96	00455	Ex. 18·128, the ratio is $\frac{6}{5} \cdot \frac{16}{19}$, or the interval by which the 19th harmonic is flatter than the minor Third
20	2025 : 2048	00490	Diaskhisma, $4Vd + 2Td = C : D\flat\flat = C_\prime\sharp : D^1\flat$, ex. 19·553
22	80 : 81	00540	Comma of Didymus, which is always meant by Comma when no qualification is added, $4Vu + Td = C : C^1$, ex. 21·506
24	524288 : 531441	00589	Pythagorean Comma, $12Vu = C : B\sharp = D\flat : C\sharp$, ex. 23·460
27	63 : 64	00684	Septimal Comma, or interval by which the 7th harmonic, 969 cents, is flatter than the minor Seventh, 996 cents, $^7B\flat : B\flat$, ex. 27·264
28	3072 : 3125	00743	Small Diësis, $Vd + 5Tu = C^\prime b : B_3\sharp$, ex. 29·614. In equal temperament this last interval would be represented (as $\|B : c$) by a Semitone of 100 cents
36	48 : 49	00896	Interval of Al Fārābī's improved Rabāb
42	125 : 128	01030	Great Diësis, the defect of 3 major Thirds from an Octave, the interval between $C\sharp$ and $D\flat$ in the meantone temperament, $3Td = C_2\sharp : D^1\flat$, ex. 41·059
44	39 : 40	01100	First interval on the Tambūr of Bagdad, the interval by which the 13th harmonic of 840·528 cents is flatter than the just major Sixth of 884·359, ex. 43·831
45	38 : 39	01128	Second interval on the Tambūr of Bagdad
46	37 : 38	01158	Third „ „ „ „
47	36 : 37	01190	Fourth „ „ „ „
49	35 : 36	01223	Fifth „ „ „ „
50	† 239 : 246	01254	Quartertone of Meshāqah, the quarter of an equal Tone, ex. ratio = 1 : 1·0293022, $\|C : Cq$
53	32 : 33	01336	33rd harmonic, interval by which the 11th harmonic exceeds the just Fourth, $F : {}^{11}F$, ex. 53·273
55	31 : 32	01379	Greek Enharmonic Quartertone, suprà, p. 265a
57	30 : 31	01424	Another Greek Enharmonic Quartertone, suprà, p. 265a
70	24 : 25	01773	Small Semitone, $Vd + 2Tu = C : C_2\sharp$, ex. 70·673
76	† 67 : 70	01908	Meantone Small Semitone, meantone $C : C\sharp$, and hence the $\sharp$ of that system, ex. 76·050, ex. ratio 1 : 1·0449
85	20 : 21	02119	Subminor Second, Greek interval, suprà, p. 264a; $A_1 : {}^7B\flat$, on the harmonical
89	19 : 20	02228	Interval from open string to second fret on the Tambūr of Bagdad
90	243 : 256	02263	Pythagorean Limma, the 'defect' of two major Tones, 408 cents, from a Fourth, 498 cents, $5Vd = C : D\flat$, ex. 90·225
92	128 : 135	02312	Larger Limma, the defect of a Fourth, 498 cents, increased by a diatonic Semitone, 112 cents (total 610 cents), from a Fifth, 702 cents, and hence the interval by which the

TABLE OF INTERVALS NOT EXCEEDING ONE OCTAVE—*continued.*

Cyclic Cents	Interval Ratios †Approximative	Logs	Name, &c.
			Fourth must be sharpened to be a diatonic Semitone below (i.e. the ' leading note' to) the Fifth, and hence the interval by which the Fourth is sharpened on modulating into the dominant, $3Vu + Tu = C : C_1\sharp = F : F_1\sharp$, ex. 92·179. This was used as the meaning of $\sharp$ in the first English edition, for which in the present 114 cents, or the Apotomē, has been substituted, to agree with Prof. Helmholtz's notation for just intonation
94	18 : 19	02348	Greek interval in Al Farabi, interval between the 18th and 19th harmonics, $d''' : e'''\flat$ on Harmonical
99	17 : 18	02480	Arabic interval
100	† 84 : 89	02509	Ex. 100·099, the nearest approximate in small numbers to the ratio of the interval of an equal Semitone, ex. ratio $1 : 1·059461$
105	16 : 17	02633	17th harmonic $= C : {}^{17}d'''\flat$ on the Harmonical, ex. 104·955
112	15 : 16	02803	Diatonic or just Semitone, ex. 111·731 cents, $Vd + Td = B : c = E : F$
114	2048 : 2187	02852	Pythagorean Apotomē, ' off-cut,' or what is left of the major Tone, 204 cents, after ' cutting off ' the Limma or 90 cents, used for $\sharp$ in this edition, $7Vu = C : C\sharp = C_1 : C_1\sharp$, ex. 113·685
117	† 100 : 107	02938	Meantone great Semitone, meantone $C : D\flat$, ex. 117·108
128	13 : 14	03219	Interval between 13th and 14th harmonics, ex. 128·298
134	25 : 27	03342	Great Limma, a Comma greater than the diatonic Semitone, 112 cents, ex. 133·237, $3Vu + 2Td = C : D'\flat$, $E_1 : F$ in the Phrygian tetrachord, suprà, p. 263d', No. 6
135	37 : 40	03386	Interval from open string to the third fret of the Tambūr of Bagdad
139	12 : 13	03476	Interval between the 12th and 13th harmonics, ex. 138·573
145	149 : 162	03633	Persian ' near the forefinger ' lute interval
146	4235 : 4608	03666	Arabic ' near the forefinger ' lute interval
150	† 221 : 241	03763	Meshāqah's 3 Quartertones, imitation of 151 cents, ex. ratio $1 : \sqrt[8]{2}^3 = 1 : 1·0905$
151	11 : 12	03779	The interval between the 11th and 12th harmonics on the trumpet scale, used in Ptolemy's equal diatonic mode, suprà, p. 264b, used by Zalzal in Arabic lute scale as 'middle finger,' see infrà, sect. K., Persia, Arabia, and Syria
155	32 : 35	03892	The 35th harmonic, septimal or submajor Second, suprà, p. 212c, ex. 155·140
165	10 : 11	04139	A trumpet interval, used in Ptolemy's equal diatonic scale, suprà, p. 264b
168	49 : 54	04219	Zalzal's ' near the forefinger ' on Arabic lute
180	59049 : 65536	04527	Abdulqadir's substitute for Zalzal's 168 cents
182	9 : 10	04576	The minor Tone of just intonation, the ' grave Second ' of the major scale, $2Vd + Tu = C : D_1$, ex. 182·404
193	† 161 : 180	04846	The mean Tone, the Tone of the meantone system, $C : D$, the mean between a major Tone of 204 cents, and a minor Tone of 182 cents, ex. ratio $1 : \tfrac{1}{2}\sqrt{5} = 1 : 1·1180340$
200	† 400 : 449	05017	An equal Tone, ex. ratio $1 : \sqrt[6]{2} = 1 : 1·22462$
204	8 : 9	05115	The 9th harmonic, major Tone, $2Vu = C : D$, ex. 203·910
224	225 : 256	05606	Diminished minor Third, $2Vd + 2Td = B_1 : D'\flat$, ex. 223·463
231	7 : 8	05799	Supersecond, or septimal Second, ${}^7B\flat : c$, on the Harmonical, ex. 231·174
240	† 74 : 85	06021	The Pentatone, or fifth part of an Octave, ex. ratio $1 : \sqrt[5]{2} = 1 : 1·1487$ in the Salendro scale, see infrà, sect. K., Java
246	125 : 144	06145	Acute diminished minor Third, $2Vu + 3Td = B_1 : D^2\flat$, ex. 244·968
250	† 200 : 231	06271	Five Quartertones, on Meshāqah's scale, ex. ratio $1 : {}^{24}\!\sqrt{2}^5 = 1 : 1·15535$
251	32 : 37	06305	The 37th harmonic, ex. 251·344
253	108 : 125	06349	Grave augmented Tone, $3Vd + 3Tu = C : D_3\sharp$, ex. 253·076
267	6 : 7	06695	Septimal or subminor Third, $G : {}^7B\flat$ on the Harmonical, Poole's minim Third, ex. 266·871
274	64 : 75	06888	The 75th harmonic, augmented Tone, $Vu + 2Tu = C : D_2\sharp$, ex. 274·583
281	17 : 20	07158	Interval on Tambūr of Bagdad
294	27 : 32	07379	Pythagorean minor Third, ancient ' middle finger ' on Arabic lute, $3Vd = C : E\flat$, ex. 294·135

TABLE OF INTERVALS NOT EXCEEDING ONE OCTAVE—*continued*.

Cyclic Cents	Interval Ratios †Approximative	Logs	Name, &c.
298	16 : 19	07463	The 19th harmonic, ex. 297·513
300 †	37 : 44	07526	The equal minor Third, $\|A : C$
303	68 : 81	07598	Persian 'middle finger' on lute
316	5 : 6	07918	Just minor Third, $Vu + Td = A_1 : C = C : E^1b$, ex. 315·641
336	14 : 17	08432	Wide or superminor Third in the chord of diminished Seventh, $'b''b : {}^{17}d'''b$ on Harmonical, ex. 336·129
342	32 : 39	08591	The 39th harmonic, ex. 342·483
345	59 : 72	08648	Arabic lute open string to string of the mean of the lengths for 204 and 498 cents, practical substitute for 355 cents
350 †	125 : 153	08780	Meshāqah's 7 Quartertones, tempered form of 355, ex. ratio 1 : $\sqrt[24]{2^7} = 1 : 1·2241$
355	22 : 27	08894	Zalzal's 'middle finger,' or *wostā*, mean length of strings for 303 and 408 cents
384	6561 : 8192	09642	Abdulqadir's substitute for 355 cents
386	4 : 5	09691	The 5th harmonic, just major Third, $Tu = C : E_1$, ex. 386·314
400 †	50 : 63	10034	Equal major Third, ex. ratio 1 : $\sqrt[3]{2} = 1 : 1·2599210$
408	64 : 81	10231	Pythagorean major Third, or Ditone, as it consists of two major Tones = 2 × 204, ex. 407·820
428	25 : 32	10721	Diminished Fourth, $2Td = E_1 : A^1b$, ex. 427·342
429	32 : 41	10763	The 41st harmonic, ex. 429·062
435	7 : 9	10914	Septimal or supermajor Third, Poole's maxim Third, ${}^7Bb : d$ on Harmonical, ex. 435·084
450 †	27 : 35	11288	Meshāqah's 9 Quartertones, ex. ratio $\sqrt[24]{2^9} = 1·297$
454	10 : 13	11394	One of Prof. Preyer's trial intervals
456	96 : 125	11464	Superfluous Third, $Vd + 3Tu = C : E_3\sharp = A^1b : C_2\sharp$, ex. 456·986
471	16 : 21	11810	The 21st harmonic, the septimal or Subfourth, $F : {}^7Bb$ on Harmonical, ex. 470·781
476	243 : 320	11954	Grave Fourth, $5Vd + Tu = C : F_1$, ex. 476·539
480 †	25 : 33	12041	Two Pentatones, the representative of the Fourth in the Salendro scale, see infrà, sect. K., Java, ex. ratio 1 : $\sqrt[5]{4}$ = 1 : 1·3195
498	3 : 4	12494	Just and Pythagorean Fourth, $Vd = C : F$, ex. 498·045
500 †	227 : 303	12543	Equal Fourth, $\|C : F$, ex. ratio 1 : $\sqrt[12]{2^5} = 1 : 1·3348$
503	80 : 107	12629	Meantone Fourth, meantone $C : F$, ex. 503·422, ex. ratio 1 : $\frac{4}{3} \times \sqrt[4]{\frac{81}{80}} = 1 : 1·3375$
512	32 : 43	12832	The 43rd harmonic, ex. 511·518
520	20 : 27	13033	Acute Fourth, $3Vu + Td = C : F^1 = A_1 : D$, ex. 519·552
550 †	500 : 687	13797	Meshāqah's 11 Quartertones, ex. ratio 1 : $\sqrt[24]{2^{11}} = 1 : 1·374$
551	8 : 11	13833	The 11th harmonic, ex. 551·318
568	18 : 25	14267	Superfluous Fourth, $2Vd + 2Tu = C : F_2\sharp$, ex. 568·718
583	5 : 7	14613	Septimal or subminor Fifth, $E : {}^7Bb$ on Harmonical, ex. 582·512
590	32 : 45	14806	The 45th harmonic, Tritone, false, sharp, augmented, or pluperfect Fourth, $2Vu + Tu = F : B_1 = C . F_1\sharp$, ex. 590·224, the Fourth $C : F$ as widened for passing into the key of the dominant, 498 + 92 = 590
600 †	99 : 140	15052	Equal Tritone, $\|F : B$, ex. ratio 1 : $\sqrt{2} = 1 : 1·4142$
610	45 : 64	15297	Diminished Fifth, $2Vd + Td = C : G^1b = F^1\sharp : c$, ex. 609·777
612	512 : 729	15346	Pythagorean Tritone, $6Vu = C : F\sharp = F : B$, ex. 611·731
628	16 : 23	15761	The 23rd harmonic, ex. 628·274
632	25 : 36	15836	Acute diminished Fifth, $2Vu + 2Td = C : G^2b = A_1 : e^1b$, ex. 631·283
650 †	90 : 131	16305	Meshāqah's 13 Quartertones, ex. ratio 1 : $\sqrt[24]{2^{13}} = 1 : 1·4556$
653	24 : 35	16486	Septimal, or Subfifth, $E^1b : {}^7Bb$ on Harmonical, ex. 653·184
666	32 : 47	16695	The 47th harmonic, ex. 665·507
666	49 : 72	16713	Arabic lute, 2nd string, a Fourth above 168 cents, ex. 666·258, and hence ·751 cents sharper than the last, the confusion with the former is due to approximations
678	177147 : 262144		Abdulqadir's substitute for 666 cents, being a Fourth above his 180 cents
680	27 : 40	17070	Grave Fifth, $3Vd + Tu = C : G_1$, ex. 680·449
697 †	109 : 163	17474	Meantone Fifth, meantone $C : G$, a quarter of a Comma too flat, ex. 696·579, ex. ratio 1 : $\frac{3}{2} \times \sqrt[4]{\frac{80}{81}} = 1 : 1·4954$
700 †	289 : 433	17560	Equal Fifth. $\|C : G$
702	2 : 3	17609	Just and Pythagorean Fifth, $Vu = C : G$, ex. 701·955

TABLE OF INTERVALS NOT EXCEEDING ONE OCTAVE—*continued.*

Cyclic Cents	Interval Ratios †Approximative	Logs	Name, &c.
720 †	95 : 144	18062	Three Pentatones, giving an acute Fifth, as in Salendro, see infrà, sect. K., Java, ex. ratio $1 : \sqrt[5]{8} = 1 : 1\cdot5157$
724	160 : 243	18149	Acute Fifth, $5Vu + Td = C : G^1$, ex. $723\cdot461$
738	32 : 49	18505	The 49th harmonic, ex. $737\cdot652$
750 †	59 : 91	18814	Meshāqah's 15 Quartertones, ex. ratio $1 : {}^{24}\!/2^{15} = 1 : 1\cdot5424$
772	16 : 25	19382	Grave superfluous Fifth, $2Tu = C : G_2\sharp$, the 25th harmonic, ex. $772\cdot627$
792	81 : 128	19873	Pythagorean minor Sixth, $4Vd = C : A\flat$, ex. $792\cdot180$
794	256 : 405	19922	Extreme sharp Fifth, $4Vu + Tu = C : G_1\sharp$, ex. $794\cdot134$
800 †	63 : 100	20069	Equal superfluous Fifth, $\|C : G\sharp$, and also equal minor Sixth, $\|C : A\flat$, the same notes differently written, ex. ratio $1 : \sqrt[4]{} = 1 : 1\cdot5874$
807	32 : 51	20242	The 51st harmonic, ex. $807\cdot304$
¶ 814	5 : 8	20412	Just minor Sixth, $Td = C : A^1\flat$, ex. $813\cdot687$
841	8 : 13	21085	The 13th harmonic, ex. $840\cdot528$
850 †	30 : 49	21323	Meshāqah's 17 Quartertones, his tempered substitute for 853 cents, ex. ratio $1 : {}^{24}\!/2^{17} = 1 : 1\cdot6339$
853	11 : 18	21388	Arabic lute, the Fourth above Zalzal's 355 cents
874	32 : 53	21913	The 53rd harmonic, ex. $873\cdot504$
882	19683 : 32768	22136	Abdulqadir's substitute for Zalzal's 853 cents, being a Fourth above 384 cents
884	3 : 5	22185	Just major Sixth, $Vd + Tu = C : A_1$, ex. $884\cdot359$
900 †	22 : 37	22577	Equal major Sixth, $\|C : A$, and also equal diminished Seventh, $\|C : B\flat\flat = A : G\flat$, ex. ratio $1 : \sqrt[4]{8} = 1 : 1\cdot6818$
906	16 : 27	22724	The 27th harmonic, Pythagorean major Sixth, $3Vu = C : A$, ex. $905\cdot865$
919	10 : 17	23045	Ratio of the 10th : 17th harmonic, the harmonic diminished Seventh, $e_,''$: $^{17}d'''$ on the Harmonical, ex. $918\cdot641$
926	75 : 128	23215	Just diminished Seventh, $Vd + 2Td = C : B^2\flat\flat = A : G^2\flat$, ex. $925\cdot416$
933	7 : 12	23408	Septimal or supermajor Sixth, $^7B\flat : g$, ex. $933\cdot129$
938	32 : 55	23521	The 55th harmonic, ex. $937\cdot632$
¶ 947	125 : 216	23754	Acute diminished Seventh, $3Vu + 3Td = C : B^3\flat\flat = A : G^3\flat$, ex. $946\cdot924$
950 †	93 : 161	23831	Meshāqah's 19 Quartertones, ex. ratio $1 : {}^{24}\!/2^{19} = 1 : 1\cdot7311$
954	72 : 125	23958	Just superfluous Sixth, $3Vd + 3Tu = C : A_3\sharp$, ex. $955\cdot031$
960 †	85 : 148	24082	Four Pentatones, the fourth note in the Salendro scale, see infrà, sect. K., Java, a close approximation to 969 cents, ex. ratio $1 : \sqrt[5]{16} = 1 : 1\cdot7411$
969	4 : 7	24304	The 7th harmonic, natural, harmonic, or subminor Seventh, $C : {}^7B\flat$ on the Harmonical, ex. $968\cdot826$
976	128 : 225	24497	Extreme sharp Sixth, $2Vu + 2Tu = C : A_2\sharp$, ex. $976\cdot537$
996	9 : 16	24988	Minor Seventh, used in the subdominant, $2Vd = C : B\flat$, ex. $996\cdot091$
999	32 : 57	25072	The 57th harmonic, ex. $999\cdot468$
1000 †	55 : 98	25086	Equal superfluous, or extreme sharp Sixth, $\|C : A\sharp$, or minor Seventh, $\|C : B\flat$
1018	5 : 9	25527	Acute minor Seventh, used in descending minor scales, $2Vu + Td = C : B^1\flat$, ex. $1017\cdot597$
1030	16 : 29	25828	The 29th harmonic, ex. $1029\cdot577$
¶ 1050 †	6 : 11	26340	Meshāqah's 21 Quartertones, ex. ratio $1 : {}^{24}\!/2^{21} = 1 : 1\cdot8340$
1059	32 : 59	26570	The 59th harmonic, ex. $1059\cdot172$
1067	27 : 50	26761	Grave major Seventh, $3Vd + 2Tu = C : B_2$, ex. $1066\cdot762$
1088	8 : 15	27300	Just major Seventh, $Vu + Tu = C : B_1$, the 15th harmonic, ex. $1088\cdot269$
1100	89 : 168	27594	Equal major Seventh
1110	128 : 243	27840	Pythagorean major Seventh, $5Vu = C : B$, ex. $1109\cdot775$
1111	10 : 19	27875	One of Prof. Preyer's trial intervals
1117	32 : 61	28018	The 61st harmonic, ex. $1116\cdot884$
1129	25 : 48	28330	Diminished Octave, $Vu + 2Td = C : C^2\flat$, ex. $1129\cdot327$
1145	16 : 31	28724	The 31st harmonic, ex. $1145\cdot036$
1150 †	35 : 68	28848	Meshāqah's 23 Quartertones, ex. ratio $1 : {}^{24}\!/2^{23} = 1 : 1\cdot943$
1158	64 : 125	29073	Superfluous Seventh, $3Tu = C : B_3\sharp$, ex. $1158\cdot941$
1173	32 : 63	29419	The 63rd harmonic, ex. $1172\cdot736$
1180	1024 : 2025	29613	The double Tritone, $4Vu + 2Tu = C : B_2\sharp$, ex. $1180\cdot447$
1200	1 : 2	30103	The Octave, $C : c$

TABLE II. *The Unevenly Numbered Harmonics of C* 66 *up to the* 63*rd.*

No.	Pitch nos.	Log	Cyclic cents	Equal notes	No.	Pitch nos.	Log	Cyclic cents	Equal notes
*1	66	0	0	C			− 1·50515	6000 +	
					33	2178	+ 1·51851	53	$c'''\sharp$ − 47
		− ·30103	1200 +		35	2310	+ 1·54407	155	d^{IV} − 45
*3	198	+ ·47712	702	g + 2	37	2442	+ 1·56820	251	$d^{IV}\sharp$ − 49
					39	2574	+ 1·59106	342	$d^{IV}\sharp$ + 42
		− ·60206	2400 +		41	2706	+ 1·61278	429	e^{IV} + 29
*5	330	+ ·69897	386	e′ − 14	43	2838	+ 1·63347	512	f^{IV} + 12
*7	462	+ ·84510	969	a′♯ − 31	45	2970	+ 1·65321	590	$f^{IV}\sharp$ − 10
					47	3102	+ 1·67210	666	g^{IV} − 34
		− ·90306	3600 +		49	3234	+ 1·69020	738	g^{IV} + 38
*9	594	+ 0·95424	204	d″ + 4	51	3366	+ 1·70757	807	$g^{IV}\sharp$ + 7
†11	726	+ 1·04139	551	f″ + 51	53	3498	+ 1·72428	874	a^{IV} − 26
‡13	858	+ 1·11394	841	a″ − 59	55	3630	+ 1·74036	938	a^{IV} + 38
*15	990	+ 1·17609	1088	b″ − 12	57	3762	+ 1·75587	999	$a^{IV}\sharp$ − 1
					59	3894	+ 1·77085	1059	b^{IV} − 41
		− 1·20412	4800 +		61	4026	+ 1·78533	1117	b^{IV} + 17
*17	1122	+ 1·23045	105	c‴♯ + 5	63	4158	+ 1·79934	1173	c^{V} − 27
*19	1254	+ 1·27875	298	d‴♯ − 2					
21	1386	+ 1·32222	471	f‴ − 29					
23	1518	+ 1·36173	628	f‴♯ + 28					
*25	1650	+ 1·39794	772	g‴♯ − 27					
27	1782	+ 1·43136	906	a‴ + 6					
*29	1914	+ 1·46240	1030	a‴♯ + 30					
31	2046	+ 1·49136	1145	b‴ + 45					

TABLE III. *Number of any Interval, not exceeding the Tritone, contained in an Octave.*

Cyclic cents in interval	Name of Interval	Number in an Octave	Cyclic cents in interval	Name of Interval	Number in an Octave
2	Skhisma	614·3	200	Equal Tone	6·0
20	Diaskhisma	61·4	204	Major Tone	5·9
22	Comma	55·8	231	Supersecond	5·2
24	Pyth. Comma	51·1	240	Pentatone	5·0
27	Septimal Comma	44·0	267	Subminor Third	4·5
28	Small Diësis	40·5	294	Pyth. minor Third	4·1
42	Great Diësis	29·2	300	Equal minor Third	4·0
50	Quartertone	24·0	316	Just minor Third	3·8
70	Small Semitone	17·0	336	Superminor Third	3·6
76	Meantone small Semitone	15·8	355	Zalzal's *wostā*	3·4
85	Subminor Second	14·2	400	Equal major Third	4·0
90	Limma	13·3	408	Pyth. major Third	2·9
92	Larger Limma	13·0	435	Supermajor Third	2·8
100	Equal Semitone	12·0	498	Just Fourth	2·4
112	Just Diatonic Semitone	10·7	500	Equal Fourth	2·4
114	Apotomē	10·6	503	Meantone Fourth	2·4
117	Meantone great Semitone	10·3	590	Just Tritone	2·0
134	Great Limma	9·0	600	Equal Tritone	2·0
182	Minor Tone	6·6	612	Pyth. Tritone	2·0
193	Mean Tone	6·2			

SECTION E.

ON MUSICAL DUODENES OR THE DEVELOPMENT OF JUST INTONATION FOR HARMONY.

(See notes pp. 208, 209, 211, 269, 272, 293, 298, 299, 301, 302, 304, 305, 306, 310, 333, 338, 345, 346, 352, and 363.)

1. *Introduction.* The following sketch is founded on my paper on the same subject in the *Proceedings of the Royal Society* for Nov. 19, 1874, vol. xxiii. p. 3. ¶ It is an attempt to develop the trichordal principles of suprà, pp. 293d and 309d. This, of course, is an inversion of what actually occurred. But the introduction of the harmonic principle has completely changed the nature of music, and its plan consequently requires reconstruction. Harmony alone is considered. Melody is made dependent on harmony. The harmony is *tertian*, that is, it includes perfect *Thirds*, major and minor; but not *septimal*, that is, it does not include the 7th harmonic of the base of a chord. But this may be superadded, see art. 23. The plan here pursued also has the advantage of showing the precise tertian relation of the notes of a chord written in the usual notation, by merely superscribing a letter, called the *duodenal*, without any change whatever in the ordinary notation itself. The notes affected by the other harmonies can then be easily indicated (art. 26)

2. The *Harmonic Elements* are the intervals of a Fifth, major and minor Third, with all their extensions, inversions, and extensions of their inversions, that is all the forms in p. 191b, c, which are here assumed. Capital letters will therefore indicate notes without regard to octave, and even allow of reduplication, or added octaves. The notation is otherwise the same as for my variation of Herr A. v. ¶ Oettingen's notation, explained on p. 277c, note *, and used through the rest of this work. The notation of intervals is used as in p. 276d, note †, so that + is 386, − is 316, ± is 702, | is 294 cents, and (...) is replaced by the proper number of cents in the interval.

3. In the *construction of the schemes* notes forming ascending Fifths are written over one another vertically; notes forming ascending major Thirds are written to the right horizontally. Against each note is written the number of cyclic cents (suprà, p. 452a) in its interval from C or the root, reduced to the same octave. A notation in Solfeggio terms (modified from that used by the Tonic Solfaists with Italian pronunciation of the vowels) is also supplied, in which *Do* stands for the root whatever the note itself may be.

4. *Harmonic Cell or Unit of Concord.*

Letter Notation.   Solfeggio Notation.

¶ $E^1\flat$ 316 G 702   *Mo* 316   *So* 702
         C o E_1 386   *Do* o   *Mi* 386

This consists of the three harmonic elements, the Fifth $C \pm G$ (or $Do \pm So$) and major Thirds $C + E_1$, $E^1\flat + G$ (or $Do + Mi$ and $Mo + So$) being placed as already explained. so that the minor Thirds $C - E^1\flat$ (or $Do - Mo$) and $E_1 - G$ (or $Mi - So$) lie obliquely upwards to the left. Such a cell is called the Cell of C (or Do) its *root*.

The student should construct such cells on any root. C has been adopted simply because it is most usual, and because it is suited to the Harmonical, on which its effect should be tried.

The cell contains therefore all forms of the major and minor chords, triad or tetrad, given in Chap. XII. above.

5. *The Harmonic Heptad or Unit of Chord Relationship.*

Letter Notation.   Solfeggio Notation.

$E^1\flat$ 316 G 702   *Mo* 316   *So* 702
$A^1\flat$ 814 C o E_1 386   *Lo* 814   *Do* o   *Mi* 386
      F 498 A_1 884   *Fa* 498   *La* 884

The heptad possesses *seven* notes, whence its name. It is formed by subjoining the cell of F (or Fa) to the cell of C (or Do), so that the Fifth of the lower cell is the root of the upper cell. This is called the Heptad of C (or Do) its central note. It contains not only the 4 *cell triads*, major $C + E_1 - G$, $F + A_1 - C$ (or $Do + Mi - So$, $Fa \perp La - Do$), minor $C - E^1\flat + G$, $F \quad A^1\flat + C$ (or $Do - Mo + So$, $Fa - Lo + Do$), but also two *union triads*, major $A^1\flat + C - E^1\flat$ (or $Lo + Do - Mo$), and minor $A_1 - C + E_1$ (or $La - Do + Mi$), which result from the union of the two cells. It therefore possesses all the six consonant chords which contain the same note C (or Do), and can hence pass readily into one another, as should be verified on the Harmonical. It possesses also the seven *condissonant triads* (p. 338, note †) containing C, the *major Trine* $A^1\flat + C + E_1$ (or $Lo + Do + Mi$) containing two major Thirds, and the *minor Trine* $A_1 - C - E^1\flat$ (or $La - Do - Mo$) containing two minor Thirds, the *pure quintal Trine* $F \pm C \pm G$ (or $Fa \pm Do \pm So$) containing two Fifths, the *major quintal Trines* $A^1\flat + C \pm G$ (or $Lo + Do \pm So$) and $F \pm C + E_1$ (or $Fa \pm Do + Mi$) consisting of a major Third and a Fifth, and the *minor quintal Trines* $A_1 - C \pm G$ (or $La - Do \pm So$) and $F \pm C - E^1\flat$ (or $Fa \pm Do - Mo$) consisting of a ¶ minor Third and a Fifth.

All of these should be studied on the Harmonical, and the readiness with which their dissonance may be removed by passing into a consonant triad containing the same note C should be felt.

6. *The Harmonic Decad or Unit of Harmony.*

Letter Notation.			Solfeggio Notation.		
$B^1\flat$ 1018	D 204		To 1018	Re 204	
$E^1\flat$ 316	G 702	B_1 1088	Mo 316	So 702	Ti 1088
$A^1\flat$ 814	C o	E_1 386	Lo 814	Do o	Mi 386
	F 498	A_1 884		Fa 498	La 884

The Decad possesses *ten* notes, whence its name. It consists of the heptad of G (or So) superimposed on the heptad of C (or Do). These two heptads have a common cell, that of C (or Do). Hence the decad of C (or Do) may also be ¶ considered as three cells, that of C (or Do) in the middle ; the dominant or that of G (or So) above, and the *subdominant* or that of F (or Fa) below. The decad of C (or Do) is the complete development of the cell of C (or Do), for the root of the upper cell is the Fifth of the root of the middle cell, while that root itself is the Fifth of the root of the lower cell.

7. *The Chords of the Decad.* The vertical axis is the column of Fifths $F \pm C \pm G \pm D$ (or $Fa \pm Do \pm So \pm Re$). Those are two horizontal axes of major Trines. The decad contains 3 cell *major* triads $F + A_1 - C$, $C + E_1 - G$, $G + B_1 - D$ (or $Fa + La - Do$, $Do + Mi - So$, $So + Ti - Re$) on the right, and 2 union *major* triads $A^1\flat + C - E^1\flat$, $E^1\flat + G - B^1\flat$ (or $Lo + Do - Mo$, $Mo + So - To$) on the left. The decad also contains 3 cell *minor* triads $F - A^1\flat + C$, $C - E^1\flat + G$, $G - B^1\flat + D$ (or $Fa - Lo + Do$, $Do - Mo + So$, $So - To + Re$) on the left, and 2 union *minor* triads $A_1 - C + E_1$, $E_1 - G + B_1$ (or $La - Do + Mi$, $Mi - So + Ti$) on the right. It has also the dissonance of the *dominant Seventh*, $G + B - D \mid F$ (or $So + Ti - Re \mid Fa$) and *minor Ninth* $G + B - D \mid F - A^1\flat$ (or $So + Ti - Re \mid Fa - Lo$), and ¶ hence of the *diminished Seventh* (the same less G or So), and of the *added Sixth* $F + A - C$ 204 D, or $F + A$ 520 D ($Fa + La - Do$ 204 Re, or $Fa + La$ 520 Re), but not the minor triad $D_1 - F + A_1$ (or $Ra - Fa + La$, see Ra in art. 11, p. 461), which is confused with it in tempered intonation. And it has also not the chords of the extreme sharp Sixth, $D^1\flat + F$ 204 $G + B_1$, p. 286*b*, or $f^1 + a$ 590 $d_1\sharp$, p. 308*b*.

8. *The Intervals of the Decad.* The relative position of the principal intervals should be observed in addition to the vertical Fifths (including Fourths), the horizontal major Thirds (including minor Sixths), and oblique minor Thirds (including major Sixths), on which the scheme is founded.

Major Tone 204, two Fifths vertically up, as C D (or Do Re).

Minor Tone 182, obliquely down to the right in the next line but *one* as G A_1 (or So La).

Defective Fifth 680, obliquely down to the right in the next line but *two* as D A_1 (or Re La).

Diatonic Semitone 112, obliquely down to the left in the next line as B_1 C (or Ti Do).

Small Semitone 70, obliquely down to the right in the next column but one, and in the next line, as $B^1\flat$ B_1 (or To Ti).

This is the smallest interval occurring in a Decad.

9. *Harmonic Trichordals.* A trichordal consists of one triad from each of the three cells of a decad. Eight such trichordals may be formed from the three major and three minor cell triads. Abbreviate the names ' major triad ' and ' minor triad ' into their first syllables, *ma.*, *mi.* (which however may, if preferred, be read more at full as ' major ' and ' minor '), and name them in the order of Subdominant, Tonic and Dominant triads, then the eight trichordals are *ma.ma.ma.*, *mi.ma.ma.*, *ma.mi.ma.*, *mi.mi.ma.*, *ma.ma.mi.*, *mi.ma.mi.*, *ma.mi.mi.*, *mi.mi.mi.* The seven tones in each trichordal reduced to a single octave constitute an *harmonic scale*, that is a scale in which each note belongs to a triad in the scale. We may begin the scale with any one of the notes of any one of the 3 generating triads. These notes may be numbered in order of sharpness when reduced to the same octave, as 4, 6, 1 in the subdominant, 1, 3, 5 in the tonic, and 5, 7, 2 in the dominant. These numbers may be simply prefixed to the trichordal they affect, to shew on what note the scale begins. We thus obtain $7 \times 8 = 56$ trichordal scales, of which 8 are ¶ generically different, each genus having 7 species. In some accounts of the modes they are all represented in fact as ma.ma.ma., differing only as to the note of the scale with which they begin. This is, of course, thoroughly erroneous. The student is recommended to make out on the Harmonical every one of these 56 scales.

10. *Principal Trichordal Scales.* The following gives a list of the principal scales thus generated referring to the places where they have been noted in the text, and the scale is noted as beginning with *C*.

The figures between two consecutive notes indicate the interval between them in cents. The number prefixed to the root of the decad indicates the note of the chord with which the scale begins, reckoned in the way just mentioned. References to the text follow. Where any one of these scales cannot be played on the Harmonical a form is given which can be played. As each note forms part of a cell triad, and mostly also of a union triad, each scale can be harmonised, and the student should therefore harmonise all of them on the Harmonical. See also p. 277, note †. No examples of the unusual VI. MI.MA.MI. are given.

I. MA.MA.MA.

¶ 1 $C = C$ 204 D 182 E_1 112 F 204 G 182 A_1 204 B_1 112 c, No. 1 of p. 274b and note. Major Mode of p. 298b. Ordinary C major.

5 $F = C$ 182 D_1 204 E_1 112 F 204 G 182 A_1 112 Bb 204 c, No. 5 of p. 275a, there called the mode of the Fourth. This must be played on the Harmonical as 5 C ma.ma.ma., which has the same intervals, namely: G 182 A_1 204 B_1 112 c 204 d 182 e_1 112 f 204 g. It is No. 5 of p. 275a, there called the mode of the Fourth.

II. MI.MA.MA.

1 $C = C$ 204 D 182 E_1 112 F 204 G 112 A^1b 274 B 112 c. The minor-major mode of pp. 305b and 309d.

III. MA.MI.MA.

1 $C = C$ 204 D 112 $E'b$ 182 F 204 G 182 A_1 204 B_1 112 c, No. 2 of p. 274b. The mode of the minor Seventh, with the leading note, or major Seventh, substituted for the minor Seventh, as in p. 303d. An ordinary form of the ascending minor scale p. 288a.

IV. MI.MI.MA.

1 $C = C$ 204 D 112 $E'b$ 182 F 204 G 112 A^1b 274 B_1 112 c. The ' instrumental ' minor scale of p. 288b; the ' modern ' ascending minor scale of p. 300d.

¶ V. MA.MA.MI.

1 $C = C$ 204 D 182 E_1 112 F 204 G 182 A_1 134 $B'b$ 182 c. Although this is called the mode of the Fourth on pp. 298b and 309d, it is a different scale from that called the mode of the Fourth on p. 275b, which is 5 F ma.ma.ma. above, under I., because the Seventh in this case is $B'b$ and in that Bb, a comma lower. See p. 277, footnote †, on this and similar confusions.

5 $F = C$ 182 D_1 134 $E'b$ 182 F 204 G 182 A_1 112 Bb 204 c. This must be played on the Harmonical as 5 $C = G$ 182 A_1 134 $B'b$ 182 c 204 d 182 e_1 112 f 204 g. This is No. 6 of p. 275b and there considered as a variant of the mode of the minor Seventh, which is really the different scale, 3 C ma.mi.mi., next immediately following.

VII. MA.MI.MI.

1 $C = C$ 204 D 112 $E'b$ 182 F 204 G 182 A_1 134 $B'b$ 182 c. This is No. 4 of p. 275a taken upwards, instead of downwards as there. The mode of the minor Seventh of p. 298b without the leading tone of p. 303c.

VIII. MI.MI.MI.

1 $C = C$ 204 D 112 $E'b$ 182 F 204 G 112 A^1b 204 B^1b 182 c, No. 3 or descending minor scale of p. 274c, the mode of the minor Third of p. 294a, No. 4.

5 $F = C$ 112 $D'b$ 204 $E'b$ 182 F 204 G 112 A^1b 182 Bb 204 c. This must be played on the

Harmonical as 5 $C = G$ 112 A^1b 204 B^1b 182 C 204 D 112 E^1b 182 F 204 g. It is No. 7 of
p. 275b, the mode of the minor Sixth of pp. 294a No. 5, 298c No. 5, 305c to 308d.

11. *The Harmonic Duodene, or Unit of Modulation.*

Letter Notation.						Solfeggio Notation.				
D^2b 134	F^1 520	A 906	C_1♯ 92	E_2♯ 478		*ru*	*fo*	*le*	*di*	*me*
G^2b 632	B^1b 1018	D 204	F_1♯ 590	A_2♯ 976		*su*	*to*	*re*	*fi*	*li*
C^2b 1130	E^1b 316	G 702	B_1 1088	D_2♯ 274		*du*	*mo*	*so*	*ti*	*ri*
F^2b 428	A^1b 814	C 0	E_1 386	G_2♯ 772		*fu*	*lo*	*do*	*mi*	*se*
B^2bb 926	D^1b 112	F 498	A_1 884	C_2♯ 70		*tu*	*ro*	*fa*	*la*	*de* ¶
E^2bb 224	G^1b 610	Bb 996	D_1 182	F_2♯ 568		*mu*	*so*	*ta*	*ra*	*fe*

In referring to the figure of the decad in art. 6 two gaps will be noticed, one to
the right at the top, the other to the left at the bottom. On filling these up according to the same laws by making F_1♯ a Fifth above B_1 and a major Third above
D, and D^1b a Fifth below A^1b and a major Third below F, we obtain the scheme
in the central rectangle of the above figures, which is called a *Duodene*, because it
consists of *twelve* notes bearing to each other the relation of the twelve notes on
the piano, which, by omitting the marks of commas $^1{}_1$, may be supposed to be
represented by the same letters. The Duodene then consists of 3 columns or
quaternions of Fifths, and four lines or major trines of Thirds, and its root, which
is that of the corresponding decad, is in the centre of the lowest line but one, so
that it is easy to construct a duodene on any note as a root.

The duodene thus completed by these extreme tones possesses two additional
union Thirds, major D^1b $+ F - A^1$b (or $Ro + Fa - Lo$) and minor $B_1 - D + F_1$♯ (or
$Ti - Re + Fi$), and consequently besides the 8 genera of scales of the decad contains ¶
the major scale of A^1b (major chords D^1b$+ F - A^1$b, A^1b$+ C - E^1$b, E^1b$+ G - B^1$b)
and the descending minor scale of E_1 (minor chords $A_1 - C + E$, $E_1 - G + B_1$, $B_1 -
D + F_1$♯), and it also gives us the chords of the extreme sharp Sixth (976 cents, a
very near approach to the 7th harmonic of 969 cents) in its three forms,

Italian Sixth A^1b 386 C 590 F_1♯ scarcely a dissonance,
French Sixth A^1b 386 C 204 D 386 F_1♯,
German Sixth A^1b 386 C 316 E^1b 274 F_1♯.

These three last chords cannot be played on the Harmonical. They arose in the days of
meantone temperament. The chord of the dominant Seventh omitting the Fifth E^1b 386
G 610 d^1b had then to be played with tempered notes as Eb 386 G 579 c♯ , because there was no
db on the instrument, and as the 7th harmonic would have been E^1b 386 G 583 $'d^1$b the effect
was so good that the chord was adopted in writing and distinguished from the chord of the
dominant Seventh, by resolving *upwards* instead of downwards.

These new notes have also introduced two new Semitones of 92 cents, F 92 F_1♯ ¶
and D^1b 92 D. But the smallest interval between any two notes remains the
small Semitone of 70 cents, A^1b 70 A_1, E^1b 70 E_1, B^1b 70 B_1.

12. *Modulation into the Dominant Duodene.*

It is obvious from the construction of the
duodene that the transition from any duodene to an adjacent one is very easy. Suppose
(see scheme in art. 11) that we omit the lowest
line D^1b $+ F + A_1$ (or $Ro + Fa + La$) and take
in the line $F^1 + A + C_1$♯ at the top, we shall
have the duodene of G, which has three lines
in common with that of C. The three new
notes introduce 2 *commas* F 22 F^1, A_1 22 A and
one *diaskhisma* C_1♯ 20 D^1b. These minute
distinctions neglected in tempered music have,
however, a powerful effect on the harmony of
justly intoned instruments. The C_1♯ is indeed

one of the extra notes and does not occur in
the decad of G or its scales. On the other
hand F_1♯, which was an extra note in the duodene of C, becomes a substantive note in the
decad, as well as duodene, of G, and we find
then that the F of the C decad becomes the
F_1♯, 92 cents higher in the G decad. This
difference is so large that it cannot be disregarded in tempered music, and it is, accordingly, there represented by an interval of 100
cents, and forms the distinguishing mark for
major scales of what is termed the *modulation
into the dominant* as just explained.

13. *Modulation into the Subdominant Duodene.*

Omit the top line $B^1\flat + D + F_1\sharp$ (or *To + Re + Fi*) of the duodene of C in scheme art. 11, and take a new line $G^1\flat + B\flat + D_1$ (or *Sa + Ta + Ra*) at the bottom to obtain the duodene of F (or *Fa*). The changes made are the reverse of those for modulations into the dominant. The notes $B^1\flat$, D are depressed by a comma of 22 cents to $B\flat$, D_1, and the extra note $F_1\sharp$ of the C duodene is raised by a

diaskhisma of 20 cents to $G^1\flat$ of the F duodene. These changes are neglected in tempered intonation. But the most important change is that B_1 (which is in the duodene, but not the decad of F) is depressed by a Semitone of 92 cents from B_1 to $B\flat$, and this being noticed in tempered music, becomes the distinguishing mark in the modulation of the major scale of C into that of the subdominant F.

14. *Modulation into the Mediant Duodene.*

Returning again to the duodene of C, we might omit the left column $D^1\flat \pm A^1\flat \pm E^1\flat + B^1\flat$ (or *ro ± lo ± mo ± to*) of the duodene of C (see scheme in art. 11) and introduce a new column on the right $C_2\sharp \pm G_2\sharp \pm D_2\sharp \pm A_2\sharp$ (or *de ± se ± ri ± ti*), in which each new note is a *great diesis* of 42 cents lower than the note in the column omitted. This differ-

ence is ignored in playing tempered music, although the distinction is preserved in writing as $D\flat$ and $C\sharp$, &c., but it is of great importance in just intonation. This is termed modulation from the root C (or *Do*) into the *Mediant*, E_1 (or *Mi*) as the root of the new duodene is the major Third or Mediant of the root C (or *Do*) of the old duodene.

15. *Modulation into the Minor Submediant Duodene.*

Similarly we might omit the right column $A_1 \pm E_1 \pm B_1 \pm F_1\sharp$ (or *la mi ti fi*) of the duodene of C (see scheme in art. 11) and in-

troduce a new column $B^2\flat\flat \pm F^2\flat \pm C^2\flat \pm G^2\flat$ (or *ta fu du su*) on the left, thus forming the duodene of $A^1\flat$ (or *Lo*) or *minor submediant*.

16. *Modulation into the Relative and Correlative Duodenes.*

But it is usual to combine these two modulations with others into the subdominant of the mediant (that is the submediant) A_1 (or *La*), generally called the *relative* on the one hand, and the dominant of the minor submediant, that is $E^1\flat$ (or *Mo*), sometimes called the *correlative*, on the other. In each case the root of the new duodene differs by a minor Third from the old root. The change for these two last modulations is considerable if we take the whole duodene into consideration, as may be seen by the schemes in art. 11, where the dotted lines mark off the new duodenes. But in the cases which occur in practice the changes are very small, especially when the difference of a comma is neglected as in tempered music. The modulation into the relative is generally from the *major* scale of the decad of C or chords $F + A_1 - C$, $C + E_1 - G$, $G + B_1 - D$ into the *minor* scale of the relative decad of A_1, either consisting of the chords $D_1 - F + A_1$, $A_1 - C + E_1$, $E_1 - G + B_1$ (in which there is only the use of D_1 for D, to indicate this modulation), or else consisting of the chords $D_1 - F + A_1$, $A_1 - C + E_1$, $E_1 + G_2\sharp - B_1$, or even $D_1 + F_2\sharp - A_1$, $A_1 - C + E$, $E_1 - G_2\sharp - B_1$, these being the three recognised forms of the minor scale. As the two latter forms are also acknowledged in

tempered intonation (which, however, confuses $F_1\sharp$ and $F_2\sharp$), the change of G into $G_2\sharp$ (or *so* into *se*, that is the sharpening of the Fifth by properly 70 cents) has generally been considered the mark of modulation from a major scale into its relative minor,—one of the commonest in music.

The modulation into the correlative is in the same way generally thought of as the change of the tonic major scale chords $F + A_1 - C$, $C + E_1 - G$, $G + B_1 - D$ (or *Fa + La - Do*, *Do + Mi - So*, *So + Ti - Re*) into the tonic minor scale chords $F - A^1\flat + C$, $C - E^1\flat + G$, $G - B^1\flat + D$ (or *Fa - Lo + Do*, *Do - Mo + So*, *So - To + Re*), in which however the chord $G + B_1 - D$ (or *So + Ti - Re*) may remain, and sometimes the chord $F + A_1 - C$ (or *Fa La Do*). This is not a modulation at all in the sense just explained, because there is no change of duodene or even decad. It is merely a change of scale within the same decad, that is, a new trichordal scale moving from 1 C ma.ma.ma. to 1 C mi.mi.mi., mi.mi.ma., or ma.mi.ma., art. 9. It would, however, be more consonant to ancient practice to restrict the term *modulation* to such changes of trichordal within the same duodene, and to use a new term for the more general operation.

17. *Duodenation.*

This is the term I propose to substitute for modulation when it means passing from one duodene to another which bears a known relation to the first. This relation may be very close, as in the cases just considered, or so remote that the two duodenes have only one note in common. Thus the duodene of $D^2\flat$ and C have only the note $B^1\flat$ in common. The annexed figure, called the *Duodenārium*, probably contains all such duodenations which

occur even in modern music, though it is impossible to be certain how far the ambiguities of tempered intonation may mislead the composer to consider as related, chords and scales which are really very far apart. It contains therefore an approximate estimate of 117, for the number of tones in an Octave which would be required to play in just intonation, and are roughly represented by the 12 tones of equal temperament.

18. *The Duodēnārĭum.*

The large figures give the cyclic cents in the interval of each note from *C*. See Table p. 440. The small figures give the corresponding numbers of the cycle of 53 with the nearest whole number of cents. See Table p. 439.

	1 −1·26	2 −·94	3 −·63	4 −·31	5 0	6 +·31	7 +·63	8 +·94	9 +1·26	
−·32	B^1♭♭ 970 / 47·974	D^3♭ 156 / 11·158	F^2 542 / 28·543	A^1 928 / 45·928	C♯ 114 / 9·113	E$_1$♯ 500 / 26·498	G$_2$♯♯ 886 / 43·883	B$_3$♯♯ 72 / 7·68	D$_4$♯♯♯ 458 / 24·453	−·32
−·27	E^1♭♭ 268 / 16·272	G^3♭ 654 / 33·657	B^2♭ 1040 / 50·1042	D^1 226 / 14·226	F♯ 612 / 31·611	A$_1$♯ 998 / 48·996	C$_2$♯♯ 184 / 12·181	E$_3$♯♯ 570 / 29·566	G$_4$♯♯♯ 956 / 40·951	−·27
−·23	A^4♭♭ 766 / 38·770	C^3♭ 1152 / 2·1155	E^2♭ 338 / 19·340	G^1 724 / 36·725	B 1110 / 53=0·1109	D$_1$♯ 296 / 17·294	F$_2$♯♯ 682 / 34·679	A$_3$♯♯ 1068 / 51·1064	C$_4$♯♯♯ 254 / 15·249	−·23 ¶
−·18	D^4♭♭ 64 / ·68	F^3♭ 450 / 24·453	A^2♭ 836 / 41·838	C^1 22 / ·22	E 408 / ·408	G$_1$♯ 794 / 39·792	B$_2$♯♯ 1180 / 3·1177	D$_3$♯♯ 366 / 20·360	F$_4$♯♯♯ 752 / 37·749	−·18 ¶
−·14	G^4♭♭ 562 / 29·566	B^3♭♭ 948 / 46·951	D^2♭ 134 / 10·136	F^1 520 / 27·521	A 906 / 44·906	C$_1$♯ 92 / 8·91	E$_2$♯ 478 / 25·475	G$_3$♯♯ 864 / 42·860	B$_4$♯♯ 50 / 6·45	−·14
−·09	C^4♭♭ 1060 / 51·1064	E^3♭♭ 246 / 15·249	G^2♭ 632 / 32·634	B^1♭ 1018 / 49·1019	D 204 / 13·204	F$_1$♯ 590 / 30·589	A$_2$♯ 976 / 47·974	C$_3$♯♯ 162 / 11·158	E$_4$♯♯ 548 / 28·543	−·09
−·05	F^4♭♭ 358 / 20·362	A^3♭♭ 744 / 37·747	C^2♭ 1130 / 1·1132	E^1♭ 316 / 18·317	G 702 / 35·702	B$_1$♯ 1088 / 52·1087	D$_2$♯ 274 / 16·272	F$_3$♯♯ 660 / 33·657	A$_4$♯♯ 1046 / 50·1042	−·05
0	B^1♭♭♭ 856 / 42·860	D^3♭♭ 42 / 6·45	F^2♭ 428 / 23·430	A^1♭ 814 / 40·815	C 0 / 0·0	E$_1$ 386 / 21·385	G$_2$♯ 772 / 38·770	B$_3$♯ 1158 / 2·1158	D$_4$♯♯ 344 / 19·340	0
+·05	E^1♭♭♭ 154 / 11·158	G^3♭♭ 540 / 28·543	B^2♭♭ 926 / 45·928	D^1♭ 112 / 9·113	F 498 / 26·498	A$_1$ 884 / 43·883	C$_2$♯ 70 / 7·68	E$_3$♯ 456 / 24·453	G$_4$♯♯ 842 / 41·838	+·05
+·09	A^4♭♭♭ 652 / 33·657	C^3♭♭ 1038 / 50·1042	E^2♭♭ 224 / 14·226	G^1♭ 610 / 31·611	B♭ 996 / 48·996	D$_1$ 182 / 12·181	F$_2$♯ 568 / 29·566	A$_3$♯ 954 / 46·951	C$_4$♯♯ 140 / 10·136	+·09
+·14	D^4♭♭♭ 1150 / 2·1155	F^3♭♭ 336 / 19·340	A^2♭♭ 722 / 36·725	C^1♭ 1108 / 53=0·1109	E♭ 294 / 17·294	G$_1$ 680 / 34·679	B$_2$ 1066 / 51·1064	D$_3$♯ 252 / 15·249	F$_4$♯♯ 638 / 32·634	+·14
+·18	G^4♭♭♭ 448 / 24·453	B^3♭♭♭ 834 / 41·838	D^2♭♭ 20 / 5·23	F^1♭ 406 / 22·408	A♭ 792 / 39·792	C$_1$ 1178 / 3·1177	E$_2$ 364 / 20·362	G$_3$♯ 750 / 37·747	B$_4$♯ 1136 / 1·1132	+·18 ¶
+·23	C^4♭♭♭ 946 / 46·951	E^3♭♭♭ 132 / 10·136	G^2♭♭ 518 / 27·521	B^1♭♭ 904 / 44·906	D♭ 90 / 8·91	F$_1$ 476 / 25·475	A$_2$ 862 / 42·860	C$_3$♯ 48 / 6·45	E$_4$♯ 434 / 23·430	+·23
	−1·26 / 1	−·94 / 2	−·63 / 3	−·31 / 4	0 / 5	+·31 / 6	+·63 / 7	+·94 / 8	+1·26 / 9	

This is the first table of modulations adapted to Just Intonation that has been constructed. But the table in Gottfried Weber's *Versuch einer geordneten Theorie der Tonsetzkunst* (Attempt at systematic theory of musical composition, 1830–2, vol. ii., § 180, p. 86), although only adapted to equal temperament, was of much assistance to me.

19. *Construction of the Duodenarium.* The arrangement is that of all the previous schemes, proceeding from bottom to top by intervals of a Fifth 702 cents (or from top to bottom by intervals of a Fourth 498 cents), and from left to right by intervals of a major Third 386 cents (or from right to left by intervals of a minor Sixth 814 cents). The number written against any note shews the cyclic intervals of the note from *C*, when all are reduced to the same Octave, see App. XX. sect. A. art. 24, xiii. p. 437b'.

But as a Fifth is 701·955 cents, and a major Third 386·314, errors of accumulation occur, and hence the cyclic numbers require corrections if the precise numbers are wanted; apply those given at top or bottom of the column, or at either end of the line containing the number. Thus F$_2$♯ ♯ 682 has the column correction + ·63, and the line correction − ·23, and its true distance from *C* is therefore 682·4 cents. On referring to the name of the note in the Table, sect. A. art. 28, p. 440, the precise number of cents to one place of decimals, the logarithm and pitch of the note will be found in addition.

The interval between any two notes, reduced to the same Octave, is the difference of the number of cyclic cents assigned; corrected if required. The number of the note by which the just note would be represented in Bosanquet's cycle of 53, is added in smaller figures under the just note, and the nearest whole number of cents is annexed. Referring to that number in the Table in sect. A. art. 27,

p. 439b, the name given by Mr. Bosanquet, the precise number of cents, the logarithm and the pitch will be found.

20. *Just Intervals Reduced to Steps of Fifths and Major Thirds.* On account of the construction by Fifths and major Thirds, we can proceed from any note to any other by taking a certain number of Fifths up or down, and then major Thirds up or down, and reducing to the same Octave. See suprà, sect. D. arts. 7 and 9, p. 452b', c, where the process is described.

21. *The Column of Fifths.*

The central column of Fifths has no superior or inferior indexes. The superior indexes ^{1}, ^{2}, ^{3}, ^{4} to the left not only serve to distinguish the columns, but indicate that the note bearing it is 1, 2, 3, 4 commas of 22 cents *sharper* or *higher* than the note of the same name in the column of Fifths. Thus Db in the column of Fifths has 90 cents, D^3b has therefore $90 + 3 \times 22 = 156$ cents as there marked. The inferior indexes $_1$, $_2$, $_3$, $_4$ to the right also not merely distinguish the columns, but indicate

that the notes are 1, 2, 3, 4 commas *flatter* or *lower* than a note of the same in the column of Fifths. Thus $C\sharp$ has 114 cents, but $C_3\sharp$ has $114 - 3 \times 22 = 48$ cents. It is thus quite easy to continue the line of Fifths up at least to $D\sharp\sharp\sharp$ 546 from the table by adding the appropriate number of commas, thus $D\sharp\sharp\sharp = D_4\sharp\sharp\sharp + 4 \times 22 = 458 + 88 = 546$ cents and down to $Cbbb$ 858 by subtracting the same as shewn by $C^1bbb - 4 \times 22 = 946 - 88 = 858$.

22. *Limits of the Duodenarium.*

These were determined thus : The central dark oblong is the duodene of C. Within the next dark oblong are all the duodenes which have at least one note in common with the duodene of C. The extremes are the duodenes of D^2b (with the note B^1b), of A^2bb (with the note D^1b), of $E_2\sharp$ (with the note $F_1\sharp$), and of B_2 (with the note A_1). Then the outer black oblong contains all the duodenes whose roots are notes in the intermediate

black oblong. Supposing the original duodene, then, to be one which had its root in the duodere of C (which may always be considered as the case), the limits allow of modulation into any duodene containing that note, and thence into duodenes which have at least one note in common with the last named. We thus obtain $9 \times 13 = 117$ notes, forming $7 \times 10 = 70$ duodenes.

23. *Introduction of the Seventh and Seventeenth Harmonics.*

If it is desired to proceed beyond tertian to *septimal* harmony to introduce the harmonic form of the chord of the dominant Seventh, with the ratios $4 : 5 : 6 : 7$ as Mr. Poole has done (see sect. F. No. 7), or even to *septendecimal* harmony to introduce the harmonic form of the chord of the minor Ninth $8 : 10 : 12 : 14 : 17$ (see p. 3;6c, note *), the number of the notes will be nearly tripled. Taking the root of the chord as C, to each minor Seventh Bb we should have to add ^7Bb, which is 27 cyclic cents flatter than this minor Seventh Bb (as shewn in the duodenary arrangement of Mr. Poole's notes, sect. F. No. 7), and to each minor Ninth as D^1b we must add $^{17}D^1b$, which is seven cyclic cents flatter than this minor Ninth. The cents in the tertian chord of the minor Ninth C E_1 G Bb D^1b are 0, 386, 702, 996, 1200+112. Hence the cents in the harmonic septimal chord of the dominant Seventh, or C E_1 G ^7Bb, are 0, 386, 702, 969, and the cents in the septendecimal chord of the minor Ninth, or

C E_1 G ^7Bb $^{17}D^1b$, are 0, 386, 702, 969, 1200 + 105. This form can be played in all its inversions on the Harmonical, see sect. F. No. 1. If the root be omitted in the chord of the minor Ninth, we obtain the chord of the diminished Seventh, which in its harmonic form is $10 : 12 : 14 : 17$, or E_1 G ^7Bb $^{17}D^1b$, in cents 0, 316, 583, 919, which can also be played on the Harmonical in all its inversions. In Mr. Bosanquet's cycle of 53, the chord of the dominant Seventh is played by the degrees 4, 21, 35, 47, or cents 0, 385, 702, 974, of which the last note is 5 cents too sharp, but the effect is good. The chord of the diminished Seventh must be played by degrees 4, 18, 30, 45, or cents 0, 317, 589, 929, the last of which is 10 cents too sharp, and the result would not be improved by taking it one degree or 23 cents flatter. Altogether it is only a slight improvement on the imitation of the tertian form, degrees 4, 18, 31, 45, or cents 0, 317, 611, 929.

24. *Need of Reduction of the Number of Just Tones.*

Of course it is quite out of the question that any attempt should be made to deal with such numbers of tones differing often by only 2 cents from each other. No ear could appreciate the multitude of distinctions. No instrument, even if once correctly tuned, would keep its intonation sufficiently well to preserve

such niceties. No keyboard could be invented for playing the notes even if they could be tuned, although, as will be seen in art. 26, it is very easy to mark a piece of ordinary music so as to indicate the precise notes to be struck. Hence some compromise is needed, such as the following.

25. *The Omission of the Skhisma. Unequally Just Intonation.*
The Cycle of 53.

The first compromise is to consider all tones differing by 2 cents (a skhisma) as identical. The dotted lines in the Duodenarium inclose $7 \times 8 = 56$ tones which differ from each other by more than 2 cents. Any note in the line just above the upper dotted line differs only by 2 cents from the note just above the lower dotted line in the preceding column. We may proceed then by perfect Fifths of 702 cents up from $D_3\sharp$ 252, the extreme note in the right-hand bottom corner of this oblong, to $D_3\sharp\sharp$ 366, at the top, and thence by an imperfect Fifth of 700 cents (the same as in the equally tempered scale) to B_2 1066 at the bottom of the next column to the left. Then again we may go by perfect Fifths to $B_2\sharp$ 1180, and then by another Fifth of 700 cents to G_1 680 $(= 1180 + 700 - 1200)$ and so on till we had by these alternating Fifths of 702 and 700 cents reached the 56th note and 55th Fifth $F^3\flat$ 450. In this way, at the 53rd Fifth or 54th note we should have reached $E^3\flat\flat$ 246, over which a short line is drawn. Now this is lower than the initial note $D_2\sharp$ 252 by only 6 cents. Hence if on the three last occasions where Fifths of 700 cents were to have been used, we had taken the perfect Fifths of 702 cents we should have made $C^1\flat = 1110$ cents, $A^2\flat\flat = 726$ cents and $F^3\flat\flat = 342$ cents, and consequently $E^3\flat\flat$ 252 cents. This would have become identical with the starting note $D_3\sharp$ 252. This mode of tuning, which if accurately executed no ear could distinguish from just intonation, forms the *unequally just intonation* mentioned in sect. A. art. 19, p. 435c. It is also the foundation of substituting for the perfect Fifth another of $31 \times 1200 \div 53 = 701 \cdot 886$ cents, so that on repeating it 53 times, and deducting 31 Octaves we should come back to the starting note. And this gives the cycle ¶ of 53 already described, sect. A. arts. 22 and 27, to which reference is made on the Duodenarium itself, shewing exactly the mode in which it can be substituted for Just Intonation without perceptible injury to the harmonic effect. For this and other less happy but more handy attempts, see sect. A. The mode of fingering this cycle is explained infrà, sect. F. Nos. 8 and 9, and of tuning it in sect. G. arts. 19 and 20.

26. *The Duodenal.* The duodenal is the letter name of the root of any duodene. *By placing it over any note or chord we indicate that that note and all which follow till a new duodenal is given are to have such values only as they would have in the duodene of which the tone indicated by the duodenal is the root.* This prevents all ambiguity by restoring in fact the notation of commas higher or lower, which alone is wanting for the representation of tertian harmony in the ordinary staff notation. If the 7th and 17th harmonics have to be introduced they ¶ will have sloping lines placed before them as in chord 17 below. The examples given are not intended as specimens of desirable harmony, but of the means of representing differences of just intonation. The first 16 chords are from *God save the Queen*; the four last are merely examples of notation.

The chords are numbered for convenience of reference, and only the treble is given for brevity. When the bass is added, the duodenal should be repeated in the bass or merely placed between treble and bass. Observe the chords 3, 9, 13, which introduce the ambiguous chord on the second. The duodenal C over chord 1 shews that we begin in the duodene of c, so that the first chord is e' 316 g' 498 c'. But G over chord 3 shews that there is a duodenation into the dominant, and that the chord is the true minor $f^{1\prime}$ 386 a' 498 d'' and not the dissonant chord of the added Sixth f' 386 a_1' 520 d''. The d'' must be retained for the voice to descend by a perfect minor Third to b_1' in chord 4, and be the true Octave of d' in that chord where C shews that the duodene of C is again reached. Hence it is not allowable to take chord 3 in the duodene of F as f' 386 a_1' 498 d_1''. The following chords 5, 6, 7, 8 are also in the duodene of C, as there is no change of duodenal. But chord 9 is in the duodene of F, because a_1' is retained from chord 8, and d_1'', f' must harmonise with it. In chord 10 the duodene of C is again reached. As purposely written in this example chord 13 is the dissonant added Sixth f'386 a_1' 520 d'', ¶ which is resolved on chord 14, but the retention of a_1' would make it more natural to take the duodene of F, as f'386 a_1' 498 d'' and then return immediately to the duodene of C. In chord 15, d'267 $^7f'$ 231 g' 386 b_1' the method is shewn by which the septimal ^7f is indicated. The duodenal C would make f without the mark before it, to be true Fourth of the root c. But this Fourth is 27 cents too sharp for the 7th harmonic of the dominant g, and hence the line sloping down to the right indicates that the Fourth has to be taken 27 cents flatter in *septimal* harmony. In ordinary tertian harmony as indicated by the duodenal only, the Fourth would remain unaltered. In chord 17, the duodenal C would shew that the $a'\flat$ must be $a^{1\prime}\flat$, the minor Sixth of the root c, or a diatonic Semitone above the dominant g. But this is 7 cents too sharp for the

17th harmonic of the dominant, and hence the line sloping down to the right indicates that it is to be 7 cents flatter. The sloping line, therefore, indicates different degrees of flattening according as it is applied to the Fourth or minor Sixth of the root expressed by the duodenal. If, therefore, we wished to have the chord e_1' 316 g'267 $'b'b$336 $^{17}d''b$ we must write the duodenal as F to get the right intonation, as in chord 19. Since the 17th harmonic of the dominant is so nearly the minor Sixth of the root, and the chord is dissonant, much of the effect would be pre-

served if this minor Sixth were used in place of the 17th harmonic provided only the 7th harmonic of the dominant were retained.

Equal temperament of course not recognising the difference of a comma, so far as sound is concerned, retains the same tempered duodene throughout, although there is a difference in writing it, as would be shewn in the Duodenarium (p. 463) if the indices were omitted. Such an omission reduces the Duodenarium to a table of modulations in any temperament which neglects the comma.

SECTION F.

¶ EXPERIMENTAL INSTRUMENTS FOR EXHIBITING THE EFFECTS OF JUST INTONATION.

(See notes pp. 6, 17, 217, 218, 222, 256, 329, and 346.)

INTRODUCTION.

At the present day ordinary musical instruments are intended to be tuned in accordance with equal temperament (see pp. 313a, 432b, art. 10; 436b, art. 22, i.; 437c, art. 25; sect. G. arts. 11 and following). The English concertina, which has 14 keys for the Octave, is still usually tuned in the older Meantone temperament ¶ (p. 433d, art. 16, and sect. G. art. 18). But neither system gives the only intervals which will allow chords in the middle part of the scale to be played without giving rise to beats. In order, then, that the ear may learn what is the meaning of 'just intonation,' it is necessary for it to have special instruments, or at least instruments specially tuned. Prof. Helmholtz has for this purpose invented a tuning for an harmonium with two rows of ordinary keys, explained on pp. 316b to 320a. Others, as Colin Brown, Liston, Poole, and Perronet Thompson, have invented harmoniums or organs with novel fingerboards; and others, as Bosanquet and J. P. White, have invented means for using the division of the Octave into 53 parts, which, as is seen in sect. E., p. 463, is practically almost identical with just intonation. A brief account of these instruments (with the exception of Prof. Helmholtz's, which is fully described in the text) will here be given. But none of them meet the wants of the student. They are all too expensive and require so much special education to use, that (with the exception of Mr. Colin Brown's) they have remained musical curiosities, some of them entirely unique. But there are two ¶ instruments which are cheap and which can be tuned so as to illustrate almost every point of theory, though they of course remain experimental instruments intended only to shew the nature of musical intervals, chords, and scales, and not to play pieces of music except especially composed exercises. These two I shall take first. They are a specially tuned harmonium and English concertina. Reed instruments are far the best for experiments, because they give sustained notes possessing a large number of powerful upper partial tones, so that any deviations from just intonation are extremely conspicuous, painfully evident indeed on any harmonium tuned in equal temperament.

1. THE HARMONICAL.

The scale of the Harmonical and the number of vibrations for every note in the first four octaves will be found on p. 17, note. The instrument has been constantly referred to in the Translator's notes to the preceding pages. It is an harmonium with one row of vibrators extending over five octaves. The tuning of the fifth octave will be explained further on.

Any one buying such an harmonium of Messrs. Moore & Moore, pianoforte and harmonium makers, 104 and 105 Bishopsgate Street, London, for 165s. net, may have it tuned as an Harmonical, by my forks, and provided with an 'harmonical bar' as presently explained, both without extra charge. I am sure that all musical students, as well as myself, must feel greatly indebted to this firm, who at the instance of Mr. H. Keatley Moore, Mus. B., a student of the first edition of this work, have so kindly undertaken to furnish this almost indispensable aid to the study of music on Helmholtz's principles at such a very moderate cost.

On the first four octaves this instrument contains all the 10 notes of the Decad of C

(p. 459b), and hence all its chords (p. 459c), and allows of playing and harmonising all the 56 trichordal scales (p. 460) contained in that decad. Its 10 notes, C D E^1b E_1 F G A^1b A_1 B^1b B_1, are placed on their usual digitals. Hence so far there is no new fingering to learn. The remaining two digitals are employed to furnish two notes of great theoretical importance, the grave Second D_1, which is of course placed on the Db or $C\sharp$ digital, and the natural or harmonic Seventh $'Bb$, which had to be placed, rather out of order, on the Gb or $F\sharp$ digital, the only one at liberty. Hence, using small letters to represent the short black keys, the keyboard for each of the first four octaves is

	d_1		e^1b			$'bb$	a^1b		b^1b		¶
C		D		E_1	F	G		A_1		B_1	C

vib. in two-foot Octave 264 $\{$ 293⅜ 297 316⅔ 330 352 462 396 422⅔ 440 475⅓ 495 528

and its scheme in the Decad form with the two additional notes is

B^1b	D	
E^1b	G	B_1
A^1b	C	E_1
	F	A_1
...	$'Bb$	D_1

In this form (...) in the second column indicates the absence of Bb, and $'Bb$ forms a column by itself. The scheme is seen filled up on p. 474c, d. The addition of $'Bb$ gives an opportunity of playing the first sixteen harmonics of C with the exception of the 11th and 13th (whence the name *Harmonical*), thus:

Note	C	c	g	c'	e_1'	g'	$'b'b$	c''	d''	e_1''	g''	$'b''b$	b''	c'''
Harmonic	1	2	3	4	5	6	7	8	9	10	12	14	15	16

By means of the 'harmonical bar' provided with the instrument, these harmonics, except the 7th and 14th, can be pressed down at the same time, and then the 7th and 14th, being on short keys, can be added with the fingers of the hands which press down the bar. The pegs which press the notes are arranged on different lines, so that the first 8 harmonics can be played by themselves, and then the effect of adding the higher Octave can be tried. It is thus possible to play the harmonics simultaneously with or without the 7th and 14th, and thus to estimate their presumed dissonant effect. To my own feeling these harmonics greatly enrich and improve the quality of the very compound tone produced.

It is evident therefore that the effects of all the intervals depending on the numbers 1 to 16 (omitting 11 and 13) can be immediately produced, and hence all the intervals on p. 212b,c, including the septimal intervals, arising from $'Bb$, which are of special importance and interest because they can be so rarely heard.

The existence of higher upper partials of the low notes can easily be made evident by beats. If we press down one of the digitals for the shortest distance that will allow the note to sound at all, we flatten it slightly, and hence put it out of tune. Keeping then C sounding fully, and slightly flattening its harmonics, one by one in this way (indicated by a prefixed grave accent) we easily obtain the beats from $C'c$, $C'g$, $C'c'$, $C'e_1'$, $C'g'$, $C''b'b$, $C'c''$, $C'd''$, $C'e_1''$, $C'g''$, $C''b''b$, $C'b''$, $C'c'''$, making evident the existence of 13 out of 15 of the upper partials of C. In the same way by slightly flattening the upper or lower notes of any of the consonant intervals, as $c : g$, we

can produce the beats which shew that the consonance has been disturbed. These are ¶ some of the most striking illustrations of Helmholtz's discoveries.

Beats between the primes of two notes are well shewn by DD_1, dd_1, $d'd_1'$, $d''d_1''$, which should beat about 9, 18, 37, 73 times in 10 seconds, the number of beats doubling for each ascent of an Octave. The very impure character of the beats of DD_1, arising from our hearing at the same time the beats of the upper pairs of notes as partials, is instructive. We can also hear the beats (all given for 10 seconds and fractions omitted) in D E^1b 50, $'Bb$ B^1b 33, E^1b E_1 33, A^1b A_1 44, B^1b B_1 50, but the higher Octaves of these notes beat too rapidly to be counted.

Combinational tones are easily heard. Any two consecutive harmonics of C give C, and by sounding two of them strongly and slightly ¶ flattening the C, the beats of this flattened $'C$ with the combinational tone may be heard, but much care and attention are necessary for this purpose. On playing b_1'' c''' the rattle of the 66 beats in a second may be heard, as well as the combinational C of 66 vib. Similarly for b_1'' and $b^1''b$ the rattle of the 39·6 beats in a second, and also the deep combinational tone E^1b of 39·6 vib. And if all three keys $b^1''b$, b_1'', and c''' be held down together, the low-pitched beat of the two combinational tones may also be heard with proper care and attention. If we play d_1' f'' we have a beat-note of 117·3 vib., very nearly B^1b. If we play d'' f'' we have the beat-note A_1 of 110 vib. If we play all three together the two beat-notes beat 73·3 times in 10 seconds. This must be carefully listened for, but the beats being so much lower in pitch cannot be confused with the higher beats of d_1'' d'', although their frequency is the same.

All the forms of the major and minor triads and tetrads on pp. 218b to 224a can be played and appreciated, and in many cases the combinational tones can be played as substantive notes with them; see my footnotes to these pages.

The effect of the analyses of dyads in my footnotes on pp. 188 to 191 can all be studied, and much of the diagrams on pp. 193 and 333 can be verified.

Most of the old Greek tetrachords on p. 263d' can be played as there pointed out.

The analysis of scales on pp. 274-278 can be illustrated.

The discords in Chap. XVII. can be mostly
¶ illustrated, as pointed out in my footnotes.

As shewn by the table on p. 17, note, the intervals 80 : 81, or comma, the minor Tone 9 : 10, and major Tone 8 : 9, the diatonic Semitone 15 : 16, and small Semitone 24 : 25, and other important intervals, can all be illustrated. Again, $'Bb : B^1b = 35 : 36$ is 49 cents, and hence almost precisely a quarter of a Tone or 50 cents, and $A_1 : {}'Bb = 20 : 21$ is 85 cents, or very nearly the Pythagorean Limma of 90 cents. The imperfect Fifth of just intonation $D : A_1$, or 680 cents, may be contrasted with the perfect Fifth $D_1 : A_1$, or 702 cents. The Pythagorean minor Third $D : F$, or 294 cents, can be contrasted with the just minor Third $D_1 : F$, or 316 cents.

But it is also necessary to note what the Harmonical cannot do. It has no Pythagorean comma, and no Pythagorean major Third, nor
¶ can it play a Pythagorean scale. It cannot play the chord of the extreme sharp Sixth, nor can it modulate into the dominant or subdominant, or relative minor (except in the descending form), but it can distinguish f 386 a_1 520 d, the chord of the added Sixth, from the minor chord f 386 a_1 498 d_1, and can modulate from C major to C minor.

It is also able to play Mr. Poole's dichordal scale $F\ A\ C,\ C\ E\ G\ 'Bb\ D$ with the peculiar minor chord $G : 'Bb : D = 6 : 7 : 9$, and the full natural chord of the major Ninth.

Method of Tuning. To be sure about the pitches, I tuned $c''528$, $a'440$, $a^{1'}b422\cdot4$, $'b'b462$ on forks with great accuracy, by means of my tuning-forks mentioned on p. 446b'. I tuned also a second set of forks each two beats flatter than the above, which I found very useful in determining the accuracy of the
¶ tuning by unisons. In fact the note of the

reed is so much more powerful than that of the fork, that the latter was quite drowned when near the unison, so that the pitch could not be determined within 3 to 5 beats in 10 seconds, and this difficulty was entirely obviated by the flat forks. After these notes then had been tuned on the Harmonical, the rest of the notes in the two-foot Octave were tuned by Fifths or Fourths, namely first $a^{1'}b$ to $e^{1'}b$, $e^{1'}b$ to $b^{1'}b$, secondly c'' to g', g' to d'', c'' to f', thirdly a_1' to e_1'', e_1'' to b_1', and a_1' to d_1'. The other notes were obtained by Octaves. The verification is by the perfect major chords FA_1C, CE_1G, GB_1D, A^1bCE^1b, E^1bGB^1b; the perfect minor chord D_1FA_1 and the perfect chord of the harmonic Seventh CE_1G^1Bb, all without beats in the two-foot Octave.

Pitch. The pitch $c''528$ was chosen to agree with the pitch adopted by Prof. Helmholtz in the text; $a'440$ was the pitch proposed by Scheibler; $a^{1'}b422\cdot4$ is within $\cdot1$ vib. of the pitch of Handel's own A fork $422\cdot5$, now in the possession of Rev. G. T. Driffield, Rector of Old, near Northampton. In the notes not tuned by forks there may be a very slight but not perceptible error, so that the Harmonical presents a series of trustworthy pitches.

Exercises. Besides numerous short airs, and special exercises, the following pieces may be played with full harmonies, and will serve to illustrate the meaning of just intonation, especially if they are contrasted with the same airs immediately afterwards played on an ordinarily tuned harmonium.

God save the Queen (in C major with its minor chord on the Second of the scale, alternating with the chord of the added Sixth).

The Heavens are telling (C major with the modulation into C minor).

Glorious Apollo (altering the brief modulation into the dominant).

The Old Hundredth (C major).

John Anderson (C minor).

Adeste Fideles (avoiding the modulation into the dominant).

Auld Lang Syne (in C major).

Dies iræ, in part (C minor modulating into C major).

Leise, leise (the prayer in *Der Freyschütz* in Poole's dichordal scale FA_1C, CE_1G^1BbD, altering the harmonies to suit the new scale).

Crudel perchè (*Nozze di Figaro,* in C minor, altered, but preserving the burst into C major).

Wanderer's Nachtlied (Schubert).

The Manly Heart (*Zauberflöte*).

So much relates to the lower four Octaves of the Harmonical, which suffice to illustrate all the principal peculiarities of just intonation. Advantage has been taken of the Fifth or 6-inch Octave to exhibit some of the higher harmonics of C 66, and to give a complete series of the first 16 harmonics of C 132, including the 11th and 13th. These notes are as follows:

Harmonics	16	17	18	19	20	22	28	24	25	26	29	30	32
Black digitals		$^{17}d'''b$		$^{19}e'''b$		$^7b'''b$			$^{25}a'''b$		$^{29}b'''b$		
White digitals	c''		d'''		e_1'''	$^{11}f'''$		g'''		^{13}a		b'''	$c^{\prime\nu}$
Pitch numbers	1056	1122	1188	1254	1320	1452	1848	1584	1650	1716	1914	1980	2112

Of course with such high pitches there has been great difficulty in tuning, and there are probably several slight errors, but none that will interfere with the general effect. I proceeded thus. The harmonics 16, 18, 20, 28, 24, 30, 32 were the Octaves of harmonics 8, 9,

10, 14, 12, 15, 16 already tuned for the lower Octave. Hence only 17, 19, 22, 25, 26, 29 remained to be treated, but they were in themselves far too high for me to tune forks for. I tuned therefore $^{17}d''b$ with 561 vib., the 17th harmonic, to which I had the Octave made by

Messrs. Valantine & Carr, music smiths, of 76 Milton Street, Sheffield. Then I tuned $^{19}e'b313\cdot5$, $^{11}f'363$, $^{23}a'b412\cdot5$, $^{13}a'429$, $^{29}b'b478\cdot5$ all harmonics of $C_{,,}16\cdot5$ and hence two Octaves too low. From these Messrs. Valantine & Carr made me forks giving the Octaves with great accuracy, and afterwards the Octaves of these forks, which so far as I could test them also appeared accurate, but it was very difficult to form an accurate judgment of the pitch of these high tuning-forks. From the forks thus made the remainder of the fifth octave was tuned. But as the tone of the reed drowned that of the fork, I had here also a second series of flatter forks constructed, beating twice in a second with the former. Of course I have not been able personally to check the tuning of all the Harmonicals, but I worked with the tuner at first and saw that he perfectly understood what was to be done, so that I confidently hope the Harmonicals he turns out will answer their purpose. One of them was exhibited in the International Inventions Exhibition of 1885, Division II., Music.

By means of this fifth octave the instru-ment has now all the first 32 harmonics of $C66$, except 6, namely 11, 13, 21, 23, 27, 31; and has all the first 16 harmonics of $c132$ without exception. There are additional loose pegs to the harmonical bar, which can be inserted, in order to play all these 16 harmonics at once, with the exception of the 7th and 14th, which, being on black digitals, must be struck with the finger as before.

The fifth octave therefore gives the trumpet scale 8, 9, 10, 11, 12, 13, 14, 15, 16, all, with exception of 14, on the white digitals. These give the peculiar intervals $10 : 11 = 165$ cents, $11 : 12 = 151$ cents, $12 : 13 = 139$ cents, $13 : 14 = 128$ cents. The inharmonic character of these intervals is, however, not well brought out, owing to the weakness of the upper partials in this region. Other intervals of interest are the approximations to the tempered Semitone, $16 : 17 = 105$ cents, $17 : 18 = 99$ cents.

The 17th harmonic allows of playing the harmonic form of the chord of the diminished Seventh in its direct form and all its inversions as

$$10 : 12 : 14 : 17 \quad = e_1'' : g'' : {}^7b''b : {}^{17}d'''b$$
$$12 : 14 : 17 : 20 \quad = g'' : {}^7b''b : {}^{17}d'''b : e_1'''$$
$$14 : 17 : 20 : 24 \quad = {}^7b''b : {}^{17}d'''b : e_1''' : g'''$$
$$17 : 20 : 24 : 28 \quad = {}^{17}d'''b : e_1''' : g''' : {}^7b'''b$$

The extreme height of the pitch of these notes, however, will prevent a due appreciation of these chords as compared with the usual forms, which can only be played at a lower pitch thus:

$$10 : 12 : 14\tfrac{2}{9} : 17\tfrac{1}{15} \quad = b_1 : d' : f' : a^Vb$$
$$12 : 14\tfrac{2}{9} : 17\tfrac{1}{15} : 20 \quad = d' : f' : a^Vb : b_1'$$
$$14\tfrac{2}{9} : 17\tfrac{1}{15} : 20 : 24 \quad = f' : a^Vb : b_1' : d''$$
$$17\tfrac{1}{15} : 20 : 24 : 28\tfrac{4}{9} = a^Vb : b_1' : d'' : f''$$

of which only the first shews the full harshness of the chord.

It is thus seen that the Harmonical is the only instrument yet tuned which brings out the full nature of just intonation for the 7th and 17th harmonics.

The difficulty in tuning the Harmonical *without* forks may be to a great extent avoided by the following means, which will enable any possessor of a cheap harmonium which he is willing to sacrifice as an experimental instrument, to get it tuned by a professional tuner. It will, however, be necessary to give up the peculiar arrangement of the fifth octave, and when it exists on any harmonium to have it tuned simply as an Octave higher than the fourth octave.

First tune the 11 notes $C\ D_1\ D\ E^1b\ E_1\ F$ $G\ A^1b\ A_1\ B^1b\ B_1$ thus. Take C to the existing pitch on the instrument. Tune the Fifths $c' : g'$, $g' : d''$, $f' : c''$ till they leave no trace of beats. Then take $c'' : e_1''$ and $a^1b : c''$ to be as perfect major Thirds without beats as the tuner can make them, verifying by the major chord $c'e_1'g'$, and minor chord $f'a^1bc''$ which should both be without beats. The combinational tones (which should be C for $c'' : e_1''$ and A^1b for $a''b : c'$) will also be a guide to the ear. But there is very little chance of perfect accuracy, the ears of tuners having been spoiled by the sharp major Thirds of equal temperament. It is best to begin by tuning these Thirds decidedly too flat, beating 10 or 20 times in 10 seconds, and then gradually to sharpen till the beats apparently vanish. The Thirds may thus remain very slightly flat, like the skhismic Thirds on p. 281d', and they will give very good results. The point is to avoid sharp major Thirds. Then tune the Fifths $e_1' : b_1'$, $a_1 : e_1'$, $d_1' : a_1'$, and $a''b : e_1''b$, $e''b : b^1b$, the necessary Octaves having been previously tuned. Verify by the major chords

$f'a_1'c''$, $c'e_1'g'$, $g'b_1'd''$; $a^1b\ c'e'b$, $e^1b\ g'b^1b$, and the minor chords $d_1'f'a_1'$, $a_1'c''e_1''$, $e_1'g'b_1'$; $f'a^Vb\ c''$, $c'e^1b\ g'$, $g'b^1b\ d''$, all of which should be perfect without sensible beats. Then only ^7bb remains to be tuned. To this we may approximate very closely thus. In the first place it is 49 cents, or say a quarter of a Tone (that is, half a tempered Semitone), flatter than b^1b which has already been tuned, and many tuners can approximate to this interval. Next, in the lowest octave B^1b and ^7Bb will beat. If the pitch happens to be $c''528$, then the beats are 33 in 10 seconds. For $c''540$, which is sharp band pitch, the beats would be not quite 34 in 10 seconds. For $c''518$, which may be taken as French pitch, the beats would be almost exactly 32 in 10 seconds. Hence by taking them as 33 in 10 seconds for any pitch, the tuner will come very near the truth. After tuning the Octaves, he will verify with the chord $c'e_1'g'^7b'b$, which should be without any sensible beats and have merely a slight roughness. Even a rough approximation to the true value of ^7Bb, as on the 53 division of the Octave, will gratify most ears.

2. The Just Harmonium.

This I used for some years. The $^7B\flat$ is sacrificed, as also the D_1; and the 12 notes are taken as in the margin, F^1 being put on the $F\sharp$ digital. It therefore contains the duodene of C, with the exception of $F_1\sharp$, and with the addition of F^1. This was tuned by an ordinary tuner in my presence in two hours from the following directions. Make the 7 major chords CE_1G, GB_1D, FA_1C, $A^1\flat CE^1\flat$, $E^1\flat GB^1\flat$, $B^1\flat DF^1$, $D^1\flat FA^1\flat$ perfect without beats.

$$
\begin{array}{ccc}
 & & F^1 \\
B^1\flat & D & \\
E^1\flat & G & B_1 \\
A^1\flat & C & E_1 \\
D^1\flat & F & A_1
\end{array}
$$

This gives more power of playing, as it contains the decad of C complete, and hence C major without the grave second, and C minor in all its forms, with all the 56 modes; $E^1\flat$ major with the grave second F, and $A^1\flat$ major without the grave second. But the harmonic Seventh $^7B\flat$, and grave second D_1 are much missed in C major. There is power of modulating from $E^1\flat$ major to its subdominant $A^1\flat$ major (without the grave second) and also into its relative minor C.
¶ Hence this plan of tuning has many advantages, especially in being easily effected by any tuner without forks.

3. The Just English Concertina.

For many years I have been in the habit of making my experiments on one of these instruments tuned thus:

Black studs $c_2\sharp$ d $d_2\sharp$ $f_1\sharp$ $g_2\sharp$ a b_1
White studs C D_1 E^1_1 F G A_1 $B\flat$

having the following duodenary arrangement:

$$
\begin{array}{ccc}
A & & \\
D & F_1\sharp & \\
G & B_1 & D_2\sharp \\
C & E_1 & G_2\sharp \\
F & A_1 & C_2\sharp \\
B\flat & D_1 &
\end{array}
$$

Hence it contains the decad of E_1 and four additional notes A, F, $B\flat$, D_1. This furnished the power of playing in G and C major with the grave seconds A_1, D_1, and in F major without the grave second G_1. Also in E_1 major without the grave second $F_2\sharp$. It can modulate perfectly from C major to the dominant G major, and from G major to its relative minor E_1. Also from E_1 major to its tonic minor E_1. But it cannot modulate perfectly from C major to its relative minor A_1, because of the absence of $F_2\sharp$ occasionally used in the subdominant. It can modulate perfectly from G major to its subdominant C major, and thence to its subdominant F major, less the grave second G_1. A considerable variety of harmony and modulation therefore lies open to it, but most pieces require special arrangement.
¶ It has been found advisable to put D_1, A_1 on the white studs and D, A on the black studs, that is D on the $D\sharp$ stud, and A on the $A\flat$ stud, and then I put $D_2\sharp$ on the $E\flat$ stud. In other respects the fingering is unaltered.

Tuning. The tuners of Lachenal's concertina factory (4 Little James Street, Bedford Row, London, W.C.) were able to tune with sufficient correctness from the following directions.

Make the 8 major chords CE_1G, GB_1D, $DF_1\sharp A$; FA_1C, $B\flat D_1F$; $A_1C_2\sharp E_1$, $E_1G_2\sharp B_1$, $B_1D_2\sharp F_1\sharp$ perfect without beats. In giving these directions I avoided using the inferior numbers except for $D_1 A_1$, which had to be distinguished from D, A. In writing music for it I generally assume the values of D and A_1 as in the key of C, and distinguish D_1 by a downstroke (as $^7b'b$ in the diagram p. 22c), and A by an upstroke (as $^{11}f'$ in the same diagram), but occasionally to prevent ambiguity I also use the up stroke to D, and the down stroke to A_1.

4. Mr. Colin Brown's Voice Harmonium.

Mr. Colin Brown, Euing Lecturer on the Science, Theory, and History of Music in the Andersonian University at Glasgow (see p. 259d, note ‡), has invented the following keyboard, a full sized model of which is in the Science Collections at the South Kensington Museum, Room Q. By the kindness of Mr. Brown I am enabled to give a perspective view of his keyboard in fig. 67.   He calls his instrument 'The Voice Harmonium.'

PLAN OF KEYBOARD.

```
                 D♯              g₂♯♯  B₁♯
             c₂♯♯  E₁♯     G♯
        C♯              f₂♯♯  A₁♯          C♯
        b₂♯  D₁♯     F♯                    b₂♯
                     e₂♯  G₁♯        B
        C₁♯      E                a₂♯  C₁♯
                d₂♯  F₁♯     A
             D              g₂♯  B₁
             c₂♯  E₁     G
        C              f₂♯  A₁          C
        b₂  D₁     F                    b₂
                  e₂  G₁        B♭
        C₁      E♭              a₂  C₁
                d₂  F₁     A♭
             D♭            g₂  B₁♭
                  E₁♭  G♭
        C♭              A₁♭            C♭
                  F♭
```

FIG. 67.

DUODENARY ARRANGEMENT.

White Digitals	Coloured Digitals	Peg Digitals
$C♯$	$E_1♯$	$g_2♯♯$
$F♯$	$A_1♯$	$c_2♯♯$
B	$D_1♯$	$f_2♯♯$
E	$G_1♯$	$b_2♯$
A	$C_1♯$	$e_2♯$
D	$F_1♯$	$a_2♯$
G	B_1	$d_2♯$
C	E_1	$g_2♯$
F	A_1	$c_2♯$
$B♭$	D_1	$f_2♯$
$E♭$	G_1	b_2
$A♭$	C_1	e_2
$D♭$	F_1	a_2
$G♭$	$B_1♭$	d_2
$C♭$	$E_1♭$	g_2

The notation used by Mr. Colin Brown, as shewn in fig. 67, is different from that used above. The column of Fifths, so far as it is there shewn, is

'$D♭$, $A♭$, $E♭$, $B♭$, F, C, G, D, A', E', B', $F''♯$.

The rest of his notation need not be specified, because it does not appear in fig. 67.

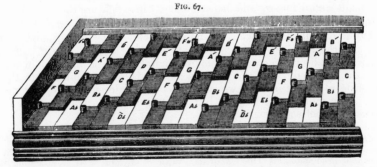

By the duodenary arrangement it is seen that the only duodenation contemplated was by Fifths up and down from the duodene of $B_1♭$ to that of $D_1♯$.   But it was not viewed in this light. It was rather considered as a series of major scales modulating into the dominant or subdominant, and also into the relative minor. The tonic minor was always taken one comma flatter than the tonic major; thus the tonic minor of C major was considered to be the relative minor of $E\flat$ major, that is C_1 minor, which, as shewn by Prof. Helmholtz's theories and the annexed duodenary arrangement, has not a single tone in common with C major.* The third column was considered merely as containing the major Sevenths and major Sixths of the relative minor scale and accounted of subordinate interest.

The arrangement of the keyboard is highly ingenious. Observing that in the major scale there are four notes in the column of Fifths, and three in the column of Thirds, it became evident that each note of the first would last during four successive modulations into the dominant, whereas each of the latter would last only through three modulations. Hence the digitals containing the former were made four parts long, and those containing the latter three parts long. In going up a series of Fifths each digital advanced one part. In the plan the longer digitals with only one note name were four parts long, and were left white. It will be seen that F is one part lower than C, C than G, G than D, and so on. Then D stands two parts higher than C. E, a major tone above D, stands also two parts above ¶ D. The long white digitals, read diagonally, give therefore the first column in the duodenary arrangement. Immediately below each is a short coloured digital, distinguished in the plan by having two names of notes on it. The lower is that of the note corresponding to the digital, and it is exactly one comma flatter than that of the long white digital above it. By this means the diagonal series of coloured digitals give the column of Thirds. Any white long digital is separated from the white digital next below it by a short coloured digital, and hence it corresponds to a rise of seven Fifths from the note of the lower white digital, and this rise gives the Pythagorean sharp, or 114 cents. Consequently the coloured digital which separates two white ones, being a comma of 22 cents flatter than the upper white one, is 92 cents sharper than the lower one. This gives the complete order, white C, coloured $C_1\sharp$ above it, and white $C\sharp$ above that; coloured C_1 below C, and white $C\flat$ below C_1. Then each digital on the right begins two parts higher, corresponding to a major Second, or two Fifths less an Octave; and the fingerboard is complete for the first two columns of the duodenary ¶ arrangement. In the fingerboard itself, as shewn in fig. 67, the lower and upper digitals are cut through at the dotted line, but they have been continued in the plan to shew the arrangement. Beginning then with any white digital as C we play the major scale in a horizontal line passing through the letters D, B_1 on the plan, and giving C, D (both white), E_1 (coloured), F, G (both white), A_1, B_1 (both coloured). The fingering is absolutely the same for all major scales whatever note is used as the white digital to commence with. The grave second D_1 is furnished by the coloured digital below the usual second.

For the relative minor, suppose the descending form with three minor chords is used; another line not quite horizontal through D_1, A_1 B_1 gives A_1 B_1 C D_1 E_1 F G A_1. To make the dominant chord major, and hence change g into $g_2\sharp$, touch the small peg which rises out of the left-hand corner of the A_1 digital, and it gives a diatonic Semitone of 112 cents below A_1, that is, the leading note to A_1. This peg is immediately to the right of the G digital. In the same way to make the subdominant chord major, and hence change f into $f_2\sharp$, use the peg $f_2\sharp$ immedi- ¶ ately to the right of the white digital F. The names of these pegs are written in small letters on the plan. We could by introducing the $c_2\sharp$ peg next to the C digital, play the complete major scale of A_1, as A_1 B_1 $c_2\sharp$ D_1 E_1 $f_2\sharp$ $g_2\sharp$ A_1, and all major scales beginning with a coloured digital would be fingered in the same way. Thus we could play the major scale of A_1 and modulate into the tonic minor scale of A_1. Similarly we could play the major scale of c_1 and modulate into the tonic minor scale of c_1. But the fingering was not intended for this, and hence it is not so convenient.

---

* Mr. Brown considers (*Music in Common Things*, p. 35) that C major and C_1 minor have one tone in common, F. This makes his descending form of the scale of C_1 minor read upwards c_1 d_1 eb f g_1 ab bb c_1', where I should use f_1, which is ready to hand on the instrument, if desired. Mr. Brown asserts (*ibid.* p. 35) that 'it is impossible to build a major and its tonic minor scale *in true key relationship*, upon the same tone of absolute pitch.' But on his own fingerboard we have the major scale of A_1, and what is, according to Prof. Helmholtz, the relative minor of C, both commencing with A_1, the one having the three minor chords $d_1 f a_1$, $a_1 c e_1$, $e_1 g b$, and the other the three corresponding major chords $d_1 f_2\sharp a_1$, $a_1 c_2\sharp e_1$, $e_1 g_2\sharp b_1$.

For the sake of perfect uniformity in fingering the difference of a skhisma is tuned, or intended so to be. Observing the two dotted lines in the duodenary arrangement, we see that no note between them differs from any other by less than a comma, but the 8 notes $A\flat$, $D\flat$, $G\flat$, $C\flat$, and C_1, F_1, $B_1\flat$, $E_1\flat$ below them, and 6 notes $D_1\sharp$, $A_1\sharp$, $E_1\sharp$ and $f_2\sharp\sharp$, $c_2\sharp\sharp$, $g_2\sharp\sharp$ above them, differ from notes between them by a skhisma only, the first eight being respectively a skhisma *flatter* than $G_1\sharp$, $C_1\sharp$, $F_1\sharp$, B_1 and $b_2\sharp$, $e_2\sharp$, $a_2\sharp$, $d_2\sharp$, and the last six respectively a skhisma *sharper* than $E\flat$, $B\flat$, F and G_1, D_1, A_1, which notes all lie between the dotted lines.

The tuning was effected, as the duodenary arrangement shews, by Fifths and major Thirds, and to overcome the difficulty of the latter the combinational tone was employed.

5. REV. HENRY LISTON'S ORGAN.

The pitch of Mr. Liston's notes has been calculated from the data furnished by ¶ Mr. Farey, *Philosophical Magazine*, vol. xxxix. p. 418. Mr. Liston's *Essay on Perfect Intonation* was published in 1812. The following is the duodenary arrangement of his notes :—

—	—	—	$C\sharp$	$E_1\sharp$	—	—
—	—	D^1	$F\sharp$	$A_1\sharp$	$C_2\sharp\sharp$	—
—	—	G^1	B	$D_1\sharp$	$F_2\sharp\sharp$	—
$F^7\flat$	$A^2\flat$	C^1	E	$G_1\sharp$	$B_2\sharp$	—
—	$D^2\flat$	F^1	A	$C_1\sharp$	$E_2\sharp$	—
$E^3\flat\flat$	$G^2\flat$	$B^1\flat$	D	$F_1\sharp$	$A_2\sharp$	$C_2\sharp\sharp$
—	$C^2\flat$	$E^1\flat$	G	B_1	$D_2\sharp$	$F_3\sharp\sharp$
—	$F^2\flat$	$A^1\flat$	C	E_1	$G_2\sharp$	$B_3\sharp$
—	$B^2\flat\flat$	$D^1\flat$	F	A_1	$C_2\sharp$	$E_3\sharp$
—	—	$G^1\flat$	$B\flat$	D_1	$F_2\sharp$	$A_3\sharp$
—	—	$C^1\flat$	$E\flat$	G_1	—	—
—	—	$F^1\flat$	$A\flat$	C_1	—	—
—	—	$B^1\flat\flat$	—	—	—	—

¶

The two isolated tones on the left were, I believe, added for the sake of tuning. It is evident that Mr. Liston contemplated considerable modulation, and provided for the tonic as well as relative minors.

6. GEN. PERRONET THOMPSON'S ENHARMONIC ORGAN.

This was the organ constructed by Robson for Gen. T. Perronet* Thompson, which I took Prof. Helmholtz to hear, as described in App. XVIII. p. 423. It had three manuals, each with a complicated fingerboard, and is completely described and figured in the General's *Principles and Practice of Just Intonation*, which is also full of curious musical information. It contained, on the whole, 40 tones to the Octave, and had considerable power of modulation, as shewn by the following table. But the gaps left indicate that the problem had not been completely grasped.

Duodenary Arrangement. ¶

C^1	E	—	—	—
F^1	A	$C_1\sharp$	—	—
$B^1\flat$	D	$F_1\sharp$	$A_2\sharp$	$F_3\sharp\sharp$
$E^1\flat$	G	B_1	$D_2\sharp$	$B_3\sharp$
$A^1\flat$	C	E_1	$G_2\sharp$	$E_3\sharp$
$D^1\flat$	F	A_1	$C_2\sharp$	$A_3\sharp$
$G^1\flat$	$B\flat$	D_1	$F_2\sharp$	$D_3\sharp$
—	$E\flat$	G_1	B_2	$G_3\sharp$
—	$A\flat$	C_1	E_2	$C_3\sharp$
—	—	F_1	A_2	—
No. of tones 7	9	9	8	7

* On the organ itself the name is painted as *Peronet* with one *r*, but the General printed and wrote his name with *rr*.

7. Mr. Henry Ward Poole's Organ.

See p. 323d, note †, and App. XVIII. p. 423a, for references to Mr. Poole's keyboards. His papers are in *Silliman's American Journal of Arts and Sciences*, 1850, vol. ix. pp. 68 83, 199–216; 1867, vol. xliv. pp. 1–22 (which contains the diagrams of his keyboard, here reproduced as figs. 68, 69, 70, p. 475, by the photographic processes of the Typo-etching Company); 1868, vol. xlv. p. 289. The first papers contain Mr. Poole's theory and an account of his Euharmonic Organ, constructed by himself and Mr. Joseph Abbey, of Newburyport, Massachusetts, having an ordinary fingerboard and a pedal to change the pipes that it affected, and playing from the major key of $D\flat$ with 5 flats to that of B with 5 sharps. The 12 digitals brought into action by each pedal produced the 7 notes of the major scale, the leading note of the relative minor scale, the perfect Seventh, and three others belonging to adjoining scales, of which only one (the grave Second) is specified.

¶ This arrangement, which was actually used in Boston, was abandoned, because it was found advisable to have all the notes under command of the hand without pedal action, and to use pedals for the bass only. But the new keyboard in its complete form does not appear from Mr. Poole's papers to have advanced beyond the stage of a cardboard model, although more recent simplifications, with 24 and 48 tones to the Octave, have been practically worked out. To these reference will be made at the conclusion of this notice.

In his theory of this keyboard, to which all subsequent remarks refer, Mr. Poole recognised 5 series of Fifths, namely those in cols. 5, 6, 7 of the Duodenarium, p. 463, and two others interposed which may be numbered as $^{7}5$ and $^{7}6$, because they contain notes which are a septimal comma (63 : 64, or 27 cents) flatter than the corresponding notes in cols. 5 and 6. These are given in the following duodenary arrangement of his notes. But instead of my superior and inferior numbers he used varieties of type, as shewn in the letterpress below fig. 70, which was photographed from the original at the same time as the fig. itself.

¶ *Duodenary Arrangement of Mr. Poole's 100 Tones.*

Cols.	5	$^{7}5$	6	$^{7}6$	7
	$E\sharp$ 522	—	—		—
	$A\sharp$ 1020	—	$C_1\sharp\sharp$ 206	—	$A_2\sharp\sharp$ 1090
	$D\sharp$ 318	—	$F_1\sharp\sharp$ 704	—	$D_2\sharp\sharp$ 388
	$G\sharp$ 816	$^{7}G\sharp$ 789	$B_1\sharp$ 2	—	$G_2\sharp\sharp$ 886
	$C\sharp$ 114	$^{7}C\sharp$ 87	$E_1\sharp$ 500	$^{7}E_1\sharp$ 473	$C_2\sharp\sharp$ 184
	$F\sharp$ 612	$^{7}F\sharp$ 585	$A_1\sharp$ 998	$^{7}A_1\sharp$ 971	$F_2\sharp\sharp$ 682
	B 1110	^{7}B 1083	$D_1\sharp$ 296	$^{7}D_1\sharp$ 269	
	E 408	^{7}E 381	$G_1\sharp$ 794	$^{7}G_1\sharp$ 767	$B_2\sharp$ 1180
	A 906	^{7}A 879	$C_1\sharp$ 92	$^{7}C_1\sharp$ 65	$E_2\sharp$ 478
	D 204	^{7}D 177	$F_1\sharp$ 590	$^{7}F_1\sharp$ 563	$A_2\sharp$ 976
	G 702	^{7}G 675	B_1 1088	$^{7}B_1$ 1061	$D_2\sharp$ 274
	C 0	^{7}C 1173	E_1 386	$^{7}E_1$ 359	$G_2\sharp$ 772
	F 498	^{7}F 471	A_1 884	$^{7}A_1$ 857	$C_2\sharp$ 70
	$B\flat$ 996	$^{7}B\flat$ 969	D_1 182	$^{7}D_1$ 155	$F_2\sharp$ 568
	$E\flat$ 294	$^{7}E\flat$ 267	G_1 680	$^{7}G_1$ 653	B_2 1066
	$A\flat$ 792	$^{7}A\flat$ 765	C_1 1178	$^{7}C_1$ 1151	E_2 364
	$D\flat$ 90	$^{7}D\flat$ 63	F_1 476	$^{7}F_1$ 449	A_2 862
	$G\flat$ 588	$^{7}G\flat$ 561	$B_1\flat$ 974	$^{7}B_1\flat$ 947	D_2 160
	$C\flat$ 1086	$^{7}C\flat$ 1059	$E_1\flat$ 272	$^{7}E_1\flat$ 245	G_2 658
	$F\flat$ 384	$^{7}F\flat$ 357	$A_1\flat$ 770	$^{7}A_1\flat$ 743	C_2 1156
	$B\flat\flat$ 882	$^{7}B\flat\flat$ 855	$D_1\flat$ 68	$^{7}D_1\flat$ 41	—
	$E\flat\flat$ 150	$^{7}E\flat\flat$ 153	$G_1\flat$ 566	$^{7}G_1\flat$ 539	—
	—	$^{7}A\flat\flat$ 652	—	$^{7}C_1\flat$ 1037	—
No. of tones	22	20	21	19	18 = 100

Thus col. 5, or 'key notes,' was represented by Roman capitals, as C, D, and had *white* digitals; col. 6, or 'Thirds,' by Roman small letters, as b, e (this was, in fact, Hauptmann's original plan in 1853, suprà, p. 276b), and had *black* digitals rising 0·4 inch; col. 7, or 'dominant Thirds, minor,' by italic small letters, as $d\sharp$, and had flat *blue* digitals rising 0·1 inch

FIG. 68.

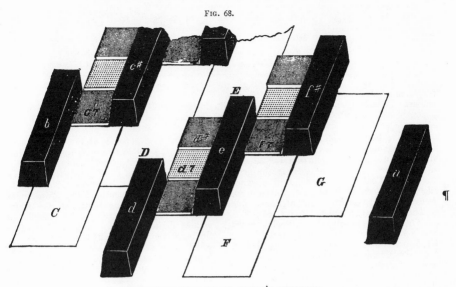

PERSPECTIVE VIEW OF MR. POOLE'S KEYBOARD.

FIG. 69.

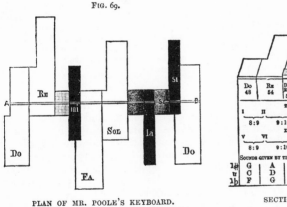

PLAN OF MR. POOLE'S KEYBOARD.

FIG. 70.

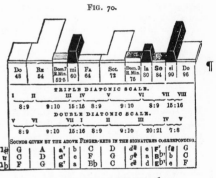

Do 48	Re 54	Dom.7 R.Min. 52·5	mi 60	Fa 64	Sol 72	Dom.3 R.Min. 75	la 80	Se 84	si 90	Do 96
TRIPLE DIATONIC SCALE.										
I	II	III	IV		V		VI	VII	VIII	
8:9		9:10	15:16	8:9		9:10		8:9	15:16	
DOUBLE DIATONIC SCALE.										
V		VI	VII	I		II		III	IV	V
8:9		9:10	15:16	8:9		9:10		20:21	7:8	
SOUNDS GIVEN BY THE ABOVE FINGER-KEYS IN THE SIGNATURES CORRESPONDING.										
1♯	G	A	a⁷	b	C	D	d♯	e	F'	f♯	G
♮	C	D	d⁷	e	F	G	g♯	a	B♭'	b	C
1♭	F	G	g⁷	a	B♭	C	c♯	d	E♭'	e	F

SECTION THROUGH A B FIG. 69.

above the white keys, marked with heraldic horizontal cross lines in fig. 68; col. ⁷5, or 'Sevenths,' by antique Roman capitals, with the index ⁷, as F⁷, and had *red* digitals, marked with heraldic vertical cross lines in fig. 68, and rising only 0·05 inch above the white digitals; col. ⁷6, or 'dominant Sevenths, minor,' had antique small letters and an index as d⁷, and *yellow* keys marked with heraldic dots in fig. 68, rising 0·15 inch above the white keys. The length of a white digital for col. 5 being taken as 4 inches, that of the black digital for col. 6 was 3 inches, and that of each of the coloured keys 1 inch. Fig. 68 gives a perspective view of this arrangement, with Mr. Poole's names of the notes on the digitals for the key of C major and some adjacent notes. Fig. 69 gives a plan of this arrangement with solfeggio names, except for two notes (pronounced with Italian vowels), to shew that it serves for *any* key beginning with a white digital and a cross line A B 'through the

centre of the third quarter (inch) of the key-note.' Fig. 70 gives the solfeggio names of the notes thus cut, with perspective views above of the remaining parts of the digitals furthest from the player. Underneath the solfeggio names are the relative number of vibrations of each note, taking *do* as 48. Below this again are the numbers of the notes in Mr. Poole's 'triple diatonic' or *trichordal* scale, with the ratios of their intervals, and also the numbers of the notes in his 'double diatonic' or *dichordal* scale (see p. 344c, note *). And finally the three last lines give the names of the notes as he writes them, supposing the first white digitals to give G (key of 1♯ or one sharp), C (natural key, where the mysterious symbol may be meant for an n—that is, 'natural,' but seems to have been reversed by the wood-engraver), and F (key of 1♭ or one flat). Interpreted into our symbols, with the interval in cents from the lowest note in each line, these will be:

rel. vib.	48	54	52½	60	64	72	75	80	84	90	96
Notes in fig. 70	g	a	$^7a_1$	b_1	c'	d'	$d_2'♯$	e_1'	$^7f'$	$f_1'♯$	g'
	c	d	$^7d_1$	e_1	f	g	$g_2♯$	a_1	^7bb	b_1	c'
	F	G	$^7G_1$	A_1	Bb	c	$c_2♯$	d_1	^7eb	e_1	f
cents	0	204	155	386	498	702	772	884	969	1088	1200
Solfeggio	Do	Re	—	Mi	Fa	Sol	—	La	Se	Si	Do
colours	white	white	yellow	black	white	white	blue	black	red	black	white

If we took only the black and white digitals, the arrangement of the keyboard would be like Mr. Colin Brown's; but this was an accident, Mr. Brown having never seen the drawings of Mr. Poole's keyboard. Both arrangements arose from the column of Fifths in the Decad (p. 459b) containing four, and the column of Thirds only three notes.

The great peculiarity of Mr. Poole's board, where Mr. Brown differs from it entirely, is in the introduction of the columns ⁷5 and 6, both containing natural Sevenths, and their amalgamation, as it were, with the col. 7 (which Mr. Brown alone uses) in three short flat digitals, placed beside the black, and hence of the same length. They are placed from front to back, as red, yellow, blue—that is, as two Sevenths and a Third, which belong to three different keys. Thus in fig. 68, to the left of the black digital for E_1 (marked e), lie three coloured digitals, 1) the red ^7Eb (of which the name is not marked in the fig.), the natural Seventh of the tonic in the key of F; 2) the yellow $^7D_1$ (marked d⁷), the natural Seventh of the dominant of the relative minor in the key of C, which is $e_1 g_2♯ b_1 {}^7d_1$; 3) the red $D_2♯$ (marked d♯), the leading note (or major Third of the dominant $b_1 d_2♯ f_1 {}^7a_1$) of the relative minor in the key of G. The situation of these digitals is such that the lowest (or red) digital gives the natural seventh of the white digital immediately adjoining (below in the figure, compare $C : {}^7C$, $F : {}^7F$). The middle (or yellow) digital gives the natural Seventh of the black digital, the right-hand top corner of which touches its left-hand bottom corner (compare $D_1 : {}^7D_1$, printed d d⁷). The uppermost (or blue) digital gives a note which is a small Semitone (24 : 25 = 70 cents) sharper than the note of the white digital on the left (compare $D : D_2♯$, marked $D : d♯$), and a diatonic Semitone (15 : 16 or 112 cents) flatter than the note of the black digital on the right (compare $D_2♯ : E_1$, marked d♯ : e). The only digital placed out of ascending order from left to right is the yellow one, which should have come between the two white digitals for so and re, but has been displaced from motives of convenience.

Mr. Poole, as Mr. Colin Brown afterwards, provided only for modulations from major keys into the dominant major, subdominant major, and relative minor. For the modulation into the tonic minor, therefore, he had to flatten by a comma. Thus C minor was considered to be the relative minor of Eb major instead of E^1b major. And although, in deference to Mr. Liston and Gen. Perronet Thompson, he also made a provision for temporarily introducing the tonic minor if desired, giving col. 4 of the Duodenarium (Silliman, vol. xliv. p. 18, art. 35), he did not require it himself. 'In the theory I have advocated,' he says, 'the major keys are based on the first series of sounds,' p. 474c, col. 5, 'and the minor keys on the Sixths of the major keys,' ibid. col. 6. 'That there must be such a relation and order is inevitable.' But this does not exclude taking minor keys also upon notes of col. 5 as a base, considered, if desired, as the Sixths of major keys on the notes of col. 4 of the Duodenarium. Otherwise tonality is destroyed by constant shifts of a comma severely felt on justly intoned instruments.

¶ Mr. Poole was also aware of the alteration by a skhisma, and of the consequent reduction of the number of pipes. He also refers to the 53 division, but he does not seem to adopt either, and is not distinct enough on these points for me to state his conclusions with certainty. In the duodenary arrangement, I have by dotted lines marked the places where the skhisma comes into play, and by affixing the cents to each note have shewn how it acts.

It will be seen that Mr. Poole had 100 notes to the octave, of which 39 arose from the harmonic Sevenths. If the skhisma were neglected there would remain only 36 tertian and 20 septimal, or in all 56 tones to the octave. The duodenary arrangement has been taken from Mr. Poole's Enharmonic table (Silliman, xliv. p. 13), consisting of 19 lines similar to the 3 at the bottom of fig. 70. He adds the following example of the fingering of chords upon his keyboard, the double numbers indicating 'that the key is touched with one finger and immediately changed for another.' The duodenals and mark of the natural Seventh are according to my notation, sect. E. art. 26, p. 465c. The upper figures refer to the Notes which follow.

'*Notes.*—1. Subdominant chord *f'* and *a₁'*.

'2. Dominant with Seventh *f'*.

'3. Same with Ninth *a'* [not *a₁'*, and hence causing a duodenation into the dominant *G*, but forming the second chord in Poole's dichordal scale of *C*. Of course, '*f'* itself is not in the duodene of *G*, but when these natural Sevenths are introduced the special marks are used. See suprà, p. 349*a*].

'4. Dominant Seventh.

'5. Dominant of the relative minor, the Seventh '*d₁* may be added [it is added here, but to secure the intonation a duodenation into the relative is marked].

'6. Subdominant with Seventh [duodenation into *F*, therefore].

'7. Grave second or Sixth of subdominant [as the duodenal gives the root *F*, the *d₁"* is sufficiently marked].

'8. The flattened note should be made natural in the next chord' [meaning 'next chord but one,' namely at 8, so that there is again a duodenation from *F* to *C* as marked].

The above examples will shew how Mr. Poole treats the chord of the dominant Seventh and the major Ninth. The three last chords are added to shew his treatment of the chord of the diminished Seventh (*Silliman*, vol. ix. pp. 78–80). He considers the first of these chords to be merely the chord of *G* with the dominant Seventh, *g' b₁' d'' ₇f'*, which is of course in the major scale of, *G*, but this lies

within the duodene of *E₁*, as I have marked it, ¶ including the ₇*F* within that duodene, as shewn in the duodenary arrangement p. 474*c*. Then he supposes that in order to resolve the chord *a₁' c'' e₁''* (the last chord), the *g'* is altered in the second chord by 'a chromatic Semitone' (that is, the small Semitone 24 : 25 = 70 cents), to *g₂'♯*, which is necessarily in the duodene of *E₁*, but this *g₂'♯* serves merely as 'a passing note' to the following *a'*, and therefore, he says, 'must be thrown out when we reckon the harmony.' But this will not explain the present use of this chord, which is now introduced without preparation, and as a means of modulation. The ratios of the chord Mr. Poole gives are 25 : 30 : 36 : 42, or taking the *g₂'♯* an Octave higher, to compare with my form, it becomes 30 : 36 : 42 : 50, that is 10 : 12 : 14 : 16⅔, or in cents 0, 316, 583, 884. Mr. J. Paul White (see below No. 9) makes the ¶ ratios of the chord 30 : 35 : 42 : 50, that is 10 : 11⅔ : 14 : 16⅔, or in cents 0, 267, 583, 884. The individual intervals in the first are 316, 267, 301, and in the second 267, 316, 301, so that the two first intervals are transposed. But in both the interval of the extreme notes is 3 : 5 = 884 cents, so that in neither have we a chord of a diminished Seventh at all, which must have 919 or 926 cents. It is only equal temperament which confuses the major Sixth and diminished Seventh together by using 900 cents for either of them.

With regard to the double diatonic or dichordal scale, which Mr. Poole always solfas as *fah sol la se do re mi fah* (where *se* is the harmonic Seventh to *do*), so that *do* is the dominant, he says that 'the most beautiful, varied, and ornate compositions are made from the elements it contains. It has the capacity in cer- ¶ tain styles of music of using with much grace accidentals, or chromatics as they are called ; for example, the *si*, the regular leading note to *do*, and the *sol♯*, a diatonic Semitone to below *la*, or the leading note to the relative minor ; these chromatics always ascending a diatonic Semitone (15 : 16) to the notes above.' In an example given he also admits *se* to be raised by 27 cents, that is to be the regular Fourth of the triple or trichordal scale, and also allows the introduction of the Sixth of this scale. Hence if we use the duodenary form and represent the dichordal scale of *F* by capitals and these permissive additions by small letters we shall have

D		
G	*b₁*	
C	*E₁*	*g₂♯*
F	*A₁*	
b♭	₇*B♭*	*d₁*

the scheme in the margin. This gives the trichordal scale of *C* major complete with its grave second, and also one form of its relative *A₁* minor complete, but both without the harmonic Seventh of the dominants, which of course he would be ready to add when the harmony in his view required it. There is also the complete trichordal scale of *F* major without the grave second. Hence his dichordal scale resolves itself into a means of bringing these three scales into close connection, chiefly by help of the chord of the Ninth $CE_1G\,{}^7B\flat D$ in the above scheme. The example that he gives of its use is the accompaniment to Figaro's *Numero quindici* from Rossini's *Barbiere*, afterwards sung as the air *Ah ! che d'amore* by Almaviva. This is written in G major. He gives the scale thus, using my notation and indicating accidentals by small letters :

DOUBLE DIATONIC SCALE IN G, WITH ACCIDENTALS.

$$G \quad A \quad a_2\sharp \quad B_1 \quad {}^7C_1 \quad c \quad c_1\sharp \quad D \quad E \quad F_1\sharp$$

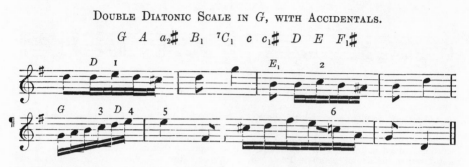

The Duodenals are mine, but as the Ninth is not in a single duodene, it can be marked only by giving the duodene containing all but the natural Seventh and indicating that by a sloping line in the usual way. The notes in inverted commas are from Mr. Poole, except the bracketed portions, which are mine.

' 1. This may be $e_1{}''$ [in that case the three first notes are in the duodene of G].

' 2 and 3. These may be c'' as well as ${}^7c''$ [in the latter case the whole run would be in D, the c'' being marked as ${}^7c''$; in the former, the run would be in G].

' 4. This may be $e_1{}''$ [this will be only if 3 is c'', so that the whole run is in G].

' 5. This note is clearly and necessarily e''. [In this case 3 certainly should be ${}^7c''$ and 4 should also be e'', but Mr. Poole does not object to $g'\,a'\,b_1{}'\,c''\,d''\,e_1{}''$ followed by e'', saying] the enharmonic change from $e_1{}''$ to e'', a rise of a comma, is often required, and I have proved that it can readily be made, for my singers, who know this change of a comma as well as others know the Tone or Semitone, will give it, even without accompaniment, with perfect accuracy, as proved by the harmony afterwards supplied as a test. All this variety within the limits of musical laws—which only forbid what is disorderly, complicated, or what the ear will not distinguish—adds to the pleasure of vocal music, and it is the exact rendering of all the melodies and harmonies which gives the charm to a good singer. [A little difficulty arises as to tonality.] When acutely perceptive of such accuracy, I had the good fortune to listen to Alboni on all the occasions when it was possible to do so. I thought her then, am still of opinion, that she was the best singer I have ever heard. It is certain that she had a wonderful exactness in executing whatever she undertook. There was no " temperament " in *her* scales, and what the strictest theory requires in intonation she understood and gave. She sang music whose analysis would alarm a student with its apparent difficulties ; but the delighted auditors perceived only a delicious and " easy " flow of melody.' [This has been quoted at length to add to the examples suprà, p. 325*a*.]

6. Mr. Poole says nothing about this c'', but I presume he would take it regularly as ${}^7c''$. In that case the whole of this would be in D major.

Mr. Poole has also devised an enharmonic keyboard pedal for the bass of his organ, but then confines himself entirely to cols. 5, ^{7}5, 6. The ' keynotes ' corresponding to the white digitals are in front in order of Fifths from left to right. Behind them, at a higher elevation, are the major Thirds lying between them. The Sevenths are in a back row behind their Fifths. This is indicated by the letters in the margin.

red	${}^7E\flat$	${}^7B\flat$	${}^7F$
black	A_1	E_1	
white	F	C	G

Mr. H. W. Poole is a native of Salem (afterwards Danvers, now Peabody), Massachusetts, U.S., and is now Professor of Public Instruction in the Government Institute of the city of Mexico, whence he kindly wrote to me on 9 March 1885, describing one of his new keyboards. From this I take the following summary and extracts :—

' I send you a stereograph of a simple form of one of them [his new manuals], which is easier to comprehend than a larger one with 48 levers to the octave. This may be called a working model, and suffices for an organ or pianoforte for instruction or study in effect of chords and fingering. It is solidly constructed in wood, ebony, and ivory, and works as freely as a common one. These 24 levers are a quarter of an inch wide, and can play a pianoforte with hammers half the common width, with single strings, but larger and lightly strained so as to yield the maximum of tone, tension nearly to breaking-point giving bad tone.' The finger-keys for each Fifth rise $\frac{1}{10}$ inch, so C is $\frac{4}{10}$ inch above $A\flat$. The white digitals have the same shape as in fig. 69, but from each projects a narrow black finger-key, with a note one comma flatter, giving the major Thirds, and fitting into the left-hand nick of the next lower white digital. Whatever white digital the player begins with, the fingering is the same, and for major scales much like that for the key of A on the usual manual. For the Seventh and Ninth of the Dichordal system separate digitals must be touched. Mr. Poole can arrange for a minor on the same tonic, but thinks it an extravagance. 'The diatonic scales with the broad ivory keys (larger than on the common board) are of first importance; next the raised ebony digitals for Thirds. The Sevenths are well provided for and convenient. The leading notes to major Thirds are introduced as diatonic Semitones below these black notes, and serve them as the black ones do the white. On my model an equal space of two measures

($\frac{1}{4} \times \frac{9}{10}$ inch each) was allotted.' They fill up the space on the left-hand side of the black Thirds, and are of the shape of the white digitals in fig. 69, only very much narrower, half the thin part being separated, and for example given to $A_2\sharp$, leading note to B_1, while the rest, including the wide part, is given to $^7A\flat$, harmonic Seventh to $B\flat$. 'My keyboard admits of equal facility in execution and in taking the chords, with the common one of 12. I think its first utility will be for teaching singing, accompanying violin players and students of harmony. For this I recommend the simple form with less outlay of money.' This form of Mr. Poole's keyboard is therefore equivalent to Mr. Colin Brown's (No. 4) with the addition of the natural Seventh. ¶

8. Mr. Bosanquet's Generalised Fingerboard and Harmonium.

Mr. R. H. M. Bosanquet's harmonium is partly described in the text, p. 328c, and its keyboard is figured and briefly explained in App. XIX., p. 429. In App. XX. sect. A. art. 27, the nature of Mr. Bosanquet's cycle of 53 and his notation, and the value of every one of his notes are explained. In App. XX. sect. E. art. 18, there is an elaborate comparison of this cycle with just intonation giving the number and pitch of every note, and, *ibid.* art. 25, it is shewn how such a cycle might have been suggested by just intonation. In sect. G. arts. 16 and 17, the methods of tuning the cycle of 53 adopted by Mr. Bosanquet and Mr. White are described. In the South Kensington Museum, Science Collections, Room Q, the harmonium itself may be inspected, Mr. Bosanquet having presented it to the Museum, as he generally employs for his own use an organ with the same fingerboard, and two sets of pipes, one set for 48 notes of the temperament advocated by Prof. Helmholtz (p. 432a), with perfect major Thirds and Fifths imperceptibly ¶ flattened by $\frac{1}{8}$ Skhisma, answering to the notes written with capital letters on the digitals in the following plan; and the other set for 36 notes of the meantone temperament, brought into separate action by a stop. The pipes are stopped, with a screw plug, so that they are more readily tuned.

It remains in this place to give the plan of the fingerboard, shewing the disposition of the notes upon it both for the 53 division and meantone temperament, and to describe its arrangement, referring especially to App. XIX., p. 429, fig. 66.

In the present plan of the keyboard, all the digitals are represented as of the same length, corresponding to that from tip to tip. This is 3 inches in the original and is here only 1 inch. At each side runs a column of figures 1 to 12 continually repeated. It will be observed that in the first column the lines terminating the oblongs come against 2, and that 2 is at the head of the column. In this case the end of each oblong gives a form of c, and in passing from one form to another, as c to c^1, we have gained a Pythagorean comma, which results from taking 12 Fifths reduced to the same Octave. In the g column headed 3 the lines are ¶ opposite 3; in the d column, headed 4, opposite 4; and so on; each Fifth corresponding to a rise of $\frac{1}{4}$ inch from tip to tip of the digitals, and to a vertical rise of $\frac{1}{12}$ inch from level to level. Hence in going from one degree to another, as c to c^1 or 4 to 5, we go backwards $12 \times \frac{1}{4} = 3$ inches, and rise $12 \times \frac{1}{12} = 1$ inch. Mr. Bosanquet says (*Mus. Int. and Temp.* p. 20) :—

'The most important practical point about the keyboard arises from its symmetry; that is to say, from the fact that every key is surrounded by the same definite arrangement of keys, and that a pair of keys in a given relative position corresponds always to the same interval. From this it follows that any passage, chord, or combination of any kind, has exactly the same form under the fingers in whatever key it is played. And more than this, a common chord, for instance, has always the same form, no matter what view be taken of its key relationship. Some simplification of this kind is a necessity if these complex phenomena are to be brought within the reach of persons of average ability; and with this particular simplification, the child or beginner finds the work reduced to the acquirement of one thing, where twelve have to be learned on the ordinary keyboard.'

PLAN OF MR. BOSANQUET'S GENERALISED KEYBOARD.

Fifths: 2　9　4　11　6　1　8　3　10　5　12　7　2　Fifths

7 to 1	12 to 6	16 to 10	21 to 15	25 to 19	30 to 24	34 to 28	38 to 32	43 to 37	47 to 41	52 to 46	3 to 50	7 to 1

The numbers in the oblongs and in the lines at the bottom of the Table are the numbers assigned to the tones of Bosanquet's cycle of 53, in sect. A. art. 27, and sect. E. art. 18, and shew their distribution on this keyboard, which was invented for playing them. The small *italic* letters under the numbers at the bottom of each oblong are the transcriptions of one of Mr. Bosanquet's names, of which all are given in sect. A. art. 27, against the number of the note. The stars preceding the numbers shew those which constitute the duodene of C. The notes of all other duodenes stand in the same relative position to their root. The capital letters are 48 out of the 56 tones between the dotted and thick lines in the Duodenarium, sect. E. art. 18, and those following the sign = are other tones differing from them by a skhisma, which are purposely identified with them, see sect. E. art. 25. These are the 48 tones used by Mr. Bosanquet for his organ, but he names them as at the bottom of each digital. The small thick Roman letters at the top of each oblong are 36 tones of the meantone temperament, and the numbers below them are the cents in the intervals from c to these notes.

This keyboard is applicable to any intonation in which all notes, by the neglect of the comma or skhisma, are reduced to one set of Fifths, no matter whether perfect or imperfect, as the flat Fifths of the meantone temperament. On referring to the plan, we see how the 53 division is placed on the notes. Mr. Bosanquet finds it convenient to use $7 \times 12 = 84$ digitals, so that there are repetitions as shewn by the figures at the bottom of the plan, each of the 12 columns containing 7 digitals. The position of the meantone notes is shewn by a small thick Roman letter at the top of the digital. Having 36 digitals at his disposal, Mr. Bosanquet has used 36 notes of the meantone scale in place of only 27. They are disposed in 12 rows with three digitals in each row. In the plan, thick lines limit the three digitals thus placed at the disposal of the meantone notes, and under the name of each note is inserted the number of cents in the interval between it and c. It will thus be seen that each note differs from the one above it in the plan by a Diĕsis of 41 cents. Thus the first row has b♯ 1159, c o = 1200, d♭♭ 41. Also each digital lies against two to the right and two to the left. The upper one to the right is 76 cents or a small meantone Semitone higher, the lower is 117 cents or a great meantone Semitone higher. The upper one to the left is, on the contrary, 117 cents lower, and the lower one to the left is 76 cents lower. The sum of the two Semitones is 193 cents $= \frac{1}{2}(204 + 182)$, a Meantone, ¶ and their difference 41 cents a Diĕsis.

The use of this fingerboard is easily acquired by any pianist, the fingering for all major keys resembling that for A major on ordinary instruments.

9. MR. J. PAUL WHITE'S HARMON.

Mr. James Paul White, of Springfield, Massachusetts, U.S. America, a tuner by profession, having been much impressed by Mr. Poole's papers in *Silliman's Journal*, cited under No. 7, determined to realise them so far as possible by means of the 53 division of the Octave. Now on examining this division by the tables in sect. A. art. 27, and sect. D. table 1, we find that the number of degrees by which any interval is represented can always be expressed by multiples, or the sums or differences of multiples, of 2, 5, 7 (which may therefore be called *indices*), as in the following table :—

Name of Interval	Its Cents	Represented in the 53 division by		
		Cents	Degrees	
Comma	22 ⎫	23	1	$= 2 \times 4 - 7$
Pythagorean Comma . .	24 ⎭			
Great Diĕsis	42	45	2	$= 2$
Small Semitone . . .	70	68	3	$= 5 \times 2 - 7$
Limma	90 ⎫	91	4	$= 2 \times 2$
Greater Limma . . .	92 ⎭			
Diatonic Semitone . . .	112	113	5	$= 5$
Minor Tone	182	181	8	$= 2 \times 4$
Major Tone	204	204	9	$= 5 + 2 \times 2$
* Supermajor Tone . . .	231	226	10	$= 5 \times 2$
† Subminor Third . . .	267	272	12	$= 7 + 5$
Pythagorean minor Third .	294	294	13	$= 7 + 2 \times 3$
Just minor Third . . .	316	317	14	$= 7 \times 2$
Just major Third . . .	386	385	17	$= 7 + 5 \times 2$
Pythagorean major Third .	408	408	18	$= 7 \times 2 + 2 \times 2$
* Supermajor Third . . .	435	430	19	$= 7 \times 2 + 5$
Fourth	498	498	22	$= 7 + 5 \times 3$
Just Tritone	590	589	26	$= 7 \times 3 + 5$
Pythagorean Tritone . .	612	611	27	$= 7 + 5 \times 4$
Grave Fifth	680	679	30	$= 5 \times 6 = 7 \times 5 - 5$
Just Fifth	702	702	31	$= 7 \times 3 + 5 \times 2$
Pythagorean minor Sixth .	792 ⎫	793	35	$= 7 \times 5$
Extreme Sharp Fifth . .	794 ⎭			
Just minor Sixth . . .	814	815	36	$= 7 \times 3 + 5 \times 3 = 53 - 5 \times 2 - 7$
Just major Sixth . . .	884	883	39	$= 7 \times 5 + 2 \times 2 = 53 - 7 \times 2$
Just diminished Seventh .	926 ⎫	928	41	$= 53 - 7 - 5$
* Supermajor Sixth . . .	933 ⎭			
Just superfluous Sixth .	954	951	42	$= 7 \times 6$
† Subminor Seventh . .	969 ⎫	974	43	$= 53 - 5 \times 2$
Extreme sharp Sixth . .	976 ⎭			
Minor Seventh . . .	996	996	44	$= 7 \times 6 + 2$
Acute minor Seventh . .	1018	1019	45	$= 5 \times 9 = 53 - 2 \times 4$
Just major Seventh . .	1088	1087	48	$= 53 - 5$
Pythagorean major Seventh .	1110	1109	49	$= 7 \times 7 = 53 - 2 \times 2$
Octave	1200	1200	53	$= 7 \times 4 + 5 \times 5$

It is really surprising how accurately the intervals of just intonation are thus represented. Only those marked * and † depending on the 7th harmonic are 5 cents too flat, or sharp respectively, which is barely perceptible.

Influenced no doubt by such a calculation as the above, Mr. Paul White conceived and executed a fingerboard of which the typographical plan below will give some conception. And this conception will be much improved by drawing pencil lines on the diagram parallel to the rows of figures sloping up (as **49** 1 6 11 16 21 26 **31** 36 41 46) and down (as 41 48 2 **9** 16 23 30 37 44) to the right. These lines will divide the plan into a number of parallelograms or irregular lozenges, each of which represents a digital of nearly the same shape, but fitting loosely into its place. These are pieces of wood all diamond-shaped and all of the same height, variously marked to assist the player, and all bearing upon them the numbers printed in the plan. The typographical plan is, of course, only approximatively correct. In reality the vertical lines are not quite vertical, and the lines parallel to the numbers differing by 12 as 39—51—10—**22**—34—46—5 at ¶ the top (which may be connected by pencil lines as just shewn) are more nearly horizontal or rather slope slightly downwards instead of up. But as the design has not been published, it was desirable to give only a conception and not an accurate plan of the arrangement with the curious slopes of the actual lines.

TYPOGRAPHICAL PLAN OF MR. J. PAUL WHITE'S FINGERBOARD.

```
                                                              51
                                         ‖39              5
                              ‖27              46              12
                    ‖15              34         53
                    22              41        =○      7
              10              29         48              14
        51         †17         36         2
¶  39         5         24       *†43        *†9
     46         12      †31         50              16
     †53         19       38              4
    =○    7         *26       45         11
     48         14      33       *52         18
        2      *21       *40       6
  43      *9         28       47       *13
     50         16      *35       1         20
      *4         23      42       8
  45      11       *30      *49       15
     52      *18       37       3
        6      25       44
¶  47      *13       32
     1         20      39
        8         27
 49         15
     3
```

The thick figures represent white digitals and serve as land-marks, forming the 53 c, 9 d, 18 e, 22 f, 27 f♯, 31 g, 40 a, 49 b.

The numbers marked * are the same as those in the plan of Mr. Bosanquet's, suprà, p. 480, and represent the duodene of note 4. The numbers marked †, or 53, 17, 31, 43, 9, represent the chord of the major Ninth $c\ e_1\ g^7b\flat\ d$.

In the columns the numbers as they proceed from top to bottom increase by 2, of course taken in the reverse direction they decrease by 2. When necessary 53 is subtracted or added here and elsewhere, as the numbers must not exceed 53.

In the lines which slope *up* to the right the numbers increase upwards by 5, and of course decrease downwards by 5.

In the lines which slope *down* to the right the numbers increase downwards by 7, and of course decrease upwards by 7.

It is thus seen that the three indices 2, 5, 7 are represented by nearly vertical and sloping lines, and it becomes easy by the preceding table to pick out any interval. Thus to take the just minor Sixth from 18 (one of the thick figures); we have by the table, $36 = 7 \times 3 + 5 \times 3$, so that we go down 3 steps on the line of 7's, and then up 3 steps on the line of 5's, and thus reach 1, the right degree for $18 + 36 = 54 = 53 + 1$. But in the table we also find $36 = 53 - 5 \times 2 - 7$, hence we may also go *down* 2 steps on the line of 5's and then *up* 1 step on the line of 7's, reaching 1 as before, but not the same 1. It is now the Octave below, and if from this new 1, we descend 4 steps on the line of 7's and then ascend 5 steps on the line of 5's, we reach the old 1, for $7 \times 4 + 5 \times 5 = 28 + 25 = 53$, or the Octave.

The body of each digital is a block of wood $2\frac{1}{2}$ inches high and not far from ¶ 1 inch square on the top. The grain of the wood is vertical so as to facilitate the action of the key on its two steel guide pins, which are driven firmly into a board as wide as the manual. The valve is opened by a pin under the key in the usual way. Of course the fingering is entirely different from ordinary fingering, but is the same in all possible keys. Contrasting his board with Mr. Bosanquet's, which he admits is admirable and would be probably regarded with more favour by musicians than his, Mr. Paul White (in a private letter to me) says 1) that his board combines the advantages of Mr. Poole's with Mr. Bosanquet's, and has digitals of a simpler construction than either, shewing also the Pythagorean tones conspicuously for every note, and having the complete cycle of tones. 2) The chords are all easy for the fingers, including those depending on the 7th harmonic. 3) Digitals differing by one comma are far apart, so that there is no danger of playing too sharp or too flat by a comma. The fingers can easily make the just chords, but to make them false by a comma is difficult. 4) This fingerboard can be made more compact than any other. The extreme width of the present instrument (the third made) is only 11 inches, or twice the ordinary width. ¶

Mr. Paul White uses only 56 digitals to the Octave, Nos. 15, 27, 39, marked ||, being the only repeats. He was kind enough to send me two photographs of his instrument, (which he calls the Harmon, and which he constructed almost entirely with his own hands,) one giving a bird's-eye view of the digitals, and the other their connection with the rods that open the valves. He has as yet not arranged any system of notation and does not himself play on his instrument from notes.

SECTION G.

ON TUNING AND INTONATION.

(See notes pp. 256, 287, 311, 325.)

Art. 1.—We have seen in sect. E. that just tertian harmony requires the discrimination of 117 different tones within the Octave. They all indeed depend upon just Fifths, Fourths, Thirds, and Sixths. But very few ears could be trusted to tune a succession of perfect Fifths and Fourths. Herr G. Appunn told me that it cost him an immense labour to tune 36 notes forming perfect Fifths and Fourths upon an experimental harmonium, and he had the finest ear for appreciating intervals that I have ever heard of. The accumulation of almost insensible into intolerable errors besets all attempts to tune by a long series of similar intervals. Even Octaves are rarely tuned accurately through the compass of a grand pianoforte. But for major Thirds and minor Sixths there is no chance at all (except by a real piece of haphazard luck) to get even one interval tuned with absolute correctness by mere appreciation of ear. Hence to attempt to tune the Duodenarium of sect. E. art. 18, p. 463, merely by Fifths and major Thirds is quite hopeless. But if we cannot tune just intervals with sufficient correctness, how can we expect to tune all the variously tempered intervals mentioned in sect. A. (and these are only a few of the most important) sufficiently well to discriminate their qualities and appreciate their merits? No ear knows *à priori* what result it has to expect, or has any means of judging whether the result obtained is correct. It follows that all attempts to tune by ear must have grievously failed, wherever they depended upon considerable alterations of just intervals, and that even the laborious ¶ and careful training of modern tuners for obtaining the very slightly altered Fifths and Fourths of equal temperament can only lead them to absolute correctness ' by accident.'

Art. 2.—To ascertain whether these theoretical views were confirmed in practice, I have made some observations on the tuning of the old meantone and the recent equal temperaments. It is easy (from the data in sect. A.) to determine the cents which should be contained in each interval, and (by measuring the actual pitch of each note with the forks described on p. 446*b'*), to find what the interval obtained in any particular case really is. For brevity I give only the names of the notes in the octave, and the interval in cents from the lowest note. But every such figure is the result of a careful observation.

Line 1 in the following Table gives the theoretical number of cents.

Line 2 gives the cents observed on a pitch-pipe of 1730 belonging to the bellfoundry Colbacchini at Padua, blown with the least ¶ force of wind possible to bring out the tone, on an organ bellows at Mr. T. Hill's, the organ-builder's.

Line 3 gives those observed on another pitch-pipe of 1780, belonging to the same and similarly blown. The mean value of *a'* on both was 425·2 vib.

Line 4 gives those observed on accurate copies of a set of tuning-forks (*b'* and *c''* missing) belonging to the bellfoundry Cavedini at Verona, supposed to be a century old, and preserved with great care, having *a'* 423·2.

Line 5 gives the cents from an octave of pipes on Green's organ at St. Katharine's, Regent's Park, from the pitches determined by me in 1878, up to which time it was one of the few organs tuned in meantone temperament. Of course in this case the tuning was modern.

Specimens of Tuning in Meantone Temperament.

Notes	C	C♯	D	E♭	E	F	F♯	G	G♯	A	B♭	B	C
1	0	76	193	310	386	503	579	697	773	890	1007	1083	1200
2	0	102	216	308	414	516	619	723	827	920	1031	1126	1232
3	0	134	236	329	443	561	626	701	814	900	982	1075	1175
4	0	117	229	301	439	507	633	733	864	934	1042	—	1175
5	0	72	198	300	382	509	586	699	785	898	1020	1096	1209

¶ Art. 3. Meantone Fifths, if properly tuned, should have 696·6 cents, and the major Thirds 386·3 cents. The old tuners did not use the Fourth in tuning, but took, for example, *c'* to *g'*, then *g'* to *d''*, and then the Octave down to *d'*, thence the Fifths *d'* to *a'*, *a'* to *e''*, and the Octave down to *e'*, so that *c'* to *e'* ought to be a just major Third. The Fifths and major Thirds actually obtained can be calculated from the Table in Art. 2 by subtraction, taking care to increase the minuend by 1200 when it is less than the subtrahend. We thus find the following values. The figures placed between the names of any two notes give the cents in the interval between them, which, neglecting decimals, should be 697 for Fifths, and 386 for major Thirds.

Line 2. Old Pitch-pipe. Fifths: *E♭* 723 *B♭*

685 *F* 684 *C* 723 *G* 693 *D* 704 *A* 694 *E* 712 *B* 693 *F♯* 683 *C♯* 725 *G♯*.

Major Thirds: *E♭* 415 *G* 403 *B*; *B♭* 385 *D* 403 *F♯*; *F* 404 *A* 382 *C♯*; *C* 414 *E* 413 *G♯*.

Line 3. Another old Pitch-pipe. Fifths: *E♭* 653 *B♭* 779 *F* 614 *C* 701 *G* 735 *D* 664 *A* 743 *E* 632 *B* 751 *F♯* 708 *C♯* 780 *G♯*.

Major Thirds: *E♭* 372 *G* 374 *B*; *B♭* 454 *D* 390 *F♯*; *F* 339 *A* 434 *C♯*; *C* 443 *E* 471 *G♯*.

Line 4. Old Forks. Fifths: *E♭* 741 *B♭* 665 *F* 693 *C* 733 *G* 696 *D* 705 *A* 705 *E*; (*B* missing); *F♯* 684 *C♯* 747 *G♯*.

Major Thirds: *E♭* 432 *G*; (*B* missing); *B♭* 387 *D* 404 *F♯*; *F* 427 *A* 383 *C♯*; *C* 439 *E* 425 *G♯*.

These old tunings are very imperfect. Both Fifths and major Thirds would make dreadful harmony. The forks are if anything worse than the pitch-pipes.

Line 5. St. Katharine's Organ. Fifths: $E\flat$ 720 $B\flat$ 689 F 700 C 699 G 699 D 700 A 684 E 714 B 690 $F\sharp$ 686 $C\sharp$ 713 $G\sharp$.
Major Thirds: $E\flat$ 399 G 397 B; $B\flat$ 378 D 388 $F\sharp$; F 389 A 374 $C\sharp$; C 382 E 403 $G\sharp$. This modern tuning is better, but not good.

The Fifths, which should be a quarter of a comma 5·4 cents flat, are often sharp; the major Thirds are unequal. But the errors here are not more than might be reasonably expected from tuning by ear.

Art. 4.—It takes a quick man three years to learn how to tune a piano well in equal temperament by estimation of ear, as I learn from Mr. A. J. Hipkins. Tuners have not time for any other method. The following are good examples:—

Line 1. The theoretical intervals, all exact hundreds of cents.
Line 2. My own piano, tuned by one of Broadwoods' usual tuners, and let stand unused for a fortnight.
Lines 3, 4, 5. Three grand pianos by Broadwoods' best tuners, prepared for examination through the kindness of Mr. A. J. Hipkins, of that house.
Line 6. An organ tuned a week previously by one of Mr. T. Hill's tuners, and used only

once, examined by the kind permission of Mr. G. Hickson, treasurer of South Place Chapel, Finsbury, where the organ stood.
Line 7. An harmonium tuned by one of Messrs. Moore & Moore's tuners, kindly prepared for my examination.
Line 8. An harmonium, used as a standard ¶ of pitch, tuned a year previously by Mr. D. J. Blaikley (p. 97d), by means of accurately counted beats, &c., with a constant blast, put at my disposal for examination by Mr. Blaikley.

Specimens of Tuning in Equal Temperament.

Notes	C	$C\sharp$	D	$D\sharp$	E	F	$F\sharp$	G	$G\sharp$	A	$A\sharp$	B	C
1	0	100	200	300	400	500	600	700	800	900	1000	1100	1200
2	0	96	197	297	392	498	590	700	797	894	990	1089	1201
3	0	99	200	305	411	497	602	707	805	902	1003	1102	1206
4	0	100	200	300	395	502	599	702	800	897	999	1100	1200
5	0	101	199	299	399	500	598	696	800	899	999	1100	1201
6	0	101	192	297	399	502	601	702	806	898	1005	1099	1201
7	0	98	200	298	396	498	599	702	800	898	999	1099	1199
8	0	100	200	300	399	499	600	700	800	900	1001	1099	1200

¶

These were all tuned by the modern way of Fifths up and Fourths down, and the object is to make the Fifth up 2 cents too close, and the Fourth down 2 cents too open. As this interval of 2 cents lies on the very boundary of perception by ear, the difficulty of tuning thus without attending to the beats is enormous. The above figures in lines 2, 3, 4, 5 shew how very close an approximation is now possible in pianofortes.
Art. 5.—The order of tuning differs in different houses. Messrs. Moore & Moore's tuners set c' by a c'' fork, and then tune in order: $c' g d' a e' b f\sharp c'\sharp g\sharp d'\sharp$. Then begin again and go on as $c' f bb e'b$. The proof of the work is that $e'b$ and $d'\sharp$ are identical. Messrs. Broadwoods' tuners also set c' from c, but then proceed thus: $c' g d' a e' b f\sharp c'\sharp g\sharp d'\sharp a\sharp f c'$, the proof being that the final agrees with the initial c'. In this case $a\sharp f$ is taken as $bb f$

or $a\sharp c\sharp$, that is a Fourth down. Observe that the tuning in both cases takes place in the Octave f to f', for which the beats of disturbed Fifths and even of disturbed Fourths are very slow. This arises from the great prominence of the second partial tone in this region on pianoforte notes. In taking the pitch of each note, I found that d', $d'\sharp$, e' taken as disturbed unisons, beat with my forks much less distinctly than f, $f\sharp$, &c., to $c'\sharp$ as disturbed Octaves. Now the above table enables us to calculate the cents in the Fifths and Fourths actually tuned, which were the intervals estimated by ear. I take only line 1 as containing the theoretical intervals, and lines 2 to 5 as being by Broadwoods' tuners, so that the order is certain. The numbers of cents placed between two notes shews the interval, all the ¶ Fourths being taken down and the Fifths up.

Pianoforte Tuning—Fourths and Fifths.

	c'500	g700	d'500	a700	e'500	b500	$f\sharp$ 700	$c'\sharp$ 500	$g\sharp$ 700	$d'\sharp$ 500	$a\sharp$ 500	f700	c'
2	500	697	503	698	503	499	706	499	700	507	497	703	
3	493	693	498	709	509	506	697	594	708	502	506	709	
4	498	698	503	698	495	501	701	500	700	501	497	698	
5	504	703	500	700	500	502	703	501	700	501	498	700	

These examples must probably be considered the best that pianoforte tuning by ear can accomplish. But even in line 5, which is the best, there are only five intervals absolutely correct, two others are only an inappreciable 1 cent in error, two are a just appreciable 2 cents wrong, two are 3 cents out, and one

wrong by the very perceptible interval of 4 cents. Now if this is the work of a clever tuner in constant practice for many hours daily for many years, in tuning one kind of temperament only, what are we to expect from those who attempt to realise new intervals?

Art. 6.—The vocalist does not, properly speaking, tune at all. It is with him a matter of ear, that is, sense of pitch, which guides the muscles to alter the tension of his vocal chords, and make them produce tones of various pitch. The ease and rapidity with which this can be done are matters of careful training, followed by long practice. They can never be acquired by those who have not the proper cerebral organisation. The extreme mobility of the voice and the difficulty of sustaining a pitch, or of exactly reaching it again after a pause, throw great impediments in the way of testing unaccompanied singers. The habit of choral singing leads to just intonation (App. XVIII.), but an accompanying instrument is quite sufficient to lead the voice astray (p. 207, note †). Hence I pass by voices altogether. The violinist apparently tunes only four strings to make three perfect Fifths, and in doing so he is assisted by an audible combinational tone, which should be just one Octave below the lower note of the pair he is tuning, and hence a Twelfth below the upper note. But he really tunes every note in the compass of his instrument g to e'''', by his method of stopping, as much as the pianoforte tuner by increasing or diminishing the tension of his strings. And according to the 'school,' he should tune the notes in equal temperament. How then does he tune? Or, to put the question in more usual language, what intonation does he use? Messrs. Cornu and Mercadier (as mentioned on p. 325d, note †) instituted a series of experiments on voices and organs by the phonautograph (*Comptes Rendus*, vol. lxviii. pp. 301 and 427), and on violins and violoncellos by means of a tin plate placed under the bridge, which was connected with a wire that conducted the vibrations to the inscribing style (see *Comptes Rendus*, 17 July 1871, vol. lxxiii. p. 178, from which, and vol. lxxiv. p. 321, vol. lxxvi. p. 432, I obtained the data which I have here reduced to cents). I give the results only for different individual players on the violin and violoncello. Some of the scales are fragmentary. I prefix the just, Pythagorean, and equal number of cents for comparison. The scales are first major and then minor. The root is omitted as unnecessary. The numbers in one line refer to a single trial.

Scale of C Major.

Notes	D	E	F	G	A	B	c
Just cents	204	386	498	702	884	1088	1200
Pyth. „	204	408	498	702	906	1110	1200
Equal „	200	400	500	700	900	1100	1200
Violin Amateurs		401	505	701			
	212	412	498	709			
	212	399	499	710			
	198	396		708			
	210	406	505	702			
	188	401		702			
	201	406	490	711			
		411	490				
	201	396	495	705			
	194	411		704			
		411					
		404		702			
Violoncello Amateurs		415		710			
		395					
	208	397	507	710			
		408					
		411					
		401		712			
		407		705			
		404					
	204	415	511	701		1109	1202
	201		511	702		1105	
	198	415	500				
	207	414	498	701		1120	1202
	202	403	499	712		1117	1201

Notes	D	E	F	G	A	B	c
Just cents	204	386	498	702	884	1088	1200
Pyth. „	204	408	498	702	906	1110	1200
Equal „	200	400	500	700	900	1100	1200
Violin Professional (M. Léonard, Belgian)		411	490				
		417	495	714			
		417					
		411					
	207				910	1128	
	216				912	1122	
						1128	
						1128	
	207	407					
		411					
		401			899		
		401					
	207	401					
	199	407		696			
	209	404					
Violoncello Professional (M. Séligmann)		393	487	702			
		407					
		419	496	716			
		411					
		399	491	702			
			496				
		414		695	892		
		407		696	896		
		403		697	907		

Scale of C Minor.

Notes	d	eb	f	g	ab	b	c'
Just cents .	204	316	498	702	814	1088	1200
Pyth. „ .	204	294	498	702	792	1110	1200
Equal „ .	200	300	500	700	800	1100	1200
Violin		292		697			1208
Amateurs		281		708			
		295		702			1210
		295					
		294					1196
		301		705			1206
		298		702			1203
	202	300		701			
	196	300	505				
	195	292	498	707			
		292					
		303					
				705	782		
					783		
				709	802	1118	1196
					802		
				696	798	1111	1209
				712	798	1118	
				704	780	1101	1197
				712	783	1102	

Notes	d	eb	f	g	ab	b	c'
Just cents .	204	316	498	702	814	1088	1200
Pyth. „ .	204	294	498	702	792	1110	1200
Equal „ .	200	300	500	700	80c	110c	1200
Violin	199	295	494	700	810		
Professional	207	311	499	701	793	1113	
(M. Ferrand, of	204	298	503	698	793	1116	1201
the Opéra	206	306	492	700	794		
Comique)	204	298					

Art. 7.—Messrs. Cornu and Mercadier conclude finally (*ibid.* vol. lxxvi. p. 434) that:—

'Musical intervals belong to at least two different systems of different values:

'1) The intervals employed in melodies which have no modulations agree with those of the Pythagorean scale.

'2) The intervals between two notes *sounded together* in chords, the basis of harmony, have for their ratios the following numbers: 2 for Octave, $\frac{3}{2}$ Fifth, $\frac{4}{3}$ Fourth, $\frac{5}{4}$ major Third, $\frac{6}{5}$ minor Third, $\frac{5}{3}$ major Sixth, $\frac{8}{5}$ minor Sixth, and $\frac{7}{4}$ Seventh, where the Fourth and Sixths were deduced from observation of the Fifth and Thirds, and the Seventh from the dominant chord.'

Thus for unaccompanied harmony of two tones (chords more than two tones were not tried) just intonation *alone* was used. For melody the major Thirds, major Sixths, and especially the major Sevenths (leading notes) were much sharpened, and the minor Thirds and minor Sixths generally much flattened. But did this arise from the custom of equal temperament? (as M. Guéroult thinks, *ibid.* 9 May 1870, vol. lxx. p. 1037, to which Messrs. Cornu and Mercadier replied, on 30 May 1870, vol. lxx. p. 1170) or really from the feeling of Fifths? The latter was impossible for the leading note, which is sometimes much sharper than in Pythagorean intonation, and the Fifths played were by no means always true. Messrs. Cornu and Mercadier say that the divergence from the mean only reaches $\frac{1}{3}$ of a comma, that is, about 7 cents; but as the Pythagorean major Sixth, major Third, and major Seventh differ from the corresponding equal tempered intervals by only 6, 8, and 10 cents respectively, this uncertainty renders it impossible to decide whether the scale played was intentionally equal or Pythagorean, or whether even it did not vary with the feeling of the moment. Taking into consideration that the pitches actually shewn in the tables vary considerably, that they very rarely repeat themselves, that the notes are sometimes flatter and sometimes sharper than either just or Pythagorean intonation, and that this uncertainty pervades even such intervals as the Fourth, Fifth, and Octave, I am inclined to adopt the hypothesis of an intentionally variable intonation. Whether founded on the feeling of Pythagorean or equal temperament, it is difficult to decide. But it is certainly not founded on any feeling of just intonation for harmony. If then these players, as Messrs. Cornu and Mercadier assert from first to last in the unmistakable terms already cited, adopt just intonation of intervals for harmony, a serious question arises as to how they treat the relations of tonality. The first part may lead and the others may be adapted to it, or the bass may determine the intonation of the other parts. In either case there would be a great variability, through which modulation, and even the adjustment of parts without much previous combined practice, would become extremely difficult; see pp. 208c and note *, 324b, c, d. But how about the return to the same key after modulations (p. 328b)? Huyghens (*Cosmotheōros,* lib. i. p. 77, as cited by Dr. Smith, *Harmonics,* 2nd ed. p. 228) suggests that as 'erring from the pitch first assumed . . . would greatly offend the ear of the musician, he naturally avoids it by his memory of pitch, and by tempering the intervals of the intermediate sounds, so as to return to it again.' But how accurately does he remember the original pitch? In some cases in the above observations of Cornu and Mercadier the Octave (and hence the original pitch) will be seen to be sharpened or flattened by more than 2 cents. In the course of a few modulations this 2 cents might easily become 22 or a comma. Where is the guarantee for remembering the original pitch? Would an alteration of a whole comma in passing through a series of modulations be noticed? We know indeed that unaccompanied singers constantly 'flatten' by much more than a comma. The Duodenarium simply shews what sounds ought to be played in modulations, and would be played on instruments with fixed tones properly arranged; not what intervals are *now* played and sung, and mere memory of ear does not suffice.

Art. 8.—Scheibler's method of tuning instruments was theoretically perfect. It consisted in tuning by his fork tonometer a series of forks 4 vib. flatter than the pitches required. The string, pipe, or reed was then tuned sharper than the fork by 4 beats in a second. (See p. 446*b* for Koenig's forks for this purpose.) This only applied to one octave, and perhaps the octave below; the others were tuned from them by estimation of Octave. The errors thus made were a minimum, but
¶ there was the obvious disadvantage of having to tune a new set of forks for each pitch desired. This Scheibler overcame by tuning auxiliary octaves on organs, and counting beats by rather a troublesome rule, entailing the need of an accurate metronome. Even then he only taught how to tune in equal temperament. For practical purposes we require not only the equal but the meantone temperament, and also the 53 division of the octave. The only sure way is by calculating the pitch number of each note, and thus determining the beats between the note and the forks of a tuning-fork tonometer. This method may be dismissed as generally impracticable. So may any method which depends upon accurately knowing the pitch number of the tuning-note. What is required is a method of tuning at any unknown pitch which an organ or piano may happen to possess at the moment within the limits of, say, $c'256$ and $c'270$ without determining exactly what that pitch really is. This would save the great trouble of entirely altering the pitch of the piano (never very certain in its results), and the still greater trouble and expense of entirely altering the pitch of an organ or harmonium.

Art. 9.—For this purpose I invented an approximative method (given on p. 785
¶ of the 1st edition of this translation and subsequently communicated to the *Musical Times*), which I here subjoin in an improved form. It is based on the result of Prof. Preyer's investigations (suprà, p. 147*d'*) that errors of $\frac{1}{5}$ or ·2 of a vibration cannot be heard in any p*r*t of the scale, so that any attempt to tune more accurately is labour thrown aw*i*y. Moreover, even at high pitches ·3, ·4, and ·5 vib. are scarcely perceptible in melody and quite inoffensive in harmony. It will be found very difficult when the beats are less than 4 in 10 seconds, that is, when the error is less than ·4 vib., to count them with any approach to accuracy. But it is only by beats that we can work effectively.

Any one who undertakes tuning should learn to estimate the meaning of 6, 10, 15, 20, 30 beats in 10 seconds. This is best done by short pendulums constructed of a piece of thread with one end tied to a curtain ring, and the other passed through a slit in a piece of firewood, round which it is ultimately tied.
¶ The stick is put under a book by the side of a table, so that the pendulum swings freely. Measure the length of the string from the centre of the ring to the beginning of the wood, which allows of an easy alteration of the length by drawing the string through the slit. Make this length $9\frac{7}{8}$ inches. The pendulum swings backwards and forwards 120 times in 60 seconds, and hence 20 times in 10 seconds. Adjust it more accurately by a seconds watch. Counting the swings one way only, there are 10 double swings in 10 seconds. By watching and counting this, say for half an hour, the tuner will learn to feel the rate of these two sets of swings. Then make another pendulum with a length of string of $4\frac{3}{8}$ inches measured as before. This vibrates much more quickly, making 180 single and hence 90 double swings in 60 seconds, and consequently 30 single and 15 double swings in 10 seconds. Finally make the length $27\frac{3}{16}$ inches, always from the centre of the ring to the stick, and the pendulum will make 6 double and 12 single vibrations in 10 seconds. Remember that if you begin counting with *one*, you will end with *seven* for 6, *eleven* for 10, *sixteen* for 15, and so on, so that you will always have to throw off one from your count.

Art. 10.—The rule has to be arranged in several forms according to the custom of tuning instruments. Harmoniums are best tuned from c' to c'', that is, in the two-foot octave. Organs are generally tuned in the principal stop, so that on touching the keys from c' to c'', the sounds are from c'' to c''', in the one-foot octave, and hence the beats are twice as fast. But for pianos it is the custom to tune from f to f' (see art. 5). Most tuners in England begin with c'', from which

c' is 'set,' and then the tuning commences. Some tuners in England and all abroad begin with a'. This makes no difference in the rule, provided the tuning octave remains the same.

Absolutely the beats arising from imperfect Fifths and Sixths vary for every difference of pitch of the lower note. As the Fifth is always too close and the Fourth too open, the reader can find the beats from the numbers in the table, p. 437c, d, by subtracting twice the larger from three times the smaller pitch number for Fifths (thus $c' : g'$ in col. ii. of the table is $258·6 : 387·6$, whence $3 \times 258·6 - 2 \times 387·6 = 775·8 - 775·2 = ·6$, giving 6 beats in 10 seconds), and four times the smaller pitch from three times the larger for Fourths (thus $g' : d' = 387·6 : 290·3$, whence $3 \times 387·6 - 4 \times 274 = 1162·8 - 1161·2 = 1·6$ or 16 in 10 seconds). But for the purposes of the rule all the beats of Fifths are supposed to be the same throughout the tuning octave, and similarly all the beats of Fourths are assumed to be the same. The errors will be found to correct each other, and in no case to exceed the permissible limits.

Art. 11.—*Rule for tuning in equal temperament at any pitch between $c'256$ and $c'270·4$.*

Tune in the following order, making the Fifths closer and the Fourths wider ¶ than perfect. The numbers between the names of the notes indicate the beats in 10 seconds.

For harmoniums :

c' 10 g' 15 d' 10 a' 15 e' 10 b' 15 $f\sharp$ 15 $c'\sharp$ 10 $g'\sharp$ 15 $d'\sharp$ 10 $a'\sharp$ 15 f'

For organs, using the metal principal, sounding thus an Octave higher than the digitals shew :

c'' 20 g'' 30 d'' 20 a'' 30 e'' 20 b'' 30 $f''\sharp$ 30 $c''\sharp$ 20 $g''\sharp$ 30 $d''\sharp$ 20 $a''\sharp$ 30 f''

For pianofortes :

c' 10 g 6 d' 10 a 6 e' 10 b 10 $f\sharp$ 6 $c'\sharp$ 10 $g\sharp$ 6 $d'\sharp$ 10 $a\sharp$ 10 f

On the piano the beats can often be heard for only 5 seconds, and then the beats will be 3 and 5 in 5 seconds, in place of 6 and 10 in 10 seconds.

In the two first cases the intervals beating 10 in 10 seconds are all Fifths up, those beating 15 in 10 seconds are Fourths down; in the last case the Fifths up beat only 6 times in 10 seconds, and the Fourths down beat 10 times in 10 seconds.

Tune each Fifth as accurately *just*, or without beats, as possible, and *then* make the interval *closer* by *flattening* the *upper* note very slightly indeed till 10 beats are heard in 10 seconds. Then from the Fifth thus reached tune a Fourth down as accurately *just*, or without beats, as possible, and *then* make the interval *opener* by *flattening* the *lower* note very slightly till 15 beats are heard in 10 seconds. From the note thus gained proceed to the next until f' is reached. The Fourth f' ¶ to c' is not tuned, as both notes have been determined. It never beats faster than 15 in 10 seconds.

If the Fifths and Fourths are not brought to be as nearly as possible just in the first instance the tuner can never be sure whether the second note of the interval is too sharp or too flat, because the error itself is too small to be judged of with accuracy on merely sounding the notes in succession, and the same number of beats would result whether we had sharpened or flattened the note, but the whole scheme would be entirely frustrated if the interval were rendered *opener* instead of *closer* or conversely.

Art. 12.—If it is preferred to commence on a', set a' to fork and proceed to e', b' up to f', &c., as in the regular scheme. Stop at f' and begin again at a', and tune the Fifth a' to d' *down*, first making the interval just, and then making it ¶ closer by *sharpening* the lower note till 10 beats in 10 seconds result. Next take d' to g' a just Fourth down, and then make the interval opener by *flattening* the lower note till 15 beats are heard in a second. Lastly, from this g' take c' a Fifth lower, and after making it just, render the interval closer by *sharpening* the lower note till 10 beats in 10 seconds are heard. This modification is merely an adaptation of the general principle that Fifths are to be closer than just, and Fourths opener than just.

The tuner should carefully familiarise himself with tuning just Fifths and Fourths, with recognising them as just by their total absence of beats in this 2-foot Octave, and by feeling how beats arise by altering either of the notes either way. On the harmonium this is easy, when the just interval has been tuned. It is only necessary to press down the digital of the upper note slightly, so as just to hear the note; this process flattens it and renders the interval *closer*; or to do the same with the lower note, which renders the interval *opener*. In either case beats ensue. As many just Fourths and Fifths exist on the Harmonical, this experiment is ready to hand.

Art. 13.—The proof of my rule consists in shewing that for $c'256$, $a'435$, $a'440$, $c'270·4$, it leads to results which no ear could distinguish from perfectly correct tuning, or which are equal at least to the very best in art. 4. For it is clear that if it holds for all these pitches, of which one is practically the lowest and one practically the highest in use, and the other two are as nearly as may be halfway between them, it must hold for all intermediate pitches. Now the results obtained by the rule are easily calculated for a given c' or a'. Using c', $c'\sharp$, d', &c., for the pitch numbers of these notes, and remembering that the rule gives 1 beat in a second for flat Fifths and 1·5 beats in a second for sharp Fourths, we have, for harmoniums or organs, supposing c' known, and the above rule accurately followed,

¶

$$2g' = 3c' - 1$$
$$4d' = 3g' - 1\cdot5$$
$$2a' = 3d' - 1$$

$$4e' = 3a' - 1\cdot5$$
$$2b' = 3e' - 1$$
$$4f'\sharp = 3b' - 1\cdot5$$
$$4c'\sharp = 3f'\sharp - 1\cdot5$$
$$2g'\sharp = 3c'\sharp - 1$$
$$4d'\sharp = 3g'\sharp - 1\cdot5$$
$$2a'\sharp = 3d'\sharp - 1$$
$$4f' = 3a'\sharp - 1\cdot5.$$

If we begin with a', then the three first equations will give

$$3d' = 2a' + 1, \quad 3g' = 4d' + 1\cdot5, \quad 3c' = 2g' + 1.$$

In the case of the pianoforte, using the Octave f to f', the equations for calculating the pitches of the notes from beats will be, if we begin with c',

$$4g = 3c' - 1$$
$$2d = 3g - \cdot6$$
$$4a = 3d - 1$$

$$2e' = 3a - \cdot6$$
$$4b = 3e' - 1$$
$$4f\sharp = 3b - 1$$
$$2c'\sharp = 3f\sharp - \cdot6$$
$$4g\sharp = 3c'\sharp - 1$$
$$2d'\sharp = 3g\sharp - \cdot6$$
$$4a\sharp = 3d'\sharp - 1$$
$$4f = 3a\sharp - 1.$$

But if we begin with a, the first three equations will give

$$3d = 4a + 1, \quad 3g = 2d + \cdot6, \quad 3c' = 4g + 1.$$

In the following calculations the rule is necessarily supposed to be carried out with perfect accuracy. Of course in practice, especially when the beats are estimated instead of being accurately counted, this is impossible. But the results will be found much more accurate than in the ordinary way of tuning entirely by estimations of ear, and the rule much more easy to manipulate than the ordinary method of tuning.

Proof of Rule for Tuning in Equal Temperament.

¶
FOR HARMONIUMS AND ORGANS.

Notes	c	$c'\sharp$	d'	$d'\sharp$	e'	f'	$f'\sharp$	g'	$g'\sharp$	a'	$a'\sharp$	b'	c''
c'256 Equal	256·0	271·2	287·4	304·4	322·6	341·7	362·0	383·6	406·4	430·5	456·1	483·3	512·0
By Rule	256·0	271·1	287·3	304·3	322·4	341·5	362·0	383·5	406·1	430·4	455·9	483·1	512·0
a'435 Equal	258·7	274·0	290·3	307·6	325·9	345·3	365·8	387·6	410·6	435·0	460·9	488·3	517·3
By Rule	258·7	273·9	290·3	307·4	325·9	345·0	365·8	387·6	410·3	435·0	460·6	488·1	517·3
a'440 Equal	261·6	277·2	293·7	311·1	329·6	349·2	370·0	392·0	415·3	440·0	466·2	493·9	523·3
By Rule	261·6	277·2	293·6	311·1	329·6	349·2	370·1	391·9	415·2	440·0	466·1	493·9	523·3
c'270·4 Equal	270·4	286·5	303·5	321·6	340·7	360·9	382·4	405·2	429·2	454·8	481·8	510·4	540·8
¶ By Rule	270·4	286·5	303·5	321·5	340·6	361·0	382·5	405·1	429·2	454·7	481·8	510·4	540·8

FOR PIANOS.

Notes	f	$f\sharp$	g	$g\sharp$	a	$a\sharp$	b	c'	$c'\sharp$	d'	$d'\sharp$	e'	f'
Equal	170·9	181·0	191·8	203·2	215·3	228·1	241·7	256	271·2	287·4	304·4	322·6	341·7
By Rule	170·9	181·0	191·8	203·2	215·2	228·2	241·7	256	271·2	287·3	304·6	322·6	341·7
Equal	180·5	191·2	202·6	214·6	227·4	240·9	255·2	270·4	286·5	303·5	321·6	340·7	361·0
By Rule	180·6	191·3	202·6	214·7	227·4	241·1	255·4	270·4	286·6	303·5	321·8	340·8	361·2

Hence it is apparent that the rule never makes an error exceeding ·3 vib., and generally keeps below this limit. Now at 256 an error of ·3 vib. amounts to 2 cents, and at 540 to less than 1 cent. The rule, therefore, properly handled will give results equal if not superior to the specimens in Art. 4.

Art. 14.—The rule applies only to one octave and gives what are known as 'the bearings,' whence the other notes must be derived by taking Octaves in the usual way.

Tuners so frequently get out in taking the Octaves that it is convenient to have a check on the estimation of ear. This is furnished by the fact that any note will beat the same number of times in a second with an imperfect Fourth below and its Octave (that is, an imperfect Fifth) above. Thus if the note have 401 vib., its imperfect Fourth below 300, and the Octave above that Fourth below 600, the beats of the Fourth are $3 \times 401 - 4 \times 300 = 3$, and the beats of the Fifth are $3 \times 401 - 2 \times 600 = 3$ also. Now the imperfect Fourths are furnished by the bearings themselves. Thus, going upwards, we have $c'\, f'\, c''$, $c''\sharp\, f'\sharp\, c''\sharp$, $d'\, g'\, d''$, &c. Going downwards, the tuner takes a Fifth and then a Fourth, as $b'\, e'\, b$, $b'\flat e'\flat b\flat$, $a'\, d'\, a$, and so on from octave to octave. Mr. Hermann Smith prefers to insert the octaves above when tuning the original bearings. Thus, if the bearings were taken in the two-foot octave c' as a' 10 d' 15 g' 10 c' 15 f' 15 $a'\sharp$ 10 $d'\sharp$ 15 $g'\sharp$ 10 $c'\sharp$ 15 $f'\sharp$ 15 b' 10 e', he would introduce the octaves in this order, $a\, a'\, d'\, d''\, g'\, g\, c'\, c''\, f'\, f\, a\sharp\, a'\sharp\, d'\sharp\, d''\sharp\, g'\sharp\, g\sharp\, c'\sharp\, c''\sharp\, f'\sharp\, f\sharp\, b\, b'\, e'\, e''\, a'$. But the method I have proposed seems simpler.

The principle of the check applies to the inversions of other imperfect intervals, and may serve as additional verifications. Thus a note beats equally with an imperfect minor Sixth below and its Octave the imperfect major Third above. Thus $500 : 801$ beat $5 \times 801 - 8 \times 500 = 8$, and $801 : 1000$ beat $5 \times 801 - 4 \times 1000 = 8$ also. Again, a note beats equally with an imperfect minor Third below and its Octave the major Sixth above. Thus $500 : 601$ beat $5 \times 601 - 6 \times 500 = 5$, and $601 : 1000$ beat $5 \times 601 - 3 \times 1000 = 5$ also. If in each case we inverted the order, we should double the beats. Thus for the Fifth, $200 : 301$ will beat $2 \times 301 - 3 \times 200 = 2$; but $301 : 400$ will beat $4 \times 301 - 3 \times 400 = 4$. For the major Third, $400 : 501$ will ¶ beat $4 \times 501 - 5 \times 400 = 4$; but $501 : 800$ will beat $8 \times 501 - 5 \times 800 = 8$. For the major Sixth, $300 : 501$ will beat $3 \times 501 - 5 \times 300 = 3$; but $501 : 600$ will beat $6 \times 501 - 5 \times 600 = 6$. The reason is obvious. The fractions expressing the minor intervals $\frac{4}{5}, \frac{8}{5}, \frac{6}{5}$ have odd denominators and even numerators, and hence their inversions reduce by dividing by 2, but this is not the case for the major intervals, $\frac{3}{2}, \frac{5}{4}, \frac{5}{3}$.

So much of the beauty of tuning pianos, harmoniums, and organs depends on the perfection of the Octave, that tuners would do well to apply the first test with the Fourth below and Fifth above, as a matter of course.

Art. 15.—*Rule for tuning in meantone temperament from c' 252·7, Handel's pitch, to c' 283·6, Father Smith's pitch for the Durham organ.*

As will be seen by the table p. 434b, c'264 or Helmholtz's pitch is a small meantone Semitone, and the Durham pitch is a meantone Tone, sharper than Handel's. The great flatness of the Fifths in the meantone intonation makes it necessary to divide the tunings into three classes, sufficiently ascertainable by a fork in Helmholtz's or even in French pitch. The first is from rather less than a Semitone to about a Quartertone flatter than French pitch; the second is French pitch and from a Quartertone flatter to a Quartertone sharper;

the third is from a Quartertone to a Semitone ¶ sharper than French.

The rule would extend to 27 notes, but on the proof it will be carried out only for 14 notes, as on the English concertina, for which this intonation is still used. And for this instrument the 'tuning octave' may be taken as c' to c''. For the few organs that still use this intonation the same Octave must be tuned, and hence, when taken on the 'principal,' the digitals must be fingered from c to c', because the beats would be otherwise too rapid to count.

Tune in the following order the numbers 25–6–7 and 40–1–2, meaning that the beats are to be 25 and 40 for the low, 26 and 41 for the medium, and 27 and 42 for the high pitch in 10 sec., according to the three grades already laid down.

c' 25–6–7 g' 40–1–2 d' 25–6–7 a' 40–1–2 e' 25–6–7 b' 40–1–2 $f'\sharp$ 40–1–2 $c'\sharp$ 25–6–7 $g'\sharp$ 40–1–2.   ¶

c' 40–1–2 f' 40–1–2 $b'\flat$ 25–6–7 $e'\flat$ 40–1–2 $a'\flat$.

The tuning is conducted in the same way as before, only in two series, from c' to $g'\sharp$ and from c' to $a'\flat$, making the Fifths and Fourths at first perfect, and then the Fifths closer and the Fourths wider. But in two cases $c'f', f'b'\flat$ the Fourth is taken upwards, and then the upper note has to be sharpened. And there is an additional verification after tuning to e', for the major Thirds $c'e', g'b', d'f'\sharp, a'c'\sharp, e'g'\sharp$, and also $f'a'$ and minor Sixths $d'b'\flat, c'a'\flat$ should be all sensibly perfect. Octaves can be verified by imperfect Fourths more easily than in equal temperament.

The equations for determining the pitch from the beats are

$2g'$	$= 3c'$	$-2 \cdot 5 - \cdot 6 - \cdot 7$
$4d'$	$= 3g'$	$-4 \cdot 0 - \cdot 1 - \cdot 2$
$2a'$	$= 3d'$	$-2 \cdot 5 - \cdot 6 - \cdot 7$
$4e'$	$= 3a'$	$-4 \cdot 0 - \cdot 1 - \cdot 2$
$2b'$	$= 3e'$	$-2 \cdot 5 - \cdot 6 - \cdot 7$
$4f'\sharp$	$= 3b$	$-4 \cdot 0 - \cdot 1 - \cdot 2$
$4c'\sharp$	$= 3f'\sharp$	$-4 \cdot 0 - \cdot 1 - \cdot 2$
$2g'\sharp$	$= 3c'\sharp$	$-2 \cdot 5 - \cdot 6 - \cdot 7$
$4d''\sharp$	$= 3g'\sharp$	$-4 \cdot 0 - \cdot 1 - \cdot 2$

$3f'$	$= 4c'$	$+4 \cdot 0 - \cdot 1 - \cdot 2$
$3b'\flat$	$= 4f'$	$+4 \cdot 0 - \cdot 1 - \cdot 2$
$3e'\flat$	$= 2b'\flat$	$+2 \cdot 5 - \cdot 6 - \cdot 7$
$3a'\flat$	$= 4e'\flat$	$+4 \cdot 0 - \cdot 1 - \cdot 2$

Proof of the Rule for Tuning in Meantone Temperament.

Notes	c'	c'♯	d'	d'♯	e'♭	e'	f'	f'♯	g'	g'♯	a'♭	a'	b'♭	b'	c''
Handel .	252·7	264·1	282·5	295·2	302·8	315·9	338·0	353·2	377·9	394·9	404·3	422·5	452·1	472·4	505·4
By Rule .	252·7	264·1	282·4	295·1	302·4	315·9	338·3	353·4	377·8	394·8	404·6	422·5	452·3	472·5	505·4
Helmholtz .	264·0	275·9	295·2	308·4	315·8	330·0	353·1	369·0	394·8	412·5	422·4	441·4	472·3	493·5	528·0
By Rule .	264·0	275·8	295·0	308·3	315·9	329·9	353·4	369·1	394·7	412·5	422·6	441·2	472·6	493·6	528·0
Durham .	283·6	296·3	317·1	331·3	339·2	354·5	379·2	396·3	424·0	443·1	453·7	474·1	507·3	530·1	567·2
By Rule .	283·6	296·6	317·0	331·6	339·2	354·6	379·5	396·9	424·1	443·5	453·7	474·2	507·4	530·6	567·2

The results are seen to be nearly as good as before, only the Durham $f'♯$ is 396·9 in place of 396·3, an error of ·6 vib. and 2·6 cents, which is scarcely perceptible.

Art. 16.—In the *Proceedings of the Musical Association* for 1874–5, vol. i. ¶ pp. 141–145, Mr. Bosanquet gives the process that he followed in tuning his cycle of 53 (see sect. F., p. 479), but it is too complicated to be abstracted with a reasonable hope of its being understood. It depended mainly on taking the beats of the differential tones in the major chord, with the major Third either in the middle or highest. In this case the Fifth was presumed to be accurately tuned to begin with. Taking the numbers in sect. A. art. 27, but doubling them for the Octave higher and supposing the Fourth to be perfect, we have

$$c''528, \quad f''704, \quad a''879\cdot32 \text{ in the cycle.}$$

Differentials   176   175·32

Beats per second   ·68   , that is, beats per minute 40·8

and the beats were apparently counted for a minute. Mr. Bosanquet, however, does not recommend the process for his harmonium because the differential tones were not distinct enough.

Art. 17.—Mr. Paul White has two methods of tuning the cycle of 53 by means ¶ of beats. We may suppose that for the given pitch of the initial *c*, all the pitch numbers have been calculated, as in sect. A, art. 27, for No. 4, Mr. Bosanquet's initial $C=264$ vib. Two places of decimals are required for this purpose, as in both methods it is necessary to rely upon the slow beats of the Thirds and Sixths and check the result at every few steps.

First method. Tune 5 minor Thirds of the cycle up, alternating with major Sixths down, to keep within the same octave. Since a minor Third has 14 degrees this gives $5 \times \frac{14}{53} = 1\frac{17}{53}$, or an Octave and 17 degrees, that is an Octave above the major Third of the cycle, which will beat slowly with the note on which we started. Thus beginning with 264 and taking first 3 cyclic minor Thirds up, and then a cyclic major Sixth down, followed by a cyclic minor Third up, we have

		up		up		up		down		up
vib.	264		317·05		380·75		457·25		274·56		329·73
beats in 10 sec.	12			14·5		17·5		10·5		12·9	

and as 264 : 329·73 is a cyclic and not a just major Third, which would be 264 : 330, ¶ it will beat 10·8 times in 10 seconds, and this will be the verification of the work. Observe that the interval of the minor Third must be too wide, and hence of the major Sixth too close, so that when we tune up, the *upper* note must be made sharp, and when we tune down, the *lower* note must be made sharp. Also that as the 5th and 6th partials are involved in the beats, the method will suit only qualities of tone, like reed-tones, with strong upper partials. Observe also that in *equal* temperament 5 equal minor Thirds are an Octave and an equal *minor* (not a major) Third. Having completed one set proceed with the next set of 5 minor Thirds (or major Sixths) until the whole cycle is complete for one octave and then tune by Octaves.

Second method, which Mr. Paul White prefers. Tune 7 cyclic major Thirds down (alternating with minor Sixths *up* to keep within the same octave). The result will be a cyclic Pythagorean minor Third of 13 degrees down, or 40 degrees up, for $3 - 7 \times \frac{17}{53} = 3 - \frac{119}{53} = 3 - 2\frac{13}{53} = 1 - \frac{13}{53} = \frac{40}{53}$. And this can be verified by 3 cyclic Fifths up, for $3 \times \frac{31}{53} = \frac{93}{53} = 1\frac{40}{53}$, such Fifths being practically perfect. Thus beginning at 528 vib. we obtain, taking major Thirds down, and minor Sixths up :

		down	down	down	up	down	down	up
vib.	528	422·74	338·47	271·00	433·95	347·44	278·18	445·45
beats in 10 sec.		17	13·9	11·2	17·5	14	11·4	17·9

		a Fourth down	another Fourth down	a Fifth up
vib.	528	396	297	445·5

The Fourths and Fifths are taken just, and the result agrees to ·05 vib. It must be remembered that the cyclic major Thirds are too close, hence in tuning down the *lower* note must be sharpened. On the contrary the cyclic minor Sixths will be too wide, and hence in tuning up, the *upper* note has to be sharpened. Having completed this set of 7 proceed to another, till the cycle is complete. This method also only suits qualities of tone, like reed-tones, with powerful 5th and 8th partials.

The process thus carried out would of course be tedious, and Mr. Paul White seems to assume a tolerably uniform beat, perhaps of 15 in 10 seconds, for he says: ‘ The beats cannot of course be made, or be made to remain uniform, but if they are nearly so, or if a few do not beat at all, the temperament is still good. I ¶ have found that the Fifths can be kept almost entirely free from beats by taking good care of the very slow beats of the Thirds. I have long been convinced that beats in the middle octave do much more good than harm in a musical cycle, for it would be impossible to tune a musical cycle of any size correctly without them. The least scratch on a reed will change a beat, while it often takes quite a scrape to cause a beat where none existed.’ The processes Mr. Paul White has worked out with the ingenious system of checks, shew that he is a thorough master of the whole art of tuning, and, a rare thing to be met with among professional tuners or even musicians, perfectly understands its rationale.

Art. 18.—A succession of just Fifths, as mentioned in art. 1, is very difficult to tune ; and one of just major Thirds is still more difficult. Hence an auxiliary stop on an organ or an auxiliary harmonium is required when just intervals have to be tuned.

It is not difficult to ascertain by ear whether a Fifth or major Third is considerably too flat. Suppose we start with c', then tune an auxiliary g' (indicated by a roman letter) decidedly flat beating 40 times in 10 seconds with c'. Then $3c'-2g'=4$, so that $\frac{3}{2}c'=g'+2$, but $\frac{3}{2}c'$ is the perfect Fifth to c', hence we must tune the required Fifth $g'=g'+2$, that is, sharper than g' by 2 beats in a second. For the next Fifth in order to remain in the same octave we should take the Fourth down. Tune the auxiliary d' so that it should be too flat, and beat 4 times in a second with the correct g'. Then $3g'-4d'=4$, and $\frac{3}{4}g'=d'+1$. But $\frac{3}{4}g'$ is the correct d', or Fourth below g'. Hence it must be tuned 1 beat in a second sharper than the auxiliary d'. And in this way by a laborious ¶ double process the succession of Fifths could be tuned with great accuracy. For the major Thirds, tune an auxiliary e' decidedly flat, and beating 4 in a second with c'. Then $5c'-4e'=4$, and true $e_1'=\frac{5}{4}c'=e'+1$. In the same way we could get $g_2'\sharp$ and $b_3'\sharp$. But for a^1b, f^2b, d^3bb we must tune auxiliary minor Sixths, which is troublesome and not feasible except on reed instruments. Tune an auxiliary $a'b$ flat, so as to beat 5 times in a second with c'. Then $8c'-5a'b=5$, and true $a^{1'}b=\frac{8}{5}c'=a'b+1$. And so on.

It appears, then, that tempered intervals which present beats of their own are more easy to tune than just intervals for which an auxiliary beating tone has to be supplied. The only satisfactory way, however, of tuning perfect and tempered intervals is by a fork tonometer, one of which suffices for every possible case that can arise, when once the pitch numbers of the notes have been calculated as in ¶ sect. A.

SECTION H.

THE HISTORY OF MUSICAL PITCH IN EUROPE.

(See note p. 16.)

Art. 1.—The *pitch number of a note* has been already defined as the number of double vibrations which the sonorous body producing it makes and communicates in one second (p. 11*a*).

Art. 2.—The *pitch number of a musical instrument*, or briefly its *musical pitch*, is taken to be the pitch number of the *tuning note* at a temperature of 59° F. = 15° C. = 12° R.

The tuning note is here assumed to be the a' of the violin, from which the pitch number of all the other notes in the scale must be calculated, or determined approximately by ear from the temperament (sect. A.) and system of tuning (sect. G.) in use. By taking a' as the tuning note, the inquiry is practically limited to European music within the last 500 years.

Art. 3.—The following passage from *Syntagmatis musici, Michaelis* PRÆTORII C., *Tomus Secundus, de Organographia*, 1619, p. 14, explains the condition of ¶ early pitch.

'In the first place it must be known that the pitch, both of organs and other musical instruments, varies greatly. Since the ancients were not accustomed to play in concert with all kinds of instruments at the same time, wind instruments were very differently made and intoned by instrument makers, some high and some low. For the higher an instrument is intoned in its own kind and manner, as trumpets, shawms, and treble viols, the more freshly it sounds and resounds. On the contrary, the deeper trombones, bassoons, bassaneldi, bombards, and bass viols are tuned, the more majestic and magnificent is their stately march. Hence when the organs, positives, clavicymbals, and other wind instruments are not in the same pitch with each other the musician is much plagued.'

Art. 4.—The authorities on whom I rely are minutely specified in my 'History of Musical Pitch' in the *Journal of the Society of Arts* for 5 March and 2 April 1880, and 7 Jan. 1881. The two last papers contain indispensable corrections and additions. In the privately printed copies there was an addendum on U.S. America from Messrs. C. R. Cross and W. T. Miller, *American Journal of Otology*, Oct. 1880.

¶ Here it must suffice to say that after learning to determine pitch to $\frac{1}{10}$ vib. (p. 444) I obtained the loan of authentic forks from the Society of Arts, Mr. A. J. Hipkins, Rev. G. T. Driffield (Handel's), Frau Naeke of Dresden, Prof. Rossetti of Padua, Mr. Blaikley, and Dr. W. H. Stone. I procured compared copies of forks in the Conservatoire at Paris, and others tuned at known temperatures to remarkable organs at Vienna, Dresden, Hamburg, Strasburg, and Seville. Then, with the assistance of many organists, I measured numerous organs in England of which the pitch had not been changed, or, with the kind help of several organ-builders, obtained untouched pipes of altered organs. When these failed I had models made of pipes of which the dimensions were given by Schlick 1511, Prætorius 1619, Mersenne 1636, Tomkins 1668, Bédos 1766, and others, which were obligingly presented to me by Mr. T. Hill, the organ-builder, on whose bellows I measured them. These constituted my own materials. Then I had recourse to the measurements and lists of Cagniard de la Tour, Cavaillé-Coll, de la Fage, Delezenne, de Prony, Euler, Fischer, French Commission on Pitch, Koenig, Lissajous, McLeod, Marpurg, Naeke, Sauveur, Scheibler, Schmahl, Dr. R. Smith, and others. From these I constructed the lists which follow. In my original papers each pitch is accompanied with full details. Here I give the smallest possible account.

Art. 5.—The pitch given is always that of a', where possible at 59° F. But this was not always the note measured. When it was not, a' was calculated on the assumption of either meantone or equal temperament. Assuming a lowest ideal pitch of a'370, which has never yet been found, I give the cents by which any other pitch exceeds this, so that the interval between any two pitches is immediately determined by subtracting the cents. I give also the date, adding occasionally *a* for *ante*, before, *p* for *post*, after, and *c* for *circa*, about; and the authority, or observer, where E. means that I am responsible for the measurement, directly or indirectly. Finally, I add a list,

classified by countries, stating the kind of pitch. I have not thought it necessary to give absolutely every fork and pitch entered in my 'History,' but have reported a large number of these entries, and especially all the most interesting of them. A complete German translation of my paper is in preparation, and will be published at Vienna.

TABLE I.—HISTORICAL PITCHES IN ORDER FROM LOWEST TO HIGHEST.

Cents	a'	Date	Observer	Place and other particulars
			1. Church Pitch, Lowest.	
000	370	—	E. . . .	Ideal lowest pitch or zero point ¶
15	373·1	—	Delezenne . .	Calculated from D.'s measurements of an open wooden pipe 1·3 metres long, taken as *c*
17	373·7	1648	E. . . .	Paris, from a model after Mersenne
19	374·2	1700a	Delezenne .	Lille, organ of l'Hospice Comtesse
31	376·6	1766	E. . . .	Paris, from a model after Bédos
33	377	1511	E. . . .	Heidelberg, from a model after Arnold Schlick (see 535 cents)
			2. Church Pitch, Low.	
66	384·3	1700c	Delezenne . .	Lille, old fork found 1854a by M. Mazingue
69	384·6	1851	,, . .	Lille, organ of St. Sauveur, rebuilt, with old pitch
100	392·2	1739	Euler . . .	St. Petersburg, a clavichord according to Marpurg, but Euler gives no particulars
104	393·2	1713	Stockhausen & E.	Strassburg Minster, great organ by A. Silbermann
114	395·2	1759	Dr. R. Smith .	Cambridge, Bernhardt Schmidt's organ at Trinity College, 1708, after being new voiced and 'shifted' in 1759
,,	,,	1720	,, .	Rome, pitch-pipes observed by Dr. R. Smith ¶
117	395·8	1789	McLeod & E. .	France, Versailles, copy of fork No. 410 at the Musée du Conservatoire, Paris, compared with the original by Cavaillé-Coll
119	396·4	1615	E. . . .	Palatinate of the Rhine, from a model of pipe described by Salomon de Caus
129	398·7	1854a	Delezenne . .	Lille, old organ of La Madeleine restored
			3. Chamber Pitch, Low.	
148	402·9	1648	E. . . .	Paris, Mersenne's Spinet, from his statement that B♮ = Bédos's 4-foot *c* (see 31 cents)
152	403·9	1730	E. . . .	Padua, from copy sent by Prof. Rossetti of the old lower *f″* fork of the bellfoundry of Colbacchini
163	406·6	1704	Sauveur . .	Paris, result of several experiments on an *e* pipe
166	407·3	1854	Delezenne . .	Lille, organ of St. Maurice repaired, old pitch kept
169	407·9	1762	Schmahl & E. .	Hamburg, organ of St. Michaelis Kirche, built by Hildebrand of Dresden, under the direction of Handel's friend, J. Mattheson (1681-1764), in the chamber pitch of the period, still preserved; now, and probably always, in equal temperament ¶
174	409	1783	Lissajous . .	Paris, Court clavecins, fork of Pascal Taskin, their tuner
178	410	—	,, .	Paris, 18th century pitch-pipe found in the cabinet of the Faculty of Sciences
184	411·4	1688	Schmahl & E. .	Hamburg, chamber pitch on the former 8-foot *Gedact* of the St. Jacobi organ (see 484 cents)
191	413·3	—	Naeke . . .	'Schneider's Oboe,' date and place unknown
196	414·4	1776	Marpurg . .	Breslau, clavichords
			4. Mean Pitch of Europe for Two Centuries.	
199	415	1754	E. . . .	Dresden, organ of the Roman Catholic Church by Gottfried Silbermann, pitch of the chained fork placed there by King August der Grechte, 1763-1827, who would not allow the pitch to be changed; the fork was lent me by Frau Naeke

TABLE I.—HISTORICAL PITCHES IN ORDER FROM LOWEST TO HIGHEST—*continued.*

Cents	a'	Date	Observer	Place and other particulars
				4. *Mean Pitch of Europe for Two Centuries*—continued.
201	415·5	1722	Naeke . . .	Dresden, organ of St. Sophie, built by G. Silbermann
211	418	1780	Euler . . .	St. Petersburg, organs ; no particulars
212	418·1	1878	E. . . .	Dresden, present pitch of the organ of the Roman Catholic Church, from a fork tuned for me there
215	419	1700c	E. . . .	London, Renatus Harris's organ at St. John's, Clerkenwell
217	419·5	1714	Naeke . .	Freiberg, Saxony, G. Silbermann's organ
218	419·6	1858	de la Fage . .	Madrid, ton de chapelle, calculated
,,	,,	1785	E. . . .	Seville, Spain, pitch of the old organ of Torje Bosch, from a fork said by the organist Don Yñiguez to be in exact unison with its a' at a mean temperature
219	419·9	1715c	E. . . .	England, rude tenor a fork, belonging to Rev. G. T. Driffield, who held it to have been made by John Shore, the inventor of tuning-forks
220	420·1	1780	E. . . .	Winchester College organ, from one of the pipes added by Green when repairing R. Harris's organ of 1681
224	421·2	1860	E. . . .	Russian Imperial Court church band from fork lent by Frau Naeke
225	421·3	1780	Naeke . . .	Vienna, fork of the Saxon organ-builder Schulze, who lived at Vienna in Mozart's time
226	421·6	1780	Naeke & E. . .	Vienna, copy of fork of Stein, who made Mozart's clavichords and pianos, lent me by Frau Naeke
229	422·3	1780	Naeke . . .	Dresden, fork of former Court organist Kirsten
,,	,,	1780c	E. . . .	Verona, from a copy of a c' fork believed to be the Roman pitch of 1780, preserved at the bell-foundry of Cavedini, procured by Prof. Rossetti of Padua
230	422·5	1751	E. . . .	England, Handel's fork belonging to Rev. G. T. Driffield. The organ at Cannons in the private chapel of the Duke of Chandos, built by Jordans, and afterwards bought by Trinity Church, Gosport, has been recently (in 1884) examined by the organist, Mr. Howlett, and found to have had in Handel's time, when he used to play on it, a $B\natural$ (now Bb) pipe of 12·3 inches long, and 1 inch in diameter; this shews that its pitch was then A423·5, or practically the same as Handel's fork
,,	,,	1820a	E. . . .	Westminster Abbey, as originally tuned by Schreider and Jordans, from indications by Mr. T. Hill, who retuned it to a'441·7. It had been altered by Greatorex to a'433·2, Smart's pitch
,,	,,	1838	E. . . .	Bath Abbey Church, as rebuilt by Smith of Bristol, from indications by Mr. T. Hill
,,	,,	1877p	E. . . .	England, Mr. J. Curwen's Tonic Solfa standard c''507, using the just a' only
,,	422·6	1790a	E. . . .	Kew Parish Church, Green's organ, untouched and in meantone temperament when measured in 1878, built as a chamber organ for George III.
,,	,,	1754c	Delezenne .	Lille, very old fork found in workshops of M. François, musical instrument maker there
,,	,,	1780c	E. . . .	Padua, from copy of the higher f'' fork of the bell-foundry of Colbacchini (see 152 cents)
231	422·7	1800c	E. . . .	England, from old fork, c''505·7, belonging to Messrs. Broadwoods
232	423	1820	McLeod & E. .	Paris, Théâtre Feydeau, Opéra Comique, from copy of fork at the Conservatoire, Paris, compared with the original by Cavaillé-Coll
233	423·2	1778	E. . . .	London, Green's organ at St. Katharine's, Regent's Park, still (when I measured it) in meantone temperament (see sect. G., p. 484c')
,,	,,	1815–1821	E. . . .	Dresden, band of the opera while C. M. von Weber (1786–1826) was conductor (*Kapellmeister*)
,,	423·3	1813	E. . . .	London, second copy of Peppercorn's fork by which the pianofortes of the Philharmonic

TABLE I.—HISTORICAL PITCHES IN ORDER FROM LOWEST TO HIGHEST—*continued.*

Cents	a'	Date	Observer	Place and other particulars
				4. Mean Pitch of Europe for Two Centuries—continued.
235	423·7	1813	E.	Society were originally tuned; this copy was prepared for the Society of Arts in 1860, and is now in the possession of Messrs. Broadwoods London, first copy of Peppercorn's fork made before 1860, belonging to Mr. Hipkins; see last entry, the original is lost, and it is impossible to say which was correct. The difference, 2 cents, is utterly insignificant
236	424·1	1740–1812	Naeke .	Eutin (18 miles N. of Lübeck), fork of Franz Anton von Weber, father of Carl Maria von Weber
237	424·2	1619	E.	Brunswick, from a model made from Prætorius's drawing of an organ pipe at a 'suitable' church pitch ¶
,,	,,	1823	Fischer	Paris, Italian Opera, mean of twenty measurements of a fork given by Spontini
,,	424·3	1750a	E.	London, old forks formerly belonging to Prof. Faraday, lent me by Mr. D. J. Blaikley
,,	,,	1749	E.	London, organ at All Hallows the Great and Less, Upper Thames Street, built by Glyn & Parker, by whom Handel's Foundling Hospital organ was built
238	424·4	1833	E.	Weimar, from a model of Töpfer's wide principal c''-pipe
239	424·6	1800c	E.	England, old fork said to have been used in Plymouth Theatre, lent me by Dr. Stainer
240	424·9	1805	E.	London, old D fork of Elliott's, by which he tuned the organ built for the Ancient Concerts at the Hanover Square Rooms, lent me by his successor, Mr. T. Hill
,,	,,	1800c	Naeke .	Germany, fork of the bassoonist Kummer
241	425·2	1730c–1780c	E.	Padua, mean of two ancient pitch-pipes belonging to the bellfoundry of Colbacchini, lent me at the request of Prof. F. Rossetti there ¶
242	425·5	1829	Lissajous	Paris, pitch of opera piano as distinct from the orchestra, verified by Monneron for de la Fage
,,	425·6	1740–1780	E.	England, Schnetzler's organ at the German Chapel Royal, St. James's Palace
,,	,,	1764	E.	Halifax, Schnetzler's organ, from indications by Mr. T. Hill
243	425·8	1824	Lissajous	Paris, pitch of opera, suddenly lowered on 31 March for Mme. Branchu, whose voice was failing. The piano for rehearsals was also lowered, and was not raised immediately when the orchestra was raised; this was called opera pitch
,,	,,	1839	de la Fage .	Bologna, Italy, pitch of fork of Tadolini, the best tuner in the town
244	425·9	1740	Tunbridge & E. .	Great Yarmouth, St. George's Chapel, by Byfield, Jordan & Bridge
246	426·5	1843	E.	Wimbledon Church, organ built by Messrs. Walker ¶
248	427	1811	Scheibler	Paris, Grand Opera
249	427·2	1878a	E.	Norwich Cathedral organ before it was altered by Bryceson, supposed to be by R. Harris
250	427·5	1877a	E.	Tonic Solfa pitch to 1877, afterwards 422·5
250	427·6	1823	Fischer	Paris, Théâtre Feydeau, fork given by Spontini
251	427·7	1696	E.	London, old organ built by R. Harris, a pipe of St. Andrew Undershaft, from Green's organ, preserved by Mr. T. Hill
,,	427·8	1788	E.	St. George's Chapel, Windsor, measured in Feb. 1880, while still in meantone temperament
255	428·7	1670	Ions .	Newcastle-on-Tyne, St. Nicholas Church organ built by Renatus Harris, frequently altered except in pitch
				5. The Compromise Pitch.
260	430	1810c	Lissajous	Paris, fork of M. Lemoine, a celebrated amateur
262	430·4	1701	E.	Fulham Parish Church organ, built by Jordans. This pitch was officially adopted in Italy in 1884

TABLE I.—HISTORICAL PITCHES IN ORDER FROM LOWEST TO HIGHEST—*continued.*

Cents	a'	Date	Observer	Place and other particulars
			5. *The Compromise Pitch*—continued.	
				as the pitch of the Italian army brass bands, giving $Bb456$, the nearest whole number to equal $Bb456\cdot13$, which would correspond to the 'arithmetical' pitch $C512$
264	431·3	1625	Lewis . . .	Lavenham (16½ miles W.N.W. of Ipswich), from a famous old tenor bell sounding $d288\cdot4$
267	431·7	1826	Fischer	Paris, Grand Opera, fork given by Spontini
269	432·2	1854a	Delezenne .	Lille, organ of St. André repaired
270	432·3	1846c	E. . . .	England, old fork which belonged to the father of Messrs. Bryceson, organ-builders, and had not been tuned since 1848, when it had been sharpened slightly
¶ 272	433	1820c	E. . . .	London, fork approved of by Sir George Smart, conductor of the Philharmonic Concerts, in possession of Mr. Hipkins, from $c''518$ using meantone temperament; if equal temperament were used it would give $a'435\cdot4$ and be a 30 years' anticipation of French pitch. Used in this way it is Broadwoods' lowest pitch. Long sold in shops as 'London Philharmonic'
273	433·2	1828	E. . . .	London, Sir G. Smart's own Philharmonic fork. Sir G. Smart considered this a' fork of his to agree with $c''518$ (see last entry). This shews that he used meantone temperament
275	433·6	1847	Byolin & E.. .	Shrewsbury, St. Mary's, built 1729 by John Harris and John Byfield, pitch altered in 1847 by Gray & Davison
276	433·9	1834	Scheibler . .	Vienna, fork I., Delezenne's Vienna minimum
,,	434	1829	Cagniard de la Tour	Paris, opera, verified by M. Montal, after the opera had recovered its pitch, the opera piano remaining at $a'425\cdot5$, which see, and also $a'425\cdot8$
¶ ,,	,,	1834c	Scheibler . .	Paris Opera, fork by Petitbout, luthier de l'opéra
278	434·3	1818	McLeod & E. .	Paris, Chapelle des Tuileries, from a copy compared by Cavaillé-Coll of fork No. 493 in the Conservatoire
,,	434·5	1869	E. . . .	Baden, fork sent officially to Society of Arts
279	434·7	—	E. . . .	London, from a model of pipe representing $b'486\cdot1$, one foot long and one inch diameter, on Renatus Harris's organ at All Hallows, Barking
280	435	1826	Naeke . . .	Dresden, opera, fork of Kapellmeister Reissiger, successor to C. M. von Weber. Naeke considers this to have been Dresden pitch from 1825 to 1830
,,	,,	1859	Fr. Com. . .	Carlsruhe, opera, the fork which determined the French *Diapason Normal*
282	435·2	1834a	Scheibler .	Paris, Conservatoire, fork made by Gand, luthier du Conservatoire
,,	435·4	1859	Koenig & E. .	Paris, the *Diapason Normal* in the Conservatoire, used extensively in Germany, officially adopted for the Belgian army in 1885. The various imperfect copies used are not cited
¶ 283	435·9	1868	Cross & Miller .	U. S. America, E. S. Ritchie's standard pitch
284	436	1802	Sarti . . .	St. Petersburg, five-foot organ pipes
,,	,,	1846p	E. . . .	London 'Philharmonic,' from Mr. Hipkins's vocal pitch, $c''518\cdot5$, which for equal temperament gives $a'436$, but on meantone temperament, for which it was first used, gave $a'433\cdot5$; the fork with which Mr. E. J. Hopkins compared the pitch of the organs at Lübeck, Hamburg, and Strassburg, see his *The Organ* ed. 1870, art. 791, p. 189
,,	,,	1878	E. . . .	London, Messrs. Bishop's standard for church organs
285	436·1	1878	E. . . .	London, fork to which Messrs. Bryceson tuned the organ at Her Majesty's Theatre
286	436·5	1834c	Scheibler .	Vienna, opera, fork II.
287	436·7	1845	Delezenne . .	Florence, fork lent by M. Marloye
,,	436·8	1740–1780	E. . . .	Dublin, Green's organ in the Refectory of Trinity College, probably sharpened
288	436·9	1869	E. . . .	Würtemberg, fork sent officially to the Society of Arts

TABLE I.—HISTORICAL PITCHES IN ORDER FROM LOWEST TO HIGHEST—*continued*.

Cents	a'	Date	Observer	Place and other particulars
				6. *Modern Orchestral Pitch, and* * *Church Pitch Medium.*
288	437	1859	Fr. Com. . .	Toulouse, Conservatoire
289	437·1	1666	E. . . .	*Worcester, cathedral organ built by Thomas and Renatus Harris, from a pipe at Mr. T. Hill's
,,	437·3	1872	Fischer . .	Berlin, from a fork furnished by Pichler, who tuned the piano of the opera
,,	437·4	1854a	Delezenne .	Paris, opera, from four forks purchased before 1854, and found to be in unison
,,	,,	1744	Streatfield & E. .	*Maidstone, Old Parish Church, built by Jordans, altered, but not in pitch, in 1878 in meantone temperament
291	437·8	1862	E. . . .	Dresden, fork given by the direction of the Court Theatre to its librarian, Herr Moritz Fürstenau, after the conference on pitch held there, by whom it was lent me to measure, meant for a'440
295	438·9	1696	E. . . .	*Boston, England, organ built by Christian Smith, from a pipe preserved by Mr. T. Hill
297	439·4	—	Delezenne .	Lille, old fork formerly belonging to the Marquis d'Aligre
,,	,,	1834c	Scheibler .	Vienna, opera, fork III.
,,	,,	1878	E. . . .	Dresden, opera pitch at date, from a fork specially prepared for me by the Court organ-builder, Jehmlich, and sent by Herr Moritz Fürstenau, librarian of the theatre
298	439·5	1812	McLeod & E. .	Paris, Conservatoire, from copy of a fork preserved there, verified by Cavaillé-Coll
,,	,,	1855	E. . . .	England, Barking, Essex, Parish Church organ (probably originally a'474·1), built by Byfield & Green, 1770, after alterations by Messrs. Walker
299	439·9	1845	Delezenne .	Turin, fork lent by Marloye
300	440	1829	Lissajous .	Paris, opera orchestra, verified by Monneron for de la Fage
,,	,,	1878	E. . . .	London, Messrs. Gray & Davison's standard pipe
301	440·2	1834	Scheibler .	Stuttgart pitch, =440 at 69° F., Lissajous measured it as 440·3 to French *Diapason Normal*, reckoned as 435, which then when corrected to 435·4 gives 440·7
,,	,,	1879	E. . . .	London, Messrs. Walker & Sons' standard pipe
,,	440·3*	1834c	Scheibler .	Vienna Opera, fork IV.
302	440·5	1878	E. . . .	London, Messrs. Bevington's standard pipe
303	440·9	1834c	Scheibler .	Paris Conservatoire, not trusted so much by Scheibler as 435·2
304	441·0	1836- 1839	Delezenne .	Paris Opera, fork of M. Leibner, who kept the pianos to pitch of orchestra, verified by Meyerbeer
,,	,,	1836	Cagniard de la Tour	Paris, Opéra Comique
,,	,,	1859	Fr. Com. . .	Dresden, fork sent to Fr. Com. by the Kapellmeister Reissiger
,,	,,	1879	E. . . .	London, church organ pitch of Messrs. Lewis & Co.
,,	441·10	1834	Scheibler	Vienna Opera, fork V., given by Prof. Blahetka as trustworthy; in 1879 this fork was found and lent to me, and then from rust and ill-treatment measured only 439·9, the greatest loss of pitch I have found in any fork
305	441·2	1878	E. . . .	London, Covent Garden Opera, fork for Messrs. Bryceson to tune the organ to
,,	441·3	1842	E. . . .	London, the equal a' corresponding to the late Dr. John Hullah's standard fork, c''524·8, purporting to be c''512; J. H. Griesbach measured it as 521·6
307	441·7	1690	E. . . .	Hampton Court Palace, Bernhardt Schmidt's organ from an original pipe, 12 inches long and 1·2 inch in diameter, giving b'b472·6
,,	,,	1660	E. . . .	Whitehall, Chapel Royal, organ by Bernhardt Schmidt, according to indications by Mr. T. Hill
,,	,,	1878	E. . . .	London, standard pipe of Messrs. Hill and Sons, from c''525·3

TABLE I.—HISTORICAL PITCHES IN ORDER FROM LOWEST TO HIGHEST—*continued.*

Cents	a′	Date	Observer	Place and other particulars
				6. *Modern Orchestral Pitch, and * Church Pitch Medium*—continued.
307	441·8	1834c	Scheibler . .	Berlin opera
310	442·5	1859	Fr. Com. . .	Toulouse opera
,,	,,	,,	,, ,,	Brussels, opera under direction of Bender
*311	442·7	1878	E. ,, . .	*Vienna, small Franciscan organ kept at modern pitch, from a fork tuned for me by the organ-builder Ullmann
312	443·0	1859	Fr. Com. . .	Bordeaux opera
,,	,,	,,	,, ,,	Stuttgart opera
,,	443·1	1815c	E. ,, . .	*Durham organ, as altered by shifting from a′474·1; a′444·7, the present pitch of new organ, is by Willis
,,	,,	1869	E. . . .	Bologna, Italy, Liceo Musicale, from fork sent officially to Soc. of Arts
313	443·2	1878	E. . . .	*Vienna, St. Stefan cathedral organ, from a fork tuned for me by organ-builder Ullmann
,,	443·3	1836	Wölfel . . .	Paris, Wölfel's pianos
,,	,,	1859	Fr. Com. . .	Gotha, opera
,,	443·4	1878	E. . . .	London, from Messrs. Bryceson's standard pipe
314	443·5	1859	Fr. Com. . .	Brunswick, opera
315	443·9	1880	Cross & Miller	U.S. America, Boston, organ of Church of the Immaculate Conception
,,	444	1860		Intended but unexecuted standard of Society of Arts to c″528
316	444·2	1880	,, ,,	U.S. America, from c″528, the 'low organ pitch' of Hutchings, Plaisted & Co.
317	444·3	1840	Cavaillé-Coll	*France, St Denis Cathedral, organ built by Cavaillé-Coll
,,	,,	1880	E. . . .	*London, Temple Church organ after rebuilding by Messrs. Forster and Andrews, who retained the pitch which they found, which was Robson's, originally built by Bernhardt Schmidt, with both Eb and D♯, and both Ab and G♯ keys, and perhaps then having a′441·7
318	444·5	1858	Lissajous . .	Madrid, Theatre Royal, fork sent to de la Fage by the Maître de Chapelle. French pitch was adopted on 18 March 1879
,,	444·6	1877	E. . . .	*London, St. Paul's, after rebuilding by Willis, from a fork belonging to Mr. Hipkins at 57°·5
,,	444·7	1879	E. . . .	*Durham Cathedral organ, rebuilt by Willis; for its original state, see a′474·1
319	444·8	1859	Fr. Com. . .	Turin opera
,,	,,	,,	,,	Weimar opera
,,	,,	,,	,,	Würtemberg concerts
,,	444·9	1857	Lissajous . .	Naples, San Carlo opera, Guillaume's fork
,,	,,	1880	Hipkins . .	London, Her Majesty's opera, fork of the theatre
320	445·0	1862	Naeke . .	Vienna, piano of Kapellmeister Proch
,,	,,	—	Schmahl . .	Hamburg 'old pitch,' date unknown
321	445·1	1834c	Scheibler .	Vienna opera, fork VI., 'a monstrous growth' (*Auswuchs*) in Scheibler's opinion
,,	445·2	1878	E. . . .	*London (from c″529·4), Mr. H. Willis's church pitch, to which he tuned the organs of the cathedrals of St. Paul's (London), Durham, Salisbury, Glasgow (established), St. Mary (Edinburgh)
322	445·4	1845	Délezenne . .	Vienna Conservatorium, fork lent by Marloye
,,	445·5	1879	Hipkins & E. .	London, Her Majesty's opera during performance
,,	445·6	,,	E. . . .	London, Covent Garden opera, fork in possession of Mr. Pitman, organist, and Sig. Vianesi, conductor. Mr. Pitman said the pitch was thus in 1878 because oboe, bassoon, and flute would not play lower
323	445·8	1867	E. . . .	London, Exeter Hall, both organs as originally built, from a pipe at the makers', Messrs. Walker; since sharpened to a′447·3
,,	,,	1856	Lissajous .	Paris opera, from the fork of M. Bodin, professor of the piano and music

TABLE I. –HISTORICAL PITCHES IN ORDER FROM LOWEST TO HIGHEST—*continued*.

Cents	a'	Date	Observer	Place and other particulars
				6. *Modern Orchestral Pitch, and * Church Pitch Medium*—continued.
323	445·9	1849–1854	E. . . .	London, from Broadwoods' original medium pitch of c″530·6, fork of the tuner Finlayson; since 1854 Messrs. Broadwoods use a'446·2 as their medium pitch. This pitch was chosen empirically
,,	446	1859	Fr. Com. .	Pesth, opera
324	446·2	1856	Lissajous	Paris, opera and Conservatoire
,,	,,	1859	Fr. Com. .	Holland, the Hague at the Conservatoire
326	446·6	1845	Delezenne .	Milan, fork lent by Marloye
327	446·8	1851	,,	Lille, festival organ, fork of the tuner Mazingue
,,	,,	1878	E. . . .	Vienna opera, from a fork sent me by the organ-builder, Ullmann, who had charge of the organ there ¶
,,	447·0	1859	Fr. Com. .	Marseilles Conservatoire
328	447·3	1879	E. . . .	London, Exeter Hall organ, from a pipe of the makers, Messrs. Walker, see 445·8
329	447·4	1856	Lissajous .	Paris, Italian opera, Bodin's fork
,,	447·5	1878	Hipkins .	London, Covent Garden opera harmonium
300	447·7	1877	E. . . .	Gloucester Festival organ, built by Messrs. Walker; from the fork to which it was tuned at 64° F., the temperature of the pipe being reduced to 59°
331	448	1854	Lissajous .	Paris, Grand Opera—also at Lyons and Liège
,,	,,	1839–1840	Schmahl .	Hamburg, opera, under Krebs
,,	448·1	1859	Fr. Com. .	Munich, opera
332	448·2	1869	E. . . .	Leipzig, Gewandhaus Concerts, from fork sent officially to the Society of Arts
333	448·4	1857	Lissajous .	Berlin, opera, fork of the conductor Taubert
,,	,,	1860	E. . . .	London, from Cramer's c″533·3, purporting to be the Society of Arts' pitch, intended for c″528
,,	448·5	1880	Cross & Miller .	Boston, Nichol's fork of Germania Orchestra, as corrected to 59° F. ¶
334	448·8	1859	Fr. Com. .	Leipzig Conservatoire
335	449	1855	Lissajous .	Paris opera, experiments by Lissajous and Ferrand, the first violin
336	449·2	1877	Hipkins .	Covent Garden Opera, pitch of the harmonium
337	449·4	1860	E. . . .	London, from Griesbach's c″534·5, tuned for the Society of Arts as c″528; he tuned a' as 445·7
338	449·7	1879	Hipkins .	London, Covent Garden opera, taken from organ a' during performance
,,	449·8	1859	Fr. Com. .	Prague, opera
339	449·9	1877	E. . . .	London, from copy of Collard's standard fork
340	450·3	1856	Lissajous .	Milan, opera
,,	450·5	1848 & 1854	Delezenne .	Lille, from forks tuned by the oboist Colin, during the performances of *Robert le Diable*, 27 April 1854, between the acts, and carefully verified
341	450·6	1877	E. . . .	Glasgow Public Halls organ, from fork settled by the organist W. T. Best and the late H. Smart, lent me by the builder Lewis
342	450·9	1880	Cross & Miller .	U.S. America, Boston Music Hall, reduced from pipe c271·2 at 70° F. ¶
345	451·5	1858	Fr. Com. .	Russian opera, from a c″ fork, probably miscalculated, as the a' from Broadwoods' c″ forks were
345	451·7	1874	E. . . .	Belgian army pitch, reduced from Koenig's 451 vib. by his old standard, and also measured from copy sent by Mahillon. On 19 March 1885 the Belgian Government adopted French pitch, A435
,,	,,	1867	Lissajous .	Milan, Scala Theatre
,,	,,	1880	Cross & Miller .	U.S. America, New York, from Chickering's c268·5 standard fork
346	451·9	1878	E. . . .	British Army regulation, from fork lent by Dr. W. H. Stone
,,	452	1885	E. . . .	The International Inventions and Music Exhibition of 1885 adopted this as the pitch of all instruments for the exhibition, being the nearest whole number to the next preceding and next following. The fork was verified by myself

TABLE I.—HISTORICAL PITCHES IN ORDER FROM LOWEST TO HIGHEST—*continued*.

Cents	*a'*	Date	Observer	Place and other particulars
				6. *Modern Orchestral Pitch, and * Church Pitch Medium*—continued.
349	452·5	1852–1874	E. . . .	London, mean of the pitch of the Philharmonic Band under the direction of Sir Michael Costa 1846-54, tuned during that period by Mr. J. Black of Broadwoods', approved by Sir Michael Costa, and recorded by Mr. Hipkins, who lent me the fork. Used as Broadwoods' highest till 1874, No. 3 of French Commission
„	„	1880	Chambers & E. .	Newcastle-on-Tyne, Schulze's Tynedock organ, from a fork tuned by Mr. Ch. Chambers, Mus. B.
350	452·9	1878	E. . . .	Kneller Hall Training School for Military Music, from a fork lent by Dr. W. H. Stone
„	453	1645	Schmahl . .	*Holstein, Glückstadt organ built 1645, improved by Schnitger 1665, measured 1879
354	453·9	1878	E. . . .	London, Willis's concert organ pitch, to which he tuned the large organs in the Albert Hall and Alexandra Palace, from pipe *c''*543·2 at 65° F., and 541·2 at 61·5° F.
„	454	1862	Naeke . .	Vienna, piano of Kapellmeister Esser, while the orchestra was at *a'*466, the regular fork at *a'*456·1, and the piano of the other Kapellmeister Proch at *a'*445
355	454·1	1877	E. . . .	Crystal Palace, from a fork *c''*540 lent by Mr. Hipkins, to which the piano for concerts was tuned
„	454·2	1715c	E. . . .	London, very old fork found at Brixton 1878 of the same make as Rev. G. T. Driffield's tenor *a*, see *a'*419·9
357	454·7	1874	E. . . .	London, from *c''*540·8, a fork representing the highest pitch of the London Philharmonic observed by Mr. Hipkins since 1874; at the suggestion of Mr. Charles Hallé, used as Broadwood's highest pitch
„	„	1879	E. . . .	London, Messrs. Steinway's London pitch
„	„	1878	E. . . .	London, Messrs. Bryceson's band pitch, to which they tuned their organ in St. Michael's, Cornhill, London
358	455·1	1877	Hipkins & E. .	London, Wagner Festival at Albert Hall, temperature probably 61·5° F., see above *a'*453·9
359	455·2	1749	Schmahl & E. .	Hamburg, old *positiv* or chamber organ, built by Lehnert, in possession of Herr Schmahl
„	455·3	1879	E. . . .	London, Erard's concert pitch, from their fork
„	455·5	1859	Fr. Com. . .	Belgium, band of Guides; probably no such fork existed. M. Bender used to give the pitch on a small clarinet, from which M. Mahillon has a fork of at least *a'*456
362	455·9	1877	E. . . .	London, fork used by one of Chappell's tuners, lent me by Dr. Stone
„	456·1	1880	Cross & Miller .	U.S. America, Cincinnati, pitch used in Thomas's orchestra. [This is said by de la Fage to have been the pitch sent by Bettini in 1857 for the London Italian opera—evidently an error]
„	„	1859a	E. . . .	Vienna, fork tuned for me by the pianoforte makers Streicher in Vienna from a fork in their possession, giving the celebrated 'sharp Vienna pitch' before the introduction of the French *Diapason Normal*. Naeke says he heard *a'*466 in the actual playing of the orchestra
366	457·2	1879	E. . . .	U.S. America, New York, from a fork obtained for me by Messrs. Steinway as representing their American pitch
369	458·0	1880	Cross & Miller .	U.S. America, New York, from a fork furnished by R. Spice as Steinway's pitch
380	460·8	„	„ „	U.S. America, highest New York pitch, from a fork furnished by R. Spice; these two last are sharper than the next, but they are put first because they belong to modern orchestral or pianoforte pitch.

TABLE I.—HISTORICAL PITCHES IN ORDER FROM LOWEST TO HIGHEST—*continued*.

Cents	a'	Date	Observer	Place and other particulars
				7. *Church Pitch, High.*
368	457·6	1640c	E. . . .	Vienna, Great Franciscan organ, stated by organ-builder Ullmann to be 240 years old in 1878, and to possess its original pitch; only used for leading the ecclesiastical chants
429	474·1	1668	E. . . .	England, in the *Pars Organica* of Tomkins's *Musica Deo Sacra* as quoted in Sir F. A. Gore Ouseley's *Collection of the Compositions of Orlando Gibbons*, 1873, makes the *f* pipe 2½ feet long
,,	,,	1683	Armes & E. . .	Durham, Bernhardt Schmidt's original organ at Durham, which had both *ab* and *g♯*. The pipe I measured in Feb. 1879 as *a*'443·1 had been shifted, and was orginally *g'♯*, which gives the above pitch. This results from an examination of the original pipes by Dr. Armes, the organist ¶
,,	,,	1708	E. . . .	Chapel Royal, St. James's, Bernhardt Schmidt's organ, now in Mercers' Hall, which I found on examination had had the pipes shifted a great Semitone. Handel played on this organ, and hence his note ordering the voice parts of an anthem written for the Chapel Royal to be transposed one Tone, and the organ part *two* Tones, referred to this organ
,,	,,	1748	E. . . .	The Jordans' organ, Botolph Lane, from indications by Mr. T. Hill
454	480·8	1879	Degenhardt & E. .	Hamburg, St. Catherine Kirche, built by Hans Stellwagen in 1543, and frequently repaired. Herr Degenhardt, the organist, declares that even at the last repairing, 1867–9, the pitch was not altered. The original pitch, however, is doubtful, and Herr Schmahl thinks it was altered formerly ¶
465	484·1	1878	Jimmerthal . .	Lübeck Cathedral, small organ, which according to the organist Jimmerthal has its *g'* in unison with the pipe on Schulze's new great organ there, which gives French *a'* in summer at 68° F.; whence the above was calculated at 59° F.
484	489·2	1688	Schmahl & E. .	Hamburg, St. Jacobi Kirche, built by Schnitger of Harburg originally in equal temperament, played on and approved by J. Sebastian Bach; pitch determined from an old pipe preserved in the organ case. Herr Schmahl the organist is accustomed to transpose all music at sight one Tone lower, which brings it to French pitch
				8. *Church Pitch, Highest.* ¶
502	494·5	1879	Schmahl & E. .	Hamburg, St. Jacobi Kirche, present pitch, used since 1866 in order to agree with Scheibler's forks, taking his *a'*440 for *g'*
506	495·5	1700	Schmahl . .	Holstein, Rendsburg, a large organ recently broken up
534	503·7	1636	E. . . .	Paris, Mersenne's *ton de chapelle* with *G*112·6 on the French four-foot pipe, this being the lowest note of his own voice
535	504·2	1511	E. . . .	Heidelberg, from a model after Arnold Schlick, who recommends that his 6½-foot Rhenish pipe, having 301·6 vib., should give *F* or *c*. If it gives *F* we have *a'*377, if it gives *c* we have the present pitch
541	505·8	1361	E. . . .	Halberstadt organ, built 1361, repaired 1495, described by Prætorius, who gives the dimensions of the largest pipe *B,,,*, whence constructing a model I arrived at the above pitch, confirmed by the four preceding pitches

TABLE I.—HISTORICAL PITCHES IN ORDER FROM HIGHEST TO LOWEST—*continued.*

Cents	a'	Date	Observer	Place and other particulars
				9. *Chamber Pitch, Highest, and Church Pitch, Extreme.*
726	563·1	1636	E. . . .	Paris, Mersenne's chamber pitch calculated from F being the pipe of 4 French feet giving 112·6 vib. See *Harmonie Universelle,* liv. 3, p. 143, but from faulty measurement Mersenne makes this pipe to have only 96 vib. But even with that assumption the pitch would be a'480·1, as at Hamburg, St. Catherine Kirche; but compare the next entry
740	567·3	1619	E. . . .	North German church pitch, called by Prætorius chamber pitch, taken as a meantone Fourth (503 cents) above Prætorius's 'suitable pitch' a'424·2, which see

TABLE II.—CLASSIFIED INDEX TO TABLE I.

The countries are arranged in alphabetical order: I. Austro-Hungary; II. Belgium; III. England, including Scotland and Ireland; IV. France; V. Germany; VI. Holland; VII. Italy; VIII. Russia; IX. Spain; X. United States of America, which for musical purposes are included in Europe.

Under each country the pitches are classified as: 1. Standards; 2. Old Forks; 3. Church Organs; 4. Concert Organs; 5. Operas; 6. Concerts, including Conservatoriums; 7. Pianofortes; 8. Military Music; 9. Other instruments.

The cents and pitch are as in the former table, to which, therefore, immediate reference can be made.

Within each division the pitches are arranged first geographically and then chronologically, but for England the organs by the same makers are generally put together.

The mark ,, means that the number or date above is to be repeated, and — that the date or place is unknown.

¶   The pitches are cited with the greatest brevity which will allow of identification.

Date	Place	Pitch	Cents	a'
		I. AUSTRO-HUNGARY.		
		3. *Church Organs.*		
1640c	Vienna	Large Franciscan organ	368	457·6
1780	,,	Organ-builder Schulz	225	421·3
1878	,,	St. Stefan	313	443·2
,,	,,	Small Franciscan organ	311	442·7
		5. *Opera.*		
1834a	,,	Scheibler, fork I.	276	433·9
,,	,,	,, ,, II.	286	436·5
,,	,,	,, ,, III.	298	439·4
,,	,,	,, ,, IV.	301	440·3
,,	,,	,, ,, V. (Blahetka) . . .	304	441·1
,,	,,	,, ,, VI. ('monstrosity') . .	321	445·1
,,	,,	Vienna Old Sharp Pitch	362	456·1
1878	,,	Ullmann	327	446·8
1859	Pesth	Fr. Com.	323	446·0
,,	Prague	,, ,, ...	338	449·8
		6. *Concerts.*		
1845	Vienna	Marloye (Conservatoire)	321	445·4
		7. *Pianofortes.*		
1780	,,	Stein, for Mozart	226	421·6
1862	,,	Esser, per Naeke	354	454·0
,,	,,	Proch, ,, ...	320	445·0
		II. BELGIUM.		
		1. *Standards.*		
1879	Brussels	Mahillon's Army Standard	345	451·7

TABLE II.—CLASSIFIED INDEX TO TABLE I.—*continued.*

Date	Place	Pitch	Cents	a'
		II. BELGIUM— *continued.*		
		5. *Opera.*		
1859	Brussels	Bender's pitch	310	442·5
		6. *Concerts.*		
,,	Liège	Conservatoire	331	448·0
		8. *Military Instruments.*		
,,	Brussels	Band of Guides (Fr. Com.)	359	455·5
		• III. ENGLAND, SCOTLAND, AND IRELAND.		¶
		1. *Standards.*		
1842	London	Hullah's c″512, really 524·8	305	441·3
1860	,,	Society of Arts intended c″528 . . .	310	444·0
,,	,,	Griesbach's attempt at c″528=534·5 . .	337	449·5
,,	,,	Griesbach's a' to his c″	322	445·7
,,	,,	Cramer's a' and c″	333	448·4
1877a	,,	Tonic Solfa College	250	427·5
1877p	,,	,, ,, ,, ...	230	422·5
		2. *Old Forks.*		
—	,,	Faraday's	237	424·3
1715c	,,	Rev. G. T. Driffield's a	219	419·9
,,	,,	Fork found buried at Brixton, a . . .	355	454·2
1751	,,	Handel's own fork	230	422·5
1800c	,,	Broadwoods' c″	231	422·7
,,	Plymouth	Dr. Stainer's a'	238	424·6
1846c	London	Bryceson's c″	270	432·3	¶
		3. *Church Organs and Bells, and Organ-builders' Church Standards.*		
1625	Lavenham	Church Bell d′288·4	265	431·3
1668a	London	Tomkins's Rule	429	474·1
		Bernhardt Schmidt :		
1660	,,	Whitehall, original	—	—
—	,,	,, altered	308	441·7
1683	Durham	Original	429	474·1
1815p	,,	(Altered)	312	443·1
1879a	,,	(New, by Willis)	318	444·7
1690	Hampton Court	Chapel	308	442·0
,,	,,	Old pipe of original	307	441·7
1708	London	St. James's Chapel Royal, original . .	429	474·1
1759	Cambridge	Trinity College, after shifting . . .	114	395·2
1683	Temple	Original	—	—
1879	,,	Altered	317	444·3
		T. & R. Harris :		
1666	Worcester	Cathedral	280	437·1	¶
		Renatus Harris :		
1670	Newcastle	St. Nicholas	255	428·7
1696	London	St. Andrew Undershaft	251	427·7
1700	,,	St. John's, Clerkenwell	215	419·0
1878a	Norwich	(?) Cathedral	249	427·2
		Green :		
1778	London	St. Katharine's, Regent's Park . .	233	423·2
1780	Winchester	Restoration of College organ . .	220	420·1
1788	Windsor	St. George's Chapel	251	427·8
1790	Kew	Parish Church	230	422·6
—	Dublin	Trinity College (altered ?) . . .	287	436·8
		Christian Smith :		
1696	Boston, Linc.	Parish Church (restored ?) . . .	295	438·9
		Glyn & Parker :		
1749	London	All Hallows the Great and Less . .	237	424·3
		Schreider & Jordans :		
1730	Westminster	Original	—	—
1820a	,,	(Altered)	230	422·5

TABLE II.—CLASSIFIED INDEX TO TABLE I.—*continued.*

Date	Place	Pitch	Cents	a'
		III. ENGLAND, SCOTLAND, AND IRELAND—*continued.*		
		3. *Church Organs, &c.*—continued.		
		Schnetzler :		
1740p	London	German Chapel Royal	242	425·6
1764	Halifax	Parish Church	,,	,,
		Byfield & Green :		
—	Barking	Original probably	429	474·1
1855	,,	(Restored by Walker)	298	439·5
		J. Byfield & J. Harris :		
—	Shrewsbury	Original	—	—
1826	,,	(Altered by Blythe)	—	—
1847	,,	(Altered by Gray & Davison) . . .	275	433·6
		Byfield, Jordan, & Bridge :		
1740	Gt. Yarmouth	St. George's Chapel	244	425·9
		Jordans :		
1744	Maidstone	Old Parish Church	289	437·4
1748	London	St. George's, Botolph Lane . . .	424	474·1
—	Fulham	Parish Church (altered ?) . . .	262	430·4
		Smith of Bristol :		
1838	Bath	Abbey Church	230	422·5
		Walker :		
1843	Wimbledon	Parish Church	246	426·5
		Bryceson :		
1878	London	St. Michael's, Cornhill . . .	357	454·7
		Schulze :		
—	Newcastle	Tynedock	350	452·8
		H. Willis :		
1879a	Salisbury	Cathedral	320	445·2
,,	Glasgow	Established Church Cathedral . . .	,,	,,
,,	Edinburgh	Episcopalian Cathedral	,,	,,
,,	London	St. Paul's, present state (like the other three at 59° F., but) at 57°·5 F. . . .	318	444·6
		Organ-builders' Standard Pipe.		
1878	,,	Bishop, c''518·5	284	436
,,	,,	Gray & Davison, c''523·2	300	440
,,	,,	Walker, c''523·6	301	440·2
,,	,,	Bevington, c''523·7	302	440·5
,,	,,	Lewis, c''524·4	304	441·0
,,	,,	Hill, c''525·3	307	441·7
,,	,,	Bryceson, c''527·3	313	443·4
,,	,,	H. Willis (church), c''529·4	320	445·2
		Experimental English 1-*foot Pipes.*		
		Diam. 1·2 inch ; wind $2\frac{1}{3}$ inch. ; vib. 477·0		
		taken as c'' gives⎫	133	398·7
		,, ,, b' ,, ⎬ in meantone temperament	247	426·6
		,, ,, b'b ,, ⎪	323	445·8
		,, ,, a' ,, ⎭	440	477·0
		Same diam. ; wind $3\frac{1}{4}$ inch. ; vib. 478·7		
		taken as c'' gives⎫	136	400·2
		,, ,, b' ,, ⎬ in meantone temperament	253	428·2
		,, ,, b'b ,, ⎪	329	447·4
		,, ,, a' ,, ⎭	446	478·7
		Bernhardt Schmidt's, same dimensions ; wind $2\frac{1}{3}$ inch. ; vib. 472·9		
		taken as c'' gives⎫	115	395·3
		,, ,, b' ,, ⎬ in meantone temperament	231	423·0
		,, ,, b'b ,, ⎪	308	442·0
		,, ,, a' ,, ⎭	425	472·9
		Diam. ·95 inch ; wind $3\frac{1}{4}$ inch. ; vib. 488·7		
		taken as c'' gives⎫	171	408·5
		,, ,, b' ,, ⎬ in meantone temperament	289	437·1
		,, ,, b'b ,, ⎪	365	456·7
		,, ,, a' ,, ⎭	482	488·7
		Diam. ·75 inch. ; wind $3\frac{1}{4}$ inch. ; vib. 498·6		
		taken as c'' gives⎫	206	416·8
		,, ,, b' ,, ⎬ in meantone temperament	323	446·0
		,, ,, b'b ,, ⎪	400	466·0
		,, ,, a' ,, ⎭	516	498·6

TABLE II.—CLASSIFIED INDEX TO TABLE I.—*continued.*

Date	Place	Pitch	Cents	a'
		III. ENGLAND, SCOTLAND, AND IRELAND—*continued.*		
		4. *Concert Organs.*		
		Elliott :		
1805	London	Ancient Concerts from d″568·3 . . .	239	429·9
		Walker :		
1867	,,	Exeter Hall, original	323	445·8
1879	,,	,, ,, sharpened . . .	328	447·3
1877	Gloucester	Festival organ	329	447·7
		Lewis :		
,,	Glasgow	Public Halls	341	450·6
		H. Willis :		
1877a	London	Concert Standard at Albert Hall and Alexandra Palace	354	453·9 ¶
1877	,,	Albert Hall, observed at 61·5° F. . .	358	455·1
		Gray & Davison :		
,,	Sydenham	Crystal Palace	355	454·1
		Bryceson :		
,,	London	Band pitch	357	454·7
		5. *Opera.*		
1857	,,	Opera, Bettini's fork (correct ?) . . .	362	456·1
		Covent Garden :		
1877	,,	Harmonium	336	449·2
1878	,,	Organ (Bryceson's fork) . . .	305	441·2
,,	,,	Harmonium	329	447·5
1879	,,	Organ (heard)	322	445·6
,,	,,	Band (performing) . . .	338	449·7
1880	,,	Theatre fork (season 1880) . .	282	435·4
		Her Majesty's :		
1878	,,	Organ	285	436·1 ¶
1879	,,	Band (performing) . . .	320	445·5
1880	,,	Theatre fork	319	444·9
		6. *Concerts.*		
		Philharmonic :		
1813–28	,,	Copy of original fork . . .	235	423·7
	,,	Another copy	233	423·3
1826	,,	Approved by Sir G. Smart . .	272	433·0
1846–1854	,,	Mean pitch while the concerts were under the direction of Sir M. Costa . .	349	452·5
1874	,,	Highest	357	454·7
1877	Sydenham	Crystal Palace band . . .	355	454·1
,,	London	Wagner Festival at Albert Hall . .	358	455·1
		7. *Pianofortes.*		
1826	,,	Broadwoods' lowest, London No. 1 of Fr. Com. .	272	433·0
1849–1854	,,	medium, London, No. 2 of Fr. Com. .	323	445·9 ¶
1854p	,,	copy now used	324	446·2
1860	,,	copy made for Society of Arts . . .	321	445·5
1852–1874	,,	highest, London No. 3 of Fr. Com. (which calculated all these forks wrongly) . .	349	452·5
1874p	,,	present highest	357	454·7
1846a	,,	Hipkins's Vocal pitch (meantone) . .	274	433·5
1846p	,,	,, ,, (equal) . . .	284	436·0
1877	,,	Collard	339	449·9
1879	,,	Erard	359	455·3
,,	,,	Steinway (in England) . . .	357	454·7
1877	,,	Chappell	362	455·9
		8. *Military Music.*		
1878	,,	British Army regulation	346	451·9
,,	,,	Kneller Hall Training School . . .	350	452·9

TABLE II.—CLASSIFIED INDEX TO TABLE I.—*continued*.

Date	Place	Pitch		Cents	a′
		IV. FRANCE.			
		1. *Standards.*			
		One French foot pipe :			
1648	Paris	Mersenne c″447	.	17	373·7
1766	,,	Dom Bédos c″450·5	.	31	376·6
1854	,,	Delezenne, c″446·4	.	15	373·1
1700p	,,	Pitch-pipe at Faculty of Sciences . .	.	178	410
1832	,,	de Prony's proposal	.	307	441·7
1834	,,	Marloye's	.	262	430·5
1858	,,	Cavaillé-Coll's proposal . . .	.	316	444·0
,,	,,	Fr. Com. ,, . . .	.	280	435·0
¶ 1859	,,	Diapason Normal, at Conservatoire . .	.	282	435·4
		2. *Old Forks.*			
1700c	Lille	Mazingue's	.	66	384·3
1754	,,	François's	.	230	422·6
1800c	,,	Cohen's	.	255	428·7
1854a	,,	Delezenne's	.	272	432·9
1859a	,,	Marquis d'Aligre's	.	297	439·4
1810c	Paris	Lemoine's	.	260	430·0
		3. *Church Organs.*			
1636	,,	Mersenne's *ton de chapelle* . . .	.	534	503·7
1700a	Lille	L'Hospice Comtesse	.	19	374·2
1789	Versailles	Palace Chapel, fork at Conservatoire .	.	117	395·8
1818	Paris	Tuileries Chapel	.	278	434·3
1840	,,	St. Denis (Cavaillé-Coll) . . .	.	317	444·3
1851a	Lille	St. Sauveur	.	69	384·6
¶ ,,	,,	La Madeleine (restored) . . .	.	129	398·7
,,	,,	St. André	.	269	432·2
		4. *Concert Organ.*			
1851	,,	Festival organ	.	327	446·8
		5. *Opera.*			
		Grand Opera :			
1811	Paris	Scheibler	.	248	427·0
1819	,,	Cagniard de la Tour . . .	.	276	434·0
1822	,,	Fischer	.	267	431·7
1824	,,	lowered for Branchu . . .	.	243	425·8
1829	,,	recovered pitch	.	276	434·0
,,	,,	orchestral pitch	.	300	440·0
1834c	,,	Scheibler's Petitbout . . .	.	276	434·0
1836–1839	,,	Delezenne's Leibner . . .	.	304	441·0
1854a	,,	,, forks . . .	.	289	437·4
1855	,,	Lissajous and Ferrand . . .	.	335	449·0
¶ 1856	,,	Bodin	.	323	445·8
1858	,,	Fr. Com.	.	331	448·0
		Italian Opera :			
1823	,,	Fischer	.	237	424·2
1856	,,	Bodin	.	329	447·4
		Opéra Comique, or Feydeau.			
1820	,,	fork at Conservatoire . . .	.	232	423·0
1823	,,	Fischer	.	250	427·6
1836	,,	Cagniard de la Tour . . .	.	304	441·0
		Provincial Opera :			
1859	Bordeaux	Fr. Com.	.	312	443·6
1838–54	Lille	Delezenne	.	340	450·5
1859	Lyons	Fr. Com.	.	331	448·0
,,	Toulouse	,, ...	.	310	442·5
		6. *Concerts.*			
1836	Paris	Mersenne's *ton de chambre* . . .	.	726	563·1
1812	,,	Conservatoire, fork there . . .	.	298	439·5

TABLE II.—CLASSIFIED INDEX TO TABLE I.—*continued.*

Date	Place	Pitch	Cents	*a'*
		IV. FRANCE *—continued.*		
		6. Concerts—continued.		
1834a	Paris	Conservatoire, Scheibler I.	282	435·3
,,	,,	,, ,, II.	303	440·9
,,	,,	,, ,, III. (Gand) . . .	282	435·2
1856	,,	,, de la Fage	324	446·2
1859	Toulouse	,, Fr. Com.	288	437·0
,,	Marseilles	,, ,, ...	327	447
		7. Pianofortes, Spinets, &c.		
1648	Paris	Mersenne's spinet	148	402·9
1713	,,	Sauveur	163	406·6
1783	,,	Pascal Taskin	174	409·0
1829	,,	Piano of opera	242	425·5
1836	,,	Wölfel's	313	443·3
		V. GERMANY.		
		1. Standards.		
1619	Brunswick	Prætorius's suitable pitch	237	424·2
1834	Stuttgard	Scheibler's pitch (reduced to 59° F.) adopted at the Congress of Physicists . . .	301	440·2
		2. Old Forks.		
1740–1812	Eutin	F. Anton von Weber's	236	424·1
1780	Dresden	Kirsten's	229	422·3
1800	—	Kummer's	239	424·9
		3. Church Organs (in order of date).		
—	N. Germany	Prætorius (called by him *chamber pitch*) highest recorded	740	567·3
1361	Saxony	Halberstadt	541	505·8
1511	Heidelberg	Schlick, high pitch	535	504·2
,,	,,	,, low pitch	33	377·0
1543	Hamburg	St. Catherine (in 1879)	454	480·8
1615	Palatinate	Salomon de Caus	119	396·4
1645	Holstein	Glückstadt	350	453·0
1688	Hamburg	St. Jacobi, low stop, old pitch . . .	184	411·4
,,	,,	,, high stops, ,, . . .	484	489·2
1700c	Holstein	Rendsburg	506	495·5
1714	Saxony	Freiberg Cathedral, Silbermann . .	217	419·5
1713–1716	Strassburg	Minster, A. Silbermann	104	393·2
1722	Saxony	Dresden, St. Sophie	201	415·5
1749	Hamburg	Lehnert's *positiv*	351	455·2
1754–1824	Dresden	Chained fork of the Roman Catholic Church .	199	415·0
1762	Hamburg	Mattheson's St. Michaelis	169	407·9
1833	Weimar	Töpfer's pipe	237	424·4
1878	Dresden	Roman Catholic Church	212	418·1
,,	Lübeck	Cathedral, old organ	465	484·1
1879	Hamburg	St. Jacobi, modern pitch	500	494·5
		5. Opera (arranged by towns).		
1822	Berlin	Fischer's Pichler's fork	289	437·3
1834	,,	Scheibler, 'trustworthy'	307	441·8
1815–1821	Dresden	Naeke's fork of Weber's time . . .	233	423·2
1859	,,	Fr. Com.	304	441·0
1878	,,	Jehmlich's fork	297	439·4
1859	Brunswick	Fr. Com.	278	443·5
,,	Carlsruhe	,, ...	280	435·0
,,	Gotha	,, ...	313	443·3
,,	Weimar	,, ...	319	444·8
,,	Stuttgard	,, ...	312	443

TABLE II.—CLASSIFIED INDEX TO TABLE I.—*continued.*

Date	Place	Pitch	Cents	*a'*
		V. GERMANY—*continued.*		
		Opera (arranged by towns)—*continued.*		
1859	Munich	Fr. Com.	332	448·1
1869	Baden	Sent to Society of Arts	278	434·5
,,	Würtemberg	,, ,,	288	436·9
		Similar forks sent from Berlin and Munich, which had adopted French pitch		
1879	Hamburg	Opera under Krebs	331	448·0
		6. *Concerts.*		
¶ —	,,	Old orchestral pitch	320	445·0
1859	Leipzig	Conservatoire Fr. Com.	334	448·8
,,	Würtemberg	Fr. Com.	319	444·8
1869	Leipzig	Gewandhaus, sent to Society of Arts . . .	332	448·2
		9. *Instruments.*		
1776	Breslau	Marpurg	196	414·4
—	—	Naeke's Schneider's oboe	191	413·3
		VI. HOLLAND.		
		3. *Church Organs.*		
—	—	The old celebrated Church organs had all been altered, and I have not succeeded in recovering their ancient pitch	—	—
		6. *Concerts.*		
1859	The Hague	Fr. Com.	334	446·2
¶		**VII. ITALY.**		
		1. *Standards.*		
1720	Rome	Pitch-pipes of Dr. R. Smith	114	395·2
1730c 1780c	Padua	Mean of pitch-pipes of the bell-foundry of Colbacchini	241	425·2
		2. *Old Forks.*		
1730c	,,	From Colbacchini's low *f''*	152	403·9
1780c	,,	,, ,, high *f''*	230	422·6
		5. *Opera.*		
1845	Florence	Marloye	287	436·7
,,	Milan	,, ...	326	446·6
,,	Turin	,, ...	299	439·4
1856	Milan	Fr. Com.	349	450·3
1857	,,	La Scala (de la Fage)	345	451·7
¶ ,,	Naples	San Carlo (Guillaume)	319	444·9
1859	Turin	Fr. Com.	319	444·8
		6. *Concerts.*		
1869	Bologna	Liceo Musicale (Society of Arts) . . .	312	443·1
		7. *Pianofortes.*		
1839	,,	Tadolini's fork	243	425·8
		VIII. RUSSIA.		
		3. *Church Organs.*		
1781	St. Petersburg	Euler	211	418·0
1860	,,	Court Church	224	421·2
		5. *Opera.*		
1802	,,	Sarti	284	436·0
1858	,,	Fr. Com. (French pitch was afterwards adopted) .	345	451·5

TABLE II.—CLASSIFIED INDEX TO TABLE I.—*continued.*

Date	Place	Pitch	Cents	a'
		IX. SPAIN.		
		3. Church Organs.		
1785	Seville	T. Bosch's organ	218	419·6
1858	Madrid	Ton de Chapelle	218	419·6
		5. Opera.		
„	„	Theatre (French pitch adopted in 1879) . . .	318	444·5
		X. UNITED STATES OF AMERICA.		
1868	New York	E. S. Ritchie's standard, and Mason & Hamlin's French pitch	283	435·9
1880	Boston	Church of Immaculate Conception	315	443·9
„	New York	Hutchings, Plaisted & Co., 'low organ pitch' .	316	444·2
„	„	Nichol's Fork, Germania orchestra	333	448·5
„	Boston	Music Hall organ (from 1863 to 1871 at French pitch)	342	450·9
„	Cincinnati	Organ tuned to Thomas's orchestra . . .	362	456·1
1879	New York	Steinway's American pitch, from a fork furnished by Steinway	366	457·2
1880	„	Steinway's, from a fork furnished by R. Spice .	369	458·0
„	„	Highest New York pitch, from a fork furnished by R. Spice	380	460·8

¶

CONCLUSIONS.

Art. 6. The two preceding tables contain the facts of the history of musical pitch in Europe since 1361, the date of the Halberstadt organ, that is for 500 years, so far as I have been able to collect information, and I have been fortunate enough to bring together such an amount of historical evidence that probably no new ¶ facts could be ascertained which would materially change the conclusions to which I have been led. These are very briefly as follows.

Art. 7. The organ was originally a mere collection of pitch-pipes, each with a fixed tone, to steady the voice of the singers of ecclesiastical chants, replacing the single pitch-pipe with a movable piston or some instrument like the flageolet (whistle) and oboe, which subsequently gave rise to the two distinct series of flue and reed pipes. But when thus collected it was necessary to fix a pitch. The guiding principles were the compass of the male voice, the rules of ecclesiastical song, the ease of the performer, to avoid introducing chromatics as much as possible (Schlick), and the standard measure or foot rule of the country. The latter suggested a whole number of feet for the length of the standard pipe, generally four feet, about the lowest note of the tenor voice, and the question thus rose what note should this tone represent? Here the answer came from ecclesiastical use,—either F or c. Schlick recommends both, thus giving pitches for any given note a whole Fourth apart. Schlick's high pitch, arising from giving a 6½-foot Rhenish pipe ¶ to c, made a'504·2. (All pitches named should be referred to in Table 1.) His low pitch arising from giving the *same* pipe to F, made a'377. These are a Meantone Fourth apart.

Art. 8.—The foot had very different lengths in different countries. If we suppose the 'scale' (or ratio of diameter to length of pipe) and the force of wind to remain the same (both in fact varied much), then the influence of the length of the foot on the pitch of the organ, supposing the four-foot or one-foot pipe to be given to the same note, may be appreciated from the table on p. 512a. In this we see a difference of more than a Tone, nearly a minor Third, between the pitch of a 1-foot pipe in France and in Saxony. The difference between the pitches of pipes of the lengths of the English foot and French foot is more than an equal Semitone. Hence probably it happened that the lowest French pitch measured, a'374·2, is a Semitone flatter than the lowest English pitch measured a'395·2. Length of foot alone would therefore account for great variety of organ-pitch, to which we must add force of wind (see the notes on experimental English 1-foot pipes, p. 506c) and different methods of voicing. The low pitches were (and still are on old organs) prevalent in France and Spain, the high pitches were at home in North Germany (see Table II.).

Names of Feet	Length	Interval
	mm.	cents
Long old French foot, or *pied de roi* . .	325	0
Long Austrian foot	316	49
Long German, or Rhenish foot . . .	314	60
ENGLISH FOOT	305	109
Old Nürnberg foot	304	116
Old Augsburg foot	296	162
Old Roman foot (medieval) . . .	295	168
Bavarian foot	292	185
Short Hamburg and Danish foot . .	286	221
Short Brunswick and Frankfurt foot . .	285	227
Short Saxon foot.	283	239

Art. 9.—The solo instruments were tuned very variously. But it became the
¶ custom to have a band to play with the organ, and the princes and petty dukes used
the same bands to play in their private apartments or ' chamber.' The very high
and very low pitch were generally found unsuitable for non-ecclesiastical music.
Hence the instruments usually adopted a pitch lower than the high and higher than
the low, and this was called ' chamber pitch,' the other being distinguished as
' church pitch.' But the same instruments had also to play with the organ.
Hence the difference had to be a definite number of degrees of the scale, a Semitone, a Tone, or a minor Third. See $a'407\text{·}9$, and especially $a'411\text{·}4$, which compare with $a'480\text{·}8$, and $a'484\text{·}1$ respectively. This was, however, not always the
case, for the very high church pitch, $a'503\text{·}7$ had a still higher chamber pitch
$a'563\text{·}1$.

Art. 10.—But this great variety occasioned much trouble, and the chamber
pitch below the high and above the low church pitch seems to have suggested
Prætorius's ' suitable pitch ' of $a'424\text{·}2$ in 1619. This was in fact a ' mean pitch,'
and as such rapidly found such favour that it spread over all Europe and, with
¶ insignificant varieties (from $a'415$ to $a'428\text{·}7$ at the extremes, an interval of 54
cents, or a quarter of a Tone), prevailed for two centuries. Handel's own fork,
$a'422\text{·}5$ in 1751, quite a common pitch at the time, and the London Philharmonic
fork, $a'423\text{·}3$ from its foundation to 1820, are conspicuous examples, but an inspection of the numerous pitches cited in Table I. sect. 4 (pp. 495d-7), will prove the
fact beyond doubt.

Art. 11.—As this was the period of the great musical masters, and as their
music is still sung, and sung frequently, it is a great pity that the pitch should
have been raised, and that Handel, Haydn, Mozart, Beethoven, and Weber, for
example, should be sung at a pitch more than a Semitone higher than they intended. The high pitch strains the voices and hence deteriorates from the effect
of the music, when applied to compositions not intended for it. Of course for
music written for a high pitch the compass of the human voice is properly studied
(see App. XX. sect. N. No. 1), and so much music has in the last fifty years been
written for a high pitch, that to perform both properly two sets of instruments
would be required. Two sets are actually in use at Dresden, one for the theatre
¶ $a'439\text{·}4$, and one for the Roman Catholic Church having $a'415$, difference 98 cents,
or about a Semitone.

Art. 12.—The rise in pitch began at the great Congress of Vienna, 1814, when
the Emperor of Russia presented new and sharper wind instruments to an Austrian
regiment of which he was colonel. The band of this regiment became noted for
the brilliancy of its tones. In 1820 another Austrian regiment received even
sharper instruments, and as the theatres were greatly dependent upon the bands
of the home regiments, they were obliged to adopt their pitch. Gradually at
Vienna, pitch rose from $a'421\text{·}6$ (Mozart's pitch) to $a'456\text{·}1$, that is, 136 cents, or
nearly three-quarters of a Tone. The mania spread throughout Europe, but at
very different rates. The pitch reached $a'448$ at the Paris Opera in 1858, and the
musical world took fright.

Art. 13.—The Emperor of the French appointed a commission to select a pitch,
and this determined on $a'435$, but made a fork called *Diapason normal*, now
found to be $a'435\text{·}4$, which is preserved at the Musée du Conservatoire, and is the
only standard pitch in the world. This pitch was widely adopted, but it is 56 cents,

or over a quarter of a Tone, sharper than Mozart's pitch, although it was 80 cents, fully three-quarters of a Semitone, flatter than the old Vienna sharp pitch a' 456·1, and 49 cents, or a quarter of a Tone, flatter than the then French opera pitch a' 448.   This pitch had been reached independently in many places, and the French commission had been twitted at taking a Carlsruhe pitch.   But it is not generally known that Sir George Smart's pitch a' 433, adopted with much hesitation for the London Philharmonic Society about 1820, and extensively sold in London as the 'London Philharmonic' for many years before the French Commission of 1859, was in fact an anticipation of the French pitch.   Both were compromises, a partial yielding to the new without entirely disregarding the old.   The pitches a' 430 to a' 436·9, therefore (interval 28 cents, or about $\frac{1}{7}$ Tone), forming Table I. Sect. 5, pp. 497-8, are termed the 'compromise pitch.'   As instruments exist for this pitch it is the only one that has a chance of being used beside the present sharp pitch of England.   Several attempts have been made to restore it, notably at Covent Garden Opera in 1880.   But the expense of new instruments for a band, about 1,000*l.*, renders any alteration extremely difficult to carry out.   The ¶ tendency in England has been to sharpen, and our orchestral and pianoforte pitch is now from a' 449·7 to a' 454·7, a difference of only 19 cents, not quite a comma. In the United States, however, the pitch has reached a' 460·8, that is 23 cents, or about a comma more.   In Germany the compromise pitch adopted was a' 440·2 as proposed by Scheibler, and it is curious that the standard pipes of the English church organ builders vary from a' 436 to a' 445·2, 36 cents, but are mostly between 440 and 441·7, an interval of only 7 cents.   The concert organs, of course, follow orchestral pitch.   (See *Postscript*, p. 555.)

Art. 14.—In England the pitch of organs varied with the note on which the four-foot or one-foot pipe was placed.   We have only one record that the one-foot pipe was placed on c'' giving a' 395·2, whereas the same pipe made to give b' produced a' 423, the mean pitch, which so long prevailed.   Put on $b'\flat$ it produced a' 442, which as a' 441·7 was Bernard Schmidt's low pitch, and is still the pitch of Mr. T. Hill, the organ-builder.   Placed on a' it gave a' 472·9, which as a' 474·1 was the highest church pitch used in England, just a Tone above mean pitch. (See p. 505*c*, III. 3, for details.)   ¶

Art. 15.—If we look into the secrets of the rise of pitch we find it always connected with wind instruments.   The first rise was from a military band, and the wind and the brass have constantly rebelled against a low pitch.   The singers have not prevailed against them except for a very short time.   The great violin school of Cremona in Italy lived in the time of mean pitch with a higher chamber pitch, and the resonance of the boxes of their violins seems to shew traces of the action of both pitches (suprà, p. 87, note *), but their great object was to insure tolerable uniformity of reinforcement, and hence they are a treasure for all time.

Art. 16.—The only possible conclusion seems to be that to sing music written for pitches different from our own, we must either transpose a Semitone (always a difficulty, and for some instruments an impossibility) or adopt a new compromise pitch, the French, already once firmly rooted in England as Sir George Smart's, and standing half-way between the extremes.   On the continent, as formerly shewn in France, and quite recently in Belgium and Italy, the government has a certain power in fixing musical pitch, by refusing to subsidise conservatories and theatres which do not adopt the pitch ordered, and commanding the regimental bands to ¶ make the change.   But beyond this their power does not extend, and the various regulations which have been made in the two countries last named shew the great difficulties that have to be overcome in introducing a new pitch even within the area under government control.   In England, however, there are no subsidised operas or musical conservatories, and even the instruments of the military bands are not provided by government.   Hence the change must be left to the gradual action of musical feeling.   We have already changed in England almost imperceptibly.   The raising of English pitch from Sir George Smart's a' 433 was to a great extent due to the individual action of the late Sir Michael Costa while conductor of the Philharmonic concerts 1846-45 (mean a' 452·5, extreme a' 454·7), to whose insistence is also due the high pitch of the Albert Hall concert organ, a' 453·9.   Perhaps a similar energetic conductor will arise to turn the tide of musical opinion in the opposite direction.

SECTION K.

NON-HARMONIC SCALES.

(See Notes, pp. 71, 95, 237, 253, 255, 257, 258, 264, 272.)

Art.	Art.
I. Introduction, p. 514.	IV. How these Divisions of the Octave may
II. Table of Non-Harmonic Scales, p. 514.	have arisen, p. 522.
III. Annotations to the Table, p. 519.	V. Results of the inquiry, p. 524.

I. INTRODUCTION.

For particulars of my researches into non-harmonic scales, see my two papers, first: 'Tonometrical Observations on some existing non-harmonic Scales' (*Proc. of the R. Society* for Nov. 20, 1884, vol. xxxvii. p. 368) and second '. On the Musical Scales of Various Nations' (*Journal of the Society of Arts* for March 27, 1885, ¶ vol. xxxiii. p. 485), in both of which I was most materially assisted by Mr. Alfred James Hipkins of Messrs. J. Broadwood and Sons'.

Properly speaking there is only one harmonic scale, that is, a scale which allows the musician to produce chords without beats, and therefore has notes with pitch numbers composed of products and multiples of the powers of 2, 3, 5, 7, 17, as shewn in Sect. E. But the term harmonic may be extended to all tempered imitations of such scales as are not worse than equal intonation. If we did not extend the use of the term thus far, we should find absolutely no harmonic scale in practical use, except by the Tonic Sol-faists when unaccompanied (App. XVIII.) Even with this extension of meaning, non-harmonic scales are greatly more numerous than harmonic. Harmony was a European discovery of a few centuries back, and it has not penetrated beyond Europe and its colonies.

¶ In order to obtain a bird's-eye glance over the scales given theoretically by ancient Greek writers (as interpreted in the text), by ancient and medieval Arabic writers (as interpreted by Professor Land); by modern Arabic theorists (as reported by Eli Smith); by Indian musicians (as reported by Rajah Sourindro Mohun Tagore); and those which I have deduced from Javese, Chinese, and Japanese instruments, with those of other countries, examined by Mr. Hipkins and myself, I have constructed the following table. The scale is represented by the numbers of cents in the interval by which any one of its notes is sharper than the lowest note, and is generally confined to one octave. The interval between any two notes in the scale is then found by subtraction. The number of the note in the scale is usually placed at the top, so that the eye can, at a glance, compare the different usages. The ratios represented by these cents may generally be found from the table in Sect. D. Each scale is numbered, and in the annotations immediately following the table, several particulars are given. It was not, however, possible to include every case in this arrangement. The complete ancient and medieval Arabian lute, Rabáb, and Tambour scales, and the complete Indian scales both in the old and modern form, and some others are therefore differently ordered, preserving, however, the expression of notes by cents as above explained. See Nos. 66 to 75.

II. TABLE OF NON-HARMONIC SCALES.

¶

Old Greek Tetrachords.

	I.	II.	III.	IV.
1. Olympos	0	112	----	498
2. Old Chromatic	0	112	182	498
3. Diatonic	0	112	316	498
4. Didymus	0	112	294	498
5. Doric	0	90	294	498
6. Phrygian	0	182	316	498
7. Lydian	0	182	386	498
8. Helmholtz	0	112	386	498
9. Soft Diatonic	0	85	267	498
10. Ptolemy's equal diatonic . .	0	151	316	498
11. Enharmonic	0	55	112	498

Greek Tetrachords after Al Farabi reported by Prof. Land.

I. *Genus molle, ordinatum.*
    a. continuum :—

12. laxum	0	386	441	498
13. mediocre	0	316	405	498
14. acre	0	267	386	498

II. TABLE OF NON-HARMONIC SCALES—*continued*.

Greek Tetrachords after Al Farabi reported by Prof. Land— continued.

	I.	II.	III.	IV.
I. *Genus molle, ordinatum*—cont.				
b. non continuum.				
15. laxum (enharmonic) . . .	o	386	460	498
16. mediocre (soft chromatic) . .	o	316	435	498
17. acre (syntonically chromatic) . .	o	267	418	498
II. *Genus forte.*				
a. duplicatum :—				
18. primum	o	231	462	498
19. secundum	o	204	408	498
20. tertium	o	182	365	498
b. conjunctum :—				
21. primum (entonically diatonic) .	o	231	435	498
22. secundum (syntonically diatonic) .	o	204	386	498
23. tertium (equally diatonic) . .	o	182	347	498
b. disjunctum :—				
24. primum (soft diatonic) . . .	o	231	413	498

Most Ancient Form of Greek Scales with 7 Tones and Octave.

	I.	II.	III.	IV.	V.	VI.	VII.	VIII.
25. Lydian	o	182	386	498	702	884	1088	1200
26. Phrygian	o	182	316	498	702	884	1018	1200
27. Doric	o	90	294	498	702	792	996	1200
28. Hypolydian	o	204	386	590	702	884	1088	1200
29. Hypophrygian (Ionic) . .	o	204	386	498	702	884	1018	1200
30. Hypodoric (Eolic) . . .	o	204	294	498	702	792	996	1200
31. Mixolydian	o	112	294	498	610	814	996	1200

Later Greek Scales with Pythagorean Intonation.

	I.	II.	III.	IV.	V.	VI.	VII.	VIII.
32. Lydian	o	204	408	498	702	906	1110	1200
33. Hypophrygian (Ionic) . .	o	204	408	498	702	906	996	1200
34. Phrygian	o	204	294	498	702	906	996	1200
35. Eolic	o	204	294	498	702	792	996	1200
36. Doric (same as No. 27) . .	o	90	294	498	702	792	996	1200
37. Mixolydian	o	90	294	498	588	792	996	1200
38. Syntonolydian	o	204	408	612	702	906	1110	1200

Al Farabi's Greek Scales as reported by Prof. Land.

	I.	II.	III.	IV.	V.	VI.	VII.	VIII.
39. Genus conjunctum medium . .	o	204	408	590	702	906	1088	1200
40. Genus duplicatum medium, or ditonum (same as No. 38) . .	o	204	408	612	702	906	1110	1200
41. Genus conjunctum primum .	o	204	435	639	702	933	1137	1200
42. Genus forte duplicatum primum .	o	204	435	666	702	933	1164	1200
43. Genus conjunctum tertium, or forte æquatum	o	204	386	551	702	884	1049	1200
44. Genus forte disjunctum primum .	o	204	435	617	702	933	1115	1200
45. Genus non continuum acre .	o	204	471	622	702	969	1120	1200
46. Genus non continuum mediocre .	o	204	520	639	702	1018	1137	1200
47. Genus non continuum laxum .	o	204	590	664	702	1088	1162	1200
48. Genus chromaticum forte . .	o	204	471	690	702	969	1088	1200
49. Genus chromat um mollissimum .	o	204	520	613	702	1018	1111	1200
50. Genus mollissin m ordinantium .	o	204	590	647	702	1088	1145	1200

Arabic and Persian Scales as reported by Prof. Land.

	I.	II.	III.	IV.	V.	VI.	VII.	VIII.
51. Zalzal, see No. 66	o	204	355	498	702	853	996	1200

Highland Bagpipe made by Macdonald of Edinburgh.

	I.	II.	III.	IV.	V.	VI.	VII.	VIII.
52. Observed	o	197	341	495	703	853	1009	1200

Modern Arabic Scale as reported by Eli Smith.

	I.	II.	III.	IV.	V.	VI.	VII.	VIII.
53. Meshāqah, theoretical . . .	o	200	350	500	700	850	1000	1200

II. Table of Non-Harmonic Scales—*continued*.

Arabic Medieval Scales as reported by Prof. Land with 7 Tones and Octave.

	I.	II.	III.	IV.	V.	VI.	VII.	VIII.
54. 'Ochaq (same as No. 33) . . .	o	204	408	498	702	906	996	1200
55. Nawa (same as No. 34) . . .	o	204	294	498	702	906	996	1200
56. Boasīlik (same as No. 37) . . .	o	90	294	498	588	792	996	1200
57. Rast	o	204	384	498	702	882	996	1200
58. Zenkouleh	o	204	380	498	678	882	996	1200
59. Rahawi	o	180	384	498	678	792	996	1200
60. Hhosaïni	o	180	294	498	678	906	996	1200
61. Hhidjazi	o	180	294	498	678	882	996	1200

Arabic Medieval Scales with 8 Tones and Octave.

	I.	II.	III.	IV.	V.	VI.	VII.	VIII.	IX.
62. 'Iraq	o	180	384	498	678	882	996	1176	1200
63. Içfahan	o	180	384	498	702	882	996	1176	1200
64. Zirafkend	o	180	294	498	678	792	882	1086	1200
65. Bouzourk	o	180	384	498	678	702	906	1086	1200

66. *Earlier Notes on the Arabic Lute as reported by Prof. Land.*

*** First, Second, &c., refer to the strings. The notes are named from the fingers—index, middle, ring, little—by which they were played.

Notes	First Octave	Second Octave	Cents Oct. 1	Cents Oct. 2 1200+
C	First : open . . .	Third : index . . .	o	o
Db	ancient near index . .	ancient middle . . .	90	90
		Persian middle . . .	—	99
	Persian near index . .		145	—
		Zalzal's middle . .	—	151
	Zalzal's near index . .	. . . .	168	—
D	index	ring	204	204
Eb	ancient middle . . .	little = Fourth : open .	294	294
	Persian middle . . .		303	—
	Zalzal's middle . . .		355	—
Fb	. . . . .	Fourth : ancient near index .	—	384
E	ring		408	—
		Persian near index . .	—	439
		Zalzal's near index . .	—	462
F	little = Second : open .	index	498	498
Gb	Second : ancient near index .	ancient middle . . .	588	588
		Persian middle . . .	—	597
	Persian near index . .		643	—
		Zalzal's middle . .	—	649
	Zalzal's near index . .	. . . .	666	—
G	index	ring	702	702
Ab	ancient middle . . .	little = Fifth : open . .	792	792
	Persian middle . . .		801	—
	Zalzal's middle . .	. . . .	853	—
Bbb	. . . . .	Fifth : ancient near index .	—	882
A	ring		906	—
		Persian near index . .	—	937
		Zalzal's near index . .	—	960
Bb	little = Third : open . .		996	—
Cb	Third : ancient near index .	Fifth : ancient middle . .	1086	1086
		Persian middle . . .	—	1095
	Persian near index . .		1141	—
		Zalzal's middle . .	—	1147
	Zalzal's near index . .	. . . .	1164	—
C	index	ring	1200	1200

II. Table of Non-Harmonic Scales—*continued*.

67. Medieval Arabic Scales as reported by Prof. Land.

*** Names of strings as in No. 66, names of notes as altered by the Arabic medieval writers.

No.	Notes	First Octave	Second Octave	Cents Oct. 1	Cents Oct. 2 1200+
1	C	First : open . . .	Third : index . . .	0	0
2	Db	remnant . . .	Persian . . .	90	90
3	Ebb	near	Zalzal . . .	180	180
4	D	index	ring . . .	204	204
5	Eb	Persian . . .	little . . .	294	294
6	F'b	Zalzal . . .	Fourth : remnant . .	384	384
7	E	ring		408	
			near (= Gbb) . . .	—	474
8	F	little	index . . .	498	498
9	Gb	Second : remnant . .	Persian . . .	588	588
10	Abb	near	Zalzal . . .	678	678
11	G	index	ring . . .	702	702
12	Ab	Persian . . .	little . . .	792	792
13	Bbb	Zalzal . . .	Fifth : remnant . .	882	882
14	A	ring		906	—
			near (= Cbb) . . .	—	972
15	Bb	little	index . . .	996	996
16	Cb	Third : remnant . .	Persian . . .	1086	1086
17	Dbb	near	Zalzal . . .	1176	1176
1'	c	index	ring . . .	1200	1200

68. Northern Tambour, or that of Khorassan, as reported by Prof. Land.

C 0, Db 90, Ebb 180, *D 204, Eb 294, F'b 384, E 408, *F 498, Gb 588, F♯ 612, *G 702, †Ab 792, G♯ 816, A 906, Bb 996, †A♯ 1020, B 1110, *c 1200, b♯ 1224, c♯ 1314, *d 1404 cents.

* Fixed tones.   † Auxiliary tones.

Rabáb or 2-stringed viol, after Prof. Land.

69. first	0	204	316	408	520	590	632	724	906
70. second	0	204	316	408	590	612	724	816	998
71. third	0	204	316	408	590	794	906	998	1180

Southern Tambour, or that of Bagdad, as reported by Prof. Land.

72. theoretical	0	44	89	135	182	231	275	320	366	413	462

Indian Chromatic Scale.

*** Arranged according to the older and more modern division as inferred from indications by Rajah Sourindro Mohun Tagore.

Degrees	1	2	3	4	5	6	7	8	9	10	11	12	13	14	15	16	17	18	19	20	21	22
Notes	C	Dbb	Db	—	D	Ebb	Eb	E	E♯	F	—	F♯	F♯♯	G	Abb	Ab	—	A	B♭b	Bb	B	B♯
73. Old .	0	51	102	153	204	264⅔	325⅓	386	442	498	549	600	651	702	753	804	855	906	966⅔	1027⅓	1088	1144
74. New .	0	49	99	151	204	259	316	374	435	498	543	589	637	685	736	787	841	896	952	1011	1070	1135

Indian Semitonic Scale as inferred from Measurement of a Madras Vina.

75. First Octave . .	0	89	178	269	373	475	596	684	781	879	996	1081	1199
Second Octave .	1199	1280	1376	1466	1567	1681	1776	1891	1984	2090	2187	2298	2398

*** The Indian partial scales enumerated by Rajah S. M. Tagore, as made up from the 19 notes in Nos. 73 or 74, 32 of them with 7 notes, 112 with 6 notes, and 160 with 5 notes each, are not given because he does not distinguish the minor variations of one degree.

Indian Partial Scales as played by Rajah Rám Pál Singh.

	I.	II.	III.	IV.	V.	VI.	VII.	VIII.
76. First	0	183	342	533	685	871	1074	1230
77. Fourth	0	174	350	477	697	908	1070	1181
78. Second	0	183	271	534	686	872	983	1232
79. Third	0	111	314	534	686	828	1017	1198
80. Fifth	0	90	366	493	707	781	1080	1087

II. TABLE OF NON-HARMONIC SCALES—*continued.*

Various Wood Harmonicons.

	I.	II.	III.	IV.	V.	VI.	VII.	VIII.
81. Balafong from *Patna*	o	187	356	526	672	856	985	1222
82. Balafong from *Singapore*	o	169	350	543	709	894	1040	1205
83. Patala from *Burmah*	o	176	350	533	707	899	1053	1246
84. Balafong from the same	o	114	350	550	687	838	1032	1196
85. Ranat from *Siam.* See p. 556	o	129	277	508	726	771	1029	1254
86. Balafong from *Western Africa*	o	152	287	533	724	890	1039	1200
¶ 87. Gen. Pitt-Rivers's Balafong from the same	o	195	289	513	686	796	1008	1209

PENTATONIC SCALES.

The Black Digitals of a Pianoforte.

	I.	II.	III.	IV.	V.	VI.
88. beginning with *C♯*	o	200	500	700	900	1200
89. ,, ,, *D♯*	o	300	500	700	1000	1200
90. ,, ,, *F♯*	o	200	400	700	900	1200
91. ,, ,, *G♯*	o	200	500	700	1000	1200
92. ,, ,, *A♯*	o	300	500	800	1000	1200

South Pacific.

	I.	II.	III.	IV.	V.	VI.
¶ 93. Balafong	o	202	370	685	903	1200

Javese Scales, as observed from Instruments and Musicians.

	I.	II.	III.	IV.	V.	VI.
94. Salendro, observed	o	228	484	728	960	1200
95. ,, assumed	o	240	480	720	960	1200

Javese Pelog, Chromatic Scale, from which the others are selected.

	I.	II.	III.	IV.	V.	VI.	VII.	VIII.
96. The seven notes	o	137	446	575	687	820	1098	1200

The Five-Note Scales Selected.

	I.	II.	III.	IV.	V.	VI.	VII.	VIII.
¶ 97. Pelog	o	—	446	575	687	—	1098	1200
98. Dangsoe (*oe* as in shoe)	o	137	—	—	687	820	1098	1200
99. Bem	o	137	—	575	687	—	1098	1200
100. Barang	o	137	—	575	687	820	—	1200
101. Miring	o	—	446	575	—	820	1098	1200
102. Menjoera (*joe* = English *you*)	o	137	446	575	—	—	1098	1200

Chinese mixed Pentatonic and Heptatonic Scales, as observed.

* Notes marked * introduced for heptatonic playing.

	I.	II.	III.	IV.	V.	VI.	VII.	VIII.
103. Flute (Ti-tsu)	o	178	*339	448	662	888	*1103	1196
104. Oboe (So-na)	o	145	297	440	637	813	1014	1216
105. Mouth-organ (Shêng)	o	210	338	498	715	908	1040	1199
106. Gong-chime (Yün-lo)	o	169	367	586	674	775	1062	1208
107. Dulcimer (Yang-chin)	o	169	*274	491	661	878	*996	1198
108. Tamboura (Sien-tsu)	o	189	386	—	702	893	—	1200
109. Balloon Guitar (p'i-p'a)	o	145	351	—	647	874	—	1195

II. Table of Non-Harmonic Scales—*continued*.

Pentatonic Scales—*continued*.

*Japanese, chiefly Pentatonic, but with extra notes marked *.*

Koto Tuning, Popular Scales.

	I.	II.	III.	IV.	V.	VI.
110. Hiradioshi, theoretical . . .	o	204	316	702	814	1200
111. „ female player . .	o	193	357	719	801	1199
112. „ music-master	o	185	337	683	790	1200
113. Akebono I., theoretical . . .	o	200	300	700	900	1200
114. Akebono II. „ . . .	o	100	500	700	800	1200
115. Kumoi I. „ . . .	o	100	500	700	800	1200
116. Kumoi II. „ . . .	o	100	500	700	800	1200
117. Han-Kumoi „ . . .	o	200	500	700	800	1200
118. Kata-Kumoi „ . . .	o	200	300	700	800	1200
119. Sakura „ . . .	o	100	500	700	800	1200
120. Iwato „ . . .	o	100	500	600	1000	1200
121. Han-Iwato „ . . .	o	100	500	700	1000	1200
122. Kata-Iwato „ . . .	o	100	500	600	1000	1200
123. Kumoi „ . . .	o	100	500	700	800	1200

Koto Tuning, Classical Scales.

	I.	II.	III.	IV.	V.	VI.
124. Ichikotsu-Chio, theoretical . .	o	200	500	700	900	1200
125. Hio-Dio „ . .	o	200	500	700	900 ,	1200
126. Sou-Dio ,, . .	o	200	500	700	900	1200
127. Wausiki-Chio „ . .	o	200	500	700	900	1200
128. Sui-Dio „ . .	o	200	500	700	1000	1200
129. Bausiki-Chio „ . .	o	300	500	700	1000	1200

Heptatonic Scales.

	I.	II.	III.	IV.	V.	VI.	VII.	VIII.
130. Classical reosen, theoretical . .	o	200	400	*600	700	900	*1100	1200
131. „ ritsusen „ . .	o	200	300	*500	700	900	*1000	1200
132. Popular I. „ . .	o	100	300	500	700	800	1000	1200
133. Popular II. „ . .	o	100	300	500	600	800	1000	1200

Japanese Biwa, Classical Instrument, Tetrachords observed on different Strings.

	I.	II.	III.	IV.	V.
134. Lowest string	o	225	332	416	512
135. Second lowest	o	223	338	429	500
136. Second highest	o	195	320	407	496
137. Highest string	o	212	321	414	503
138. Mean	o	214	328	416	503
139. Theoretically assumed as . .	o	200	300	400	500

III. Annotations to the Table.

Nos. 1 to 11 are those given in the text, pp. 262-5, but No. 8 was merely suggested by Prof. Helmholtz.

Nos. 12 to 24 are from App. III. to Prof. Land's paper, *Over de Toonladders der Arabische Musiek* (on the Arabic musical scales), and contain his corrections of the very faulty MS. of Al Farabi; the numbers are also given by Kosegarten, p. 55. After 24, the numbers suddenly cease in the MS.

Nos. 25 to 31 are the old theoretical form of the Greek scales with the old tetrachords, see suprà, p. 268 *c, d'*.

Nos. 32 to 38 are taken from suprà, p. 269 *a*.

Nos. 39 to 50 are from Prof. Land, *ibid*. p. 38, corrected from the MS. at Leyden; for No. 45 the copyist had repeated No. 44, and Prof. Land has supplied the numbers by analogy.

No. 51 is inferred from the complete set of notes used on the Arabic lute at different times as shewn in No. 66.

No. 52. The Highland bagpipe representing that scale has been inserted immediately afterwards to show its practical identity. It was played to Mr. Hipkins and myself by Mr. C. Keene, the artist.

No. 53. The very modern survival of the same scale has been put next. It is described, suprà, p. 264 note **. In practice each note might be sharpened by one or more quartertones.

Nos. 54 to 65 are the twelve scales given, suprà, p. 284, from Prof. Land, but the four which employ 8 notes are now placed last. No. 61, *Hhidjazi*, is, in fact, more harmonic than the usual equal temperament. If we begin on the note vii., and reckon the intervals from it through an octave, afterwards subtracting 996, it gives the scale 0 204 384 498 702 882 1086 1200, and if 384 882 1086 were each increased by 2 cents, this would be our just major scale. The difference is not felt even in chords, as I have ascertained by actually playing them on a properly tuned concertina.

No. 66 is the complete collection of the notes on the old Arabic lute, as used at different times, reported by Prof. Land. Of course the Persian and Zalzal's notes could not be used together, and when Zalzal's 355 and 853 were used, both 294 and 408, and also both 792 and 906 had to be discontinued, producing No. 51.

No. 67 gives the complete 17 mediæval Arabic notes as determined by Prof. Land, with the ¶ 2 extra ones which appear in the second Octave. Villoteau (op. cit. suprà, p. 257, note ‡, ed. 1809, folio, vol. i.) declared, as is well known, that the most generally received Arabic division of the Octave is into thirds of a Tone (op. cit. p. 613). Prof. Land has demonstrated (Gamme Arabe, p. 62) that this is not the case. Villoteau, an excellent musician, sent to Egypt by the French Government to study the native music, had every facility given to him, and had native musicians at his beck and call. How did he arrive at this opinion? After an attentive study of his book I consider the following hypothesis probable. The greater number of theorists gave 17 notes to the Octave. This was the medieval Arabic scale, No. 67. Villoteau was not used to just intervals, and he was a very ¶ poor arithmetician (see the remarkable note op. cit. p. 668). He was used to the 'musicians' cycle' of 55 degrees (suprà, p. 436d, viii.), in which the Tone contained 9, the major Semitone 5, and the minor 4 degrees (op. cit. pp. 667, 678). When he heard the scale of Rast played (see p. 284 and No. 57, for the medieval form 0 204 384 498 702 882 996 1200 cents), which was the principal Egyptian scale, he tried to sing it as A B $C\sharp$, &c., in his 55 degrees, but was immediately told that his $C\sharp = 392$ cents was too sharp. This would hardly have been the case if the true medieval 384 had been played to him. He next tried A B $C = 305$ cents, but then C was too flat. Now the interval C to $C\sharp = 87$ cents was his minor Semitone of 4 degrees. Hence he concluded that 3 degrees = 65 cents in his tempera- ¶ ment would be right, and that satisfied the natives (op. cit. p. 679). This, however, was one third of his Tone. But he found also 17 tabaqat or transpositions of each scale, proceeding by Fourths, called by him perfect, but as he really considered 17 of these Fourths to make up 7 Octaves, he arrived at a cycle of 17 degrees, each having 71 cents (ex. 70·588233) or being almost precisely a small Semitone 24 : 25 (=ex. 70·673 cents). To the nearest cent, using his symbols, where × means increased by one third, and ♯ increased by two thirds of a Tone, and ↳, ♭ mean diminished by the same amount, the notes of this cycle in cents were, 1 A 0, 2 $A \times = B\flat$ 71, 3 $A\sharp = B$↳ 141, 4 B 212, 5 C 282, 6 $C \times = D\flat$ 353, 7 $C\sharp = D$↳ 424, 8 D 494, 9 $D \times = E\flat$ 565, 10 $D\sharp = E$↳ 635, 11 E 706, 12 F 776, 13 $F \times = G\flat$ 847, 14

$F\sharp = G$↳ 918, 15 G 988, 16 $G \times = A\flat$ 1058, 17 $G\sharp = A$↳ 1129, 1' A 1200. Then he writes the Rast as he heard it, as A B $C \times$ D E $F \times$ G A, which therefore gave the cents 0 212 353 494 706 847 988 1200. Now it is difficult to conceive that he could have heard the medieval Rast in this way, even though the intervals were determined purely by estimation of ear, apparently his only method of estimation. But the probability is that medieval Rast, like the other medieval scales, had become a thing of the past, and that what Villoteau heard in Egypt in 1800 was what Eli Smith in 1849 tells us Meshâqah (suprà, p. 264b) laid down at Damascus, namely No. 53, that is, the normal scale 0 200 350 500 700 850 1000 1200 cents, a survival of Zalzal's with the neutral Third and Sixth, and this is very accurately represented by the above scale of Rast, as Villoteau notes it. At the very outset Villoteau says that some divide the Octave into Tones, Semitones, and Quartertones (op. cit. p. 613). This shews that the 24 division was even acknowledged. But Villoteau was perfectly ignorant of equal temperament, and hence paid no attention to this. On the other hand he found the 17 divisions in the theorists, and made them equal, because he thus seemed to reconcile theory and practice. But he only obtained an outline thus, as is evidently shewn by his speaking (op. cit. p. 612) of 'les divisions et subdivisions des tons de la musique arabe en intervalles si petits et si peu naturels, que l'ouïe ne peut jamais les saisir avec une précision exacte, ni la voix les entonner avec une parfaite justesse.' It was evident there were many other Tones (the Quartertones) not in his list of 17. Indeed he says (op. cit. p. 673) : 'Ils savent aussi qu'il y a d'autres degrés intermédiaires aux précédents, et ils en font usage même assez fréquemment, mais ils ne sauraient dire au juste quelle est la nature et l'étendue de l'intervalle qui sépare ces degrés les uns des autres.' These were possibly the Quartertones (very uncertainly produced) by which we learn from Eli Smith that the Arabs, like the Indians, continually varied their scale. If the 17 thirds of tones of Villoteau, just given in his notation, be read as 1 C, 2 $D\flat$, 3 $E\flat\flat$, 4 D, 5 $E\flat$, 6 $F\flat$, 7 E, 8 F, 9 $G\flat$, 10 $A\flat\flat$, 11 G, 12 $A\flat$, 13 $B\flat\flat$, 14 A, 15 $B\flat$, 16 $C\flat$, 17 $D\flat\flat$, as in No. 67, his 12 scales will be found to correspond precisely in names of the notes with those given by Prof. Land (suprà, p. 284), but the whole of the intervals, which were originally of 90 or 24 cents, are now equalised as 71 cents by this confused temperament of Villoteau's in which medieval Arabic music seems to have been intended, but the modern form was really misrepresented. To shut up 24 Quartertones into 17 thirds of Tones, at least two must be given to one note on seven occasions. Thus (in cents) Villoteau's 71 was 50 or 100, his 282 was 250 or 300, his 424 was 400 or 450, his 565 was 550 or 600, his 776 was 750 or 800, his 918 was 900 or 950, and his 1129 was 1100 or 1150. This would fully account for the indistinctness complained of.

No. 68 is a complete Arabic medieval scale,

with additional intervals, 612, 816, 1110, 1200 + 24, 1200 + 114, played on a tambour. These very long-necked guitars allow of minute subdivision of the string.

Nos. 69 to 71 are the various notes produced by the Rabáb, according to the three methods of tuning the second string as 316, 408, or 590 cents, the intervals between pairs of notes on both strings being identical. In No. 69 the note 520 cents is played on the second string, but is here inserted in order of pitch.

No. 72 gives the most extraordinary and most limited scale known, produced by using only the open string and 39, 38, 37, 36, and 35 fortieths of it; the open second string being tuned in unison with the sharpest note of the first string. It is valuable as showing a primitive method of obtaining scales and a division of one-eighth of the keyboard into 5 equal parts.

Nos. 73 and 74 are an attempt to represent the Indian Chromatic Scale from indications in Rajah Sourindro Mohun Tagore's *Musical Scales of the Hindus*, Calcutta, 1884, and the *Annuaire du Conservatoire de Bruxelles*, 1878, pp. 161–169, the latter having been drawn up by Mons. V. Mahillon from information furnished by the Rajah. As regards the 7 *fixed* notes (*prakrita*) of the C scale (*sharja gráma*), C, D, E, F, G, A (a comma sharper than our A_1), B, there seems to be no doubt of the theoretical values. As to the 12 *changing* notes (*vikrita*), the values given can be considered only as approximate. The division of the intervals of a major Tone of 204 cents into 4 degrees (*s'rutis*); of a minor Tone of 182 cents into 3 degrees; and of a Semitone of 112 cents into 2 degrees, as indicated by the superscribed numbers, is also certain. But whether the 4 parts of a whole Tone were equal and each 51 cents, and the three parts of a minor Tone were also equal and each equal to $60\frac{2}{3}$ cents, and the two parts of a Semitone were also equal and each therefore 56 cents, is quite uncertain. This, however, was assumed to be the case in calculating No. 73, and the results are probably not much out. Nor is it likely that the alterations by degrees (produced on increasing the tension of the string by pressing behind high frets, or deflecting the string along low frets, or by arranging the movable frets) were even approximately constant. In No. 74 we have the modern Bengali division of the finger-board referred to in the above books. It seems that the string is first divided into half and a quarter, giving the Octave and Fourth (theoretically). Then the distance from the First to the Fourth on the finger-board is divided into 9 equal parts, and that from the Fourth to the Octave into 13 equal parts, and each distance represents the interval of a degree (*s'ruti*). From these data the values of No. 74 have been calculated. It will be seen by subtraction that the first 9 degrees thus found vary from 49 to 63 cents, and the last 13 from 45 to 64 cents. The cents found, however, from the inverse ratio of the lengths will differ slightly from those used practically. The ♯ and ♭ (*tibra* and *komala*) are used in this scale for deviations of two degrees when a

major tone is divided, and for deviations of one degree when a minor Tone or Semitone is divided. This is done by the Rajah in his translations into ordinary notation. In addition I have taken the liberty to use ♭♭ to represent only *one degree* flatter than the single ♭, and wish it to be read 'very flat' (*ati-komala*); similarly ♯♯ is one degree sharper than ♯, and should be read 'very sharp' (*ati-tibra*). The Rajah not having distinguished the very flat and very sharp notes from the simply flat and sharp ones in his 304 scales, I have avoided citing them at length. Similarly I have not been able satisfactorily to find the cents for the F scale or *mad'hyama gráma* (usually represented as our just major scale in which A should be one *comma* instead of one *degree* flatter than in the C scale, as it ¶ would appear to be), or for the E scale or *gand'hára gráma* (in which D and A appear to be one degree flatter, and B one degree sharper than in the normal C scale or *sharja gráma*), and hence I have not given them in the table. But, using numbers before the notes for degrees, they may possibly be for the F scale, 1C, 5D, 8E, 10F, 14G, 17A, 21B, which would use degree 17, and the corresponding note may be called grave A, and written A'. For the E scale we may possibly have 1C, 4D, 8E, 10F, 14G, 17A, 22B, where 4D is now utilised, and becomes grave D'. But these A', D', are not our A_1, D_1, and hence the scales are different from ours. This is, however, pure conjecture.

No. 75, for the old national Indian instrument, a Vina from Madras, in the South ¶ Kensington Museum, gives the value of 24 notes by measuring vibrating lengths of string from fret to bridge, and is, of course, very uncertain. It will be found, however, that the notes agree with No. 74 better than with No. 73, for the scale C D♭ D E♭ E F F♯ G A♭ A B♭ B c. The other degrees could be easily produced by pressing the string behind the frets, which were about one inch in height.

Nos. 76 to 80 are five observations of scales played by Rajah Rám Pál Singh, and observed with forks. These were set by altering the movable frets of a sitár. The first and fourth are placed together in the table, as they are believed to have been meant for the same scale, and differed only because they were set on different days. They seem to be meant for Rajah Sourindro Mohun ¶ Tagore's first scale. The second setting seems meant for his 13th, the third for his 29th, and the fifth for his 9th.

No. 81. A wood harmonicon in the South Kensington Museum, stated to have come from Patna, but probably arrived from some hill tribes. Its scale resembles one which I deduced by measurements of strings from a Tár of Cashmere, which was 0 175 354 512 720 896 1062 1237, but I thought this scale too uncertain to put in the table.

No. 82. A wood harmonicon sent direct from Singapore to Mr. A. J. Hipkins, taking the central Octave.

Nos. 83 to 87. Wood harmonicons in South Kensington Museum, of which the last belonged to General Pitt-Rivers.

Nos. 88 to 92 are inserted because they are the

examples of pentatonic scales usually given. They are at any rate now used for pentatonic Scotch music. See suprà, p. 259d. They are not, however, by any means usual forms.

No. 93. A balafong from the South Pacific belonging to General Pitt-Rivers, seems to be intended for No. 90, but 370, 685 cents for 400 and 700 cents are both rather flat.

Nos. 94 to 102. Javese instruments examined at the Aquarium, London, in 1882, and pitch of notes determined by forks. In Nos. 96 to 100 the order of these notes settled by Mr. W. Stephen Mitchell, and the order confirmed by information received through Prof. Land of Leyden. Nos. 101 and 102 were inferred from information of missionaries obtained by Prof. Land. The assumption in No. 95 of a division of the Octave into ¶ five equal parts was confirmed by other measurements communicated by Prof. Land.

Nos. 103 to 109. These seven scales were taken from the playing of Chinese musicians at the International Health Exhibition, 1885, during four private interviews. No. 106 had two additional tones of 497 and 797 cents above the lowest note; these were omitted in playing by the musician. There was a second y'in-lo at the South Kensington Museum, also with 10 tones, which gave 0 52 240 266 418 437 586 589 712 738, but these form no scale. They may be a fund out of which scales are constructed. The four following may be among such:—To a sharp Fifth, 0 240 418 589 712; to a flat Fifth, 0 188 366 534 686; to a flat Fifth again, 0 214 385 537 686; to a fourth, 0 197 349 498. The last is a Meshâqah tetrachord, ¶ see No. 53. A chime of four small bells, belonging to Mr. Hermann Smith, gave 0 312 480 724.

Nos. 110 to 139. Japanese scales. Accepting the statement of native musicians that the intervals are those of equal temperament, or at least so near them that Japanese ears do not perceive the difference, then the theory gives Nos. 110 and Nos. 113 to 128 for 'popular' koto tunings. There are 13 strings to the koto, but only 5 give the scale,

the rest being Octaves or unisons. Nos. 111 and 112, heard at the 'Japanese Village' Knightsbridge, shew how practice sometimes overrides theory. Several of the scales seem to be identical, but they are at different pitches, and hence no more identical than our major scales for different keys.

Nos. 124 to 129 are 'classical' koto tunings of which 124 126 128 are classed as *riosen* or analogous to our major scales, and 125 127 129 as *ritsusen* or analogous to our minor scales. This must have been effected by sharpening some of the notes by pressure on the string behind the bridges which limit their vibrating length.

Nos. 130 and 131 are both 'classical' heptatonic scales, owing to the introduction of the notes marked *.

Nos. 132 and 133 are both 'popular' heptatonic scales, in which however the introduced notes are not pointed out in Mr. Isawa's Report on Japanese music at the Health Exhibition of 1885, Educational Division, from which the scales No. 110 and Nos. 113 to 133 have been taken.

Nos. 134 to 139 result from an examination of the Biwa, a classical instrument, closely resembling the Arabic lute, fretted only as far as the Fourth. The strings are tuned to one another in six different ways, and hence produce a great variety of notes. By touching the strings on the frets (taking care not to press behind them) and determining the pitch of the notes sounded, Mr. Hipkins and I found that the tetrachord produced differed according to the string employed, as shewn in Nos. 134 to 137. The mean of the intervals thus determined is given in No. 138, and accepting equal temperament, as in the whole of Mr. Isawa's report, the corresponding divisions are given in No. 139. These divisions were in so far assumed to be correct that the three Semitones, though materially differing (having the mean values of 114, 89, and 86 cents), are, in Mr. Isawa's account of the notes, considered as alike, and exactly half of the value of first Tone, which had a mean value of 214 cents.

IV. How these Divisions of the Octave may have arisen.

It is impossible to trace such scales to their germs. Singing and playing on pipes were probably the first music. Striking of bone, wooden, and metal bars was ¶ probably also a very early form, and as their notes are tolerably persistent they are valuable for determining scales (see Nos. 81 to 87 and 93 to 102). But scales themselves are a great development, and two or three notes, varied rhythmically, probably long preceded them. We must be content to commence with strings, certainly a very late form of musical instrument, on which, however, the chief work of the older theorists was expended. After the examples in the latter part of the Table we have no right to assume an accurate musical 'ear,' or appreciation of just intervals. Even in Europe it requires much practice for the majority to sing accurately in tune or to appreciate small errors. (See suprà, p. 147d.) We now know that on any stringed instrument such as the violoncello or guitar, where, to prevent jarring, the string has to rise further and further from the finger board as the finger proceeds from the nut towards the bridge, the pressure of the finger in 'stopping' the string either on the board or fret increases the tension of the string and hence makes the note sharper than it would be if the string could be stopped at its natural height, though even then, as we have seen (p. 442a), the results are not absolutely trustworthy. The law that the number of vibrations is inversely proportional to the length of the string holds but very roughly for such instruments. Thus stopping a violoncello string exactly at its middle gives a note sharper than the Octave, hence the finger has to be placed sensibly nearer the nut. Moreover, the amount of error depends on the nature of the string. The examples Nos. 134 to 137 in the table are very instructive. The determination with which those differences are overlooked (see Annotations to these Nos.) is equally instructive. It is evident that Euclid in his Canon for the Pythagorean notes No. 32, and Abdulqadir in giving his rule for obtaining the 17 notes of his scale No. 67, considered the division to be perfect, and Prof. Land in calculating the value of the notes had, of course, to assume that it was so. The old intervals were, therefore, not so accurately tuned as was supposed, and hence when we take them to be accurately tuned we are ourselves inaccurate.

The fact was strongly impressed on me in making an instrument on which I could play any scale expressed in cents. I had a Dichord, that is, a double monochord, constructed with wires 1,200 millimètres long, diameter ·3 mm., height of nut 7 mm., of bridge 24mm., from sound board. Then having a number of laths 5 mm. thick, I used them as moveable finger-boards, and marked on one the place of the notes of the just scale as determined by the theory of inverse ratios. Trying, by my Harmonical, I found every place much too sharp, and it was only by marking the places which gave unisons that I was able to correct the error. If any one constructs such an instrument, I recommend his setting off the place where he should stop for each semitone for two octaves, by a well-tuned pianoforte, and then dividing each distance representing a semitone into 10 parts. Each of these parts will represent 10 cents with quite sufficient accuracy, and can be subdivided by the eye. Thus a geometrical scale can be constructed by means of which the places to touch the string can be marked off on a new lath, and the scale played. This geometrical scale was placed under one string and the finger-board to be played from under the other. I found it best for accuracy to stop the string with the side of my thumb-nail. All the principal scales above given were thus realised.

Now assuming the usual law of division, suppose a string divided in half, giving the Octave, and each half subdivided in half, giving the Fourth and double Octave. This division being the simplest possible would naturally give a preponderance to the Fourth, whence would arise the tetrachords, the foundation of Greek, and hence of European, and of Persian and Arabic music. The Fourth is also recognised in India; but in pentatonic regions, especially in Java, where the string is not in use (the rabáb they use is Arabic in name and origin, Nos. 69 to 71), the Fourth is not correct. It is clearly, therefore, not a primitive interval, and the quartering of the string may really have much to do with its adoption.

The interval was, however, too wide, and it was necessary to subdivide it. The most obvious plan was again to quarter it. Thus, the additional distances of $\frac{1}{16}$, $\frac{2}{16} = \frac{1}{8}$, $\frac{3}{16}$ of the string from the nut would be obtained, giving the vibrating lengths $\frac{15}{16}$, $\frac{7}{8}$, $\frac{13}{16}$, the ratio. The first gives the diatonic Semitone of 112 cents, and its defect from the Fourth, 498 cents, the major Third of 386 cents, see No. 1. This $\frac{15}{16}$ of the string = 112 cents is conspicuous in Nos. 1 to 4, some of the oldest forms. The $\frac{7}{8}$, = 231 cents, we find in No. 72. The $\frac{13}{16}$ = 360 cents did not

come into use, but it is practically Zalzal's 355 cents, No. 51.

Continuing this simplest of all subdivisions by two, we have half of $\frac{1}{16} = \frac{1}{32}$ of the string from the nut, giving the vibrating length $\frac{31}{32}$ of the string = 55 cents. Hence we obtain the enharmonic division No. 11. At the same time my observations on p. 265, note *, hold good, for the errors in coming so near the nut as $\frac{1}{32}$ of the string, would be too great to obtain anything like accurate results by measurement. On my dichord I found it impossible to take less than a semitone of 100 cents with any degree of certainty. It is interesting to observe that this $\frac{31}{32}$ of the string gives very nearly 50 cents or the Quartertone, and still more nearly $54\frac{6}{11}$, the 22nd part of an Octave, corresponding to the Indian degree (Nos. 73 and 74 Annotation), and is really the commencement of the variation of notes by about a Quartertone.

Then the divisions attempted in the Southern Tambour No. 72, and also in forming the Persian middle-finger note, 303 cents (see No. 66), by taking a place half-way between that for 294 cents and 408 cents, and again for Zalzal's middle-finger note of 355 cents, by taking a place half-way between 303 and 408 cents, and finally the modern Bengali division of the distance occupied by a Fourth on the finger-board into 9 parts (see No. 74) and that for the following Fifth into 13 parts, suggest that the attempts were made to divide the $\frac{1}{4}$ of the string from the nut to the Fourth, by other simple numbers beside 2. The division into three parts would be more difficult, but might be done very fairly by guess. Now the distances of the stopping-place from the nut $\frac{1}{12}$, $\frac{2}{12} = \frac{1}{6}$ of the string or the vibrating lengths $\frac{11}{12}$ and $\frac{5}{6}$ of the string corresponding to 151 cents and 316 cents, the first, the Threequartertone, which is the real parent of all the neutral intervals to be considered presently, and the second the minor Third. Both occur in No. 10, and the minor Third occurs also in Nos. 3 and 6.

The division of the whole string into thirds could hardly have taken place, but $\frac{1}{3}$ of the string from nut gives the vibrating length $\frac{2}{3}$ of the string = 702 cents, the Fifth, which, as exceeding the Fourth of 498 cents, would not be regarded till the tetrachord had been extended to the Octave. The defect of a Fourth from a Fifth gave the major Tone, one of the most important intervals, also obtained directly by taking $\frac{1}{3}$ of $\frac{1}{3}$, or $\frac{1}{9}$ of the string from the nut, giving $\frac{8}{9}$ of the string as the vibrating length = 204 cents. This interval finally absorbed all the others, except the Fourth, especially after the observation that it was also the defect of two Fourths from an Octave. Its direct use is apparent in No. 19, where 204 cents corresponds to $\frac{1}{9}$ string from nut, and 408 cents to $\frac{1}{5}$ of the vibrating length of the 204 cents from the stop for 204 cents. Thus if the nut be called A, and the stopping places for 204, 408, and 498 be B, C, D, and the bridge be Z, we shall have $AD = \frac{1}{4} AZ$, whence $DZ = \frac{3}{4} AZ$; also $AB = \frac{1}{9} AZ$, whence $BZ = \frac{8}{9} AZ$; also $BC = \frac{1}{9} BZ$, whence $CZ = \frac{8}{9} BZ = \frac{64}{81} AZ$. If we suppose the whole length AZ divided into 324 parts, then $AB = 36$, $BC = 32$, $CD = 13$. But the whole division is obtained by taking halves and thirds. No. 5 is the reverse of No. 19. Let the stopping-places for 90 and 294 be E and F. Then $DZ = \frac{3}{4} AZ$, ¶ $FZ = \frac{9}{8} DZ = DZ + \frac{1}{8} DZ$, so that F is found from DZ by adding $\frac{1}{8} DZ$, which is obtained by continual halving. Again $EZ = \frac{9}{8} FZ = FZ + \frac{1}{8} FZ$, so that E is found from FZ by adding $\frac{1}{8} FZ$, which is again obtained by continual halving. This made it easier to produce No. 5 than No. 19. The complicated value of $AE = \frac{13}{256} AZ$ would be thus altogether avoided. Nevertheless it is most probable that Nos. 5 and 19 were both obtained simply 'by ear,' and that they were never exactly 'in tune.'

The next division of the Fourth to be expected is by 5, as in No. 72, and this is therefore advanced by Prof. Land as a probable means of obtaining the Arabic tetrachord. This gives the lengths of the string from the nut, $\frac{1}{20}, \frac{2}{20} = \frac{1}{10}, \frac{3}{20}, \frac{4}{20} = \frac{1}{5}, \frac{5}{20} = \frac{1}{4}$, and the vibrating lengths $\frac{19}{20}, \frac{9}{10}, \frac{17}{20}, \frac{4}{5}, \frac{3}{4}$ of the string giving inter- ¶ vals of 89, 182, 281, 386, 489 cents. Of these 89 is a fair representative of 90 in No. 5,

which has just been otherwise obtained. The minor Tone, 182 cents, occurs direct in Nos. 6 and 7, and also possibly in No. 2, where it would be simpler to obtain it from the open string as $\frac{9}{10}$ its length than as $\frac{24}{25}$ of the vibrating length of 112 cents, that is, as $\frac{24}{25}$ of $\frac{15}{16}$ of the string. The 281 cents appears not to have been used, but it approximates to 294 cents, which may have been introduced from Greece in place of it. This is, however, mere conjecture. The 386, or major Third, is in No. 7 only (for No. 8 is not ancient), and there is very little probability that it was tuned direct. It might have been got as $182 + 204$ or as $498 - 112$. Both the 182 and 386 cents were certainly lost at an early time in 204 and 408 cents, so that it is difficult to suppose that 386 at least was ever obtained directly. It was indirectly produced in Nos. 1 and 2, where, judging from Japanese habits, the tuner tried to get the Semitone by 'feeling,' and left the major Third to arise as the defect of the Semitone from the Fourth.

The division into 7 parts belongs to a much more advanced stage, and never seems to have come into use. But we may understand No. 9 thus. Stopping at $\frac{1}{7}$ the string, we obtain a vibrating length of $\frac{6}{7} = 267$ cents. Then taking $\frac{1}{3}$ of $\frac{1}{7}$, or $\frac{1}{21}$ the length for the stopping place, we obtain a vibrating length of $\frac{20}{21}$ string = 85 cents, and thus find both intervals in No. 9. Of course when the ball had been set rolling, and there was no harmony to check the fancies of dividers or musicians, such forms as Nos. 12 to 24 could be produced. But the ancient Nos. 1 to 7 and 9 to 11 (No. 8 was not ancient) are sufficient to have traced.

V. RESULTS OF THE INQUIRY.

The chief points of interest which the exhibition of these scales affords appear to be the following :

1. The predominance of the Fourth, and mere evolution of the Fifth, in Greece, Arabia, India, and Japan.

These may be only different forms of some original system. The Chinese may have imported the principle, but on this the extreme uncertainty pervading all exhibitions of Chinese scales hitherto made (including Van Aalst's treatise on *Chinese Music*, 1884) renders it difficult to judge. The Fourths actually heard are uncertain, see Nos. 103 to 109. But they seem at home in Japan, where they are ¶ used in tuning, but may have been imported, and they are occasionally absent. In the important Javese scales, Nos. 94 to 102, they are never in tune. Even in Arabia and India they are apt to be altered. The specimens of ruder

music, Nos. 81 to 87, are not favourable to the Fourth. The Fifth never had the same predominance. It is constantly too sharp or too flat. In modern India generally it is too flat. In one set of scales in Java it is too sharp (No. 94); in the other set as flat, as in India (Nos. 74 to 79). These differences probably pervaded also the scales of other countries as actually used, but we know Greece and Arabia from theory only. That the Fifth is true on the bagpipe (No. 52) depends apparently on the use of the drone, which would produce frightful beats if it were as much out of tune as in the other cases cited.

2. The use of Tones and Semitones of about 200 and 100 cents depends upon the Greek tetrachordal system as modified by Pythagorean intonation.

In Zalzal's scale (No. 51), and even in the medieval Arabic scales, Nos. 57 to 65 (that is, omitting the three Nos. 54 to 56, which are identical with the Greek, and the exceptional scales, Nos. 68 to 72), they do not both exist. In the Indian scales they are overridden by the system of 22 degrees, and only undesignedly come close to our equal intonation. In the ruder scales Nos. 82 to 87 they cannot be traced. In Java they do not exist actually.

In China great difficulty was felt by the native musicians of the Health Exhibition of 1884 in respect to the Semitones. The Tones were variable, and in some cases there seemed to be a liking for the minor Tone as in tuning the Tamboura, No. 108. In Japan Semitones and Tones play a great part theoretically, but in the only practical cases I have been able to observe, Nos. 111 and 112 both were very uncertain, and the Fourth was absent.

3. Neutral intervals, each lying between two European intervals, and having the character of neither, but serving for either, abound.

The earliest instance is 151 cents in No. 10, which is the Threequartertone between the Semitone of 90 or 112 cents and the tone of 182 or 204 cents; thus $\frac{1}{2} \times (90 + 204) = 147 = \frac{1}{2} \times (112 + 182)$. In Zalzal's, No. 51, it was $355 - 204 = 151$, and $498 - 355 = 143$ cents; in the bagpipe observed, No. 52, it was $341 - 197 = 144$, and $495 - 341 = 154$, $853 - 703 = 150$, $1009 - 853 = 146$ cents, merely variants of tuning; Meshāqah's theory gave 150 cents. In bagpipe music it serves indifferently for what would be a Tone or a Semitone in music for another instrument. In the Japanese cases, No. 111 represents the two theoretical Semitones by $357 - 193 = 164$, and $801 - 719 = 82$ cents, and No. 112 by $337 - 185 = 152$, and $790 - 683 = 107$ cents.

Zalzal's neutral Third of 355 cents, No. 51, is so truly neutral between the just minor and major Thirds, 316 and 386 cents, that Mr. Hipkins was quite unable to determine to which it most nearly approached in character, but for 345 and 365 cents, as tried on my dichord (p. 522b), the minor and major characters were slightly but decidedly felt. In the observed bagpipe this interval was 341 cents. In Meshāqah's Quartertone temperament it was 350 cents, which may be taken as its usual tempered form. It is the 374 of the New Indian, No. 74, as shown in Nos. 76, 77, and 80. Compare also Nos. 81 to 84, 103, 105, 106, and 109. The correlative neutral Sixth arises similarly.

The neutral Tritone, 550 cents, is also ¶ sometimes found, but it is rare, and as the Tritone, 600 cents, is itself rare, this neutral form is not easily observed. See Nos. 73, 74, 76, 78, 79, 82, 83, 84, and perhaps 96.

4. Modern Arabic and Indian scales have changing or alternative intervals, produced by varying the pitch of one or more of the regular notes in any one scale by a Quartertone or Degree.

To a smaller extent alternative tones are known in Europe. Thus the just major scale borrows its occasional grave second from the subdominant key, where the difference is only a comma of 22 cents. Nos. 62, 63, 65 have each two tones which differ by only a comma of 24 cents, namely 1176 and 1200 in Nos. 62 and 63, and 678 and 702 in No. 65. The scales consequently have 8 tones, as our just major scale of C would have if we inserted both D and D_1. This, however, disappears in tempered intonation. Again our just ascending minor scale of A_1 has two alternative notes, $F_2\sharp$ and $G_2\sharp$, as well as F and G, and these remain in tempered intonation. If we inserted these, we might say that our minor scale had 9 tones. Similarly No. 64 has 8 tones, with three intervals of 114 cents between 180 and 294, 678 and 792, 1086 and 1200, and one interval of 90 cents between 792 and 882. We might suppose that 792 and 882 are alternative notes, and that we might play either

o 180 294 498 678 882 1086 1200, or else
o 180 294 498 678 792 1086 1200,

the three final intervals in the first case being 204 204 114, and in the second 114 294 114. This, however, is only an illustration, and is not the point raised. Meshāqah's complete scale consists of 24 Quartertones to the Octave, for two Octaves, each tone having its own individual name. Of these, only 7 are selected to form the normal scale, namely No. 53. But any one of these 7 notes may be raised (or also probably depressed) by one or two Quartertones. And so freely is this variation of the scale employed, that of the 95 snatches of melodies which Eli Smith reports from Meshāqah, there are only 7 in which some change is not occasionally made. Sometimes the change is in ascending and not in descending or conversely. Thus in the air called *Remel*, I find 9 notes, the alternatives 950 and 1100 being introduced so that the scale, instead of ending 700 850 1000 1200 ends as 700 850 950 1000 ¶ 1100 1200. Something of the kind occurs in bagpipe-playing at the present day, owing to the system known as 'crossfingering,' which gives nominally two ways of fingering the same g'', but actually produces two slightly different notes, the sharper being used in ascending passages. This was observed by Mr. Bosanquet at a bagpipe competition, and has been confirmed on inquiry by Messrs. Glen, the great bagpipe makers of Edinburgh.

A similar thing apparently occurs in Indian scales, where some of the notes may be depressed one, two, or three degrees, and others raised by similar amounts as shewn in Nos. 73 and 74. And there seem to be other alterations of the kind not written, but conditioned by the *rāgini* or modelet in which the musician is playing. This is a point which greatly requires elucidation. ¶

These tones changing by a degree are made by pressing the string behind the fret or deflecting it along the fret. A similar thing occurs in playing the Japanese koto. The player is constantly pressing slightly or heavily on the string beyond the bridge, or pulling the string towards the bridge, and thus more or less sharpening or flattening the pitch of the note.

5. Scales of five tones may be formed by omission from scales of seven tones, but on the other hand many scales of five tones seem to be entirely independent of tones of seven tones, neither generating them nor being generated by them.

Though some of Chinese pentatonic scales, as Nos. 103 and 107, seem to be derived from heptatonic or conversely, yet all the Javese pentatonic scales are thoroughly independent of any heptatonic form. No. 94 and Nos. 97 to 102 could not be expressed as parts of even our chromatic scale of 12 semitones. European musicians, indeed, persist in hearing and writing the *Salêndro* scales,

properly o   240   480   720   960   1200,
    as o   200   500   700   900   1200,
    or o   300   500   700   1000   1200

but this must arise from their not appreciating 240 cents, which is almost a neutral interval between a Tone, 200, and a minor Third, 300, and is hence mistaken by European ears sometimes for one and sometimes for the other. As for the Pelog scales, I cannot find that any one has ventured to put their airs into a European dress, the intervals No. 96 are so strange. I have however tried to appreciate them by having a concertina tuned with the white studs in Salêndro (tuning E and F and ¶ also B and C as unisons), and the black studs in Pelog, and writing them with the notes

which would belong to the stud used in the ordinary tuning. For the Salêndro I used the airs in Raffles's Java, and in Crawford's paper in the Tagore collection. For the Pelog I had to invent airs myself. The characters of the two sets are quite unlike. But pentatonic Scotch airs played with the Salêndro scale are quite recognisable. Whether this scale is the primitive pentatonic scale it is quite impossible to say.

In Japan the koto tunings are all pentatonic, and according to theory the 'popular' have intervals of a Semitone, a Tone, and a *major Third*, whereas the 'classical' have intervals only of a Tone and a *minor Third*, just as on the commonly received black digitals of a piano; but the classical is said to come from China. In practice probably these intervals are varied as in Nos. 111 and 112.

6. Pentatonic scales do not necessarily arise from inability to appreciate Semitones.

This is shewn by the Javese Pelog notes, which contain intervals of 137, 129, 112, 133, and 102 cents, and by the Japanese popular koto tunings, No. 110 and Nos. 113 to 123, all pentatonic, and theoretically founded on the diatonic Semitone. If in practice the diatonic Semitone sometimes grows to a Threequarter-tone, it also sinks to a small Semitone, see Nos. 111 and 112.

8. There is an entire absence of tonality in our sense of the term and of any attempt at harmony.

There is regard to the final cadence, at least in Meshāqah's scales, and probably in all. There is in the Indian a *ruler* note (*vádi*), ¶ see p. 243*c'*, and *minister* notes (*samvádi*), which function as our tonic, dominant and subdominant in certain respects, and Prof. Helmholtz thinks he discovers a reference to a tonic in Aristotle (p. 241). But the European feeling of tonality is one of very late growth, and in non-harmonic scales must have been something quite different, and if we refer it to the same feeling as our own, it is from want of power to appreciate the feeling of those who use non-harmonic scales. This is parallel to what constantly happens in appreciating the intervals of these scales.

There is plenty of *ensemble* playing with notes of very different qualities of tone, but but they regularly proceed in unisons and Octaves. In the Indian instruments there are sympathetic and secondary strings. The former have their partials evoked by the notes played. The latter, generally tuned in rela- ¶ tions of an Octave Fourth or Fifth, are occasionally thrummed. But there is nothing like a chord, or a tissue of harmony. It would not be possible with the notes at command. There is also *discant* playing as in the old polyphony before harmony proper was invented. Prof. Land, speaking of the Gamelan, or band of Javese musicians sent by the independent prince of Solo to the Arnheim Industrial Exhibition in 1879, says: 'The musical treatment is this. The rabáb plays the tune in the character of leader' [at the Aquarium, the player of the gambang (wooden bar harmonium) seemed to be leader]; 'the others play the same tune, but figured, and each for himself and in his own way; the sáron (metal bar harmonium) resumes the motive or tune. All this is accompanied by a sort of *basso ostinato*, and a rhythmical movement of the drum,

and the whole is divided into regular sections and subsections by the periodical strokes of the gongs and kenongs [kettles]. The variations of the same tune by the different instruments produce a sort of barbarous harmony, which has, however, its lucid moments, when the beautiful tone of the instruments yields a wonderful effect. But the principal charm is in the quality of the sound, and the rhythmical accuracy of the playing. The players know by heart a couple of hundred pieces, so as to be able to take any of the instruments in turn.'

In his report on Japanese music, Mr. S. Isawa, director of the Musical Institute at Tokio, distinctly claims a species of harmony for Japan, and gives an arrangement of the Greek 'Hymn to Apollo' (Chappell, p. 174), which he had directed 'a Court musician, and a member of the [Musical] Institute, to harmonise purely according to the principles of Japanese classical music.' It was set for five instruments, the Riuteki (fuye), Hichiriki, Sho, Koto, and Biwa. I possess the copy of the music in European notation, sent to the Educational Section of the Health Exhibition in 1884. Though much was in Octaves, the koto played a figured form, with dissonances, followed by consonances. A non-professional Japanese gentleman, a student of physics, acquainted with European music, in answering my questions, says: 'Anything like European [harmony] cannot be heard in Japan. If it exist, it is of the rudest possible description. We have certainly *ensemble* playing with many instruments of different sorts; but it seems to me that we have no idea of such things as chords. . . . We go generally parallel in Octaves and in Fifths, rarely in Fourths, but there are cases where two different tones, not belonging to the three consonances, are sounded, but they are not *harmonic*, but what

Helmholtz calls *polyphonic*. We have many *figures* for accompaniment. . . . In popular music, we meet with cases where two instruments play Octaves or Fifths. With singing this would also hold, but it is very rare that people ever sing chorus.'

At the same time, as the Japanese use a system of twelve notes to the Octave, which they do not seem to distinguish from European equally tempered notes, and which will probably be soon reduced to that form by the labour of Mr. Isawa, there seems to be no reason why harmony should not be naturalised, like so many European customs, in the wonderfully progressive country of Japan.

It may be added, although it cannot appear from the table of the scales, that in listening to native Javese, Chinese, and Japanese performers, there seemed to be a total absence of what we term expression. There was no piano and forte, no shading or *nuance*, merely a hard playing of the notes, as on street mechanical pianos. They appeared to depend principally on gongs, clacks, or accumulation of various instruments to give rhythm and spirit to the music. But so far as I could judge by the very little India music I heard from Rajah Rám Pál Singh, it seems to have some expression, as it certainly has an extremely varied rhythm, sounding very strange to European ears. (See Siamese scales, *Postscript*, p. 556.) ¶

SECTION L.

RECENT WORK ON BEATS AND COMBINATIONAL TONES.

(See notes throughout Part II., pp. 152–233, and especially pp. 43, 55, 126, 151, 152, 155, 156, 157, 159, 167, 199, 202, 204, 205, 226, 229, 231, and 420. The reader is particularly requested to defer any reference to this Section L until he has studied Part II., and become familiar with the whole phenomenon of beats and combinational tones, and with Prof. Helmholtz's theories respecting their origin. Until such familiarity has been gained, much of what follows will be unintelligible.)

Art. 1. *Papers considered.*

The papers here considered are

i. R. KOENIG. *Ueber den Zusammenklang Zweier Töne* (On the sounding of two tones at the same time). Pogg. *Annal.* Feb. 1876, vol. 157, pp. 177–237.

This paper appeared a year before the 4th German edition of Helmholtz's *Tonempfindungen*, and is cursorily referred to, suprà, p. 159b. The other papers of Koenig here mentioned appeared subsequent to Prof. Helmholtz's 4th German edition. But this paper is placed first because it commenced the new investigations. A translation appeared in the *Philosophical Magazine*, June 1876, and supplement of the same date, pp. 417–446, 511– 525, under the title 'On the Simultaneous Sounding of Two Notes,' and communicated by the late W. Spottiswoode, President R. S., who also read a paper on 'Beats and Combination Tones' before the Musical Association on May 5, 1879 (*Proceedings of Mus. A.*, 1878-9, pp. 118-130), when he exhibited K.'s apparatus and repeated several of his experiments.

ii. R. KOENIG. *Ueber die Erregung harmonischer Obertöne durch Schwingungen eines Grundtones* (On the excitement of harmonic upper partials by the vibrations of a fundamental tone). Wiedemann, *Annal.* 1880, vol. xi., pp. 857–870.

This treats of a subject incidentally mentioned, suprà, p. 159a, in reference to combinational tones.

iii. R. KOENIG. *Ueber den Ursprung der Stösse und Stosstöne bei harmonischen Intervallen* (On the origin of beats and beat-tones for harmonic intervals). Wiedemann, *Vnnal.* 1881, vol. xii. pp. 335–349, introducing an entirely new method of experimenting by means of the wave-siren.

iv. R. KOENIG. *Beschreibung eines Stosstöneapparates für Vorlesungsversuche* (Description of a beat-tone apparatus for lecture-room experiments), Wiedem., 1881, immediately after the last paper, vol. xii. pp. 350–353.

v. R. KOENIG. *Bemerkungen über die Klangfarbe* (Remarks on quality of tone) ¶ Wied., 1881, vol. xiv., pp. 369–393, more fully describing the wave siren of No. iii.

In drawing up this notice I made use solely of the original German papers just cited. But I find that Dr. Koenig has republished the whole of his 16 acoustical papers, of which those just cited form the 9th, 14th, 10th, 11th, and 16th, respectively, in the French language, in one volume, with beautifully printed wood-engravings, under the title of *Quelques Ex-*périences *d'Acoustique*, 1882, to be had at his present establishment, 27 Quai d'Anjou, Paris, and I have made some use of the additional notes then added. I cordially recommend this collection as a valuable and almost indispensable supplement to Prof. Helmholtz's work.

vi. R. H. M. BOSANQUET. *On the Beats of Consonances of the Form h : 1.* Proceedings of the Physical Society of London, vol. iv., Aug. 1880 to Dec. 1881, pp. 221–256. This was written before B. had seen No. iii. and iv. above.

vii. W. PREYER. *Ueber die Grenzen der Tonwahrnehmung* (On the limits of the perception of tone) containing the sections. I. The Lower Tones. II. The Highest Tones. III. Sensitiveness for Difference of Pitch. IV. Sensitiveness ¶ for the Sensation of Interval. V. Sensation of Silence. Forming the first part of the first series of *Physiologische Abhandlungen* (Physiological Essays) edited by W. Preyer, M.D. and Ph.D., Prof. of Physiology and Director of the Physiological Institute at Jena, 1876, the year before the publication of the 4th German edition of Helmholtz, who quotes it several times. It is here inserted for completeness.

viii. W. PREYER. *Akustische Untersuchungen* (Acoustical Investigations) in the same collection, second series, fourth part, containing I. Deepest Tones without upper partial Tones (suprà, see footnote, pp. 176–7). II. Combinational Tones and upper partial Tones of Tuning Forks. III. Contributions to the Theory of Consonance. IV. Notice on the Perception of the smallest differences of Pitch, Jena, 1879.

These will be cited by the initial of the author, K. or B. or P. followed by the number of the paper, and generally by the page, which in the case of P. will refer to the separate editions of vii. and viii. Prof. Helmholtz will be cited as H., generally followed by the page of this edition.

¶   Art. 2. *Koenig's Simple Tones.*

(*a*) *Simple Tones of Forks.* The tones dealt with by K. are as simple as K. could make them. 'The forks that I used with resonators,' says K. iii. 337, 'had no recognisable harmonic upper partials at all. The occurrence of harmonic upper partials in tuning-forks depends not so much on the lowness of their pitch and the amplitude of their vibrations as on the relation of the amplitude to the thickness of the prongs.'

From a *c* fork (128 d. v.) with prongs 7 mm. (= ·28 inch) thick, K. obtained as many as 4 partials. From another *c* fork with prongs 15 mm. (= ·59 inch) thick and 20 mm. (= ·79 inch) wide, only 2 partials were generally obtained, but extremely violent blows brought out a 3rd partial. With prongs 29 mm. (1·14 inch) thick and 40 mm. (= 1·57 inch) wide, it was not possible to hear even a faint Octave, and a Twelfth, except when the opening of the resonator tuned to them, almost touched the prongs of the fork. The pitch of forks varies directly as the thickness, and inversely as the square of the length, of their prongs (K. iii. 338). K. proceeds to mention that the forks he used, even the largest, when placed before properly tuned resonators, had no detectable upper partials. Subsequently B. repeated K.'s observations in part with the stopped organ-pipes of B.'s experimental organ, in which only the Twelfth or 3rd partial was perceptible and could be allowed for. Also, afterwards, K. iii. 342, used stopped organ-pipes and tuning-forks. B. used tones

of moderate force, and K. also used weak tones with the pipe, and not the strong tones of his tuning-forks mentioned in K. i.

(b) *Simple Tones of the Wave Siren.* however, K. invented the wave siren.

To avoid the suspicion of upper partials,

An harmonic curve constructed on a large scale and reduced by photography was cut on the edge of a wheel. The wheel revolved under a narrow slit, placed exactly in the position of a radius of the wheel, through which wind was driven as the wheel rotated. The curve alternately cut off and let pass the stream of air, and produced a perfectly simple tone, the pitch of which depended on the rapidity of rotation. Forms of this wave siren are figured in K. iii. 346, 347, and K. v. 386. The last shews 16 harmonic curves which may be made to act in any groups, producing all the combinations of perfectly simple tones,

of which the ratio numbers lie between 1 and 16. Hence, although upper partials are found on most tuning-forks, and especially on certain of K.'s forks, it would be wrong to assume (as P. ii. 38 apparently assumes) that in all K.'s cases, at least the Octave was audible. This was not the case with stopped organ-pipes used by both K. and B., and still less so with the tones of the wave siren. K.'s results therefore cannot be explained by upper partial tones. But when we are dealing with compound tones, *each* pair of partials forms a ¶ combination of simple tones to which K.'s observations apply.

Art. 3. *The Phenomena which arise when two notes are sounded together, according to Koenig and Bosanquet.*

(a) *The facts as distinct from theory.* We must distinguish the phenomena from any theoretical explanation of them that may be proposed. The phenomena described by such an acoustician as K., so careful in experiments, so amply provided with the most exact instruments, will, I presume, be generally accepted. The theory by which he seeks to account for them is a matter for discussion. The following relates to two simple tones only, and this must be carefully borne in mind, because H. 159b apparently imagined that the tones used really had upper partials.

(b) *Upper and Lower Beats and Beat-notes.*

If two simple tones of either very slightly or greatly different pitches, called generators, be sounded together, then the upper pitch number necessarily lies ¶ between two multiples of the lower pitch number, one smaller and the other greater, and the differences between these multiples of the pitch number of the lower generator and the pitch number of the upper generator give two numbers which either determine the frequency of the two sets of beats which may be heard or the pitch of the two beat-notes which may be heard in their place. The term ' beat-notes ' is here used *without any theory as the origin of such tones,* but only to shew that they are tones having the same frequency as the beats, which are sometimes heard simultaneously.

Referring to the tables in the Translator's footnote to p. 191 suprà, which relate to compound tones, and therefore contain multiples of the pitch numbers (or of the numbers which give the interval ratios) of two generators, we see from the minor Tenth 5 : 12, that the prime 12 of the upper generator lies between 10 and 15, the 2nd and 3rd multiples of the lower generator, and hence the beat or beat-note frequencies would be $12 - 10 = 2$, and $15 - 12 = 3$. If, then, the two generators are low enough, say, having the pitch-numbers $5 \times 6 = 30$ and $12 \times 6 = 72$, the beats heard would be $2 \times 6 = 12$ and $3 \times 6 = 18$, which would be plainly distinguishable as beats.

But if they were higher, as $5 \times 20 = 100$ and $12 \times 20 = 240$, the beats would be $2 \times 20 = 40$, and $3 \times 20 = 60$, which, though far too rapid to be counted, would be clearly heard as beats, and *at the same time* the beat-notes of 40 and 60 vib. would *also* be audible. If, however, they were much higher, as $5 \times 100 = 500$ and $12 \times 100 = 1200$, then only the beat-notes of $2 \times 100 = 200$ and $3 \times 100 = 300$ vib. would ¶ be heard. The beats heard are then the same *as if* the upper generator were simple and the lower generator compound; but it must be remembered that *both* generators are really simple.

The frequency arising from the lower multiple of the lower generator is called the frequency of the *lower* beat or *lower* beat-note, that arising from the higher multiple is called the frequency of the *higher* beat or beat-note, without at all implying that one set of beats should be greater or less than the other, or that one beat-note should be sharper or flatter than the other. They are in reality sometimes one way and sometimes the other.

(c) *Limits within which either one or both Beat-Notes are heard.* Both sets of beats, or both beat-notes, are not usually heard at the same time. If we divide the intervals examined into groups (1) from 1 : 1 to 1 : 2, (2) from 1 : 2 to 1 : 3, (3) from 1 : 3 to 1 : 4, (4) from 1 : 4 to 1 : 5, and so on, the lower beats and beat-tones extend over little more than the lower half of each group, and the upper beats and beat-tones over little more than the upper half. For a short distance in the middle of each period both sets of beats, or both beat-notes, are audible, and these beat-notes beat with each other, forming secondary beats, or are replaced by new or secondary beat-notes.

(d) *Beat-Notes and Differential Tones.* The lower beats, as long as they are distinctly audible, and refer to an interval less than 5 : 6, or a minor Third, agree with the beats of H. 171a, and when they have a greater frequency than from 16 to 20 there is also heard the beat-note, which then coincides in pitch with the differential tone of H. 153a. Above a minor Third, H. 171d says the beats are practically inaudible. K. however hears them—and B. vi. 235–237 also heard them—passing over into a roll and a confused rattle, as far as the major Sixth $(C : A_1)$ for the lower beats.

¶ In other respects K.'s beat-notes are different from H.'s differential tones. Thus for the minor Tenth 5 : 12, our first example, the beat-notes are 2 and 3, as just shewn, but the differential tone is $12 - 5 = 7$, which is not obtained by K. i. 216, who says : 'These intervals, which are formed by high tones, allow the beat-notes to be heard quite loudly, but give no trace of differential tones. Thus $c''' : b'''$ (8 : 15) gives only 1 and no trace of 7, $c''' : d'''$ (4 : 9) gives only 1 and nothing of e''' (5); $c''' : f''$ (3 : 8) only f and f'' and no a''' (5) at all, hence the differential tones must be extraordinarily weaker than the beat-notes. But I was able to establish the actual existence of these differential tones with certainty by forming the above intervals with deeper notes, which, lasting longer, allowed me by means of auxiliary forks to get a definite number of beats with the differential tones in ¶ question.' This experiment I have repeated several times. I made the tone of the generating forks as loud as possible by holding them over resonance jars. The auxiliary fork had to be held at a considerable distance from its jar in order to reduce its loudness to about that of the differential tone, to allow the beats to be counted. Thus the mistuned minor Tenth 223·77 : 539·18 gave the differential tone 315·41, which, although inaudible, beat with the fork 319·59 audibly 4·18, which was counted as 4·2. And so on in other cases.

In the case of a mistuned Octave, it appears to me that the lower fork acts as this auxiliary fork to catch the differential tone. Thus the mistuned Octave 223·77 : 451·14 gives the differential tone 227·37, which would possibly have been quite inaudible if it had not been caught by the lower fork, with which it made ¶ 3·6 beats in a second, as I myself counted. In this case I had to hold the higher fork far above the resonance jar. The beats were heard as low beats at the pitch of the lower fork. Also in this case, on continuing to hold the high fork over the resonance jar of the upper fork to weaken its sound, but bringing the low fork over the higher resonance jar as closely as possible, the higher Octave of that fork, or 447·54, was produced, which beat with the higher fork also 3·6 times; but now the beat was clearly and distinctly at the pitch of the upper fork. This was a beat of the 2nd partial of the lower fork with the upper fork, and was altogether distinct in character from the lower beat. Hence the beats could not be confused when heard separately, although the frequency was the same. K., so far as I can see, does not anywhere mention the pitch of the beats he heard, but B., vi. 237–9, says that in all cases he has observed, when the required partials have been removed, 'the beats . . . consist entirely of variations of intensity of the lower note,' and adds that, 'as he (K.) does not analyse the beats, we cannot tell whether the variations of the lower note were produced in his experiments.' Mr. Blaikley (*Proc. Mus. Assn.* 1881–2, p. 25) relates, however, that when K. exhibited the beats to him in Paris, K. said : 'You hear distinctly—there can be no doubt about it—that the beating note is the lower one.' This gives K.'s opinion, which Mr. Blaikley did not share.

Observe that K. does not deny the existence of tones having the pitch of differential (or summational) tones, but, as in this case, he shews their existence, and that they are distinct from his beat-notes, having frequently a different, and only occasionally the same, pitch. When, therefore, B., vi. 239, talks of second, third, and fourth combinational tones having been demonstrated directly by K., he seems to have identified beat-notes and differential tones, which, however, K. distinguishes.

The above observations of K. on differential tones are taken from the German edition of K.'s paper in 1876. In the French republication in 1882 they still appear, but in parentheses, and with a long note (ibid. p. 130), in which he states that subsequent investigations have induced him to change his opinion, as he finds that even very wide harmonic intervals between extremely weak tones may produce distinct beats. Hence (for the case where an auxiliary fork produced beats with two forks having the ratio 8 : 15, shewing a very weak tone 7, that might have arisen from 'the tone of the lower beats of 8 and 15'), in his German summary of results, K. i. 236, paragraph III. 6, admitted the actual existence of differential tones, though 'extraordinarily weaker than beat-notes;' but in his French republication (p. 147) he has altered this paragraph to : 'No experiment has yet proved with certainty the existence of differential and summational tones.' Observe that the existence of tones with the pitch of differential tones is not disputed. It is only the theoretical origin of such tones that is called in question. At present it seems impossible to decide that point.

(e) *Bosanquet's summary of the phenomena.* B. vi. 228. 'As two notes of equal amplitudes separate from unison, they are at first received by the ear in the manner of resultant displacements, consisting of the beats of a note whose frequency is midway between the primaries. When the interval reaches about two commas [say 43 or 50 cents], the ear begins to resolve the resultant displacements, and the primary notes step in beside the beats. When the interval reaches a minor Third in the ordinary parts of the scale, neither the beats nor the intermediate pitch of the resultant note are any longer audible, at least as matter of ordinary perception ; but the resultant displacement which reaches the ear is decomposed, and produces the sensation of the two primary notes, perfectly distinct from each other : that is to say, Ohm's law has set in, and is true, for ordinary perceptions and in the ordinary regions of the scale, for the minor Third and all greater intervals.' These phenomena are not mentioned by Koenig, and in my own observations I feel a difficulty in appreciating them.

Art. 4.—*Objective Beats and Subjective Beats, Beat-Notes and Differential Tones.*

(*a*) *Objective Beats.* Beats of a disturbed unison exist objectively as disturbances in the air before it reaches the ear. They are reinforced by resonators, they ¶ disturb sand, &c. In the case of the beats of harmonium reeds in Appunn's tonometer, they strongly shook the box containing the reeds. Other beats, beat-notes, and combinational tones appear not to exist externally to the ear.

(*b*) *Subjective Beats and Notes.* K. i. 221 says : ' Neither these combinational tones nor the beat-notes already described are reinforced by resonators.' B. vi. 233-4, after describing his improved resonator, by means of which he can effectually block up both ears against any sound but that coming from a resonance jar (see p. 43*d'*, note ‡), says : ' By means of these arrangements I some time ago examined the nature of the ordinary first difference-tone, and convinced myself that it is not capable of exciting a resonator. In short, the difference-tone of H., or first [lower] beat-note of K., as ordinarily heard, is not objective in its character. . . . When the nipples of the resonator-attachment fitted tightly into the ears, nothing reached the ear but the uniform vibrations of the resonator sounding *C*. But if there was the slightest looseness between the nipple and the passage of either ear, the second note (*c*) of the combination got in, and gave rise to the subjective difference-tone (first [lower] beat-note of K.), by the interference of which with the ¶ *C* I explain the beats on that note. *These beats are therefore subjective.*'

This expression is not meant to imply that they are the product of the imagination, but that they do not exist externally to the ear. Hence, when H. 157*c* says that they are at least partly objective, although he admits that the greater part of the strength of combinational tones arises only within the ear, and again says, H. 216*a*, that he has ' always been able to hear the deeper combinational tones of the second order, when the tones have been played on the harmonium and the ear was assisted by proper resonators,' he had possibly not succeeded in blocking both ears properly against the outer air.

(*c*) *Preyer's experiments to shew the subjectivity of Differential Tones.*—P. viii. II. had seven tuning forks of extraordinary delicacy constructed, giving *f* 170⅔, *c'* 256, *f'* 341⅓, *a'* 426⅔, *c''* 512, *f''* 682⅔, *g''* 768 vibrations, and hence having the ratios 2 : 3 : 4 : 5 : 6 : 8 : 9, which were so ready to vibrate on the slightest excitement that they could be experimented on at night only. The three lowest forks had the following partials.

Fork *f* had the 2nd *f'* strong, the 3rd *c''* strong, and the 4th *f''* weak.
Fork *c'* had the 2nd *c''* strong and the 3rd *g''* strong.   ¶
Fork *f'* had the 2nd *f''* strong.
Sounding these forks in pairs to get the differential tones,

c'' & *f* gave *f'* or 6−2=4 ; *f''* & *f* gave *c''* or 8−2=6 ; *g''* & *c'* gave *c''* or 9−3=6 ;

and that these tones were objective enough was shewn by their making the forks *f'*, *c''* vibrate sympathetically. But we see that *f'* and *c''* are partials of *f* and *c'*, which existed already strongly on those forks, and if the forks *f* and *c'* were sounded separately, they also made the forks *f'*, *c''* vibrate sympathetically. Hence these results did not prove the objective existence of their differential duplicates. On the other hand, the pairs of forks giving the audible differential tones—

$$f''-c'' = f \text{ or } 8-6 = 2, \quad a'-c' = f \text{ or } 5-3 = 2, \quad c''-f'=f \text{ or } 6-4=2,$$
$$g' -c'' = c' \text{ or } 9-6 = 3, \quad a'-f = c' \text{ or } 5-2 = 3, \quad f''-a'=c' \text{ or } 8-5 = 3,$$
$$g''-a' =f' \text{ or } 9-5 = 4, \quad g''-f' = a' \text{ or } 9-4 = 5, \quad f''-c' =a' \text{ or } 8-3 = 5,$$

utterly failed to produce the slightest effect on the forks having the same pitch.

(*d*) *Subjectivity of Summational Tones.* Again, for summational tones the combined forks

$$f +f'=c'' \text{ or } 2 + 4 = 6, \quad f + c'' =f'' \text{ or } 2 + 6 = 8, \quad c' + c'' = g'' \text{ or } 3 + 6 = 9$$

gave tones perfectly objective, but then these tones c'', f'', g'' already existed as partials of one of the two forks excited. On the other hand,

$$f +c'=a' \text{ or } 2 + 3 = 5, \quad c' + a' =f'' \text{ or } 3 + 5 = 8, \quad f' + a' = g'' \text{ or } 4 + 5 = 9$$

were inaudible, that is, neither existed without nor within the ear. ' Perhaps,' says P., ' they might be made audible after properly arming the forks by means of resonance boxes *while* sounding. But the observation would not be easy.' Just as H. could hear the cases he cites (Pogg. *Ann.* vol. xc. 1856, p. 519) ' only with great difficulty.' But the forks tried by him each possessed the Octave, as he states (*ibid.* pp. 506, 510).

¶ Now, when Octaves exist, and in case of the siren other partials are strongly developed, these summational tones — as G. Appunn pointed out to P.—could be conceived as differential tones of the second order—that is, differential tones arising from the first differential acting on the partial, if such action is admitted. Thus H. found $b +f''= d$ or $2 + 3 = 5$, but then b and f' included the Octaves b' and f'' or 4 and 6, and we had, the first differential $f' - b = B$ or $3 - 2 = 1$, and the second $f'' - B = d''$ or $6 - 1 = 5$, or, without using B in the formula, $f'' - (f' - b) = d''$ or $6 - (3 - 2) = 5$. In this way P. proceeds to shew that all the cases recorded can be explained. Hence he concludes that the summational tone, if not existing as a partial on one of the tones, is entirely generated within the ear. Thus, according to K., i. 220, from $c' : g' = 2$ he heard clearly $5 = e''$, $7 = e'' + c'$, $8 = e'' + g'$, $9 = d'''$, $10 = e'''$ ¶ and 11, the last by auxiliary forks which beat with the required tones. But there were in this case the partials 2, 4, 6, $8 = c'$, c'', g'', e''', and 3, 6, 9, $12 = g'$, g'', d''', g'''. Hence from the summational tones we have 8 and 9 as partials, while $5 = 8 - 3$, $7 = 9 - 2$, $10 = 12 - 2$, $11 = 12 - (3 - 2)$, and so on. ' Therefore,' adds

P., ' even the comprehensive investigations of Koenig do not make the [external objective] existence of summational tones probable.' Hence, like the differential tones, they must be generated within the ear.

K., in the French edition of his papers (note, p. 127), says : ' This explanation is not admissible, because it assumes that two sounds always generate a differential tone, which is not correct. For example, take two tones corresponding to the fundamental tones c' and e_1' [my notation], or $256 : 320 = 4 : 5$. They give the beat-note $c' = 1$; but this sound 1 does not form the sound 7 with the Octave of c', that is, with $c'' = 8$; nor does it form the sound 9 with the Octave of e_1' or $e_1'' = 10$, as we can be convinced by sounding at the same time the primary sounds C and $c'' = 1 : 8$, or C and $e_1'' = 1 : 10$, even when these latter are much stronger than the beat-note in question and than the two Octaves of the primary sounds c' and e_1'.' The explanation by differentials of the second order given by P. is an adoption of a theory of Appunn ; and, of course, until the reality of the differentials assumed can be proved, remains a merely theoretical explanation.

Art. 5.—*Theory of Beats, Beat-Notes, and Combinational Tones.*

(*a*) *Origin of Beats.* ' How do the beats of mistuned consonances arise ? ' asks B. vi. 228, and replies : ' They may be regarded as springing from interference of new notes, which arise by transformation, in the passage of the resultant forms through the transmitting mechanism of the ear, before the analysis of the sensorium.'

The theory of beats of a disturbed unison on the hypothesis of interference is given in H. 164. The theory of differential and summational tones is given in H. 159*a* and App. XII. pp. 411–413. This, however, extends only to the first differential and first summational tone. But H. 158*b*, *c*, gives a theory for the ¶ generation of such tones within the ear owing to the non-symmetrical structure of its drumskin and the looseness of the joint between the hammer and anvil within the drum. And B. vi. 242–8, by means of some perhaps rather hazardous assumptions, succeeds in shewing that the asymmetry of the drumskin acting upon the waves of air coming to them would, as he terms it in the above extract, ' transform ' the result into one for which the displacement is not relatively infinitesimal, but in which its higher terms must be taken in consideration. Then proceeding to the fourth order of displacement, he ultimately obtains six summational and six differential tones ' produced by direct transformation of the primaries ' (B. vi. 246), so that he avoids the introduction of numerous differential tones of various orders (H. 200–203, B. vi. 241), which H. seems to have borrowed from Scheibler, who, although he did great things with the beats of tuning-forks, was not a physical authority. Calculation based on the introduction of these entirely hypothetical, because always inaudible, tones leads, as K. i. 200 shews, to the right number of beats ; but, as he says, ' we are compelled continually to assume the existence of tones which have not only not been heard themselves, but which are supposed actually to generate and be generated by other likewise inaudible tones.' We have an example in the first differential of a tone which, when it does not coincide with the lower beat-note, is appreciable only by beats with an auxiliary tone, and is hence very faint indeed in respect to the generators; and yet these are supposed to be the progenitors of others relatively weaker, till at last they produce one strong enough to be well heard (K. i. 186). The difficulty is surmounted by B., so far as the existence of the ultimate tone, without assuming the action of hypo- thetical intermediate generators. But we know nothing of the strength of these ultimate tones as determined by the formula, and we are constrained to believe that what depends upon the higher powers of the displacement, when the latter is not infinitesimal in respect to the length of the wave, must be extremely small, not at all comparable with the beat-notes actually heard, and hence must be in- sufficient to explain them. That is, we may admit all the differential and summa- tional tones of H. and B. without having approached a satisfactory explanation of the main phenomenon, the beat-note.

(b) *Can Beats generate Tones ? First, Beats of Intermittence.* Now, the obvious ¶ hypothesis is that the beats coming within the frequency of musical notes are heard as tones. H. 156c, in mentioning this, states three objections, of which P. viii. 27 says that not one is at present tenable. They are : (1) that this hypo- thesis does not explain summational but only differential tones. On which P. remarks that summational tones, which have been heard only when at least the second partial of the generators was audible, can be explained as differential tones of the second order, as noted suprà, p. 532 b. (2) That 'under certain conditions the combinational tones exist objectively,' which is against art. 4, p. 531 ; and P. viii. 25 especially observes that the only experiment which H. has cited (Pogg. *Annal.* vol. xcix. p. 539) to prove the objective existence of summational tones by sand strewed on a membrane cannot be critically examined because the two generators are not specified. (3) That 'the only tones which the ear hears correspond to pendular vibrations of the air.' P. considers this to be disproved by the inter- mittence tones obtained by K., who rotated a disc perforated with 128 holes before tuning forks of different pitches, and obtained the same tone of intermittence what- ever was the pitch of the fork. This tone was accompanied by two variant tones ¶ having pitch numbers equal to the sum and difference of the frequencies of the fork and intermittences.

K. i. 230, varied the experiment by con- structing a disc 'with three circles, each with 96 equidistant holes, the diameters of which increased and diminished on the first circle 16 times from 1 to 6 mm. (= ·04 to ·24 inch), on the second 12, and on the third 8 times. On blowing through a tube 6 mm. (= ·24 inch) in diameter, and revolving the disc slowly, the separate periods of holes on each circle gave separate beats. On revolving continually more quickly, the 16 periods of the first, then the 12 of the second, and finally the 8 of the third circle, passed over into a musical tone. Finally, when the high tone of the 96 holes on revolv- ing 8 times in a second had reached g″ with 768 d. vib., the deep tones c, G, C—answering to the numbers of the periods 128, 96, 64 d. vib.—could be heard loudly and powerfully at the same time as g″.'

(c) *Can Beats generate Tones ? Secondly, Beats of Interference.* Now, in reference to these tones of intermittence, K. i. 231 remarks that although they show great similarity to beating combinations, as proving the possibility of separate maxima of intensity passing over into a continuous tone, they were in reality very ¶ different from such combinations, because in the case of beats there was a change of sign, a maximum of condensation being followed by a maximum of rarefaction.

This was precisely the objection made by Lord Rayleigh when Mr. Spottiswoode gave his account of Koenig's experiments (*Proc. Mus. Ass.* 1878-9, p. 128), and he in conse- quence could not understand how beats could generate tone. B. (*ibid.* p. 129) raised the same objection, which he developed in B. vi. 223-5. He there shews that in the case of two tones of equal strength, less than two commas from a unison, 'the resultant dis- placement' would produce a tone 'whose fre- quency is the arithmetical mean between the frequencies of the two primaries, and having oscillations of intensity whose frequency is defined by a pendulum vibration of frequency equal to half the difference of the frequencies of the primaries.' Then B. says that if the law held for widely separated notes, as for the 'Fifth (4 : 6), the note heard would be the major Third, which would beat rapidly . . . but as a matter of fact the note 5 is not heard at all in the above case.' Further, 'supposing that in some unexplained way the beats whose speed is' *half* the difference of the frequencies of the primaries, as just stated, 'gave rise to a note as supposed by K., then the speed of that note does not agree with that required for K.'s first [lower] beat-note, which has the same speed as H.'s difference-tone,' or the *whole* in- stead of *half* the difference of the frequencies.

Now this objection was fully realised by K. i. 232–3, which paper was before B. when he wrote the passage just cited, but was possibly overlooked by him. K. i. 232 says : 'If two tones of 80 and 96 d. vib. are sounded together, they generate a tone of $\frac{1}{2}$. (80 + 96) = 88 vibrations with an intensity increasing and diminishing 16 times, and at each passage from one beat to another there is a change of sign, so that the maximum of compression of the first vibration of the following beat is half a vibration behind the maximum of compression of the last vibration of the preceding beat.' To meet this case he made two experiments. In the first he divided a circle into 176 parts, and in the five points 1, 3, 5, 7, 9 he drilled five holes, gradually increasing and then ¶ diminishing in size. Similarly in the points 12, 14, 16, 18, 20, and then in the points 23, 25, 27, 29, and 30, and so on. 'When such a disc was blown upon through a pipe with the diameter of the largest opening, in addition to the tone 88 and the very powerful tone of the period 16 both of the tones 80 and 96 could be heard, but they were very weak, and, on account of the roughness of the deep tone, difficult to observe.' In this case the phase was the same throughout. To imitate the change of phase, K. i. 233 divided each of two concentric circles, running parallel to each other, into 88 parts, and 'disposed the holes which were to represent the successive beats alternately on each. As 88 holes and 16 periods give $5\frac{1}{2}$ holes to each period, K. took two periods together, and pierced on the first circle the divisional points 1, 2, 3, 4, 5, 6, and ¶ on the second 6, 7, 8, 9, 10, 11, then again on the first 12, 13, 14, 15, 16, 17, and on the second 17, 18, 19, 20, 21, 22, and so forth.'

The sizes of the holes were alternately increasing and diminishing to represent beats. 'When these circles of holes were blown upon at the same time through two pipes of the diameter of the largest opening, and placed on the same radius, one circle from above and the other from below, then at each revolution of the disc there were created 88 isochronous impulses, varying 16 times in intensity, which changed sign on each transition from one period of intensity to the other. In this experiment the two tones 80 and 96 were more distinct than in the first experiment, where the circles of holes were blown upon from one side only.'

On B.'s objections just quoted (supra, p. 533), K. observes (French edition, p. 143, note): 'The change of phase of the separate vibrations of a variable amplitude, forming the beats, does not cause these maxima of intensity to be produced in contrary directions. Besides, these maxima remain isochronous, and consequently fulfil the conditions under which primary impulses are combined to form sounds. The only influence which the change of phase in question exerts on the disposition of the waves consists in these maxima of intensity not standing apart by a whole number of complete vibrations, but by an odd number of half-vibrations. The disc of the siren in which the resultant compressions of all the successive vibrations of the complex sound are represented by holes of a proper size, and, still better, the disc that has its rim cut out according to the curve of a series of successive beats [art. 5 (e) below], render this mechanism readily apprehensible, and allow of shewing that, notwithstanding the change of phase, the beat-note must always have the same frequency as the beats.'

(d) *Would a Tone generated by Beats be louder than the Primaries?* K. i. 234 then proceeded to meet Tyndall's objection (*On Sound*, 3rd ed. p. 350) that if the resultant tones (as he calls them) were formed from the beats of the primaries they would be heard when the primaries were weak, which is not the case. K. observes that beats would always be more powerful than their primary tones, 'provided that equal amplitudes of vibration produced equal intensities for all tones,' and proceeds to shew by experiment that this is not the case, and that 'deep tones must have much larger amplitudes of vibration than high tones in order to exhibit the same intensity.'

(e) *Experiments with the Wave Siren.* Thus the question was left till 1881, when K. applied his Wave Siren, originally exhibited in the London International Exhibition of 1872, already partly described in art. 2 (b), p. 529, to solve the question experimentally. The complete form (K. v. 386) was of course applicable to ¶ any pairs of tones with ratios expressible by numbers not exceeding 16. But the simplest method was to draw out the two harmonic curves, and the result of their combination, as is done above (H. 30 b, c) on a very large scale, and then reduce the drawing by photography to the required dimensions. Then the compound curve thus drawn was inverted, so that the high parts became low and the low high, cut, and affixed to the rim of the wave siren. The reason for inversion in this case was that the heights on the curve represented greater intensities, but on the siren would give less intensities.

K. iii. 345 then says : 'When a disc with such a rim is rotated before a slit fixed over it in the direction of the radius, and of a length at least equal to the greatest height of the curve, the slit will be periodically shortened and lengthened according to the law of the curve ; and if wind is blown through the slit, a motion in the air must be generated corresponding to the same law. And this motion must be pre-

cisely the same as that produced by the simultaneous sounding of two really simple tones without any admixture of upper partials.' The beauty of this arrangement thus consists in our knowing precisely what tones act, and that they are undoubtedly simple. The result is thus described : 'The discs for different intervals, when the rotation was slow, gave beats, and when it was more rapid, beat-notes, exactly corresponding to those observed when two tuning-forks were sounded together. Thus the major Second 8 : 9 produced the lower beat-note 1 ; the major Seventh 8 : 15, the upper beat-note 1 ; the disturbed Twelfth 8 : 23, the upper beat-note of the second period, which is again = 1, loudly and distinctly. In the same way the ratios 8 : 11 and 8 : 13 gave quite distinctly and at the same time the upper and lower beat-notes 3 and 5 for the first, and 5 and 3 for the second :' $11-8=3, 2 \times 8-11=5$, and $13-8=5, 2 \times 8-13=3$.

(f) Beats and Beat-Notes heard together. The preceding experiment shews the gradual passage of beats into tones, the transitional part being where both beats and tones are heard together. This occurs where the rotation is sufficiently quick to generate a tone (see H. 174–9, and especially footnote † to p. 176), but not so fast as to destroy the distinct perceptions of beat.

To this I drew attention in a footnote to p. 231 of the 1st edition of this translation, now reproduced in a modified form (suprà, p. 153c', note). This hearing of the two phenomena K. i. 227 explains by a theory of H. (contained on pp. 217-8 of the 1st English ed., but omitted in this 2nd English ed., because it was struck out in the 4th German ed., H. having altered his opinion) that tones are heard in the cochlea and noises in other parts of the ear. In the additions to the 4th German ed. (suprà, pp. 150b to 151d) H. attributes the hearing of both musical tone and noise to the cochlea, and reserves the labyrinth for the sensation of revolution of the head, thus agreeing with Exner. P. viii. 29-33, thinks there are many reasons why we should not accept the theory that all perceptions of noise are due to the cochlea. If so, he says, 'animals without a cochlea would be deaf. Fishes certainly are mostly dumb, and do not hear acutely, as anglers well know, but they are not *deaf*.' On examining Exner's paper (suprà, p. 151d, note *), and especially Anna Tomaszewicz's ' Contributions to the Physiology of the Labyrinth of the Ear,' (*Beiträge zur Physiologie des Ohrlabyrinths*, Medic. Inaugural Dissertation, Zurich, 1877), with other phenomena, he comes to the conclusion that the cochlea hears only musical tones with a pitch-number not less than about 16 (the lowest audible musical tone as usually produced), and that separate noises are heard by other parts of the ear—if not in the vestibule, then in the sacculus. He considers it probable, as others have also thought, that the function of the semicircular canals is rather to give a sensation of the direction whence sound comes. The point is, however, still undecided.

K., in the French republication of his paper (p. 137), says :—' At all events the simultaneous perception of separate beats and the sound which results from their succession is no more in contradiction with the new hypothesis than with the old, for we can very well suppose that, beside the general excitement of the basilar membrane due to each separate beat, the particular parts of this membrane, whose proper tones correspond to the period of the impulses, are more strongly shaken, and execute lasting vibrations giving the perception of sound.'

Lord Rayleigh, in his Presidential Address to the British Association meeting at Montreal, Canada, in Aug. 1884, says :—' Every day we are in the habit of recognising, without much difficulty, the quarter from which a sound proceeds, but by what step we attain that end has not yet been satisfactorily explained. It has been proved that, when proper precautions are taken, we are unable to distinguish whether a pure tone (as from a vibrating tuning-fork held over a suitable resonator) comes to us from in front or from behind. This is what might have been expected from an *à priori* point of view ; but what would not have been expected is, that with almost any other sort of sound the discrimination is not only possible, but easy and instinctive. In these cases it does not appear how the possession of two ears helps us, though there is some evidence that it does ; and even when sounds come to us from the right or left, the explanation of the ready discrimination which is then possible with pure tones is not so easy as might at first appear. We should be inclined to think that the sound was heard much more loudly with the ear that is turned towards than with the ear which is turned from it, and that in this way the direction was recognised. But if we try the experiment we find that—at any rate with notes near the middle of the musical scale—the difference of loudness is by no means so very great. The wave-lengths of such notes are long enough, in relation to the dimensions of the head, to forbid the formation of anything like a sound-shadow in which the averted ear might be sheltered.'

(g) Beat-Notes and Beat-Tones. After K.'s final experiment (p. 532d') on the passage of beats into tones, we might perhaps disuse the interim term ' beat-note,' which implied no theory as to its origin, but only a statement as to its frequency, and use K.'s term ' beat-tone,' implying that the tone is *generated* by beats. But just because ' beat-note ' does *not* imply a theory, and because no theory has been at present generally accepted, nor is sufficiently supported by proofs to be so, it will be convenient to continue the use of the word ' beat-note,' which simply states that the frequency of the beat is identical with that of the note. At the same time we must not disuse the terms ' differential and summational tones, of various orders,' because if they really exist they are a decidedly different phenomenon from beat-notes, and only in the most frequently observed case coincide in pitch (but not in intensity) with beat-notes. P. viii. 29, however, decides to identify the two. K., on the other hand, considers the existence of differential and summational tones not proved.

Messrs. Preece and Stroh, referring to their machines for the synthesis of vowels, noticed infrà, sect. M. art. 2, p. 542d, say (Proc. R. S. 27 Feb. 1879, vol. xxviii. p. 366): 'The curves arrived at synthetically do not differ materially from those arrived at analytically by H. They principally differ in the prominence of the prime. But the prime can be dispensed with altogether. Curves produced by the synthetic machine, compounded of the different partials without their prime, shew that there exist *beats* or resultant sounds. A vowel sound of the pitch of the prime may be produced by certain partials alone, without sounding the prime at all. The beat, in fact, becomes the prime. This point is clearly illustrated by the automatic phonograph, and graphically by the sketch drawn by the synthetic curve machine. In fact, every two partials of numbers indivisible by any common multiple [divisor?], if sounded alone, reproduce by their beats the prime itself. Thus, the 3rd and 5th partials, or the 2nd and 3rd, &c., will result in the reproduction of the prime.' Observe that this gives the beat-note, not the differential tone. The differential tone of 3 and 5 is 2, but the beat-note is $2 \times 3 - 5 = 1$. 'In fact, the figure illustrates not only this, but it shews that when the number of partials introduced is increased, the beats become more and more pronounced.' Mr. Stroh from his own experience considers the the beat-notes thus produced to be generated in the same way as K. supposes.

¶   (*h*) *Koenig's explanation of Summational Tones.* In the matter of summational tones, P. (see p. 532b) explains them as differential tones of the second order. K. i. 217–8 thinks that they arise as beat-notes from upper partials. But P. viii. 24 notes that this explanation fails when very high partials would be required.

Thus, to get the summational tone 64 from 31 : 33 we should require the 32nd partial, which is not heard. So, from the reed tones 496 and 528, P. heard 1024. The 32nd partials would be 16896 and 15872, difference 1024. But such partials are inaudible, 'whereas every term of the acoustical equations

$$2 \times 528 - \quad (528 - 496) = 1056 - \quad 32 = 1024$$
$$3 \times 528 - (2 \cdot 528 - 496) = 1584 - 560 = 1024$$

is easily proved.' The tones mentioned certainly exist; the question is only, are they
¶ powerful enough to produce the result?

To the above remarks K. replies in the French edition, p. 127, note, continuing the passage already quoted (p. 532b'): 'M. Preyer cites in favour of his views that on sounding together free reeds of 496 and 528 d. vib. = 31 : 33, he heard the sound 1024 d. vib. = 64, and he thinks that we cannot assume that the reeds had the 32nd partial, 16896 and 15872 d. vib. If the sound really observed was 64, and not the Octave of 31 or 33, we might be really astonished that the 32nd partials were sufficiently strong in these tones to produce it; but the explanation proposed by M. Preyer is absolutely inadmissible, for 496 and 528 d. vib., even when they have considerable force, give 32 beats, which do not as yet allow the deep tone C_i to be heard, so that at any rate such tone must be extremely weak. Now the Octave of 528 (or 1056) is the 33rd harmonic of this excessively weak sound. But two primary sounds of 32 and 1056 d. vib., even when extremely powerful, never produce a sound of 1024 d. vib. The second manner in which M. Preyer thinks the sound might have been produced is equally opposed to all that has been directly observed when two primary tones sound together. Thus he makes the Octave of 528 (i.e. 1056) produce with 496 d. vib. a differential tone of 560 d. vib., and then makes this tone 560 produce with the Twelfth of 528 (i.e. 1584) a new differential of 1024. But these two sounds of 496 and 1056 ($= 2 \times 496 + 64$) give the beat-note 64, and not 560; and if the sound 560 really existed it would give with 1584 ($= 2 \times 560 + 464 = 3 \times 560 - 96$) the beat-note 96, and also more faintly 464, but not 1024.'

(*i*) *Koenig's theory for the origin of Beat-Notes.* K. i. 186 gives the following theory for the origin of tones from beats. He says that 'the beats of the harmonic intervals, as well as of the unison, should be deduced directly from the composition of waves of sound, and we should assume that they arise from the periodically
¶ alternating coincidences of similar maxima of the generating tones, and of the maxima with opposite signs. The similar maxima for these harmonic intervals, as in the case of unisons, will either exactly coincide, or else there will be maxima of condensation in the higher tone lying between two successive vibrations of the fundamental tone, slightly preceding one and slightly following the other; but in both cases the effect on the ear will be the same, for a beat (fluctuation) is no instantaneous phenomenon, but arises from a gradual increase and diminution of the intensity of tone.' Then he adds some drawings of the compounded vibrations of two tuning-forks, one of which bore a piece of smoked glass and the other a style. These are almost precisely the same as the curves drawn by means of Donkin's harmonograph, and inserted at the end of B. vi., opposite p. 256. That is, both K. and B., who are strongly opposed in opinion, refer to practically identical curves in support of their own views. This serves to shew the extremely difficult and delicate nature of the investigation.

(*k*) *Lecture Demonstration of Beat-Notes.* In the beat-notes produced by the wave siren, K. had the great advantage of producing tones which could be continued for any length of time, whereas those from tuning-forks vanished so rapidly that they could be with difficulty recognised. But this did not suffice for lecture demonstrations. Hence K. invented a machine which produces beat-tones audible over a whole lecture room.

This consists (K. iv.) of pairs of glass tubes adjusted so as to give notes with definite intervals by longitudinal vibrations. These are held at the node by two clamps against the surface of a wheel bearing a thick cloth tire, which continually dips into a trough of water, and thus rubs the tubes sufficiently to produce loud tones and either one or both of the beat-notes continuously, and loud enough to be appreciated by the whole audience. A piece of paper wrapped round the node and bearing the number of the relative pitch enables the glass tubes to be selected and changed with the greatest rapidity.

Art. 6.—*Influence of difference of Phase on Quality of Tone.*

H. p. 126a finds that difference of phase has no effect on quality of tone. But, on p. 127c, H. points out 'an apparent exception,' on which K. v. 376 remarks that if quality depends on the relative intensity of the harmonic upper partials, ¶ and this relative intensity is really altered by difference of phase, the influence of this difference is 'actual, and not merely apparent.' Then observing on the difficulties attendant on H.'s rule for finding the differences of phase (suprà, p. 124c), he proceeds to describe his new experiments with the wave siren (for which reason they are mentioned in this place), which certainly admit of very much more precision. They were conducted thus: K. compounded harmonic curves of various pitches, and with various assumptions of amplitudes, under four varieties of phase: (1) the beginning of all the waves coinciding; (2) the first quarter, (3) the halves, and (4) the third quarter of each wave coinciding; briefly said to have a difference of phase of $0, \frac{1}{4}, \frac{1}{2}, \frac{3}{4}$. These were reduced by photography, inverted, and placed on the rim of the disc of a wave siren, and then made to speak. He gives the remarkable curves which resulted in a few cases, and instructions for repeating the experiments. The following are his conclusions (K. v. 391):—

'The composition of a number of harmonic tones, including both the evenly and unevenly numbered partials, generates in all cases, quite independently of the ¶ relative intensity of these tones, the strongest and acutest quality tone for the $\frac{1}{4}$ difference of phase, and the weakest and softest for $\frac{3}{4}$ difference of phase, while the difference 0 and $\frac{1}{2}$ lie between the others, both as regards intensity and acuteness.

'When, unevenly numbered partials only are compounded, the differences of phase $\frac{1}{4}$ and $\frac{3}{4}$ give the same quality of tone, as do also the differences 0 and $\frac{1}{2}$; but the former is stronger and acuter than the latter.

'Hence, although the quality of tone principally depends on the number and relative intensity of the harmonic tones compounded, the influence of difference of tone is not by any means so insignificant as to be entirely negligible. We may say, in general terms, that the differences in the number and relative intensity of the harmonic tones compounded produces those differences in the quality of tone which are remarked in musical instruments of different families, or in the human voice uttering different vowels. But the alteration of phase between these harmonic tones can excite at least such differences of quality of tone as are observed in musical instruments of the same family, or in different voices singing the same ¶ vowel.'

Of course, as K. v. 392 observes, the complete wave siren figured on K. v. 386 is applicable to numerous other investigations.

Art. 7.—*Influence of Combinational Tones on the consonance of Simple Tones.*

This is a brief notice of P. viii. III. It would appear from H.'s theory of consonance (see especially suprà, pp. 200d and 205b) that, if there were no upper partial or combinational tones, dissonance and consonance could not be distinguished—in the Thirds for example. P.'s experiments rendered this doubtful. He had a series of 11 forks made, very accurately tuned to—

Vib.	1000	1100	1200	1300	1400	1500	1600	1700	1800	1900	2000
Cents		165	151	138	129	119	112	105	99	93	89
Sums	0	165	316	454	583	702	814	919	1018	1111	1200

where the upper line gives the numbers of vibrations, the second the cents in the intervals between two successive forks, and the bottom the sums of those cents, or the cents in the intervals between any fork and the lowest, from which the cents in the intervals between any two forks can be immediately deduced by subtraction ; and by a reference to the table in Sect. D. the names of such intervals can be found. P. selected the difference of 100 vib. because it was small enough to allow of a sensation of roughness when two successive forks were sounded together. And he selected the pitches 1000 and 2000 because they precluded hearing upper partials, while the frequency 1000 to 2000 not being sufficient to have any effect on distinguishing consonance, the absence of power to distinguish it could not be ascribed to the high pitch.

¶ Both practised and unpractised ears immediately recognised on them that successive forks were dissonant to each other ; this was due to the small difference of 100 vib. Almost all other intervals of these 11 forks, when the forks were not too loud, were frequently considered consonant, especially by musicians—such as 10 : 13, 11 : 13, 12 : 19, 17 : 20, &c. Also the ratios expressible by small numbers (except 8 : 9 and 9 : 10), namely, 5 : 7, 5 : 9, 6 : 7, 7 : 8, 7 : 9, 7 : 10, often passed as consonances ; and though the 1 : 2, 2 : 3, 3 : 4, 3 : 5, 4 : 5, 5 : 6, 5 : 8 were generally preferred, some observers found 6 : 7, 15 : 19, 11 : 13, &c. more harmonious than the pure Thirds and Sixths, 4 : 5, 5 : 6, 5 : 8, 3 : 5, especially than the minor Sixth 5 : 8. The listener was always kept in ignorance of the numerical ratios, and only one person was tried at a time. The sum of the different judgments was therefore :

' After all upper partials and combinational tones have been eliminated from a dissonant pair of sounds, it loses the disagreeable effect of dissonance.'

The Octave and Fifth were generally recognised with certainty, probably from long practice. This appears to be an excellent proof of H.'s theory. And the less care there was taken to exclude upper partials and combinational tones, the more unpleasant became the dissonance, and the easier it was for the ear alone to determine the interval immediately. But this is not all. H.'s theory that dissonances should be recognised only by beats of the partials or combinational tones implies that, if these were too far distant in pitch to produce beats, there would be no roughness, and hence no beats. This did not prove to be the case. The pair 1400 : 1600 vib. formed a dissonance, although all partials and combinational tones differed by 200. The ratio 8 : 9 was universally called a cutting dissonance, even in the 4 times and 8 times accented Octave.

¶ The explanation of the above phenomena seemed to require a remodelling of H.'s theory, and P.'s conclusions are stated thus (P. viii. 58) :—

' (1) The larger the least two numbers required to express the ratio between two tones, the greater the number of combinational tones, which always form an arithmetical series, and arise, whether upper partials be present or not (H. 155c).

' (2) The greater the number of simple tones which affect the ear simultaneously, the less distinct is each single tone.

' (3) The more coincidences there are between the tones which might be and are generated by any interval, the more pleasing is the sensation ; and the fewer the coincidences the more confusing, and hence unpleasant, the impression.'

And as these conclusions hold for tones which, on account of their own distance from each other and the distance of their partials and combinational tones, cannot generate sensible beats, P. considers that this is both a formal and an actual extension of H.'s theory of consonance. But if, with K., we consider these differential tones absolutely insensible, it would be difficult to see how they would affect the result, and the facts noted would still require explanation.
¶ The whole subject of combinational tones and beats evidently requires much more examination.

SECTION M.

ANALYSIS AND SYNTHESIS OF VOWEL SOUNDS.

(See Notes, pp. 75, 118, 124.)

Art. 1.—*Analysis of Vowel Sounds by means of the Phonograph.* The following is a brief account of a paper by Prof. Fleeming Jenkin, F.R.SS. L. and E., and Mr. J. A. Ewing, B.Sc., F.R.S.E., *On the Harmonic Analysis of certain Vowel Sounds,* 'Transactions of the Royal Society of Edinburgh,' vol. xxviii.

pp. 745-777, plates 34-40, communicated June 3 and July 1, 1878, and published with additions to July 19, 1878—that is, subsequently to the appearance of the 4th German ed. of this work.

Messrs. Jenkin and Ewing make use of a variety of Mr. T. A. Edison's phonograph which, by means of a style affixed to a vibrating disc against which words are spoken or sung, impresses the amplitude of vibration at any time on a piece of tinfoil passed beneath it by machinery. On repassing the style over these indentations the vibrations are recommunicated to the disc, and the sounds reproduced sufficiently, on the form of the instrument used by these gentlemen, for listeners to understand sentences impressed during their absence. Then the indented foil was passed under another style in communication with a system of delicate levers, ending in one of Sir W. Thompson's electrical squirting recorder tubes, which magnified the depth of the indentations 400 times, and squirted their form, without friction, on to a telegraph-paper band wound round a cylinder revolving at such a speed as to magnify the length of the indentations 7 times. Perfect records of the vibrations registered by the phonograph were thus obtained, ¶ of sufficient size to be measured. The amplitudes of the compound vibrations of the curves were measured to the 200th part of an inch ('005 inch). Then, as the apparatus could not properly determine high partials, the curves were assumed to be compounded of six partials, and the ordinates or amplitudes had to be determined by Fourier's formula—

$$y = A_0 + A_1 \sin x + A_2 \sin 2x + \ldots + A_n \sin nx + \ldots$$
$$+ B_1 \cos x + B_2 \cos 2x + \ldots + B_n \cos nx + \ldots$$

The period was taken as the length between two minima of ordinates, and divided into 12 equal parts for successive values of x, and then the corresponding values of y were measured. The 12 resulting simultaneous equations, giving the values A_0 to A_6 and B_1 to B_5, were then solved by Professor Tait's formulæ (given in the paper), and thus the amplitudes of the six partials for any length of the ordinate were determined. The Authors say :— ¶

'The experiments were chiefly directed to the two sounds ō and ū (the vowels in oh! and food). Several different voices were employed. Voice No. 1 was a powerful baritone with a considerable range and good musical training. No. 2 was a high set and somewhat harsh voice of limited range and without musical training. No. 3 was a rich and well-trained bass voice of a man of eighty. Nos. 4 and 5 were somewhat alike, being voices of moderate range and power and with some musical training. No. 6 was a powerful bass. Generally the vowels were sung in tune with notes given by a piano,' the pitch of which was supposed to be c' 256, but was probably much higher.

Photo-lithographs of the records of the vibratory curves are given in the paper, and ingeniously arranged tables are added shewing the maximum amplitudes of the partials for each pitch of the prime. Of these, the following is Table VII. p. 761 slightly re-arranged, with the names of the upper partials inserted :—

VOWEL SOUND ō ('OH'). ¶

VOICE	PITCHES AND AMPLITUDES OF THE FIRST SIX PARTIALS					
No.	I.	II.	III.	IV.	V.	VI.
2	$f'\sharp$ 44	$f''\sharp$ 32	$c'''\sharp$ 6	$f'''\sharp$ 0	$a'''\sharp$ 4	$c''''\sharp$ 2
1 5	f' 121 53	f'' 71 19	c''' 7 6	f''' 1 2	a''' 5 3	c''' 4 1
1 2 3 4 5	e' 105 51 53 55 52	e'' 69 30 18 34 53	b'' 7 5 3 7 5	e''' 3 2 1 2 6	$g'''\sharp$ 2 1 2 2 5	b''' 3 1 1 0 2

Vowel Sound ō ('OH')—*continued.*

VOICE No.	PITCHES AND AMPLITUDES OF THE FIRST SIX PARTIALS					
	I.	II.	III.	IV.	V.	VI.
	d'	*d''*	*a''*	*d'''*	*f'''*♯	*a'''*
1	119	76	5	1	3	2
2	66	40	4	0	3	2
5	27	42	6	4	2	1
	c'	*c''*	*g''*	*c'''*	*e'''*	*g'''*
1	110	160	15	10	10	7
3	37	30	1	4	1	1
4	54	25	0	3	2	1
5	47	41	5	3	3	2
	b	*b'*	*f''*♯	*b''*	*d'''*♯	*f'''*♯
1	70	126	15	14	6	1
2	45	66	7	4	6	2
3	36	31	2	4	1	0
4	45	43	4	2	3	0
5	47	61	2	14	8	2
	b♭	*b'*♭	*f''*	*b''*♭	*d'''*	*f'''*
1	75	185	13	8	11	1
2	49	104	18	6	4	2
4	25	82	5	7	1	2
5	48	70	13	7	1	3
	a	*a'*	*e''*	*a''*	*c'''*♯	*e'''*
1	125	190	25	22	5	2
2	32	58	6	8	6	2
3	40	36	4	4	3	2
4	16	54	4	4	1	1
5	40	68	10	8	3	0
	g	*g'*	*d''*	*g''*	*b''*	*d'''*
1	69	103	27	6	2	2
2	23	51	14	3	2	2
3	46	29	2	2	2	1
4	33	44	7	2	1	2
5	32	50	3	6	1	2
	f♯	*f'*♯	*c''*♯	*f''*♯	*a''*♯	*c'''*♯
2	18	58	15	3	1	0
4	28	62	10	2	3	2
	f	*f'*	*c''*	*f''*	*a''*	*c'''*
1	55	140	45	4	8	1
3	25	37	11	3	4	2
5	70	109	58	13	12	5
	e	*e'*	*b'*	*e''*	*g''*♯	*b''*
1	72	131	73	7	10	4
3	40	67	35	5	5	2
4	25	49	21	6	6	0
5	41	88	64	13	5	2
	d	*d'*	*a'*	*d''*	*f''*♯	*a''*
1	44	134	82	16	22	10
3	27	61	38	19	4	2
4	20	46	21	3	4	1
5	33	72	56	5	7	2
	c	*c'*	*g'*	*c''*	*e''*	*g''*
1	18	95	61	33	3	0
3	19	48	33	18	2	4

VOWEL SOUND ō ('OH')—*continued.*

VOICE No.	PITCHES AND AMPLITUDES OF THE FIRST SIX PARTIALS					
	I.	II.	III.	IV.	V.	VI.
	B	*b*	*f'*‡	*b'*	*d''*‡	*f''*‡
1	25	15	28	31	6	5
3	21	46	29	28	10	0
4	12	34	23	10	4	1
5	6	38	23	25	6	3
	B♭	*b*♭	*f'*	*b'*♭	*d''*	*f''*
1	37	58	61	47	11	0
3	28	41	25	36	9	0
5	18	26	15	15	2	2
6	18	22	32	75	9	2
	A	*a*	*e'*	*a'*	*c''*‡	*e''*
1	15	15	18	29	6	3
3	26	41	35	39	18	2
5	15	8	22	21	4	0
6	9	46	44	80	12	4
	G	*g*	*d'*	*g'*	*b'*	*d''*
1	13	0	15	40	8	4
6	34	30	8	45	9	7
	F	*f*	*c'*	*f'*	*a'*	*c''*
6	22	10	15	8	34	1

¶

The following is a table of the results for *ū* for voices 1 and 5 only, where, for brevity, I give only the pitches of the primes, for the pitches of the partials are given in the preceding table, and the numbering of the partials is sufficient to shew the great peculiarity of the jump from one reinforced partial to two, the second being then by far the most prominent, and the different pitches at which different voices make the change. Voice 5 could not get out a clear *ū* at the pitch *a*. To these are added the results obtained from voice 5 for the vowels *ā°* ('awe') and *ā* ('ah').

VOWEL *ū* ('oo'), VOICE 5 — AMPLITUDES OF PARTIALS

PITCH OF PRIME	I.	II.	III.	IV.	V.	VI.
e'	136	6	2	4	3	1
c'	85	3	0	3	2	2
b	287	26	12	3	8	0
b♭	250	8	11	1	1	2
g	38	128	9	4	10	11
f	23	135	5	10	3	5
e	34	148	18	7	6	3
d	33	107	14	4	3	0
c	31	74	41	4	6	3
B	18	50	28	3	5	1

VOWEL *ū* ('oo'); VOICE 1 — AMPLITUDES OF PARTIALS

PITCH OF PRIME	I.	II.	III.	IV.	V.	VI.
d'	94	7	0	2	3	2
b♭	22	189	12	38	2	8
a	13	120	12	12	4	0
g	22	136	6	16	3	1
f	21	108	7	13	6	2
e	27	127	12	14	2	2

¶

VOWEL *ā°* ('AWE'), VOICE 5

PITCH OF PRIME	I.	II.	III.	IV.	V.	VI.
a	41	48	48	3	6	2
g	24	44	32	15	3	0
f	14	18	14	23	2	0
e	23	39	32	40	6	2
d	19	26	20	30	7	3

VOWEL *ā* ('AH'), VOICE 5

PITCH OF PRIME	I.	II.	III.	IV.	V.	VI.
d'	9	22	14	3	2	2
c'	20	48	58	15	10	0
b	20	51	56	8	3	2
a	37	62	46	20	4	3
g	23	35	24	25	6	0
f	24	29	12	24	1	1
e	12	23	11	15	9	3

On the first table the authors remark :—

'At the pitches ordinarily used in speech, the vowel ō consists almost wholly of the two constituents—a prime and its Octave—the ratio of whose amplitudes may vary widely. But when the range is extended so as to reach lower pitches, higher partials successively appear in such a way as to allow the highest strongly reinforced partial to remain in the neighbourhood of $b'b$. . . . Generally we may say, . . . that there is a wide range of reinforcement, extending over about two Octaves (from f or g to f''), within which all tones are more or less strongly reinforced, and that there is a specially strong reinforcement at the pitch $b'b$.'

But this last did not appear for the artificial ō's produced by Prof. Crum Brown's instrument (described in the paper), and recognised by the ear as ō's. They also draw attention to the sudden alteration of amplitude of the 4th partial with voice 6, and also of the 5th partial, for pitches Bb, A, G, as compared with F; and to similar sudden alterations in the 3rd and also 4th partial with voices ¶ 1, 3, 5.

After discussing the results condensed in these tables, the authors review former vowel theories and give their own conclusions, of which the following may be noted ; but the whole paper requires careful examination, as a most original and laborious study of a very difficult subject.

'In distinguishing vowels the ear is guided by two factors, one depending on the harmony or group of relative partials, and the other on the absolute pitch of the reinforced constituents. It seems not a little singular that the ear should attach so distinct a unity to sounds made up of such very various groups of constituents as we have obtained from different voices and at different pitches, so as to recognise all these sounds as some one particular vowel. We are forced to the conclusion . . . that the ear recognises *the kind of oral cavity* by which the reinforcement is produced. . . . The vowel-producing resonance cavities are clearly distinguished in virtue of two properties—first, the absolute pitch at which they produce a maximum reinforcement; and, second, the area of pitch over which reinforcement acts. The latter property, when it is extensive, is very probably due to the existence of subordinate proper tones not far from each other in pitch. . . . We ¶ should . . . describe the ū cavity as an adjustable cavity, with a very limited range of resonance, whose effect is to reinforce strongly only one partial above the pitch of a. . . . If we assume that the ō cavity is absolutely constant, we must describe it as a cavity reinforcing tones throughout nearly two Octaves, or from g to f''. . . . We are disposed to regard it as more probable, that in human voices the ō cavity is slightly tuned or modified according to the pitch on which the vowel is sung; . . . the genuine character of ō is given by a cavity reinforcing tones over rather more than one Octave, with an upper proper tone never far from $b'b$. . . . It is very satisfactory to find that the ō's given by the human voices which we have experimented with are marked by the strong resonance on $b'b$ which Helmholtz has noticed by quite different methods of observation. It tends to shew that our ō was essentially the same vowel sound as his, and to give us confidence in the mode of experiment we have adopted.' (*Ibid.* pp. 772–775.)

Art. 2.—*Synthetical Production of Vowel Sounds.* A most ingenious method of producing artificial vowels was invented, and is explained by Messrs. W. H. Preece and A. Stroh in their paper entitled *Studies in Acoustics : On the Synthetic Examination of Vowel Sounds,* 'Proceedings of the Royal Society,' Feb. 27, 1879, vol. ¶ xxviii. No. 193, pp. 358–67. Mr. Stroh, to whom all the machinery is due, was kind enough, on May 29, 1884, for the purposes of this Appendix, to shew me the machines in action, and to reproduce the results many times over in order that I should be able to judge of them. Essentially there are four machines. First, one to produce the curve resulting from compounding 8 harmonic curves, representing partials, with maximum amplitudes decreasing inversely as the number of the partial increased, but with arrangements for altering the amplitudes and phases of composition. The resulting figures are extremely beautiful. Secondly, a machine for cutting the curve thus produced, but on a reduced scale, on the edge of a brass disc, so that 30 periods were included in one circumference of this disc, the curves being automatically transferred from the first machine. Third, a machine by which an axis on which 8 of the discs thus cut were placed, representing 8 partials. These discs by springs could be brought into action in any combinations, and could convey the resulting vibration to a style working against a sensitive disc like that of the telephone. The sensitive disc on vibrating produced the corresponding sound audibly. Not being satisfied with these results, Mr. Stroh took the combinations and amplitudes which these experiments shewed were likely to succeed best, made the corresponding compound curves by the synthetic machine, cut them by the second machine, mounted them as in the third, and then in a fourth or vowel machine conducted the vibrations from each compound curve to a disc, which spoke them. The details and drawings of the first and fourth machine, the speaking disc, and various compound curves are given in the paper. The curves are also compared with those resulting from my table of Prof. Helmholtz's results (suprà, p. 124c, d, footnote), which had also appeared in the 1st edition of this translation, p. 181. The table of the intensities of the partials given in the paper (on its p. 365)—though I am not quite sure that they agreed with those I heard—are as follows, the pitch of the prime being $B\flat$:—

Vowels	1	2	3	4	5	6	8	16
U	ff	mf	pp					
O	mf	f	mf	p				
A	p	p	p	mf	mf	p	p	
E	mf		mf			ff		
I	mf	p				p		mf

¶

The effect of these vowels on my ear was not like that of human vowels, hence I found it extremely difficult to place them anywhere in the human vowel scale. Roughly, I felt that—

    U was a sort of *oo*, tending towards *oh !*

    O was more like the word *awe* than *oh !*

    A was a very high *ah*, tending to the long sound of English *a* in *fat*.

    E was very imperfect, and had the effect of a hollow low French *ê* mixed with English *u* in *but*.

    I, was the worst vowel. It had none of the character of *ai* in *air*, but was far from *ee*. The sound ' tootled.'

When taken in rapid succession, the ear at once recognised that these sounds were meant for *oo*, *oh*, *ah*, and perhaps *ay*, *ee* ; but on prolonging the sound of any one, the character of the vowel became lost, as indeed is frequently the case in singing. Curious effects resulted from raising and lowering the pitch. The *O* ¶ flattened became a very decent *oo* (in *boot*), and the *A* flattened almost a good *oh*. The effect of taking all an Octave higher was not so successful.

The synthesis of Prof. Helmholtz and that of Messrs. Preece and Stroh, together with the analysis of Messrs. Fleeming Jenkin and Ewing, in art. 1, prove distinctly that difference in the quality of tone, taking only harmonic partials, is the foundation of vowels, and also that difference of phase has, so far as they could observe, no effect on the ear. (But see suprà, App. XX. Sect. L, art. vi. p. 536.) Both, however, also prove that there is much more yet to be learned before we can satisfactorily imitate spoken vowels. Each of these methods of synthesis necessarily relates to sung vowels, which are quite distinct from spoken vowels, and indeed never satisfactorily imitate them. It appears to me that the mode of vibration of the vocal chords is a most important element of vowel character, and that the resulting effect is modified by the resonance in the ventricles of Morgagni, in the cartilagenous larynx more or less covered by the epiglottis (acting, possibly, like the cup mouthpiece of brass instruments, see suprà, p. 98d, note), in the ¶ pharynx, and between the pillars of the velum, before it reaches the larger resonance cavities of nose and mouth, with which we are almost solely able to deal. By the original mode of vibration of the vocal chords for spoken vowels many inharmonic proper tones are probably produced, which are overcome in singing, and this is possibly one of the many differences between speaking and singing. Also, we should bear in mind that each speaker has his personal quality of ' voice ' (that is, mainly, of vowel sound), by which he would be recognised in the dark, and that in each individual the feeling of the moment varies the pitch and the characteristic quality of his vowels ; so that there are really millions of different qualities of tone all recognised generically as the same vowel. And yet in the artificial vowels just considered I could not recognise any exact form of human vowel with which I was acquainted, although I have made speech sounds an especial study for more than forty years. We have an analogy in the multiform presentment of the human countenance, which is nevertheless unhesitatingly recognised as distinct from that of the anthropoid ape.

SECTION N.

MISCELLANEOUS NOTES.

(See pp. 78, 179.)

¶ ### 1. *Compass of the Human Voice.*

Instruments can be tuned or manufactured at almost any required pitch. The human voice is born, not manufactured. Although by skilful training its compass can generally be somewhat extended, both upwards and downwards, yet it must in general be considered to be an instrument beyond human control. The usages of Europe have, however, made it the principal instrument, and, when it is present, have reduced all others to an accompaniment. Hence it is necessary that these other instruments should have their compass and pitch regulated by that of the human voice. Now the voice, like the viol family, represents at least four different instruments—soprano, alto, tenor, and bass, with two intermediate ones, ¶ mezzo-soprano, between soprano and alto, and barytone, between tenor and bass. It is therefore as necessary to determine the average and exceptional compass of these species of voice as it is to know the compass of any other instrument, in order that composers may be certain as to what sounds can be reproduced, and not demand any other. To do this, the precise acoustic meaning of each written musical note should be ascertained. The difficulty of determining it has been shown by the preceding history of musical pitch (pp. 494–513), from which, combined with the tables of meantone and equal intonation (pp. 434 and 437), it is evident that Handel's sustained a'' in the *Hallelujah* chorus had 845 vib., but would now be sung to 904 vib.; and that Mozart's f''' in the *Zauberflöte* would have meant 1349 ¶ vib., but would now have to be sung at 1455 vib. The strain that this would put upon voices is evident, and no composer who wished his music to be well represented would think of making such demands on his singers. It appeared, therefore, necessary to ascertain more precisely than had been hitherto done, and to express in numbers of vibrations, the limits of the different kinds of voices. If the composer will then only translate his written notes into numbers of vibrations, by the table on p. 437, according to the pitch he employs, he will avoid all danger of straining singers.

Through the kindness and liberality of the choir conductors Messrs. Henry Leslie, W. G. McNaught, J. Proudman, Ebenezer Prout, L. C. & G. J. Venables, and 542 members of the choirs they conducted, I was able to examine a sufficient number of singers, in January, 1880, to arrive at something like a trustworthy account of the compass of the voice. I gave each singer a paper with the words *do re mi fa sol la ti do* printed on them in four columns up and down to the requisite extent, and then started them on *do* in 4 different pitches, 507, 522·5, 528, 540·7 vib. (representing the just c'' corresponding to a' 422·5 Handel's pitch, a' 435·4 the French pitch, a' 440 Scheibler's pitch, and the equal c'' of a' 454·7, the highest Philharmonic pitch of 1874, respectively). I got them to sing up and down in chorus under the direction of the conductor, and to mark with a pencil the highest and lowest note each one could reach, first *easily*, or secondly *by an effort* (falsetto of male voices being excluded in the first case, but not in the last). From these papers I determined by calculation, on the assumption of just intonation (as being most probable for unaccompanied singers), the numbers of vibrations in the limiting notes. These are contained in the following table, together with the mean height and depth of all the voices. The extreme highest limit for male voices, as it included falsetto, is a mere curiosity. For writing music, the mean should not be assumed as the limit, for perhaps half the chorus could not reach it. But it would be perfectly safe to write from the highest low easy limit to the lowest high easy limit. Thus, for sopranos it would not do to write up to b'' 993 and down to f 180, but it would be quite safe to write up to f'' 704 and down to b 253. Viewed in this way, my results agree more nearly with Randegger's, which I add for comparison. These last are given in a staff-notation form in his primer on *Singing* (Novello, 1879); and as he politely informed me that he assumed Broadwood's medium pitch a' 446·2, I was able to calculate the vibrations. All the numbers of vibrations are given to the nearest integer only, and it is to these numbers that attention should be especially paid, the names of the notes being merely guides. Those letters preceded by a turned period relate to high pitch in the column 'Actual,' and those not so preceded relate to a medium pitch, as French or German. But in the column 'Mean' no precise system at all could be selected. In Randegger's, of course, Broadwood's medium pitch is intended. If, however, the notes be played on any ordinary piano, they will seldom be in error to the extent of a quarter of a Tone.

MEAN AND ACTUAL COMPASS OF THE HUMAN VOICE.

VOICES OBSERVED	EASY LOWER LIMIT		VOICES OBSERVED	EXTREME LOWER LIMIT	
	Mean	Actual		Mean	Actual
146 Sopranos	f 180	·b 253 to ·c 135	173 Sopranos	eb 162	·g 203 to c 130
91 Altos	eb 161	ab 211 to c 132	108 Altos	d 147	g 198 to B 124
107 Tenors	G 98	e 163 to ·D 76	114 Tenors	E 85	·B 127 to C 66
125 Basses	E ·81	Ab 106 to C 66	140 Basses	C♯ 72	·F 90 to ·A, 56

VOICES OBSERVED	EASY HIGHER LIMIT		VOICES OBSERVED	EXTREME HIGHER LIMIT	
	Mean	Actual		Mean	Actual
145 Sopranos	b″ 993	f‴ 1408 to f″ 704	173 Sopranos	c‴♯ 1124	·a‴b 1690 to ·g″ 811
83 Altos	g″♯ 836	·d‴ 1216 to ·e″ 676	105 Altos	b″b 952	g″ 1584 to ·f″ 721
114 Tenors	c″ 521	d′ 608 to ·e′b 317	112 Tenors	d″ 617	·g″ 811 to g′ 396
120 Basses	f′♯ 375	·c″ 541 to d′ 294	139 Basses	b′b 483	·c‴ 1081 to e′ 330

RANDEGGER'S STATEMENT OF LIMITING TONES.

VOICES	REGULAR		VOICES	EXCEPTIONAL	
	Lower Limit	Upper Limit		Lower Limit	Upper Limit
Soprano	bb 236	c‴ 1061	Soprano	bb 236	f‴ 1417
Mezzo Soprano	g 199	b″b 945	Mezzo Soprano	g 199	c‴ 1061
Alto	e 167	f″ 708	Alto	e 167	g″ 795
Tenor	c 133	b′b 473	Tenor	c 133	c″♯ 562
Barytone	Ab 105	f′ 354	Barytone	F 87	g′ 398
Bass	F 89	e′b 316	Bass	D 75	f′ 354

2. *Harmonics and Partials of a Pianoforte String struck at one-eighth of its length.*

On p. 77, note *, will be found Mr. Hipkins's observations on the striking-point of pianoforte strings, shewing that one-seventh of the length, which seemed to be assumed as usual by Prof. Helmholtz, was not in use generally, or (p. 76d′) at Steinways'. Prof. Helmholtz conceived that the origin of this presumed custom was to get rid of the 7th partial, which he also considered likely to injure the quality of tone. Mr. Hipkins's experiments were therefore made with the object of determining whether when the striking-place was one of the nodes the corresponding partial disappeared, as results from the mathematical formula (12a) suprà, p. 383b.

Mr. Hipkins's first experiments are detailed in his paper entitled 'Observations on the Harmonics of a String, struck at one-eighth its length' (*Proc. Royal Society,* 20 Nov. 1884, vol. xxxvii. p. 363). The main facts are given suprà, p. 78d. The results were all witnessed by Dr. Huggins, F.R.S., and myself. The string was exactly 45 inches long, and was struck at precisely one-eighth its length from the wrestplank-bridge (that nearest the player). When it was touched with a piece of felt at 5·63, 16·88, and 28·13 inches from the belly-bridge (that farthest from the player), which

are three positions of the nodes for the 8th partial or third Octave higher, selected to avoid errors (as not being positions of the nodes of the 2nd or 4th partials), in each case the 8th harmonic was well heard. It was not so strong as the 4th, 5th, 6th, 7th, and 9th, all of which were heard, but quite unmistakable, and was heard better on removing the felt immediately after the note had been produced. The 16th partial was also heard when the string was touched at its nodes 2·81 and 8·44 inches from the belly-bridge, which are nodes of the 16th but not of the 8th partial.

What was heard was the *harmonic,* not the simple partial tone, and it was suggested, that perhaps touching the string at the node *coerced* the string and obliged it to vibrate with these nodes, notwithstanding that it was struck in one of the series of such nodes. Mr. Hipkins, therefore, at my suggestion made a new series of experiments, detailed in his paper entitled 'Observations on the Upper Partial Tones of a Pianoforte String, struck at one-eighth its length' (*Proc. of the Royal Society,* 15 Jan. 1885, vol. xxxviii. p. 83). These experiments I also witnessed. The object was to leave the string perfectly uncoerced, and to avoid the use of resonators, on which some suspicion had been (wrongly, as I believe) cast by at least one observer. Calculating the pitch of the partials, which would be the same as that of the harmonics, and the interval which the tempered notes of the piano would make with them (as in Table II., suprà, p. 457), a string of the corresponding note was slackened (or tightened, as convenient; sometimes both alternately), while the other unison strings were damped with the usual tuners' wedges; and in the same way only one of the three lower strings was allowed to vibrate. Then the low note and the high note were struck simultaneously. It is evident that the high note being slightly out of unison with the upper partial of the lower note, beats would ensue if such a partial existed. Now, for the 5th, 6th, ¶ 7th, 8th, 9th, 10th, and 11th partial such beats were perfectly audible, but their duration for the 11th partial with 1487 vib. was so short that higher ones were not tried. For the 8th partial the beats were quite distinct, and, on removing the wedges that damped the unison strings for the high note, and striking the three high strings without the lower note, it was evident that the beats heard were the same in rapidity and character as when the single string was sounded with the low one.

The fact, therefore, that the 8th partial existed was conclusively proved. Various causes have been assigned. On p. 383d, note *, I have suggested that if terms omitted by the hypothesis named were introduced, perhaps there would be a residuum which would account for hearing the 8th partial. This partial was ¶ really much weaker than the 7th. The last, indeed, was quite clear and ringing, so that it did not seem affected by striking the string so near its node.

It is curious that when the nodes do not lie very close the harmonic could be brought out by touching the string somewhat near the proper place. Thus for the 2nd harmonic, node at 22·5 inches from the belly-bridge; the next nodes were 1·2 inch nearer and 1·2 inch farther from that spot, and on trial the 2nd harmonic or Octave came out when the string was touched between 22·1 and 22·95 inches

from the belly-bridge, but not at 22·05 and 23·0, so there was a 'play' of ·85 inch. For the 3rd harmonic there was similarly a 'play' of ·65 inch. A very remarkable fact was that by stopping within 1·5 inch of the belly-bridge, the simple prime or lowest partial came out unaccompanied by any other audible partials. This was tested by beats of forks, shewing that the 2nd and 4th partials did not exist.

Various causes for the sounding of the 8th harmonic have been suggested. One of these was that the hammer of the pianoforte, being round and soft, did not strike at one point, and so excited the string on each side of the node. To avoid this action the much harder hammer of the highest note (A in the 3-inch octave) was used in supplementary experiments made on 2 April 1885. The width of the part of the hammer that came in contact with the string did not exceed $\frac{1}{10}$ inch. And again, an ivory edge, not more than $\frac{1}{20}$ inch in width, was used instead of the felt covering of the hammer. The 8th partial, tried by beats, in both cases came out much stronger than before, and the beats could be distinctly heard 10 or 12 feet off. Again, it was supposed that the string might not be uniform, and that if the striking-place were slightly moved from the theoretical node, an actual node would be reached, and the partial quenched. Hence the ivory head of the hammer was shifted so as to strike up to $\frac{1}{10}$ inch away from the node on either side. The partial was heard strongly, but the sound of the note was not so pleasant as when the string was struck at the actual node.

It has been also suggested that the string moved the points of support, but that we had no means of testing. The phenomenon, therefore, remains unexplained; but thanks to Mr. Hipkins, who had the resources of Broadwoods' establishment and the assistance of experienced tuners at command, there is no doubt whatever of the fact, that a pianoforte string when struck at a node by a hard or soft hammer does not lose the corresponding partial, and does not materially enfeeble the partials with adjacent nodes.

3. History of Meantone Temperament.

This is the temperament usually, but wrongly, known as 'unequal' (suprà, p. 434a'), which prevailed so long over Europe, and is ¶ not yet entirely extruded.

Arnold Schlick, Spiegel der Orgelmacher vñ Organisten (Mirror of Organ-builders and Organists), 1511, chap. viii., orders the Fifths FC, CG, GD, DA to be tuned as flat as the ear could bear, so as to make the major Third FA decent. Then he tunes AE, BD in the same manner. Beginning again with F, he tunes the Fifths down FBb, Bb Eb, sharpening the lower note for the same reason. Then he tunes Eb Ab and makes Ab 'not sharp, but somewhat flatter than the Fifth requires, on account of the proof (vmb das brifen), although, however, the G♯ thus made is never a good Third or perfect Sixth to the Fifth E and B♮ for cadences in A.' He prides himself, however, on the Ab or G♯, and shews how to disguise inaccuracies. And he refutes those who would make G♯ good for cadences in A in the chord E G♯ B, by saying this produces weak-

ness, and takes away the effect of good and strange consonances. For the rest, he tunes B F♯ with the upper note flat, and apparently F♯ C♯ in the same way. This was really an unequal temperament, and looks very like the meantone temperament spoiled, but that system was not yet discovered. Schlick's editor (Rob. Eitner, in the Monatshefte der Musik-Geschichte, monthly parts of the History of Music, part I., 1869) says that what Schlick claims for Ab and G♯ was supposed to be the invention of Barth. Fritz of Brunswick in 1756, 245 years later.

Giuseppe Zarlino of Chioggia, 50 years after Schlick, in his Le istitvtioni harmoniche (Institutes of Harmony), Venice, 1562, speaks of alcuni (some people) who seemed to think that the interval of the comma should be distributed among the two nearest intervals, and the others left in their natural form (cap. 43, p. 128). This would give a meantone for the second of the major scale = $\frac{1}{2} \times (204 + 182) =$ 193 cents, but leave the others very dissonant, and to this Zarlino rightly objects. It would give the major scale C_0, D_{193}, E_{386}, F_{498}, G_{702}, A_{884}, B_{1088}, c_{1200} cents, so that the Fourth $D : G$ would have 509 cents, and the Fifth $D : A$ would have 691 cents, which coming in the midst of just intervals would be intolerable, and beyond the natural key it fails entirely. Zarlino's remedy (chap. 42, p. 126) is to diminish every Fifth by two-sevenths of a comma, and he proceeds to shew how this affects the tuning. It preserves the small Semitone $24 : 25 = 70\cdot673$ cents. He says, p. 127, 'although in instruments thus tempered consonances cannot be given in their perfect—that is, their true and natural form— yet they can be used when the chords have to be given in their true and natural proportions. I say this,' he adds, 'because I have frequently made the experiment on an instrument which I had made for the purpose, and the effect may be tried on any other instrument, especially the harpsichord and clavichord, which are well adapted for the purpose.' Then, in chap. 43, he proceeds to shew that this temperament is rationally constructed, and that no other is so (*che per altro modo non si possa fare*, that is, *ragioneuolmente*). It is quite clear, then, that Zarlino, as has often been asserted, did not invent the meantone temperament, and did not consider equal temperament worth mentioning, even if he was acquainted with it.

Francis Salinas of Burgos in Spain, born 1513, died 1590, blind from infancy, Professor of Music in the University of Salamanca, Abbé of St. Pancras de Rocca Scalegno, in the kingdom of Naples, in 1577 published his *De Musica libri septem*, of which a very imperfect, and, as respects temperament, incorrect account is given in Burney & Hawkins. Salinas says (lib. iii. cap. xv. p. 143, I translate his Latin) :—

'From what has been said, in order that Tones should be rendered equal, the minor must be increased and the major diminished. It must be observed that this can be done in several ways, because the comma, by which they differ, may be divided in many ways. Of these, *three* have been thought out up to this time, which seem to me most suitable (*aptissimi*). Hence arise three ways of tempering imperfect instruments. The *first* is to divide the comma into three proportional parts, giving *one* to the minor Tone, and taking *two* from the major Tone. This gives a new Tone, larger than the minor and smaller than the major. The decrement is twice the increment, and through the maximum inequality the tone becomes equal.'

The comma has $21\cdot506$ cents, hence $\frac{1}{3}$ comma has $7\cdot169$, and $\frac{2}{3}$ comma has $14\cdot388$ cents. Then $182\cdot404 + 7\cdot169 = 189\cdot572 = 203\cdot910 - 14\cdot338$ cents, which is what the above statement comes to, giving $189\cdot572$ cents for the new Tone. Notwithstanding this very precise statement, Salinas ought to mean *precisely the reverse*. His object was to make the Tritone perfect, and to make it consist of three new Tones. Now a Tritone $F : B$ consists of 2 major Tones and 1 minor Tone, —that is, 3 minor Tones and 2 commas, or $590\cdot224$ cents, $\frac{1}{3}$ of which is $196\cdot741$ cents, which is $182\cdot404 + 14\cdot338$ and $203\cdot910 - 7\cdot169$

—that is, the reverse of the former result. By a singular error perpetuated in a figure (which, of course, being blind he could not see), Salinas makes the Tritone in this place consist of 2 minor Tones and 1 major Tone—that is, 3 minor Tones and 1 comma, having the ratio $18 : 25$, or cents $568\cdot718$, which is not the Tritone, but the superfluous Fourth, and may here be called the *false* Tritone. This mistake seems to have arisen thus. The Octave, as he rightly says, has 6 minor Tones, 2 commas, and a great Diësis. 'The comma being divided into 3 proportional parts, if one is added to each minor Tone, 2 commas will be added to the six Tones, and one to three, equally distributed among them. From which distribution it will follow in this constitution of the temperament that the Tritone consists of three minor Tones and one comma, or ¶ 2 minor Tones and one major,' whence he deduces the ratio $18 : 25$. But he thus altogether loses sight of the great Diësis, and considers a Tritone to be half of an Octave after it has been diminished by a Diësis. On p. 155 he again notices the Tritone as $32 : 45$, the correct ratio. The *false* Tritonic temperament therefore makes the Tone $189\cdot572$, the Fifth $694\cdot786$, and the false Tritone $568\cdot718$ cents. But the true Tritonic system gives the Tone of $196\cdot74$, the Fifth of $698\cdot37$ and the true Tritone $590\cdot22$ cents.

Salinas continues his account of the three temperaments thus : 'The second [temperament] divides the comma into 7 proportional parts, giving 3 to the minor and taking 4 from the major Tone.' This is Zarlino's temperament already described, and preserves the ¶ small Semitone $24 : 25 = 70\cdot673$ cents. 'The third will arise from halving the comma, giving half to the minor and taking half from the major Tone.' Then he adds (p. 164) : 'Wherefore any one of these three temperaments seems most suitable for artificial instruments ; nor have any more been as yet thought out (*neque plura adhuc excogitata sunt*) ;' that is, Salinas, like Zarlino, utterly ignores the equal temperament. 'The first, so far as I know, has been laid down by no one.' From which it is to be inferred that it was his own invention. 'The second I have also found in the harmonic institutions of Joseph Zarlino of Chioggia,' as already given. 'The third was commenced, but not perfected, by Luigi or Ludovico Folliano of Módena,' who must have been Zarlino's 'some people' (*alcuni*). 'And ¶ Joseph Zarlino has properly considered it in his harmonic demonstrations. But no one has previously acknowledged all three, nor observed upon their relation and mutual order.'

It was Salinas who finished Folliano's work, and in chaps. 22 to 25 he describes the result thoroughly. As, therefore, we consider Watt, and not the Marquis of Worcester, to have invented the steam engine, we must consider Salinas, and not Folliano, to have invented the meantone temperament. I give a comparative table of all three schemes in cents to the nearest integer, from $E\flat$ to $G\sharp$, distinguishing the true and false Tritonic and adding the Equal, which will shew the real relations of these three temperaments to each other.

Notes	True Tritonic	False Tritonic	Zarlino	Meantone	Equal
C	o	o	o	o	o
C♯	89	64	71	76	100
D	197	190	192	193	200
E♭	305	316	313	310	300
E	393	379	383	386	400
F	502	505	504	503	500
F♯	590	569	575	580	600
G	698	695	696	697	700
G♯	787	758	766	773	800
A	895	884	887	890	900
B♭	1003	1010	1009	1007	1000
B	1092	1074	1079	1083	1100
c	1200	1200	1200	1200	1200

¶ The true Tritonic, making the Tritone 590·22 cents, necessarily differs very slightly from equal temperament, which makes it 600 cents, while the false Tritonic, making the Tritone 568·716, or a comma too flat, approaches very near to Zarlino's and the Mean-tone, so that I think Salinas must have intended to use this one, which he lays down so clearly, and that he accidentally made a mistake of a comma in estimating the Tritone, by hastily neglecting the Diësis. For later usages see suprà, pp. 320, 321.

4. *The History of Equal Temperament.*

When once the Pythagorean division of the Octave had been settled, and it had been observed that 12 Fifths exceeded 7 Octaves by the small interval of a Pythagorean comma (p. 432, art. 9), the idea of distributing this error among the 12 Fifths was obvious. Aristoxenus, a pupil of Aristotle, the son of a musician and a writer on music, is said to have advocated this. At any rate he stated that the Fourth consisted of two Tones and a ¶ half, which is exactly true only in equal temperament. Amiot reports equal temperament from China long previously even to Pythagoras. In later times Mersenne (*Harmonie Universelle*, 1636) gives the correct numbers for the ratios of equal temperament, and says (Livre 3, prop. xii. 'Des genres de la musique') of equal temperament that it 'est *le plus usité* et le plus commode, et que tous les practiciens avoüent que la division de l'Octave en 12 demitons leur est plus facile pour toucher les instruments.' This should imply that there were numerous instruments in equal temperament, but I have not been able to find any noticed. Bédos (*L'Art du facteur d'Orgues*, 1766) knows only meantone temperament, which he gives directions for tuning. In Germany, Werckmeister (*Orgelprobe*, 2nd edit. 1698) says that he can only recommend equal ¶ temperament, and Schnitger of Harburg in Hanover, and afterwards of Hamburg, an admirer of Werckmeister, built the organ of St. Jacobi-Kirche in Hamburg in 1688, and tuned it in intentionally equal temperament. Herr Schmahl, who had been the organist there since 1838, never knew it otherwise tuned, and could find no record of any change of intonation in the archives of the church, and he also could not recollect having ever heard of any other intonation in North Germany. His master, Demuth (died 1848) of St. Catharinen-Kirche, whose memory extended backwards to 1810, also knew of no other tuning in North Germany. Of course the temperament never was thoroughly equal, so that when Herr Schmahl practised on the St. Catharine's organ, the usual keys C and G were not so good as the unusual keys F♯ and

D♯. Dr. Robert Smith, 1759, must have heard equal temperament, or else he could hardly have spoken of 'that inharmonious system of 12 hemitones' producing a 'harmony extremely coarse and disagreeable' (*Harmonics*, 2nd ed. pp. 166-7), but it may have been only an experimental instrument of his own.

As regards the recent introduction of equal temperament into England, Mr. James Broadwood, in the *New Monthly Magazine*, 1 Sept. 1811, proposed it, and gave the error of the Fifths as $\frac{1}{40}$ Semitone ($= 2\frac{1}{2}$ cents), which was to him the smallest sensible interval. On 1 Oct. 1811, Mr. John Farey, sen., shewed that this was too much (it should be 1·954, or about 2 cents—that is, about $\frac{1}{50}$ Semitone), and referred to the article 'Equal Temperament' in Rees's *Cyclopedia*. Hereupon, on 1 Nov. 1811, Mr. James Broadwood rejoined that he gave merely a practical method of producing equal temperament, 'from its being in most general use, and because of the various systems it has been pronounced the best deserving that appellation by Haydn, Mozart, and other masters of harmony.' Unfortunately he gives no references, and consequently this assertion can be taken only as an unverified impression. Haydn died 1808, Mozart 1791, but the Hamburg organs had equal temperament long before that time. Sebastian Bach (died 1750) is generally credited with introducing equal temperament, but M. Bosanquet says 'there is no direct evidence that he ever played upon an *organ* tuned according to equal temperament' (*Musical Intervals and Temperament*, 1876, p. 31). Bitter, however, states, in his life of Sebastian Bach, that he once played on the St. Jacobi organ at Hamburg, and expressed his approval of the tuning, and even applied for the post of organist. The *wohl temperirtes Clavier*, or well-tuned clavichord, the notes of which are very fugitive, was the instrument mentioned by Carl Philip Emanuel Bach, who died 1788.

As regards Mr. James Broadwood's statement that equal temperament was in 1811 'in most general use' presumably in England—

Mr. Hipkins has been at some pains to ascertain how far that was the case, and from him I learn that Mr. Peppercorn, who tuned originally for the Philharmonic Society, was concert tuner at Broadwoods', and a great favourite of Mr. James Broadwood. His son writes to Mr. Hipkins that his father 'always tuned so that all keys can be played in, and neither he nor I [neither father nor son] ever held with making some keys sweet and others sour.' Mr. Bailey, however, who succeeded Mr. Peppercorn as concert tuner, and tuned Mr. James Broadwood's own piano at Lyne, his country house, used the meantone temperament to Mr. Hipkins's own knowledge, and no other. Not one of the old tuners Mr. Hipkins knew (and some had been favourite tuners of Mr. James Broadwood) tuned anything like equal temperament. Collard, the Wilkies, Challenger, Seymour, all tuned the meantone temperament, except that, like Arnold Schlick, 1511 (see p. 546d), they raised the $G\sharp$ somewhat to mitigate the 'wolf' resulting from the Fifth $E\flat : G\sharp$ in place of $E\flat : A\flat$. Hence Mr. James Broadwood did not succeed in introducing equal temperament permanently even into his own establishment, and all tradition of it died out long ago. So far runs Mr. Hipkins's interesting information.

In 1812 Dr. Crotch (*Elements of Musical Composition*, pp. 134-5) gives the proper figures for equal temperament, shews how it arose, that its Fifths are too flat and its major Thirds too sharp, adding 'this will render all keys equally imperfect,' but says nothing to recommend it. Yet in 1840 Dr. Crotch (who died in 1847) had his own chamber organ tuned in equal temperament, as I have been informed by Mr. E. J. Hopkins, author of *The Organ, &c.*

It is one thing to propose equal temperament, to calculate its ratios, and to have trial instruments approximately tuned in accordance with it, and another thing to use it commercially in all instruments sold. For pianos in England it did not become a trade usage till 1846, at about which time it was introduced into Broadwoods' under the superintendence of Mr. Hipkins himself. At least eight years more elapsed before equal temperament was generally used for organs, on which its defects are more apparent, although not to such an extent as on the harmonium.

In 1851, at the Great Exhibition, no English organ was tuned in equal temperament, but the only German organ exhibited (Schulze's) was so tuned.

In July 1852 Messrs. J. W. Walker & Sons put their Exeter Hall organ into equal temperament, but it was not used publicly till November of that year. Meanwhile, in Sept., Mr. George Herbert, a barrister and amateur, then in charge of the organ in the Roman Catholic Church in Farm Street, Berkeley Square, London, had that organ tuned equally by Mr. Hill, the builder. Though much opposed, it was visited and approved by many, and among others by Mr. Cooper, who had the organ in the hall of Christ's Hospital (the Bluecoat School) tuned equally in 1853. ¶

In 1854 the first organ built and tuned originally in equal temperament, by Messrs. Gray & Davison, was made for Dr. Fraser's Congregational Chapel at Blackburn (both chapel and organ have since been burned). In the same year Messrs. Walker and Mr. Willis sent out their first equally tempered organs. This must therefore be considered as the commercial date for equal temperament on new organs in England. On old organs meantone temperament lingered much later. In 1880, when I wrote my *History of Musical Pitch*, from which most of these particulars have been taken, I found meantone temperament still general in Spain, and used in England on Greene's three organs, at St. George's Chapel, Windsor (since altered), at St. Katharine's, Regent's Park (see p. 484c'), and at Kew Parish Church; and while many others ¶ had only recently been altered, one (Jordan's at Maidstone Old Parish Church) was being altered when I visited it in that year. Hence, in England, equal temperament, though now (1885) firmly established, is not yet quite 40 years old on the pianoforte, and only 30 years old on the organ.

The difficulty of tuning in equal temperament led to the invention of Scheibler's tuning-fork tonometer. In Sect. G., art. 11, p. 489, will be found a practical rule for tuning in sensibly equal temperament at all usual pitches.

5. *Professor Mayer's Analysis of Compound Tones and Harmonic Curves.*

The following are two of the numerous acoustical contrivances of Mr. Alfred M. Mayer, Ph.D., Member of the National (American) Academy of Sciences, and Professor of Physics in the Stevens Institute of Technology, Hoboken, New Jersey, United States (see suprà, p. 417c).

1. *New Objective Analysis of Compound Sounds.* The analysis of compound sounds by resonators has two disadvantages: first, that it is subjective, inasmuch as but one observer at a time is capable of hearing the results; and, secondly, that the range of pitch reinforceable by a resonator is too great for extreme accuracy in the estimation of the actual component sound present. Both of these disadvantages were thus overcome (*Phil. Mag.* Oct. 1874, vol. xlviii. pp. 271-3, with a figure).

A Grenié's free-reed pipe, of the pitch $C = 128$, had part of its wooden chamber removed and replaced by morocco leather, at one point of which 8 silk cocoon fibres were attached, having their opposite extremities attached to 8 tuning-forks tuned to C, c, g, c', ¶ e' g', $^{\flat}b'b$, c'', at the point of the upper node in each where it divides into segments when giving its upper harmonic, so that this harmonic was eliminated. The cocoon fibres were stretched till they made no visible ventral segments when vibrating. The reed was tuned accurately to the C fork (of 64 vibrations) by means of the g fork. The forks were placed on proper resonance boxes. When the reed was sounded each fork 'sang out' loudly, but if the prongs of any fork were only slightly loaded the fork was mute, and was so rapidly affected that Prof. Mayer estimates (same vol. p. 519) that the effect of intervals such as 2000 : 2001 (or ·87 or not quite 1 cent) can be rendered sensible to the ear. On ceasing to sound the reed the forks continued to sound, and produced a tone of so nearly the same

quality as that of the reed that it was easy to feel that the difference was due to the absence of partials higher than the eighth. By this means, then, the analysis and synthesis of a compound tone can be shewn to a large audience at once, and all doubt as to its objective reality removed. At the same time the air in the resonance chamber of the reed acts on the leather cover as in hearing it would have acted on the drumskin of the ear, and the conduction of that vibration by the cocoon fibres replaces the complicated arrangements in the interior of the drum and the fluid of the labyrinth of the ear, while the forks themselves serve as the organs in the cochlea. Prof. Helmholtz's physiological theory of audition is thus perfectly exhibited in a ' working model.'
¶ The action of the resonance chambers of the forks is simply to make the effects heard at a distance.

2. *Harmonic Curves* (see p. 387*d*). In the *Philosophical Magazine*, Supplement for January 1875, vol. xlviii. pp. 520–525, Prof. Mayer gives curves compounded by six curves of sines (p. 23*d'*), representing six partial tones, where, for convenience, the amplitudes are taken to vary as the wave lengths, and to have the same initial phase. They are combined by taking the algebraical sum of their ordinates, which law would of course not hold true for the amplitudes chosen (about one-third of the length of the wave). The resulting figure bears a most remarkable resemblance to fig. 25, suprà, p. 84*b*. There is the same sudden rise on the left and step-like descent on the right, but the steps are more rounded, and the upper crest more pointed, and there are five steps in addition to the crest. Prof. Mayer then combines two such compound curves, and thus produces the resulting curves of two compound tones forming an Octave (with one high and one low crest, and also one high and one low trough, and the steps uneven and reduced in number), a Fifth (with four crests, two with long and two with short descents, the shorter having only one step), and a major Third (with eight crests, two extremely small, two moderate, three intermediate between the two last and one high, the ascents being abrupt as before, the descents rather wavy than stepped). These had all been drawn on a large scale with several hundred ordinates, and were reduced photographically. They form an excellent practical illustration of the nature of harmonic motions.

6. *The presumed different Characters of Keys, both Major and Minor.*

See suprà, p. 310*c* to 311*c*. It is first necessary to know what is the presumed phenomenon to account for. In the discussion of my paper ' On the Measurement and Settle-¶ ment of Musical Pitch ' (*Journal of the Society of Arts*, 25 May 1877, p. 686), Prof. (now Sir George) Macfarren, Principal of the Royal Academy of Music, spoke of ' the difficulty of representing the compositions of different eras, which had been written for different standards of pitch,' and added ' it was a marvellous fact that, while the pitch was felt to be changed, the impression of the character of the keys seemed to remain with reference to the nominal key, not to the number of vibrations of each particular note. Thus the key of *D* at the present day represented the same effect as was produced by the same key according to one's earliest recollections; it did not sound like the key of *E*♭, although it might be of the same pitch. If Mozart's symphony in *C* were to be played a Semitone lower, to bring it to the original pitch, it would not sound at all the ¶ same. How far this result was subjective — how much depended on the imagination of the hearer, and how much on the physical facts—was a deep, perhaps an insoluble question; but it was one which really ought to be considered.' The Chairman (Mr. William Pole, F.R.S., Mus. D. Oxon), on the contrary, said: ' In a practical point of view the French did an exceedingly good thing when they fixed on one pitch, . . . and they had practically done so, not only for France, but the Continent generally. He had the gratification some time ago of hearing Beethoven's *Sinfonia Eroica* played at a Conservatoire concert in *E*♭, as it should be, but he could not get rid of the idea, when he heard it played at the Philharmonic concerts, that it was in *E*♮.'

The mention of the performance of symphonies by Sir G. Macfarren and Dr. Pole takes the whole question out of the action of individual instruments, in which there is no doubt of considerable variety depending on the tonic, but this can be traced in every case to some defect of the instrument itself, as has been considered in the text (*loc. cit.*). The point of the long and short keys on a pianoforte, spoken of by Prof. Helmholtz, has been well worked out by Mr. G. Johnstone Stoney, D.Sc., F.R.S. Dublin, in a paper read before the Royal Dublin Society on 16 March 1883 (*Scientific Proc. R. Dublin Soc.* p. 59), who calls attention to the fact that in *A* all the Fifths are on white digitals and the major Thirds on black digitals, while in *A*♭ the Fifths are on black and the major Thirds on white digitals, and argues that this must make considerable difference in playing. Mr. Hipkins, however, gives it as his opinion that it is impossible to tell in the performance of a first-rate player whether he is striking a white or a black key. That would relegate the difference to the degree of skill of the player. But this does not at all affect the organ, the harmonium, or the voice. And by reference to symphonies we are constrained to consider the question independently of any particular instrument, as a simple acoustical fact.

In order to ascertain what that fact is supposed to be, according to recognised musicians of high standing, I give a condensed re-arrangement of the characters attributed to different keys in Mr. Ernst Pauer's *Elements of the Beautiful in Music* (Novello), p. 23, placing the major and minor keys in opposite columns, and proceeding by intervals of a Semitone.

Presumed characters of

Major Keys.

C. Expressive of feeling in a pure, certain, and decisive manner, of innocence, powerful resolve, manly earnestness, deep religious feeling.

C♯. Scarcely used; as *D♭* it has fulness of tone, sonorousness, and euphony.

D. Expressive of majesty, grandeur, pomp, triumph, festivity, stateliness.

E♭. Greatest variety of expression; eminently masculine, serious and solemn; expressive of courage and determination, brilliant, firm, dignified.

E. Expressive of joy, magnificence, splendour, and highest brilliancy; brightest and most powerful key.

F. Expressive of peace and joy, also of light passing regret and religious sentiment.

F♯. Brilliant and very clear; as *G♭* expresses softness and richness.

G. Favourite key of youth; expresses sincerity of faith, quiet love, calm meditation, simple grace, pastoral life, and a certain humour and brightness.

A♭. Full of feeling and dreamy expression.

A. Full of confidence and hope, radiant with love, redolent of genuine cheerfulness; especially expresses sincerity.

B♭. Has an open, frank, clear, and bright character, admitting of the expression of quiet contemplation; favourite classical key.

B. Expresses boldness and pride in fortissimo, purity and perfect clearness in pianissimo; seldom used.

Minor Keys.

C. Expressive of softness, longing, sadness, earnestness and passionate intensity, and of the supernatural.

D♭. The most intensely melancholy key.

D. Expressive of subdued melancholy, grief, anxiety, and solemnity.

E♭. Darkest and most sombre key of all; rarely used.

E. Expressive of grief, mournfulness, and restlessness of spirit.   ¶

F. Harrowing, full of melancholy, at times rising into passion.

F♯. Dark, mysterious, spectral, and full of passion.

G. Expresses sometimes sadness, at others quiet and sedate joy, with gentle grace or a slight touch of dreamy melancholy, occasionally rising to a romantic elevation.

A♭. Fit for funeral marches; full of sad, heartrending expression, as of an oppressed and sorrowing heart.

A. Expresses tender womanly feeling, especially the quiet melancholy sentiment of Northern nations; also fit for Boleros and Mauresque serenades; and finally for sentiments of devotion mingled with pious resignation.

B♭. Full of gloomy and sombre feeling, like *E♭*; seldom used.   ¶

B. Very melancholy; tells of quiet expectation and patient hope.

In reading over this Table it is impossible not to feel that the character, often contradictory, arises from the reminiscence of pieces of music in those keys, as the author indeed admits (*ib.* p. 22). Such a distinction as that made between *F♯* and *G♭*, which, in equal temperament, is a mere matter of notation, but is here made to yield incompatible results, shews that the writer was thinking more of treatment than of actual sound. This is confirmed by his saying (*ib.* p. 26) : ' We shall often find that the general character of a key may be changed by peculiarities and idiosyncrasies of the composer; and thus a key may appear to possess a cheerful character in the hands of one writer, whilst another composer infuses into it a melancholy expression; all depends on the treatment, on the individual feeling of the composer, and on his acute understanding of the characteristic qualities of the key he employs.' The writer then goes on to consider the effect of rhythm and time, and the different characters which he assigns to their varieties, independently of the key employed, clash so much with the preceding that it is difficult to know what is supposed to belong to one and what to the other.

Now the acoustical facts, independently of any particular instrument or temperament or any errors of tuning or performance (both numerous but variable), are these. Whether we take just intonation, or that of any uniform linear or cyclic temperament, carried on to a sufficient number of tones to prevent the occurrence of 'wolves' within the piece of music performed, the one thing aimed at is to have the intervals between the same notes of the same scale precisely the same, at whatever pitch they are played, or however they may be conventionally noted. If there is any difference between the scales of, say, just A_1 and $A^1♭$, which have a difference of 70 cents, or equal *A* and *A♭*, which differ by 100 cents, or meantone *A* and *A♭*, which differ by 76 ¶ cents, or Pythagorean *A* and *A♭*, which differ by 114 cents, this difference must be due solely to pitch. There is no doubt that on the piano, the organ, and each instrument of the orchestra, the difference will be considerable and very appreciable, but that does not enter into consideration. What effect does simple difference of pitch in the tonic produce? In the human voice and in all instruments quality of tone varies together with the pitch. A change of tonic implies a different pitch for the most frequently returning sounds, and those most important to the nature of the key. Hence it produces a different quality of tone, with a variation in the range of partials possessed, and consequently affects the distinctness of the delimitation of the principal consonances and dissonances of the keys, and by that means alters their audible effect. For intervals so small as we have supposed this difference must necessarily be small, whereas the difference of the keys of A and $A\flat$ is said to be great. If so, it can only arise from errors of intonation or performance. In the days of the old meantone temperament in its defective state of 12 notes only to the Octave, there was a vast difference between the keys of A and $A\flat$; the first had all its chords correct (supposing $G\sharp$ were not sharpened, supra, p. 546d), the latter had all the chords involving $A\flat$ and $D\flat$ (which had to be represented by $G\sharp$ and $C\sharp$) frightfully erroneous.

¶ It seems to me that the feeling of a difference in the character of the keys whose tonics vary but slightly in pitch was established at this time (in Sir George Macfarren's younger days, b. 1813). Any difference so slight as a Semitone would have been strongly felt (except in passing from A to $B\flat$, the two extreme good keys). Whereas for differences of a Fifth there would generally not have been such violent distinctions (except at the extremes A to E and $B\flat$ to $E\flat$). It would appear that these mechanical distinctions partly influenced composers in their choice of a key, and produced what has become an hereditary prejudice, for which there is no longer any ground, and which never ought to have existed in just unaccompanied singing. But even at that time we have composers ignoring the difference. Handel composed his dead march in *Saul* in C, and having written an anthem ('O sing unto the Lord a new song') for St. ¶ James' Chapel Royal, in which the organ had

a pitch one Tone higher than his own (see sect. H. p. 503b, under a' 474·1), he directed the singers, as the voice parts were too high, to take them a Tone lower, and the accompanying organ to play *two* Tones lower. Had Handel any idea of the innate difference of the character of keys? and, if he had not, what does it all amount to? Beyond those differences inevitable to varieties of pitch, already pointed out, and easily perceived by slowly playing up the major scale on any instrument from its lowest note, or singing it on any voice from the lowest note it can reach easily (see p. 544d)—beyond such differences, *all* seems to be subjective, or due to hereditary feeling created by former defective temperaments, or at present to mechanical errors of tuning, stopping, or blowing, especially in unusual keys. Possibly a composer at the present day would write a piece of a totally different character, as pointed out in the table, according as he made the signature $F\sharp$ or $G\flat$, but that must have been a reaction on his own mind, for the tones he would play would be precisely the same in both cases.

That tuners of the piano sometimes still intentionally tune unequally, and hence make the effect of A and $A\flat$ really very different, has nothing whatever to do with the matter. Those who do so have not learned their profession. Similarly for players on a pianoforte who cannot equalise the effect of the long and short keys. But the singer knows very well when a piece of music falls upon his bad notes, and ruthlessly transposes the key, quite reckless of these presumed varieties of key-character.

7. *Dr. W. H. Stone's Restoration of* 16-*foot C, to the Orchestra.*

See p. 175c on the deepest tones which can be heard. The following is condensed from information furnished me by Dr. Stone.

Dr. Stone has for some years endeavoured to restore to the orchestra the lower notes of the 16-foot Octave, which appear to have been neglected of late. It seems to him a contradiction that, while the organ possessed that Octave, and another, the 32-foot, below it, and while even an instrument of so comparatively feeble a tone as the pianoforte could obtain these deep notes, they should be absent from the full band. Most of the great composers have employed them, especially Beethoven and ¶ Onslow. Many passages of their compositions had to be partially transposed, often (as in the C minor Symphony of Beethoven) much to the detriment of the general effect. In the Trio of this great work a scale passage occurs several times for the double basses alone, beginning on the 16-foot G_i. But this note being entirely absent on the ordinary three-string basses, as used in England, it was there customary to take it either altogether or in part in the Octave above. Some players, indeed, were in the habit of letting down the A_i or lowest string, by a Tone to G_i, for this special passage; but the resonance of a string thus slackened was far inferior to what could be obtained by more legitimate means. The fine part for the contrafagotto in the same symphony, descending to C_i, was usually omitted, or played an Octave higher by the ophicleide. The *Pastoral Symphony* likewise

frequently contains F_i natural, a note quite unattainable except on the four-string basses, whose lowest note is E_i (p. 18c).

It was obvious, in attempting to remedy this defect, that of the three modes by which vibrations in a stretched string may be slackened, two, namely, length and thickness, were inadmissible. The first renders the instrument so large as to be unwieldy and out of the reach of an ordinary arm. The second was found to cause rotation of the string under the impulse of the bow acting at its periphery, and thus to generate false notes. The third remained, in increasing the specific gravity of the string without enlarging its diameter. This was satisfactorily accomplished by covering a gut string with heavy copper wire such as is used for the lowest strings of pianos. The note C_i was obtained, and an instrument thus strung was exhibited in London in the International Exhibition of 1872.

But it became clear that to give the new notes full power, and to prevent the danger of shaking the instrument to pieces, a means of strengthening the belly in the direction of strain was required, which should not unduly increase the weight of the sound-board. This requisite was ingeniously fulfilled by Mr. Meeson:--Four strips of white pine are glued on to the back of the belly, running its whole length, one on one side of the ordinary bass-bar, and three on its other side, thus corresponding in number, and to a certain degree in position, to the increased number of strings.

Two of them cross and intercept the usual *f*-shaped sound-holes, thus removing a weak place in the belly, and causing it to vibrate more homogeneously. They appeared on trial to add great power to the instrument throughout, and to remove the inequality and varying intensity of vibration which exists on most old instruments even by celebrated makers, and which musicians usually designate by the term 'wolf.' The bars are curved to an elliptical shape to fit the hollow of the belly, and to give the greatest resistance to compression with the smallest quantity of material. Even in a double bass the quantity of wood required is very small, and from its lightness when perfectly dry it hardly exceeds an ounce in weight. From their shape and function they are termed *elliptical tension bars.*

It appeared from subsequent experiments that the same system was applicable to the smaller members of the viol and violin family, giving an increased sonority and firmness to the tone. It succeeds best with violins of sweet but feeble quality, and in some of the older Italian instruments, where the progress of decay had to a certain extent diminished the volume of sound.

The contrafagotto or double bassoon as made on Dr. Stone's designs by Herr Haseneier of Coblenz, consists of a tube 16 feet 4 inches long, truly conical in its bore, and enlarging from ¼-inch diameter at the reed to 4 inches at the bell or lower extremity. It is curved on itself for convenience of manipulation, so that in actual length it is about equal to the ordinary bassoon. Its extreme compass is from $C_{\prime}$ to c', but its ordinary range is to g only, the other notes being difficult to bring out. Haydn gives a part to such an instrument in his *Creation*, Mozart uses it occasionally, Beethoven frequently, Mendelssohn sometimes. ¶

Both of these instruments I have heard in use. Their tone is not perfectly continuous, but is very good, and when played in conjunction with other instruments, musically effective.

8. *On the Action of Reeds.*

(See pp. 95 to 100.) Knowing the long, patient, and practical attention which Mr. Hermann Smith had paid to the action of reeds, I requested him to furnish me with an account of the results of his experience. He obligingly sent me a series of elaborate and extensive notes, which the space at my command utterly precludes me from giving at proper length, and which I am therefore forced to represent by the meagrest possible outline, with the hope that they will appear elsewhere in suitable detail. Only passages in inverted commas contain Mr. Hermann Smith's actual words, the rest is my own necessarily imperfect attempt to condense his statements.

Reeds may be classed as, 1. *single*, whether *striking* (in clarinet of cane, and reed-pipe of organ of metal) or *free* (harmonium and American organ, both of metal); 2. *paired* (bassoon and oboe, both of cane, in action compressible; horns and larynx, both membranous, in action extensible); 3. *streaming* (flutes, flageolets, flue-pipes of organ, all of rushing air, in action abstracting). The last kind has been partly considered, suprà, pp. 396-7.

In the *clarinet* the reed is straight and very thin at the tip, but the edge of the section of the wooden tube against which it strikes presents a slight curve, whereas in the organ pipe the reed is curved and the edge against which it strikes is straight (p. 96c). The time of vibration consists (1) of the time of forward motion, which may vary slightly; (2) of the time of rest, which may vary greatly; (3) of the time of recoil, which does not vary 'When the reed is placed in the mouth, the air on both surfaces of the reed is of equal pressure, and on increase of strength in the wind, the tendency would be to separate the reed still further from the edge of the mouthpiece, were it not that a current of air quickly passing into a tube exercises suction at the orifice of entry; therefore the elastic reed yields in the direction of the place of suction, so that it is held there. The current of wind having been sent forward with impetus, leaves behind it a partial vacuum, which is strongest close upon the inner face of the tongue. There, then, is a region of least pressure, which continues to exist during the transit of the pulse of compressed air to the first found point of outlet. When that point is reached, the external air rushes in and restores equilibrium, and in doing so causes the shock of arrested motion in which the reed recoils,' and forthwith the action commences as before.' ¶

The *clarinet* should not be described as a stopped pipe, though both have unevenly numbered partials and give similar pitch for similar length, because 'in the clarinet there is a propulsive current going through the pipe; in the stopped pipe, on the contrary, there is an abstracting current acting outside by suction.' Too much has been attributed to the cylindrical bore for producing only the unevenly numbered partials. 'An oboe reed fixed on the clarinet tube gives oboe pitch of tone and oboe partials.' The Japanese *Hichi-riki* has an inverted conical bore, that is, the diameter is the smallest at the point furthest from the reed. 'Like the clarinet, it gives notes which are an Octave lower in pitch than would be calculated from its length. The first note after the fundamental is the Twelfth. The reed is broad, not single as on ¶ the clarinet, but double and as the bassoon reed, differing, however, in having an enlarged base where it fixed into the tube. . . . On substituting an oboe reed, the pitches of the notes correspond to those of the oboe and the first tone after is the Octave. . . . If the end of the pipe is placed full within the mouth, and is blown through without the use of any reed whatever (and without any action from the lips), clear and powerful sounds are elicited, varying as the openings of the holes are varied, provided one of the upper holes is left open . . . it is indifferent whether the end of the wide diameter or that of the narrow is taken into the mouth, either way sounds are in this manner readily produced.' This effect Mr. Hermann Smith attributed to a stream reed from the open hole.

Bassoons and *oboes* have paired reeds, which touch down their outer edge, and do not vibrate length-wise but cross-wise, so that a transverse section through them has alternately the form of the outer lines in fig. 63, p. 387c, when they are open, and of two parallel lines when they are closed. The reeds are sections from a small hollow reed-plant of particular growth (*arundo donax* or *sativa*), made very thin at the tip, and rendered supple by the moisture of the mouth. The player's lip restricts the size of the oval in notes of high pitch. 'The pressure of wind would keep the pair of reeds apart but for the influence of the suction when the current is thrown through. The vibration therefore is produced by the same kind of action as in the clarinet, but there is a new mechanical method for bringing it about.'

¶ The membranous reeds formed by the lips and vocal chords are reeds of *extension*, beginning to vibrate from a state of *closure*, contrary to all other reeds. In *horns* the cup acts as an exhaust chamber, and when it is too large the upper notes cannot be well produced; 'that is to say, the necessary degree of vacuum cannot be brought about in time to coincide with the reciprocating return of the column of air in the tube.' In the larynx the ventricles of Morgagni between the true and false chords probably act as exhaust chambers.

The *stream reeds* have been already considered (p. 396c'), but Mr. Hermann Smith has developed his theory of *displacement action*, or the actual tone of air under cleavage, deduced originally from observations of the different sounds of wind sweeping through the branches of leafless trees, in which tone is ¶ produced without a vibrating agent. Mr. Hermann Smith found the common doctrine of friction unsatisfactory. In 1870 he had made a series of rods about 5 ft. long, the sides of each having same smoothness and of the uniform width of 1¼ inches, with a V-shaped or triangular section, and these he swept swiftly through the air like swords, sharp edge first. He found that, although the friction surface was similar on each, they developed different notes, which he discovered to be according to the thickness of the back, the pitches being inversely as the thickness. *C* 528 vib. requires a thickness somewhat less than half an inch. 'Covering irregularly the tips of the rods did not affect the sound. Half an inch thickness of spongy felt fixed on the back of the rod, the same width being preserved, lowered the pitch a Fourth. The felt entangled air in ¶ its pores, so that the vacuum by suction was less perfect.... With less speed of stroke the pitch is again lowered. ... In organ pipes, &c., in all wind instruments, a sudden displacement of air must take place immediately near the agent of vibration, in which space a right degree of vacuum is requisite, else the right note will not follow. To this result all the devices of mouthpieces tend. This work of displacement in the origination of sound raises a question distinct from the transmission of sounds in waves.'

The free reed is supposed to have been adopted from China. But the European and Chinese forms are different. The Chinese reed is stamped out in the same piece as the frame, with which it lies level. The reeds act upon tubes which (agreeing with M. W. Weber's law, though made long before its discovery)

are three-quarters of the half-wave length. The harmonium reed is placed above the frame, and the end turns up from it. If it is set level with the frame it will not vibrate. To produce vibration a stream of air must pass between the tongue and the frame, producing a partial vacuum on the underside of the reed, and the amount of suction thus caused must be properly graduated or there will be no action. The chief peculiarity of the free reed is that the pitch is only slightly affected by the cavities with which it is associated, but these boxes or cavities, according to their dimensions, and governed by the operation of partial occlusion, mainly determine the *quality of tone*.

The reed is not properly compared to a vibrating rod, because the reed actually in use is not uniform. In a series, the low reeds are thickest at tip and thinnest at root; high reeds thinnest at tip and thickest at root. Hence they are affected differently by different pressures of wind, and alter their pitch differently. Expressive playing therefore becomes playing out of tune. Only with a constant blast will a free reed maintain a constant pitch. The stronger blast flattens deep reeds and sharpens high ones, and from this cause arises much of the painful dissonance of series of chords played on these instruments.

To remedy this defect the action of the wind on the tongue in the American organ is limited by making the frame very thin, which is 'dished out' underneath till the edge passed by the tongue is barely thicker than the tongue, instead of being 8 or 10 times as thick as on the harmonium. The suction in the harmonium is longer in time and stronger in degree, the tongue moves a greater distance, and more intensity of tone is produced. 'The American reed cannot make a deep excursion, for the suction is spent as soon as the tongue gets below the edge of the frame.' It is therefore not suited for expression, but produces a smooth and flexible kind of tone.

Mr. Hermann Smith considers the harshness of the free reed, resulting from its large number of partials, to be chiefly due to its proportions. 'The tongue is inordinately long in proportion to its width, and hence under the stress of the wind (which necessarily shifts its incidence during the movements of the tongue and its re-course) there is developed a diagonal strain or torsion from one corner at the tip of the tongue to the opposite corner of the root, so that a lateral irregular motion is set up accompanying the longitudinal vibration.' Hence he concludes that 'the long reed is a wrong reed,' and that, 'in view of its liability to lateral torsion, the rectangular form is about the worst.... Trial of various reeds shews that a long rectangular reed is strident in tone, that increase of width in reeds brings in increasing proportion smoothness of tone, and that the width may be increased till it equals the length.'

In 'voicing,' a bend is made across the tongue, turning the point upwards. This somewhat checks lateral vibration. By an early plan of his, 'reeds were yoked together by a bar across the middle of the length, and the improved quality of tone was obviously due to the fact that the two reeds were equivalent to one broad reed, and that the bar across hindered the operation of any diagonal strain or twist during vibration.... When a reed is much curved it is slow in speech, and a great amount of wind passes wastefully, compensated only by the smoother tone.' Mr. Hermann Smith says his ' best toned reeds have been series in which, according to his design, the openings made by the curve given to the reed were filled up at the sides by arched blocks added to the top of the frame, following the line of curve. The discontinuity was therefore sharply defined, yet the tones were mellow and rich. A reed mounted on wood surface may have its quality greatly changed by an interposed pad of leather or felt between the reed frame and the wood, the extreme harshness disappears and the tone is smoother altogether, shewing how much that is unpleasant is due to the jarring from arrested motion.'

Mr. Hermann Smith's ' conclusions are that, in the making of free reed instruments, broad reeds should be used, and with broad channels or boxes or cavities of varied shape; that within the large chamber small suction chambers should be placed below or beyond the reed tongue, in imitation of the ventricles of the larynx; and if these afford areas and cavities suitable for the displacement accordant with the pitches, the speech will be quickened and firmness given to the tones, or, in other words, the mechanical motion of air and reed will be steadied. The rectangular form of reed, except when stops of hard metallic quality are required, should be abandoned, and broad shapes substituted, having tips semicircular, semi-oval, or ovate, or shapes to ensure a central line of strain. Weighting the tips of reeds should be avoided as much as possible.'

Mr. Hermann Smith's latest device in the treatment of reeds is designed to overcome the difficulties of inordinately long or weighted reeds. His plan is ' to use metal or material of uniform thickness, and to get the degrees of flattening by drilling out or excising portions of material at or near the root of the reed, and then to fill such spaces as are thus made with other fixed pieces of metal that are neutral, and do not enter into vibration. Thus the sides of the reed tongues remain with the fibre intact, unweakened by thinning or scraping, which takes the best vigour out of the reed. Any degree of flattening may be attained according as the excision is made to extend up the tongue. A like treatment of the half of the tongue forming the tip will produce the opposite effect, sharpening pitch by lightening the tip ; the spaces left by the excised portions are then covered with lighter material, such as goldbeater's skin.'

Mr. Hermann Smith states that his device of ' adding to the large cavities small exhaust chambers or cup-like cavities, fixed just below the tongue of the reed, causes the most unmanageable reeds, even those in the 32-foot Octave, when made broad on the above plan, to render good musical service, to be free in speech, and to produce a full pervading quality of tone, devoid of the harshness of long reeds having heavily weighted tips.' Mr. Augustus Stroh (see Sect. M. No. 2, p. 542*d*) informs me that he has been led to a similar contrivance in a machine he has recently constructed.

For further details of Mr. Hermann Smith's inventions respecting reeds, see Specifications to his Patents 1878, No. 227 and No. 4942 ; 1880, No. 68 ; and 1884, No. 7777. A large amount of varied information may also be found in his treatise entitled ' In the Organ and in the Orchestra,' now (1884–5) publishing in *Musical Opinion*, a monthly magazine. Of this treatise 25 chapters have already appeared full of interesting elucidations of instrumental difficulties.

9. *Postscript.*

Standard Musical Pitch in England.—In consequence of a communication from our Foreign Office, due to the Belgian change of pitch (p. 501*d*), Sir G. A. Macfarren, Principal of the Royal Academy of Music, convened a public meeting of musicians, theorists, instrument-makers, and their friends, ' to consider the desirability of a standard musical pitch for the United Kingdom.' It took place on 20 June 1885, and was well attended. Three resolutions were passed: (1) declaring uniformity desirable: (2) recommending French pitch ; (3) declaring it advisable to take steps for its adoption in civil and military bands. A committee of 4 theorists, 15 musicians, and 4 instrument-makers was appointed to carry out the resolutions. See the *Times*, 22 June 1885, p. 7, col. 2, and p. 9, col. 3, and *Musical Opinion*, 1 July 1885, p. 493.

Addenda to History of Pitch. In 1845 the pitch of the Philharmonic Society was *a*′ 447·1, according to a fork tuned in the orchestra at that date by Mr. R. S. Rockstro.

Herr Eduard Strauss, of Vienna, performing at the Inventions Exhibition 1885, used *a*′ 452·5, but the bandsmen said the pitch of the opera was nearly a Quartertone flatter, say *a*′ 447.

The band of the Pomeranian (Blücher) Hussars performing at the same Exhibition used *a*′ 460·8, or, as the bandmaster said, ' exactly a Semitone sharper than French pitch.'

Effect of Rust, &c., on Tuning-forks. See p. 445 *d*′. Mr. Rockstro possesses a fork which in 1859 was at French pitch, and now through rust (and possibly bad treatment) shews only *a*′ 424·5, that is, has gone down by 42 cents, far exceeding any other fork examined.

Flute Intonation. Mr. R. S. Rockstro, in June 1885, kindly brought me an eight-keyed flute, about 40 years old, an excellent instrument of its kind. This he played so as to preserve its natural intonation without correction. Before striking any new note it was ascertained that *g*′ 404 vib. remained constant. The *a*′ 461 vib. was a quarter of a Tone too sharp. The result gave in cents, reckoning *g*′ as 1200=0, *c*′ 488, *c*′♯ 564, *d*′ 678, *d*′♯ 763, *e*′ 906, *f*′ 1000, *f*′♯ 1098, *g*′ 1200=0, *g*′♯ 92, *a*′ 229, *a*′♯ 292, *b*′ 431, *c*″ 513, *c*″♯ 566. This seems meant for meantone intonation, sharpening many of the sharps to pass as the flats above them. But even in this case the intonation was imperfect, and Mr. Rockstro thought it was rather due to a series of compromises.

Mr. R. S. Rockstro brought at the same time the ' Rockstro model ' flute, invented by himself, to have a more correct equal intonation than Boehm's. This was blown in the same way, but in this case *a*′ was always brought to 452 vib. Result in cents, reckoning a' as 1200 = 0, was c' 312, $c'\sharp$ 406, d' 506, $d'\sharp$ 601, e' 697, f' 801, $f'\sharp$ 895, g' 1001, $g'\sharp$ 1097, a' 1200 = 0, $a''\sharp$ 102, b' 201, c'' 304, $c''\sharp$ 404. This is very good, and may be better than the above numbers shew, as, on account of the difficulty of sustaining notes on the flute without variation, it was not possible to determine the pitch of each note within less than 1 vib. in a second. With regard to the lowest c' from the open end of the flute, Mr. Rockstro says he leaves it purposely too sharp in relation to a', because it is easy, by management of lip, to blow it in tune, but if it were originally in tune, it would not sound sharp enough in very soft passages.

Siamese Scales. The King of Siam sent over his Court Band with their instruments to ¶ the London Inventions Exhibition 1885, and the Siamese minister obligingly allowed Mr. Hipkins and myself to determine the musical scale. Prince Prisdang told us that the intention was to divide the Octave into 7 equal intervals, each of which would then have 171·43 cents. Hence the following comparison. The scales are given as usual in cents from the lowest note.

Theoretical scale :—0, 171, 343, 514, 686, 857, 1029, 1200 cents, having a neutral Second 171 lying between 100 and 200, a neutral Third 343 lying between 300 and 400, a slightly sharpened Fourth 514 for 500, a slightly flattened Fifth 686 for 700, a neutral Sixth 857 lying between 800 and 900, and a neutral Seventh 1029 lying between 1000 and 1100, but much nearer the former. As there is no harmonic interval but the Octave, and as the ¶ Siamese seem to tune by Octaves and single degrees, there is room for much variation from the ideal intonation, as shewn in the following observed scales.

Ranat ek or wood harmonicon, first Octave 0, 208, 326, 537, 698, 883, 1048, 1208, second Octave (pitch of lowest note 382·6 vib.), 0, 200, 359, 537, 711, 883, 1057, 1222, third (incomplete) Octave 0, 193, 347, 549, 698 (two more bars, too high to measure). This instrument is tuned by lumps of wax mixed with some heavy substance stuck to the underpart of the bar. The tuning lump having fallen from the second bar of the first two Octaves, it was quite out of tune, and its proper pitch (registered above) was determined by a comparison with other instruments. In the Ranat, p. 518, No. 85, all the lumps had been removed, hence, it was entirely out of tune.

¶ Ranat t'hong or brass harmonicon (pitch of lowest note 382·6 vib.) scale 0, 200, 340, 537, 699, 881, 1043, 1207.

Ranat lek or steel harmonicon, first Octave (the second bar absent), 0, 327, 519, 679, 856, 1075, 1202, second Octave (pitch of lowest note, 385·5 vib.), 0, 150, 299, 447, 614, 743, 960, 1179, third (incomplete) Octave 0, 90, 222, 430, 609.

Tak'hay or crocodile, a three-stringed instrument with high frets, played with a conical plectrum, 0, 198, 362, 528, 720, 890, 1080, 1250.

Hence 52 single degrees were examined, each of which should have had theoretically 171·43 cents. In reality 5 were less than 132, 8 between 140 and 159, 12 between 160 and 167, 9 between 170 and 179, 3 between 180 and 185, 6 between 190 and 198, and 9 between 200 and 219. Hence only 15 approached

to equal Tones, and only 2 approached to equal Semitones, both sets being clearly erroneous, while the 21 between 160 and 179 were tolerably close approximations to the ideal. Bearing these variations in mind, it is probable that p. 518, Nos. 81, 82, and 83, at least, belonged to this system of 7 intentionally equal heptatones, as they may be called. And this confirms the conception that Salendro, p. 518, Nos. 94 and 95, consists ideally of 5 equal pentatones.

The instruments were beautifully and artistically ornamented, the execution by the musicians was florid and musicianly in accurate and varied rhythm, there was an observance of light and shade, together with a clear conception of melody, but none of harmony. Besides the harmonicons there were kettles or gongs (k'hong), a three-stringed viol (saw t'hai), a two-stringed fiddle (saw Chine), the three-stringed crocodile (tak'hay) ; reed instruments, flutes, and drums.

Japanese Scales, see pp. 519 and 522, Nos. 110 to 139. In July 1885, Mr. Isawa, Director of the Musical Institute, Tokio, Japan, sent to the Inventions Exhibition several tuning-forks and tables. From the tables it appeared that the *classical* 12 Ritsu or Semitones resulted from tuning 11 perfect Fifths up (or Fourths down), and then a Fifth too flat by a Pythagorean comma, giving the scale : 0, 114, 204, 318, 408, 522, 612, 702, 816, 906, 1020, 1110, 1200 cents. But the 13 forks sent had the following pitch (as determined by the Translator), the number and name of the Ritsu and the name of the nearest European note at French pitch being prefixed : –I. Ichikotsu d' 292·7, II. Tangin $d'\sharp$ 305·6, III. Hiyōjō e' 326·2, IV. Shōretsu f' 343·1, V. Shimomu $f'\sharp$ 365·7, VI. Sōjō g' 391·5, VII. Fushō $g'\sharp$ 410·1, VIII. Waushiki a' 437, IX. Rankei $a'\sharp$ 460, X. Banshiki b' 491·5, XI. Shinsen c'' 517·3, XII. Kamimu $c''\sharp$ 549·5, I'. Ichikotsu d'' 585·4, this gives the scale in cents : 0, 75, 188, 275, 385, 503, 583, 693, 782, 897, 986, 1091, 1200.

Mr. Isawa also sent forks for tuning the *popular* scale Hiradioshi (p. 519a, Nos. 110 to 112) in two forms, old and new, both different from those already given. Old style in cents 0, 102, 502, 706, 809, 1197, evidently meant for just 0, 112, 498, 702, 814, 1200. New style in cents 0, 85, 502, 708, 793, 1200, for Pythagorean 0, 90, 498, 702, 792, 1200.

Mr. Isawa also sent a Standard Tuning-fork giving d 145·45 at 52° F. French pitch d = 145·2 vib.

There was also a monochord on which many scales were indicated, and two sets of reed pitch-pipes, which cannot be described for want of space.

Modern Greek Scale. According to Meshaqah, in Eli Smith (op. cit. p. 264, note §), the modern Greeks divided the Octave into 4 × 17 = 68 parts, and form the scale by 12, 9, 7, 12, 9, 7, 12 of these divisions. Since, then, 1200 ÷ 68 = 17·65 cents, the scale in cents will be 0, 212, 371, 494, 706, 865, 988, 1200, which again has a neutral Third and Sixth, 371, 865. If the scale had consisted of 12, 8, 8, 12, 8, 8, 12 of these divisions, we should have got the precise scale of *Villoteau* (p. 520a' l. 5), which is a singular additional justification of his division of the Octave into 17 equal parts.

INDEX.